AF774006

The Minerals, Metals & Materials Series

Marc André Meyers · Hector Alfredo Calderon Benavides · Sonia P. Brühl
Henry A. Colorado · Elvi Dalgaard
Carlos Nelson Elias · Roberto B. Figueiredo
Omar Garcia-Rincon · Megumi Kawasaki
Terence G. Langdon · R.V. Mangalaraja
Mery Cecilia Gomez Marroquin
Adriana da Cunha Rocha
Julie M. Schoenung · Andre Costa e Silva
Mary Wells · Wen Yang
Editors

Proceedings of the 3rd Pan American Materials Congress

TMS

Springer

Editors
Marc André Meyers
University of California-San Diego
La Jolla, CA
USA

Hector Alfredo Calderon Benavides
ESFM-IPN
Mexico City
Mexico

Sonia P. Brühl
UTN-National University of Technology
Buenos Aires
Argentina

Henry A. Colorado
Universidad de Antioquia
Medellín
Colombia

Elvi Dalgaard
Pratt & Whitney Canada
Longueuil
Canada

Carlos Nelson Elias
Military Institute of Engineering
Rio de Janerio
Brazil

Roberto B. Figueiredo
Universidade Federal de Minas Gerais
Belo Horizonte
Brazil

Omar Garcia-Rincon
Ternium Mexico SA de CV
San Nicolas de los Garza
Mexico

Megumi Kawasaki
Hanyang University
Seoul
Korea (Republic of)

Terence G. Langdon
University of Southern California
Los Angeles, CA
USA

R.V. Mangalaraja
University of Concepción
Concepción
Chile

Mery Cecilia Gomez Marroquin
Universidad Nacional de Ingeniería
Lima
Peru

Adriana da Cunha Rocha
Federal University of Rio de Janeiro
Rio de Janeiro
Brazil

Julie M. Schoenung
University of California, Irvine
Irvine, CA
USA

Andre Costa e Silva
Universidade Federal Fluminense
Rio de Janeiro
Brazil

Mary Wells
University of Waterloo
Waterloo, ON
Canada

Wen Yang
ETH Zurich
Zürich
Switzerland

ISSN 2367-1181 ISSN 2367-1696 (electronic)
The Minerals, Metals & Materials Series
ISBN 978-3-319-52131-2 ISBN 978-3-319-52132-9 (eBook)
DOI 10.1007/978-3-319-52132-9

TMS owns copyright; Springer has full publishing rights

Library of Congress Control Number: 2016962043

© The Minerals, Metals & Materials Society 2017
This work is subject to copyright. All rights are reserved by the Publisher, whether the whole or part of the material is concerned, specifically the rights of translation, reprinting, reuse of illustrations, recitation, broadcasting, reproduction on microfilms or in any other physical way, and transmission or information storage and retrieval, electronic adaptation, computer software, or by similar or dissimilar methodology now known or hereafter developed.
The use of general descriptive names, registered names, trademarks, service marks, etc. in this publication does not imply, even in the absence of a specific statement, that such names are exempt from the relevant protective laws and regulations and therefore free for general use.
The publisher, the authors and the editors are safe to assume that the advice and information in this book are believed to be true and accurate at the date of publication. Neither the publisher nor the authors or the editors give a warranty, express or implied, with respect to the material contained herein or for any errors or omissions that may have been made. The publisher remains neutral with regard to jurisdictional claims in published maps and institutional affiliations.

Printed on acid-free paper

This Springer imprint is published by Springer Nature
The registered company is Springer International Publishing AG
The registered company address is: Gewerbestrasse 11, 6330 Cham, Switzerland

Preface

The Pan American Materials Congress (PAMC) is in its third iteration and was originally initiated from a partnership between the Associação Brasileira de Metalurgia, Materiais e Mineração (ABM) located in Brazil and The Minerals, Metals & Materials Society (TMS) located in the United States. This partnership produced two previously successful materials science and engineering conferences, titled "Pan American Materials Congress" occurring in 2010 and 2014 and held in conjunction with ABM's large annual conference. These events were co-chaired by Prof. Sergio Neves Monteiro, ABM's incoming President. The 3rd PAMC, hosted by TMS, includes nine participating professional societies, and is co-located with the TMS 2017 Annual Meeting & Exhibition. It is the first time that this international materials science and engineering conference is held in North America, with TMS in the role of host society. A program covering a variety of materials science topics has been created based on the input from leading scientists and engineers representing eight countries and nine international materials, metals, and minerals societies listed below:

1. Argentina: Asociación Argentina de Materiales (SAM)
2. Brazil: Associação Brasileira de Metalurgia, Materiais e Mineração (ABM)
3. Peru: Asociación Peruana de Metalurgia, Materiales Y Minerales (APMMM)
4. Colombia: Colombian Materials Society
5. Chile: Instituto de Ingenieros de Minas de Chile (IIMCh)
6. Canada: Metallurgy and Materials Society (MetSoc), Canadian Institute of Mining, Metallurgy, and Petroleum (CIM)
7. Chile: Sociedad Chilena de Metalurgia y Materiales (SOCHIM)
8. Mexico: Sociedad Mexicana de Materiales (SMM)
9. United States: The Minerals, Metals & Materials Society (TMS; Host Society)

The participation of additional materials societies throughout the Americas is being sought and is under discussion. The organizers of this congress seek to provide an international Pan American focused program to address the needs of the materials science and engineering communities as they relate to government, academic, and industrial institutions, while providing an intimate setting for professionals to interact with and form strategic partnerships with their peers. Student participation is strongly encouraged and is a focus for the lead organizers of this event. Additionally, as far as we are aware, this is the only international materials science conference where the emphasis is exclusively on North and South America.

The 3rd PAMC technical programming encompasses a wide range of materials, metals, and minerals with applications specific to the international communities that are represented, including symposia on materials for transportation and infrastructure, materials for the oil and gas industry, and minerals extraction and processing. These proceedings contain the following sections, which correspond to the themes of the conference:

Advanced Biomaterials
Advanced Manufacturing
Materials for Green Energy
Materials for Infrastructure
Materials for the Oil and Gas Industry
Materials for Transportation and Lightweighing
Minerals Extraction and Processing
Nanocrystalline & Ultra-fine Grain Materials & Bulk Metallic Glasses
Steels

From this program, it is expected that rich discussions and collaborative opportunities will result, heavily focused on the Americas. The congress is scheduled to run for three consecutive days, with sessions in both the morning and afternoon. Special attention has been paid to communications and complementary planning between the congress organizers and TMS staff, and TMS 2017 symposia organizers and volunteers, to ensure that the sessions are synergistic and not duplicative of the TMS 2017 Annual Meeting & Exhibition programming.

Marc André Meyers
Hector Alfredo Calderon Benavides
Sonia P. Brühl
Henry A. Colorado
Elvi Dalgaard
Carlos Nelson Elias
Roberto B. Figueiredo
Omar Garcia-Rincon
Megumi Kawasaki
Terence G. Langdon
R.V. Mangalaraja
Mery Cecilia Gomez Marroquin
Adriana da Cunha Rocha
Julie M. Schoenung
Andre Costa e Silva
Mary Wells
Wen Yang

Contents

About the Editors

Marc André Meyers is Distinguished Professor in the University of California, San Diego. This is the highest professorial level in the UC system and represents an honor that is reserved for only a small fraction of the tenured faculty. He has had visiting professorships at the University of Karlsruhe (Institute of Technology), University of Metz, and Cambridge University (Cavendish Laboratory). He is also a life member of Clare Hall, Cambridge. He is currently supported by the major U.S. funding organizations: National Science Foundation, the Office of Naval Research (MURI), Lawrence Livermore National Laboratory, University of California Office of the President, and DARPA.

Throughout his career, he received a number of important awards. The most prestigious of these, the Acta Materialia Materials and Society Award, which was bestowed in 2010, has a most distinguished list of recipients that includes global leaders in the materials science field. Other awards are from Europe (Humboldt Society Senior Scientist Award in Metal Physics, Germany, Heyn Medal, German Materials Society, J.S. Rinehart Award from the DYMAT Association, and *Materials Science and Engineering: A* Journal Prize), China (Lee Hsun Lecture Award from the Institute of Metal Research, Chinese Academy of Sciences) and U.S. (ASM Barrett Silver Medal, TMS-Educator Award, ASM Albert Sauveur Accomplishment Award, Albert White Educator Award, SMD/TMS Distinguished Scientist and Distinguished Service Awards). He is a Fellow of TMS, APS, and ASM International. In 2011 he was elected Corresponding Member of the Brazilian Academy of Sciences. Marc Meyers is the author of more than 400 papers, four technical books, as well as the editor of eight books.

Hector Alfredo Calderon Benavides is Professor of Applied Physics and Electron Microscopy at the Physics Department at ESFM-IPN (National Polytechnic Institute). He has a Ph.D. in Materials Science and Engineering from Northwestern University since 1985. He joined the Institut für Angewandte Physic from 1986 to 1994 at the ETH-Zürich. Since 1995 his work has been done at IPN in Mexico. His research focuses on theory and application of aberration-corrected and high resolution analytical electron microscopy, with particular emphasis on the application of aberration-corrected low dose electron microscopy. Dr. Calderon studies the structure and properties of materials, nanowires and nanoparticles as related to the application of nanomaterials to sustainable energy technologies. The main trust of his research is the development of devices to transform solar energy into a fuel. He has authored numerous publications on synthesis and characterization of materials and chapters of books.

Sonia P. Brühl has a doctoral degree in Physics from National University of Rosario, Argentina (1995). She is currently working as Full Professor at the National University of Technology (UTN), Faculty of Concepción del Uruguay, in Argentina. She has also been head of the Surface Engineering Group since 1999.

Dr. Brühl is an experienced researcher in the field of Plasma Surface Engineering and she has a vast teaching experience in physics and surface engineering, in both graduate and postgraduate engineering programs. Today, her R&D work is devoted to CVD and PVD coatings over stainless steel to achieve wear and corrosion resistance, and plasma diffusion treatments such ion nitriding, nitrocarburizing or carburizing on steels. She is actively involved in surface characterization, adhesion and wear tests, and corrosion resistance of modified layers and coatings. She is an active participant in national and international R&D programs, adviser in HR training for R&D, and also leader in program coordination. As such, she regularly attends international materials conferences and publishes in peer-reviewed journals, such as *Surface and Coatings Technology*, among others, where she has also taken part as invited editor. Dr. Brühl has experience in the organization of congresses and workshops and at present she is involved in the directive board of the Argentine Materials Association (SAM), as well as in the coordination of the doctoral program in materials science and engineering of UTN.

Henry A. Colorado is a full-time faculty member from the School of Engineering at the Universidad de Antioquia at Medellin, Colombia. He is a founder member and vice-president of the recently created Colombian Materials Society.

His research mainly includes new materials and composites fabricated with hazardous wastes, traditional ceramics and additive manufacturing for structural materials. Dr. Colorado is the director of CCComposites Lab at Universidad de Antioquia. Other area of research includes materials in arts, with several projects going on in collaboration between Engineering and the Arts Schools at Universidad de Antioquia.

Dr. Colorado obtained a B.S. and M.S. in Mechanical Science and Engineering from Universidad Nacional de Colombia (with honors) in 2005, and M.S. from the University of California, Los Angeles. He finished his Ph.D. in Materials Sciences at UCLA in June 2013. Other experiences include 5 months in Argonne National Lab working in hazardous wastes, 5 months in Centro Atomico Bariloche (Argentina), and training in nuclear materials at Idaho National Lab and Oak Ridge National Lab. Dr. Colorado has currently active collaboration with faculty from several schools abroad the world. He also is founder and a business partner of I+D Recycling Solutions, a start-up company located in Colombia for the processing of hazardous wastes.

Elvi Dalgaard holds a B.Eng. in Metallurgical Engineering from McGill University, as well as a Master's in Materials Engineering focusing on automotive alloy development and a Ph.D. in Materials Engineering with the research focused on the solid state welding of aerospace alloys, from the same institution. She has extensive experience in the characterization of advanced materials, including SEM, EDS, EBSD, TEM and XRD techniques. Her recent work includes initiating and managing research projects in the aerospace industry in the areas of metals, alloys, joining, thermoplastic composites, coatings, and advanced manufacturing including additive manufacturing. She is very active in Canadian and international professional societies, holding a board position in both ASM International (Montreal Chapter) and the Metallurgical Society of Canada (MetSoc). Dr. Dalgaard has a strong interest in encouraging young people, especially young women, in the scientific and engineering fields, and has participated in numerous mentoring and teaching activities such as being a mentor with the Chaire Marianne Mareschal (promotion du genie auprès des femmes), and organizing the ASM Materials Mini-Camp for high school students.

Carlos Nelson Elias

Roberto B. Figueiredo is Assistant Professor at the Federal University of Minas Gerais, in Brazil, where he graduated in mechanical engineering. He obtained his Ph.D. at the University of Southern California with a CAPES-Fulbright fellowship. He was a postdoctoral research associate at the University of Southampton and a postdoctoral visiting researcher at the University of Southern California. He joined the faculty of the Federal University of Minas Gerais in 2011. He has worked extensively on severe plastic deformation processing, both experimentally and using computer modeling, and on characterization of structure and mechanical properties of ultrafine-grained metallic materials. He received the NanoSPD Young Researcher Award in 2014.

Omar Garcia-Rincon earned a B.S. degree in Metallurgy (1997), M.Sc. in Materials Science (1999) both from Universidad Autonoma de Nuevo Leon (Mexico), and Ph.D. in Materials Science (2006) from the University of Sheffield (U.K.). Currently he is a product technologist in TERNIUM MEXICO SA de CV since 2008 where his work has focused on the optimization and design of steel products for different end use applications including automotive, appliances, construction, oil and gas among others. Main research focus in physical and mechanical metallurgy, phase transformation and relationships between microstructure and properties.

Megumi Kawasaki is Associate Professor in the Division of Materials Science & Engineering at Hanyang University, South Korea, where she joined the faculty as Assistant Professor in 2012. At present, she holds an adjunct research associate professorship in aerospace and mechanical engineering at USC.

Megumi obtained her B.Eng. degree in Metallurgy and Materials Science from Osaka Prefecture University in Japan in 2002 and then received an M.S. degree in 2004 and Ph.D. degree in 2007 in Materials Science at the University of Southern California, USA.

Her research interests lie in the area of the synthesis and characterizing unique properties of hybrid UFG metals and nanocomposites processed by HPT. She has collaborated actively with many researchers around the world and has published over 100 papers in peer-reviewed journals in the last 10 years. She has received an Early Career Award from the Korean Institute of Metals and Materials in April 2016.

Terence G. Langdon is Professor of Engineering Emeritus at the University of Southern California and Professor of Materials Science at the University of Southampton in the U.K. He graduated in physics from the University of Bristol, obtained his Ph.D. in Metallurgy from Imperial College of the University of London and received a D.Sc. in Physics from the University of Bristol. He is a fellow of the Royal Academy of Engineering, the European Academy of Sciences and Academia Europaea. He received honorary doctorates from the Russian Academy of Sciences in 2003 and Peter the Great Saint Petersburg Polytechnic University in 2016. He was awarded the Blaise Pascal Medal by the European Academy of Sciences in 2008, the Lee Hsun Award from the Chinese Academy of Sciences in 2009, the Honorary Medal "De Scientia et Humanitate Optime Meritis" from the Academy of Sciences of the Czech Republic in 2009, the Acta Materialia Gold Medal in 2012 and the State Prize in Science and Technology of the Republic of Bashkortostan (Russia) in 2015. He is a fellow of TMS, ASM, MRS, AAAS and the American Ceramic Society and an honorary member of the Japan Institute of Metals. He is listed on Web of Science with more than 1000 peer-reviewed papers, more than 40,000 citations, and an h-index of 103.

R. V. Mangalaraja is Professor in the Department of Materials Engineering, University of Concepción at Concepcion, Chile. He served as the director of the Department of Materials Engineering from 2010 to 2016. He is the current president of the Chilean Metallurgy and Materials Society (SOCHIMM). He is a life member of Indian Ceramic Society (ICS) and Indian Institute of Ceramics (IIC). Dr. Mangalaraja earned his B.Sc. degree in Physics from Madurai Kamaraj University in 1993 and received his M.Sc. degree in Materials Science in 1995 and M.Tech. in Ceramic Technology and Ph.D. in Technology in 2003 from Anna University-Chennai, India. He has postdoctoral experiences from AIST, Nagoya, Japan; Lulea University of Technology, Lulea, Sweden; and Anna University, Chennai, India. His primary research interests are focused in the areas of nanotechnology-driven advanced ceramics (functional ceramics for energy and environmental applications) and micro–nano-structure–property relationship of ceramics and metallic materials. He is the author or coauthor of more than 120 refereed journal publications, four book chapters, and more than 110 research presentations in several international conferences.

Mery Cecilia Gomez Marroquin

Adriana da Cunha Rocha obtained a Bachelor of Science degree in Materials Sciences and Metallurgical Engineering from the Pontificia Universidade Catolica do Rio de Janeiro (PUC-Rio) and both M.S. and Ph.D. degrees in Materials Sciences at the same university. During the years 2003–2005 she held a postdoctoral appointment at the Materials Sciences Division of the Lawrence Berkeley National Laboratory (LBNL) in Berkeley, California, working in the areas of high temperature corrosion and fuel cells. She is currently adjunct professor at the Federal University of Rio de Janeiro (COPPE/UFRJ) at the Metallurgical and Materials Engineering Program (PEMM) and a collaborator researcher of the Laboratory of Non-Destructive Testing, Welding and Corrosion (LNDC—COPPE/UFRJ), which is dedicated to the research of materials for the oil and gas industry.

Julie M. Schoenung is Professor of Materials Science and engineering at UC Irvine. She received her Ph.D. in 1987 and M.S. in 1985 in materials engineering from the Massachusetts Institute of Technology, and a B.S. in ceramic engineering from the University of Illinois, Urbana-Champaign. Dr. Schoenung was recently selected to be the recipient of the 2018 ASM Edward DeMille Campbell Memorial Lectureship and the 2016 *Acta Materialia* Holloman Award for Materials & Society. She was appointed as a co-editor-in-chief for the *Journal of Sustainable Metallurgy*, and has served for many years as a key reader for *Metallurgical and Materials Transactions A* (2003–present). In 2016, she was elected as a fellow of Alpha Sigma Mu, the materials honor society. In 2012, she was elected as an ASM Fellow and selected as a recipient of the Chime Bell Award, Hubei Province, China. In 2002, she became an AT&T industrial-ecology faculty fellow. Dr. Schoenung's research activities seek to provide fundamental insight into structure–processing–property mechanistic relationships in material systems for a variety of applications. Innovative synthesis and consolidation processes are combined to fabricate material systems that exhibit unique behavior, thereby providing new knowledge into the mechanisms that govern the observed behavior. Dr. Schoenung also has many years of experience in studying the materials-selection process in a variety of applications. She also conducts research into the analysis of factors that guide the materials-selection decision-making process, such as economics, environmental impact and toxicity, cost-performance trade-offs, and market potential. Dr. Schoenung uses tools and datasets from several disciplines, including management theory, process economics, life cycle assessment, and environmental economics in her research approach.

Andre Costa e Silva

Mary Wells is Associate Dean of Outreach and Professor with the Faculty of Engineering, University of Waterloo, and Professor of Materials in the Department of Mechanical and Mechatronics Engineering at the same university. Dr. Wells's research interests include mathematical modeling of industrial metallurgical operations including casting, extrusion, forging, and computation modeling of the interdependency between processing, microstructure, and properties.

Wen Yang obtained her Ph.D. from Northeastern University, China, in 2010, working in the Shenyang National Laboratory for Materials Science, which is operated by Institute of Metal Research of the Chinese Academy of Sciences. She was one of the first researchers studying the biological seashells in China.

From 2011 to 2013, Dr. Yang worked in the United States as a postdoctoral researcher on the mechanical behavior of biological materials in the group of Prof. Marc Meyers and Prof. Joanna McKittrick at UC San Diego, collaborating with Prof. Robert Ritchie at the Lawrence Berkeley National Laboratory and Markus J. Buehler at MIT. Since 2014, she has been a postdoctoral researcher in the Complex Materials group at ETH Zürich. Her interests include the characterization and mechanical behavior of biological and bio-inspired structural materials, collaborating with different groups and institutes including Max Plank Institute, Cambridge University, Imperial College, and institutes in the Chinese Academy of Sciences.

Since 2010, Dr. Yang has been a member of the Biomaterials Committee of TMS. She has co-organized two symposia for TMS (2016 and 2017) as well as helped organize the *6th International Conference on Mechanics of Biomaterials and Tissues* (*ICMOBT*) in 2015. Dr. Yang is on the editorial board of *Journal of Mechanical Behavior of Biomedical Materials* (*JMBBM*) and Austin Biomolecules. She is invited editor of two special issues of *JMBBM* and an independent reviewer for seven journals. She was evaluated as a valued reviewer in *Acta Biomaterialia* and *Materials Science and Engineering C*. She has published 25 papers with 436 citations and has an h-index of 12.

Part I
Advanced Biomaterials

Analysis of Biomimetic Surgical Clip Using Finite Element Modeling for Geometry Improvement and Biomaterials Selection

Thays Obando Brito, Bianca Bastos Dos Santos, Leonardo Sales Araújo, Luiz Henrique De Almeida and Marysilvia Ferreira Da Costa

Abstract An absorbable suture clip (MU9102934-1) for surgical applications was designed based on the bite mechanism of ant *Atta laevigata*. In order to emulate the behavior of the ant mandible, the clip was designed to naturally fall after some time, relieving the inconvenience of the clip removal process. The structure consists of a metallic handle and an absorbable polymeric tip. This study aims to optimize its geometry and select the best biomaterials to the handle structure, by analyzing its mechanical performance using the finite element method (FEM). The biomaterials selected for the simulations of the handle were AISI 316L and AISI 420 stainless steels. FEM analysis was performed using ANSYS FE software. The stress and strain distributions for each material and geometry changes were analyzed. From the analysis performed, the clip was optimized in order to be applied in a less traumatic form.

Keywords Surgical clip · Biomaterials · *Atta laevigata* · FEM

Introduction

Suture is used to help the healing process of the epithelial tissues, leading to a better and faster result, through the approximation of the edges of an injury, sealing blood vessels [1–3]. The approximation of the skin edges can be done by suture yarn, adhesive or metal clip [2]. Despite the constant improvement of the suture modes, they still have limitations concerning, due to the material used, the high cost, the shape, the placement and release mechanisms or the traumatic placing and healing process.

---

bolsista de IC (doutorado) da FAPEAM, 55 92 981864039

T.O. Brito (✉) · B.B. Dos Santos · L.S. Araújo · L.H. De Almeida · M.F. Da Costa
Programa de Pós-Graduação em Engenharia Metalúrgica e de Materiais,
PEMM-COPPE-UFRJ, CP 68505, Rio de Janeiro, RJ 21941-972, Brazil
e-mail: thaysdesigner@hotmail.com

© The Minerals, Metals & Materials Society 2017

M.A. Meyers et al. (eds.), *Proceedings of the 3rd Pan American Materials Congress*,
The Minerals, Metals & Materials Series, DOI 10.1007/978-3-319-52132-9_1

About 1000 years B.C. the use of jaws of ants for suture was reported in the Indian medical text *Charaka Samhita* [2]. The use of an ant jaw as a suture component permits the approximation of the edges and, during healing period, the jaw degrades and the remaining parts fall from the skin [4]. This technique is illustrated in Fig. 1.

According to the technique based on this bioinspired mechanism of suture using an ant mandible, a bioabsorbable surgical clip was developed based on the *Atta laevigata* bite mechanism, which resulted in technological patent type utility model [5], as shown in Fig. 2.

The innovative aspect of this clip is that there is no need for the patient to return to the hospital for the removal of the stitches, leading to a less traumatic and more comfortable technique for both the patient and the healthcare professional. This is because the product has a subsystem as an absorbable biomaterial. The absorbable surgical clip design is composed by the subsystems: handle structure, approach structure and cover, Fig. 2b. The handle structure is composed by a metallic material and a bioabsorbable polymer attachment structure. Figure 3 shows the main dimensions of the absorbable suture clip. The handle has to be capable of having a dimensional span to open and close the approach structure by means of the elastic properties of the metallic material, at the same time having high elastic modulus to provide enough for to permit the penetration of the approach structure in the skin.

Fig. 1 Suture performed with jaw ant *Atta*. Ant jaw approximating the tissue edges (**b**). Ant jaw as suture points (**c**). Adapted from [7]

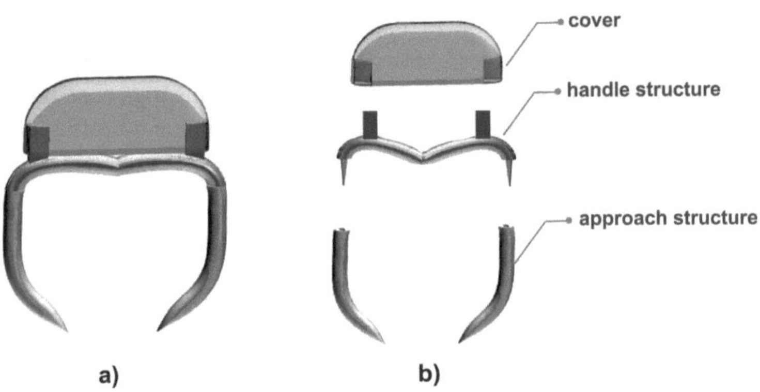

Fig. 2 Design of bioabsorbable suture clip (**a**) and their subsystems (**b**). Adapted from [8, 5]

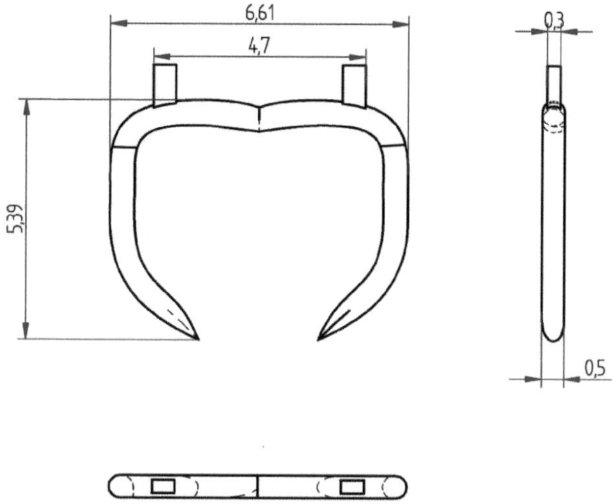

Fig. 3 The technical design of bioabsorbable suture clip. Adapted from [5]

The aim of this research is to optimize the geometry, dimensions and analysis the biomaterials, polymer and metal, for the surgical clip, estimating their mechanical performance using the Finite Element Method (FEM).

Materials and Methods

Two biomaterials, already used as medical applications, were selected for this analyses: stainless steel AISI 316L (austenitic) and AISI 420 (martensitic). The clip was modeled by Solid Edge software by two pieces of metal and polymer. The opening mechanism of the clip was simulated using the software ANSYS 17.2®. The element used is SOLID185 with 4-nodes tetrahedral configuration.

The primary analysis was geometric to prescribe displacement. After that, nonlinear quasi static analysis was made at the handle structure of the surgical clip to verify the stress and strain distribution in the surgical clip. According to these, the design of clip and selection of the best biomaterial for the perfect functioning of that can be optimized.

Results and Discussion

The geometric analysis was performed on the handle structure and the approach structure to prescribe large displacement. An angle of 15° was set from the calculation of height of the handle structure and the clip design of the plane of symmetry, defining the opening of the clip. It results in the displacement of 0.05 mm for X and Y directions, equivalent to a force of 30 N. Figure 4 shows the 15° angle and the resultant opening span of the clip.

After defining the opening span for the clip, FE simulations were performed using the selected metallic materials for the handle (AISI 316L and AISI 420). An elastic modulus of 200 GPa was set for AISI 420 and 192 GPa [6] was used for AISI 316L. The Poisson's ratio used for both steels was 0.3. Figure 5 shows the location of prescribed offset of 0.05 mm and the stress distribution in the handle structure of AISI 316L stainless steel material.

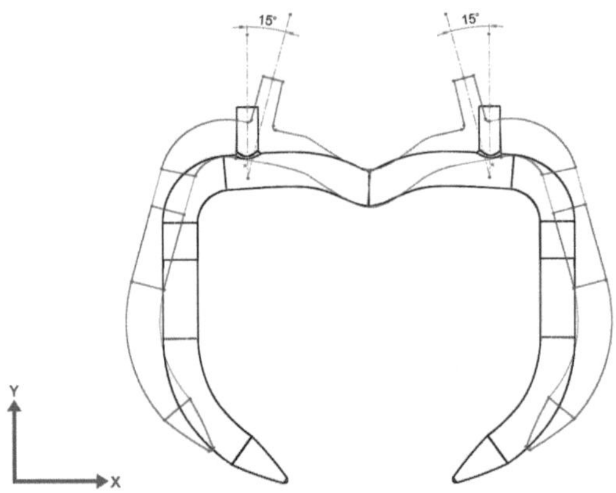

Fig. 4 Schematic of the clip opening geometry with a 15° angle

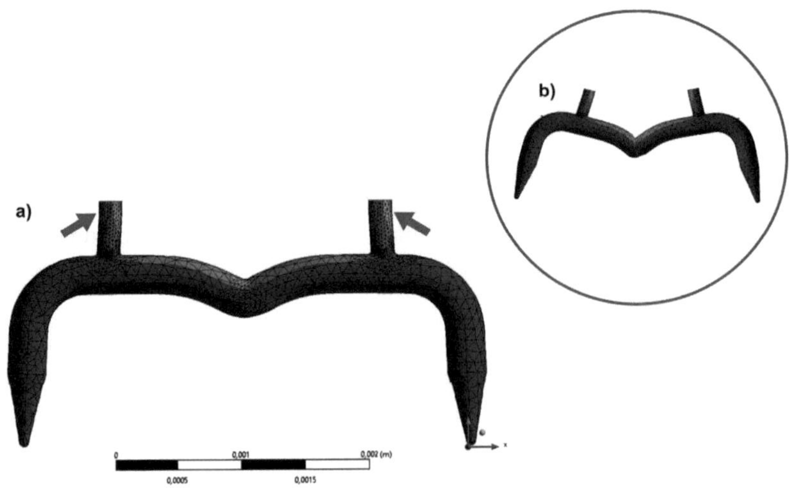

Fig. 5 Identification of boundary conditions (prescribe displacement) on the of handle structure (**a**) and open clip (b)

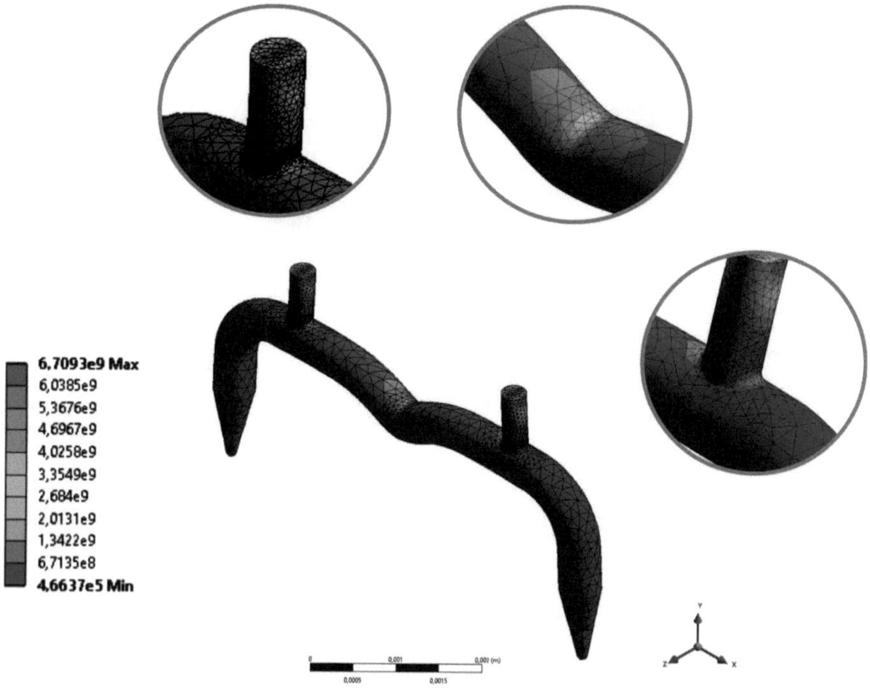

Fig. 6 Identification of stress concentration on the handle structure of the AISI 316L surgical clip

For the design considered, the more intense stress concentrations were located at point of contact between the opening tool (at the upper part of the handle) and at the joint responsible for the elastic opening of the handle. The stress concentration at these points are far above the yield strength and ultimate tensile strength of both steels. For AISI 316L the relevant stress concentration is identified as shown in Fig. 6, approaching 3 GPa. This indicates that plastic deformation and possible rupture of the material will occur, especially at the joint, where there is an angle of 110°.

It has also been established the same displacement to the handle opening of stainless steel AISI 420. Figure 7 shows the clip simulation AISI 420 with the tension distribution to identifying regions of clip handle structure.

The stress concentration in the steel AISI 420 handle structure was also predominant in the central region, as identified in Fig. 7, reaching a pressure of about 3GPa. This also far exceeds its strength, equivalent to 967 MPa [6], indicating that fracture should occur in this region.

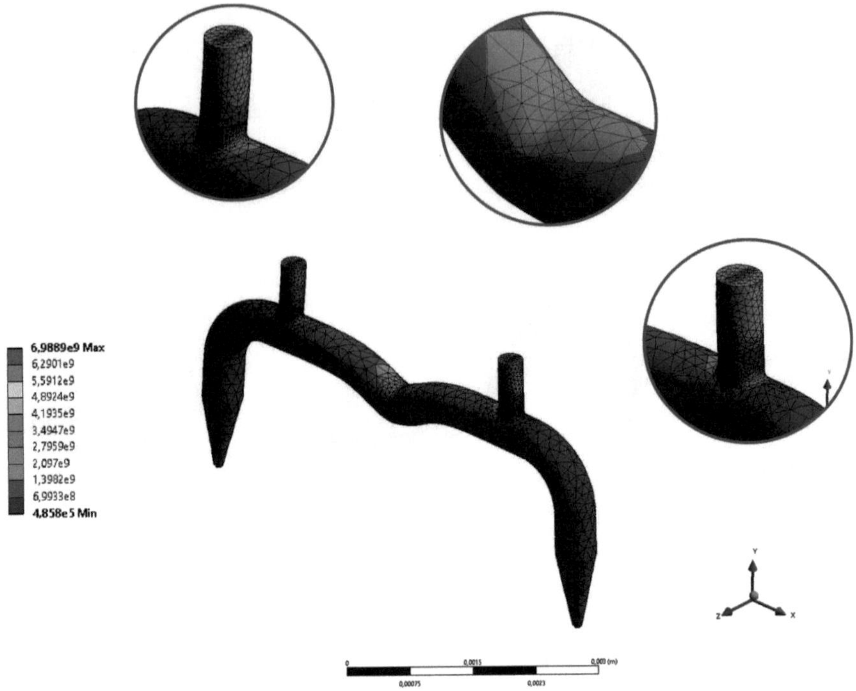

Fig. 7 Identification of stress concentration on the handle structure of the surgical clip of stainless steel AISI 420

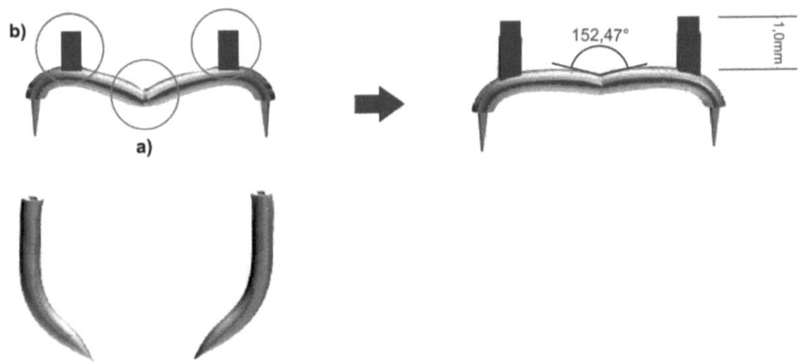

Fig. 8 Identification of regions of the handle structure that need to be optimized based on the analysis in ANSYS. Adjust at the angle of the center (**a**) and increase of the levers (**b**) in the handle structure of the clip

Conclusion

For the specific design initially defined, it is observed that, the stress concentrations on specific points were far above the strength of the metallic biomaterials selected. This indicates that the geometry of the clip must be optimized, as shown in Fig. 8, since the clip may not have any plastic deformation, relying solely on a efficient spring effect of the handle of the surgical clip structure.

Another alternative is to select ferritic and martensitic stainless steels, already in use in the medical field, to obtain an efficient spring effect in the handle structure of the surgical clip.

Acknowledgements The authors would like to acknowledge The Center for Nonlinear Mechanics (MECANON) from Federal University of Rio de Janeiro, as well as the National Council for Scientific and Technology Development (CNPq) and Foundation support of the State of Amazonas Research (FAPEAM) for the financial support.

References

1. Goffi, F. S. (2004). *Técnica Cirúrgica: bases anatômicas, fisiopatológicas e técnicas cirúrgicas* (4ª ed). São Paulo: Atheneu.
2. Hering, F., Gabor, S., & Rosenberg, D. (1993). *Bases técnicas e teóricas de fios e suturas*. Roca: São Paulo.
3. Magalhães, H. P. (1989). *Técnica cirúrgica e cirurgia experimental*. Ed. Sarvier, São Paulo.
4. Mota, P. (2008, July). *A latitude do Olhar: Expedição Amazônica* (Revista Perspectiva, Matosinhos - PT), n. 13, pp. 6–7.
5. INPI. (2012). Instituto Nacional de Pesquisa da Amazônia (Manaus, AM). *Grampo bioabsorvível*. Brasil patente BR n. MU9102934-1.
6. Handbook. (1990). *Properties and selection*. Ohio, ASM International. 10th ed., V. 2.
7. Brito, T. O. (2014). *Biônica aplicada no estudo da mandíbula da formiga Atta laevigata (f. smith 1858) para desenvolvimento de grampos de sutura na área médica*. Dissertação, Universidade Federal do Rio Grande do Sul, Porto Alegre.
8. Brito, T. O. (2009). *Sistema de sutura a partir da técnica biônica*. Monografia, Universidade Federal do Amazonas, Manaus.

Chemical Composition Effect of Sol-Gel Derived Bioactive Glass Over Bioactivity Behavior

L.A. Quintero and D.M. Escobar

Abstract Bioactive glasses (BG) are a group of inorganic materials widely used in Bone Tissue Engineering (BTE). These biomaterials react with body fluids resulting in the formation of bone like apatite layer. In this study, sol-gel derived bioactive glass was synthesized in the SiO_2-CaO-P_2O_5 system according to augmented constrained mixture experimental design, with percentage restrictions for each oxide as follows: $58 \leq SiO_2 \leq 70$; $6 \leq P_2O_5 \leq 9$ and $24 \leq CaO \leq 34$. BG were conformed into short-bulk cylinders and immersed in Simulated Body Fluid (SBF) solution for 7 and 14 days in order to carry out bioactivity tests. Apatite layer formation was confirmed by Scanning Electron Microscopy (SEM) and Energy Dispersive X-ray Analysis (EDX). The results showed apatite layer formation depended on BG chemical composition proved with *p*-values from ANOVA analysis below 0.05 indicating factors significance over the response. The formed apatite layer presented a Ca/P ratio similar to bone apatite, this result is appropriate for biomaterials used in BTE.

Keywords Bioactive glasses · Apatite layer · Bone tissue engineering · SBF

Introduction

Bioactive glasses (BG) are a group of bio-ceramics based on silicon oxide which have been used due to their excellent bioactivity that increases apatite layer formation allowing a suitable and safe chemical bond between the material and living bone [1]. The formed apatite layer on the surface of BG in presence of simulated body fluid (SBF) is similar to that of the mineral phase present in the bone [2]. It has been proved that BG dissolution products could stimulate cell proliferation and

L.A. Quintero (✉) · D.M. Escobar
Biomaterials Research Group, Engineering Faculty,
University of Antioquia, St 70 N° 52-21, Medellin, Colombia
e-mail: lindsey.alejandra@gmail.com

© The Minerals, Metals & Materials Society 2017
M.A. Meyers et al. (eds.), *Proceedings of the 3rd Pan American Materials Congress*,
The Minerals, Metals & Materials Series, DOI 10.1007/978-3-319-52132-9_2

differentiation that eventually can promote new bone formation and, according to silicon oxide quantity, bone formation is faster with a BG implant than synthetic hydroxyapatite [3, 4].

Most used BG is 45S5 Bioglass® first reported by Larry Hench [5] and obtained through melting-quenching process. However, this process requires higher temperatures and limits porosity, particle size and specific surface of final powder [6]. An alternative synthesis route is sol-gel process, a technique used since early 90s. Sol-gel synthesis provides powder with lower crystallinity, different morphologies and is more versatile than melt method [2].

Within the glass and glass-ceramic systems, SiO_2-CaO-P_2O_5 has been considered as a base system to include new formulations and molar proportions of oxides [7]. These proportions have been conventionally defined through laborious practices as trial-and-error methods, in which many resources are expended. Therefore, the use of more sophisticated statistical methodologies as Mixture Experimental Designs (MED) could be a useful tool to determinate a suitable bioactivity behavior from the control of main oxides for the BG synthesis. Statistical softwares provide the experimental region conformed to provided restrictions consistent with constrained mixture design (CMD) theoretical equations [8].

Bioactivity behavior, it can be measured through bioactive tests according to reported by Kokubo et al. [9] by submerging the material in SBF. BG in presence of SBF ions can form a bone like apatite layer, especially those obtained by sol-gel techniques due to OH^- groups on their surface, which are able to induce Hydroxyapatite nucleation [2]. There are diverse chemical and physical procedures to evaluate the formed apatite layer, such as Fourier Transformed InfraRed (FTIR), X-ray Diffraction (XRD), Scanning Electron Microscopy (SEM), Energy Dispersive X-ray (EDX), among others.

In this study, BG was synthetized according to an augmented constrained mixture design to establish the best conditions in a highly bioactive BG for BTE applications. The percentage of formed apatite layer was measured through SEM analysis and the stoichiometric ratios of that layer, namely Ca/P ratio near to natural ratio, was measured with EDX several times in similar places to determine a point of comparison.

Materials and Methods

Materials

Tetraethylorthosilicate (TEOS: $C_8H_{20}O_4Si$) and Calcium Acetate (CaAc: Ca $(CH_3COO)_2$) were purchased from Merck Inc; Ammonium Dihydrogen Phosphate (ADP: $NH_4H_2PO_4$) was purchased from Carlo Erba Reagent; Ethanol was purchased from Panreac. All chemicals from SBF preparation were provided from Sigma-Aldrich, Germany.

Mixture Experimental Design

In this study an analysis of variance (ANOVA) based on an augmented CMD was used to evaluate the effect of molecular fraction of SiO_2 (X1), P_2O_5 (X2) and CaO (X3) on quantity and quality of formed apatite layer. Based on CMD, the samples were prepared by mixing the three components at different ratios with the summation of the proportions (X1 + X2 + X3) one. The restrictions for each oxide were selected according to percentages reported by different authors [7, 10] ($58 \leq SiO_2 \leq 70$; $6 \leq P_2O_5 \leq 9$ and $24 \leq CaO \leq 34$). Samples codes and molecular fraction of each experimental point as 10 combinations are showed in Table 1. Experimental points of compositions were defined through R software with mixexp package, which allows to determine the experimental region showed in Fig. 1. The experiments were performed in triplicates of the mixtures, for a total of 30 samples evaluated.

The selected model for ANOVA test was a special cubic model presented in Eq. 1, where *Y* is the response, βi are the constants for each interaction and *Xi* are the variables

$$Y = \beta 1X1 + \beta 2X2 + \beta 3X3 + \beta 12X1X2 + \beta 13X1X3 + \beta 23X2X3 + \beta 123X1X2X3 \quad (1)$$

Fisher's LEAST SIGNIFICANCE DIFFERENCE (LSD) test was performed to determine significant effects of the variables. The chosen significance level was 5% which correspond *p*-value = 0.05. The *p*-values below 0.05 indicated the statistical significance of the factors.

Table 1 Molecular fraction of BG components based on CMD

Sample code	Molecular fraction		
	SiO_2	P_2O_5	CaO
BG1	0.67	0.09	0.24
BG2	0.68	0.08	0.24
BG3	0.58	0.09	0.33
BG4	0.64	0.07	0.29
BG5	0.58	0.08	0.34
BG6	0.70	0.06	0.24
BG7	0.60	0.06	0.34
BG8	0.63	0.09	0.28
BG9	0.59	0.07	0.34
BG10	0.65	0.06	0.29

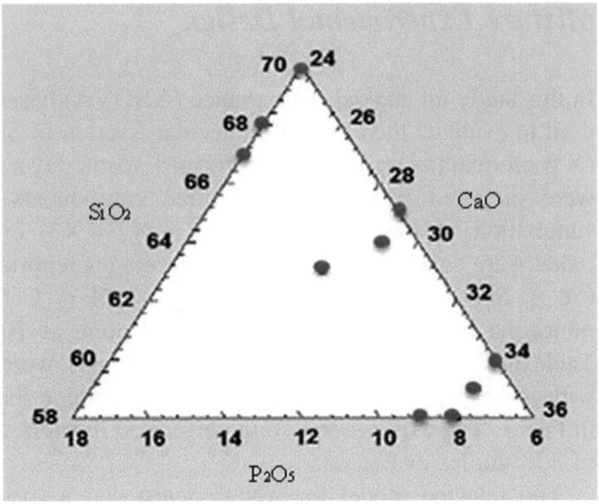

Fig. 1 Experimental region of CMD obtained with R software

Bioactive Glass Synthesis

Sol-gel derived BG was synthesized by dissolving TEOS into an ethanol/distilled water solution, with a molar ratio 1:4 both TEOS/distilled water and TEOS/ethanol according to Vaid et al. [10]. The mixture was left in agitation. ADP and ethanol/distilled water solution with a molar ratio 1:4 both ADP/distilled water and ADP/ethanol were added after 1 h to react completely in stirring for another 45 min. Subsequently, calcium acetate was added to react for 11 min. Finally, an acetic acid/distilled water solution (6:1) was added into mixture (distilled water/BG 1:4). The final solution was kept in agitation until gel point.

BG gel was kept in a container for 3 days at room temperature. The gel was heated at 120 °C for 2 days to remove all water content. Followed by a final mash to obtain a dry powder.

Sample Characterization

Every synthesized BG was analyzed by X-ray diffraction (XRD) with XPert PANalytical Empyrean Series II diffractometer after thermal treatment at a final temperature of 1050 °C. This instrument worked with voltage and current settings of 45 kV and 40 mA respectively and used Cu-Kα radiation (1.5405980 Å).

In Vitro Bioactivity Study

The SBF solution was prepared according to Kokubo et al. [9]. BG powder was uniaxially pressed at 30 MPa for 2 min in order to obtain cylindrical disks. The cylinders were immersed in the SBF solution and incubated at 37 °C in close tubes for 7 and 14 days. Afterwards, the cylinders were removed from the SBF solution and washed with distilled water to removed undissolved salts. Finally they were analyzed by Scanning Electron Microscopy (SEM) using a microscope JOEL-JSM 6490 LV that operated at the acceleration voltage of 20 kV. Energy dispersive X-ray (EDX) was performed on the same equipment.

Results and Discussion

XRD Analysis of the Synthesized BG

Figure 2 presents XRD results for all 10 different BG with narrow and differentiable peaks due to augmented crystallinity with thermal treatment. Figure 2a shows from BG1 to BG5 and Fig. 2b from BG6 to BG10. Primary and secondary peaks in all synthesized BG present consistencies regarding angle and intensity according with reported for comparable systems [4, 11]. Peaks at $2\theta = 30.9$ and 21.3 are primary and secondary peaks for SiO_2, respectively. Peaks at $2\theta = 34.4$ and 31.3 are primary and secondary for CaO; and peaks at $2\theta = 27.7$ and 25.7 are primary and secondary for P_2O_5. Those peaks are present in all DRX spectrum; however, BG1, BG2, BG3, BG4, BG8 and BG10 spectra show some other peaks (more than two) possibly due to the formation of undesirable phases, as for BG5, BG6 and BG9 spectra have one or two peaks of other phases, possible pseudo-wollastonites which tend to form with same oxides. BG7 shows no undesirable peaks.

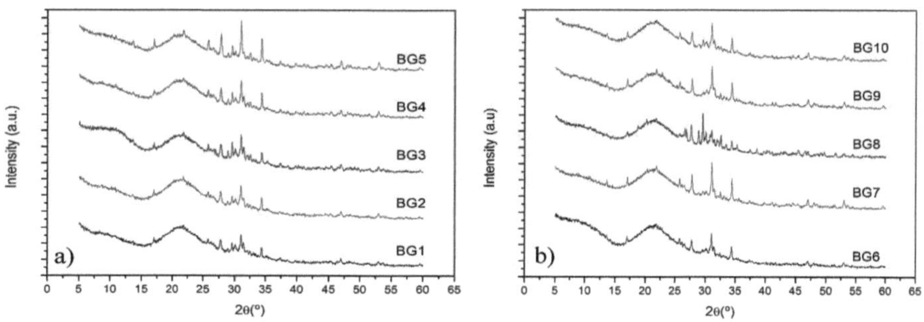

Fig. 2 XRD patterns of synthesized BG after 1050 °C treatment. **a** BG1 to BG5; **b** BG6 to BG10

Sample Characterization After In Vitro Tests

Apatite layer formed on the surface of BG cylinders after in vitro tests were analyzed with SEM images and EDX analysis both for 7 and 14 days of study. Figure 3 shows SEM micrographs for each BG, it can be seen that for some of the cylinders the formed apatite layer on their surface decreased with time, which means that the layer is unstable and not desirable for BTE.

On the other hand, the formed layer exhibits a small morphology size as shown in Fig. 4 for the apatite layer on the surface of BG7 at 14 days of study. A small particle size is advantageous since the natural apatite is micro and nano-size [12, 13].

Figures 5 and 6 summarizes Ca/P ratio results and apatite layer percentage (ALP) for synthesized BG after in vitro test. ALP was obtained through image analysis with ImageJ software. Data were collected in triplicate and those results were used to perform statistical analysis with R software.

Statistical results showed that the selected model fits the data (R^2 = 0.98), this means that it is possible to analyze the behavior of apatite layer percentage knowing the BG composition.

The Ca/P ratio results showed that according to BG compositions, the formed apatite layer can be similar of natural hydroxyapatite, which varies between 1.37 and 1.87 [13] indicating the appropriate results for BG5 and BG7 obtained after 14 days of SBF immersion. From Fig. 6 is observed two different tendencies for

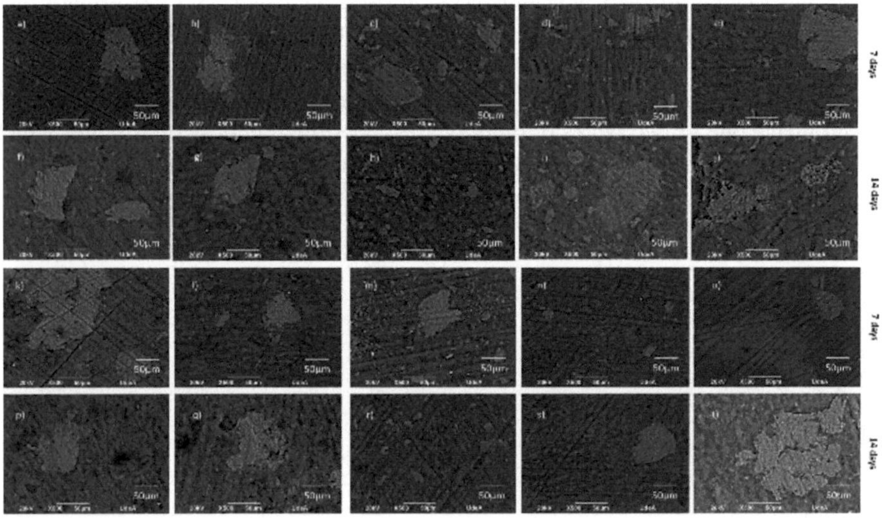

Fig. 3 SEM micrographs of synthesized BG after SBF immersion. **a** and **f** are SEM images for BG1 after 7 (**a**) and 14 (**f**) of immersion; **b** and **g** are for BG2; **c** and **h** are for BG3; **d** and **i** are for BG4; **e** and **j** are for BG5; **k** and **p** are for BG6; **l** and **q** are for BG7; **m** and **r** are for BG8; **n** and **s** are for BG9; **o** and **t** are for BG10

Fig. 4 SEM micrograph of formed apatite layer for BG7 after 14 days of immersion in SBF

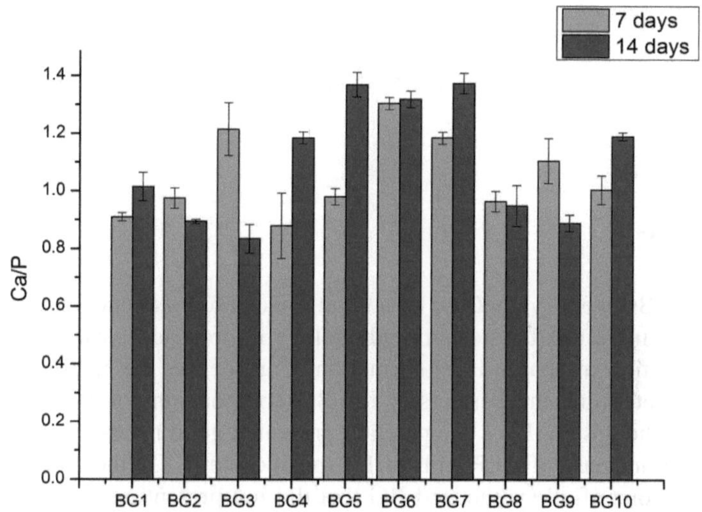

Fig. 5 Ca/P ratio of the apatite formed for each BG after in vitro test at 7 and 14 days

ALP at 7 and 14 days. Some of them (BG1, BG2, BG3, BG5 and BG6) decreased ALP with SBF exposition time, whereas the opposite behavior was exhibited for the remains BG (BG4, BG7, BG8, BG9 and BG10). These trends are attributed to the different silicon composition employed in each synthetized BG. From CMD, the obtained p-value <0.05 indicate that silicon represents the highest significance variable on response (growing apatite layer), which is in agreement with previous reported by Hench [5, 14]. That means, large concentrations of silicon reduces any kind of tissue bonding, whereas lower proportions the bonding is promoted highly with soft tissues [5]. In this way, the growing of apatite layer is observed to middle

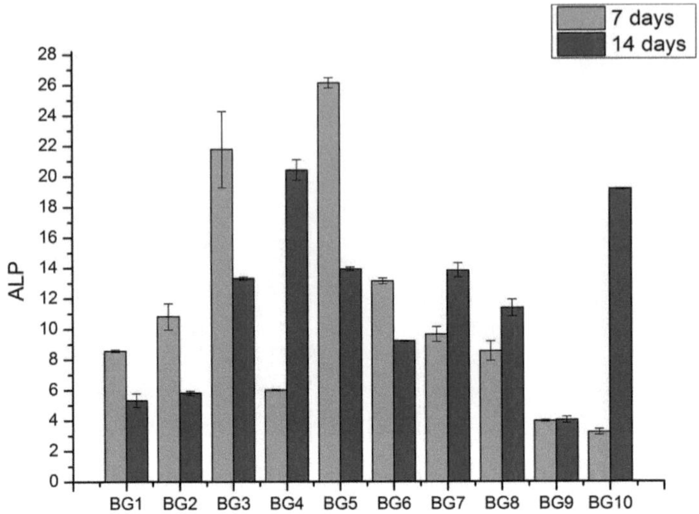

Fig. 6 Apatite layer percentage of the apatite formed for each BG after in vitro test at 7 and 14 days

combinations of CMD and reduction of this layer for boundary of experimental design.

Conclusions

In this study, BG were synthetized by sol-gel route at room temperature in a ternary system following a mixture design methodology in order to investigate the role of oxides proportion in apatite layer formation. Considering the results, it has been possible to determine the best result for BG composition for BTE applications according to the design. BG7 composition presented good behavior during in vitro tests with the increase of ALP with SBF exposition time. On the other hand, BG7 was the only one of all synthetized BG that did not present undesirable peaks for some other phases like pseudo-wollastonites on XRD analysis. This confirms that BG7 has the BG typical phase on its structure.

CMD was a useful tool to determine suitable conditions for BG synthesis in order to develop a bioceramic for BTE application with beneficial performance to promote bone like apatite layer formation.

Acknowledgements The authors are thankful with Biomaterials Research Group for providing the necessary reagents and studies during the development of this project.

Conflict of Interest The authors declare that they have no conflict of interest.

References

1. Peter, M., Binulal, N. S., Nair, S. V., Selvamurugan, N., Tamura, H., & Jayakumar, R. (2010). Novel biodegradable chitosan–gelatin/nano-bioactive glass ceramic composite scaffolds for alveolar bone tissue engineering. *Chemical Engineering Journal, 158*(2), 353–361. doi:10.1016/j.cej.2010.02.003
2. Catauro, M., Bollino, F., Renella, R. A., & Papale, F. (2015). Sol-gel synthesis of SiO_2-CaO-P2O_5 glasses: Influence of the heat treatment on their bioactivity and biocompatibility. *Ceramic International, 41*(10), 12578–12588. doi:10.1016/j.ceramint.2015.06.075
3. Jones, J. R. (2013). Review of bioactive glass: From Hench to hybrids. *Acta Biomaterialia, 9* (1), 4457–4486. doi:10.1016/j.actbio.2012.08.023
4. Dziadek, M., Zagrajczuk, B., Menaszek, E., Wegrzynowicz, A., Pawlik, J., & Cholewa-Kowalska, K. (2016). Gel-derived SiO_2–CaO–P2O_5 bioactive glasses and glass-ceramics modified by SrO addition. *Ceramic International, 42*(5), 5842–5857. doi:10.1016/j.ceramint.2015.12.128
5. Hench, L. L. (2006). The story of bioglass. *Journal of Materials Science: Materials in Medicine, 17*(11), 967–978. doi:10.1007/s10856-006-0432-z
6. Faure, J., Drevet, R., Lemelle, A., Ben Jaber, N., Tara, A., El Btaouri, H., et al. (2015). A new sol-gel synthesis of 45S5 bioactive glass using an organic acid as catalyst. *Materials Science and Engineering C, 47*, 407–412. doi:10.1016/j.msec.2014.11.045
7. Siqueira, R. L., & Zanotto, E. D. (2013). The influence of phosphorus precursors on the synthesis and bioactivity of SiO_2-CaO-P2O_5 sol-gel glasses and glass-ceramics. *Journal of Materials Science: Materials in Medicine, 24*(2), 365–379. doi:10.1007/s10856-012-4797-x
8. Cornell, J. A. (2002). *Experiments with mixtures: Designs, models, and the analysis of mixture data* (3rd ed.). Hoboken, NJ, USA: Wiley.
9. Kokubo, T., & Takadama, H. (2006). How useful is SBF in predicting in vivo bone bioactivity? *Biomaterials, 27*(15), 2907–2915. doi:10.1016/j.biomaterials.2006.01.017
10. Vaid, C., & Murugavel, S. (2013). Alkali oxide containing mesoporous bioactive glasses: Synthesis, characterization and in vitro bioactivity. *Materials Science and Engineering C, 33* (2), 959–968. doi:10.1016/j.msec.2012.11.028
11. Desogus, L., Cuccu, A., Montinaro, S., Orrù, R., Cao, G., Bellucci, D., et al. (2015). Classical Bioglass® and innovative CaO-rich bioglass powders processed by Spark Plasma Sintering: A comparative study. *Journal of the European Ceramic Society, 35*(15), 4277–4285. doi:10.1016/j.jeurceramsoc.2015.07.023
12. Saltzman, W. M. (2009). Biomechanics. In *Biomedical engineering. Bridging medicine and technology* (p. 656). Cambridge University Press.
13. Wu, S., Liu, X., Yeung, K. W. K., Liu, C., & Yang, X. (2014). Biomimetic porous scaffolds for bone tissue engineering. *Materials Science and Engineering R: Reports, 80*(1), 1–36. doi:10.1016/j.mser.2014.04.001
14. Hench, L. L., Splinter, R. J., Allen, W. C., & Greenlee, T. K. (1971). Bonding mechanisms at the interface of ceramic prosthetic materials. *Journal of Biomedical Materials Research, 5*(6), 117–141. doi:10.1002/jbm.820050611

Effects of *Dialium guineense* Based Zinc Nanoparticle Material on the Inhibition of Microbes Inducing Microbiologically Influenced Corrosion

Joshua Olusegun Okeniyi, Gbadebo Samuel John, Taiwo Felicia Owoeye, Elizabeth Toyin Okeniyi, Deborah Kehinde Akinlabu, Olugbenga Samson Taiwo, Olufisayo Adebola Awotoye, Ojo Joseph Ige and Yemisi Dorcas Obafemi

Abstract This paper investigates the effects of *Dialium guineense* based zinc nanoparticle material on the inhibition of microbes inducing microbiologically influenced corrosion (MIC) in metals. Extract of leaf from the natural plant were used as precursor for zinc nanoparticle material, which was characterized by scanning electron microscopy and energy dispersive spectroscopy (SEM + EDS) instrument. Sensitivity of the developed zinc bio-nanoparticle material from this on different strains of microbes that are known to induce microbiologically influenced corrosion, in metallic materials, was then studied and compared with that obtained from a commercial antibiotic employed as control. Results showed that the bio-material capped nanoparticle exhibited inhibited growth of the studied different MIC inducing microbes. Zones of inhibition, the sensitivity measure of the biosynthesized material against the microbial strains either surpassed or compared well with the zones of inhibition from the commercial antibiotic (control). These

J.O. Okeniyi (✉) · G.S. John
Mechanical Engineering Department, Covenant University, Ota 112001, Nigeria
e-mail: joshua.okeniyi@covenantuniversity.edu.ng

T.F. Owoeye
Department of Chemistry, Covenant University, Ota 112001, Nigeria

E.T. Okeniyi
Petroleum Engineering Department, Covenant University, Ota 112001, Nigeria

D.K. Akinlabu
Biochemistry Programme, Department of Biological Sciences, Covenant University, Ota 112001, Nigeria

O.S. Taiwo · O.A. Awotoye · O.J. Ige · Y.D. Obafemi
Applied Biology and Biotechnology Unit, Department of Biological Sciences, Covenant University, Ota 112001, Nigeria

© The Minerals, Metals & Materials Society 2017
M.A. Meyers et al. (eds.), *Proceedings of the 3rd Pan American Materials Congress*, The Minerals, Metals & Materials Series, DOI 10.1007/978-3-319-52132-9_3

results engender implication on the prospects of the zinc bio-nanoparticle usages in corrosion inhibition and protection system for metals in microbial corrosion influencing environment.

Keywords Biomaterial-based nanoparticle · *Dialium guineense* leaf-extract · Microbiologically-influenced-corrosion inducing microbes · Microbial-growth inhibition · Material characterisation

Introduction

Activities of microbes on industrial installation facilities are largely traceable to metallic corrosion attacks in industrial systems, including chemical processing, oil and gas exploration and production as well as power generation industries [1–3]. Usually, microbes that could be found on metal surfaces may not directly cause a unique form of metallic corrosion [1]. However, the physical presence of their colony, in physiological types that may include sulphate reducing/sulphur oxidising bacteria, Methanogen, nitrate reducing bacteria and Enterobacteria, can promote and accelerate corrosion processes through by-products of their metabolic activities [2, 3]. For instance, interactions of microbes with corrosion products would prevent the normal characteristics of corrosion products to form films that could have prevented further metallic corrosion attacks. Also, further acceleration of corrosion reactions could be enhanced through microbial induced reduction reaction that facilitates necessary conditions for initiation and continuance of corrosion forms such as pitting, selective de-alloying, under-deposit, crevice, galvanic and erosion corrosion attacks. Any of these forms of microbiologically influenced corrosion (MIC) attacks degrades metallic materials in industrial installations by converting the metals into oxides, salts or other compounds of much lesser structural integrity [1]. Treatments for avoiding these bio-corrosion induced failures gulp millions of dollars annually, while cost of other related crises could range into billions of dollars, in many countries [4]. In physical terms the cost of corrosion increases annually, thus necessitating cost-saving corrosion control practices that will be attended with individual safety and other environmental benign consequences [5, 6].

Methods for mitigating MIC degradation of metals employ strategies that include physical control, chemical control, bioelectric antimicrobial effect, cathodic protection and biological control applications [7, 8]. Chemical control method takes the form of using oxidising and non-oxidising biocides for bio-corrosion inhibition [1, 8]. However, many of the known biocides exhibits specific activities of effectiveness on selected types of bio-corrosion inducing microbes while they may not be effective against other types [8], thus necessitating needs for more studies on applicable means of curbing activities, and consequently bio-corrosion effects, of MIC inducing microbes [6].

Adequate understanding of knowledge involving not only metallurgical and electrochemical but also microbiological and chemical interactions on micro-organism induced corrosion has been proposed for tackling difficulties of MIC processes [8]. For instance, while chemical types of biocides and the types of microbes against which they are effective were reported in [8], other non-corrosion related studies have also shown that extracts from natural plants exhibit antimicrobial growth effects [9–12]. These anticorrosion effects of plant extracts are well corroborated by several experimental studies on metallic corrosion protection in microbial simulating environments where extracts from natural plants have exhibited corrosion inhibition effectiveness [13–22]. Although, a recent study had deliberated on the antimicrobial effect of plant extract capped silver nanoparticle [23], there is still dearth of studies on the effect of biosynthesized nanoparticle materials on microbes inducing MIC. This is in spite of reported works that have shown that physico-chemically synthesized metallic nanoparticle [24] and micro-organism synthesized (i.e. biosynthesized) nanoparticle [25] inhibited growth of microbes that are known to induce MIC. More especially, motivation for this study was drawn from the report in [26], wherein bactericidal activity, against clinical pathogens, by the stem extract from *Dialium guineense* (*D. guineense*) Leguminosae was investigated, with positive results. From the reported studies, it could be observed that some of the pathogens from antimicrobial studies using plant extracts [9–12, 26] or using plant extract capped silver nanoparticle materials [23] find similarities with microbes that were identified from biofilms of MIC deterioration of metals in industrial facilities [1–3]. These include microbial strains such as *Escherichia coli*, *Micrococcus varians*, *Serratia* spp., *Staphylococcus aureus* and *Pseudomonas aeruginosa*.

Apart from these materials, zinc based nanoparticle material have also been identified with very useful and distinct antimicrobial activity, which have made this metallic nanoparticle material find usefulness in biocides coatings in water treatment, paint and cosmetic applications [27]. Details from studies indicate that Zn based nanoparticle combines the special advantages of exhibiting minimal effect on human cells with strong toxicity against wide range of micro-organism, including those identified in biofilms inducing MIC, excellent stability and long shelf life [27–30]. These biocidal effectiveness of nanoparticle materials, even against microbial strains that could be resistant to other antibiotic materials, is due to combination of their small size and consequent high surface-to-volume ratio that facilitate intimate interaction with the membranes of the microbial organisms [31]. However, biosynthesis of nanoparticle materials, using plant extract precursors as reducing agents, is attracting research interest due to its economic and eco-friendly advantages over the chemical and physical methods of nanoparticle synthesis [31, 32]. While studies have employed extracts from other plants for the biosynthesis of Zn based nanoparticle with attendant antimicrobial effects [31, 32], there is dearth of such study on *Dialium guineense* mediated Zn nanoparticle material. It is due to these considerations that the objective of this study was to investigate the effects of zinc nanoparticle that was biosynthesized from the use of *Dialium guineense* (*D. guineense*) leaf-extract as precursor on the inhibition of the growth of microbes known to induce microbiologically influenced corrosion (MIC).

Experimental Methods

Extraction from Dialium guineense *Leaf Material*

By following procedures described in studies [21, 33], fresh leaves of *D. guineense* was collected, dried under cover maintained at 20 °C and blended into powder. The blended leaf powder was then wrapped in Whatman filter paper and placed in a condenser equipped soxhlet extractor having CH_3OH (methanol) as the solvent. The *D. guineense* extract solution obtained from this processing was then concentrated over water bath to a pasty remainder of the extraction.

Biosynthesis of D. guineense *Based Nanoparticle Material*

From the extract paste of the *D. guineense* plant material, 25 mg was dissolved and made up to a volume of 1000 ml using 1 M $Z_2SO_4{\cdot}7H_2O$ (Zinc sulphate heptahydrate), which was obtained from Sigma Aldrich®. Sample was taken from this after 48 h and was centrifuged, using Laboratory Centrifuge, Model SM-80-2 obtained from Surgifield (England), at 3500 rpm for 15 min. The supernatant was then decanted and the biosynthesized nanoparticle was transferred to a watch glass where it was air dried before being collected in Eppendorf tubes and stored at room temperature for further use as the biosynthesized *D. guineense* based zinc nanoparticle in the study.

Characterisation of the Biosynthesized Zinc Nanoparticle

For characterisation of the biosynthesized zinc nanoparticle material, selected sample of it was placed on a sample stub. This sample stub was then loaded onto the stage of a Quorum Sputter Coater, Model Q150R ES obtained from Quorum Technologies Limited® (England), on which the specimen of biosynthesized zinc nanoparticle material was coated with gold. The sample stub was then removed and placed on standard sample holder that was mounted onto a Pro X PHENOM™, Model 800-07334 obtained from Phenomworld® (Netherlands), for scanning electron microscopy and energy dispersive spectroscopy (SEM+EDS) analyses.

MIC Inducing Microbes for the Study

Bacterial isolates of *Escherichia coli* (*E. coli*), *Micrococcus varians* (*M. varians*), *Serratia* spp., *Staphylococcus aureus* (*S. aureus*) and *Pseudomonas aeruginosa* (*P. aeruginosa*) were obtained from the culture collection centre in the Biotechnology Unit of Department of Applied Biological Sciences, Covenant University, Ota, Ogun State, Nigeria. The test isolates were maintained on nutrient broth and incubated overnight at 37 °C between 18 and 20 h [2, 12]. From the overnight culture of each organism, 2 ml was obtained into sterile tube and was made up with sterile distilled H_2O to match the turbidity of 0.5% Mcfarland turbidity standard [26].

Inhibition Study on MIC Inducing Microbes by **D. guineense** *Based Zinc Nanoparticle*

Sterile nutrient agar plates were seeded with the mixture of test organism that was obtained from the previously described procedure, using agar well diffusion method. Sterile cork borer of 9 mm was used to bore well on the already seeded nutrient agar.

The biosynthesized zinc nanoparticle of 1 g by mass was dissolved and thoroughly mixed in 10 ml of C_2H_6OS (Dimethyl sulfoxide; DMSO). Sterile pipette were then used to disperse 0.2 ml of the dissolved biosynthesized nanoparticle material into the already bored well on the agar plate. The plates were then incubated at 37 °C for 24 h, after which zones of inhibition were measured in mm unit [23].

For this MIC inducing microbe inhibition study, the positive control of 10 μg Gentamicin, obtained from Abtec Biological Limited™ (Liverpool, UK), was used for comparison with the inhibition results obtained through use of the zinc bio-nanoparticle.

Results and Discussion

Characterised Results of the Biosynthesized Zinc Nanoparticle

The SEM images of the biosynthesized zinc nanoparticle material are as shown in Fig. 1. In the figure, the scanned electron microscopic image at 1200× magnification is shown in Fig. 1a and the ParticleMetric® rendering of the image at this same magnification is as shown in Fig. 1b. From the ParticleMetric® analysis by the

SEM facility, circle equivalent diameters ranging in μm unit from (median = 16.6: average = 20.5) to (median = 8.28: average = 9.56) was obtained from 233 clusters of particles indicated by coloured identification in Fig. 1b. However, it could be observed from this image that much smaller particles than those in the indicated coloured ranges were not among the identified particles by the SEM system, which therefore laid credence to the fact that particles in the nano-scale range was attained from the biosynthesis experiment.

Result of the energy dispersive spectroscopic analysis of the biosynthesized zinc nanoparticle, by the Pro X PHENOM™ facility, is presented as the spectral plot in Fig. 2. From the 97, 658 counts made in 30 s for the EDS test-results, spikes of zinc element could be observed after major spikes indicating carbon, nitrogen, oxygen and chromium. Other major spikes, identifiable from the EDS analysis of the

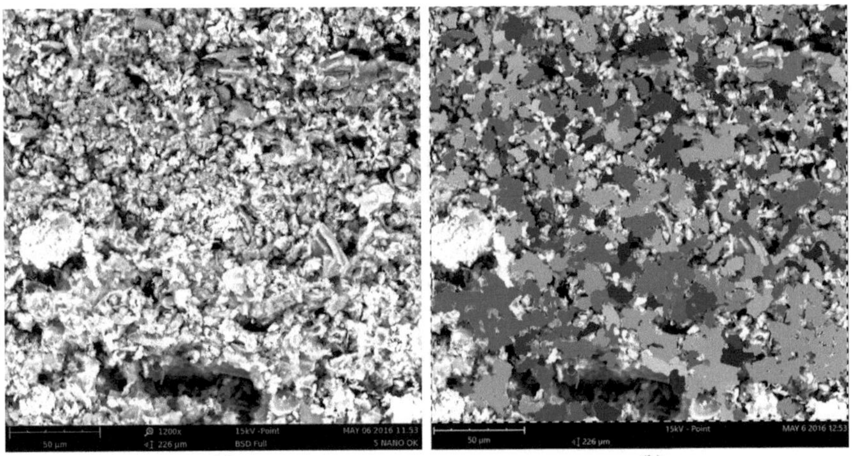

(a) (b)

Fig. 1 SEM images of the *D. guineense* based zinc nanoparticle **a** scanned electron microscopic image **b** ParticleMetric® rendering of the scanned microscopic image

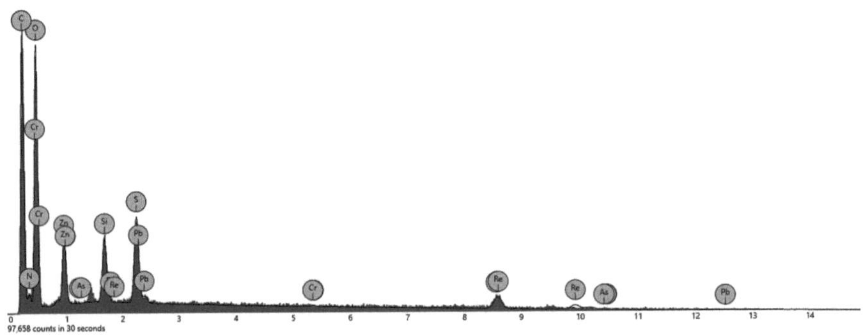

Fig. 2 Results from EDS analysis for the *D. guineense* capped zinc nanoparticle material

biosynthesized nanoparticle, include silicon and sulphur among other traces of elements. From the reported works on plant extract biochemical characterisation [34] and on "green", plant extract dependent nanoparticle synthesis [35], it could be understood that the other identified elemental spikes from the EDS analysis proceed from the biochemical constituent of the *D. guineense* plant extract. The bio-constituents from which these other elemental elements are obtained find usefulness both as reducing agent and stabilizing organic "coat" at the nano-scale level for the biosynthesized zinc nanoparticle material.

D. guineense *Based Zinc Nanoparticle Material Inhibition of Microbes Inducing MIC*

Results of the study of the inhibition effect on the growth of microbes inducing MIC by the *D. guineense* capped zinc nanoparticle material are presented in Fig. 3. In this figure, the inhibition effect on the growth of microbes inducing MIC by the Gentamicin, control antibiotic, is also presented.

The results in Fig. 3 show that the *D. guineense* based Zn nanoparticle material exhibited higher sensitivities than Gentamicin, the control antibiotic, in four out of the five tested microbes that are known to induce MIC. The biomaterial exhibited sensitivity activity against *S. aureus* with 22.00 ± 4.24 mm zone of inhibition compare to the control antibiotic to which the microbial strain was resistive with 10.50 ± 3.54 mm zone of inhibition. In similar manner, the biosynthesized nanoparticle material was sensitively active against *P. aeruginosa* with 26.00 ± 00.00 mm zone of inhibition in comparison to the antibiotic to which the *P. aeruginosa* strain was resistive with 15.0 ± 7.07 mm zone of inhibition. The

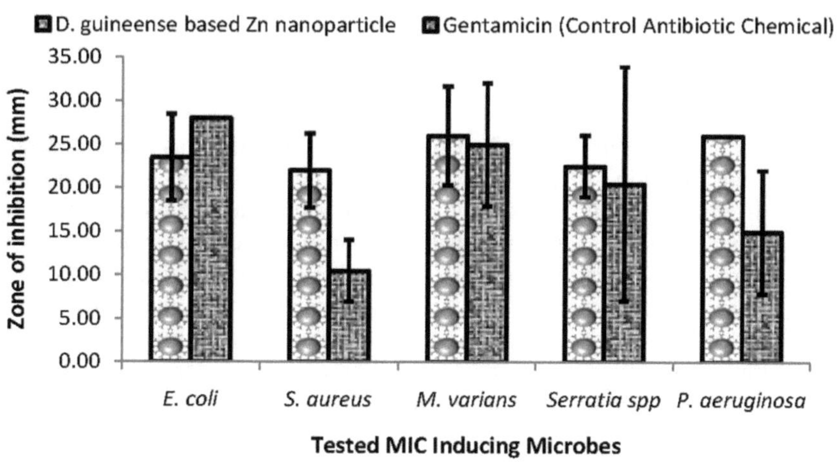

Fig. 3 Plots of measured zones of inhibition of tested MIC inducing microbes

resistance of *P. aeruginosa* to the antibiotic drug finds agreements with reports from antimicrobial research by Olajubu et al. [26] in which *P. aeruginosa* had been identified as "multiple drug-resistant strains". In that study also, *P. aeruginosa* was resistant to Gentamicin but was sensitive to the crude extract from the stem of *D. guineense*, the same type of natural plant having its leaf extract being used as precursor for the biosynthesized Zn nanoparticle in this study. The zone of inhibition obtained for *P. aeruginosa* in this study just surpassed, by 1 mm, the optimally effective zone of inhibition obtained from *Syzygium cumini* based silver nanoparticle reported in [23]. *M. varians* and *Serratia* spp. both exhibited sensitivity activities of 26 $\pm$ 5.66 and 22.50 $\pm$ 3.54 mm, respectively, to the biosynthesized Zn nanoparticle, which compares well to the sensitivities of the microbes to Gentamicin at respective 25.00 $\pm$ 7.07 and 20.50 $\pm$ 13.44 mm zones of inhibition. It was only against *E. coli* that Gentamicin exhibited the higher sensitivity of 28.00 $\pm$ 0.00 mm in this study, than the *D. guineense* based Zn nanoparticle material that exhibited the lower but yet comparable sensitivity activity of 23.50 $\pm$ 4.95 mm zone of inhibition against the microbial strain. This zone of inhibition against *E. coli* from the biosynthesized Zn nanoparticle material in this study was also within the zone of inhibition range, of 20–26 mm obtained from the use of different concentrations of *Syzygium cumini* capped silver nanoparticle, against this same microbial strain in [23].

Implications from these results include the fact that the developed *D. guineense* based Zn nanoparticle could find usefulness for the development of either new or the improvement of existing anti-MIC protection coating system that are predicated on nanotechnological methods [36]. From such system, the release of Zn ion by the biomaterial capped nanoparticle into the cells of microbes inducing MIC deterioration of metal could instigate mechanisms that would lead to the cell death of the microbial strains [23] and consequently protect metals from the causes of MIC.

Conclusions

In this paper, *D. guineense* based Zn nanoparticle biomaterial was developed and its effects on the inhibition of microbes that are known to induce microbiologically influenced corrosion (MIC) attacks of metals in industrial facilities. SEM + EDS analyses from the study suggested that the biomaterial capped nanoparticle was constituted of biochemical constituents that could act as stabilizing organic "coat" for the biosynthesized material. It was also established in the study that the biosynthesized Zn nanoparticle material exhibited sensitivity to the growth of the different microbial strains investigated, including *E. coli*, *M. varians*, *Serratia* spp., *S. aureus* and *P. aeruginosa*. The results of the zones of inhibition against the growth of the microbes studied compare well with that obtained from the control of Gentamicin, a commercial antibiotic, as well as the results obtained from other studies that were deliberated on antimicrobial investigations. These results engender positive implications on the suitability of *D. guineense* based zinc nanoparticle in nanotechnological methods of coating system for protecting metals from MIC attacks in industrial facilities. Due to the results from this work also, recommendation is being proposed for further studies on the inhibition effect and MIC protection mechanism on metallic materials that could be obtained from Zn nanoparticle material developed from the usage of *D. guineense* leaf-extract as precursor/reducing agent.

References

1. Akpan, G. U., & Iliyasu, M. (2015). Fungal populations inhabiting biofilms of corroded oil pipelines in the Niger Delta region of Nigeria. *Sky Journal of Microbiology Research, 3,* 36–40.
2. Akpan, G. U., Abah, G., & Akpan, B. D. (2013). Correlation between microbial populations isolated from biofilms of oil pipelines and corrosion rates. *The International Journal of Engineering and Science, 2,* 39–45.
3. Babu, B. R., Maruthamuthu, S., Rajasekar, A., Muthukumar, N., & Palaniswamy, N. (2006). Microbiologically influenced corrosion in dairy effluent. *International Journal of Environmental Science and Technology, 3,* 159–166.
4. Adesina, A. Y., Aliyu, I. K., & Al-Abbas, F. M. (2015). Microbiologically influenced corrosion (MIC) challenges in unconventional gas fields. In *CORROSION 2015*. Houston, TX: NACE International, Paper No. 5725.
5. NACE International. (2016). *International measures of prevention, application, and economics of corrosion technologies (IMPACT) study.* Houston, TX: NACE International.
6. Schmitt, G. (2009). *Global needs for knowledge dissemination, research, and development in materials deterioration and corrosion control.* World Corrosion Organization.
7. Lin, J., & Ballim, R. (2012). Biocorrosion control: Current strategies and promising alternatives. *African Journal of Biotechnology, 11,* 15736–15747.
8. Videla, H. A., & Herrera, L. K. (2005). Microbiologically influenced corrosion: Looking to the future. *International Microbiology, 8,* 169–180.
9. Kanyanga, C. R., Munduku, K. C., Ehata, T. M., Lumpu, N. S., Maya, M. B., Manienga, K., et al. (2014). Antibacterial and antifungal screening of extracts from six medicinal plants collected in Kinshasa-Democratic Republic of Congo against clinical isolate pathogens. *Journal of Pharmacognosy and Phytotherapy, 6,* 24–32.
10. Okere, O. S., Sangodele, J. O., Adams, M. D., Ogunwole, E., & Shafe, M. O. (2014). Antibacterial and antifungal activity of methanolic leaf extract of *Allium sativum* on selected pathogenic strains. *International Journal of Tropical Disease & Health, 4,* 1146–1152.
11. Kavitha, K. S., & Satish, S. (2013). Evaluation of antimicrobial and antioxidant activities from *Toona ciliata* Roemer. *Journal of Analytical Science and Technology, 4,* 1–7. doi:10.1186/2093-3371-4-23
12. Mahesh, B., & Satish, S. (2008). Antimicrobial activity of some important medicinal plant against plant and human pathogens. *World Journal of Agricultural Sciences, 4,* 839–843.
13. Okeniyi, J. O., Loto, C. A., & Popoola, A. P. I. (2016). Anticorrosion performance of *Anthocleista djalonensis* on steel-reinforced concrete in a sulphuric-acid medium. *HKIE Transactions, 26,* 138–149. doi:10.1080/1023697X.2016.1201437.
14. Okeniyi, J. O., Loto, C. A., & Popoola, A. P. I. (2016). Total-corrosion effects of *Anthocleista djalonensis* and $Na_2Cr_2O_7$ on steel-rebar in H_2SO_4: Sustainable corrosion-protection prospects in microbial/industrial environment. In: R. E. Kirchain, B. Blanpain, C. Meskers, E. Olivetti, D. Apelian, J. Howarter, A. Kvithyld, B. Mishra, N. R. Neelameggham, & J. Spangenberger (Eds.), *REWAS 2016: Towards materials resource sustainability* (pp. 187–192). Cham, Switzerland: Springer. doi:10.1007/978-3-319-48768-7_27

15. Okeniyi, J. O., Omotosho, O. A., Okeniyi, E. T., & Ogbiye, A. S. (2016). Anticorrosion performance of *Solanum aethiopicum* on steel-reinforcement in concrete immersed in industrial/microbial simulating-environment. In: *TMS2016 Supplemental Proceedings* (pp. 409–416). Cham, Switzerland: Springer. doi:10.1007/978-3-319-48254-5_49
16. Okeniyi, J. O., Loto, C. A., & Popoola, A. P. I. (2015). Evaluation and analyses of *Rhizophora mangle* L. leaf-extract corrosion-mechanism on reinforcing steel in concrete immersed in industrial/microbial simulating-environment. *Journal of Applied Sciences, 15,* 1083–1092.
17. Okeniyi, J. O., Loto, C. A., & Popoola, A. P. I. (2015). Inhibition of steel-rebar corrosion in industrial/microbial simulating-environment by *Morinda lucida*. *Solid State Phenomena, 227,* 281–285.
18. Okeniyi, J. O., Loto, C. A., Popoola, A. P. I., Omotosho, O. A. (2015). Performance of *Rhizophora mangle* L. leaf-extract and sodium dichromate synergies on steel-reinforcement corrosion in 0.5 M H_2SO_4-immersed concrete. In: *Corrosion 2015 Conference & Expo*. Houston, TX: NACE International, Paper No. 5636.
19. Okeniyi, J. O., Omotosho, O. A., Ogunlana, O. O., Okeniyi, E. T., Owoeye, T. F., Ogbiye, A. S., et al. (2015). Investigating prospects of *Phyllanthus muellerianus* as eco-friendly/sustainable material for reducing concrete steel-reinforcement corrosion in industrial/microbial environment. *Energy Procedia, 74,* 1274–1281.
20. Okeniyi, J. O., Loto, C. A., & Popoola, A. P. I. (2014). Corrosion inhibition performance of *Rhizophora mangle* L. bark-extract on concrete steel-reinforcement in industrial/microbial simulating-environment. *International Journal of Electrochemical Science, 9,* 4205–4216.
21. Okeniyi, J. O., Loto, C. A., & Popoola, A. P. I. (2014). Electrochemical performance of *Phyllanthus muellerianus* on the corrosion of concrete steel-reinforcement in industrial/microbial simulating-environment. *Portugaliae Electrochimica Acta, 32,* 199–211.
22. Loto, C. A., & Popoola, A. P. I. (2011). Effect of tobacco and kola tree extracts on the corrosion inhibition of mild steel in acid chloride. *International Journal of Electrochemical Science, 6,* 3264–3276.
23. Logeswari, P., Silambarasan, S., & Abraham, J. (2015). Synthesis of silver nanoparticles using plants extract and analysis of their antimicrobial property. *Journal of Saudi Chemical Society, 19,* 311–317.
24. Naik, K., & Kowshik, M. (2014). Anti-biofilm efficacy of low temperature processed AgCl–TiO_2 nanocomposite coating. *Materials Science and Engineering C, 34,* 62–68.
25. Zarasvand, K. A., & Rai, V. R. (2016). Inhibition of a sulfate reducing bacterium, Desulfovibrio marinisediminis GSR3, by biosynthesized copper oxide nanoparticles. *3 Biotech, 6,* 1–7.
26. Olajubu, F. A., Akpan, I., Ojo, D. A., & Oluwalana, S. A. (2012). Antimicrobial potential of *Dialium guineense* (Wild.) stem bark on some clinical isolates in Nigeria. *International Journal of Applied and Basic Medical Research, 2,* 58. http://dx.doi.org/10.4103%2F2229-516X.96811
27. Navale, G. R., Thripuranthaka, M., Late, D. J., & Shinde, S. S. (2015). Antimicrobial activity of ZnO nanoparticles against pathogenic bacteria and fungi. *JSM Nanotechnology & Nanomedicine, 3,* 2–9.
28. Reddy, L. S., Nisha, M. M., Joice, M., & Shilpa, P. N. (2014). Antimicrobial activity of zinc oxide (ZnO) nanoparticle against *Klebsiella pneumoniae*. *Pharmaceutical Biology, 52,* 1388–1397.
29. Meruvu, H., Vangalapati, M., Chippada, S. C., & Bammidi, S. R. (2011). Synthesis and characterization of zinc oxide nanoparticles and its antimicrobial activity against *Bacillus subtilis* and *Escherichia coli*. *Rasayan Journal of Chemistry, 4,* 217–222.
30. Liu, Y., He, L., Mustapha, A., Li, H., Hu, Z. Q., & Lin, M. (2009). Antibacterial activities of zinc oxide nanoparticles against *Escherichia coli* O157: H7. *Journal of Applied Microbiology, 107,* 1193–1201.

31. Salem, W., Leitner, D. R., Zingl, F. G., Schratter, G., Prassl, R., Goessler, W., et al. (2015). Antibacterial activity of silver and zinc nanoparticles against *Vibrio cholerae* and enterotoxic *Escherichia coli. International Journal of Medical Microbiology, 305,* 85–95.
32. Azizi, S., Ahmad, M. B., Namvar, F., & Mohamad, R. (2014). Green biosynthesis and characterization of zinc oxide nanoparticles using brown marine macroalga *Sargassum muticum* aqueous extract. *Materials Letters, 116,* 275–277.
33. Okeniyi, J. O., Loto, C. A., & Popoola, A. P. I. (2014). Electrochemical performance of *Anthocleista djalonensis* on steel-reinforcement corrosion in concrete immersed in saline/marine simulating-environment. *Transactions of the Indian Institute of Metals, 67,* 959–969.
34. Okeniyi, J. O., Ogunlana, O. O., Ogunlana, O. E., Owoeye, T. F., & Okeniyi, E. T. (2015). Biochemical characterisation of the leaf of *Morinda lucida*: Prospects for environmentally-friendly steel-rebar corrosion-protection in aggressive medium. In: *TMS2015 Supplemental Proceedings* (pp. 635–644). Cham, Switzerland: Springer. doi:10.1007/978-3-319-48127-2_78
35. Makarov, V. V., Love, A. J., Sinitsyna, O. V., Makarova, S. S., Yaminsky, I. V., Taliansky, M. E., et al. (2014). "Green" nanotechnologies: Synthesis of metal nanoparticles using plants. *Acta Naturae, 6,* 35–44.
36. Fuerbeth, W. (2015). New coatings for corrosion protection using nanoparticles or nanocapsules. In: *CORROSION 2015.* Houston, TX: NACE International, Paper No. 5554.

Injectability Evaluation of Bone-Graft Substitutes Based on Carrageenan and Hydroxyapatite Nanorods

J.I. González and C.P.O. Ossa

Abstract The first injectable bone substitutes were introduced for orthopedic trauma applications since more than a decade, and over recent years the number of commercial products has dramatically increased. These substitutes can be injected into a fracture space for augmentation as an alternative to bone graft, or around a screw for augmentation if the bone is weak, so the injectability of the substitute must be optimum with a good behavior within and our of syringe. The aim of this work was to study the injectability of substitutes based on carrageenan CG with 1, 1.5, 2.5 and 60 wt% hydroxyapatite HA nanorods. Initially carrageenan and hydroxyapatite were characterized and then injectability tests were performed with the syringe between the compression plates of a testing machine. The material also was characterized by scanning electron microscopy. The results revealed that none of the samples had phases separation and they did not exceed 300 N of force (97.08, 107.84 and 149 N to each material), that the injectability was 95.71, 93.69 and 90.63% and the CG was a good vehicle for HA nanorods. Therefore, the substitutes are adequate for manual handling.

Keywords Injectable bone substitute · Carrageenan · Hydroxyapatite · Injectability · Nanorods

Introduction

Millions of patients worldwide are affected by bone defects caused by bone disorder or injury, large bone defects resulting from malignancy, trauma, congenital diseases and infection commonly occur in orthopedic and craniofacial surgery. The number

J.I. González (✉) · C.P.O. Ossa
Biomaterials Research Group, Engineering Faculty,
University of Antioquia, St 70 N° 52-21, Medellin, Colombia
e-mail: jazmin.gonzalez@udea.edu.co

C.P.O. Ossa
e-mail: Claudia.ossa@udea.edu.co

© The Minerals, Metals & Materials Society 2017

M.A. Meyers et al. (eds.), *Proceedings of the 3rd Pan American Materials Congress*,
The Minerals, Metals & Materials Series, DOI 10.1007/978-3-319-52132-9_4

of bone fractures has dramatically increased in the most industrialized countries where it raised to a public healthcare issue [1, 2]. In this context, there is an increasing demand in the development of bone substitutes. Clinically, those defects are reconstructed using autologous bone grafts, allografts and biocompatible materials. Autografts are the gold standard for treatment but they have problems such as donor site morbidity, limited availability and increased surgery time. Allografts are expensive, carry the risk of disease transmission and adverse host immune response [2]. These problems have increased the need for synthetic bone substitute materials (BSM) since orthopedic surgery often requires filling bone defects, prosthesis fixation, and fracture immobilization [3–5].

Reconstruction of irregular shaped defects can be problematic when the implant cannot be readily formed prior to surgery using the available tools. Furthermore, implantation of a preformed bone graft requires an invasive surgical procedure [1, 6, 7]. In comparison, an injectable bone substitutes (IBS) is capable of implantation using minimally invasive surgical techniques, offers rapid and effective bone repair at low cost, can provide a better fit to the defect, limits postoperative infectious risk, has better bone-biomaterial interface even for defects with complex geometrical shapes and enables a fast transfer to industrial scale. Consequently, the use of IBS has the potential to reduce pain to the patient, risk of infection, treatment cost and scared tissue. IBS are used in several clinical conditions, as osteoporotic and osteoarthritis fracture, congenital deformity correction, tumor resection and reconstruction, generic infection, pathological degenerative bone destruction, and other degenerative diseases [4, 6–8].

In general, IBS have been designed to fill open voids or gaps in a macroenvironment under little to no pressure, IBS have structures that serve as a bridge for bone formation whose main functions are to give a pre-adjustment of the desired anatomical shape, be support for the periosteum, accelerating bone remodeling and carrying antibiotics or growth factors [5, 7]. Most injectable ceramic materials consist of micro or nanoparticles suspended in appropriate vehicles, which must have a good viscosity for a better performance and handling. indeed, the system should consist in a suspension of solid material into a suitable liquid vehicle like a polymer, so it can be easily injected and placed into complex defect sites or cavities [4, 8, 9].

Synthetic bone substitutes, mainly based on nano hydroxyapatite (nHA), have been developed as treatment for bone defects due to its biocompatibility, bioactivity and strong ion exchange capacity [2]. Synthetic HA ($Ca_{10}(PO_4)_6(OH)_2$) exhibits strong affinity to host hard tissue due to the chemical similarity between HA and the mineral phase of human bone tissue. Formation of chemical bonds with the host tissue offers HA a greater advantage in clinical applications than most other bone substitutes such as allografts or metallic implants [10, 11]. This research used nHA as ceramic phase in the IBS and the vehicle was *k*-carrageenan (CG), which is a natural polymer used far more widely than agar as emulsifier and gelling, thickening and stabilizing agent in pharmaceutical and industrial formulations. Its relatively low cost promotes its use in environmental and commercial applications as well [12, 13]. This hydrophilic polysaccharide has a backbone composition of sulphated disaccharides (galactose and anhydrogalactose units) linked by glycosidic unions. CG is obtained from certain species of red seaweeds, the most important types of CG from commercial point of view are *kappa*, *iota* and *lambda*, and this classification depends on the substitution degree that occurs on their free hydroxyl groups, the substitutions are generally either the addition of ester sulfate or the presence of the 3, 6-anhydride on the 4-linked residue [12, 14].

The aim of this study was to investigate the injectability of IBS based in nHA and CG. According to Bohner and Baroud [15] the injectability of a paste is defined as its capacity to stay homogeneous during injection, independently of the injection force. However, several protocols can be found in the literature and in all cases a syringe system is used with or without a cannula, and authors measure the amount or volume of paste extruded at a constant force and/or the pressure/force required to extrude the paste [1, 7, 9, 16]. The evaluated formulation with 1, 1.5 and 2.5 wt% of CG and 60 wt% of nHA showed an excellent injectability and they did not exceed 300 N of force, none of the samples had phases separation.

Materials and Methods

Synthesis and Characterization of Carrageenan and Hydroxyapatite Nanorods

All of the reagents used in this study were of analytical grade and they were used without any further purification. Carrageenan was purchased from Caisson Labs (USA) and was characterized using Fourier transform infrared (FTIR) espectroscopy Perkín Elmer model Spectrum one detector DTGS, with eight canning and a wave number between 4000 and 400 cm^{-1}.

The raw materials to synthesize nHA were calcium nitrate tetrahydrate $Ca(NO_3)\cdot 4H_2O$ 99.8% (Panreac, Spain) and diammonium hydrogen phosphate $(NH_4)_2HPO_4$ (Carlo Erba, Spain). The synthesis was carried out by chemical precipitation and hydrothermal treatment. It was performed by mixing stoichiometric amounts of $Ca(NO_3)_2\cdot 4H_2O$ and $(NH_4)_2HPO_4$ (Ca/P = 1.67), pH was increased to ~10 in both solution with the addition of ammonia, then the phosphate solution was slowly dropped into nitrate solution with continuous stirring. After that, the as-obtained mixed solution was transferred to a Teflon-lined stainless steel autoclave with 50 ml capacity. At that point the autoclave underwent hydrothermal treatment at 180 °C for 24 h. After the hydrothermal treatment, the autoclave was cooled down naturally and the resulting product was purified by a six cycles of washing process with distilled water. Finally, purified product was dried in an oven at 75 °C for 24 h and then macerated.

nHA powder was characterized using X-ray diffraction Rigaku using a Cu source with $\lambda = 1.5818$ Å at 2θ angle from 0° to 60°, filed emission scanning electron microscopy (FESEM) JEOL 5900 and transmission electron microscopy

(TEM) FEI model Tecnai G2. Nanorods were deagglomerated at an ultrasonic processor with a probe of 6 mm model Vibra Cell. The mean size (including diameter and length) is derived from statistical analysis of TEM image using ImageJ software.

Preparation of Injectable Bone Substitute (NHA/CG)

The injectable bone substitute (IBS) were prepared with different percentages with 1, 1.5 and 2.5 wt% of CG, 60 wt% of nHA and 5 wt% of cross-linking agent (Glutaraldehyde 25%, from Sigma Aldrich) according to CG amount. Each quantity of CG was dissolved in distilled water at 75 °C for 10 min with constant agitation, then nHA and glutaraldehyde were added into the polymer solution at 75 °C for 5 min. After, the solution final was manually mixed at 25 °C during 5 min and finally it was stored in a syringe to further characterization. The samples were named 1CG, 1.5CG and 2.5CG for 1, 1.5 and 2.5 wt% of carrageenan.

Injectability Assay and Characterization

The IBS injectability was tested with the material placed into a 10 ml syringe with a cannula 14G (diameter of opening = 1.6 mm) and extruded by applying a force at a crosshead speed of 15 mm/min using mechanical testing machine (Digimess) with a load cell of 50 kgf, as it is shown in Fig. 1. The force during the injection was recorded. The injectability was evaluated as the percentage of the mass of the paste extruded from the syringe divided by the original mass of the paste inside the syringe [9].

The surfaces of IBS specimens were observed using a scanning electron microscope (JEOL model JSM-6490LV) equipped with an energy-dispersive X-ray (EDX) microanalyzer. The test was performed after the samples were dried at room temperature for 2 days.

Statistical Analysis

Each measurement was carried out in triplicate and the results are expressed as a mean ± standard deviation (SD). Statistical analysis was carried out using a one-way analysis of variance with 95% confidence intervals (ANOVA). Fisher's Least Significant Difference (LSD) were used as post hoc analysis with the R free software, the differences were considered to be significant at a level of $p < 0.05$.

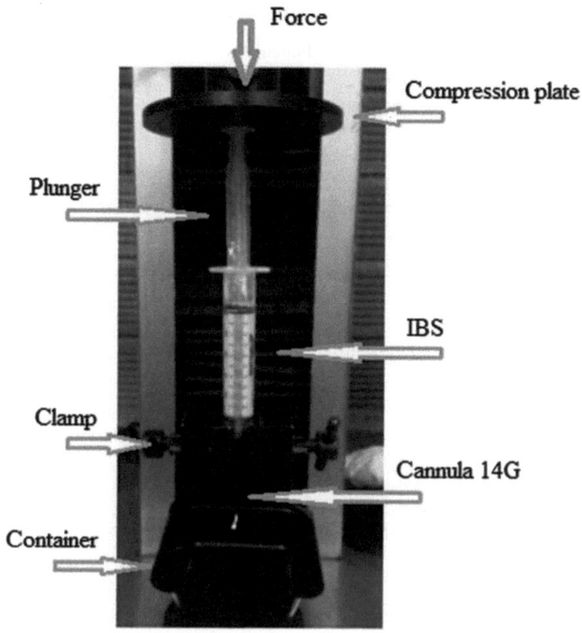

Fig. 1 Assembly to injectability testing

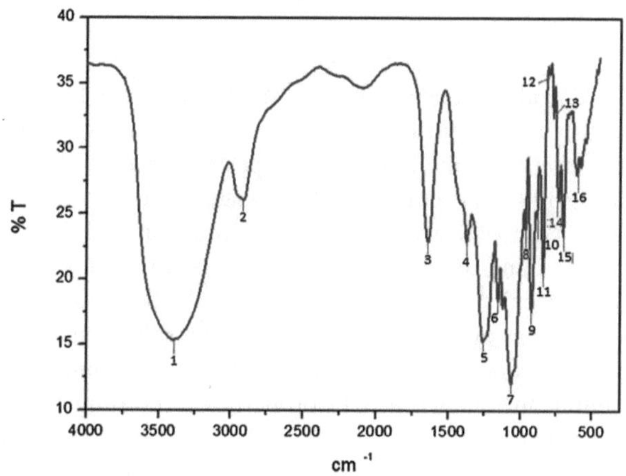

Fig. 2 FTIR spectrum of analytical carrageenan CG powder

Results and Discussion

Characterization of Carrageenan

Figure 2 shows FTIR spectroscopy pattern applied to the carrageenan powder. In literature the wave number range between 1400 and 600 cm^{-1} is mentioned as the most interesting region of the three main types of carrageenans: *kappa, lambda and iota.*

Table 1 Listing of the detected absorbance peaks in carrageenan powder

#	Bands (cm^{-1})	Value in literature	Functional group	Kappa	Iota	Lambda	References
1	3399.48	3400–3000	O–H (stretching)	X	X	X	[19]
2	2915.82	2920	C–H (stretching)	X	X	X	[20]
3	1638.49	1650	Amides I and II				[19]
4	1376.79	1380–1355	Sulphates	X	X	X	[17]
5	1263.18	1260–1210	O=S=O (asymmetric stretching)	X	X	X	[17]
6	1159.54	1160–1155	C–O–C (asymmetric stretching)	X	X	X	[19]
7	1067.28	1080–1040	C–O + C–OH	X	X	X	[17]
8	968.09	970–965	Glycosidic bonds (asymmetric stretching)	X	X	X	[17]
9	927.84	930	C–O–C (3,6-anhydrogalactose)	X	X	X	[14]
10	890.70	900–890	C_6 β-D-galactose	X	X	X	[14]
11	846.11	850–840	C_4–O–S galactose	X	X		[14]
12	800.41	805–800	C_2–O–S (3,6-anhydrogalactose)		X		[14]
13	772.74		Unidentified				
14	735.49	740–725	C–O–C α(1,3) (stretching)	X	X		[17]
15	703.70	705	Sulphates in C_4, galactose				[17]
16	608.83	615–608	O=S=O (bending)	X	X	X	[17]

The absorbance peaks are shown in Table 1. Most of them could be associated with values from literature that identify corresponding functional groups and their occurrence in the three main types of carrageenans. The broad band between 3700 and 3000 cm^{-1} corresponds to water molecules and is not related to the contents of each type of carrageenan [17]. Bands around 1650 cm^{-1} can be assigned to the amide I and amide II bands, proposed for identification of proteins [18]. The band 772.74 cm^{-1} is unidentified and could be associated to some reagent used in the extraction process of CG [13], the other bands shown in Table 1 belong to *k*-carrageenan with some traces of *iota* and some protein. This does not represent a significant change in CG properties, according to the three ways classification of carrageenans (based on the number and position of ester sulfate groups, family type and specific properties), *kappa* and *iota* CG belong to the same family, besides they belong to group of carrageenans with gelling properties [12].

Characterization of HA Nanorods

Figure 3 shows the XRD patterns of the as-synthesized powders. They were identified as HA (JCPDS card: No. 73-0294). It could be seen that there were numerous sharp peaks and low background in the XRD pattern of the HA powders.

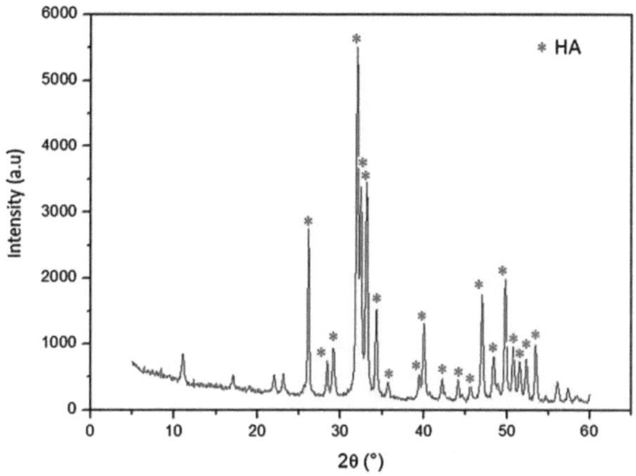

Fig. 3 XRD pattern of the synthesized powder nHA

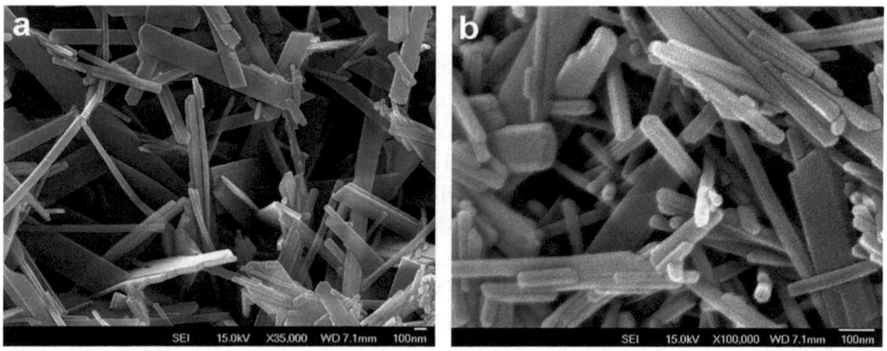

Fig. 4 FESEM images of nHA prepared by hydrothermal reaction. **a** 35,000×. **b** 220,000×

The main characteristic peaks are observed in 2 theta = 31.7°, 32.2° and 33°, there is also the presence of some secondary peaks in 2 theta = 26°, 40.5°, 46.5° and 49.4° and other peaks with lower intensity in 2 theta = 29° and 53.2°. Meanwhile, the XRD pattern of the sample did not reveal any other phases than HA. The shape of the strong diffraction peaks indicates that the samples are fairly well crystallized. The crystallinity of the HA powders synthesized via hydro-thermal method is much higher than those synthesized via normal chemical precipitation, sol–gel or normal micro-emulsion methods [21].

The FESEM image (Fig. 4) shows micrographs of HA powder synthesized by hydrothermal method, it is possible to observe structures in nanorods with several agglomerates.

In all above images the morphology is in rods, such morphology is similar to HA crystals in mineral phase from bones, the crystal size is very small with 2–5 nm thick by 20–80 nm in length or width, even when there are forming of agglomerates

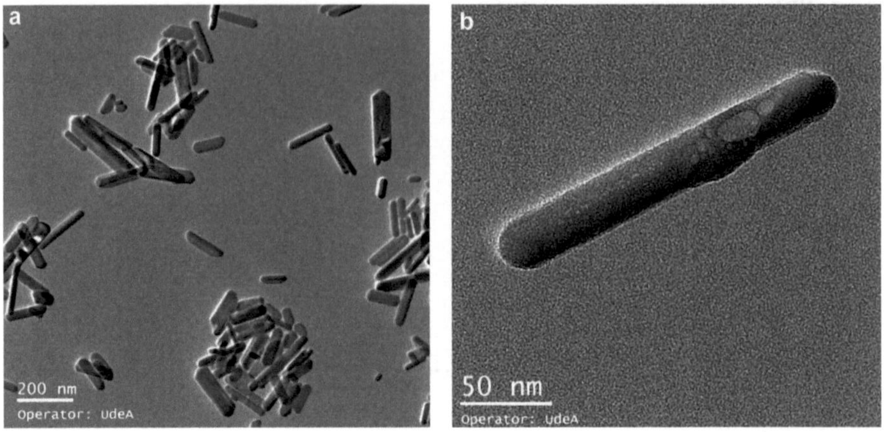

Fig. 5 TEM images of nHA prepared by hydrothermal synthesis. **a** Several nanorods. **b** Single nanorods

the size remains in the order of nanometers, which can have serious repercussions on the mechanical properties of bone, such as its strength and elastic modulus [22, 23]. Furthermore, the morphology of nHA has a significant effect on surface chemistry and bioactivity of the material [24].

Figure 5 shows nHA TEM images with more details since they are deagglomerated and thus can be observed individually, confirming the similarity with the nanocrystals present in bone apatite.

According to observed through FESEM and TEM images, powder morphology is in nanorods, typical structure of powders obtained by the hydrothermal method, which can be achieved since nanoparticles to nanostructures 3D, depending on the variation of parameters such as temperature, time and pH [10, 21, 24, 25].

The size of the nanorods was measured from several TEM images. The nanorod length was 100.56 nm (less than 190 nm), value reported in the literature [21] and the synthesized nanorods in this research are closer to size of apatite crystals in bone (80 nm) [25]. The diameter was 40.67 nm which is similar to reported in the literature (40 nm) [21], but higher that crystal bone (20 nm) [25]. Finally, the aspect ratio calculated was 3.33 similar to reported in literature and the size of crystal bone 4 [21, 26].

HA is considered nano with they have 100 nm or less in at least one direction [25]. Studies have shown that HA based biomaterials nano exhibit better resorption and higher bioactivity than those using micro size, since the release of calcium ions from a nHA is similar to that presented in the biological apatite. Also, it has been found that the nano HA has the ability to decrease apoptotic cell death and thus enhances cell proliferation and activity related to bone growth. This could be due to the high surface area and roughness present in the nHA, resulting in better adhesion and cell-matrix interactions [25].

Injectability Test and Characterization of IBS

Figure 6 shows the IBS behavior in the injectability tests depicting the evolution of the applied force during the material movement through the syringe; here also it shows a control of only nHA and water without CG. The overshoot at the beginning of the curve is caused by the yield stress, which is the critical force that must be applied to make the paste start to flow. The subsequent plateau is related to the load needed to maintain the paste flow, as desired. Both the yield stress and the extrusion force (in plateau stage) rise with an increase in amount of carrageenan CG. At the end of the extrusion, the force increases steeply until reaching the maximal extrusion force (97.08 ± 5.31 N, 107.84 ± 5.16 N and 149 ± 0.01 N for 1CG, 1.5CG and 2.5CG samples, respectively), which is caused by the mechanical contact between the syringe plunger and the syrige's bottom when all the material has been extruded. According to ANOVA test a p value <0.05 was found whereby the amount of CG has a statistically significant effect on the extrusion force.

Similarly, Liu et al. [27] obtained a paste with homogeneous and mouldable shape, this characteristics allows its handling and the possibility to take a desired geometry. Zhang et al. [16] reported maximum values of extrusion force below to 110 N for injectable materials with silanized-hydroxypropyl methylcellulose (Si-HPMC) and calcium phosphates particles and Neves et al. [8] obtained values near to 100 N for bone substitutes compounds of alginate, strontium and HA; these values are similar to reported in this study and they remain under the force range than it is possible applied manually [28]. This fact demonstrates excellent handling properties: such as injectability and cohesion. The results for nHA-carrageenan substitutes are far from those found by Morais et al. [2], who obtained values lower than 8 N for bone substitutes based in alginate and chitosan polymers. It exists a strong influence of amount ceramic added, because this material increases the flow opposition during injectability.

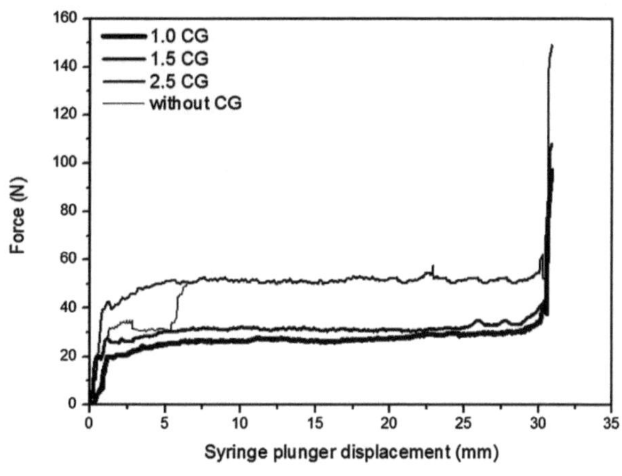

Fig. 6 Injectability curves obtained with several amounts of CG

The previous behavior could be caused by high viscosity in the paste, since the viscosity is higher in the sample 2.5CG the friction between the particles is increased and therefore injectability is decreased. Bohner and Baroud [15] showed in their research that in some studies the polymers had a positive effect on the injectability of cements, which could be explained by the decrease in viscosity since presence of polymer between grains decreases friction among them. However, in this study opposite happens, perhaps could be necessary an optimum amount of polymer, in which all nanorods are uniformly distributed and can give easier flow of the paste, in cases where the percentage of polymer exceeds the optimal amount, the nanorods could form agglomerates more localized and thus hindering the flow.

The injectability expressed here as the extruded fraction was decreased by an increase of the amount of CG as is shown in Fig. 7. In control the injectability was 34.53 ± 1.22%, IBS with 1CG the injectability was 95.71 ± 1.19%, while to 1.5CG was 93.69 ± 0.45% and 2.5 CG has an injectability of 90.63 ± 2.08%.

According to ANOVA test a $p < 0.05$ was found whereby the amount of CG has a statistically significant effect on the injectability percentage.

The injectability of a chitosan bone substitute with particles mixtures of tetra calcium phosphate (TTCP)-dicalcium phosphate anhydrous (DCPA) with diameter between 1 and 50 μm [6] was greater than 90%, this value is similar to obtained in this study. Meanwhile, Dorati et al. [4] evaluated an injectable material compound of chitosan and bovine bone particles in microscale sizes, between 141 and 1340 μm, they presented an injectability of 70%. Values in injectability are related to the particle size, therefore a valid strategy to increase this property is to change the particle size [15], the interaction between the polymer and ceramic components is greater in smaller scale (nanometer) and this interaction generate greater injectability degree. Furthermore, the material achieved more easily remain in a single phase, during the injectability test it not only released the solvent (water) due to the material is extruded together, water, CG and nHA, but control had a syneresis effect.

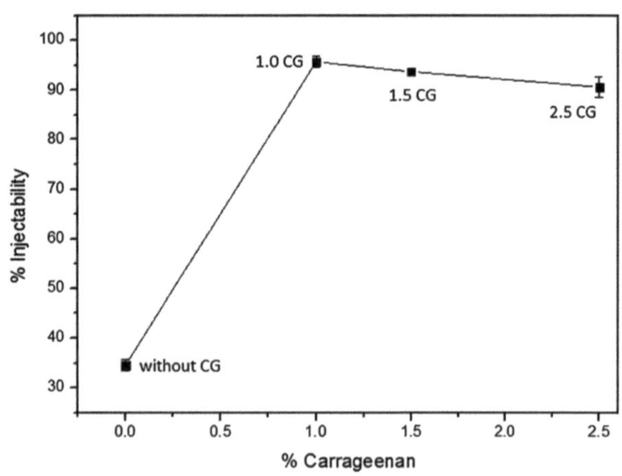

Fig. 7 Extruded fraction of the IBS as a function of the amounts of CG

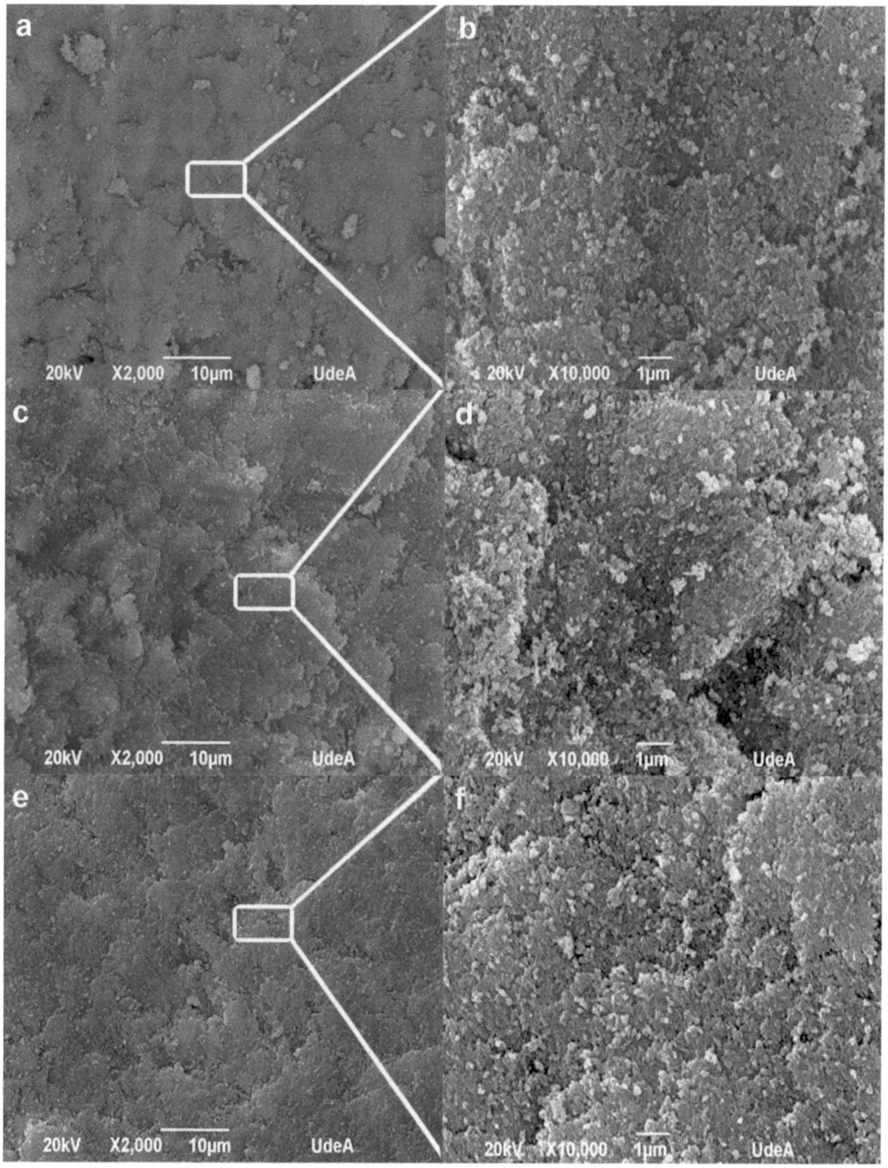

Fig. 8 SEM images of the three IBS. **a** and **b** 1CG. **c** and **d** 1.5CG. **e** and **f** 2.5CG

IBS were additionally characterized by SEM as is shown in Fig. 8. Images a and b corresponding to IBS with 1% of CG with different magnifications (2000 and 10,000× respectively), c and d corresponding to IBS with 1.5% of CG with different magnifications too and e and f corresponding to IBS with 2.5% of CG.

Similar result was obtained by Cui et al. [1] 150–157, how get a injectability closer to 95% for injectable material with 1 and 1,5% de glass particles and below value 84% for 2%.

The IBS morphology reveals that the CG can involve and aggregate the nHA very well. In greater detail, it is possible to observe, in Fig. 7 b, d, f, agglomerates of powder enveloped by polymeric material. In all cases the amount of CG was appropriate to suspend the nanorods and thus be a suitable vehicle for substitutes [2].

Conclusions

The handling properties of the IBS fabricated in this study presented low but different maximum extrusion forces, due to the low amount of CG used. According to the results, the CG is a potential vehicle carry nHA particles, since it is able of properly aggregatethem, maintaining the injectable system totality and allowing a good handling of the IBS with low and stable injectability forces. The amount of material extruded showed that all IBS had an efficient performance and the SEM micrographs revealed that the amount of CG used was appropriate to suspend the nHA and thus be a suitable vehicle for substitutes. In future, a biological follow-up study must be conducted to ascertain the biomedical application of the developed IBS.

Acknowledgements The authors are thankful with Biomaterials Research Group and Colciencias (2016-257 project) for providing the necessary reagents and studies during the development of this project, also they wish to thank to Diego Giraldo from GIPIMME Research Group of University of Antioquia for allowing the use of mechanical testing machine.

Conflict of Interest The authors declare that they have no conflict of interest.

References

1. Cui, X., Zhang, Y., Wang, H., Gu, Y., Li, L., Zhou, J., et al. (2016). An injectable borate bioactive glass cement for bone repair: Preparation, bioactivity and setting mechanism. *Journal of Non-Crystalline Solids, 432,* 150.
2. Morais, D. S., Rodrigues, M. A., Silva, T. I., Lopes, M. A., Santos, M., Santos, J. D., et al. (2013). Development and characterization of novel alginate-based hydrogels as vehicles for bone substitutes. *Carbohydrate polymers*, *95*, 134.
3. Colon, D. A., Yoon, B. J. V., Russell, T. A., Cammisa, F. P., & Abjornson, C. (2015). Assessment of the injection behavior of commercially available bone BSMs for Subchondroplasty® procedures. *The Knee, 22,* 597.
4. Dorati, R., Colonna, C., Genta, I., De Trizio, A., Modena, T., Klöss, H., et al. (2015). In vitro characterization of an injectable in situ forming composite system for bone reconstruction. *Polymer Degradation and Stability*, *119*, 151.

5. Tulyaganov, D. U., Reddy, A. A., Siegel, R., Ionescu, E., Riedel, R., & Ferreira, J. M. F. (2015). Synthesis and in vitro bioactivity assessment of injectable bioglass−organic pastes for bone tissue repair. *Ceramics International*.
6. Song, H. Y., Rahman, A. E., & Lee, B. T. (2009). Fabrication of calcium phosphate-calcium sulfate injectable bone substitute using chitosan and citric acid. *Journal of Materials Science: Materials in Medicine, 20*, 935.
7. Liu, W., Zhang, J., Rethore, G., Khairoun, K., Pilet, P., Tancret, F., et al. (2014). A novel injectable, cohesive and toughened Si-HPMC (silanized-hydroxypropyl methylcellulose) composite calcium phosphate cement for bone substitution. *Acta Biomaterialia, 10,* 3335.
8. Neves, N., Campos, B. B., Almeida, I. F., Costa, P. C., Cabral, A. T., Barbosa, M. A., et al. (2016). Strontium-rich injectable hybrid system for bone regeneration. *Materials Science and Engineering: C, 59*, 818.
9. Liu, W., Zhang, J., Weiss, P., Tancret, F., & Bouler, J. M. (2013). The influence of different cellulose ethers on both the handling and mechanical properties of calcium phosphate cements for bone substitution. *Acta Biomaterialia, 9,* 5740.
10. Jin, X., Zhuang, J., Zhang, Z., Guo, H., & Tan, J. (2015). Hydrothermal synthesis of hydroxyapatite nanorods in the presence of sodium citrate and its aqueous colloidal stability evaluation in neutral pH. *Journal of Colloid and Interface Science, 443,* 125.
11. Sadat-Shojai, M., Atai, M., Nodehi, A., & Khanlar, L. N. (2010). Hydroxyapatite nanorods as novel fillers for improving the properties of dental adhesives: Synthesis and application. *Dental Materials, 26,* 471.
12. Prajapati, V. D., Maheriya, P. M., Jani, G. K., & Solanki, H. K. (2014). Carrageenan: A natural seaweed polysaccharide and its applications. *Carbohydrate Polymers, 105,* 97.
13. Li, L., Ni, R., Shao, Y., & Mao, S. (2014). Carrageenan and its applications in drug delivery. *Carbohydrate Polymers, 103,* 1.
14. Campo, V. L., Kawano, D. F., Da Silva, D. B., & Carvalho, I. (2009). Carrageenans: Biological properties, chemical modifications and structural analysis—A review. *Carbohydrate Polymers, 77,* 167.
15. Bohner, M., & Baroud, G. (2005). Injectability of calcium phosphate pastes. *Biomaterials, 26,* 1553.
16. Zhang, J., Liu, W., Gauthier, O., Sourice, S., Pilet, P., Rethore, G., et al. (2016). A simple and effective approach to prepare injectable macroporous calcium phosphate cement for bone repair: Syringe-foaming using a viscous hydrophilic polymeric solution. *Acta Biomaterialia, 31,* 326.
17. Prado-Fernández, J., Rodrıguez-Vázquez, J. A., Tojo, E., & Andrade, J. M. (2003). Quantitation of κ-, ι-and λ-carrageenans by mid-infrared spectroscopy and PLS regression. *Analytica Chimica Acta, 480*, 23.
18. Gómez-Ordóñez, E., & Rupérez, P. (2011). FTIR-ATR spectroscopy as a tool for polysaccharide identification in edible brown and red seaweeds. *Food Hydrocolloids, 25,* 1514.
19. Chang, M. C., & Tanaka, J. (2002). FT-IR study for hydroxyapatite/collagen nanocomposite cross-linked by glutaraldehyde. *Biomaterials, 23,* 4811.
20. Mansur, H. S., Sadahira, C. M., Souza, A. N., & Mansur, A. A. P. (2008). FTIR spectroscopy characterization of poly (vinyl alcohol) hydrogel with different hydrolysis degree and chemically crosslinked with glutaraldehyde. *Materials Science and Engineering C, 28,* 539.
21. Lin, K., Chang, J., Cheng, R., & Ruan, M. (2007). Hydrothermal microemulsion synthesis of stoichiometric single crystal hydroxyapatite nanorods with mono-dispersion and narrow-size distribution. *Materials Letters, 61,* 1683.
22. Jiménez, E. B. M. (2010). Espumas Inyectables de Hidroxiapatita Obtenidas Por El Método de Espumado de La Fase Líquida de Un Cemento de Fosfato Tricálcico Alfa, Universidad Politécnica de Cataluña.
23. Navarro, M. E. (2005). Desarrollo Y Caracterización de Materiales Biodegradables Para Regeneración Ósea, Universidad Politécnica de Cataluña.

24. Swain, S. K., & Sarkar, D. (2011). A comparative study: Hydroxyapatite spherical nanopowders and elongated nanorods. *Ceramic International, 37,* 2927.
25. Sadat-Shojai, M., Khorasani, M.-T., Dinpanah-Khoshdargi, E., & Jamshidi, A. (2013). A comparative study: Hydroxyapatite spherical nanopowders and elongated nanorods. *Acta Biomaterialia, 9,* 7591.
26. Tan, J., Chen, M., & Xia, J. (2009). Water-dispersible hydroxyapatite nanorods synthesized by a facile method. *Applied Surface Science, 255,* 8774.
27. Liu, H., Li, H., Cheng, W., Yang, Y., Zhu, M., & Zhou, C. (2006). Novel injectable calcium phosphate/chitosan composites for bone substitute materials. *Acta Biomaterialia, 2,* 557.
28. Krebs, J., Ferguson, S. J., Bohner, M., Baroud, G., Steffen, T., & Heini, P. F. (2005). Clinical measurements of cement injection pressure during vertebroplasty. *Spine, 30,* E118 (Phila. Pa. 1976).

Synthesis and Characterization of $Ni_{0.5}Zn_{0.5}Fe_2O_4$@$mSiO_2$ Core Shell Nanocarrier for Drug Delivery Applications

Mohd Qasim, Khushnuma Asghar and Dibakar Das

Abstract $Ni_{0.5}Zn_{0.5}Fe_2O_4$@mesoporousSiO_2 (NZF@$mSiO_2$) core shell nanocarrier was synthesized by sol–gel method using tetraethyl orthosilicate (TEOS) and cetyltrimethylammonium bromide (CTAB) and characterized for different physicochemical properties. The structural and morphological properties were studied by X-ray powder diffraction (XRD), transmission electron microscope (TEM), and field emission scanning electron microscope (FESEM) techniques. XRD pattern and TEM micrographs confirm the coexistence of $Ni_{0.5}Zn_{0.5}Fe_2O_4$ and SiO_2 phases in the nanocomposites. Average crystallite size of $Ni_{0.5}Zn_{0.5}Fe_2O_4$ NPs was found to be around ~21 nm. Particles size of NZF@$mSiO_2$ measured by TEM and FESEM are found to be ~200–400 nm. High-resolution transmission electron microscopy (HRTEM) results confirm successful formation of NZF@$mSiO_2$ core shell nanocomposites having well symmetric structure and ellipsoidal shape. HRTEM analysis confirmed the presence of pores (5–10 nm) on the surface of SiO_2 nanosphere. Magnetic properties of NZF@$mSiO_2$ nanocarriers were studied by vibrating sample magnetometer (VSM) technique. NZF@$mSiO_2$ nanocarriers were found to be super-paramagnetic in nature with negligible coercivity and remanent magnetization. The M_s value for NZF@$mSiO_2$ was found to be 9.5 emu/gm.

Keywords Magnetic nanoparticles · Targeted drug delivery · Nanocarriers

Introduction

Recently, nanomaterials are increasingly being used in different fields because of their unique properties compared to their bulk counterpart [1, 2]. Among them, magnetic nanoparticles (MNP) are attracting significant attention because of their

M. Qasim · K. Asghar · D. Das (✉)
School of Engineering Sciences and Technology, University of Hyderabad, Hyderabad 500046, India
e-mail: ddse@uohyd.ernet.in

© The Minerals, Metals & Materials Society 2017
M.A. Meyers et al. (eds.), *Proceedings of the 3rd Pan American Materials Congress*, The Minerals, Metals & Materials Series, DOI 10.1007/978-3-319-52132-9_5

applications in biology and nano-medicine. MNP shows superparamagnetic behavior below certain critical size, which makes them more suitable for biomedical applications. They are being used in various biomedical fields such as targeted drug delivery, hyperthermia, imaging, magnetic resonance imaging (MRI), magnetic separation etc. The unique properties of being able to be guided by an external magnetic field make them very useful core materials for nano-carrier design. For a biomedical application such for drug delivery, the MNP needs to be stable in aqueous solutions, which can be achieved by the surface modification of MNP and thus by making core-shell nanoparticles. The coated shell material simultaneously provides several functions such as oxidation resistance of MNP core, prevention of agglomeration, provide biocompatibility, hydrophilicity, and functionality to hold drug molecules. Various types of coating materials including metals (Gold, Silver etc.), metal oxide (silica, calcium carbonate etc.) and polymers (polyethylene glycol, PLGA, PNIPAM etc.) are being used to design core-shell nanocarriers for drug delivery applications [3, 4]. Among them, impervious silica layer is usually used to coat the magnetic core to achieve improved biocompatibility, stability and dispersibility. Mesoporous silica nanoparticles (MSNs) have been used as carriers for drug/gene/antimicrobial agents due to its biocompatibility, hydrophilicity, stability and drug loading ability (porosity) [2]. CTAB based templating technique has been generally used to synthesize MSNs with high surface areas, and tunable pore sizes. Pure MSN based nanocarriers (without any surface functionalization) lacks in tumor targeting ability which minimizes delivery of drug at a tumor site. Magnetic targeting ability can be achieved in nanocarriers by incorporating MNP within MSN matrix. Magnetic tumor targeting is simple and inexpensive than antibody based targeting [4]. Recently, mesoporous silica based magnetic nanocarriers are getting increasing attention in biomedical research such as, targeted drug delivery due to its biocompatibility, high surface area, and high porosity [5–9]. A primary aim of magnetic nanocarrier based drug delivery system is to develop a platform that effectively reduces systemic toxicity of drugs while retaining their pharmacological activity. Magnetic nanocarrier based drug delivery systems offer several advantages over the administration of molecular free drugs viz. specific targeting ability, enhanced permeability and retention (EPR), controlled drug release, improved solubility and stability of drugs, low toxicity to normal cell, low clearance, long circulation time etc., which contribute to enhanced tumor cell death [4]. Soft magnetic Ni–Zn ferrites nanoparticles are one of the most versatile magnetic nanoparticles as they have high saturation magnetization, high Curie temperature, superparamagnetic nature, chemical stability, low coercivity and biodegradability. Recently, anticancer activity of Ni–Zn ferrite was conducted and it was reported that Ni–Zn ferrites nanoparticles have potential cytotoxicity against studied cancer cells—HT29, MCF-7, and HepG2 cells [10, 11]. In our previous work, we have reported albumen coated $Ni_{0.5}Zn_{0.5}Fe_2O_4$ (NZF) nanoparticles and subsequent deposition of $Zn_{0.95}Ni_{0.05}O$ on egg albumen stabilized $Ni_{0.5}Zn_{0.5}Fe_2O_4$ nanoparticles by simple green and environmental friendly sol gel route [1]. In the present work, we have combined superparamagnetic $Ni_{0.5}Zn_{0.5}Fe_2O_4$ and MSN together in one entity to fabricate multifunctional nanocarrier for anticancer drug delivery application. In this study, $Ni_{0.5}Zn_{0.5}Fe_2O_4$@mesoporousSiO_2 (NZF@$mSiO_2$) core shell nanocarrier was synthesized by sol–gel method using pre-synthesized NZF NPs, TEOS (as a silica source) and CTAB (as sacrificial template) and characterized for different physicochemical properties. These NZF@$mSiO_2$ NPs could have potential applications in targeted drug delivery due to its magnetic tunability and biocompatibility.

Materials and Methods

All reagents used in this synthesis were of analytical grade. Tetraethyl orthosilicate (99.99% TEOS), Zinc nitrate (>98% $Zn(NO_3)_2{\cdot}6H_2O$), nickel nitrate (>98% $Ni(NO_3)_2{\cdot}6H_2O$), iron nitrate (>98% $Fe(NO_3)_3{\cdot}9H_2O$) from Sigma Aldrich and citric acid, ammonia from SRL, India were used in this study without any purification. Cetyltrimethyl ammonium bromide (CTAB) (C19H42BrN purchased from SDFCL was used as template.

$Ni_{0.5}Zn_{0.5}Fe_2O_4$ NPs were prepared following our earlier reported gel-combustion method [1, 12]. NZF@$mSiO_2$ nanocarrier was prepared by in situ deposition of silica via hydrolysis-condensation of TEOS in presence of NZF NPs and CTAB [2]. For the synthesis of NZF@$mSiO_2$ nanocarrier, first colloidal NZF NPs were obtained by dispersing NZF NPs and CTAB in an alkaline mixture of water. The mixture was then subjected to magnetic stirring at 50 °C. Subsequently, 2.5 mL TEOS was added to the above suspension and was stirred for 12 h at room temperature. Resultant product was magnetically separated, dried and calcined at 500 °C to obtain NZF@$mSiO_2$ nanocarrier.

The crystalline structure of samples was analyzed by a Bruker D8 Advance X-ray diffractometer. Particle sizes and morphology of samples were analyzed by using TEM (FEI Tecnai T20G2 S TWIN TEM) and FESEM (Carl Zeiss Ultra 55 FESEM). A drop of aqueous dispersion of NZF@$mSiO_2$ nano-carrier was placed on carbon coated copper grid and grid was examined under TEM. A bit of powered sample on carbon tape was used for FESEM measurement. The magnetic property of NZF@$mSiO_2$ nano-carrier was studied using a Lakeshore (Model 7407) VSM in magnetic fields up to 1.5 T at ambient temperature. A pellet of 1 mm thickness was used for VSM measurement.

Results and Discussion

NZF@$mSiO_2$ nanocarrier was prepared by a two step process. First step involves preparation of superparamagnetic NZF NPs. CTAB serves not only as the stabilizing surfactant for the transfer of hydrophobic NZF to the aqueous phase but also as the organic template for the formation of mesopores in the sol–gel reaction [6]. When NZF NPs and CTAB are dispersed in water, CTAB tends to form soft assembly of micellar rod holding NZF NPs due to hydrophobic interaction between NZF NPs and non-polar tells of CTAB. On TEOS (silica source) addition to this medium hydrolysis-condensation of TEOS occurs and presence of these soft templates containing NZF NP provides heterogeneous nucleation sites for silica [4, 6]. Continuous deposition of silica on CTAB template gives rise to three dimensional meso-structure of silica network. CTAB can be removed by washing with suitable solvent (acidic ethanol solution, pH 1.4), or by calcinations above its decomposition temperature.

The phase and crystalline structure of powdered NZF@mSiO_2 nano-carrier were studied by XRD using CuKα radiation (CuKα = 1.540598 Å) at room temperature. Powder XRD pattern of NZF@mSiO_2 nano-carrier and pure mesoporous silica are shown in Fig. 1a, b, respectively. XRD pattern of NZF@mSiO_2 nano-carrier shows a broad hump at 22° with few intense sharp diffraction peaks centered around 18°, 30°, 36°, 37°, 43°, 53°, 57°, 63°, and 74°. The broad peak around 22° can be attributed to the amorphous nature of silica [2]. All the sharp diffraction peaks can be attributed to the presence of cubic NZF NPs with (111), (220), (311), (222), (400), (422), (511), (440), and (533) reflection planes, respectively [1]. The observed sharp diffraction pattern matches well with the standard diffraction pattern of the spinel cubic structure of $Ni_{0.5}Zn_{0.5}Fe_2O_4$ {JCPDS card No. 520278}. XRD pattern of pure mesoporous silica shows presence of a broad peak around 22° without any sharp diffraction peaks. Appearance of both broad and sharp characteristic diffraction peaks of silica and NZF in XRD pattern of NZF@mSiO_2 nano-carrier confirm successful formation of nanocomposite and also suggests that the structure of NZF nanocrystals does not alter after mSiO_2 coating. The crystallite size of NZF NPs was estimated by Debye–Sherrer formula using most intense peak (311) and found to be ~21 nm.

Particle size and morphology of NZF NPs and NZF@mSiO_2 nano-carrier have been studied using TEM and FESEM. TEM micrographs of NZF NPs and NZF@mSiO_2 nano-carrier are shown in Fig. 2a, b, respectively. It was observed

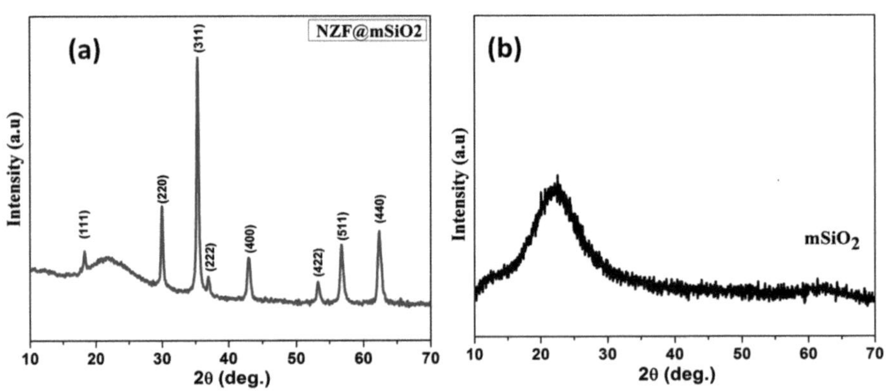

Fig. 1 X-ray diffraction pattern of NZF@mSiO_2 (**a**) and pure mesoporous silica (**b**)

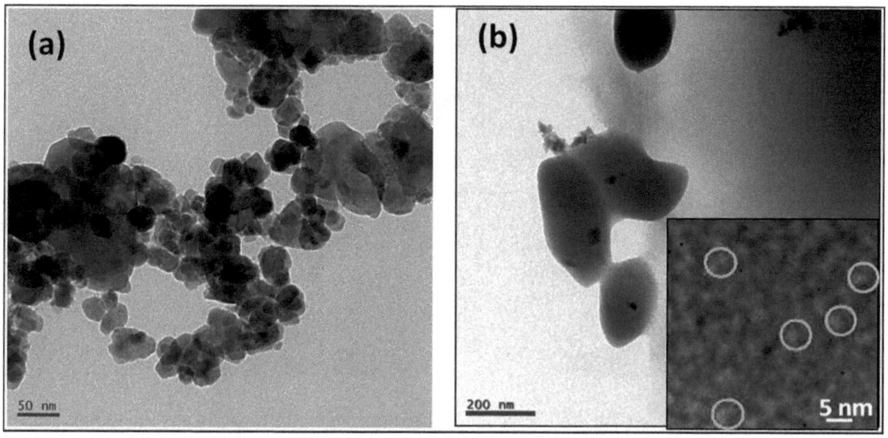

Fig. 2 TEM micrographs of pure NZF NPs (**a**) and NZF@mSiO_2 (**b**). *Inset* of figure (**b**) is HRTEM image of outer mesoporous silica which shows presence of pores (*red circles*) (Color figure online)

from TEM micrograph that prepared NZF NPs was nearly spherical in shape. Particles size of NZF NPs was estimated to be ~20–25 nm. This value is in good agreement with the values obtained from XRD data, which suggest particles to be made up of single crystal. Formation of mesoporous NZF@mSiO_2 with core shell type of nanocomposites was confirmed by TEM analysis. It was observed from TEM micrographs that particles are having well symmetric structure with ellipsoidal shape. Dark black NZF NPs of size 20–25 nm can clearly be seen in the center of mSiO_2 nanocarrier. Since ferrite is magnetic in nature it absorbs more electron than silica hence appears darker than silica in the TEM image [1]. Mesoporous nature (less dense network) of silica also make it appear lighter (grey) in TEM image. Dense silica nanoparticles of similar size are expected to appear relatively darker in TEM images due to high density. Appearance of brighter grey color of silica is due to its mesoporous nature. Further existence of pores on the surface and formation of mesoporous silica were examined by HRTEM analysis, which clearly shows presence of pores (red circle) of diameter 5–10 nm (inset of Fig. 2b). These pores are formed by removal of CTAB templates on calcinations at 500 °C. In our previous work we have shown hydrolysis and condensation of TEOS in the absence of CTAB give solid silica particle [13]. The diameter of pores depends on the diameter of formed sacrificial micellar rod due to hydrophobic interaction between nonpolar chains of surfactant molecules. Particles size of NZF@mSiO_2 nano-carrier was found to ~200–400 nm. Similar TEM results were obtained by Kim et al. [6] for iron oxide-mesoporous silica core shell nanoparticles.

FESEM images of NZF@mSiO_2 nano-carrier are shown in Fig. 3a, b. It was observed that particles are ellipsoidal in shape as observed under TEM. Enlarge FESEM images of single NZF@mSiO_2 particle is shown in Fig. 3b. Particles size of NZF@mSiO_2 nano-carrier observed from FESEM micrograph is

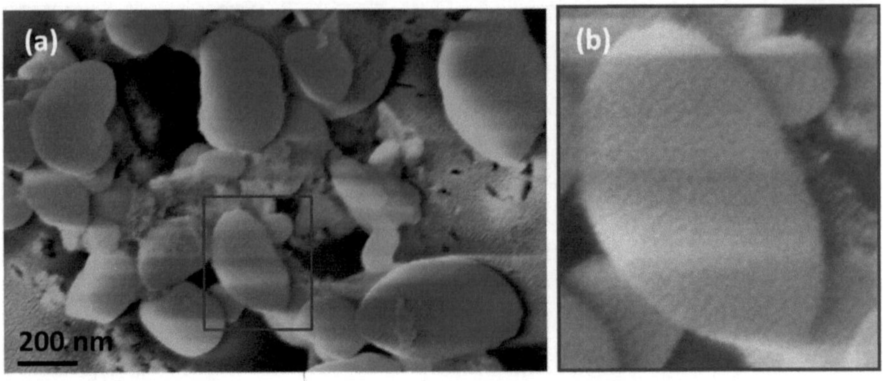

Fig. 3 (a) FESEM micrograph of NZF@mSiO_2 nanocarriers. Enlarge image of single particle (*red* frame of **a**) is shown in image (**b**) (Color figure online)

Fig. 4 Room temperature magnetization-hysteresis (M-H) curve of NZF@mSiO_2 nanocarriers

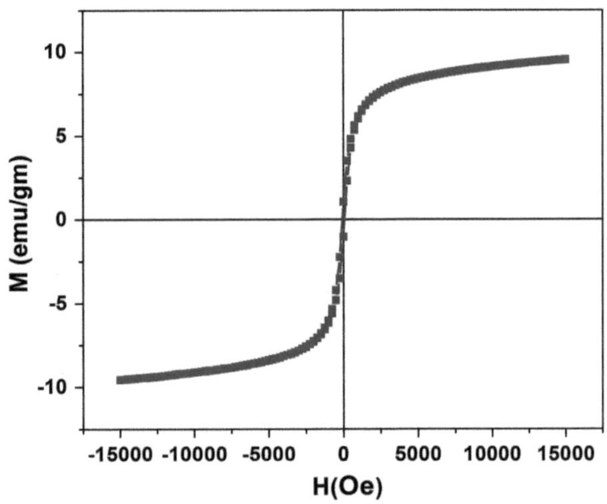

found to 200–400 nm. Appearance of rough surface of NZF@mSiO_2 nano-carrier in FESEM image suggests porous nature of mSiO_2. NZF NPs were not seen due to uniform coating of mSiO_2 on them. Obtained results are similar to the results obtained by TEM.

Magnetic property has been studied by VSM techniques and obtained room temperature M-H curve of NZF@mSiO_2 nano-carrier is shown in Fig. 4. Prepared NZF@mSiO_2 nano-carrier is found to be superparamagnetic in nature with very low remanent magnetization and coercivity. The superparamagnetic behavior of NZF@mSiO_2 nano-carrier can be attributed to the smaller NZF NPs. Magnetic controllability of MNPs depends upon its saturation magnetization (M_s) value. The M_s value of NZF@mSiO_2 nano-carrier is found to be 9.5 emu/gm. Saturation magnetization (M_s) value of magnetic nanocomposites is proportionally dependent on magnetic content of the nanocomposite. In the present case low content of magnetic NZF NPs in the NZF@$mSiO_2$ nano-carrier is the reason for low M_s value of NZF@$mSiO_2$ nano-carrier. The M_s value of present system is sufficient for targeted drug delivery applications.

This NZF@$mSiO_2$ nano-carrier could have potential to be used as biocompatible, inexpensive magnetic nano-carrier for anti-cancer targeted drug delivery. In depth study of NZF@$mSiO_2$ nano-carrier for drug delivery applications is underway which will help to identify the true potential of NZF@$mSiO_2$ nano-carrier for biomedical applications.

Conclusions

$Ni_{0.5}Zn_{0.5}Fe_2O_4$@mesoporousSiO_2 (NZF@$mSiO_2$) core shell nanocarrier was synthesized by sol–gel method using TEOS and CTAB and characterized for different physicochemical properties.

XRD pattern and TEM micrographs confirm the formation of nanocomposite. Particles size of NZF@$mSiO_2$ measured by TEM and FESEM are found to be ~200–400 nm. Morphology of the particles is found to be ellipsoidal in shape. NZF@$mSiO_2$ nanocarriers were found to be super-paramagnetic in nature with negligible coercivity and remanent magnetization. Due to biocompatible, porous, superparamagnetic and hydrophilic nature, NZF@$mSiO_2$ nanocarriers are very suitable for targeted drug delivery applications.

Acknowledgements M. Qasim greatly acknowledges the financial support obtained from University Grants Commission (UGC) in the form of MANF fellowship in carrying out this research work (2012-13/MANF-2012-13-MUS-UTT-15733). The technical support received from the School of Engineering Sciences & Technology (SEST), Central Instruments laboratory (CIL) and Centre for Nanotechnology at the University of Hyderabad is greatly appreciated.

References

1. Qasim, M., Asghar, K., Singh, B. R., Prathapani, S., Khan, W., Naqvi, A. H., et al. (2015). Magnetically recyclable $Ni_{0.5}Zn_{0.5}Fe_2O_4/Zn_{0.95}Ni_{0.05}O$ nano-photocatalyst: Structural, optical, magnetic and photocatalytic properties. *Spectrochimica Acta Part A: Molecular and Biomolecular Spectroscopy, 137,* 1348–1356.
2. Qasim, M., Singh, B. R., Naqvi, A. H., Paik, P., & Das, D. (2015). Silver nanoparticles embedded mesoporous SiO_2 nanosphere: An effective anticandidal agent against Candida albicans 077. *Nanotechnology, 26,* 285102.
3. Mura, S., Nicolas, J., & Couvreur, P. (2013). Stimuli-responsive nanocarriers for drug delivery. *Nature Materials, 12,* 991–1003.
4. Tian, Y., Jiang, X., Chen, X., Shao, Z., & Yang, W. (2014). Doxorubicin-loaded magnetic silk fibroin nanoparticles for targeted therapy of multidrug-resistant cancer. *Advanced Materials, 26*(43), 7393–7398.

5. Kim, J., Kim, H. S., Lee, N., Kim, T., Kim, H., et al. (2008). Multifunctional uniform nanoparticles composed of a magnetite nano-crystal core and a mesoporous silica shell for magnetic resonance and Fluorescence Imaging and for Drug Delivery. *Angewandte Chemie, 120,* 8566–8569.
6. Kim, J., Lee, J. E., Lee, J., Yu, J. H., Kim, B. C., An, K., et al. (2006). Magnetic fluorescent delivery vehicle using uniform mesoporous silica spheres embedded with monodisperse magnetic and semiconductor nanocrystals. *Journal of the American Chemical Society, 128*(3), 688–689.
7. Lee, J. E., Lee, N., Kim, T., Kim, J., & Hyeon, T. (2011). Multifunctional mesoporous silica nanocomposite nanoparticles for theranostic applications. *Accounts of Chemical Research, 44* (10), 893–902.
8. Chen, W. H., Luo, G. F., Lei, Q., Cao, F. Y., Fan, J. X., Qiu, W. X., et al. (2016). Rational design of multifunctional magnetic mesoporous silica nanoparticle for tumor-targeted magnetic resonance imaging and precise therapy. *Biomaterials, 76,* 87–101.
9. Alvarez-Berríos, M. P., Sosa-Cintron, N., Rodriguez-Lugo, M., Juneja, R., & Vivero-Escoto, J. L. (2016). Hybrid nanomaterials based on iron oxide nanoparticles and mesoporous silica nanoparticles: Overcoming challenges in current cancer treatments. *Journal of Chemistry, 2016,* 1–15.
10. Sadiq Al-Qubaisi, M., et al. (2013). Cytotoxicity of nickel zinc ferrite nanoparticles on cancer cells of epithelial origin. *International Journal of Nanomedicine, 8,* 2497–2508.
11. Sadiq Al-Qubaisi, M., et al. (2013). Induction of apoptosis in cancer cells by NiZn ferrite nanoparticles through mitochondrial cytochrome C release. *International Journal of Nanomedicine, 8,* 4115–4130.
12. Khan, J. A., Qasim, M., Singh, B. R., Khan, W., Das, D., & Naqvi, A. H. (2014). Polyaniline/$CoFe_2O_4$ nanocomposite inhibits the growth of Candida albicans 077 by ROS production. *Comptes Rendus Chimie, 17,* 91–102.
13. Qasim, M., Ananthaiah, J., Dhara, S., Paik, P., & Das, D. (2014). Synthesis and characterization of ultra-fine colloidal silica nanoparticles. *Advanced Science, Engineering and Medicine, 6,* 965–973.

Part II
Advanced Manufacturing

Carbon Based Coatings Deposited on Nitrided Stainless Steel: Study of Thermal Degradation

Eugenia L. Dalibón, Naureen Ghafoor, Lina Rogström, Vladimir J. Trava-Airoldi, Magnus Odén and Sonia P. Brühl

Abstract Amorphous hydrogenated carbon (DLC) coatings have a high hardness depending on the relative amount of sp^3/sp^2 bondings. They also exhibit an extremely low friction coefficient and are chemically inert. However, these coatings have some disadvantages which limit their applications. For instance, adhesion is poor when they are deposited on metallic substrates and they are also unstable at high temperatures, degrading into graphite and loosing hardness. In this work, DLC coatings were deposited on precipitation hardening stainless steel (PH Corrax) which was plasma nitrided before the coating deposition. The samples were submitted to annealing treatments for an hour at different temperatures from 200 to 600 °C, together with a control group, which was only coated but not nitrided. After each annealing cycle, Raman Spectroscopy, nanoindentation and microscopy were used to check film properties. It was demonstrated that the nitriding pre treatment improved not only adhesion but also the thermal stability of the DLC, slowing degradation and preventing delamination.

E.L. Dalibón · S.P. Brühl (✉)
Surface Engineering Group, Universidad Tecnológica Nacional
(UTN-FRCU), Ing. Pereira 676, E3264BTD Concepción del Uruguay, Argentina
e-mail: sonia@frcu.utn.edu.ar; sbruhl@gmail.com

E.L. Dalibón
e-mail: dalibone@frcu.utn.edu.ar

N. Ghafoor · L. Rogström · M. Odén
Nanostructured Materials, Department of Physics, Chemistry and Biology (IFM),
Linköping University, 581 83 Linköping, Sweden
e-mail: naugh@ifm.liu.se

L. Rogström
e-mail: linro@ifm.liu.se

M. Odén
e-mail: magod@ifm.liu.se

V.J. Trava-Airoldi
Instituto Nacional de Pesquisas Espaciáis (INPE), Av. dos Astronautas 1758,
12.227-010 São José dos Campos, SP, Brazil
e-mail: vladimir@las.inpe.br

© The Minerals, Metals & Materials Society 2017
M.A. Meyers et al. (eds.), *Proceedings of the 3rd Pan American Materials Congress*,
The Minerals, Metals & Materials Series, DOI 10.1007/978-3-319-52132-9_6

Keywords DLC films · Thermal degradation · Duplex treatment · Plasma nitriding

Introduction

The DLC coatings are characterized by high hardness, low friction coefficient, good wear resistance and chemical inertia. They are used in different industrial applications such as tool coating or mechanical parts that can reach high temperatures during operation. They are frequently used for various sliding situations where they can be exposed to localized heating caused by friction, such as engine components. For these reasons, thermal stability is a very important property for different applications of DLC coatings [1, 2].

Some studies have been published about the thermal stability of DLC coatings in which it was reported that the DLC films could not retain their diamond- like properties at high temperature. This was due to the changes in their structure, because of the conversion from sp^3 bonds to sp^2 bonds [1, 3].

In order to improve the thermal stability of these coatings, either dopant elements such as Si or N or nanoparticles-dispersed composites have been added [2, 4, 5].

Another disadvantage of these films is that they present adhesion problems when they are deposited on metallic substrates. Among other reasons, this is because the carbon diffuses into the metal delaying the DLC nucleation. In addition, the iron has a catalytic effect that leads to the formation of graphite and finally, the thermal expansion coefficients of the coatings and the steels are not compatible, causing poor adhesion and high residual stresses [6].

In order to overcome these problems different interlayers between the substrate and the DLC film (e.g. Si) have been studied as well as diffusion treatments on the substrate, like plasma nitriding [6–8].

This work deals with the thermal stability of DLC coatings deposited on a particular stainless steel, the precipitation hardening type (PH), over nitrided and non nitrided substrate.

Experimental

Corrax® PH (Uddeholm) precipitation hardening stainless steel was used as base material. The samples were 6 mm high and 24 mm in diameter. The chemical composition in mass percentage of Corrax® is 0.03% C, 12% Cr, 1.4% Mo, 0.3% Mn, 0.3% Si, 9.2% Ni, 1.6% Al and Fe as balance. All samples were aged at 530 °C for 2 h according to supplier recommendations to increase hardness. Nitriding was carried out for 10 h in an industrial facility at 390 °C using a gas mixture composed of 20% N_2 and 80% H_2. The DLC coatings were deposited by the Plasma Assisted Chemical Vapor Deposition technique (PACVD) with an asymmetrical bipolar DC pulsed discharge, using methane as the precursor gas. The process lasted 2 h and it was at 150 °C. Previously, a thin amorphous silicon interlayer was deposited using silane gas as precursor.

The DLC coatings were deposited on only aged PH stainless steel (named coated samples) and on nitrided steel (named duplex samples).

The annealing was carried out at 200, 300, 400, 500, 600 °C for 1 h on coated and duplex samples in vacuum following a method already reported in the literature [3]. The nanohardness of the films was measured employing a nano-indenter with 15 mN load before and after the annealing process.

The coatings were characterized by EDS and Raman spectroscopy before and after the thermal process as well. The microstructure of coating was analyzed by OM and SEM-FIB. The surface of coatings was observed by OM after annealing.

Results and Discussion

Microstructure of DLC Films and the Nitrided Layers

The Raman spectra for DLC coating on duplex and coated samples without annealing presented two overlapping bands known as the D and G. The D band appeared approximately at 1389 cm^{-1} and the G band, approximately at 1550 cm^{-1}. The intensity ratio of the D and G bands, I_D/I_G, was about 0.8 (Fig. 1) which indicates a low percent of sp^3 C–C bonds.

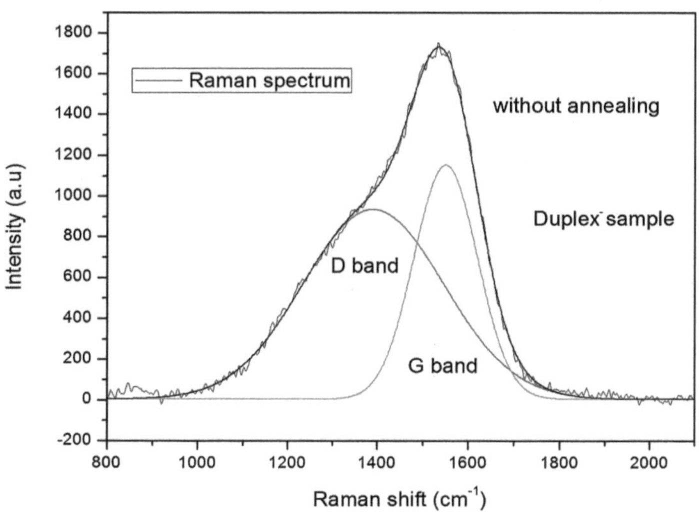

Fig. 1 Raman spectrum of duplex sample without annealing

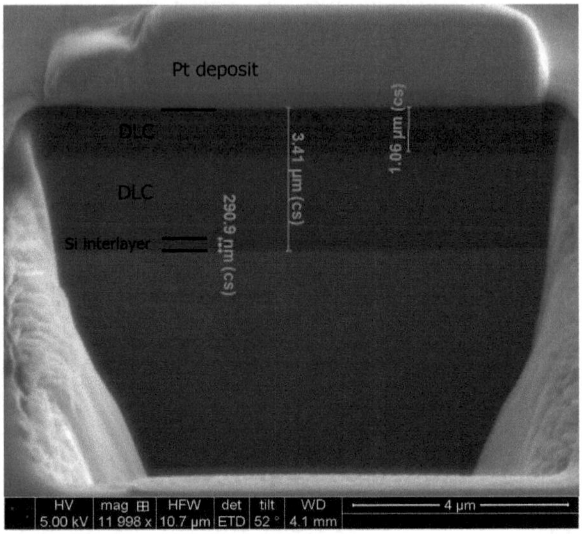

Fig. 2 SEM image of coating on duplex sample

Moreover, according to the I_D/I_G ratio and the Full Width at Half Maximum (FWHM) of the D band, it could be inferred that the coating is largely amorphous with a cluster size smaller than 2 nm [9, 10].

The hydrogen content was about 12%, which was estimated from the slope of the fitted line to the base of the original spectrum [10].

In the EDS spectrum (not shown), C and Si were detected as expected. The coating thickness was about 3.5 µm with a two layers structure, as it can be observed in the cross-section (Fig. 2) which was analyzed by SEM/FIB. Prior to the milling process, a platinum protective layer was deposited onto the region of interest in order to prevent the erosion of the surface by the ion beam. In a previous work, the two DLC layers were analyzed and it was found that the closest layer to the surface is rich in carbon rather than silicon. The pure silicon interlayer cannot be well distinguished but it is approximately 0.3 µm. So, the EDS Si signal corresponds to the interlayer and the deeper DLC layer. The coating presented a well-defined interface with the substrate both in the duplex sample and the coated one.

The nitrided layer thickness was about 14 µm, which corresponds to a region of nitrogen in solid solution. It looked white after etching with Vilella reagent in the optical micrograph (Fig. 3). Dark regions were not observed, which normally indicates chromium nitrides precipitation and the formation of second phases.

Moreover, the XRD analysis on the duplex sample revealed α_N and γ_N peaks that correspond to nitrogen supersaturated martensite (expanded martensite) and expanded austenite respectively. There was no sign of Fe or Cr nitrides as it was reported by some of the authors on a prior publication [11].

Fig. 3 Optical micrograph of the nitrided layer

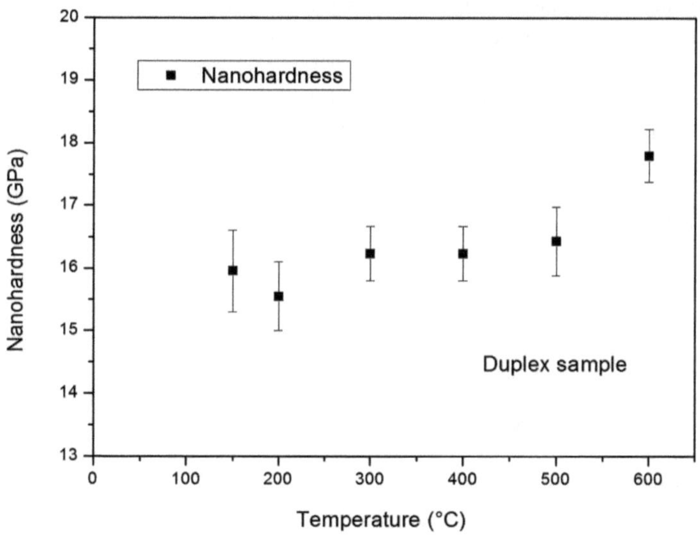

Fig. 4 Relation between nanohardness and annealing temperature of DLC coating (duplex sample)

Characterization of Coatings After Annealing

The hardness was about 16 GPa for duplex and coated samples before annealing. The results of nanohardness after the annealing and as-deposited (PACVD, 150 °C) of the duplex sample are presented in Fig. 4. Each point is the average of twenty measurements and the error was determined using the standard deviation. The hardness values obtained in both samples correspond purely to the coating hardness, because the penetration depth was less than 1/10 of the film thickness.

In the duplex sample, it can be observed that the nanohardness did not decrease; on the contrary, it remained approximately constant until 500 °C, because the variation of the hardness values is within error. At 600 °C, the hardness changed probably due to the modification of structure and morphology that the coating suffered, as it is shown below.

The results of nanohardness after the annealing and as-deposited (150 °C) of the coated sample are shown in Fig. 5.

In the coated sample, the hardness decreased at 300 °C. At 600 °C, it was not possible to obtain a value for the nanohardness because the dispersion was very large. This could be due to different factors. On one hand the coating detached in several regions (as it will be shown below) and on the other hand it is probable that the coating was transformed in some areas because of some of the hardness values were very low as it was reported in the literature for graphitization in DLC films [3]. In addition, although the hardness was measured in an area where the coating was present, the indentation could have been influenced by an adjacent uncoated area, due to the fact that plastic deformation is induced around it.

Regarding the film structure, in both samples, the Raman spectra presented similar features after annealing at different temperatures as it can be observed in Figs. 6 and 7. In order to compare both figures and show clearly the difference between spectra, the scale is the same on the "y" axis for both figures. Therefore, in the duplex sample graph (Fig. 7), the Raman spectrum corresponding to the annealing temperature of 400 °C does not appear because its background was very high and it is outside the range.

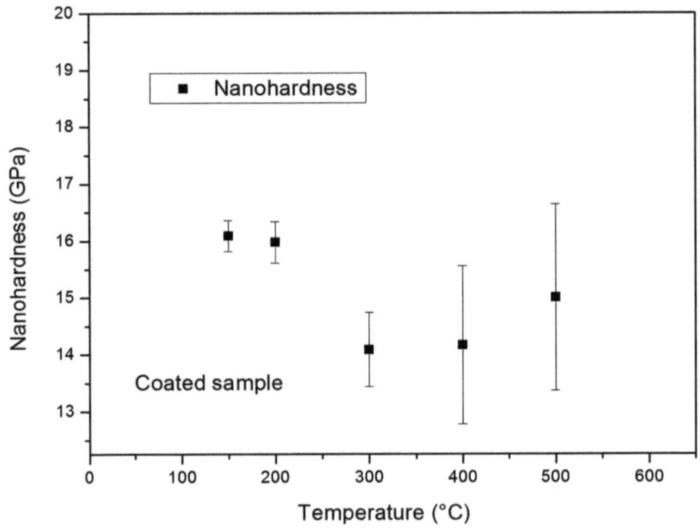

Fig. 5 Relation between nanohardness and annealing temperature of DLC coating (coated sample)

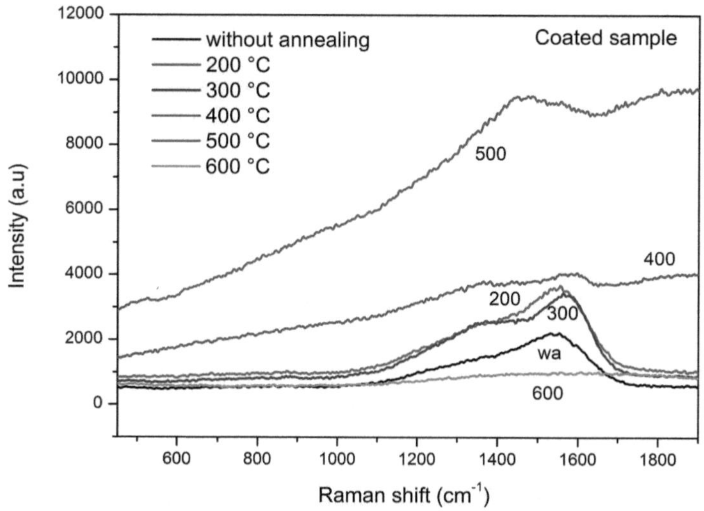

Fig. 6 Raman spectrum of coated sample for annealing at different temperatures

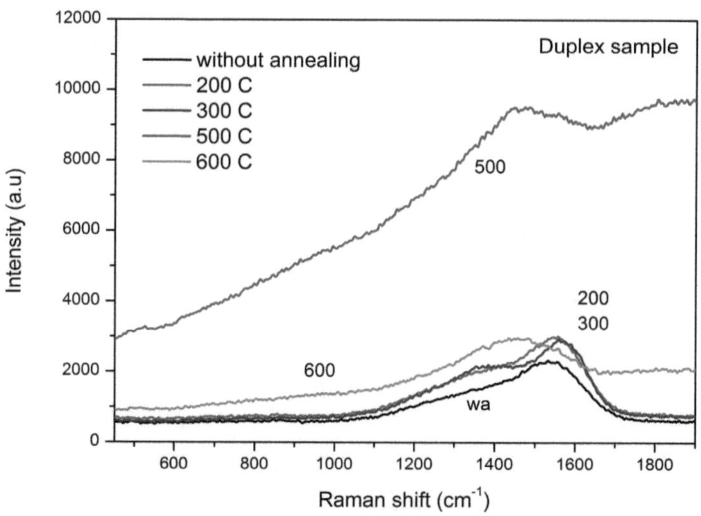

Fig. 7 Raman spectrum of duplex sample for annealing at different temperatures

Without annealing the Raman spectrum presented two overlapping bands known as the D and G bands, which were well positioned, indicating a good quality film as it was mentioned above (Fig. 2).

With annealing at 200 °C, a slight D peak position shift can be observed, and at 300 °C, it can be said that the film starts to transform (sp^3 to sp^2).

At 400 °C, a very strong fluorescence appears indicating a degradation of the diamond-like structural properties of the films.

At 500 °C, there is no DLC coating structure at all and the graphite-like component increases. A small band in the region of 500 cm^{-1} can also be detected which could correspond to the silicon interlayer.

At 600 °C, this transformation effect is more evident and the graphite band appears to be displaced.

With respect to the changes on the coating surface, a change could be observed after the annealing process at 600 °C in the duplex sample as it can be seen in Fig. 8. Large cracks were produced but there was not detachment of the coating.

In the coated sample, different features could be detected. The transformation of the coating could be observed at lower temperature, 400 °C (Fig. 9), where part of the coating was detached. After annealing at 600 °C, the transformation was completed and some cracks appeared on the surface.

According to the results of hardness and the surface images, it could be concluded that in the duplex sample the transformation of the DLC coating was produced at a higher temperature (about 600 °C) than in the coated sample. In fact, the change of color is a sign of coating degradation, the zone where the coating was vanished is observed in red.

Taking into account the Raman spectra, degradation of the coatings was detected in both samples, but probably the rate of graphitization in the duplex coating sample was lower than in the coated sample and consequently the effect of the transformation was less noticeable and it did not have any influence on the coating properties.

The adhesion was better in the duplex sample than in the coated sample. This could be due to the difference between the elastic moduli. This difference is smaller between the coating and the nitrided layer than between the coating and the stainless steel. Consequently, the stress gradient is lower, and the coating was not detached in the duplex sample.

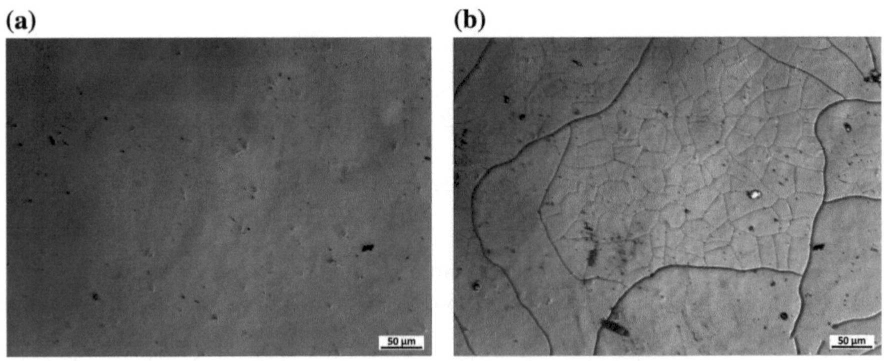

Fig. 8 Optical micrograph of duplex sample, with annealing: **a** at 400 °C, **b** at 600 °C

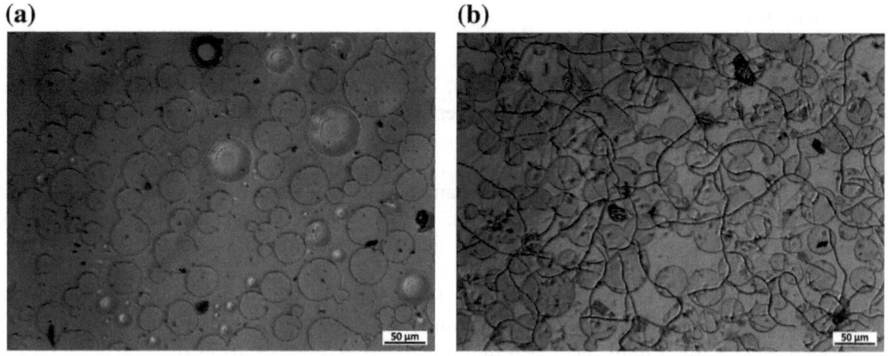

Fig. 9 Optical micrograph of coated sample, with annealing: **a** at 400 °C, **b** at 600 °C

On the other hand, during the annealing, thermal stresses are induced. Probably, these stresses were higher in the coated sample than in the duplex sample because of the presence of the silicon layer (with thermal coefficient similar to DLC) and the nitrided layer that modifies the thermal coefficient compared to stainless steel without any treatment [12, 13]. Moreover, in this sample, as there is chemical affinity between silicon of the interlayer and nitrogen of the nitrided layer, the silicon reacts with the nitrogen forming SiN as it was detected by XPS analyses, which were made by some of the authors recently. The presence of silicon nitride reduces the thermal stresses since its thermal expansion coefficient is close to the silicon thermal expansion coefficient of the interlayer [14–16].

Conclusion

The DLC coating deposited over nitrided PH steel presented better thermal stability than the same coating over stainless steel without treatment. In the duplex sample, the hardness did not decrease, the morphology of the surface did not change until 600 °C and the coating did not detach. This could be due to following: (i) the duplex samples had better adhesion, (ii) the similarity of the thermal coefficients of the nitrided layer and the DLC coating during the thermal treatment.

Acknowledgements The authors would like to thank to research scholars of GIS group (UTN) for their help in preparing of samples and tests.

References

1. Li, H., Xu, T., Wang, C., Chen, J., Zhou, H., & Liu, H. (2006). Annealing effect on the structure, mechanical and tribological properties of hydrogenated diamond-like carbon films. *Thin Solid Films, 515,* 2153–2160.
2. Zou, Y. S., Wu, Y. F., Huang, R. F., Sun, C., & Wen, L. S. (2009). Mechanical properties and thermal stability of nitrogen incorporated diamond-like carbon films. *Vacuum, 83,* 1406–1410.
3. Zhang, S., Bui, X. L., & Li, X. (2006). Thermal stability and oxidation properties of magnetron sputtered diamond-like carbon and its nanocomposite coatings. *Diamond and Related Materials, 15,* 972–976.
4. Choi, H. W., Gage, D. M., Dauskardt, R. H., Lee, K.-R., & Oh, K. H. (2009). Effects of thermal annealing and Si incorporation on bonding structure and fracture properties of diamond-like carbon films. *Diamond and Related Materials, 18,* 615–619.
5. Er, K., & So, M. (2010). Thermal annealing behavior of Si-doped diamond like-carbon films deposited by reactive sputtering. *Journal of Ceramic Processing Research, 11,* 760–764.
6. Borges, C. F. M., Pfender, E., & Heberlein, J. (2001). Influence of nitrided and carbonitrided interlayers on enhanced nucleation of diamond on stainless steel 304. *Diamond and Related Materials, 10,* 1983–1990.
7. Trava-Airoldi, V. J., Bonetti, L. F., Fernandes, J. A., Blando, E., Hübler, R., Radia, P. A., et al. (2007). DLC film properties obtained by a low cost and modified pulsed-DC discharge. *Thin Solid Films, 516,* 272–276.
8. Choi, J., Soejima, K., Kato, T., Kawaguchi, M., & Lee, W. (2012). Nitriding of high speed steel by bipolar PBII for improvement in adhesion strength of DLC films. *Nuclear Instruments and Methods in Physics Research, Section B: Beam Interactions with Materials and Atoms, 272,* 357–360.
9. Casiraghi, C., Ferrari, A. C., & Robertson, J. (2005). Raman spectroscopy of hydrogenated amorphous carbons. *Physical Review B, 72,* 1–13.
10. Ferrari, A. C., & Robertson, J. (2000). Interpretation of Raman spectra of disordered and amorphous carbon. *Physical Review B, 61,* 14095–14107.
11. Dalibon, E. L., Trava-Airoldi, V., Pereira, L. A., Cabo, A., & Brühl, S. P. (2014). Evaluation of the mechanical behaviour of a DLC film on plasma nitrided AISI 420 with different surface finishing. *Surface and Coatings Technology, 255,* 22–27.
12. Li, X.-Y., Sun, Y., & Bell, T. (1999). Stability of nitrogen S-phase in austenitic stainless steel. *Materials Research Advance, 90,* 901–990.
13. Wei, Ch., & Yen, J.-Y. (2007). Effect of film thickness and interlayer on the adhesion strength of diamond like carbon films on different substrate. *Diamond and Related Materials, 16,* 1325–1330.
14. Azzi, M., Amirault, P., Paquette, M., Klemberg-Sapieha, J. E., & Martinu, L. (2010). Corrosion performance and mechanical stability of 316L/DLC coating system: Role of interlayers. *Surface and Coatings Technology, 204,* 3986–3994.
15. Chen, C. W., Huanga, C. C., Lina, Y. Y., Chen, L. C., & Chen, K. H. (2005). The affinity of Si–N and Si–C bonding in amorphous silicon carbon nitride (a-SiCN) thin film. *Diamond and Related Materials, 14,* 1126–1130.
16. Tien, C. L., & Lin, T. W. (2012). Thermal expansion coefficient and thermomechanical properties of SiN(x) thin films prepared by plasma-enhanced chemical vapor deposition. *Applied Optics, 51,* 7229–7235.

Comparative Mechanical Analysis Between Epoxy Composite Reinforced with Random Short Cuarua Fibers and Aligned Long Curaua Fibers

Natália O.R. de Maciel, Carolina G.D. Ribeiro, Jordana Ferreira, Janaina S. da Vieira, Carlos Maurício Vieira, Frederico M. Margem and Sergio N. Monteiro

Abstract Synthetic fibers have been used for many years to attend the demands required by the most technological fields, but their use have been questioned due to the impact of them on the environment. In this way, the natural fibers have received considerable attention because of some their characteristics, besides low cost, they are flexible, viable, renewable and considered abundant substitutes. Thus, this paper is about curaua fibers, which belong to the family *Bromeliaceae* and it intends to compare the mechanical strength between epoxy composite reinforced with random short curaua fibers and aligned long curaua fibers. The results show that aligned long curaua has higher mechanical strength.

Keywords Curaua fibers · Composites · Environment · Mechanical strength

Introduction

Composites are defined as engineering materials made from two or more constituents with different physical, chemical or mechanical properties. Composites are typically stronger than the individual materials [1].

Composites reinforced with fibers offer many benefits such as high strength, high durability, low weight, water resistance, chemical resistance, electrical resistance, fire

N.O.R. de Maciel (✉) · C.G.D. Ribeiro · J. Ferreira · J.S. da Vieira · C.M. Vieira
UENF, Advanced Materials Laboratory, LAMAV, State University of the Northern Rio de Janeiro, Av. Alberto Lamego, 2000, 28013-602 Campos dos Goytacazes, RJ, Brazil
e-mail: nataliaoliveiraroque@gmail.com

F.M. Margem
Redentor, BR 356, 25, 28300-000 Itaperuna, RJ, Brazil

S.N. Monteiro
Military Institute of Engineering, IME, Praça General Tibúrcio, 80, 22291-270 Praia Vermelha, RJ, Brazil

© The Minerals, Metals & Materials Society 2017
M.A. Meyers et al. (eds.), *Proceedings of the 3rd Pan American Materials Congress*, The Minerals, Metals & Materials Series, DOI 10.1007/978-3-319-52132-9_7

resistance and corrosion resistance. For example, materials as glass fiber and carbon fiber (synthetic fibers) have been widely used in automotive and aircraft industries. However, these materials have serious drawbacks they are non-biodegradable, non-recyclable, from non-renewable resources, present high energy consumption in the manufacturing process and health risk when inhaled [2, 3].

In contrast, natural fibers show a lot of advantages such as specific strength properties, low cost, low density, good thermal properties and mainly enhanced energy recovery and biodegradability. Moreover, plant fibers are abundant, non-toxic, non-corrosive, non-irritable to skin, eyes or respiratory system [4]. Because of all advantages, natural fiber composites (NFCs) have emerged as more environmental friendly and a realistic alternative to substitute synthetic fibers [5]. So, natural fibers composites are seen as potential materials for many engineering applications [6].

Despite the natural fibers have several advantages, they present some disadvantageous factors such as: (1) the chemical composition and fiber properties of plant tissue are also different at different stages of the growing season further, (2) incompatibility between the fiber and the matrix and (3) natural fibers absorb water from the air and direct contact from the environment. This absorption deforms the surface of the composites by swelling—decreasing strength—and creating voids—increase in mass. These factors currently limit the large scale production of natural fibers composites [7, 8]. However, studies have been done because the advantages of using natural fibers in polymeric composites may overcome the disadvantages.

Thus, the aim of this work was to evaluate and compare the tensile properties of composites with epoxy matrix reinforced with random short curaua fibers (RSCFs) and aligned long curaua fibers (ALCFs).

Experimental Procedure

The curaua fiber is obtained from a plant that grows in the Amazon region whose scientific name is *Ananas Erectifolius*, the firm that commercializes this natural fibers is called Amazon paper. The typical aspect of curaua plant and a bundle of soft fibers are showed in Fig. 1. In this work the fibers were used (Fig. 1).

The fibers were used in the as-received condition, without any surface treatment, after a room temperature drying procedure.

The measured distribution of 100 fibers revealed a dispersion interval in length from 150 to 164 mm (Fig. 2), with an average of 157 mm and diameter from 0.044 to 0.193 mm, with an average of 0.098 mm (Fig. 3), i.e. the average length is more than 15 times its critical length of 3 mm, obtained in epoxy pullout tests [9]. The density was calculated as approximately 960 kg/m^3.

For carrying out the work 13 tensile specimens were made in a rectangular dog-bone shape with 5.8 $\times$ 4.5 mm of reduced gage dimensions for each fiber

Fig. 1 **a** "*White* curaua" and "*purple* curaua" plant respectively **b** bundle curaua fiber [6] (Color figure online)

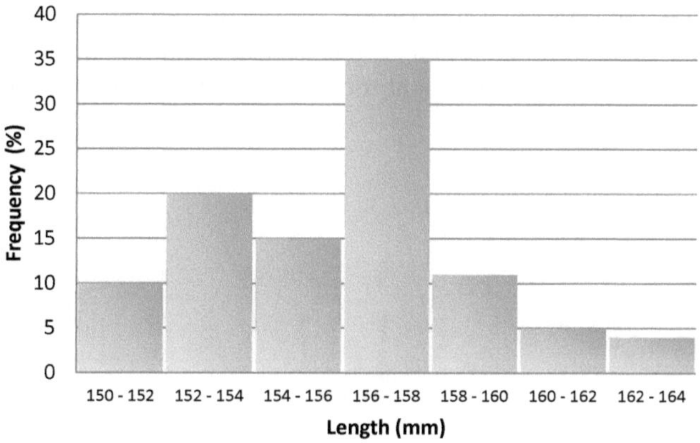

Fig. 2 Histogram for the statistical distribution of length of the supplied curaua fibers

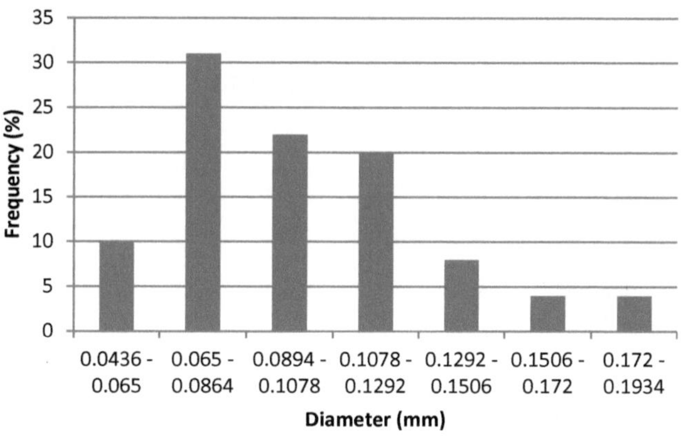

Fig. 3 Histogram for the statistical distribution of diameter of the supplied curaua fibers

concentration (ASTM D790): 0% fiber, 30% aligned long curaua fiber and 30% random short curaua fiber in polymeric matrix of epoxy.

The specimens with 0% fiber were made in silicone mold. In the case of the samples with 30% fiber, the fibers were cut in specimen size for composites of ALCFs (aligned long curaua fibers) and in the critical length (≤ 1 cm) for composites of RSCFs (random short curaua fibers). For next step, long and short fibers were inserted, separately, in the metal mold in alignment (ALCFs) or randomly (RSCFs) and then were bathed in epoxy resin. All specimens were manufactured at room temperature and that were manufactured in the metal matrix were subjected to press to 2.0 t per 24 h during the curing process. This fabrication process is known as direct method and it has low cost and demands short time. However, it is difficult to control the axial arrangement of fibers and this fact can compromise the tensile strength of unidirectional composites [10].

After curing, the samples were grinded, measured with the aid of a caliper rule, numbered and then were subjected to tensile test using an Instron Model 5582 machine with a speed of 2 mm/min at room temperature. Figure 4 shows the steps described above.

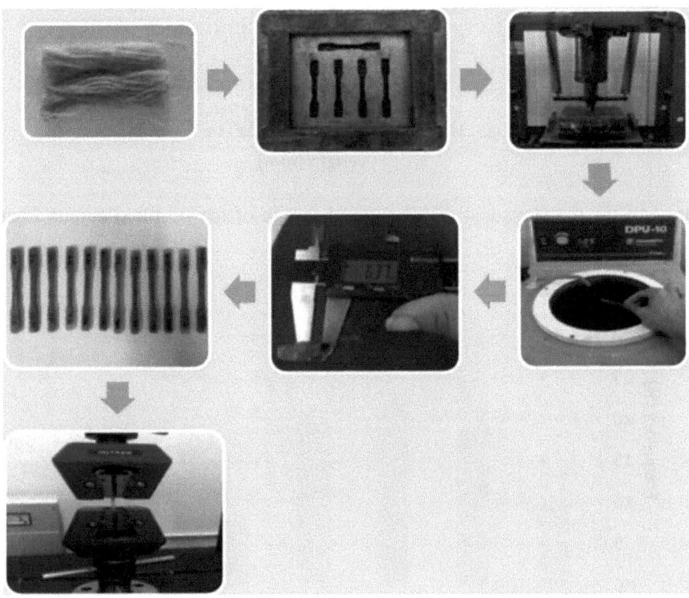

Fig. 4 Procedure to the tensile test

Results and Discussion

Scanning Electron Microscopy—SEM

The fracture analyses of the tensile test specimens were performed by SEM, microscopic analysis and macroscopic observation.

Figures 5 and 6 show a typical SEM fractography of a tensile rupture specimen for epoxy composite reinforced with 0 and 30% curaua fiber respectively. In Fig. 5, it is possible to conclude that it is a brittle material once the surface of the fracture is perpendicular to the direction of the tractive tension; moreover, the fracture features river plates and a single direction.

In Fig. 6, it is possible to see the fibrils characteristics of the fiber and its regularity. It can also note the traction stress on the fiber that has spread to the fiber-matrix interface and caused the rupture. Furthermore, it is possible to observe the brittle fracture in the matrix already explained in the previous Figure.

Figure 7 illustrates the aspect of the tensile ruptured specimens corresponding to each volume fraction of curaua fiber. In the first specimen it is possible to see that the crack occurred transversely to the tensile axis with no participation of the fibers

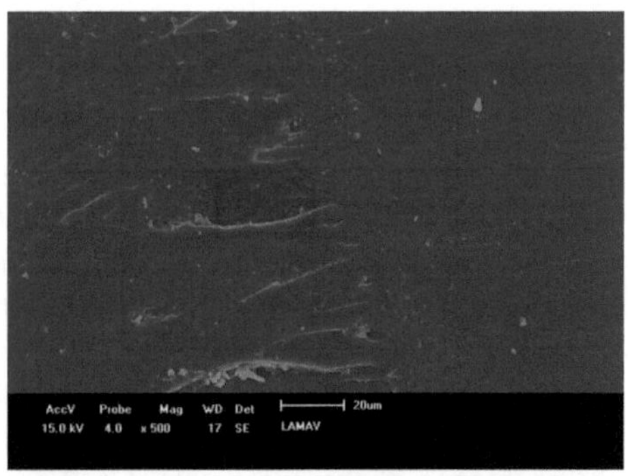

Fig. 5 Tensile fracture surface of pure epoxy specimen (0% fiber) with magnification (×500)

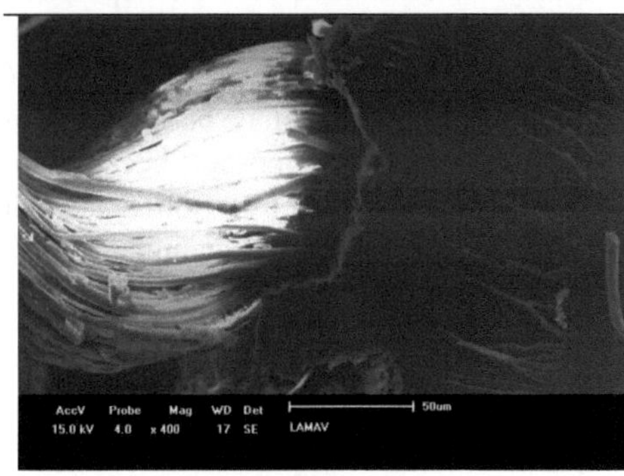

Fig. 6 Tensile fracture surface of epoxy composite reinforced with 30% of curaua fibers with magnification (×400)

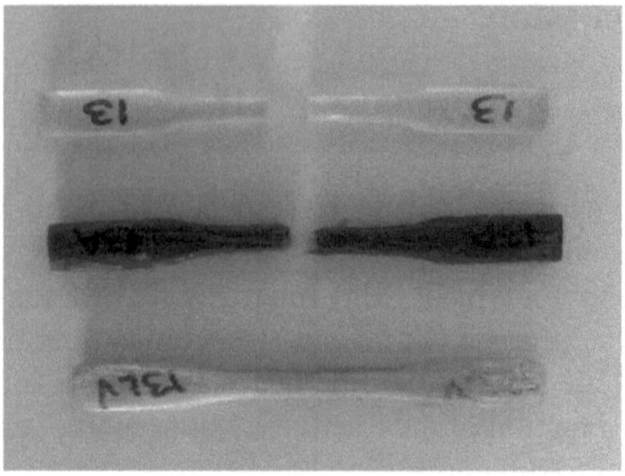

Fig. 7 Tensile ruptured specimens for each volume fraction of curaua fiber incorporated into the epoxy matrix

because the matrix is a brittle material, In the other specimens with curaua fibers, the macroscopic aspects reveals an increase of non-uniform rupture due the insertion of the fibers. However, the crack continuous occurred transversely to the tensile axis.

Tensile Test

Figure 8 exemplifies the typical load vs. elongation curves for each composites 0% curaua fibers, 30% RSCFs and 30% ALCFs respectively. These curves revealed that the composites reinforced with curaua fiber apparently present limited plastic deformation and sudden rupture characterizes a brittle material.

In Fig. 9 the graphic with the stress value of each specimen was plotted. It is possible to realize that the composites with 30% RSCFs is similar to 0% of curaua fiber composite, that is, the increase in volume of RSCFs practically did not change the resistance of the composite. On the other hand, adding ALCFs increased considerably the resistance of the composite.

The average stresses for the samples with 0% fiber, 30% random short fibers and 30% aligned long fibers were 31, 34 and 61 MPa approximately.

Another factor that may be available is the coefficient of variation or relative dispersion, which is important because this statistical tool apparently permits the comparison of varieties free from scales effects [11]. In this case, the coefficient of variation is equal 34% to specimen with 0% of fiber; 31% to RSCFs and 23% to ALCFs.

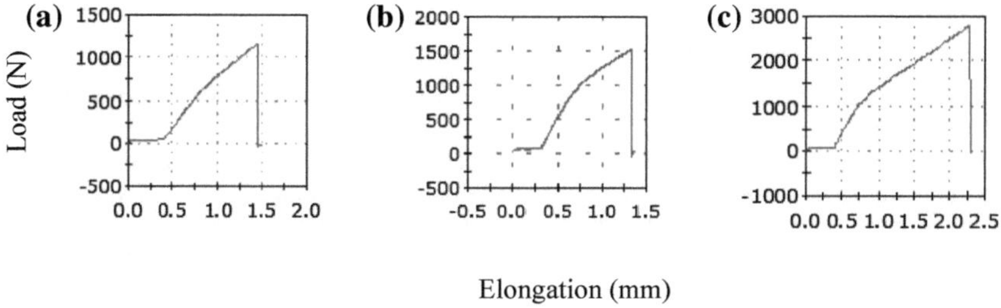

Fig. 8 Load versus elongation curves for epoxy composites reinforced with **a** 0%, **b** 30% random short and **c** 30% aligned long in volume fraction of curaua fibers

Fig. 9 Comparative graph of stress among specimens with 0% fiber, 30% RSCFs and 30% ALCFs

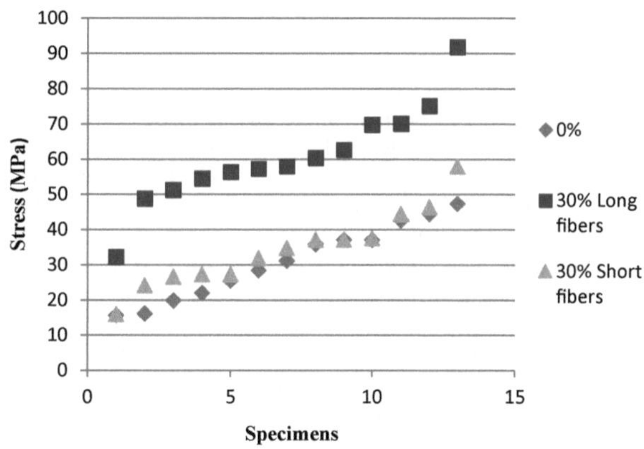

Conclusions

- The incorporation of aligned long curaua fiber significantly increases (100%) the tensile strength in epoxy matrix composites;
- The increase in volume of random short fibers did not change the tensile strength results, so it is not feasible to use random short fibers even if it is with the same concentration of aligned long fibers;
- Besides the incorporation of aligned ling curaua fiber increase the tensile strength, the coefficient of variation was less than 30% and lowest among the three specimens.

Acknowledgements The authors thank the support to this investigation by the Brazilian agencies: CNPq, CAPES, FAPERJ and TECNORTE/FENORTE.

References

1. Thakur, V. K. (2013). *Green composites from natural resources* (p. 406). Boca Raton, FL: CRC Press.
2. Ticoalu, A., Aravinthan, T., & Cardona, F. (2010). A review of current development in natural fiber composites for structural and infrastructure applications. In *Southern Region Engineering Conference (SREC 2010)*, Toowoomba, Australia.
3. Cheung, H., Ho, M., Lau, K., Cardona, F., & Hui, D. (2009). Natural fiber-reinforced composites for bioengineering and environmental engineering applications. *Composites Part B Engineering, 40*(7), 655–663.
4. Shalwan, A., & Yousif, B. F. (2013). In state of art: Mechanical and tribological behavior of polymeric composites based on natural fibers. *Materials and Design, 48,* 14–24.
5. Cao, Y., Shibata, S., & Fukumoto, I. (2006). Mechanical properties of biodegradable composites reinforced with bagasse fiber before and after alkali treatments. *Composites Part A: Applied Science and Manufacturing, 37*(3), 423–429.
6. Pickering, K. (2008). *Properties and performance of natural-fiber composites* (p. 576). Amsterdam: Elsevier.
7. Rowell, R. M., Han, J. S., & Bisen, S. S. (1997). Changes in fiber properties during the growing season. In R. M. Rowell, R. A. Young & J. K. Rowell (Eds.), *Paper and composites from agro-based resources* (pp. 23–38). Boca Raton, FL: CRC Lewis Publishers.
8. Westman, M. P., Fifield, L. S., Simmons, K. L., Laddha, S. G., & Kafentzis, T. A. (2010). *Natural fiber composites: A review.* Pacific Northwest National Laboratory.
9. Monteiro, S. N., Ferreira, A. S., & Lopes, F. P. D. (2009). Pullout tests of curaua fibers in epoxy matrix for evaluation of interfacial strength. In *Proceedings of Characterization of Minerals, Metals and Materials—TMS Conference* (pp. 1–7). San Francisco, USA.
10. Monteiro, S. N., Ferreira, A. S., & Lopes, F. P. D. (2009). Tensile properties of epoxy composites reinforced with continuous curaua fibers. In *Proceedings of Characterization of Minerals, Metals and Materials—TMS Conference* (pp. 1–7). San Francisco, USA.
11. Brown, D. C. E. (1998). Coefficient of variation. In *Applied Multivariate Statistics in Geohydrology and Related Sciences* (pp. 155–157) [Internet]. Berlin: Springer [citado 12 de setembro de 2016]. Disponível em: http://link.springer.com/chapter/10.1007/978-3-642-80328-4_13

On Tool Wear in Rotary Tool Micro-Ultrasonic Machining

Sandeep Kumar, Akshay Dvivedi and Pradeep Kumar

Abstract Micro-ultrasonic machining (micro-USM) is used to fabricate complex micro-features in brittle and hard materials. In micro-USM, both dimensional and form accuracy of machined component depend mainly on the shape of the tool. However, tool wear is an unavoidable phenomenon of this process, which affect the accuracy of micro-feature. The tool suffers by three types of wear (longitudinal, lateral and edge wear) in micro-USM. Accumulation of micro-chips and abrasives in the machining gap (between tool and workpiece) are the main reasons responsible for severe tool wear in micro-USM. This article reports on a new method named as rotary tool micro-USM to reduce tool wear. The rotary tool micro-USM involves abrasive slurry with providing simultaneous rotation and vibration to the tool. Rotation of the tool helped abrasives to replenish from the machining gap easily. Micro-channels were fabricated and characterized by using stereo microscope. From the results, it was found that rotary tool micro-USM resulted in very less tool wear and as a result of that micro-channels of better dimensional and form accuracy were developed.

Keywords Micro-USM · Tool rotation · Micro-channel · Tool wear · Form accuracy

S. Kumar (✉) · A. Dvivedi · P. Kumar
Department of Mechanical and Industrial Engineering,
Indian Institute of Technology Roorkee, Roorkee 247667, India
e-mail: sandeepkumar71@gmail.com

A. Dvivedi
e-mail: akshaydvivedi@gmail.com

P. Kumar
e-mail: kumarfme@gmail.com

© The Minerals, Metals & Materials Society 2017
M.A. Meyers et al. (eds.), *Proceedings of the 3rd Pan American Materials Congress*,
The Minerals, Metals & Materials Series, DOI 10.1007/978-3-319-52132-9_8

Introduction

Ultrasonic machining (USM) is a promising technique used to fabricate complex features in all types of brittle and hard materials viz. quartz, glass, ceramics, silicon, titanium alloys etc. [1]. In USM, material is removed in the form of small chips from workpiece surface by the impact of abrasive particle [2, 3]. Micro-USM is downscaled from macro-USM and it can generate 3D complex micro-feature using simple or complex shaped micro-tools. The final shape of the cavity depends entirely on the shape and size of the tool. The tool wear is an unavoidable phenomenon of micro-USM which affect the accuracy of the micro-feature. Therefore, accuracy of micro-features can be maintained by controlling the shape, size and material of the tool as well. As per literature available, researchers have already demonstrated the feasibility of micro-USM to machine micro-components [4, 5]. Adithan [6] stated that machining rate decreased with decrease in the length as well the weight of the tool. M. Adithan also revealed that poor circulation of abrasive slurry was main the cause of low machining rate and tool wear in USM. Yu et al. [7] reported accumulation of debris in the machining gap results in low machining efficiency as well as high tool wear. Pei et al. [8] reported that the profile of machined surface is greatly influenced by abrasive particles present in the machining gap (between tool and workpiece). Yu and Rajurkar [9] applied a uniform wear method integrated with CAD/CAM software. Uniform wear method resulted in regaining of the shape of the tool while machining of 3D micro-shapes. Cheema et al. [10] reported that tool suffered three types of wear, namely longitudinal wear (face wear), lateral or side wear, and edge wear in micro-USM. Face wear of tool affects the micro-feature depth, whereas side and edge wear results in the taper on micro-feature wall and rounded bottom corners of the feature respectively. To compensate the tool wear in micro-USM. A mathematical model to measure the tool wear in micro-USM without tool rotation was presented by Cheema et al. [11].

In micro-USM tool wear depends on many factors such as frequency of vibration, tool material, workpiece material, amplitude, abrasive material, abrasive size, concentration of abrasive slurry and static load. Some researchers have reported experimental investigations on tool wear by considering above mentioned process parameters. The accumulation of abrasives in the machining gap is one of the main causes of excessive tool wear and poor form accuracy of micro-features in case of micro-USM. Therefore, this study reports on a new method named as rotary tool micro-USM to reduce tool wear. The rotary tool micro-USM involves abrasive slurry and simultaneous rotation and vibration to the tool. Microchannels were fabricated using rotary tool micro-USM. Further the effect of tool rotation speed on tool wear as response characteristics was investigated.

Experimental Setup

Materials and Methods

Microchannels were fabricated on borosilicate glass using a layer-by-layer machining approach [10]. An in-house designed and developed setup of rotary tool micro-USM was used to fabricate microchannels. Schematic representation of experimental setup is illustrated in Fig. 1. The rotary tool micro-USM setup consists of a power supply of 250 W, frequency of 25 ± 1 kHz, range of tool rotation speed from 0 to 2000 rpm and amplitude of 15 μm respectively. Vertical feed was provided to the workpiece with the help of counter weight balance mechanism (Fig. 1). Silicon carbide was used as abrasive material. The tool material selected was tungsten carbide with circular cross section (Ø 600 μm) for the current investigation.

Tool Wear in Micro-USM

Tool wear is an unavoidable phenomenon of micro-USM and it cannot be completely eliminated. However, it can be minimized by controlling input process parameters. In micro-USM, the tool suffers by three types of wear, namely longitudinal wear, lateral wear (face wear) and edge wear. Longitudinal wear is the reduction in the length of the tool which affect the depth of micro-feature, lateral wear is the taper formation around the tool surface which results in taper on side walls of micro-feature. The edge wear is the reduction in the diameter of the tool face and it results in rounded corners at bottom of micro-features [10]. The schematic representation of different types of tool wear is shown in Fig. 2.

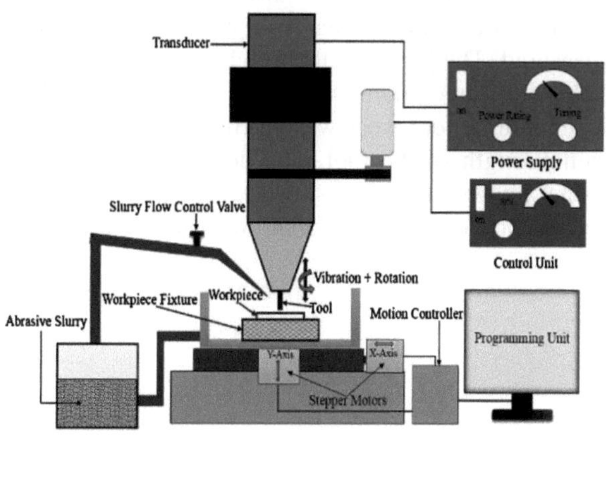

Fig. 1 Schematic representation of rotary tool micro-USM setup [12]

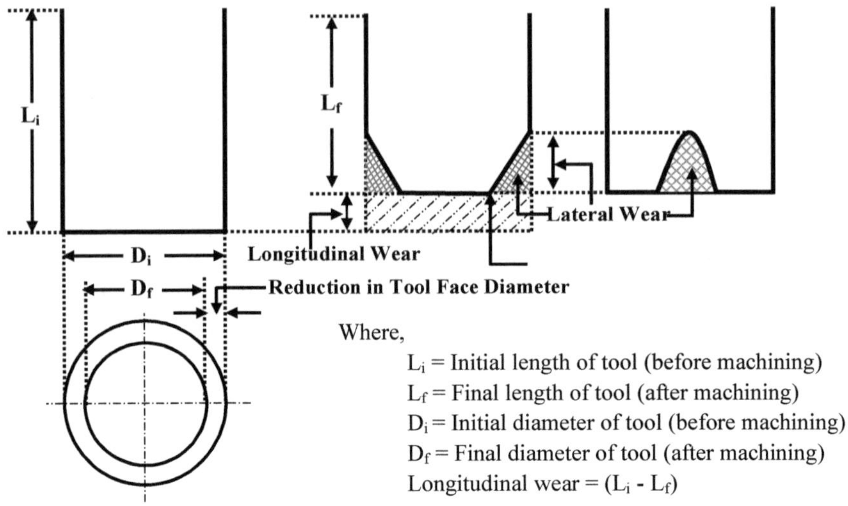

Fig. 2 Schematic representation of tool wear in micro-USM

Process Parameters

The objective of current the investigation was to reveal the effect of tool rotation during micro-USM. Thus, experiments were conducted by varying tool rotation speeds from 100 to 400 rpm. In order to avoid any possible lateral vibrations at higher tool rotation speeds, the tool rotation speed was limited to 400 rpm. The value of other process parameters such as power rating, abrasive size, workpiece feed rate, concentration of slurry and static load were based on published literature [13] and they were kept constant throughout experimentation. Cheema et al. [10] emphasized in there study, that longitudinal tool wear has the largest contribution (86–90) in overall tool wear followed by lateral wear (7–8%) and edge wear (2–3%). Consequently, only longitudinal tool wear was selected as response characteristic in the present study. The values of constant process parameters are given in Table 1. Each experiment was repeated three times and the average value was measured. The length of the tool before and after machining was measured with the help of a stereo zoom microscope (Nikon: SMZ-745T) at a magnification of 50×. Longitudinal tool wear was calculated by subtracting the final length of tool from initial length of the tool. The depth and width of microchannels were measured using image analysis using Dewinter DMI premium optical microscope. The micrograph of cross sectional view of microchannel is shown in Fig. 2.

Table 1 Constant process parameter and their values

S. no.	Process parameters	Values/machine setting
1	Static load (gm)	75
2	Frequency (kHz)	25 ± 1
3	Abrasive size (mesh)	#1000
4	Workpiece feed rate (mm/min)	15
5	Slurry concentration (%)	20
6	Power rating (%)	40
8	Tool diameter (μm)	600
9	Abrasive material	Silicon carbide (SiC)
10	Liquid medium for slurry	Tap water
11	Tool material	Tungsten carbide (WC)
12	Workpiece material	Borosilicate glass

Results and Discussion

The microscopic view of the tool after machining, fabricated microchannel and cross sectional view of microchannels are shown in Fig. 3. In Fig. 3a–d (iii), hatched portion denotes the desired profile of microchannel and double dash chain line denotes the actual profile of microchannels.

Effect of Tool Rotation Speed on Tool Wear, Width and Depth of Microchannel

At tool rotation speed of 100 rpm, it was observed that the tool wear and depth of microchannel were low (as it is shown in Fig. 4) whereas the overcut was large [as shown in Fig. 3a (iii)]. This can be attributed to the fact that at 100 rpm, tool rotation helped the trapped abrasive particles to move away (outward) from the machining gap (between tool and workpiece) but at the same time due to the low rotation speed (100 rpm) of the tool, the circulation of abrasives was not effective. This resulted in a low depth of the microchannel (Fig. 4). Moreover, multiple layers of abrasives kept on flowing in the lateral gap (between tool face and workpiece wall) due to which increased width of microchannels was obtained [Fig. 3a (ii)].

After increasing the tool rotation speed from 100 to 200 rpm, both depth of microchannel and longitudinal tool wear were increased (Fig. 4) whereas the width of the microchannel decreased [Fig. 3b (ii)]. Due to increased rotation speed, circulation of abrasives was improved and more number of abrasives came in contact with the rotating and vibrating tool. This caused an increase in tool wear. Further, the improved circulation of abrasives resulted in a decreased width of the microchannel. Fresh abrasives came into the working gap as the rotation speed was

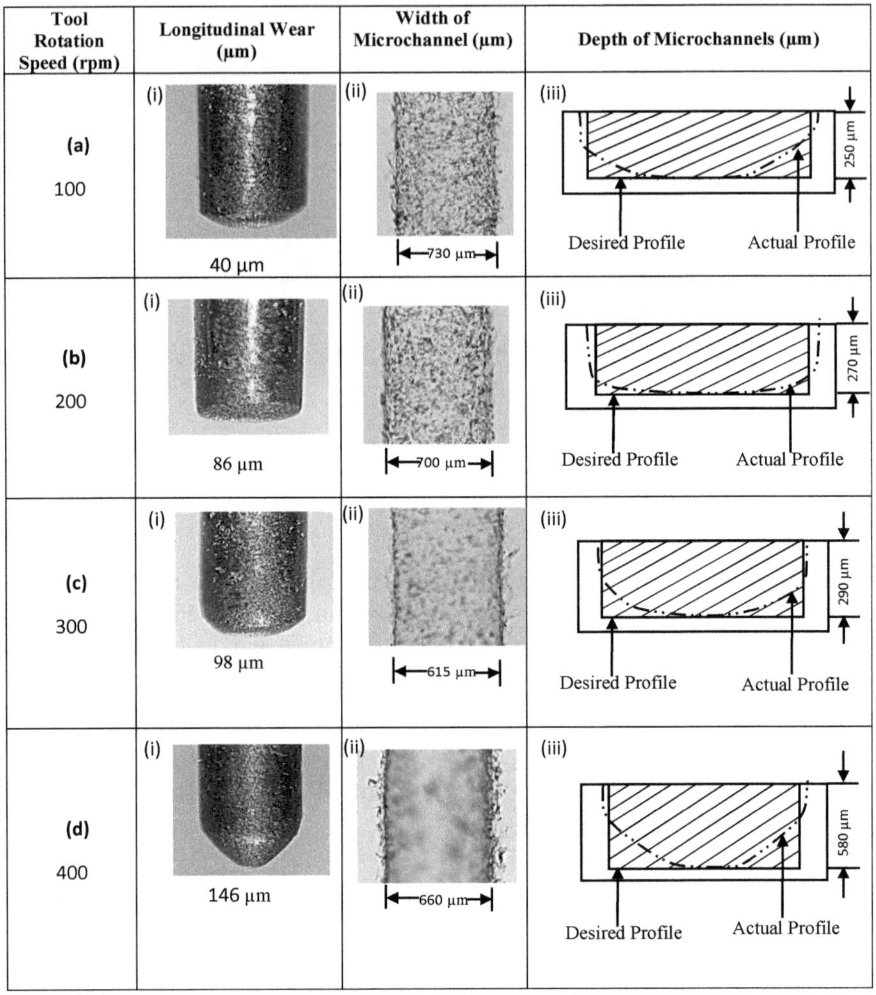

Fig. 3 Microscopic images (45×) of tool [**a–d** (i)], microchannel [**a–d** (ii)] and cross sectional view of microchannel [**a–d** (iii)]

increased. These abrasives further impacted on workpiece surface resulting in more removal of material. Thus, the depth of microchannel was increased.

With further increase in tool rotation speed from 200 to 300 rpm, longitudinal wear and depth of microchannel were slightly increased (Fig. 4) whereas the width of microchannel was decreased [Fig. 3 c (iii)]. This can be explained with the fact that interaction between rotating tool and fresh abrasives was more due to increase in tool rotation speed, which in turn resulted in increasing the tool wear at 300 rpm. Furthermore, a uniform laminar layer of abrasives was observed in between the tool and workpiece. Due to this layer, a good acoustic bond was formed. Subsequently,

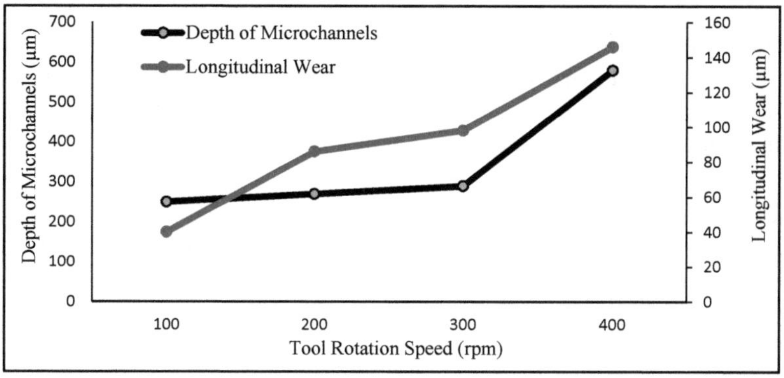

Fig. 4 Effect of tool rotation speed on depth of microchannels and longitudinal wear

the maximum energy was transferred from the abrasive to the workpiece. Thus width of microchannel was decreased and depth of microchannel was increased. This can be evidenced from Figs. 3c (ii) and 4.

On further increase in tool rotation speed from 300 to 400 rpm, tool wear, depth and width of microchannels were increased (Figs. 3d (i), (ii) and 4). It seems that at 400 rpm, the tool exerted high centrifugal force on the abrasives and as a result, less abrasives impacted on workpiece surface. This led to the direct contact of tool and workpiece. Due to the tool and workpiece interaction, lateral wear and edge wear were also dominated at 400 rpm [Fig. 3d (i)]. Therefore, overall tool wear was increased at 400 rpm (Fig. 4). This phenomenon (direct contact between tool and workpiece) was also reported in the literature [10]. Further, the direct contact of tool and workpiece damaged the edges of microchannel as shown in Fig. 3d (ii). The worn tool penetrated inside the workpiece. Consequently, the depth of microchannel was increased and microchannel of deteriorated form accuracy (tapered wall and curved surface) was formed [Fig. 3d (iii)].

Conclusions

The present study focused on the tool wear in rotary tool micro-USM. Microchannels were successfully fabricated at different tool rotation speeds. The following conclusions can be drawn from this study:

- Rotary tool micro-USM possesses the flexibility to fabricate 3D microchannels in all types of brittle and hard materials.
- Tool rotation speed played an important role in deciding the tool wear and form accuracy in rotary tool micro-USM.
- Tool rotation speed of 100 and 400 rpm resulted in large width and poor form accuracy of microchannels.

- The better form accuracy as well as lowest tool wear was obtained at a tool rotation speed of 300 rpm.
- Further, it is necessary to investigate the effect of other process parameters (power rating, abrasive size, workpiece feed rate, concentration) also for desired form accuracy, depth of microchannel and even lower tool wear.

Acknowledgements The authors highly acknowledge the financial support from Department of Science and Technology (DST), Government of India, Grant no. SB/FTP/ETA/207/2012.

References

1. Thoe, T. B., Aspinwall, D. K., & Wise, M. L. H. (1998). Review on ultrasonic machining. *International Journal of Machine Tools and Manufacture, 38,* 239–255.
2. Benedict, G. F. (1987). *Non-traditional manufacturing processes* (pp. 67–86). New York: Marcel Dekker Inc.
3. Pandey, P. C., & Shan, H. S. (1980). *Modern machining processes* (pp. 7–30). India: Tata McGraw-Hill Education Pvt. Ltd.
4. Sun, X. Q., Masuzawa, T., & Fujino, M. (1996). Micro ultrasonic machining and its applications in MEMS. *Sensor and Actuator A—Physics, 57,* 159–164.
5. Egashira, K., & Masuzawa, T. (1999). Microultrasonic machining by the application of workpiece vibration. *CIRP of Annals, 48,* 131–134.
6. Adithan, M. (1981). Tool wear characteristics in ultrasonic drilling. *Tribology International, 14,* 351–354.
7. Yu, Z., Hu, X., & Rajurkar, K. P. (2006). Influence of debris accumulation on material removal and surface roughness in micro ultrasonic machining of silicon. *CIRP of Annals, 55,* 201–204.
8. Pei, W., Yu, Z., Li, J., Ma, C., Xu, W., Wang, X., et al. (2013). Influence of abrasive particle movement in micro-USM. *Procedia CIRP, 6,* 551–556.
9. Yu, Z. Y., & Rajurkar, K. P. (2004). Study of 3D micro-ultrasonic machining. *Journal of Manufacturing Science and Engineering, Transactions of the ASME, 126,* 727–732.
10. Cheema, M. S., Dvivedi, A., & Sharma, A. K. (2015). Tool wear study in fabrication of microchannels in ultrasonic micromachining. *Ultrasonics, 57,* 57–64.
11. Cheema, M. S., Singh, P. K., Tyagi, O., Dvivedi, A., & Sharma, A. K. (2016). Tool wear and form accuracy in ultrasonically machined microchannels. *Measurement, 81,* 85–94.
12. Kumar, S., & Dvivedi, A. (2016). Fabrication of 3D complex micro-features used in bio-medical applications. *International Conference on Nanotechnology for Better Living, 3,* 255.
13. Cheema, M. S. (2015). *An ultrasonic micromachining approach to fabricate microchannels on borosilicate glass.* Ph.D. Thesis, Department of Mechanical and Industrial Engineering, IIT Roorkee.

Numerical Modeling of High-Velocity Impact Welding

Ali Nassiri, Shunyi Zhang, Tim Abke, Anupam Vivek, Brad Kinsey and Glenn Daehn

Abstract To support the lightweighting aim in the automotive industry, High-Velocity Impact Welding (HVIW) can be used to join dissimilar metals. The manufacturing industry often relies on numerical simulations to reduce the number of trial-and-error iterations required during the process development to reduce costs. However, this can be difficult in high strain rate manufacturing processes where extremely high plastic strain regions develop. Thus, a traditional Lagrangian analysis is not able to accurately model the process due to excessive element distortion. In order to further understand the science behind HVIW processes and benefits of various numerical simulation methodologies, two methods were utilized to simulate Al/Fe bimetallic system which is of interest for the automotive industry. First, a Smoothed Particle Hydrodynamics (SPH) model of two impacting plates was created. Using SPH method, metal jet emission was investigated which previously was impossible. The results then were compared with an Arbitrary Lagrangian-Eulerian (ALE) method. Finally, the numerical results were compared with experimental tests using a Vaporizing Foil Actuator Welding process.

Keywords Numerical modeling · Welding · Smoothed particle hydrodynamics · Arbitrary Lagrangian-Eulerian

A. Nassiri (✉) · A. Vivek · G. Daehn
Center for Design and Manufacturing Excellence (CDME),
The Ohio State University, 1314 Kinnear Rd., Columbus, OH 43212, USA
e-mail: nassiri.3@osu.edu

A. Nassiri · A. Vivek · G. Daehn
Department of Materials Science and Engineering,
The Ohio State University, 2041 College Rd., Columbus, OH 43210, USA

S. Zhang · B. Kinsey
Department of Mechanical Engineering, University of New Hampshire,
33 Academic Way, Durham, NH 03824, USA

T. Abke
Honda R&D, North America, 21001 State Route 739, Raymond, OH 43067, USA

© The Minerals, Metals & Materials Society 2017

M.A. Meyers et al. (eds.), *Proceedings of the 3rd Pan American Materials Congress*,
The Minerals, Metals & Materials Series, DOI 10.1007/978-3-319-52132-9_9

Introduction

Car bodies are traditionally made with stamped and spot welded steel. But owing to increasing cost of energy as well as the increasing ambitions to reduce CO_2 emission, there is a growing interest in creating lightweight vehicles. One way to achieve mass reductions is using multi-material assemblies. Creation of such structures requires joining of dissimilar metals. Due to differences in melting points and the tendency to form brittle intermetallic compounds (IMCs), many dissimilar metal pairs cannot be joined with traditional fusion-based techniques such as resistance spot welding and Metal Inert Gas (MIG) welding.

One way to join dissimilar metals is through HVIW. HVIW is a solid state welding technique which contains a class of methods including EXplosive Welding (EXW), Magnetic Pulsed Welding (MPW), Laser Impact Welding (LIW), and Vaporizing Foil Actuator Welding (VFAW). In all these methods, the metals are joined by first accelerating a flier workpiece, either a metal tube or a metal plate, to a high velocity with a significant kinetic energy before it strikes a stationary workpiece. The methods differ in the precise mechanism by which the flier workpiece is accelerated [1]. In addition, the length scale of the weld is different in these method. High precision, short cycle time, less wrinkling and uniform strain distribution are among the outstanding advantages of HVIW.

A remarkable feature of metals joined using HVIW is the emergence of a characteristic wavy morphology at the interface between the two welded components (see Fig. 1a–e). Although the origin of the wavy morphology has been the subject of comprehensive investigation both from mechanical and metallurgical points of view, a unique fundamental physical theory for this phenomenon is not yet widely accepted in the literature. Shear instability generated by high velocity collision [2, 3], jet indentation mechanism [4, 5], vortex shedding mechanism [6], and stress wave

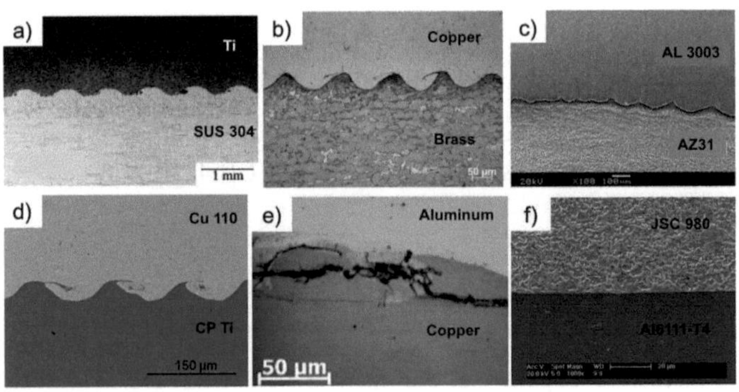

Fig. 1 Wavy morphology observed in: **a** titanium-stainless steel [9], **b** copper-brass [10], **c** aluminum-magnesium [11], **d** copper-titanium [12], **e** aluminum-copper [13], and **f** dual phase steel-aluminum

mechanism [7] are among the proposed mechanisms. Generally, a successful weld has been characterized by a wavy morphology with no IMCs. But depending on material properties, crystal structures of the metal pairs to be joined, and impact conditions, the welding interface may also consist of defects characterized by an abrupt failure due to the excessive temperature and formation of discontinuous voids along the interface (see Fig. 1e) and the onset of welding characterized by a waveless interface that looks like an adhesive bond (see Fig. 1f) [8].

To perceive a clear correlation between the emergence of wavy morphology and quality of the weld, numerical simulation provides an alternative tool of scientific investigation, instead of conducting expensive, time-consuming experiments in laboratories. Also predicting process parameters, especially temperature at the interface will reduce the number of trial and error experimental tests in a production setting. However, due to the highly coupled and dynamic nature of the HVIW processes, numerical simulations are very challenging, especially near the welding zone where large deformations occur. Thus, a traditional Lagrangian method where the mesh is fixed to the component geometry fails as a results of excessive element distortion.

An Arbitrary Lagrangian-Eulerian (ALE) method is an alternative approach. This method is a technique that combines the features of Lagrangian and Eulerian analyses within the same mesh to maintain a high quality during simulations involving large deformation. The ALE method has been used in many research efforts including the study of a solid–fluid interaction [14], high strain rate process [1], and metal cutting-forming process [15].

Despite the great success, mesh-based numerical methods suffer from difficulties in some aspects which limit their applications in many complex problems. For example, employing a mesh can lead to difficulties in dealing with problems with deformable boundary, moving interface, and extremely large deformation. With the advancement in hardware capability and parallel processing, the next generation of computational methods, so called *meshless* methods, have received significant attention. Several meshless methods have been proposed for different applications. Although these methods may share common features, they are different in the function approximation, approach for creating discretized system equations, and detailed implementation process [16]. Among all meshless methods, Smoothed Particle Hydrodynamics (SPH) has received significant attention. In SPH, a collection of particles is used to represent a continuum body. These particles have a spatial location in the numerical model. SPH has been used in many fields including ballistic impact [17], non-Newtonian fluids [18], and high-velocity impact welding [19].

In this study, in order to further understand the science behind HVIW and to compare the drawback of various numerical simulation methodologies, the impact between Al6111-T4 and high strength steel (JSC980) was simulated using the SPH and ALE methods. The benefits of SPH modeling with respect to capturing the jetting effect and voids at the interface were demonstrated. To experimentally validate the numerical efforts, VFAW was implemented. In the experimental test, Photon Doppler Velocimetry (PDV) was used to measure the impact velocity.

Scanning Electron Microscope (SEM) was used to investigate for evidence of melting and IMCs. Good agreement between the numerical simulations and experimental results provided confidence in the numerical modeling.

Numerical Simulation Model

The 2D numerical simulations were carried out to simulate an impact between two Al6111-T4 and JSC980 plates. The dimensions of the flier (i.e., Al6111-T4) and base plate (i.e., JSC980) are 20 and 1 mm and 20 and 1.4 mm, respectively. The simulations were conducted by imposing a flier plate velocity normal to the interface (i.e., impact velocity, V_F = 727 m/s) and fixing the initial angle between the flier and base plate (ß = 10°) as initial conditions. The bottom of the base plate was fixed as a boundary condition (see Fig. 2). The material behavior was modelled using a Johnson-Cook constitutive relation:

$$\sigma = (A + B\varepsilon_{eff}^{n})(1 + Cln\dot{\varepsilon})(1 - T^{*m}) \quad (1)$$

where σ is the flow stress, ε_{eff} is the effective plastic strain, $\dot{\varepsilon} = \frac{\dot{\varepsilon}_{eff}}{\dot{\varepsilon}_0}$ is the plastic strain rate, $T^* = \frac{T - T_{room}}{T_{melt} - T_{room}}$ is the homologous temperature, and A, B, C, n and m are material parameters. The Johnson-Cook parameters for Al6111-T4 and JSC980 are given in Table 1 [20, 21]. Due to the high strain rate, adiabatic heating was assumed with an inelastic heat fraction of 0.9 [22]. The pressure at the contact point, which is incorporated in the stress tensor, was computed using Gruneisen state equation [23].

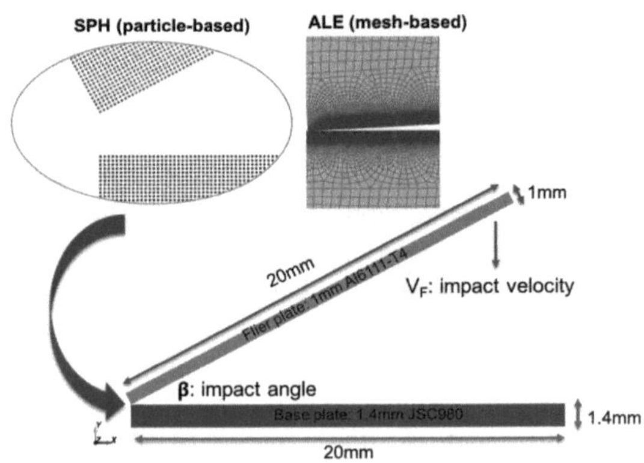

Fig. 2 FEA model setup

Table 1 Material properties, Gruneisen state equation constant and Johnson-Cook parameters for Al6111-T4 and JSC980

Material	ρ (kg/m^3)	E (GPa)	v	k (W/mk)	C_p (J/kg K)				
Al6111-T4	2710	70	0.33	154	900				
JSC980	7860	214	0.3	30	680				
	C (m/s)	S1	S2	S3	γ_0				
Al6111-T4	5451	1.259	0	0	2.17				
JSC980	4569	1.49	0	0	2.17				
	A (MPa)	B (MPa)	C	n	m	$\dot{\varepsilon}_0$	T_{melt} (K)	T_{room} (K)	
Al6111-T4	153	528	0.0026	0.5176	3.09	1	923	294	
JSC980	980	2000	0.0026	0.83	1.4	1	1643	292	

SPH Model

SPH simulation was carried out using the LS-DYNA software package. Total of 1,932,966 particles were generated in the model. The main advantage of the SPH methods is to bypass the requirement for a numerical grid to calculate spatial derivatives. This avoids severe problems associated with mesh tangling and distortion which usually occur in classical Lagrangian analysis involving high strain rate events. The method converts a set of partial differential equations (PDE) into a set of ordinary differential equations (ODE) [16].

ALE Model

ALE simulation was carried out using the Abaqus Explicit software package. ALE consists of two fundamental tasks at each time increment: (1) creating a new mesh and (2) remapping the solution variables from the old mesh to the new mesh. A structured quad mesh (i.e., 10 μm × 10 μm) was used in near the contact region. To avoid a spurious solution due to the potentially flexible elements, hourglassing was controlled through a viscous approach. To control the adaptive remeshing process, the smoothing algorithm was set to improve the aspect ratio in each iteration based on the evolving geometry. The frequency of remeshing was set to 1 and adaptive remeshing was executed with 40 sweeps per increment. Using these two parameters, a new mesh was found by sweeping iteratively over the adaptive mesh domain and moving nodes to smooth the mesh.

Experimental Test

To validate the numerical simulations, experimental tests with VFAW process were conducted. Figure 3 illustrates the setup for VFAW. This method was invented by Daehn's research group at the Ohio State University [24]. In terms of the length scale of the weld, this technique is very close to MPW but unlike MPW, which tends to work best only for highly conductive materials, the efficiency of VFAW does not depend on the conductivity of the flier material. In addition, the foil actuator in VFAW is a consumable of each cycle, while MPW commonly uses a permanent solenoid actuator (see Refs. [1, 12] for further information about VFAW).

The horizontal distance between two stand-offs (D) was set to 30.8 mm to create an oblique collision of 10°. The aluminum dog-bone shaped foil was connected to the terminals of the capacitor bank that acts as the current source. When the capacitor bank is discharged, the foil is electrically vaporized within tens of microseconds and the expanding gas from the rapid vaporization of the foil provides the driving force to repel the flier plate into a high speed collision with the base plate. Photon Doppler Velocimetry (PDV) was used to measure the impact velocity [1, 12] and used as input into the numerical simulations.

Results

Figure 4 shows the comparison between the experimental observation and numerical simulations. Both SPH and ALE methods were able to capture the wavy morphology. The average predicted wavelength by the SPH and ALE methods were 52 and 70 μm, respectively, while the experimental test shows that the average wavelength was 59 μm. As is clear, the predicted wave by the ALE method the joint was weaker than the SPH method. This is due to the severe deformation at the interface that degrades the mesh quality. The flattened element which is caused by

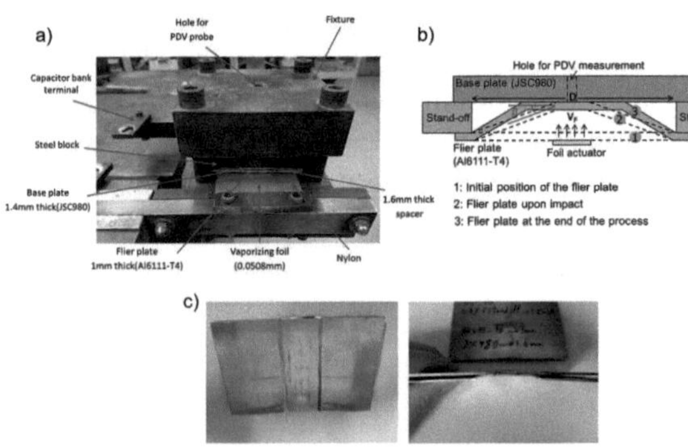

Fig. 3 **a** Experimental setup, **b** schematic of the VFAW process, and **c** successful welded sample

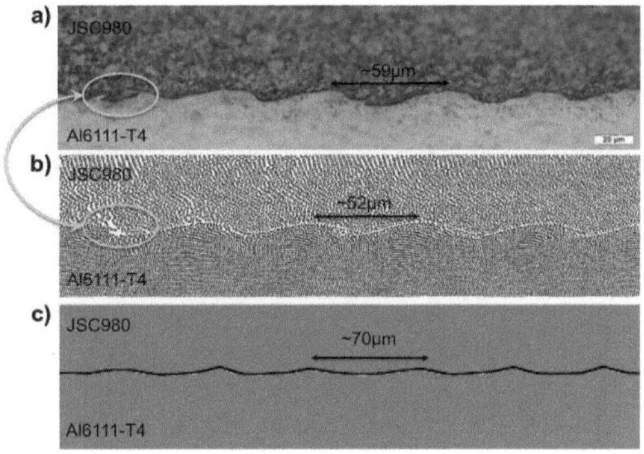

Fig. 4 Interface between Al6111-T4 and JSC980 **a** optical micrograph, **b** SPH numerical simulation, and **c** ALE numerical simulation

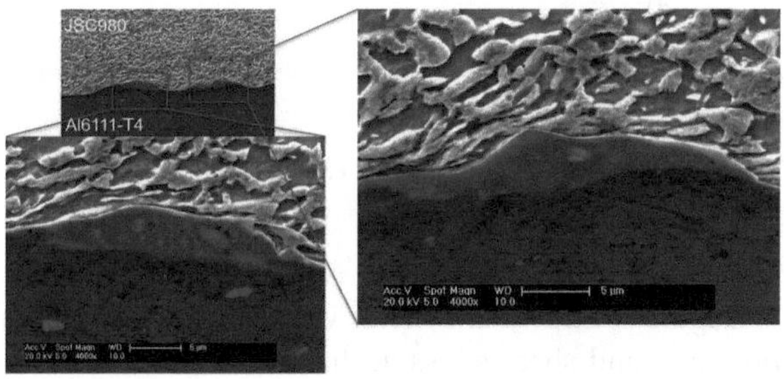

Fig. 5 SEM images of the interface between Al6111-T4 and JSC980 at a wavy region

Fig. 6 SEM images of the interface between Al6111-T4 and JSC980 at a straight region

significant impact pressure and intense interfacial shearing could be a reason limiting the ALE computation despite the adaptive remeshing during the evolution of the interface. To investigate for evidence of melting and IMCs along the cross section, Scanning Electron Microscope (SEM) was used. As shown in Figs. 5 and 6, continuous IMCs were found along the interface both in the wavy and straight regions. Although the formation of IMCs could be detrimental to the weld quality [25], microstructural analyses show that the thickness of the IMCs layer along the interface was mainly less than 5 μm. Such a narrow region could provide an inter material-bonding role for the Al/Fe system and behave less as an embrittling phase.

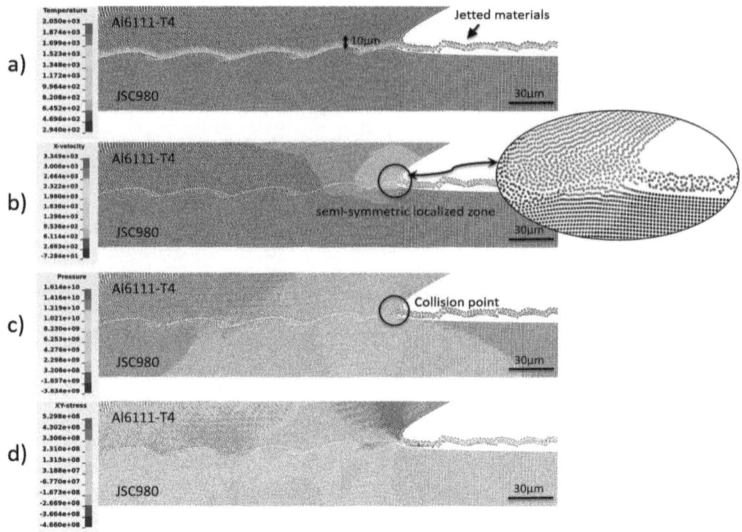

Fig. 7 Predicted **a** temperature, **b** collision velocity, **c** pressure, and **d** shear stress based on SPH method

Figure 7 shows the field variables including temperature, collision velocity, pressure, and shear stress at the interface between the two welded materials based on the SPH method. As is clear in Fig. 7a, the predicted temperature was highly localized near the interface where the plastic deformation developed. Since the process occurs over a short time scale, there is no time for heat transfer away from the interface. Thus any potential melting and IMCs formation will be localized. Numerical modeling shows that the thickness of this region is approximately less than 10 μm. Figure 7b shows the collision velocity field near the contact zone. As is evident, a semi-symmetric localized high velocity zone (with collision point velocity of ~2300 m/s) was developed in the Aluminum side which has less density and shear modulus. This may suggest that for this material combination, the jetted materials were mainly composed of Al6111-T4 rather than JSC980. Figure 7c shows that the pressure at the collision point is approximately ~16 GPa.

Figure 7d shows the shear stress where different signs were produced at the interface. The interface with opposite shear stresses on both sides of the interface may confirm a successful bond because the same orientation shearing prevents bonding.

Figure 8 shows the similar field variables based on the ALE method. As is clear, the predicted temperature in ALE was much less (~730 K) than the SPH method (>1000 K consistently at the interface with local particles up to ~2000 K). The predicted collision velocity in ALE was matched with the SPH method, while the pressure and shear stress in the ALE were less and higher, respectively, compared to the SPH results. This is likely due to element distortion.

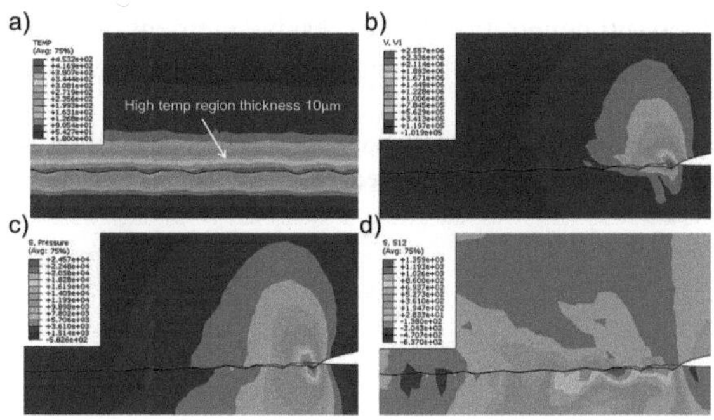

Fig. 8 Predicted **a** temperature, **b** collision velocity, **c** pressure, and **d** shear stress based on ALE method

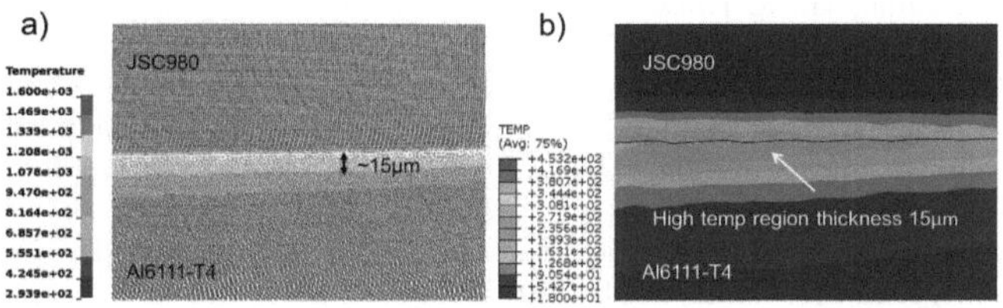

Fig. 9 Predicted temperature profile in the straight region in **a** SPH simulation, **b** ALE simulation

Figure 9 shows the temperature field in the straight region based on both SPH and ALE methods. As is clear, both methods confirm that the high temperature region with a thickness of approximately 15 μm was localized close to the interface, although the predicted temperature based on the ALE method was much less than the SPH method.

Conclusion

In this study, the numerical simulations were conducted for the HVIW process by two different approaches: a SPH method and an ALE method. The accuracy of each method was corroborated by comparisons with experimental observations. The SPH numerical results were in good agreement with the experimental tests including the capturing of the characteristic wavy morphology, temperature prediction and interfacial melting observed through microstructural analysis. The ALE simulation results were capable of predicting the wavelength, but it most likely can produce only a weakly shaped interface in case of dissimilar impact welding. Also, using SPH method, metal jet emission was investigated which previously was impossible. To see the effect of IMCs on the strength of the weld, tension and hardness tests will be conducted. In addition, the jetted materials will be collected during the experimental test and characterized in terms of composition and compared with the SPH results. Finally, to avoid penetration and excessive element distortion, a new contact algorithm will be defined and implemented for the ALE method.

Acknowledgments Funding from the U.S. National Science Foundation (CMII-1537471) and Honda Motor Company are gratefully acknowledged. Many thanks to Genevieve Lee for helping to acquire SEM images.

References

1. Nassiri, A., Chini, G., Vivek, A., Daehn, G., & Kinsey, B. (2015). Arbitrary Lagrangian-Eulerian finite element simulation and experimental investigation of wavy interfacial morphology during high velocity impact welding. *Materials and Design, 88,* 245–358.
2. Nassiri, A., Chini, G., & Kinsey, B. (2014). Spatial stability analysis of emergent wavy interfacial patterns in magnetic pulsed welding. *CIRP Annals-Manufacturing Technology, 63*(1), 245–248.
3. Nassiri, A., Kinsey, B., & Chini, G. (2016). Shear instability of plastically-deforming metals in high-velocity impact welding. *Journal of the Mechanics and Physics of Solids, 95,* 351–373.
4. Abrahamson, G. R. (1961). Permanent periodic surface deformations due to a traveling jet. *Journal of Applied Mechanics, 28*(4), 519–528.
5. Bahrani, A. S., Black, T. J., & Crossland, B. (1967). The mechanics of wave formation in explosive welding. *Proceedings of the Royal Society of London Series A. Mathematical and Physical Sciences, 296*(1445), 123–136.
6. Cowan, G. R., Bergmann, O. R., & Holtzman, A. H. (1971). Mechanism of bond zone wave formation in explosion-clad metals. *Metallurgical and Materials Transactions B, 2*(11), 3145–3155.
7. El-Sobky, H., & Blazynski, T. Z. (1975). Experimental investigation of the mechanics of explosive welding by means of a liquid analogue. In *Proceeding of 5th International Conference on "High Energy Rate fabrication", Denver, Colorado* (pp. 1–21).
8. Sapanathan, T., Raoelison, R. N., Padayodi, E., Buiron, N., & Rachik, M. (2016). Depiction of interfacial characteristic changes during impact welding using computational methods: Comparison between Arbitrary Lagrangian-Eulerian and Eulerian simulations. *Materials and Design, 102,* 303–312.
9. Manikandan, P., Hokamoto, K., Deribas, A. A., Raghukandan, K., & Tomoshige, R. (2006). Explosive welding of titanium/stainless steel by controlling energetic conditions. *Materials Transactions, 47*(8), 2049–2055.
10. Faes, K., Baaten, T., De Waele, W., & Debroux, N. (2010). Joining of Copper to Brass using magnetic pulse welding. In *Proceedings of the 4th International conference on High Speed Forming, Columbus, OH* (pp. 84–96).
11. Kore, S. D., Imbert, J., Worswick, M. J., & Zhou, Y. (2009). Electromagnetic impact welding of Mg to Al sheets. *Science and Technology of Welding and Joining, 14*(6), 549–553.
12. Vivek, A., Hansen, S. R., Liu, B. C., & Daehn, G. S. (2013). Vaporizing foil actuator: A tool for collision welding. *Journal of Materials Processing Technology, 213*(12), 2304–2311.
13. Raoelison, R. N., Sapanathan, T., Buiron, N., & Rachik, M. (2015). Magnetic pulse welding of Al/Al and Al/Cu metal pairs: Consequences of the dissimilar combination on the interfacial behavior during the welding process. *Journal of Manufacturing Processes, 20,* 112–127.
14. Sarrate, J., Huerta, A., & Donea, J. (2001). Arbitrary Lagrangian-Eulerian formulation for fluid–rigid body interaction. *Computer Methods in Applied Mechanics and Engineering, 190* (24), 3171–3188.

15. Wen, Q., Guo, Y. B., & Todd, B. A. (2006). An adaptive FEA method to predict surface quality in hard machining. *Journal of Materials Processing Technology, 173*(1), 21–28.
16. Liu, G. R., & Liu, M. B. (2003). *Smoothed particle hydrodynamics: A meshfree particle method.* Singapore: World Scientific publisher.
17. Hayhurst, C. J., & Clegg, R. A. (1997). Cylindrically symmetric SPH simulations of hypervelocity impacts on thin plates. *International Journal of Impact Engineering, 20*(1), 337–348.
18. Xu, X., Ouyang, J., Yang, B., & Liu, Z. (2013). SPH simulations of three-dimensional non-Newtonian free surface flows. *Computer Methods in Applied Mechanics and Engineering, 256,* 101–116.
19. Nassiri, A., & Kinsey, B. (2016). Numerical studies on high-velocity impact welding: smoothed particle hydrodynamics (SPH) and arbitrary Lagrangian–Eulerian (ALE). *Journal of Manufacturing Processes.* 10.1016/j.jmapro.2016.06.017
20. Smerd, R. (2005). Constitutive behavior of aluminum alloy sheet at high strain rates. Ph.D. dissertation, Waterloo, Canada.
21. Banerjee, A., Dhar, S., Acharyya, S., Datta, D., & Nayak, N. (2015). Determination of Johnson cook material and failure model constants and numerical modelling of Charpy impact test of armour steel. *Materials Science and Engineering A, 640,* 200–209.
22. Rittel, D. (1999). On the conversion of plastic work to heat during high strain rate deformation of glassy polymers. *Mechanics of Materials, 31*(2), 131–139.
23. Hallquist, J. O. (2006). *LS-DYNA theory manual.* Livermore: Livermore Software Technology Corporation.
24. Johnson, J. R., Taber, G., Vivek, A., Zhang, Y., Golowin, S., Banik, K., et al. (2009). Coupling experiment and simulation in electromagnetic forming using photon doppler velocimetry. *Steel Research International, 80*(5), 359–365.
25. Nassiri, A. (2015). *Investigation of wavy interfacial morphology in magnetic pulsed welding: Mathematical modeling, numerical simulations and experimental tests* (p. 167). University of New Hampshire.

Selective Laser Sintering of Polyamide/Hydroxyapatite Scaffolds

Frederic Dabbas, Steferson Luiz Stares, José Maria Mascheroni, Dachamir Hotza and Gean Vitor Salmoria

Abstract Selective Laser Sintering (SLS) is an additive manufacturing technique that enables final products to be processed without additional machining. SLS permits the fabrication of implants and scaffolds with complex geometry for biomedical applications. In this study, composite scaffolds of polyamide (PA2200) filled with particles of hydroxyapatite (HA) were fabricated using SLS. The microstructure and mechanical properties were characterized. The effects of SLS processing parameters, including particle content and laser power, were investigated. Particle content and laser energy play a key role in the final density and mechanical properties of the sintered components. This study demonstrated that HA-reinforced PA can be successfully manufactured by SLS with controlled porosity features.

Keywords Additive manufacturing · Selective laser sintering · Polyamide · Hydroxyapatite · Scaffolds

---

F. Dabbas · G.V. Salmoria
Laboratory of Innovation on Additive Manufacturing and Molding (NIMMA), Federal University of Santa Catarina (UFSC), Florianópolis, SC 88040-900, Brazil
e-mail: f.dabbas@gmail.com

G.V. Salmoria
e-mail: gean.salmoria@ufsc.br

S.L. Stares · D. Hotza (✉)
Interdisciplinary Laboratory for the Development of Nanostructures (LINDEN), Federal University of Santa Catarina (UFSC), Florianópolis, SC 88040-900, Brazil
e-mail: d.hotza@ufsc.br

S.L. Stares
e-mail: steferson.stares@ufsc.br

J.M. Mascheroni
Alkimat Technology, São José, SC, Brazil
e-mail: jose@alkimat.com.br

© The Minerals, Metals & Materials Society 2017
M.A. Meyers et al. (eds.), *Proceedings of the 3rd Pan American Materials Congress*, The Minerals, Metals & Materials Series, DOI 10.1007/978-3-319-52132-9_10

Introduction

The ultimate goal in the area of reconstructive medicine is the structural and functional restoration of tissue to its natural state. Bone tissue engineering offers an alternative approach for repairing bone defects. Scaffolds, which are one of the key for tissue engineering, should possess sufficient mechanical strength to provide structural support and porous structure to guide new bone tissue in-growth. Moreover, the scaffolds should also possess good biocompatibility and osteoconductivity [1].

Bioceramics like hydroxyapatite and polymers like polyamide have been given a lot of attention as candidate scaffolds materials since they possess highly desirable characteristics. A composite material may be prepared in such a way that the final product may acquire some excellent properties that cannot be found individually in either material [2, 3].

Selective laser sintering (SLS) is an additive manufacturing process that can be used for the fabrication of composite components. The process starts by applying a thin layer of the powder materials to the building platform. A powerful laser beam then fuses the powder at exactly the points defined by the computer-generated component design data. The platform is lowered and another layer of powder is applied. Then the material is fused to bond with the layer below at the predefined points. The process can be used to create a wide variety of shapes with tailored macro- and microscopic porosities for a broad field of applications [4, 5].

In the present work, scaffolds have been developed with the goal of fabricating bioactive structures to use in bone tissue engineering. Sintering behavior and microstructure were studied as well as the strength and elastic modulus.

Experimental

Materials

The powders used in this study were commercial polyamide (PA2200, Eosint) with average particle size of 60 μm, and hydroxyapatite (HA, Fluka, 90%) with a mean particle size of 5 μm. The HA content in the composites were 0, 5, 10 and 20 wt%. The composites were prepared with a Y mixer for a period of 2 h.

Selective Laser Sintering

The composite specimens (dimensions: 35 mm × 5 mm × 1.4 mm) were processed by Selective laser sintering (SLS) using a RF-excited CO_2 laser, with a wavelength of 10.6 μm, laser beam diameter of 250 μm, scan speed of 57 mm s^{-1}

Table 1 SLS process parameters used in manufacturing of PA2200/HA scaffolds

PA2200/HA (wt/wt)	100/0	95/5	90/10	80/20
Laser power (W)	4		5	
Energy density (W mm^{-2})	0.281		0.351	
Layer thickness (μm)	150			
Powder bed temperature (°C)	140			

and chamber temperature of 110 °C. The further processing parameters used are listed in Table 1.

Mechanical Tests

Ultimate strength and elastic modulus (bending) were performed on a TA Instruments analyzer, model Q800, with single cantilever mode. Stress–strain curves were obtained at a strain rate of 2 mm min^{-1} and 30 °C. The mean values and standard deviations were calculated.

Scanning Electron Microscopy (SEM)

The composite scaffolds were observed under a scanning electron microscope (SEM, Phillips XL30) in order to investigate the fracture surface, particle aspects and microstructure. The scaffolds were coated with gold (Bal-Tec Sputter Coater SCD005).

Density and Open Porosity

The theoretical density (TD, g cm^{-3}) was determined from weight and volume measurements. The apparent density (AD, g cm^{-3}) was calculated according to the rule of mixture. From these values, the open porosity (OP, %) of composite specimens was calculated by Eq. (1),

$$OP = \frac{TD - AD}{TD} \tag{1}$$

Results and Discussion

The PA2200/HA scaffolds consist of 9 selective laser sintered layers with 150 μm layer thickness of powder deposition. Table 2 summarizes the properties of the manufactured specimens. During the sintering of composite scaffolds a significant laser energy absorption by materials was observed, making sintering possible at low laser power. In a previous work [6], FTIR evaluations showed that both PA2200 and HA materials, used in this work, have a significant absorption at a wavelength of 10.6 μm. Moreover, preheating (140 °C) of the powder helps to lower the energy required from laser and ensures the stability of the processed parts during the sintering process.

A larger amount of hydroxyapatite resulted in more laser power for sintering. This indicates that the amount of HA greatly influences the degree of sintering. This can be explained by the change in the specific heat capacity of the material [7]. The specific heat capacity is the amount of energy required by a material specimen to raise the material temperature by 1 K. At a larger HA ratio, the specific heat capacity of the material is reduced, indicating that the amount of heat that the material requires to raise the material's temperature by 1 K was lower. Hence, there would be less stored heat within the sintering system, resulting in lower conduction of heat during sintering and less fusion between particles. Savalani and coauthors [8] have shown that the specific heat capacity value of PA-HA was the half than that of pure PA, which suggested that materials with high HA content may absorb a great amount of the energy supplied by the laser power in comparison to pure PA. As a result, the effect of laser power on the material was less significant.

Among the various sintering parameters, the laser power and laser scan speed directly determine the amount of energy imparted on the parts in the SLS process and are directly related to the specific energy density, which is the most important parameter in SLS. Beal et al. [9] evaluated the effect of the variation of energy density, by varying laser power and laser scan speed, on the mechanical properties of a polymeric material. It was found that laser power had more influence on density and mechanical properties of the samples than laser scan speed. In this work, the variation in power from 4 to 5 W resulted in the increase in the apparent density. This can be explained by the fact that higher energy density levels cause a better fusion of the powder particles, resulting in a more solid part being formed [10]. Moreover, since the density of HA (3.15 g cm^{-3}) was higher than that of the PA2200 (0.54 g cm^{-3}), the composite density increases with increase in HA

Table 2 Composition, process parameters and properties of PA2200/HA composites manufactured by SLS

PA/HA (wt/wt)	100/0	95/5	90/10	80/20
Laser power (W)	4		5	
Energy density (W mm^{-2})	0.281		0.351	
Theoretical density (g cm^{-3})	1.020	1.031	1.044	1.075
Apparent density (g cm^{-3})	0.538	0.505	0.572	0.632
Porosity (%)	47	51	45	41

content at constant volume. The theoretical density was calculated according to the weight and volume measurements. Based on the calculations, the open porosity ranges varied from 41 to 51%. Savalani et al. also showed that the apparent density of laser sintered polyamide and hydroxyapatite powder increased with increasing energy density [11].

In tissue engineering, a number of architectural characteristics, including porosity, pore size, and permeability are significant parameters in biological delivery and tissue regeneration. The ability to control scaffold architecture can provide significant insights into how scaffold architecture and material affect tissue regeneration. Previous studies have found that, for bone scaffolds, both macro and microporosity play an important role in osteogenesis and bone regeneration [12, 13]. Scaffolds with high porosity enable effective release of biofactors such as proteins, genes, or cells and provide good substrates for nutrient exchange. However, the mechanical property that is important in maintaining the structural stability of the biomaterial is often compromised as the result of increased porosity [14]. Previous works have shown that scaffolds with level of porosity from 40 to 65% were able to promote the connection and proliferation of cells in biological tests [15–17].

Figure 1 shows representative SEM microstructure images of PA2200/HA of scaffolds after SLS. The particles presented a good interparticle bonding and the

Fig. 1 SEM images of PA2200/HA composites (240×): **a** 100/0, **b** 95/5, **c** 90/10 and **d** 80/20 PA/HA (weight proportion)

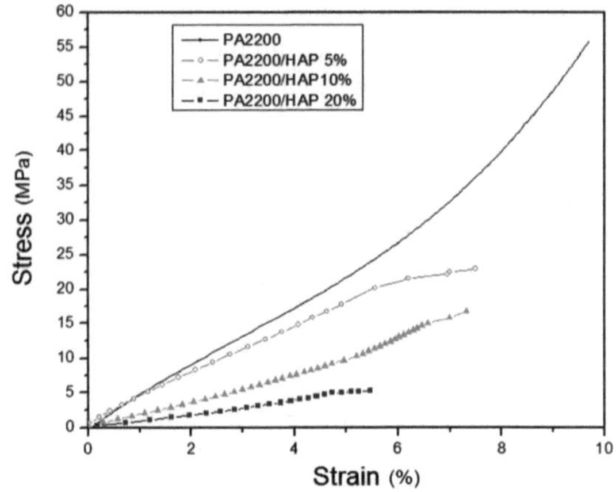

Fig. 2 Stress versus strain curves for pure PA2200 and PA2200/HA composites manufactured by SLS

Table 3 Elastic moduli and ultimate strength values of PA2200/HA composites manufactured by SLS

PA/HA (wt/wt)	100/0	95/5	90/10	80/20
Elastic modulus (GPa)	0.38 ± 0.05	0.31 ± 0.08	0.17 ± 0.03	0.12 ± 0.01
Ultimate strength (MPa)	55.7 ± 10	22.2 ± 9	17.7 ± 3	5.2 ± 1

HA particles are homogeneously distributed between the PA2200 matrix. It is also possible to observe the presence of pores with sizes ranging from 20 to 100 μm. For bone scaffolds, in general, pore sizes in the range 20–60 μm are ideal for cell spreading and proliferation. However, a minimum pore size of approximately 100 μm is considered sufficient to enable bone tissue regeneration to occur [18–20].

Figure 2 shows representative curves of the stress versus strain behavior for pure PA2200 and PA2200/HA sintered composites (80/20, 90/10, and 95/5) obtained by flexural mechanical analysis. Table 3 presents the average value for the elastic modulus and the ultimate strength of specimens.

As expected, sintered composites had a lower value for the elastic modulus and ultimate strength than the sintered PA2200 pure specimen in the flexural tests [21]. The low values for the ultimate strength and elastic modulus of the 80/20 and 90/10 composite specimens indicate the low chemical affinity between the PA2200 and HA phases. Higher weight percentage of HA means less binder and consequently less material to fuse and facilitate composite particulate bonding. On the other side, the higher ultimate strength and elastic modulus of 95/5 specimen is because the PA phase provides greater strength and stiffness. In contrast to the literature [22, 23], in this work, the increase of laser power did not promote the increase on the mechanical properties of sintered scaffolds, as expected by the fact that higher energy density levels cause a better fusion of the powder particles, resulting in a more solid part being formed. The effect of HA content was more significant that the effect of laser power, resulting on decrease of mechanical properties.

The values calculated for the standard deviation of the elastic modulus and ultimate strength results for all specimens indicated the presence of defects due to the particle packing process, influencing the mechanical behavior of the material. The complex relationship between the effects of the laser powder sintering fabrication process and the composites composition on the microstructure formation (sintering degree, co-continuous phases and porosity) plays an important role in the specimen properties, and can lead to high standard deviation values for the elastic modulus and ultimate strength of the composite specimens [24].

About mechanical properties, the bone structure is dynamic in time and becomes fit to the imposed stresses in an adaptive manner, the values of their mechanical properties are not constant from bone to bone, from individual to individual, and therefore, only mean data are obtained. For the cortical human bone, for example, ultimate strength values in the order of 26–83 MPa and elastic modulus values of 1.6–4.4 GPa depending on the density and orientation [25, 26]. The strength values of PA2200/HA 100/0, 95/5 and 90/10 obtained in this work are in the range of strength of the cortical bone, which can make them suitable for use in bone tissue engineering.

Conclusions

In the present work, porous PA2200/HA structures were fabricated by Selective Laser Sintering (SLS) for use in bone reconstruction. Particle content and laser energy play a key role on the final density and mechanical properties of the sintered scaffolds. These structures show porosity of 41–51%, ultimate strength of 5.2–55.7 MPa and elastic modulus of 0.12–0.38 GPa. The properties of scaffolds produced are closely within the range of the reported experimental data for human cortical bone. Further analysis to evaluate the connectivity, biocompatibility and kinetics of degradation, are in course to that make the composites suitable for applications as scaffolds for repairing cortical defects.

References

1. Feng, P., et al. (2014). A novel two-step sintering for nano-hydroxyapatite scaffolds for bone tissue engineering. *Scientific Reports, 4,* 5599–5609.
2. Neuendor, R. E., et al. (2008). Adhesion between biodegradable polymers and hydroxyapatite: Relevance to synthetic bone-like materials and tissue engineering scaffolds. *Acta Biomaterialia, 4,* 1288–1296.
3. Supova, M. (2009). Problem of hydroxyapatite dispersion in polymer matrices: A review. *Journal of Materials Science Materials in Medicine, 20,* 1201–1213.
4. Liu, F. H. (2014). Synthesis of biomedical composite scaffolds by laser sintering: Mechanical properties and in vitro bioactivity evaluation. *Applied Surface Science, 297,* 1–8.

5. Liu, F. H., Shen, Y. K., & Lee, J. L. (2012). Selective laser sintering of a hydroxyapatite silica scaffold on cultured MG63 osteoblasts in vitro. *International Journal of Precision Engineering and Manufacturing, 13,* 439–444.
6. Dabbas, F. (2006). Selective laser sintering of composite polyamide/hydroxyapatite and polyethylene/hydroxyapatite with applicable functional gradients in tissue engineering (M.Sc. Dissertation, Federal University of Santa Catarina, 2006), 85–95.
7. Childs, T. H. C., et al. (1999). Selective laser sintering of an amorphous polymer—Simulations and experiments. *Proceedings of the Institution of Mechanical Engineers, Part B, Journal of Engineering Manufacture, 213,* 333–349.
8. Savalani, M. M., et al. (2007). Fabrication of porous bioactive structures using the selective laser sintering technique. *Proceedings of the Institution of Mechanical Engineers Part H: Journal of Engineering in Medicine, 221-H*(8), s873–886.
9. Beal, V. E., et al. (2009). Statistical evaluation of laser energy density effect on mechanical properties of polyamide parts manufactured by selective laser sintering. *Journal of Applied Polymer Science, 5,* 2910–2919.
10. Caulfield, B., McHugh, P. E., & Lohfeld, S. (2007). Dependence of mechanical properties of polyamide components on build parameters in the SLS process. *Journal of Materials Processing Technology, 182,* 477–488.
11. Savalani, M. M., et al. (2012). The effects and interactions of fabrication parameters on the properties of selective laser sintered hydroxyapatite polyamide composite biomaterials. *Rapid Prototyping Journal, 18,* 16–27.
12. Karageorgiou, V., & Kaplan, D. (2005). Porosity of 3D biomaterial scaffolds and osteogenesis. *Biomaterials, 26,* 5474–5491.
13. Hulbert, S. F., Morrison, S. J., & Klawitte, J. J. (1972). Tissue reaction to 3 ceramics of porous and non-porous structures. *Journal of Biomedical Materials Research, 6,* 347–374.
14. Hollister, S. (2005). Porous scaffold design for tissue engineering. *Nature Materials, 4,* 518–524.
15. Liao, H. T., et al. (2011). Fabrication of tissue engineered PCL scaffold by selective laser-sintered machine for osteogenesis of adipose-derived stem cells: The research has proven that a bone tissue engineered scaffold can be made using the selective laser sintering method. *Virtual and Physical Prototyping, 6,* 57–60.
16. Sudarmadji, N., et al. (2011). Investigation of the mechanical properties and porosity relationships in selective laser-sintered polyhedral for functionally graded scaffolds. *Acta Biomaterialia, 7,* 530–537.
17. Shuai, C., et al. (2013). Optimization of TCP/HAP ratio for better properties of calcium phosphate scaffold via selective laser sintering. *Materials Characterization, 77,* 23–31.
18. Yang, S. F., et al. (2001). The design of scaffolds for use in tissue engineering. Part 1. Traditional factors. *Tissue Engineering, 7,* 679–689.
19. Hao, L., Lawrence, J., & Chian, K. S. (2004). On the effects of CO_2 laser irradiation on the surface properties of a magnesia partially stabilised zirconia (MgO-PSZ) bioceramic and the subsequent improvements in sin human osteoblast cell adhesion. *Journal of Biomaterials Applications, 19,* 81–105.
20. Hollister, S. J., Maddox, R. D., & Taboas, J. M. (2002). Optimal design and fabrication of scaffolds to mimic tissue properties and satisfy biological constraints. *Biomaterials, 23,* 4095–4103.
21. Stares, S. L., et al. (2013). PLLA/HA composite laminates. *Advanced Engineering Materials, 15,* 1122–1124.
22. Kruth, J. P., et al. (2007). Consolidation phenomena in laser and powder-bed based layered manufacturing. *Manufacturing Technology, 56,* 730–759.
23. Goodridge, R. D., et al. (2012). Laser sintering of polyamides and other polymers. *Progress in Materials Science, 57,* 229–267.

24. Salmoria, G. V., et al. (2008). Selective laser sintering of PA12/HDPE blends: Effect of components on elastic/plastic behavior. *Polymer Testing, 27,* 654–659.
25. Cowin, S. C. (2001). *Bone mechanics handbook* (2nd ed., pp. 459–496). Boca Raton, FL: CRC Press.
26. Jameson, J. R. (2014). *Characterization of bone material properties and microstructure in osteogenesis Imperfecta/Brittle bone disease* (Ph.D. Thesis, Marquette University), pp. 101–117.

Part III
Materials for Green Energy

A Comparison Between Recycled Spent Zeolite and Calcite Limestone for Manganese Removal

Adarlene M. Silva, Rodrigo S. Figueiredo and Versiane A. Leao

Abstract The current work initially investigated the manganese removal by sorption in a recycled spent zeolite (faujasite) used in the oil industry. Equilibrium adsorption was described by the Langmuir isotherm ($r^2 > 0.99$) with a maximum loading of 10.9 mg/g-zeolite. The performance of the spent zeolite was compared with that of limestone (a low cost sorbent). In this latter case, equilibrium sorption was also modelled using the Langmuir isotherms with 1.03 mg/g-limestone as maximum uptake. Fixed-bed sorption on both materials was also studied and solid loadings increased with the aqueous metal concentration. The Thomas model was selected to describe the breakthrough curves and showed good correlation with the experimental data for both samples and indicated a maximum solid uptake of 0.33 mg/g-solid in faujasite and 0.03 mg/g in limestone.

Keywords Spent zeolite · Limestone · Manganese · Fixed-bed · Sorption

Introduction

Manganese and sulphate removal is currently considered to be one of the main challenges in the water management in the mineral industry. Because its water chemistry is more complex than most transition metals, manganese is not easily removed by simply increasing the solution pH (the practice most commonly used in the industry) and then a different approach must be selected to treat manganese-laden mine waters.

Manganese is an essential nutrient for the human body and thus there is no restriction on water-containing manganese for human consumption at low to moderate concentrations. Notwithstanding, it may cause gray or black staining and

A.M. Silva · R.S. Figueiredo · V.A. Leao (✉)
Universidade Federal de Ouro Preto, Campus Morro do Cruzeiro S/N Bauxita, Ouro Preto, MG, Brazil
e-mail: versiane@demet.em.ufop.br

© The Minerals, Metals & Materials Society 2017
M.A. Meyers et al. (eds.), *Proceedings of the 3rd Pan American Materials Congress*,
The Minerals, Metals & Materials Series, DOI 10.1007/978-3-319-52132-9_11

bitter metallic taste to drinking water and thus the USEPA defines a secondary drinking water standard of 50 μg/L. In addition, its concentration in wastewaters is also controlled with countries setting maximum limits around 1 mg/L [1].

Kothari [2] observed that manganese removal can be achieved by various methods depending on the water to be treated, but usually oxidation and precipitation by air or other oxidizing agents (chlorine, hypochlorite, potassium permanganate, ozone) is usually applied. Air is used to oxidize Mn^{2+} (as encountered in aqueous solutions) to Mn^{4+}, which then precipitates as MnO_2, but Mn^{2+} oxidation by air is kinetically slow at pH values below 8.5. Sorption process (adsorption or ion exchange) can also be applied to remove manganese.

Zeolites can be an interesting option for metal removal from mine waters through sorption. For instance, Jimenez et al. [3] assessed the effects of temperature and pH on the sorption of single metal solutions containing either Cr^{3+}, Cd^{2+}, Ni^{2+} or Mn^{2+} in the zeolite scolecite and a greater manganese removal at pH 6 was reported. This was justified by the low H^+ content at this pH, which did not compete with the Mn^{2+} ions for the sorption sites. Likewise, Omri and Benzina [4] studied the manganese sorption in activated carbon. The equilibrium data were fitted to Langmuir, Freundlich and Temkin isotherms. The sorption isotherm was best described by the Langmuir equation with 172 mg/g as maximum uptake at pH 4.0.

Taffarel and Rubio [5] investigated the manganese removal by a Chilean zeolite composed of mordenite and clinoptilolite at 25 °C and pH 6.0–6.8. An increase in pH resulted in larger sorption capacities, which was also credited to a reduced competition between protons and manganese by the sorption sites. The sorption isotherm was described by the Langmuir model suggesting the formation of a Mn^{2+} monolayer on the zeolite surface with an adsorption capacity of 20.9 mg/g for the NaOH-activated zeolite. Subsequently, Taffarel and Rubio [6] analyzed Mn^{2+} adsorption on Mn-oxide coated zeolite after activation with NaCl at pH 6.0 and the loading capacity increased to 29.7 mg/g.

Aziz et al. [7] studied the batch sorption of 2 mg/L single solutions of Cu, Zn, Pb, Cd, Ni and Cr on limestone at pH 8.5. The sorption data were fitted to the Langmuir isotherm, and nickel showed the highest loading (38 mg/kg-limestone) as compared with the other metals. In another study, a 3 g/L limestone pulp (particle size < 210 μm) was produced from a mixed metal solution (10 mg/L of Pb, Cu, Zn and Cd) at 25 °C with a pH of 6.0, in which the metals such as copper and lead showed the highest affinity for limestone [8].

Spent zeolites and other adsorbents are usually sent to landfilling, which requires a previous stabilization step. Therefore different strategies are proposed to dispose these kind of materials in such a way to use in the cement and ceramic industries [9] or to recycle as adsorbent [10]. Therefore, the approach selected in the current work was to compare manganese sorption on a spent zeolite from the oil industry with that of calcite limestone (a low cost sorbent) in both batch and fixed bed experiments, applying the Thomas model to describe the fixed-bed sorption on both materials.

The Thomas Model

Thomas model assumes that sorption can be described by the Langmuir kinetic equation [11]:

$$\frac{\partial_q}{\partial_t} = K_{T1}C(q_m - q) - K_{T2}q \tag{1}$$

In Eq. 1, q_m is the maximum sorption capacity (mg/g), K_{T1} is the second-order forward rate constant K_{T2} is the first-order reverse rate constant whereas C and q are respectively the aqueous concentration and the solid loading (mg/g) at time t.

A mass balance for fixed-bed results:

$$v\frac{\partial C_t}{\partial Z} + \frac{\partial C_t}{\partial t} + \frac{1-\varepsilon}{\varepsilon}\rho_p\frac{\partial q_t}{\partial t} = 0 \tag{2}$$

where v represents the interstitial velocity (cm/s) and ε bed porosity, C_t is the liquid phase concentration, z is the axial coordinate, ρ_p is the adsorbent density.

Thomas proposed the following solution for Eqs. 1 and 2:

$$\frac{C_t}{C_0} = \frac{J(\frac{n}{r}, nT)}{J(\frac{n}{r}, nT) + \{1 - J(n, \frac{nT}{r})\}\exp\{(1-\frac{1}{r})(n - nT)\}} \tag{3}$$

With $r = 1 + bC_0$, $n = \frac{\rho_p qkZ(1-\varepsilon)}{\varepsilon v}$, $T = \frac{\varepsilon(\frac{1}{b} + C_0)}{\rho_p q(1-\varepsilon)}\left(\frac{vt}{Z} - 1\right)$;

The J function is given by Eq. 4, with I_0 being the zero-order Bessel function of the first kind.

$$J(x, y) = 1 - \int_0^x \exp(-y - \tau)I_0(2y\tau)d\tau \tag{4}$$

Equation 5 is usually used as an approximation of Eq. 4 to reduce the computational work [12]:

$$J(x, y) = \frac{1}{2}\{1 - \mathrm{erf}(\sqrt{x} - \sqrt{y}) + \frac{\exp(-(\sqrt{x} - \sqrt{y})^2}{\sqrt{\pi}(\sqrt{y} + (xy)^{0.25})}\} \tag{5}$$

Equations 3–5 were solved using Mathematica 9.0 package to find the best values for k and q with minimizes the difference between experimental and predicted values of C/C_0 at different times.

Materials and Methods

A calcite limestone sample (2.68 ton/m^3, 40.3% calcium and 1.23% magnesium) was kindly provided by Brazilian producers. The −850 + 590 μm (0.569 m^2/g specific surface area) size fraction was used for the fixed-bed manganese sorption tests. The second adsorbent was a spent rare earth-doped zeolite used in hydrocarbon cracking. After washing with sodium hydroxide, the particle size fraction −104 + 74 μm (282.3 m^2/g surface area) was selected for the experiments.

Sorption isotherm was produced from the batch data using synthetic solutions. The procedure comprised transferring 120 mL of a limestone pulp (3 g/L solids) to 250 mL Erlenmeyer flasks. Manganese concentrations varied from 1.5 to 25 mg/L at pH 8.6–8.9. Furthermore, 0.5 g of zeolite was mixed with 100 mL of a manganese sulfate solution assaying from 50 to 120 mg/L Mn^{2+} at pH 6.5. The flasks containing these pulps were stirred at 600 cycles min^{-1} until equilibrium. At the end of the experiment, the loaded material was separated from the solution by simple filtration. The final and the initial solutions had their manganese levels determined by ICP-OES. Solid loading was determined by mass balance and such experiments were carried out at 23 ± 2 °C.

The fixed-bed experiments were performed with both the solid adsorbents, at 23 ± 2 °C and the experimental conditions were listed in Table 1. Solid particles were transferred to a glass column and glass wood felts were placed on the top and bottom of the fixed bed to prevent loss of the solid particles. After loading, distilled water was passed through the bed to remove any fine particles that could have been loaded in the column. The column was fed upward by a peristaltic pump (Milan) to either avoid or minimize low-resistance pathways for the solution, and the flow rate was varied from 1 to 3 mL/min. Samples of the column effluent were collected as a function of time and their manganese content was analysed in a ICP-OES (Varian, 725). Solid loading was determined by a mass balance on manganese.

Table 1 Experimental conditions applied in the experiments with either limestone or spent zeolite

Parameter	Spent zeolite	Limestone
Particle size	−104 + 75 μm	−850 + 590 μm
Mn conc. (mg/L)	10, 25, 50	1.5 and 15
Solution pH	6.5	6.5–8.0
bed depth (cm)	6	20
Column diameter (cm)	1.0	2.0
Flow rate (mL/min)	1 and 2	1, 2 and 3

Results and Discussion

In the current work, a comparison between the calcite limestone and the spent zeolite during manganese removal was undertaken. The Thomas equation was applied to model fixed-bed experiments whereas the sorption isotherm was described by the Langmuir equation.

The Thomas model was developed with the premise that the sorption isotherm follows the Langmuir equation and Fig. 1 shows the equilibrium sorption data fitted to such equation.

Figure 1 revealed that Mn^{2+} sorption in both the materials can be described by the Langmuir isotherm in which both had quite different affinity for the adsorbate. In limestone, the maximum manganese loading (q_{max}) was 1.03 mg/g with 10.92 mg/g for the spent zeolite. The spent zeolite approached a rectangular isotherm which indicates a high affinity for the sorbent and this reflected in the observed high b value (3.39). Assis et al. [7] observed the limestone loadings varying from 0.012 (Zn) to 0.038 mg/g (Ni) while studying Cu, Zn, Ni, Pb, Cd and Cr on limestone. Similarly, the maximum loading was 7.1 mg/g in a Chilean zeolite [5].

Subsequently fixed-bed experiments were performed so that the continuous Mn^{2+} sorption by both materials (limestone and spent zeolite) was investigated. These experiments were performed to study the effects of flow rate and initial manganese concentration on manganese sorption.

The Thomas equation reproduced the sorption process fairly accurately at low manganese concentrations (Fig. 2). According to Fig. 2, it is implied that an increase in aqueous manganese concentration caused a reduction in the time required to reach column saturation, which means that at higher metal concentrations, the volume of solution treated is smaller at the same flow rate. Similar results were observed during the adsorption of Congo-red dye in rice hulls [13]. This outcome was ascribed to the changes in the concentration gradient which affected

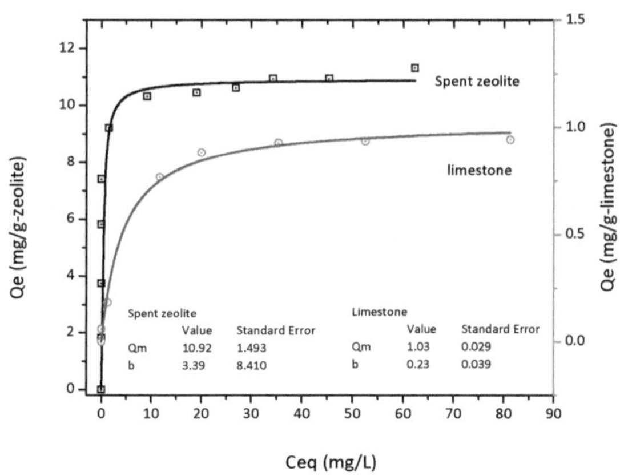

Fig. 1 Isotherms describing manganese sorption on either limestone or spent zeolites at 23 ± 2 °C

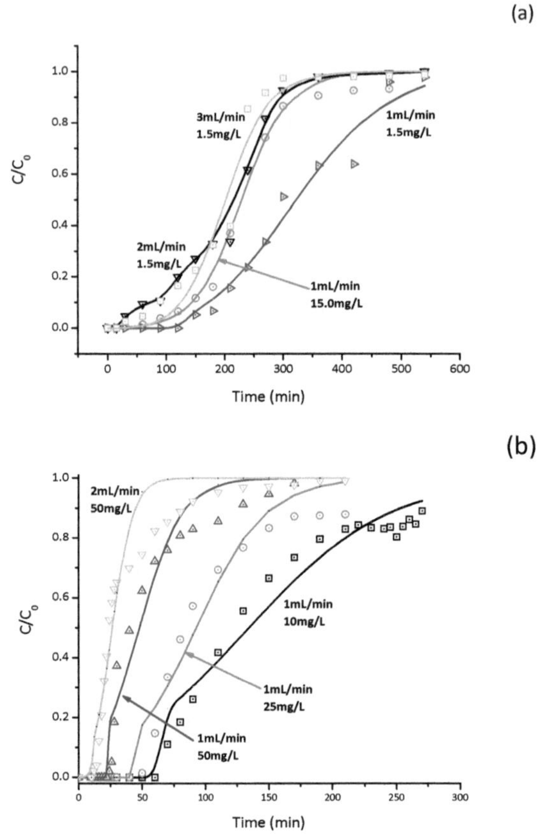

Fig. 2 Breakthrough curves for manganese sorption on limestone (**a**) and the spent zeolite (**b**). The experimental conditions are listed in Table 1

the saturation rate. Similarly, a study of lead adsorption on clinoptilolite revealed that an increase in the concentration (from 200 to 500 mg/L) resulted in faster breakthrough [14].

Fitting the experimental data to the Thomas model produced solid phase (q_0) and bed loading (N_0) values, which increased with the flow-rate at the same initial concentration, particularly in the case of manganese sorption on limestone. With the spent zeolite such effect was not so remarkable because the bed capacity slightly increased from 438.3 (1 mL/min) to 461.5 mgMn/L-bed (2 mL/min) as the flow rate was increased. Furthermore, the higher inlet concentration (15.0 mg/L) enabled a better utilization of the limestone sorption capacity as higher bed and particle loadings were achieved as compared with the 1.5 mg/L-manganese water. Conversely particle and bed loadings showed a slightly reduction when the initial manganese concentration was increased from 10 (518.8 mg/g) to 50 mg/L (438.3 mg/g) using the spent zeolite.

Additionally, the rate constant (k_T) was remarkably affected by the flow rate. For instance it varied from 7.0 to 16.0 mL/mg min as the flow rate increased from 1 to 3 mL/min for Mn^{2+} sorption on limestone. Likewise it changed from 1.27 to

Table 2 Parameters produced during fitting of the Thomas model to manganese sorption on limestone and spent zeolite

Z (cm)	Q (cm^3/min)	C_0 (mg/L)	k_T (mL/mg min)	q_0 (mg/g)	N_0 (mg/L-bed)	SSE
20	1	1.5	6.99	0.009	14.72	3.70E-02
20	2	1.5	9.80	0.012	19.48	2.16E-02
20	3	1.5	15.99	0.017	29.28	6.06E-02
20	1	15.0	1.74	0.031	53.01	1.84E-02
6	1	10.0	1.91	0.333	518.8	7.41E-02
6	1	25.0	1.44	0.289	449.1	8.01E-02
6	1	50.0	1.27	0.282	438.3	1.31E-01
6	2	50.0	2.77	0.297	461.5	2.05E-01

2.77 mL/mg min as the flow rate was increased in the experiments with the spent zeolite. This outcome suggested that the film diffusion seemed to be the rate-limiting step during manganese uptake by both materials [15] (Table 2).

Conclusions

The results showed a new application for a spent zeolite of the oil industry (faujasite) after its washing with a sodium hydroxide solution. The zeolite was able to load soluble manganese at pH 6.5 producing a maximum loading of 10.9 mg/g. This sorption capacity was 10 times higher than that observed in the low cost calcite limestone (1.03 mg/g). Further experiments are required to assess the manganese sorption capacity in the presence of other ions in solution and to establish the competitive effects of metals such as calcium and magnesium on the manganese uptake in such zeolite.

Acknowledgements Funding provided by Vale and the agencies FINEP, CAPES, CNPq and FAPEMIG are acknowledged.

References

1. Silva, A. M., Cruz, F. L. S., Lima, R. M. F., Teixeira, M. C., & Leão, V. A. (2010). Manganese and limestone interactions during mine water treatment. *Journal of Hazardous Materials, 181*(1–3), 514.
2. Kothari, N. (1988). Groundwater, iron and manganese—An unwelcome trio. *Water-Engineering & Management, 135*(2), 25–26.
3. Jimenez, R. S., Bosco, S. M. D., Carvalho, W. A. (2004). Remoção de metais pesados de efluentes aquosos pela zeólita escolecita - Influência da temperatura e do pH na adsorção em sistemas monoelementares. *Química Nova* 27, 734–738.

4. Omri, A., & Benzina, M. (2012). Removal of manganese(II) ions from aqueous solutions by adsorption on activated carbon derived a new precursor: *Ziziphus spina-christi* seeds. *Alexandria Engineering Journal, 51*(4), 343–350.
5. Taffarel, S. R., & Rubio, J. (2009). On the removal of Mn^{2+} ions by adsorption onto natural and activated Chilean zeolites. *Minerals Engineering, 22*(4), 336–343.
6. Taffarel, S. R., & Rubio, J. (2010). Removal of Mn^{2+} from aqueous solution by manganese oxide coated zeolite. *Minerals Engineering, 23*(14), 1131–1138.
7. Aziz, H. A., Adlan, M. N., & Ariffin, K. S. (2008). Heavy metals (Cd, Pb, Zn, Ni, Cu and Cr (III)) removal from water in Malaysia: Post treatment by high quality limestone. *Bioresource Technology, 99*(6), 1578–1583.
8. Sdiri, A., & Higashi, T. (2013). Simultaneous removal of heavy metals from aqueous solution by natural limestones. *Applied Water Science, 3*(1), 29–39.
9. Verbinnen, B., Block, C., Van Caneghem, J., & Vandecasteele, C. (2015). Recycling of spent adsorbents for oxyanions and heavy metal ions in the production of ceramics. *Waste Management, 45,* 407–411.
10. Pokonova, Y. V. (2012). Adsorbents from spent zeolites. *Solid Fuel Chemistry, 46*(3), 179–184.
11. Chu, K. H. (2010). Fixed bed sorption: Setting the record straight on the Bohart-Adams and Thomas models. *Journal of Hazardous Materials, 177*(1–3), 1006–1012.
12. Silva, A. M., Cordeiro, F. C. M., Cunha, E. C., & Leão, V. A. (2012). Fixed-bed and stirred-tank studies of manganese sorption by calcite limestone. *Industrial and Engineering Chemistry Research, 51*(38), 12421–12429.
13. Han, X., Wang, W., & Ma, X. (2011). Adsorption characteristics of methylene blue onto low cost biomass material lotus leaf. *Chemical Engineering Journal, 171*(1), 1–8.
14. Vukojević Medvidović, N., Perić, J., Trgo, M., & Mužek, M. N. (2007). Removal of lead ions by fixed bed of clinoptilolite—The effect of flow rate. *Microporous and Mesoporous Materials,* 105(3), 298–304.
15. Cooney, D. O. (1999). *Adsorption design for wastewater treatment* (p. 190). Boca Raton: Lewis Publishers.

Effect of Mo^{6+} Substitution on Microstructure and Lithium Ionic Conductivity of Garnet-Type $Li_7La_3Zr_2O_{12}$ Solid Electrolytes by Field Assisted Sintering Technology

Fei Chen, Junyang Li, Yanhua Zhang, Dunjie Yang, Qiang Shen and Lianmeng Zhang

Abstract Advanced rechargeable lithium batteries with high energy density are required as power sources for electric or modern storage systems. Solid-state batteries with non-flammable inorganic solid electrolytes are one candidate to replace the currently-used liquid electrolyte. In recent years, a novel class of inorganic ceramic solid electrolyte with garnet structure $Li_7La_3Zr_2O_{12}$, known as LLZO, has been developed, which has several superior advantages, such as high Li-ion conductivity, high chemical and electrochemical stability in air. However, Li-ion conductivity for LLZO solid electrolyte is still too low to be applied in the industry. In order to obtain high Li-ion conductivity of LLZO, the priority is to stabilize its cubic phase, because the conductivity of tetragonal phase is two orders of magnitude lower than that of the cubic phase. Previous studies indicated that elemental doping was an effective means to stabilize the cubic phase as well as increase the density of ceramic samples, especially with Al^{3+}. It can not only substitutes the corresponding element in the lattice but also helps expel pores through low melting-point phase formed at grain boundaries, which leads to a good connection between the cubic grains. In addition, Mo (normally Mo^{6+}) substitution for Zr^{4+} may lead to more Li vacancies in LLZO, which is beneficial to the enhancement of ionic conductivity. In this study, Mo^{6+} doped $Li_{6.5}La_3Zr_{1.75}Mo_{0.25}O_{12}$ (LLZM) solid electrolytes are successfully prepared via field assisted sintering technology (FAST). The effect of sintering temperature on the microstructure and lithium ionic conductivity is mainly investigated. The results show that pure cubic phase LLZM can be obtained at the range of temperatures from 1050 to 1150 °C for no more than 10 min. For the sample sintered at 1150 °C, a maximum relative density of >95% with a total ionic conductivity as high as 1.3×10^{-4} S cm^{-1} are obtained at room temperature. The change of ionic conductivity is ascribed to the smaller ionic size

F. Chen (✉) · J. Li · Y. Zhang · D. Yang · Q. Shen · L. Zhang
State Key Laboratory of Advanced Technology
for Materials Synthesis and Processing, Wuhan University of Technology,
Wuhan 430070, People's Republic of China
e-mail: chenfei027@gmail.com

© The Minerals, Metals & Materials Society 2017
M.A. Meyers et al. (eds.), *Proceedings of the 3rd Pan American Materials Congress*,
The Minerals, Metals & Materials Series, DOI 10.1007/978-3-319-52132-9_12

of Mo^{6+} (0.62 Å) to Zr^{4+} (0.72 Å). The higher valence of Mo^{6+} to Zr^{4+} can reduce the Li^+ concentration and stabilize the cubic phase.

Keywords Solid electrolytes · Field assisted sintering technology · Lithium ionic conductivity

Introduction

Recently, solid electrolytes with high relative density are believed to be a promising alternate for liquid electrolyte to solve safety issues caused by lithium dendrites and the usage of flammable organic electrolytes in the lithium ion batteries (LIBs), which have greatly hindered its development [1–4]. Inorganic oxides known as $Li_7La_3Zr_2O_{12}$ (LLZO) with a garnet-type structure is regarded as a candidate material due to its high room temperature ionic conductivity (10^{-4} S cm^{-1}), good electrochemical stability [5, 6] and mechanical strength [7] since it was first reported by Weppner et al. [5].

There were two phases of garnets: tetragonal phase and cubic phase [8, 9]. The garnet framework structure is consist of dodecahedral LaO_8 and octahedral ZrO_6 as shown in Fig. 1a. By these two polyhedron connecting with each other, a three-dimensional network of lithium migration pathway which consist of tetrahedral (Li_1, 24d) and octahedral (Li_2, 96h) site is formed (see Fig. 1b) [9, 10]. The cubic phase is formed by a disordered occupation of lithium atoms, while the tetragonal phase is formed by a more ordered occupation, and this makes the single lithium atom migration in cubic phase much easier than the fully synchronous migration in tetragonal phase [8, 9, 11]. This differences is also reflected in the ionic

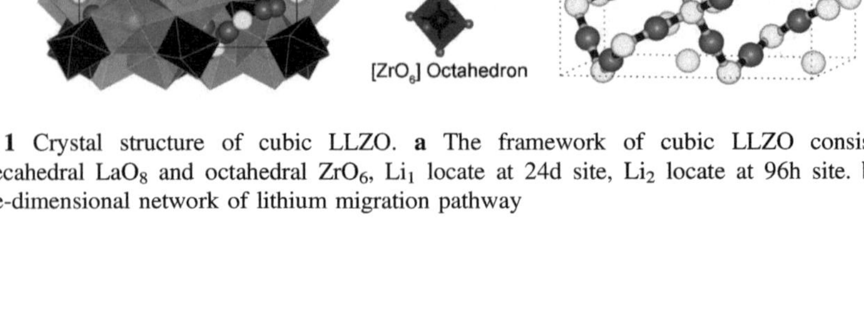

Fig. 1 Crystal structure of cubic LLZO. **a** The framework of cubic LLZO consists of dodecahedral LaO_8 and octahedral ZrO_6, Li_1 locate at 24d site, Li_2 locate at 96h site. **b** The three-dimensional network of lithium migration pathway

conductivity of the cubic phase (10^{-4} S cm^{-1}) and the tetragonal phase (10^{-6} S cm^{-1}). Thus the priority of obtaining LLZO with high ionic conductivity is to stabilize its cubic phase.

Previous studies indicate that elemental doping is an efficient method to stabilize its cubic phase as well as improve the ionic conductivity of ceramics. Aluminum was widely used as substituted element in recent research [12–16]. Rangasamy et al. [12] and Tsai et al. [13] reported their work on Al-substituted LLZO with high ionic conductivity of 4×10^{-4} and 3×10^{-4} S cm^{-1} respectively. They believed that Al substituting for Li creates charge compensating Li vacancies which determine phase stability. Zhang et al. [14, 15] prepared Al-doped LLZO with high grain ionic conductivity of 1.35×10^{-3} S cm^{-1} by controlling lithium ion concentration to 6.35 mol in LLZO and found that lithium ion concentration can affect ionic conductivity via changing lithium ion distribution in the grain. After summing up recent research on Al-doped LLZO, we can conclude that Aluminum is located at Li_1 site and its higher valence than Li^+ can reduce lithium ion concentration as well as stabilize the cubic phase.

The 3-D lithium migration pathway is formed by the superposition of LaO_8 dodecahedron and ZrO_6 octahedron. Use amounts of element with different ionic size to substitute La^{3+} or Zr^{4+}, the ionic conductivity can be significantly improved. Liu et al. [17] and Janani et al. [18] used Ta^{5+} (Ionic radius = 0.64 Å) to substitute Zr^{4+} and the ionic conductivity of LLZ-Ta can be improved to 7.24–7.63 $\times 10^{-4}$ and 3.7×10^{-4} S cm^{-1} respectively. Dhivya and Murugan [19] used Y^{3+} (Ionic radius = 0.9 Å) and Ta^{5+} to substitute La^{3+} (Ionic radius = 1.032 Å) and Zr^{4+} respectively and the ionic conductivity of LLZ-Ta can be improved to 4.36×10^{-4} S cm^{-1}. Deviannapoorani et al. [20] used Te^{6+} (Ionic radius = 0.56 Å) to substitute Zr^{4+} and the ionic conductivity of LLZ-Te can be improved to 1.02×10^{-3} S cm^{-1}. Huang et al. [21] substituted Zr^{4+} with Si^{4+} (Ionic radius = 0.4 Å), In^{3+} (Ionic radius = 0.8 Å) and Ge^{4+} (Ionic radius = 0.53 Å) and the experimental result showed that Si and In doping deteriorated electrical properties, while Ge doping led to an improvement of ionic conductivity of 7.63×10^{-4} S cm^{-1}. In short, element whose ionic radius located at a proper range (0.53–0.64 Å for current study) doping was proved to be favorable to improve ionic conductivity, while the rest of them deteriorated the performance. Density functional theory calculation had also been applied by Miara et al. [22] to search for all possible dopants and their site preference for LLZO by analyzing the defect energy of doped LLZO. As expected, for Zr site, cations doping prefer to be stable ($E_{defect} < 1\,\mathrm{eV}$) when differences of ionic radius between doping element and Zr were less than about 0.2 Å. Thus, for a stable doping, the ionic radius of doping element should be in the range of 0.52–0.9 Å. For all these reasons, the molybdenum is chosen to be doped in LLZO in this work. The higher valence of Mo^{6+} can introduce more lithium vacancies into garnet crystal to stabilize the cubic phase. The ionic radius of Mo^{6+} is 0.62 Å, thus Mo-LLZO is stable and is also expected to obtain high ionic conductivity by modifying the garnet framework.

Field assisted sintering technology (FAST) is a combination of hot-pressing and electric field activation sintering technology, which can efficiently decrease the sintering temperature and time, and get a dense simple. In this study, Mo^{6+} doped $Li_{6.5}La_3Zr_{1.75}Mo_{0.25}O_{12}$ (LLZM) solid electrolytes was successfully prepared via field assisted sintering technology (FAST). The effect of sintering temperature on the microstructure and lithium ionic conductivity were mainly investigated.

Experimental Procedures

Cubic $Li_{6.5}La_3Zr_{1.75}Mo_{0.25}O_{12}$ was prepared by a solid-state reaction from the stoichiometric amount of high purity Li_2O (purity of 99.99%, particle size of ~10 μm, produced by Aladdin), La_2O_3 (purity of 99.99%, ~10 μm, produced by Sinopharm Chemical Reagent Co., Ltd., China), ZrO_2 (purity of 99.9%, ~5 μm, produced by Guangdong Orient Zirconic Ind. Sci. & Tech. Co., Ltd., China), MoO_3 (purity of 99.9%, ~10 μm, produced by Aladdin) powders with excess Li_2O of 10 wt% and 1.5 wt% γ-Al_2O_3 (purity of 99%, ~20 nm, produced by Aladdin, China) were added to compensate the loss of lithium caused by volatilization during high temperature sintering and stabilize the cubic phase respectively. The powders were ball milled using zirconia balls for 12 h and dried in a vacuum oven at 80 °C for 6 h. After that, a graphite die filled with the dried powders was put into a plasma activated sintering system (ED–PASIII, Elenix Ltd., Japan). A 30 s activation step was carried out with a pulse current of 30 V/10A before the sintering. The sintering temperature was varied from 1050 to 1200 °C with a constant uniaxial pressure of 10 MPa and a holding time for 3 min. Ar was chosen as protect gas to stabilize the pressure and prevent the loss of lithium. After the sintering, the sample was naturally cooled down to room temperature.

X-ray diffraction (XRD, Rigaku Ultima III) with Cu Kα radiation of (40 kV, 40 mA) over the range of 10–90° (2θ) and field-emission scanning electron microscopy (FESEM, FEI-Quanta-250) were used to characterize the phase structure and microstructure of the LLZM ceramics. The density of the ceramics were measured by the Archimedes' method. The Li ionic conductivity was measured using an electrochemical workstation (Autolab PGSTAT 30) in the frequency range from 10 Hz to 10 MHz at room temperature and an impedance analyzer (Agilent E4980A) in the frequency range from 20 Hz to 2 MHz. A thin film of Au was sputtered on both sides of the pellets as the experimental blocking electrodes. Impedances are determined from the intercepts of the relevant capacitive arcs at the real axes in the Nyquist plots and conductivities are calculated using the following equation:

$$\sigma = \frac{1}{Z} \times \frac{L}{A} \tag{1}$$

where σ is the ionic conductivity, Z is the impedance, L is the pellet thickness and A is the pellet area. Typical dimensions of the pellets were 1.5–2.0 mm in thickness, and 10 mm in diameter.

Result and Discussion

XRD patterns of samples sintered at different temperature varied from 1050 to 1200 °C are shown in Fig. 2. The pure cubic garnet pattern at the bottom comes from pdf#45-0109, which is the pattern of garnet ($Li_5La_3Nb_2O_{12}$). It can be seen that a cubic garnet structure appears at a sintering temperature of 1050 °C, but the broad peaks mean that 1050 °C is the lowest limit of sintering temperature for LLZM. The sharp peaks of simples sintered at 1100 and 1150 °C suggest the well formation of a garnet structure. When the sintering temperature increase to 1200 °C, the disappearance of garnet structure and the only existence of $La_2Zr_2O_7$ phase reveal that LLZM can be synthesized in the temperature range of 1050–1150 °C. Previous research on Al-doped LLZO using FAST method showed that cubic phase appeared at a wide temperature range of 900–1230 °C, and pure cubic phase could be obtained at the range of 1100–1180 °C. Compared with Al-LLZO, garnet structure could only be obtained in a small temperature range in LLZM. When the sintering temperature rose above 1150 °C, garnet structure disappear, which revealed that Mo-doping could reduce the sinterability of LLZO.

SEM micrographs of the cross sections of the samples sintered at various temperature are shown in Fig. 3. For the LLZM pellet sintered at 1050 °C (See Fig. 3c), particles are composed of incompletely grown grains and amounts of pores formed from the incompact contact between grains. For the pellet sintered at 1100 °C (Fig. 3b), the pores are decreasing with the grain size getting bigger compared with that of 1050 °C. When the temperature increases to 1150 °C, grains are closely

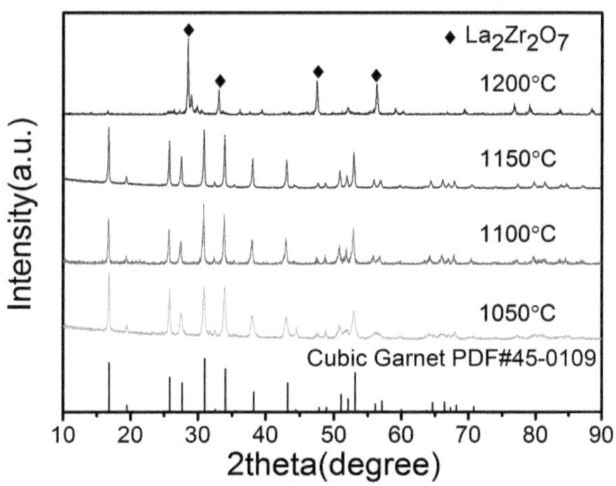

Fig. 2 XRD patterns of the samples sintered at various temperature

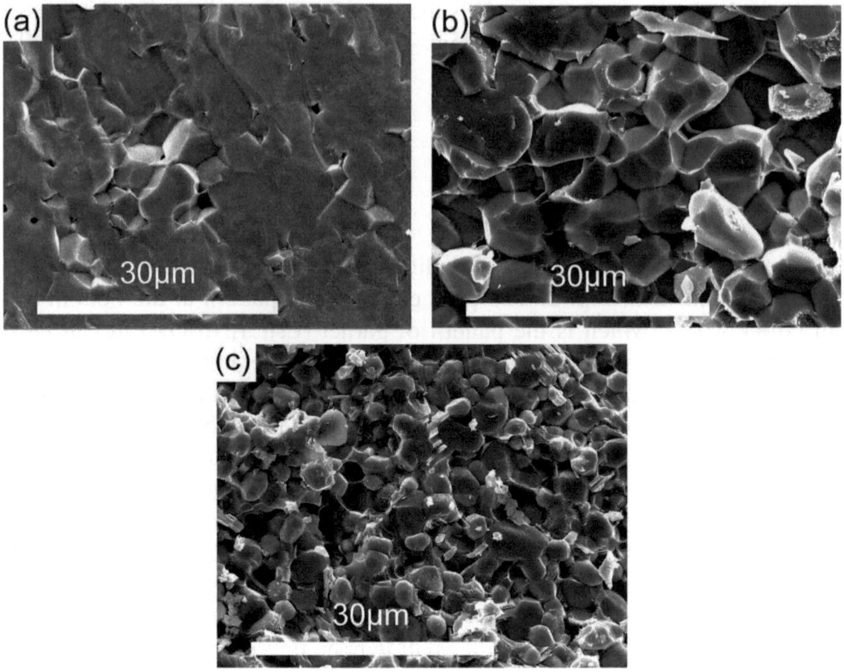

Fig. 3 SEM micrographs of cross section of samples sintered at 1150 °C (a), 1100 °C (b), 1050 °C (c)

contact with each other with tiny pores inside, and the fracture is transgranular (Fig. 3a). In short, the rate of mass transfer between grains increases when the sintering temperature raises, the differences on microstructure are thus emerged.

The microstructure of sample in different sintering temperature are much different, which cause significant influence on ionic conductivity. Nyquist plots of the impedance spectrum for LLZM pellets with a sintering temperature range of 1050–1150 °C measured at room temperature are shown in Fig. 3a. For the sample sintered at 1150 °C, the semicircle in high frequencies can't be clearly observed, which should be caused by the almost completely removing of the grain boundary resistance. The ionic conductivity can be calculated with the Z′ axis coordinate value of the semicircle's termination point and the sample dimensions. Table 1 summarizes the Ionic conductivity and relative density of simples in various sintering temperature. When the sintering temperature raise up to 1150 °C, the relative density changes from 80.5 to 95.2% and the ionic conductivity increases by 2 orders of magnitude. It means that the relative density is a main factor that influence the ionic conductivity. For the sample sintered at 1150 °C, a total ionic conductivity with the highest value of 1.3×10^{-4} S cm^{-1} at room temperature compared with that of 1050 and 1100 °C is obtained. A high relative density comes from the compact contact of grains which contributed to a high ionic conductivity. So we can say that the effect of the relative density on ionic conductivity is significant.

Table 1 Relative density and ionic conductivity in room temperature of LLZM simples sintered at various temperature

Temperature (°C)	1050	1100	1150
Relative density (%)	80.5	87.1	95.2
Ionic conductivity (S cm^{-1})	1.3×10^{-6}	1.74×10^{-5}	1.3×10^{-4}

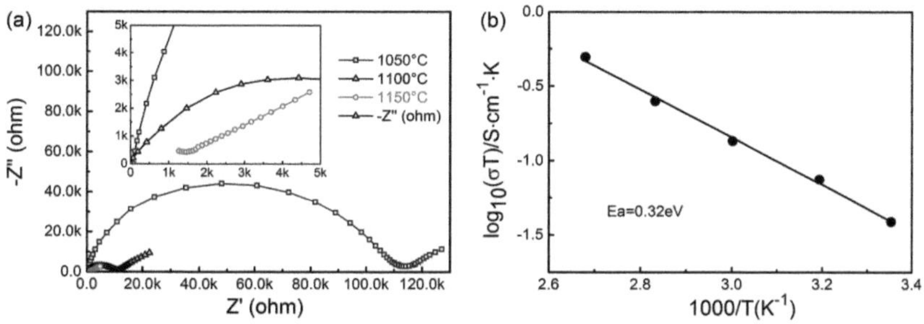

Fig. 4 **a** Typical AC impedance plots (frequency range: 10 Hz–10 MHz) of the samples sintered at the range of 1050–1050 °C measured at room temperature. **b** Arrhenius plot for total conductivity of cubic LLZM simples sintered at 1150 °C

The temperature dependence of total ionic conductivity can be defined as the Arrhenius equation:

$$\sigma = \frac{A}{T}\exp\left(\frac{-E_a}{K_B T}\right) \tag{2}$$

where σ is the total ionic conductivity, A is the pre-exponential constant, T is the absolute temperature, E_a is the activation energy, K_B is the Boltzmann constant. Figure 4b shows Arrhenius plot for total conductivity of cubic LLZM simples sintered at 1150 °C. The determined activation energy from the slope of Arrhenius plot is 0.32 eV, which is slightly larger than previous reports of 0.30 eV. The disparity of activation energy should be caused by the tiny pores on LLZM pellets observed from Fig. 2a, which leads to a harder migration for lithium ions between grains.

Conclusions

The molybdenum doped garnet $Li_{6.5}La_3Zr_{1.75}Mo_{0.25}O_{12}$ with cubic phase was successfully prepared via the FAST method. XRD result showed that a sintering temperature range of 1050–1150 °C was required to obtain high ionic conductivity and dense pure cubic phase in LLZM. The maximum total ionic conductivity of 1.3×10^{-4} S cm^{-1} observed in the sample sintered at 1150 °C along with a high relative density of more than 95% proved that doping with Mo could efficiently increase the stability of cubic phase by reducing lithium ion concentration and the ionic conductivity by modifying the garnet framework structure. The samples sintered by FAST exhibited high relative density with a constant uniaxial pressure of 10 MPa which led to an easy transport of lithium ions between grains. Thus we can confirm that the relative density is a main factor that influence the ionic conductivity. Compared with the best performance of other elements doped-LLZO (10^{-3} S cm^{-1}), the ionic conductivity of LLZM doesn't reach our expectations. Pores remain exist in the pellets with a relative density of 95.2% and the influence of lithium ion concentration on LLZM is barely investigated. By optimizing both of those points, LLZM still has potential to reach a higher performance.

Acknowledgements This work is financially supported by the Fundamental Research Funds for the Central Universities, the National Natural Science Foundation of China (Nos. 51202171 and 51472188) and the "111" project (No. B13035).

References

1. Deviannapoorani, C., Dhivya, L., Ramakumar, S., & Murugan, R. (2012). Synthesis of garnet structured Li_{7+x} $La_3Y_xZr_{2-x}O_{12}$ (x=0-0.4) by modified sol-gel method. *Journal of Sol-Gel Science and Technology, 64*(2), 510–514.
2. Kamaya, N., Homma, K., Yamakawa, Y., et al. (2011). A lithium superionic conductor. *Nature Materials, 10*(9), 682–686.
3. Knauth, P. (2009). Inorganic solid Li ion conductors: An overview. *Solid State Ionics, 180* (14–16), 911–916.
4. Tsai, C. L., Roddatis, V., Chandran, C. V., et al. (2016). $Li_7La_3Zr_2O_{12}$ interface modification for Li dendrite prevention. *ACS Applied Materials & Interfaces, 8*(16), 10617–10626.
5. Murugan, R., Thangadurai, V., & Weppner, W. (2007). Fast lithium ion conduction in garnet-type $Li_7La_3Zr_2O_{12}$. *Angewandte Chemie International Edition England, 46*(41), 7778–7781.
6. Kotobuki, M., Kanamura, K., Sato, Y., & Yoshida, T. (2011). Fabrication of all-solid-state lithium battery with lithium metal anode using Al_2O_3-added $Li_7La_3Zr_2O_{12}$ solid electrolyte. *Journal of Power Sources, 196*(18), 7750–7754.
7. Yu, S., Schmidt, R. D., Garcia-Mendez, R., et al. (2016). Elastic properties of the solid electrolyte $Li_7La_3Zr_2O_{12}$ (LLZO). *Chemistry of Materials, 28*(1):197–206.
8. Awaka, J., Takashima, A., Kataoka, K., Kijima, N., Idemoto, Y., & Akimoto, J. (2011). Crystal structure of fast lithium-ion-conducting cubic $Li_7La_3Zr_2O_{12}$. *Chemistry Letters, 40*(1), 60–62.
9. Meier, K., Laino, T., & Curioni, A. (2014). Solid-state electrolytes: Revealing the mechanisms of Li-ion conduction in tetragonal and cubic LLZO by first-principles calculations. *Journal of Physical Chemistry C, 118*(13), 6668–6679.
10. Chen, Y., Rangasamy, E., Liang, C., & An, K. (2015). Origin of high Li^+ conduction in doped $Li_7La_3Zr_2O_{12}$ garnets. *Chemistry of Materials, 27*(16), 5491–5494.
11. Xu, M., Park, M. S., Lee, J. M., Kim, T. Y., Park, Y. S., & Ma, E. (2012). Mechanisms of Li^+ transport in garnet-type cubic $Li_{3+x}La_3M_2O_{12}$ (M=Te, Nb, Zr). *Physical Review B, 85*(5), 052301.
12. Rangasamy, E., Wolfenstine, J., & Sakamoto, J. (2012). The role of Al and Li concentration on the formation of cubic garnet solid electrolyte of nominal composition $Li_7La_3Zr_2O_{12}$. *Solid State Ionics, 206*, 28–32.

13. Tsai, C. L., Dashjav, E., Hammer, E. M., et al. (2015). High conductivity of mixed phase Al-substituted $Li_7La_3Zr_2O_{12}$. *Journal of Electroceramics, 35*(1–4), 25–32.
14. Zhang, Y., Chen, F., Tu, R., Shen, Q., & Zhang, L. (2014). Field assisted sintering of dense Al-substituted cubic phase $Li_7La_3Zr_2O_{12}$ solid electrolytes. *Journal of Power Sources, 268*, 960–964.
15. Zhang, Y., Chen, F., Tu, R., Shen, Q., Zhang, X., & Zhang, L. (2016). Effect of lithium ion concentration on the microstructure evolution and its association with the ionic conductivity of cubic garnet-type nominal $Li_7Al_{0.25}La_3Zr_2O_{12}$ solid electrolytes. *Solid State Ionics, 284*, 53–60.
16. Suzuki, Y., Kami, K., Watanabe, K., et al. (2015). Transparent cubic garnet-type solid electrolyte of Al_2O_3-doped $Li_7La_3Zr_2O_{12}$. *Solid State Ionics, 278*, 172–176.
17. Liu, T., Ren, Y. Y., Shen, Y., Zhao, S. X., Lin, Y. H., & Nan, C. W. (2016). Achieving high capacity in bulk-type solid-state lithium ion battery based on $Li_{6.75}La_3Zr_{1.75}Ta_{0.25}O_{12}$ electrolyte: Interfacial resistance. *Journal of Power Sources, 324*, 349–357.
18. Janani, N., Ramakumar, S., Kannan, S., & Murugan, R. (2015). Optimization of lithium content and sintering aid for maximized Li^+ conductivity and density in Ta-doped $Li_7La_3Zr_2O_{12}$. *Journal of the American Ceramic Society, 98*(7), 2039–2046.
19. Dhivya, L., & Murugan, R. (2014). Effect of simultaneous substitution of Y and Ta on the stabilization of cubic phase, microstructure, and Li^+ conductivity of $Li_7La_3Zr_2O_{12}$ lithium garnet. *ACS Applied Materials & Interfaces, 6*(20), 17606–17615.
20. Deviannapoorani, C., Dhivya, L., Ramakumar, S., & Murugan, R. (2013). Lithium ion transport properties of high conductive tellurium substituted $Li_7La_3Zr_2O_{12}$ cubic lithium garnets. *Journal of Power Sources, 240*, 18–25.
21. Huang, M., Dumon, A., & Nan, C. W. (2012). Effect of Si, In and Ge doping on high ionic conductivity of $Li_7La_3Zr_2O_{12}$. *Electrochemistry Communications, 21*, 62–64.
22. Miara, L. J., Richards, W. D., Wang, Y. E., & Ceder, G. (2015). First-principles studies on cation dopants and electrolyte| cathode interphases for lithium garnets. *Chemistry of Materials, 27*(11), 4040–4047.

Electromagnetic Levitation Refining of Silicon–Iron Alloys for Generation of Solar Grade Silicon

Katherine Le, Yindong Yang, Mansoor Barati and Alex McLean

Abstract At present, expensive semiconductor grade silicon (SEG-Si) is used for the manufacture of cells to convert solar energy into electricity. This results in a high cost for photovoltaic electricity compared to electricity derived from conventional sources. The processing of inexpensive metallurgical silicon, or ferrosilicon alloys, is a potentially economical refining route to produce photovoltaic silicon. With phosphorus being one of the most difficult impurities to remove by conventional techniques, this project investigated the use of electromagnetic levitation for dephosphorization of silicon–iron alloy droplets exposed to hydrogen–argon gas mixtures. The effects of time, temperature, hydrogen partial pressure, iron content in the alloy, and initial phosphorus concentration were evaluated.

Keywords Dephosphorization · Electromagnetic levitation · Ferrosilicon · Solar grade silicon · Photovoltaic

Introduction

Solar energy is a promising renewable energy resource for the future with potential to satisfy the growing global energy demand. Silicon is the second most abundant element found in the Earth's crust (27.7%) [1], and the most dominant photovoltaic (PV) material in the industry, comprising the base component of nearly 90% of cells produced [2]. The development of solar grade silicon is hindered by costly and energy intensive production methods, which result in silicon material comprising approximately 45–50% of the total cost of a typical crystalline-silicon module [3]. Current speculation regarding solar energy technology attaining grid parity with

K. Le · Y. Yang (✉) · M. Barati · A. McLean
Department of Materials Science and Engineering, University of Toronto,
184 College St., Toronto, ON M5S 3E4, Canada
e-mail: yindongyang@yahoo.com

© The Minerals, Metals & Materials Society 2017

M.A. Meyers et al. (eds.), *Proceedings of the 3rd Pan American Materials Congress*,
The Minerals, Metals & Materials Series, DOI 10.1007/978-3-319-52132-9_13

existing electricity technologies, as a result of the recent price decline in solar modules has been partly due to the reduced usage and substitution for silicon materials [3].

The upgrading of inexpensive ferrosilicon alloys or metallurgical grade silicon to solar grade silicon is a potentially economical refining route and several research studies have been carried out. Numerous techniques have been developed to target and remove specific impurity elements in metallurgical silicon alloys, which include: slag refining, acid leaching, solvent refining, solidification refining, and vacuum melting. The investigated methods have been found to be selective, with no single method able to remove all the impurity elements [2]. Phosphorus and boron are deemed as the two most difficult impurity elements to remove from metallurgical silicon by investigated techniques, and have been the focus of several recent studies. In addition to reductions in production costs of SoG-Si, effective removal of the aforementioned impurity elements is essential for producing silicon through metallurgical means [4].

Electromagnetic levitation (EML) refining is an excellent technique for containerless processing of liquid metals. Advantages include the lack of metal contamination from refractory containers, inductive stirring, rapid melting, and approximate spherical geometry of the metal droplet during levitation. The objective of this work is to apply EML to study dephosphorization of ferrosilicon alloys under a hydrogen–argon atmosphere. The effects of time, temperature, hydrogen partial pressure, iron content in the alloy, and initial phosphorus concentration were investigated. Upon achieving effective dephosphorization, the partially refined alloy can be processed by a combination of previously investigated metallurgical techniques into solar grade silicon.

Theoretical Aspects

Phosphorus Removal from Si–Fe Alloy Melts

Evaporative losses of phosphorus (P) in iron (Fe) melts follows the reaction:

$$[\mathrm{P}]_{\mathrm{Fe}} \Leftrightarrow 1/2\mathrm{P}_2(\mathrm{g})$$

Ueda et al. determined the activity coefficient of phosphorus in Si–Fe alloys at a controlled phosphorus partial pressure, showing a maximum value at a certain composition due to strong interaction between silicon and iron. This effect resulted in decreased phosphorus solubility in the alloy. Consequently, the possibility for dephosphorization of silicon alloyed with iron was demonstrated. A value of 0.199 was obtained for the first order interaction parameter e_P^{Fe} for the effect of iron on the activity coefficient of phosphorus [5].

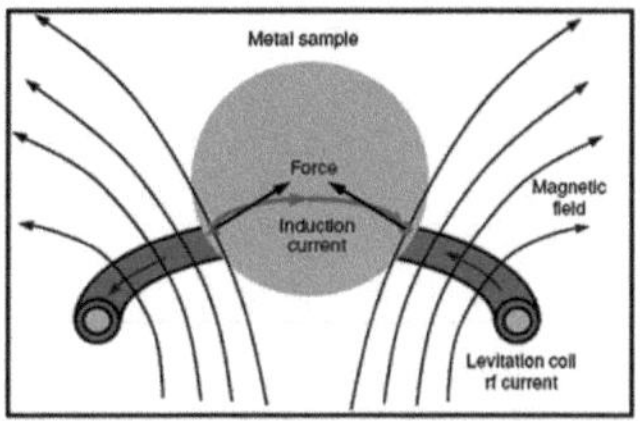

Fig. 1 Electromagnetic levitation principle [9]

From steelmaking data, the interaction parameter for the effect of hydrogen on phosphorus dissolved in an iron melt is $e_P^H = 0.33$ [6]. By using a hydrogen–argon gas mixture, the argon can be considered to produce a vacuum effect, accelerating phosphorus evaporation, while hydrogen in the system generates a reducing atmosphere and increases phosphorus activity in the Si–Fe melt, thus promoting phosphorus removal.

Electromagnetic Levitation Refining

Electromagnetic levitation (EML) is a process that can be utilized for refining molten metals in a containerless environment [7–9]. This prevents potential contamination from impurities while generating inductive stirring [8]. The EML process involves passing an AC electric current through a coil, inducing an eddy current in the conductive sample, which in turn, is melted by heating from the induced current [8]. The electromagnetic force resulting from the interaction of the EM field, and inductive eddy current lifts the sample [8]. Figure 1 depicts the principle of electromagnetic levitation.

Experimental Aspects

Design of Levitation Equipment

The electromagnetic levitation facility used in this work is shown in Fig. 2. The apparatus consists of a quartz tube chamber (15 mm O.D., 13 mm I.D., 304 mm length) a copper levitation coil wound using 1/8 in. diameter tubing, a rotatable platform housing a copper mold and alumina rod. Power is provided to the coil using an Ameritherm-Ambrell high frequency induction heating system with a rated terminal output of 10 kW and a frequency range from 150 to 400 kHz. Adjustment in applied current allows for the manipulation of the vertical position of the droplet within the levitation coil, and accordingly the amount of heating provided to the sample based on interaction with field flux lines. A minimum working current to support the droplet in levitation is dependent on sample material, mass, gas flow rate, and coil design.

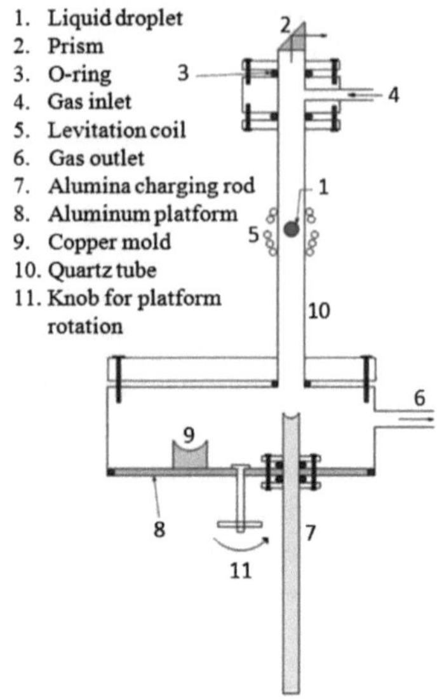

Fig. 2 Schematic diagram of EML apparatus

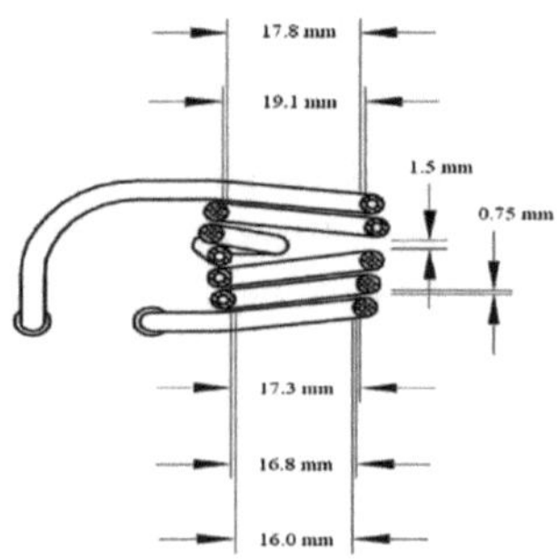

Fig. 3 Schematic diagram of levitation coil

Figure 3 shows a schematic of the five-turn levitation coil used in this work [10]. The coil is a low-angle conical configuration comprising two sections: a lower three-turn primary cone, and an upper inverted cone with two reverse turns. With this configuration, the lower cone serves to provide sufficient lifting force and droplet heating to appropriate temperature, while the upper inverted cone provides vertical and lateral stability of the droplet.

Temperature Measurement

Droplet temperatures were measured using a CHINO IR-CA Q3088 two-color pyrometer with an uncertainty of ±15 °C. The pyrometer is aimed on the droplet surface through the viewing window at the top of the levitation chamber and

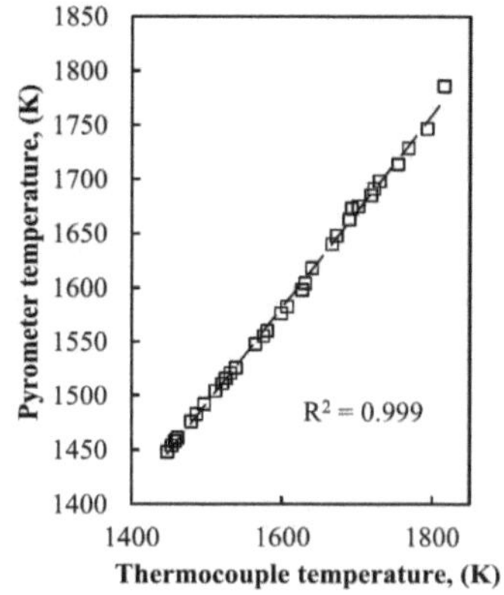

Fig. 4 Pyrometer calibration for molten iron by comparison with a submerged thermocouple

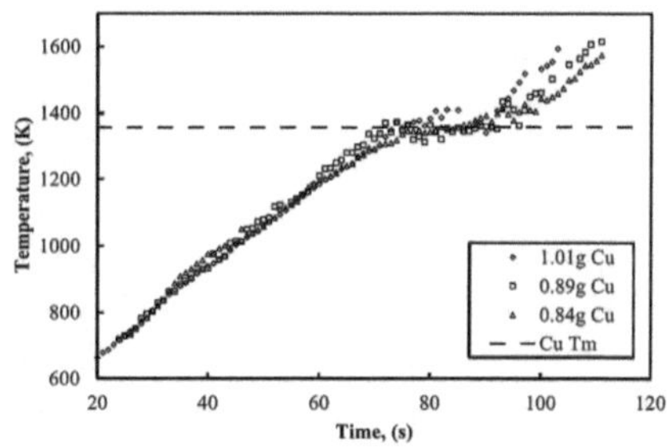

Fig. 5 Pyrometer calibration using levitated copper droplets of different mass

continuous temperature readings obtained. The optical viewing window serves to minimize spectral losses during measurements. Two separate pyrometer calibration tests were carried out. For the first test, measurements were taken from a thermocouple submerged in molten iron contained in a crucible located within an induction furnace under a hydrogen–argon atmosphere [10]. The melting point of iron was compared with the recorded measurements from both the thermocouple and the pyrometer (Fig. 4). For the second calibration test, the levitation furnace was used with copper droplets. The pyrometer calibration was verified by measuring the recorded melting temperatures of levitated copper droplets of different mass as shown in Fig. 5 [10].

Levitation of Silicon–Iron Alloys

Silicon–iron alloy samples were cut into 0.6–0.7 g (±0.01) sections. A specimen is placed on the alumina charging rod, and raised into the sealed levitation chamber to a position between the upper and lower cones of the coil. The chamber is purged with the reaction/inert gas mixture for 1–2 min and checked for leaks. With application of electrical power to the coil, the solid specimen levitates, generally at temperatures of 1173–1273 K. The charging rod is retracted as soon as stable levitation of the solid specimen is obtained. The specimen usually melts within 1 min, and the temperature is adjusted to the required value by control of the applied current. After a given reaction time, the rotatable platform is turned so that the copper mold is positioned below the coil. When the power is turned off, the droplet falls and is quenched in the copper mold. The specimen is then recovered for subsequent analysis by inductively coupled plasma optical emission spectroscopy.

Results and Discussion

Effect of Time and Temperature

The results in Fig. 6 shows that phosphorus removal increases with both time and temperature. The rate of removal appears to follow first-order reaction kinetics. Steady state conditions are achieved after approximately 45 min, beyond which the concentration of phosphorus remains essentially constant. The higher processing temperature of 1993 K results in a lower final phosphorus content, with 72% of phosphorus removal achieved after levitation of a 25 wt%Fe–Si alloy with an initial P content of 0.025 wt% for 40 min in a 50%H_2–Ar mixture at 0.5 L/min flow rate. However processing at high temperatures is unfavorable from an energy standpoint. Furthermore, with the vapour pressure of silicon known to increase significantly at temperatures above the melting point, processing at higher temperature would result in silicon vaporization losses [11].

Effect of Hydrogen Partial Pressure

The effect of hydrogen concentration on dephosphorization behavior is shown in Fig. 7 for flow rates of 0.5 and 1.2 L/min. In both cases, a positive correlation

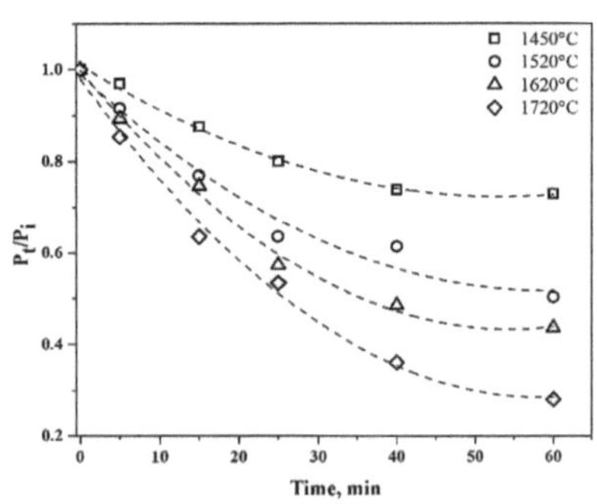

Fig. 6 Effect of time and temperature on dephosphorization with 50% H_2–Ar

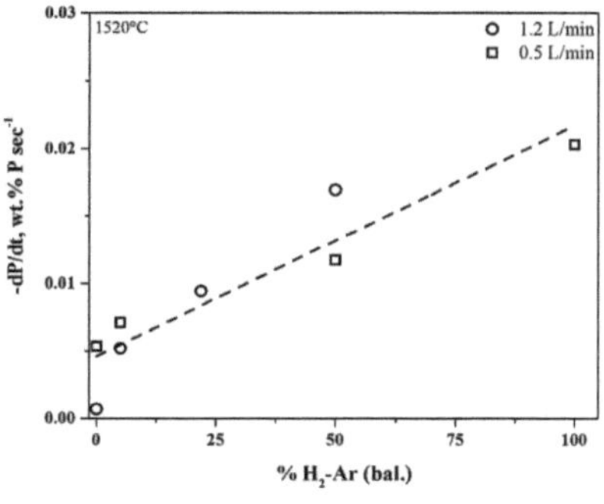

Fig. 7 Effect of H_2 concentration on dephosphorization

between a higher partial pressure of hydrogen and increased rate of dephosphorization is observed. The effect of gas flow rate for the range investigated does not appear to be significant. Further work is required to identify and quantify the reaction products in order to confirm the mechanism of phosphorus removal from silicon–iron alloys.

Effect of Iron Alloying

Based on the results shown in Fig. 8, increasing Fe content is observed to impede phosphorus removal. This effect is likely due to the stronger affinity of Fe for P, than Si for P, as indicated by the calculated iso-activities of P in the ternary Si–Fe–P system, Fig. 9, and work by Khajavi et al. [12]. In Fig. 9, the activity of P displays a large negative deviation from ideality. An increase in Fe content, or decrease in Si content is shown to increase the deviation of P activity from ideality, which is consistent with the observed affinity of P for Fe in Si–Fe melts.

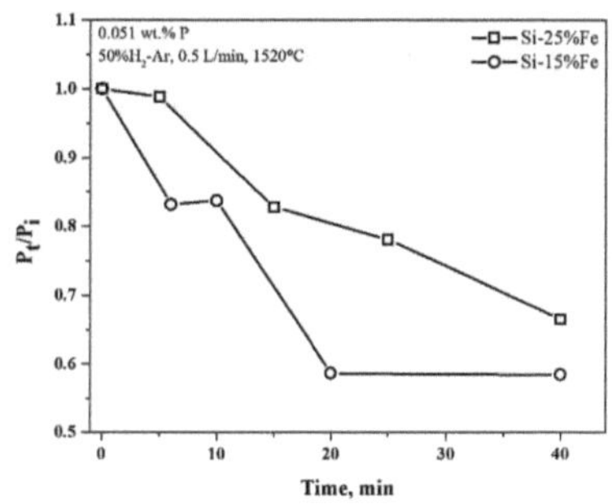

Fig. 8 Effect of Fe on dephosphorization

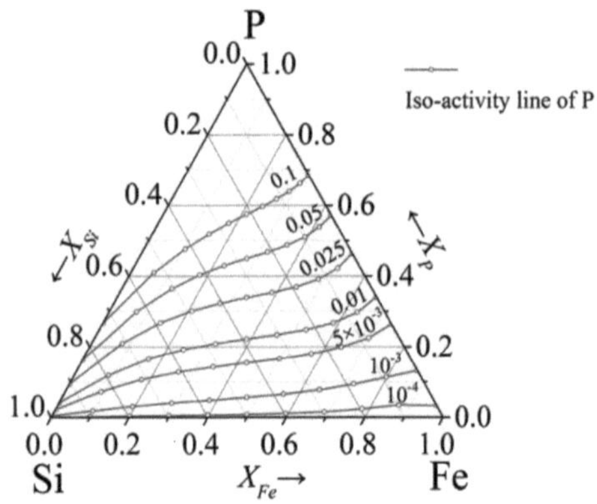

Fig. 9 Iso-activity curves of [P] in Si–Fe–P alloys

Effect of Initial Phosphorus Concentration

It is observed that a higher initial phosphorus concentration in the silicon alloy droplets results in a greater dephosphorization (Fig. 10). At lower initial phosphorus concentrations, and phosphorus depletion in the alloy melt as the reaction proceeds, the overall rate is expected to be controlled by mass transfer in the liquid metal. It is noteworthy that the dephosphorization rate for the higher initial phosphorus concentrations has promising implications for the refinement of high alloy steels. In such operations, strict control of alloying element contents including chromium, nickel, and silicon is critical. Under oxidizing conditions, the risk of alloying element loss is significant due to preferential oxidation compared with phosphorus. Accordingly, as demonstrated during the present work, refining of high-alloyed steels under a reducing environment would prevent alloy loss due to be oxidation.

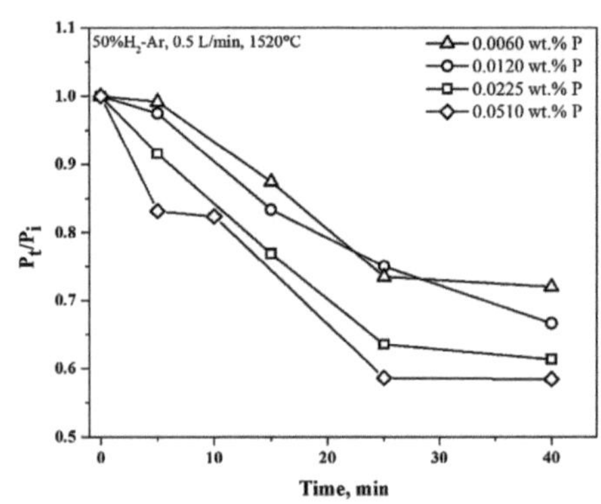

Fig. 10 Effect of initial phosphorus concentration on dephosphorization

Conclusions

Dephosphorization was achieved by electromagnetic levitation refining of silicon–iron alloys in H_2–Ar gas mixtures. The degree of dephosphorization was found to increase with reaction time, with 72% of phosphorus removed after 40 min at 1993 K in a 50%H_2–Ar mixture. However, processing at lower temperatures is expected to reduce evaporation losses of silicon and increase the process efficiency. Higher concentrations of hydrogen in the gas mixture as well as higher initial phosphorus concentration in the alloy were observed to increase dephosphorization. Increase of iron content in the silicon–iron alloys was found to lower the rate of phosphorus removal. Further work based on these preliminary findings could facilitate the development of an innovative technique for refining metallurgical grade silicon alloys and ultimately contribute to a cost effective route for the production of solar grade silicon.

Acknowledgements Appreciation is expressed to the Natural Sciences and Engineering Research Council of Canada for providing funding for this project through a Strategic Research Grant.

References

1. Monroe, J., & Wickander, R. (2006). *Essentials of geology* (4th ed.). Belmont, USA: The Thomson Corp.
2. Lynch, D. (2009). Winning the global race for solar silicon. *JOM Journal of the Minerals Metals and Materials Society, 61,* 41–48.
3. The International Renewable Energy Agency. (2013). Solar technology brief.
4. Johnston, M., Khajavi, L., Li, M., Sokhanvaran, S., & Barati, M. (2012). High-temperature refining of metallurgical-grade silicon: A review. *JOM Journal of the Minerals Metals and Materials Society, 64,* 935–944.
5. Ueda, S., Morita, K., & Sano, N. (1997). Thermodynamics of phosphorus in molten Si-Fe and Si-Mn alloys. *Metallurgical and Materials Transactions B, 28B,* 1151–1155.
6. Hino, M., & Ito, K. (2010). *Thermodynamic data for steelmaking*. Sendai, Japan: Tohoku University Press.
7. Hectors, D., Van Reusel, K., & Diesen, J. (2008). Experimental validation of electromagnectic-thermal coupled modelling of levitation melting. *Przeglad Elektrotechniczny, 84,* 140–143.
8. Asakuma, Y., Sakai, Y., Hahn, S., Tsukada, T., Matsumoto, T., Fujii, H., et al. (2000). Equilibrium shape of a molten silicon drop in an electromagnetic levitator in microgravity environment. *Metallurgical and Materials Transactions B, 31,* 327–329.
9. Popa, M. (2010). Study of an electromagnetic levitation system. *Nonconventional Technologies Review, 1,* 34–38.
10. Wu, P., Yang, Y., Barati, M., & McLean, A. (2014). Electromagnetic levitation of silicon and silicon-iron alloy droplets. *High Temperature Materials and Processes, 33*(1), 1–7.
11. Wei, K. X., Ma, W. H., Dai, Y. N., Yang, B., Liu, D. C., & Wang, J. F. (2007). Vacuum distillation refining of metallurgical grade silicon (I)—Thermodynamics on removal of phosphorus from metallurgical grade silicon. *Transactions of Nonferrous Metals Society of China, 17,* 1022–1025.
12. Khajavi, L. T., & Barati, M. (2012). Thermodynamics of phosphorus removal from silicon in solvent refining of silicon. *High Temperature Material Process, 31,* 627–631.

Green Extract of Mate Tea as Corrosion Inhibitor of Copper and Aluminum

Ana M. Derna, Claudia M. Méndez, Liliana M. Gassa and Alicia E. Ares

Abstract Many of the corrosion inhibitors in current use are expensive and toxic so the current trend is to study compounds that are environmentally friendly and efficient. Inhibitors are easy to use and they have the additional advantage that can be applied in situ and without causing significant disruption in the process. In the present work we investigate the influence of the aqueous extract of yerba mate as corrosion inhibitor of copper and aluminum in 0.5 M NaCl and 1 M HCl solutions, using different electrochemical techniques. The following inhibition efficiencies: 56% for copper in 0.5 M NaCl with 25% concentration of extract, 17% for copper in 1 M HCl with 5% concentration of extract, and 99% for aluminum in 0.5 M NaCl for 1% concentration of extract were obtained by potentiodynamic polarization.

Keywords Inhibitors · Mate tea · Copper · Aluminum · Corrosion parameters

Introduction

The use of inhibitors to control the corrosion of metals in contact with the aggressive environment is an important accepted practice [1]. For corrosion inhibitors it is important that they are inexpensive and safe, so plant extracts are environmentally acceptable, available and renewable sources. These represent a

A.M. Derna · C.M. Méndez · A.E. Ares
Faculty of Sciences, National University of Misiones,
1552 Azara Street, 3300 Posadas, Misiones, Argentina

L.M. Gassa · A.E. Ares (✉)
CIC of the National Research Council (CONICET) of Argentina,
Buenos Aires, Argentina
e-mail: aares@fceqyn.unam.edu.ar

L.M. Gassa
Instituto de Investigaciones Fisicoquímicas Teóricas y Aplicadas,
INIFTA, Calle 64 Diag. 113, 1900 La Plata, Buenos Aires, Argentina

© The Minerals, Metals & Materials Society 2017

M.A. Meyers et al. (eds.), *Proceedings of the 3rd Pan American Materials Congress*,
The Minerals, Metals & Materials Series, DOI 10.1007/978-3-319-52132-9_14

rich source of ingredients, which have a high inhibition efficiency [2]. A corrosion inhibitor is a chemical that can lessen the effect of corrosion by reducing the average velocity of attack on the metal. Inhibition mechanisms are complex. It is accepted as likely, that the inhibitor is adsorbed on the metal surface, the inhibitor reacts with the metal in a thin, compact and protective layer, or the inhibitor forms a precipitate on the metal surface. Some inhibitors act on the cathodic reaction (cathodic inhibitors), or the anodic reaction (anodic inhibitors) or on both (mixed inhibitor).

There are some investigations reported in literature regarding different natural products as potential corrosion inhibitors for copper, aluminum, other metals and alloys, in aqueous solutions at alkaline or neutral pH values. There is little information for acid environments [3, 9]. Generally, an organic compound containing a functional group with heteroatoms performs well as a corrosion inhibitor. The strength of the coordination bonds of parallel heteroatoms ($O < N < S < P$) can provide an idea of their potential effectiveness as corrosion inhibitor [10]. The mate tea is a Neotropical tree species, native to the basins of Alto Paraná and Alto Uruguay and some tributaries of the Paraguay River [11, 12].

For this work, mate tea is used as raw material (ground mixture of leaves and sticks). A commercial brand marketed from the Province of Misiones, Argentina has been chosen.

Experimental Procedure

Obtaining the Extract

The aqueous extract was obtained by taking 50 g of the mate tea mixture and placed in a container of about 200 ml. This was connected to a light bulb by a hose to a vacuum flask and a vacuum trap, in order to suction the extract into the bulb. The dimensions of the container, the water temperature and the characteristics of the bulb were those defined by the Argentinian norm IRAM 20540-1 [13].

Approximately 20 ml of demineralized water at 100 °C were poured on the solid and then allowed to stand for 20 s. The vacuum trap was switched on, producing suction for 20 s, by passing the solution to the kitset. Then, hot water was added and the operation was repeated until the 500 ml of water were used.

The final obtained volume was 375 ml due to retention of liquid in the solids. The obtained extract was left to cool to room temperature (25 °C). Then a centrifuge was used to separate the suspended solids. Relatively small volumes (approximately 10 ml) were processed in 5 min. Finally, the solids were discarded and the mate extract was kept in the refrigerator for up to 48 h before using it as an inhibitor.

Electrochemical Measurements

Measurements were carried out in a conventional three-electrode cell in a Reference Gamry 600® potentiostat. As a working electrode, copper and commercial purity aluminum were used in the form of rectangular areas (2.08 and 1.10 cm^2, respectively).

The exposed surface of each working electrode was roughened using sandpaper of different grain sizes from #600 to #1500, rinsed with demineralized water and dried with soft paper. A platinum electrode was used as counter electrode and saturated calomel as reference electrode (E = − 0.242 V vs. <ENH). The solutions in use included 0.5 M NaCl (pH = 5) and 1 M HCl (pH = 0) prepared from analytical grade reagents with different concentrations of mate tea extract. Concentrations of 1, 5 and 25% vol. of extract were used for potentiodynamic polarization experiments and concentrations of 0.66 and 6.67% vol. were used for cyclic voltammetric experiments. All measurement were done at room temperature and in an atmosphere saturated with N_2.

Potentiodynamic Polarization

The polarization curves were obtained after 1 h to stabilize the electrode potential at open circuit, sweeping between −0.3 and 0.3 V with respect to the open circuit potential, at a rate of 0.166 mV/s.

Cyclic Voltammetry

Two types of experiments were conducted.

In the first type of experiments, the working electrode was allowed to dip in mate tea extract during different times (1, 12 and 24 h). The electrode was then placed in the cell containing 0.5 M NaCl without the presence of the extract under study. The voltammograms were obtained between −0.800 and 0.500 V, using a sweep rate of 25 mV/s.

In the second type of experiments, the extract was added to 0.5 M NaCl solution, to obtain concentrations of inhibitor of 0.66 and 6.67% (v/v). Before each test the system was maintained at −0.850 V for 300 s for stabilization. The voltammograms were acquired between −0.850 and 0.850 V, using two sweep velocities between 5 and 25 mV/s.

Results and Discussion

Potentiodynamic Polarization

Potentiodynamic polarization curves obtained using copper in 0.5 M NaCl solution in the absence and presence of different concentrations of aqueous extract of mate tea studied as inhibitor at 25 °C can be seen in Fig. 1. The corrosion parameters are summarized in Table 1.

Figure 1 shows that as the concentration of extract increases, the current densities are slightly modified. The data in Table 1 shows that the aqueous extract of yerba mate inhibits the corrosion process in the range of the investigated concentrations. At the same time efficiency increases with concentration, reaching a maximum value of 56% for a 25% vol. of inhibitor concentration.

Potentiodynamic polarization curves obtained with copper in 1 M HCl solution, in the absence and presence of different concentrations of aqueous extract of mate tea studied as inhibitor at 25 °C can be seen in Fig. 2. The corrosion parameters are summarized in Table 2.

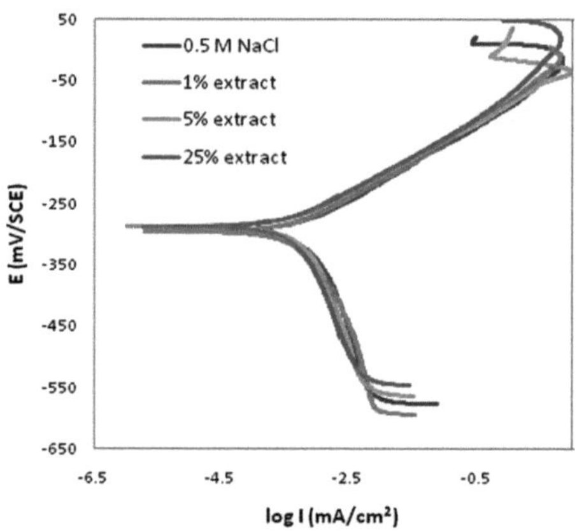

Fig. 1 Potentiodynamic polarization curves of copper in 0.5 M NaCl with the presence of different concentrations of aqueous extract of mate

Table 1 Copper electrochemical parameters at different concentrations of extract of mate tea in 0.5 M NaCl solution

Concentration (%)	E_{corr} (mV)	Log I (mA/cm^2)	I_{corr} (mA)	Efficiency (%)
0	−295	−3.2218	0.00060	–
1	−296	−3.4111	0.00039	35
5	−287	−3.4621	0.00035	42
25	−289	−3.5758	0.00027	56

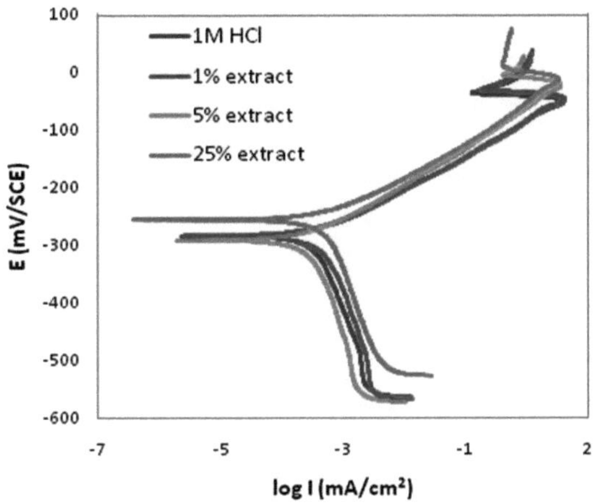

Fig. 2 Potentiodynamic polarization curves on copper in 1 M HCl presence of different concentrations of aqueous extract of mate tea

Table 2 Copper electrochemical parameters at different concentrations of extract of mate tea in 1 M HCl solution

Concentration (%)	E_{corr} (mV)	Log I (mA/cm^2)	I_{corr} (mA)	Efficiency (%)
0	−285	0	1	–
1	−286	−0.0213	0.9521	5
5	−292	−0.0799	0.8319	17
25	−257	0.2860	1.9319	–

For copper in hydrochloric acid, it is observed that the efficiency reaches 17% in the concentration of 5% of mate extract in the medium, while increasing the concentration the current density increases, so it no longer acts as an inhibitor but potency the corrosion process of copper using 1 M HCl with 25% extract.

Potentiodynamic polarization curves of aluminum in 0.5 M NaCl solution, in the absence and presence of different concentrations of aqueous extract of mate studied as inhibitor at 25 °C can be seen in Fig. 3. The corrosion parameters are summarized in Table 3.

For aluminum in 0.5 M NaCl (Table 3) the highest percentage of efficiency is achieved for the concentration of yerba mate extract of 1%. As the concentration increases, the efficiency decreases. The extract with a concentration of 25% vol. no longer inhibits the corrosion process. The inhibition efficiency was calculated according to Eq. (1).

$$E = 1 - \frac{I'corr}{Icorr} * 100 \quad (1)$$

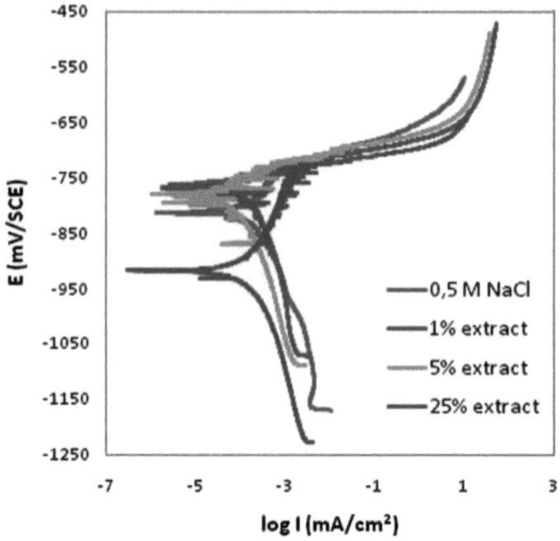

Fig. 3 Potentiodynamic polarization curves of aluminum in 0.5 M NaCl solution in the presence of different concentrations of aqueous extract of mate tea

Table 3 Electrochemical parameters of aluminum at different concentrations of mate extract in 0.5 M NaCl solution

Concentration (%)	E_{corr} (mV)	Log I (mA/cm^2)	I_{corr} (mA)	Efficiency (%)
0	−814	−3.8430	0.0001435	–
1	−768	−5.7475	0.0000018	99
5	−964	−5.6158	0.0000024	98
25	−918	−3.8018	0.0001578	–

where E is the efficiency in % of inhibition, and Icorr and I′corr are the current density in the absence and presence of mate extract in the medium, respectively, in mA/cm^2. Both of them were obtained by using the Tafel extrapolation method. Figures 4, 5 and 6 show micrographs of the surface of the copper and aluminum electrodes, in 0.5 M NaCl and 1 M HCl in absence and presence of aqueous extract of mate tea, respectively, after potentiodynamic sweep.

In Fig. 4a, it is observed that brown dark corrosion products have covered the surface, almost entirely. While in Fig. 4b, corresponding to the electrode tested in 0.5 M NaCl with 25% extract of mate tea. The surface is covered by corrosion products to a lesser extent compared to what can be seen in Fig. 4a, there are even

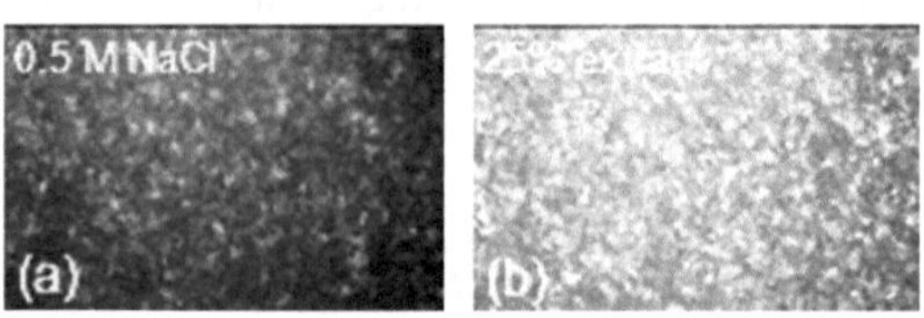

Fig. 4 Micrograph (25×) of the copper surface, after potentiodynamic polarization testing in 0.5 M NaCl, **a** in the absence and **b** in the presence of 25% aqueous extract of mate tea

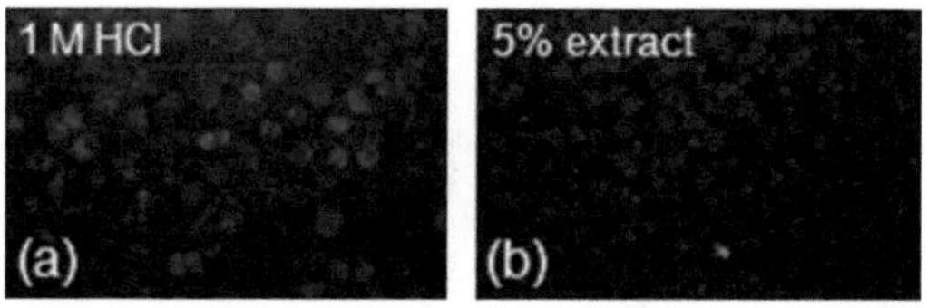

Fig. 5 Micrograph (25×) of the copper surface, after potentiodynamic polarization testing in 1 M NaCl, **a** in the absence and **b** in the presence of 5% aqueous extract of mate tea

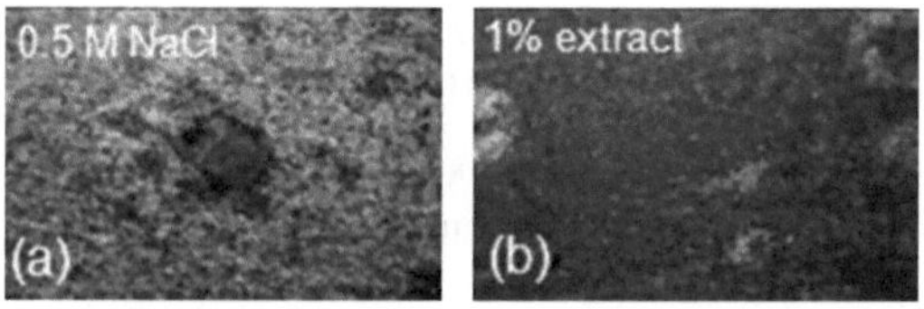

Fig. 6 Micrograph (25×) of the copper surface, after potentiodynamic polarization testing in 1 M NaCl, **a** in the absence and **b** in the presence of 1% aqueous extract of mate tea

significative oxide uncoated areas. This is consistent with the value of inhibition efficiency of 56% for 25% of mate extract in the medium (Table 1). In Fig. 5a, corresponding to the surface of the copper electrode after the potentiodynamic polarization test in 1 M HCl solution, the area covered by corrosion products is dark brown and on this layer, the presence of gray spots distributed throughout the surface can be observed. Similarly, in Fig. 5b, corresponding to the electrode that was tested in 1 M HCl solution with 5% extract of mate tea, the surface is covered by a layer similar to those characteristics observed in Fig. 5a but in more compact or homogeneous form. In this case, the value of the inhibition efficiency obtained was 17–5% of mate extract in the medium (Table 2). In Fig. 6a, corresponding to the surface of the aluminum electrode after potentiodynamic polarization test in 0.5 M NaCl solution, pitting was observed on the entire surface.

While in Fig. 6b, corresponding to the electrode that was tested in 0.5 M NaCl solution with 1% of extract of mate, the surface was covered by a homogeneous layer of green color throughout its length, which could be attributed to the formation of a complex compound between the corrosion products and the extract of yerba mate. In this case the value obtained corresponding to the inhibition efficiency reached of 99% for 1% to the extract in the medium (Table 3).

Cyclic Voltammetry

Figure 7 shows the copper voltammogram obtained in a 0.5 M NaCl solution with different immersion ex situ times (1, 12 and 24 h) in the concentrated extract of mate tea (30 g/l) at 25 mV/s of sweep rate.

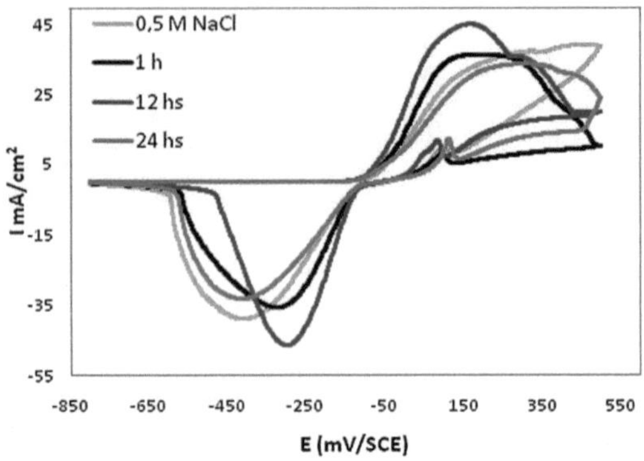

Fig. 7 Cyclic voltammogram for copper in 0.5 M NaCl solution with immersion times of 0, 1, 12 and 24 h in concentrated aqueous of extract of mate tea (30 g/l)

The voltammogram shows a broad area of anodic current from 0.050 to 0.410 V, which could be associated with the oxidation of Cu to Cu(I) [10]. In the reverse sweep a cathodic peak is observed at −0.380 V, which can be attributed to the reduction of soluble species [10]. For 1 h, immersion is defined in the anodic sweep a width anodic peak. During cathodic sweep, a reactivation peak is observed in 0.100 V and the reduction peak shifts to potentials that are more positive. After 12 h of immersion, the anodic and cathodic peaks are clearly defined, with current values higher than those obtained with the target. By increasing the immersion time in the inhibitor (24 h) can be seen a decrease in the values of currents peaks, probably due to better anchor of the inhibitor on the surface, to remain longer in contact with it before being exposed to aggressive medium. It reversing the potential sweep direction towards the cathodic direction, it is observed that the reactivation process occurs, in which species had formed anodically, now reduced giving a cathodic peak [14]. In Fig. 8 is observed micrographs of the surface of the copper electrodes after voltammetry test.

In Fig. 8a, corresponding to the surface of the copper electrode, after the cyclic voltammetry test (in 0.5 M NaCl solution), after 1 h of immersion, ex situ, in concentrated mate extract (30 g/l), it is observed that a homogeneous layer covers the surface. In Fig. 8b, corresponding to the electrode remained submerged for a

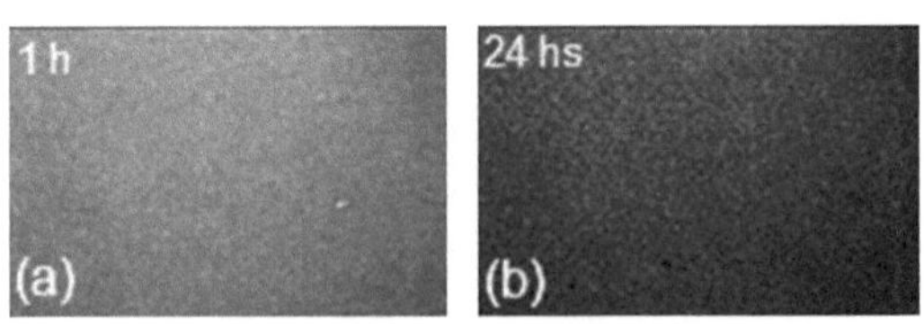

Fig. 8 Micrograph (25×) of the copper surface, after the cyclic voltammetry testing in 0.5 M NaCl, with immersion time of **a** 1 h and **b** 24 h in concentrated extract of mate (30 g/l)

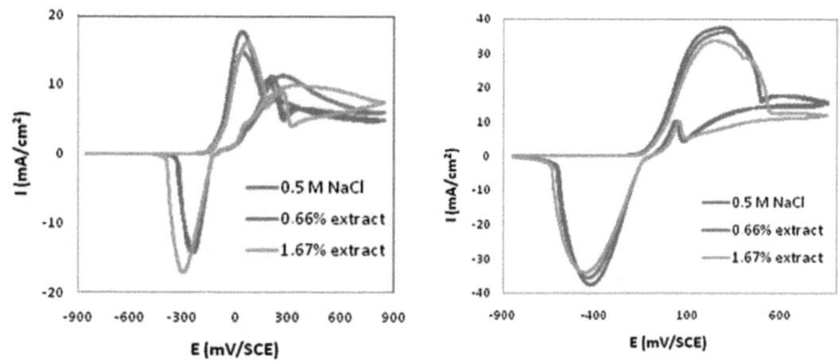

Fig. 9 Cyclic voltammogram for copper in 0.5 M NaCl solution with 0.66 and 1.67% of aqueous extract of mate tea, at: **a** at 5 mV/s and **b** 25 mV/s

period of 24 h in the extract of mate tea, the surface is covered by a homogeneous layer and has black spots distributed evenly. In Fig. 9 the voltammograms for copper in 0.5 M NaCl solution and 0.5 M NaCl solution with 0.66 and 1.67% aqueous extract of mate tea, for two different sweep rates, are showed. It is observed that for low sweep speeds are better defined the peaks of oxidation and reduction of copper, and current values are approximately the half of peak of those obtained at higher inhibitor concentrations for both velocities. This is probably due to a slow sweep velocities, the inhibitor has more time to interact with the metal surface.

Conclusions

The aqueous extract of mate tea has promising properties for use as a corrosion inhibitor of copper and aluminum in 0.5 M NaCl solution and 1 M HCl, and the efficiencies have been determined in the present research.

The efficiencies of corrosion inhibition were determined by potentiodynamic polarization, but should still be optimized the tests to achieve higher inhibition values using lower concentrations of mate tea extract.

The cyclic voltammetry studies show a complex response of metals studied in presence of mate tea extract in function of time of immersion and ex situ at different concentrations added to the aggressive solution. This makes it necessary to determine the mechanism by which this natural inhibitor is acting on the substrates used.

Acknowledgements This work was supported by Consejo Nacional de Investigaciones Científicas y Técnicas (CONICET) and Agencia Nacional de Promoción Científica y Tecnológica (ANPCyT), from Argentina.

References

1. Buchweishaija, J. (2009). Phytochemicals as green corrosion inhibitors in various corrosive media: A review. *Tanzania Journal of Science, 35,* 77–92.
2. Raja, P. B., & Sethuraman, M. G. (2008). Natural products as corrosion inhibitor for metals in corrosive media—A review. *Materials Letters, 62,* 113–116.
3. Al-Otaibi, M. S., Al-Mayouf, A. M., Khan, M., Mousa, A. A., Al-Mazroa, S. A., Alkhathlan, H. Z. (2014). Corrosion inhibitory action of some plant extracts on the corrosion of mild steel in acidic media. *Arabian Journal of Chemistry, 7,* 340–346.
4. Singh, A., Ahamad, I., & Quraishi, M. A. (2016). Piper longum extract as green corrosion inhibitor for aluminium in NaOH solution. *Arabian Journal of Chemistry, 9,* S1584–S1589.
5. Halambek, J., Berkovic, K., & Vorkapic-Furac, J. (2013). *Laurus nobilis* L. oil as green corrosion inhibitor for aluminium and AA5754 aluminium alloy in 3% NaCl solution. *Materials Chemistry and Physics, 137,* 788–795.
6. Singh, A., Singh, V. K., & Quraishi, M. A. (2010). Aqueous extract of Kalmegh (*Andrographis paniculata*) leaves as green inhibitor for mild steel in hydrochloric acid solution. *International Journal of Corrosion.* Article ID 275983, 10p. doi:10.1155/2010/275983
7. Hazwan Hussin, M., Jain Kassim, M., Razali, N. N., Dahon, N. H., & Nasshorudin, D. (2016). The effect of *Tinospora crispa* extracts as a natural mild steel corrosion inhibitor in 1 M HCl solution. *Arabian Journal of Chemistry, 9,* S616–S624.
8. Khadraoui, A., Khelifa, A., & Hamitouche, H. (2014). Inhibitive effect by extract of *Mentha rotundifolia* leaves on the corrosion of steel in 1 M HCl solution. *Research on Chemical Intermediates, 40,* 961–972.
9. Rosliza, R., & Wan Nick, W. B. (2010). Improvement of corrosion resistance of AA6061 alloy by tapioca starch in seawater. *Current Applied Physics, 10,* 221–229.
10. Baeza, H., Guzmán, M., Ortega, P., & Vera, L. (2003). Corrosion inhibition of copper in 0.5 M hydrochloric acid by 1,3,4-thiadiazole-2,5-dithiol. *Journal of the Chilean Chemical Society, 48,* 23–28.
11. Ramallo, L. A., Smorczewski, M., Valdez, E., Paredes, A. M., & Schmalko, M. E. (1997). Composicion quimica del extracto acuoso de la yerba mate. I Congreso Sul-Americano da Erva-Mate, II ReuniaoTecnica do Cone Sul Sobre a Cultura da Erva-Mate. *Curitiba Brasil,* 411.
12. Sabbatella, O. P. Pokolenko, J. J., & Schmalko, M. E. (2009). Influencia de la Composición en la Extracción de los Solubles de la Yerba Mate, Revista Ciencia y Tecnología 11.
13. Instituto Argentino de Racionalización de Materiales. (1997). Norma 20540–1: Yerba Mate: Materiales y procedimientos a utilizar en la determinación de los caracteres organolépticos de la yerba mate, bajo forma de mate.
14. Rosatto, S. S., Cabot, P. L., Sumodjo, P. T. A., & Benedetti, A. V. (2001). Electrochemical studies of copper-aluminum-silver alloys in 0.5 M H2SO4. *Electrochimica Acta, 46,* 1043–1051.

Platinum Salt Synthesis as Precursor to Get Heterogeneous Catalyst for Biofuels Production

Adriana I. Martínez-Montalvo, Sherly C. Acosta-Beltrán, Jonathan F. Sierra-Cantor and Carlos A. Guerrero-Fajardo

Abstract The trend in the industry is aimed at developing sustainable processes thus the search for alternative fuels such as biofuels can be the best alternative to replace fossil fuels but they are not yet economically competitive. This research shows a chemical route for the synthesis of tetraammineplatinum (II) chloride salt which needs several stages. Each one production step was evaluated through platinum determination by atomic absorption and their products were characterized by XRD, IR and XRF analyses. Taking into account that the overall yield to tetraammineplatinum (II) chloride was of 61% a pilot plant was proposed and evaluated. This salt is used in the preparation of precursors compounds for getting catalysts supported on structured pore size allowing impregnation of catalytic species and promote the transfer of materials triglyceride molecules, as well as the evaluation of the scaling up of this process to industrial scale.

Keywords Chemical synthesis · Tetraammineplatinum (II) chloride salt · Biofuels · Catalyst precursor

---

A.I. Martínez-Montalvo (✉)
Chemistry and Environmental Engineering,
Universidad Nacional de Colombia, Bogotá, Colombia
e-mail: aimatinezm@unal.edu.co

S.C. Acosta-Beltrán · J.F. Sierra-Cantor · C.A. Guerrero-Fajardo
Department of Chemistry, Universidad Nacional de Colombia, Bogotá, Colombia
e-mail: scacostab@unal.edu.co

J.F. Sierra-Cantor
e-mail: jfsathelo@gmail.com

C.A. Guerrero-Fajardo
e-mail: caguerrerofa@unal.edu.co

© The Minerals, Metals & Materials Society 2017
M.A. Meyers et al. (eds.), *Proceedings of the 3rd Pan American Materials Congress*,
The Minerals, Metals & Materials Series, DOI 10.1007/978-3-319-52132-9_15

Introduction

The trend in the industry is currently focused at developing of sustainable processes, and investigations are centered on biofuels considered one of the best alternatives for replacing fossil fuels such as oil. Between biofuels is included the biodiesel defined as "*monoalkyl fatty acid esters long chain derived from renewable lipids such as vegetable oils used in engines compression ignition (diesel engines) or heating boilers*" (ASTM), this is produced from the transesterification reaction between a compound lipid and an alcohol to produce an ester and a bioproduct, glycerol [1]. Taking in account the growing concern generated by the global warming, biofuels such as biodiesel have attracted attention even they are not yet economically viable because of, among other things, the use of homogeneous catalysts which are difficult to separate from the reaction product. For this reason a large portion of the research in this field is aimed on the development of heterogeneous catalysts, which could be separated in a simple way from the product with unit operations such as filtration; however, these catalysts have a low conversion even compared with the rate showed by the homogeneous catalysts [2, 3].

The tetraammineplatinum (II) chloride salt, (see Fig. 1), is used as a catalyst in chemical synthesis and precursor of platinum catalysts.

It is also used in the pharmaceutical field. This salt is synthesized from platinum which has presented an increasing in its production over 1.28 metric tons on average over the last ten years in Colombia [4]. This production is mainly associated with the strip mining of gold and silver [5]. Figure 2 can be seen the data production of platinum in the last 10 years in the country.

It can be seen in Fig. 2 that in 2013 is displayed a remarked increasing of platinum production, mainly because of there were a big number of dredges operating in the department of Choco—Colombia, where 99% of the country's

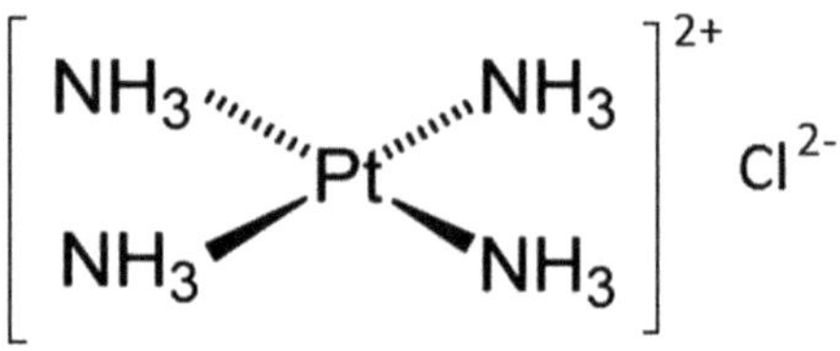

Fig. 1 Tetraammineplatinum (II) chloride salt structure. Taken from: https://pubchem.ncbi.nlm.nih.gov/compound/Tetraammineplatinum_II__chloride#section=Top

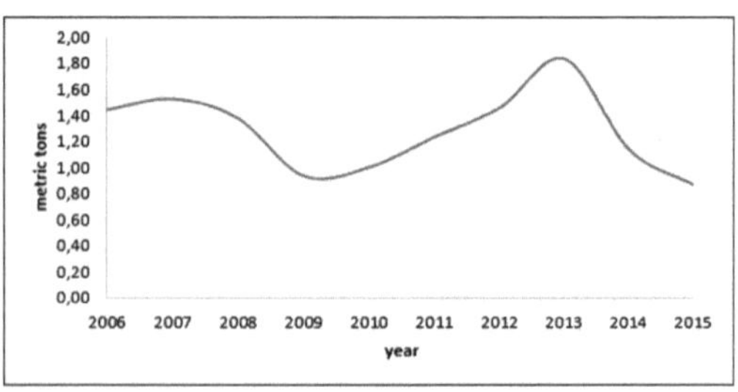

Fig. 2 Platinum production. Historic change in the last 10 years. The data shows a maximum of 1:28 tons at 2013. Data taken from: http://www.simco.gov.co/?TabId=121

platinum is produced. Due to Colombia does not have enough technology to produce final products from platinum, the great majority of domestic production is exported. In this work will be presented a methodology for the synthesis of tetraammineplatinum (II) chloride salt on a laboratory scale showing an alternative for the preparation of Platinum catalytic materials impregnating zeolites, this last used as catalyst for the transesterification reaction for biodiesel production. Finally, it is proposed a reaction system for the pilot production of platinum salt on scale.

Synthesis of Tetraammineplatinum (II) Chloride Salt

Methodology

From pure platinum, we start with the synthesis of chloroplatinic acid, H_2PtCl_6, subsequently we reduce to chloroplatinous acid, H_2PtCl_4 and finally the complex tetraammine platinum chloride (II) is prepared (Fig. 3).

First Stage

According to [6], chloroplatinic acid (H_2PtCl_6) is synthesized by dissolving of platinum wires in aqua regia during 4 weeks keeping a temperature constant of 98 °C with a pressure of 462.08 mmHg and a constant agitation (200 rpm). This procedure consume 156 mL of HCl and 92 mL of HNO_3, once dissolved entirely platinum, the excess of HNO_3 is removed with consecutive evaporation and adding totally 80 mL HCl. On the first solution the H_2PtCl_6 was taken to solid state by evaporating water from the solution and the efficiency was also determined. During the repetition of the process, the acid was left within solution and we proceeded to the second stage (Fig. 4).

$$Pt + 4HNO_3 + 6HCl \rightarrow H_2PtCl_6 + 4NO_2 + 4H_2O \tag{1}$$

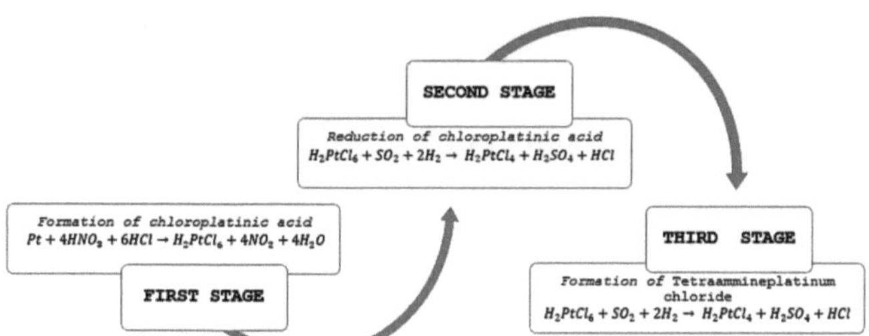

Fig. 3 Schematic diagram for the experimental methodology. *Source* Authors, 2016

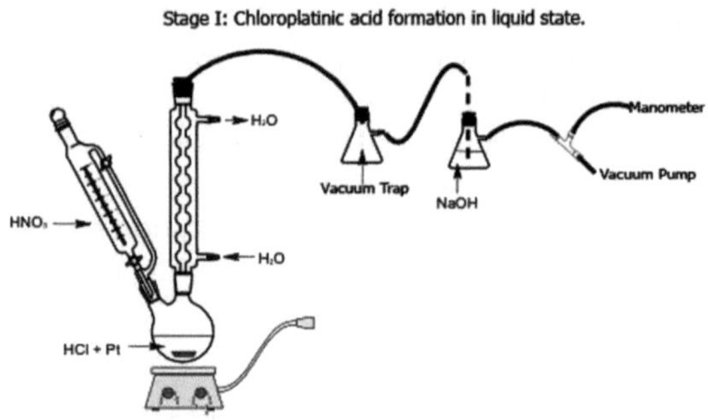

Fig. 4 Experimental assembly for the formation of chloroplatinic acid. *Source* Authors, 2016

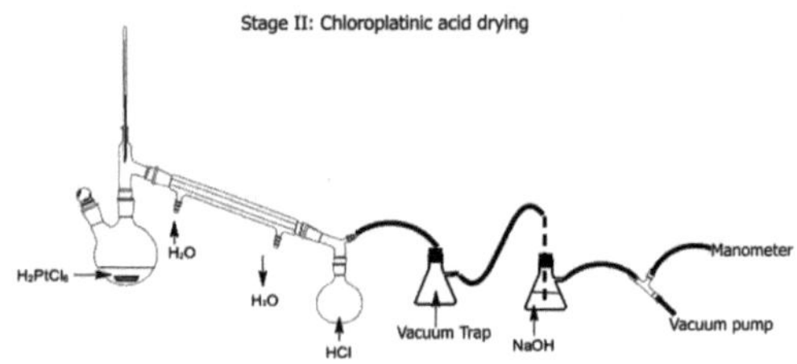

Fig. 5 Experimental assembly for the chloroplatinic acid drying. *Source* authors, 2016

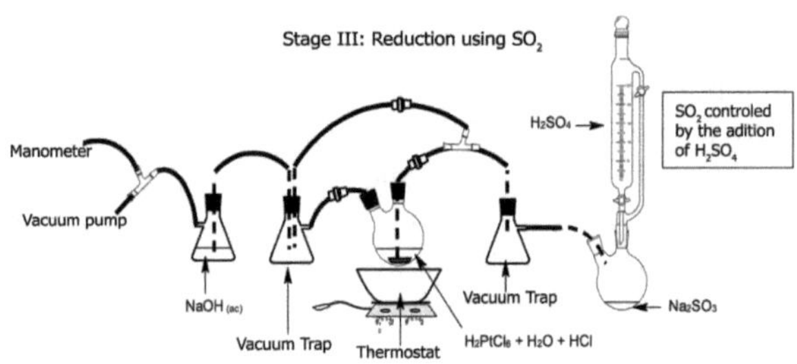

Fig. 6 Experimental assembly for the chloroplatinic acid reduction. *Source* Authors, 2016

Second Stage

As it is suggested in [7], a SO_2 gas is prepared from H_2SO_4 and $Na_2S_2O_5$ and this gas is circulated within the H_2PtCl_6 solution with a constant temperature of 90 °C and stirring. A equimolar process was made such that all the chloroplatinic acid would be reduced to chloroplatinous acid, H_2PtCl_4, decreasing in this way the formation of complexes sulfate (Figs. 5 and 6).

$$H_2PtCl_6 + SO_2 + 2H_2O \rightarrow H_2PtCl_4 + H_2SO_4 + HCl \qquad (2)$$

Third Stage

With the conditions of 82 °C and 44.5 mmHg, 50 mL of water were added, slowly and under constant stirring should be added concentrated NH_3 until a yellow solution is obtained which correspond to the salt. The temperature is maintained until the gases are colorless, then an excess of HCl is added and the acid pH is checked. We added 100 mL of acetone and 50 mL of methanol, the reaction conditions were maintained during 20 h with constant agitation (200 rpm). After that, the obtained solid is filtered and washed with a mixture of alcohol-acetone until test of sulphates is negative. Finally, it is washed with acetone. The salt obtained was dried under ambient conditions.

$$H_2PtCl_4 + 6NH_3 \rightarrow [Pt(NH_3)_4]Cl_2 + 2NH_4Cl \quad (3)$$

Analysis, Discussion and Results

The chloroplatinic acid obtained in the first solution was taken to solid state to perform the mass balance (Table 1).

Table 2 shows a porcentual error of 1.77% obtained in this test. It could be due to a loss of mass in the reaction system during the process which increases the expected value of chloroplatinic acid (Fig. 7).

To corroborate reduction from H_2PtCl_6 to H_2PtCl_4 a cyclic voltamperometry is performed (see Fig. 8), with a glassy carbon electrode to a sample of 500 mL

Table 1 Initial and final conditions for the first stage

	Initial	Final
Pt mass (g)	0.883	0.0

Table 2 Results obtained from the chloroplatinic acid

	Theoretical	Experimental	% error
Chloroplatinic acid mass (g)	1.856	1.889	1.77

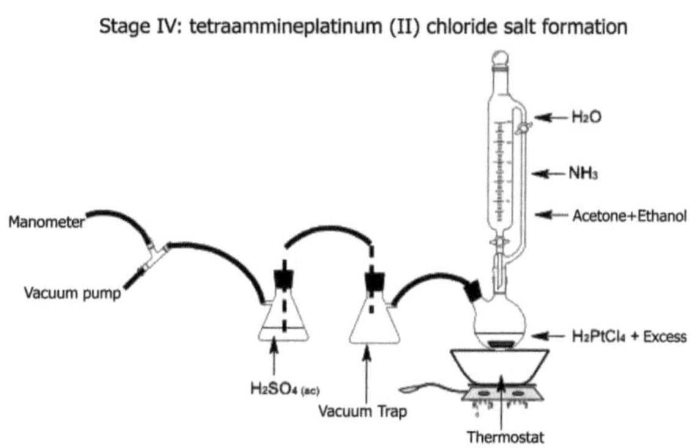

Fig. 7 Experimental assembly for the tetraammineplatinum (II) chloride salt formation. *Source* Authors, 2016

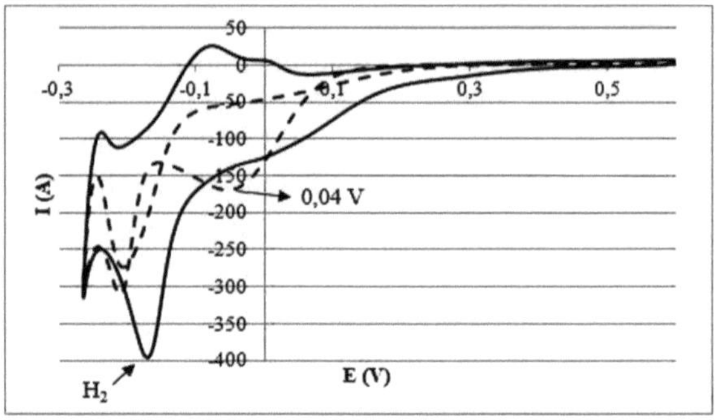

Fig. 8 Cyclic voltamperogram from the reduced solution with SO_2. The first scan is displayed by the *dotted line* and the fifth scan is represented by the *solid line*

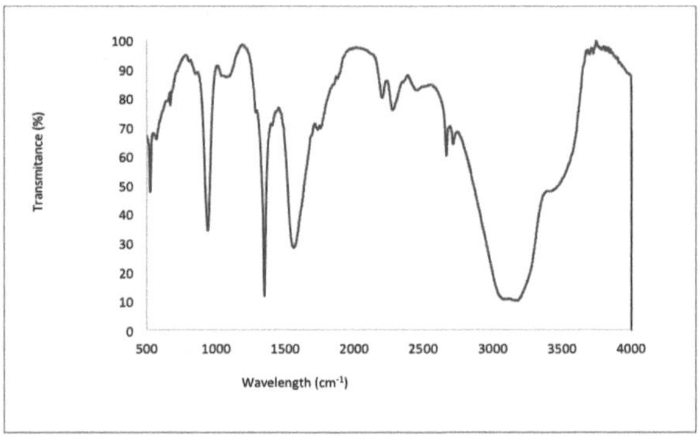

Fig. 9 Infrared spectra for the $[Pt(NH_3)_4]Cl_2$ obtained during the process

solution of H_2PtCl_4 0.05 M approximately, with a scan rate of 0.05 V/s and this solution was left stabilize for 10 s. With this, it is observed in 0.04 V the reduction peak of a single species of platinum along with the corresponding hydrogen absorption peak, due to acid medium in which the sample is. Not observed oxidation peak due to platinum does not oxidize naturally, in the sweep 5 decreases the intensity of reduction peak, which is characteristic of platinum due to its nature. In the final stage of the process are obtained 1.023 g of Tetraammineplatinum chloride unpurified. The overall mass balance is obtained that the platinum in the input must be equal to that present in the salt, which represents 58% of the total mass of this. According to the mass of platinum that fed to the second synthesis (0.961 g), mass expected salt is 1.65 g of what follows that the overall yield of the process is 61%.

In the characterization by infrared, (see Fig. 9) the following bands were obtained:

3431 cm^1 (Medium intensity, broadband) stretch O–H bond of waters of hydration, 3106 cm^1 (Intense, broadband) stretch bond N–H the ammoniac.

1560 cm^1 (intense) in-plane bending H–N–H of ammonia.

1348 and 939 cm^1 (intense and acute) Bands associated with the presence of acetone used to wash the final product.

524 cm^1 (low intensity, acute band) possible stretching-Cl.

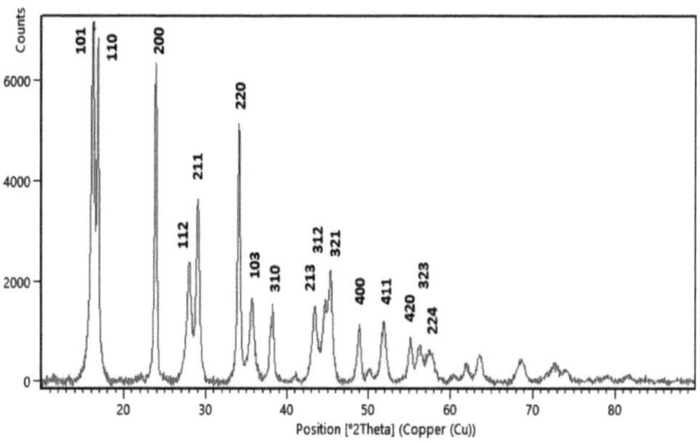

Fig. 10 DRX of $[Pt(NH_3)_4]Cl_2$ obtained

Analysis by XRD, (see Fig. 10) was performed with the program HighScore Plus 2.0, cell parameters were found are, a = b = 7.442 Å; c = 8.060 Å and a = b = g = 90°. With these parameters and reviewing the Miller indices, it was found that the unit cell has a simple tetragonal structure. It can see that peaks presents widening which can be explained by two factors; first, effect of the solvent within the crystal formed and second, formation of nanoparticles of tetraammineplatinum chloride salt that are ideal for further use as catalyst precursor.

Proposed Pilot Scale Reactor

Because this salt is classified as a fine chemical product, a batch process is proposed for the scaling process. Figure 11 displays a simplified diagram for the observed process. It is proposed firstly a type reactor stirred tank, as different stages of the process require a high residence time and continuous agitation, such as chloro-platinic acid production that is the longest stage of the process. It is anticipated that the use of a single reactor is sufficient to carry out all reaction steps, ending with a crystallization step in which the salt is obtained. Reactor are connected to storage tanks the reagents of different stages, the addition of these is determined by the concentration of species into the reactor, also necessary to install a system for measuring tank level, and another to control the pressure and temperature. It is process in vacuum as this decreases the boiling point of the reactants and thus the risk of chloroplatinic acid decomposition in the first stage.

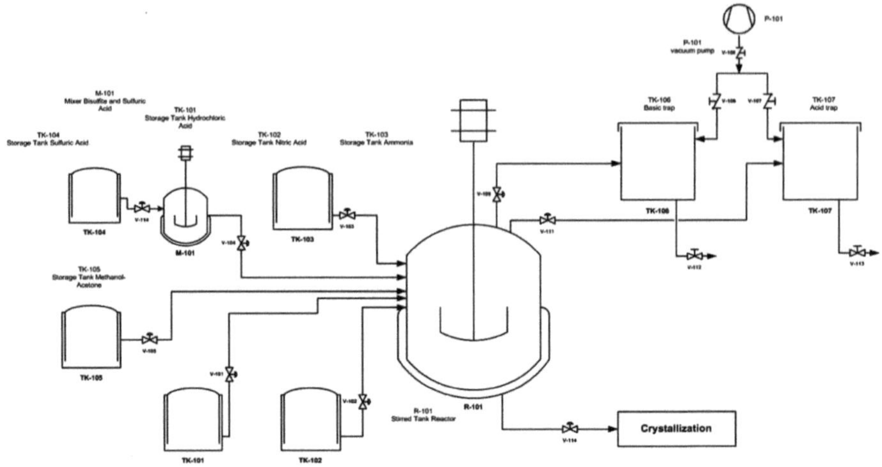

Fig. 11 Simple schematic assembly for the reaction system proposed by the work team. Design made by the authors, 2016

Conclusions

Reduction from H_2PtCl_6 to H_2PtCl_4 is confirmed using cyclic voltammetry, since the reduction of a single species of platinum it is shown, which in this case is considered as the chloroplatinous acid.

During the first synthesis step of chloroplatinic acid production an error rate of 1.77% was obtained, because a mass loss occurs in the reaction system, as data showed, the balloon mass before and after the reaction, to avoid this should use a reaction system with thicker walls for subsequent synthesis.

An overall yield of 61% in the synthesis of $[Pt(NH_3)_4]Cl_2$ was obtained, this performance although not as low, shows that there are significant losses in the process that may be associated mainly washings carried out in the last stage of the synthesis, since in this is used alcohol and salt is very soluble in these compounds.

Acknowledgements Special gratitude to the research group: Energy Recovery of Natural Resources—APRENA, Department of Chemistry from National University of Colombia for the contribution to this research.

References

1. Lizana, D. Antecedentes Generales sobre Biodiesel. In *Ecodesarrollo.cl*. Available: http://www.ecodesarrollo.cl/descargas/Antecedentes_Biodiesel_D.pdf. Accessed September 5, 2016.
2. Yang, L., Zhang, A., & Zheng, X. (2009). Shrimp shell catalyst for biodiesel production. *Energy and Fuels, 23*, 3859–3865.

3. Helwania, Z., Othman, M., Aziz, N., Fernando, W., & Kim, J. (2009). Technologies for production of biodiesel focusing on green catalytic techniques: A review. *Fuel Processing Technology*, *90*, 1502–1514.
4. Sistema de informacion minero Colombiano. In *Unidad de planeación minero energética*. Available: http://www.upme.gov.co/generadorconsultas/Consulta_Series.aspx?idModulo=4&tipoSerie=118&grupo=361. Accessed September 5, 2016.
5. Unidad de Planeacion minero energetica, "SIMCO", 2005. In *Sistema de infromación minero colombiano*. Available: http://www.simco.gov.co/Portals/0/estadisticas/Imagen/platino.pdf. Accessed September 5, 2016.
6. Schweizer, E., & Kerr, G. T. (1978). Therma decomposition of hexachloroplatinic acid. *Inorganic Chemistry, 17*(8), 2326–2327.
7. Rochow, E. G. (1960). *Inorganic syntheses* (Vol. 6, pp. 250–253). Mc-Graw-Hill book Company.
8. ASTM sets the standard for biodiesel. In *Standarization News*. Available: https://www.astm.org/SNEWS/JF_2009/nelson_jf09.html. Accessed September 5, 2016.

Technical and Environmental Assessment of an Alternative Binder for Low Traffic Roads with LCA Methodology

Alejandra Balaguera Quintero, Diana Gómez Cano, Gloria Carvajal Peláez and Yhan Paul Arias

Abstract Currently, low traffic roads in most countries are made up of unpaved roads; therefore, to increase the bearing capacity and durability of soils, using stabilizers such as lime and portland cement is required. In this paper, the results obtained from the addition of alternative binder materials based on industrial by products such as alkali activated coal ashes that work as soil stabilizers with sustainability criteria and are assessed through Life Cycle Assessment (LCA); this process is approached from the preparation, packaging and storage of binder material, its activation and finally the application in test sections obtaining unconfined compressive strengths of the order of 2 MPa; which represented an increase in resistance above 300% for the same soil without stabilization.

Keywords Alternative binder · Life cycle assessment · Soil stabilization

Introduction

Alternative binders are formed from two components, the powdery materials of alumina-silicates nature and alkaline activator, usually sodium or potassium hydroxide comprise materials with microstructures similar to Ordinary Portland Cement (OPC). These material have been known since the last three decades as geopolymer [1–3] or alkali cements [4, 5], and its use in construction industry ranges from cements, mortars and concretes [6, 7]; among others. Alkali cements produced from industrial by-products like fly ash, blast furnace slag, thermally modified clays, and others such as metakaolin, are used due to their chemical

A.B. Quintero · D.G. Cano · G.C. Peláez
Facultad de Ingeniería, Universidad de Medellín, Carrera, 87 N° 30-65, Medellín, Colombia

Y.P. Arias (✉)
Facultad de Arquitectura, Universidad Nacional de Colombia, Cl. 59a #63-20, Medellín, Colombia
e-mail: ypariasj@unal.edu.co

© The Minerals, Metals & Materials Society 2017

M.A. Meyers et al. (eds.), *Proceedings of the 3rd Pan American Materials Congress*, The Minerals, Metals & Materials Series, DOI 10.1007/978-3-319-52132-9_16

composition and reactivity as precursors for the manufacture of unconventional cementitious materials [8]. Materials with high contents of alumina Al_2O_3 and reactive silica SiO_2 or fly ashes, have been used successfully for decades as a partial replacement of OPC, thus contributing to the improvement of physical-mechanical properties and durability compared to traditional materials [9, 10].

The production of a ton of OPC generates approximately between 0.87 and 0.97 t of carbon dioxide gas CO_2 [1, 11, 12] where the construction industry accounts for 7% of anthropogenic emissions worldwide, this due to the use of fossil fuels [13]. According to *The Colombian National Administrative Department of Statistics*, the cement production is on the rise, in 2009 6.97 Million Metric ton (MMt) and in 2015 13.15 MMt were produced, generating an increment of nearly double the production in six years. This increase is linked to population growth and strong demand in the construction industry.

In 2012 it was found that coal is the second source of main energy due to its abundance and the policies adopted by industrialized countries, which produce about 80% of the energy required from fossil fuels; coal production worldwide exceeds 6185.85 MMt per year; Colombia particularly presented an annual coal production from 34 to 77 MMt, being the tenth largest producer of coal, with 74.35 MMt [14]. Fly ashes from coal combustion are generated in large volumes, approximately 10,000 t per year [15]. In addition, in the production of palm oil Colombia ranks first in America and fourth worldwide [16], with a production of 5.4 MMt per year, this shows that 5% ashes are a product of biomass processing. Therefore, the use and reuse of these materials represent significant advantages, since their production and use have a positive environmental impact by reducing anthropogenic CO_2 emissions over conventional cementitious materials that have in their production a high consumption of energy and natural resources [17].

In most developing countries the roads are not paved and have high levels of deterioration due to the susceptibility of the soil and the water regime to which they are exposed, hindering proper operation, especially during the rainy season [18]. To this problem it is added the financial inability to pave the entire network of this type of roads, which means the need for rehabilitation and maintenance in tertiary network with techniques that contribute to their stability and proper functioning.

In Colombia the *research network and technological innovation in new materials and construction processes for road infrastructure INNOVIAL* has integrated coal ash as an alternative material in soil stabilization for low traffic roads, showing significant improvements not only technical but also environmental and social for the sustainable development in the construction business [19]; however, the material conditions for stabilizing the soils and the processes associated with their manufacture and initial processing of raw materials for their production is a subject without enough research, because it is not considered a commercial product yet and there are not systems for optimal storage to preserve their properties. The materials used for packaging and subsequent storage in addition to moisture conditions and storage times directly influence the physical, chemical and mechanical properties of the stabilizer and therefore stabilized soils; from the economic point of view it is evidenced a high cost in the types of packaging used for storage, which is revealed in the increase of environmental impacts at the time of production and marketing. This statement implies exploring into new stabilizer products for unpaved roads from industrial waste management from the manufacturing process, packaging, storage and environmental assessment.

Materials and Methodology

Materials

Silty-clay soil from Colombia that comes from a low traffic road unpaved was extracted for stabilization with fly ash and Na(OH). The used soil had a plastic index of 17.5% with characteristics AASHTO A-7-5, and contents of SiO_2 and Al $(OH)_3$. Modified proctor tests were performed in the soil and the stabilized soil to determine the optimal soil moisture. See Table 1.

The stabilization or improving of the mechanical properties of the soils that are the study object were performed with the use of fly ashes product of the coal combustion in a thermoelectric process of a Colombian textile industry.

The ashes used had mass contents of SiO_2 + Al_2O_3 + Fe_2O_3 around 82.35% and CaO of 4.46%, which makes it an F type ash. The quantity of burned coal was 3.9%, see Table 2.

Through DRX Rietveld analysis [20], it was determined that 72.25% in mass of the coal ash corresponds to a vitreous phase, the rest of the mineral components of crystalline character represent 27.75% of the ashes, see Table 3.

The ashes particle size is described in Fig. 1, with a d_{80} of 45 μm, which allows classifying it as an F Type fly ash.

Table 1 Materials

Soil (% in mass)	Fly ash (% in mass)	Optimum moisture (%)	Maximum density (kN/m^3)
100.00	0.00	25.75	14.66
86.00	14.00	22.50	14.70

Table 2 FRX of fly ash (%) in mass

SiO_2	Al_2O_3	Fe_2O_3	CaO	MgO	Na_2O	SO_3	Loss on ignition 110–1000 °C
46.2	31.4	4.75	4.46	1.82	3.63	0.68	3.9

Table 3 Mineralogical phases for fly ash (%) in mass

% mullite	% quartz	% hematite	% vitreous phase
17.49	9.01	1.59	72.25

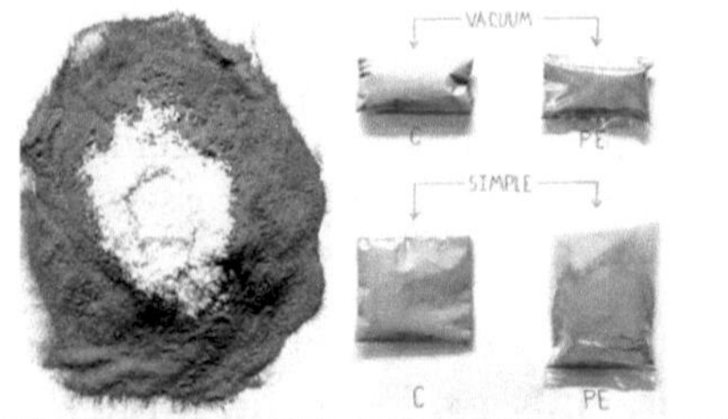

a) Stabilizer Mixture of NaOH and fly ash b) Packed Stabilizer Product

Fig. 1 Stabilizer product

Alkaline Activator

Activation of the fly ash, it has been made whit solid flakes of Na(OH) for the necessary concentration of 4 M. NaOH 98% purity was used.

Methodology

A design of experiment (DOE) methodology was proposed to assess factors that may modify the final conditions of the stabilizing packaged product. The packed stabilizer corresponds to the mixture of fly ash and Na(OH) solid flakes, see Fig. 1.

The experimental matrix is presented Table 6, where:

F I: Packaging Material
F II: Type of seal
F III: Baling Moisture
F IV: Storage humidity.

F I with two levels: *cellulose* (C) and *polyethylene* (PE). Vacuum sealed (*Vacuum*) and (*simple*) sealed without vacuum were considered for *F II*. Moisture packaging material *F III* included kiln (*dried*) fly ash and (*standard*) humidity conditions. Finally, the *F IV* was evaluated by storing the packed stabilizer to (*standard*) and above 90% humidity. For all measures of the experimental matrix 14% mass of stabilizer product and 86% of soil was used. In each experimental run the stabilizer packaged product was stored for 14 days and the effectiveness of the soil stabilizer was measured by its response to *Unconfined Compressive Strength* (UCS). Similarly, it was performed at 28 days of storage. To assess the response to UCS test pieces of 50 mm of diameter and 100 mm high were used, cured in a sealed container for 7 days at 25 °C. The compressive test of the samples was conducted in 3000 Humboldt equipment with a load cell of 45 kN.

Finally, for environmental valuation the life-cycle assessment (LCA) methodology was used through the following stages, according to the ISO-14044 standard.

Stage 1. Goal Definition and Scoping

The quantification of environmental impacts generated during the processes of stabilizer preparation, activation, packaging and storage, was from gate to gate LCA, because the analysis was carried out from the acquisition of raw materials for manufacturing until obtaining the stabilizing product, using as a functional unit 25.0 kg of stabilizing product manufactured and stored. Table 4 shows the impact categories considered.

Stage 2. Life Cycle Inventory

Resources consumption and waste generation and emissions attributable to product life-cycle were identified, according to information gathered, depending on the input and output variables of the evaluated processes (see Table 5).

Stage 3. Impact Assessment and Interpretation

According to ReCiPe methodology, the evaluation was carried out in three consecutive phases: *classification* of the impacts for each impact category; *characterization*, involving the assessment of the real impact of each category by characterization factors; and finally, *normalization* where multiplicative weighting factors.

Table 4 Definition of impact categories

Categoría de impacto
Use of fossil fuel (MJ)
Energy consumption (MJ)
Breathable inorganics (SO_2)
Climate change (CO_2)

Table 5 Variable definition of input and output

Variables de Entrada		Variables de Salida	
Distance	Lengths traveled for transporting materials	Emissions (CO_2–SO_2)	Generation of emissions in relation to fuel emission factors (ACPM—Diesel) and energy (kWh)
Fuel/energy	Type and amount of consumption in relation to performance of vehicle/equipment and usage times		

Results and Discussion

Technical Assessment of the Stabilizer

The technical results of the stabilizer were performed by *analysis of variance* ANOVA. The null hypothesis indicates that the factors evaluated have no difference or a significant effect on UCS, i.e. arise in changes in the properties of the stabilizers do not arise. The confidence level (β) is defined as 95%, therefore, the significance level (α) is 5% ($\alpha = 0.05$). Under these conditions if the p-value is less than α, the null hypothesis is rejected (see Table 6). Experimental matrix and results are presented.

The p-values obtained for 14 days and 28 days of storage age, indicated that none of the factors or double interactions between them influenced the response variable. See Table 7.

Although the results Fig. 2 reported an average increase of 400–600% of the UCS on the soil with stabilizing product with respect to the unstabilized soil. It was observed that there is a variation of properties depending on the time of stabilizer storage.

Table 6 Comparing UCS means

Measure	Main factors				UCS (MPa) for age store		t_0	*Hypothesis*
	I	*II*	*III*	*IV*	14 days	28 days	$t_{\alpha/2} = 4.3$	*Ho: Ux = Uy*
1	C	Vacuum	Dray	Std.	1.05	1.48	11.22	Rejected
2	PE	Vacuum	Dray	>90%	1.36	1.58	0.82	Accepted
3	C	Vacuum	Std.	>90%	1.85	1.76	0.63	Accepted
4	PE	Simple	Dray	Std.	2.38	1.79	2.30	Accepted
5	PE	Simple	Dray	>90%	2.05	0.98	7.16	Rejected
6	C	Vacuum	Std.	Std.	1.57	0.89	4.50	Rejected
7	PE	Simple	Std.	Std.	2.78	0.95	2.73	Accepted
8	C	Vacuum	Dray	>90%	2.42	1.43	3.01	Accepted
9	C	Simple	Std.	Std.	2.61	1.64	9.92	Rejected
10	C	Simple	Std.	>90%	1.49	1.52	0.15	Accepted
11	PE	Simple	Dray	Std.	3.07	1.42	23.32	Rejected
12	PE	Vacuum	Dray	Std.	2.71	1.35	64.50	Rejected
13	PE	Vacuum	Std.	>90%	1.98	1.21	14.80	Rejected
14	PE	Simple	Std.	>90%	2.13	1.92	1.31	Accepted
15	C	Simple	Dray	>90%	3.02	1.92	58.19	Rejected
16	PE	Vacuum	Std.	Std.	2.57	1.26	10.78	Rejected

Table 7 ANOVA

Factor	Value—*p*	
Main factors	*14 days*	*28 days*
Packaging material (A)	0.321	0.268
Type of seal (B)	0.124	0.189
Baling moisture (C)	0.774	0.733
Storage humidity (D)	0.425	0.278
Double integration		
Packaging material*type of seal (A*B)	0.741	0.852
Packaging material*baling moisture (A*C)	0.444	0.064
Packaging material*storage humidity (A*D)	0.124	0.190
Type of seal*baling moisture (B*C)	0.537	0.083
Type of seal*storage humidity (B*D)	0.556	0.127
Packaging material*storage humidity (C*D)	0.600	0.761

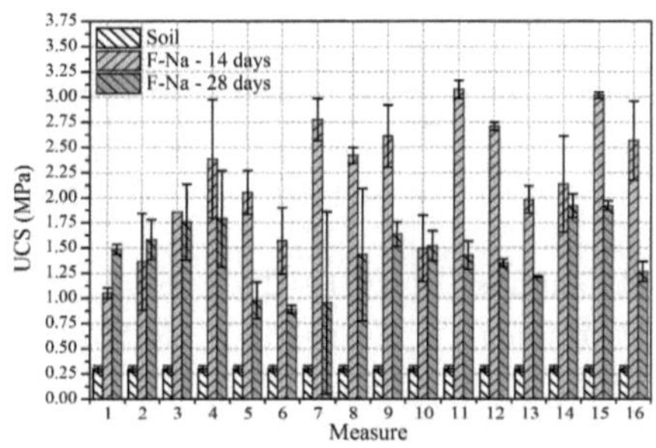

Fig. 2 Unconfined compression strength

Environmental Assessment of the Stabilizer

The results of the environmental assessment are summarized in two stages of the product life-cycle: *preliminary* and manufacturing. Table 8 shows the overall results during the life cycle, obtained by processing input and output variables. In fuel consumption and emissions, a considerable excess was found for the manufacturing stage; because at this stage it was required higher power consumption unlike the preliminary stage, where only transportation was needed to move the materials to the gathering site and using local material.

In Table 9, the contribution by life cycle stage for each impact category is showed. It is notably highlighted the effect generated by the manufacturing stage with a contribution in all cases greater than 90%, this represents a contribution of 92.21% of the total impact generated throughout the life cycle evaluated. For the preliminary stage of the contribution only reached 1%. Moreover, the impact category that has affected the most is the availability of fossil fuels mainly in relation to the activities carried out for the manufacturing stage, while other categories had lower involvement close to 0%.

Table 8 Life cycle consumption and emissions

Stage	Activity	Consumption fuel vehicle (gal)	Emissions CO_2 vehicle (kg)	Emissions SO_2 vehicle (kg)
Preliminary	Transport materials	1.28E+00	3.88E−01	9.78E−04
	Transport packaging	2.91E+01	8.74E+00	2.22E−02
Stage	Activity	Power consumption equipment (kWh)	Emissions CO_2 equipment (kg)	Emissions SO_2 equipment (kg)
Manufacturing	Preparation of materials	1.36E+02	3.51E+01	5.32E−01
	Dosage	4.44E+01	1.50E+02	2.26E+00
	Mixed and packed	9.90E+00	8.72E+01	1.32E+00
	Storage	5.24E+02	1.42E+02	2.15E+00

Table 9 LCA results of the impact categories

Etapa	Fossil fuel use (MJ)	Global Warming CO_2 (DALY)	Inorganic respirable SO_2 (DALY)	Health damage (DALY)
Preliminaries	4.63E+02	6.46E−16	2.90E−11	1.74E−11
Fabricación	5.49E+03	6.45E−12	1.73E−06	1.04E−06

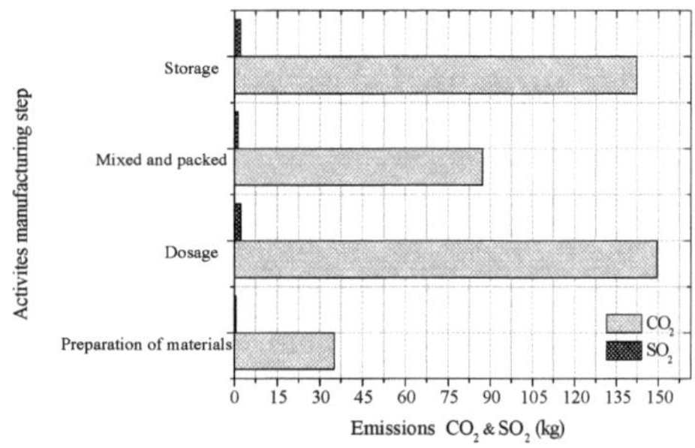

Fig. 3 Contribution of CO_2 and SO_2

Finally, it was found that the greatest impact for all categories in the manufacturing stage was caused by the composition and storage activities representing 98% more CO_2 emissions compared to SO_2 for all the activities (see Fig. 3).

Conclusions

The factors: **F I**, **F II**, **F II** and **F IV** have no statistical influence on the response variable, this indicates that the condition of fly ash material with a moisture of 0.15% is suitable to design the stabilizing product, allowing to decrease the energy required by the processes like drying.

It is recommended for Kraft paper and high density polyethylene packaging a simple or vacuum sealed and a moisture of fly ash below 0.15% so that the stabilizer product can be stored at humidities above 90%.

For environmental purposes laminated kraft paper and biodegradable represents the best packaging material for the stabilizing product, although it should be noted that the storage time tested was less than 30 days. It is recommended to assess it for longer periods.

By using the LCA methodology for assessing environmental impacts, it became clear that the packaging process has high impacts due to the energy required during the manufacturing stage associated with composition and storage. From the foregoing, it is proposed as future research, standardize processes, using alternative energy to optimize electricity consumption.

Acknowledgements Thanks are given to "INNOVIAL", "Universidad de Medellín" and "Universidad Nacional de Colombia".

References

1. Duxson, J. L., Fernández-Jiménez, P., & Provis, A. (2007). Geopolymer technology: The current state of the art. *Materials Science, 42*(6), 2917–2933.
2. Duxson, P., Provis, J. L., Lukey, G. C., Mallicoat, S. W., Kriven, W. M., & Van Deventer, J. S. J. (2005). Understanding the relationship between geopolymer composition, microstructure and mechanical properties. *Colloids and Surfaces A: Physicochemical and Engineering Aspects, 269*(1–3), 47–58.
3. Yip, C. K., Lukey, G. C., & Van Deventer, J. S. J. (2005). The coexistence of geopolymeric gel and calcium silicate hydrate at the early stage of alkaline activation. *Cement and Concrete Research, 35*(9), 1688–1697.
4. Palomo, A., Grutzeck, M. W., & Blanco, M. T. (1999). Alkali-activated fly ashes: A cement for the future. *Cement and Concrete Research, 29*(8), 1323–1329.
5. Fernández-Jiménez, A., & Palomo, A. (2005). Composition and microstructure of alkali activated fly ash binder: Effect of the activator. *Cement and Concrete Research, 35*(10), 1984–1992.
6. Criado, M., Palomo, A., & Fernández-Jiménez, A. (2005). Alkali activation of fly ashes. Part 1: Effect of curing conditions on the carbonation of the reaction products. *Fuel, 84*(16), 2048–2054.
7. Sanz, M. C. (2007). Nuevos materiales cementantes basados en la activación alcalina de cenizas volantes. caracterización de geles N-A-S-H en función del contenido de sílice soluble. efecto del Na_2 So_4. Universidad Autónoma de Madrid (pp. 1–356).
8. Billong, N., Melo, U. C., Njopwouo, D., Louvet, F., & Bonnet, J. P. (2009). Effect of mixture constituents on properties of slaked lime-metakaolin-sand mortars containing sodium hydroxide. *Cement and Concrete Composites, 31*(9), 658–662.

9. Fernández-Jiménez, A., & Palomo, A. (2003). Characterisation of fly ashes. Potential reactivity as alkaline cements. *Fuel, 82*(18), 2259–2265.
10. Rodriguez Martinez, E. D. (2009). Eficiencia de activadores alcalinos basados en diferentes fuentes de silice para laproducción de sistemas geopoliméricos de ceniza volante (M.Sc. thesis, Universidad Autonoma de Valencia, pp. 1–72).
11. Hasanbeigi, A., Price, L., & Lin, E. (2012). Emerging energy-efficiency and CO_2 emission-reduction technologies for cement and concrete production: A technical review. *Renewable and Sustainable Energy Reviews, 16*(8), 6220–6238.
12. Pacheco-Torgal, F. (2015). Introduction to handbook of alkali-activated cements, mortars and concretes. In *Handbook of alkali-activated cements, mortars and concretes* (pp. 1–16).
13. Garcia-Lodeiro, A. F.-J. I., & Palomo, A. (2015). An overview of the chemistry of alkali-activated cement-based binders. In *Handbook of alkali-activated cements, mortars and concretes* (pp. 19–47).
14. Candil, N. A. N., Moreno, J. R., Castañeda, J. F. F., Villazón, R. A., & Galvis, J. J. M. (2012). *La Cadena del Carbón* (Report MinMinas, UPME).
15. Velásquez Vallejo, L. F., De La Cruz Morales, J. F., & Sánchez Morales, J. F. (2007). Remoción de carbón inquemado de las cenizas volantes producidas en el proceso de combstión de carbón. In *Energética* (pp. 107–112).
16. www.portafolio.co/economia/finanzas/colombia-cuarto-productor-aceite-palma-mundo-59140 Colombia, cuarto productor de aceite de palma en el mundo, 2014.
17. Rashad, A. M. (2014). A comprehensive overview about the influence of different admixtures and additives on the properties of alkali-activated fly ash. *Materials and Design, 53,* 1005–1025.
18. Fredy Alexander, M. C., & Yhan Paul, A. J. (2015). Evaluación del polvo de ladrillo como estabilizante de suelo perteneciente a vías terciarias (pp. 1–10).
19. Zuluaga, D. M., & Arias Jaramillo, Y. P. (2015). Valoracion de las cenizas de carbón para la estabilización de suelos mediante activación alcalina y su uso en vías no pavimentadas. (Bachelor thesis, Universidad de Medellín, pp. 1–80).
20. Arias, Y. P., Londoño, D., & Tobón, J. I. (2011). Correlación entre DRX de polvo usando el método Rietveld y técnica disolución selectiva en la determinación de las fases cristalina y amorfa en cenizas de carbón para uso en cementos alternativos (Memoryof I Reunión Latinoamericana de Cristalografía, Argentina, Vol. 41).

Part IV
Materials for Infrastructure

Anticorrosion and Adsorption Mechanism of *Rhizophora mangle* L. Leaf-Extract on Steel-Reinforcement in 3.5% NaCl-Immersed Concrete

Joshua Olusegun Okeniyi, Olugbenga Adeshola Omotosho, Cleophas Akintoye Loto and Abimbola Patricia Idowu Popoola

Abstract This paper studies anticorrosion and adsorption mechanism of *Rhizophora mangle* L. leaf-extract on steel-reinforcement in concrete immersed in 3.5% NaCl test-environment. Open circuit potential, macrocell current and corrosion rate measurements were obtained from steel-reinforced concrete samples, into which different concentrations of the leaf-extract was admixed during casting, and which were immersed in the saline/marine simulating-environment. Corrosion noise resistance was modelled as the ratio of standard deviation of the corrosion potential to the standard deviation of the corrosion current. Analyses of these test-results showed that the corrosion rate from linear polarization resistance exhibited very good correlation with the noise resistance model for the leaf-extract concentrations studied. Further analyses identified *Rhizophora mangle* L. leaf-extract concentration with excellent corrosion inhibition efficiency performance on steel-reinforcement. Also data of anticorrosion performance followed the Langmuir adsorption isotherm model, which indicated physisorption as the prevalent corrosion-protection mechanism by the *Rhizophora mangle* L. leaf-extract on steel-rebar in concrete immersed in the 3.5% NaCl, simulating saline/marine environment.

Keywords *Rhizophora mangle* L. leaf-extract · Steel-reinforcement in concrete · Corrosion noise resistance · Correlation fitting model · Adsorption isotherm · Saline/marine simulating-environment

---

J.O. Okeniyi (✉) · O.A. Omotosho · C.A. Loto
Mechanical Engineering Department, Covenant University, Ota 112001, Nigeria
e-mail: joshua.okeniyi@covenantuniversity.edu.ng

C.A. Loto · A.P.I. Popoola
Chemical, Metallurgical and Materials Engineering Department,
Tshwane University of Technology, Pretoria 0001, South Africa

© The Minerals, Metals & Materials Society 2017
M.A. Meyers et al. (eds.), *Proceedings of the 3rd Pan American Materials Congress*,
The Minerals, Metals & Materials Series, DOI 10.1007/978-3-319-52132-9_17

Introduction

Corrosion protection methods are necessary applications for maintaining structural integrity and attaining service life sustenance of steel reinforced concretes designed for aggressive environments where the steel-rebar is susceptible to corrosion attacks [1–5]. However, ascertaining effectiveness of an applied corrosion protection technique requires adequate monitoring of corrosion condition and interpretation of measured corrosion data. Nevertheless, problems persists because interpretation of measured data from non-destructive testing of concrete steel-reinforcement corrosion portends its own challenges and difficulties, apart from the technical skill and financial necessities of periodic survey that are required for corrosion monitoring [5, 6]. Inherent challenges and difficulties attending interpretation of corrosion test-results often originate from the fact that corrosion test-measurements exhibit more scatter, of stochastic/random deviations from the expected values prevailing in the given corrosive test-system, than other types of tests [6, 7]. These randomized deviations could be due to variety of factors in the material or in its environment that might be unaccounted for in the corrosion testing design but that are potent at masking underlying fundamental conditions in corrosive systems. Complexities from these render transformation of corrosion test-results unusable for real life service application, in the stead of applicability of such test-results for realistically and reliably addressing industrial corrosion problems [7].

According to recommendations from literature [2, 5, 8], reliability of corrosion test-data could be improved through the combinations of corrosion testing methods for complementing one another. In addition to this, use of statistical modeling and analyses [9] has been prescribed by standard specifications, including ASTM G16-95 R04 [6] and ASTM C876-91 R99 [10], and the literature [7] for tackling difficulties from stochastic corrosion test-data interpretation. While there are several steel-rebar corrosion protection techniques each of which could require monitoring for effectiveness assessment, the use of corrosion inhibitor remains a preferred method because of its advantages of lower cost and ease of applicability [11, 12]. Also concerns for eco-friendliness and non-toxicity for sustainable corrosion protection is gearing concurrent research interests towards the study of novel inhibitors from natural organic sources [13–15]. Study have shown that leaf-extract obtained from *Rhizophora mangle* L. constitute an example of natural organic material that is non-toxic [16] and which had been used individually [17] and in synergies [18] with positive results for the inhibition of steel-reinforcement corrosion in acidic medium. In spite of this, there is paucity of study deliberating on the mechanism by which *Rhizophora mangle* L. leaf-extract exhibit corrosion protection on concrete steel-reinforcement in NaCl medium. For examples, such analyses of corrosion protection mechanism on steel-rebar has been reported for *Anthocleista djalonensis* leaf-extract admixture in steel-reinforced concrete immersed in NaCl medium [14] and for *Phyllanthus muellerianus* leaf-extract in steel-reinforced concrete immersed in H_2SO_4 environment [15]. Therefore, this paper deliberates on the analyses of electrochemical test-data from steel-reinforced concrete admixed with *Rhizophora mangle* L. leaf-extract for modeling anticorrosive and adsorption mechanism of the leaf-extract in steel-reinforced concrete immersed in 3.5% NaCl, simulating saline/marine environment.

Experimental

Steel-reinforced concrete specimens, into which were admixed different concentrations of *Rhizophora mangle* L. (*R. mangle* L.) *Rhizophoraceae* leaf-extract, are prepared by standard procedure prescribed in [19, 20] and detailed in [17, 18]. The Ø12 mm steel-rebar was cut into 190 mm lengths of steel rods. Each of the rods were subjected to uniform surface preparations as prescribed in [20], before 150 mm of the steel-rebar were embedded centrally in the 100 × 100 × 200 mm concrete, during the concrete casting. These steel-reinforced concrete were cast in duplicates (Dup), i.e. having similar concentrations of *R. mangle* L. leaf-extract admixture [21], which varies from 0 g/L (control specimens) in increments of 1.6667 up to 6.6667 g/L of concrete mixing water. By this admixture design, the total number of steel-reinforced concrete specimens equals 10 (ten) for this study. From these ten specimens, electrochemical measurements were obtained, using techniques detailed in [22], first in five days interval for the first 40 days and thereafter in seven days interval for the next six weeks. These measurements constitute $n = 15$ data-points of corrosion test-measurements for each of the electrochemical test-techniques employed in the study. These electrochemical test-methods for the study include corrosion potential versus $Cu/CuSO_4$ electrode (CSE) (Tinker & Rasor) as per ASTM C876-91 R99 [23], by high impedance digital multimeter [24–26], corrosion current by zero resistance ammeter (Corrosion Service®) [2, 27, 28], and corrosion rate by linear polarization resistance instrument (Metal Samples®) [29, 30].

Each of the three measured variables of electrochemical test-data was subjected to statistical distribution modeling analyses, as prescribed in [6] and detailed in [14], of the Normal and Weibull distribution with data compatibility to each distribution tested by Kolmogorov-Smirnov goodness-of-fit (K-S GoF) statistics [7, 9, 31–33]. By this distribution fitting, the Normal mean, μ_N, and standard deviation, σ_N, were estimated from the respective formula [14]:

$$\mu_N = \frac{\sum_{i=1}^{n} x_i}{n} \tag{1}$$

$$\sigma_N = \sqrt{\frac{\sum_{i=1}^{n} (x_i - \mu_N)^2}{n-1}} \tag{2}$$

In these equations, x_i is the requisite measurement of electrochemical test-data of the corrosion potential or corrosion current or corrosion rate and $n = 15$, the number of data points of the corrosion test-measurements in the 82 days experimental period. Similarly, the Weibull mean, μ_W, and standard deviation, σ_W, were estimated, respectively, from the relationships [14, 31, 34, 35]:

$$\mu_W = c\Gamma\left(1 + \frac{1}{k}\right) \tag{3}$$

$$\sigma_w = \sqrt{c^2\left\{\Gamma\left(1 + \frac{2}{k}\right) - \left[\Gamma\left(1 + \frac{1}{k}\right)\right]^2\right\}} \tag{4}$$

where k and c are the Weibull shape parameter and the Weibull scale parameter, respectively, and $\Gamma(\xi)$ is the gamma function of (ξ), which for the arbitrary term ξ, is given by:

$$\Gamma(\xi) = \int_0^\infty e^{-t} t^{(\xi-1)} dt \tag{5}$$

By the foregoing analyses, the distribution model followed by the corrosion potential and corrosion current find usefulness for the evaluation of noise resistance, R_n, as the ratio of the standard deviation, σ_V, of corrosion potential to that of the corrosion current, σ_I [2, 14, 15, 34, 36]:

$$R_n = \frac{\sigma_V}{\sigma_I} \tag{6}$$

The *R. mangle* L. leaf-extract admixture concentration, ρ, the noise resistance, R_n, and the statistical distribution model of corrosion rate, *CR*, were then subjected to correlation modeling with requisite modeling efficiency analyses of correlation coefficient, Nash-Sutcliffe efficiency and analysis of variance [11]. Experimental and correlated prediction of corrosion rate models obtained from that analysis were then subjected to anticorrosive inhibition efficiency, η, estimation and surface coverage, θ, modeling that was employed for adsorption isotherm and, subsequently, adsorption mechanism modeling, using [2, 14, 15, 37]:

$$\eta = \frac{CR_{\text{control specimen}} - CR_{\text{admixed specimen}}}{CR_{\text{control specimen}}} \times 100 \tag{7}$$

$$\theta = \frac{CR_{\text{control specimen}} - CR_{\text{admixed specimen}}}{CR_{\text{control specimen}}} \tag{8}$$

Results and Discussion

The plots of the modelled results from the statistical distribution analyses of corrosion potential and corrosion current are presented in Fig. 1. Included in these plots include the mean ± standard deviation for corrosion potential along with the linear plots of ASTM C876-91 R99 criteria in Fig. 1a and the mean ± standard deviation for the corrosion current in Fig. 1b.

The plots in Fig. 1 showed that the Normal and the Weibull models of corrosion potential, Fig. 1a, and of corrosion current, Fig. 1b exhibited similar patterns of result models for all the specimens studied. Despite these similarities, the goodness-of-fit analyses by the Kolmogorov-Smirnov test-statistics, the results of which are plotted in Fig. 2, showed that the corrosion potential data followed the Normal more than the Weibull distribution. Specifically, datasets of corrosion potential of the 1.6667_Dup and of the 3.3333_Dup specimens were not scattered like the Weibull distribution, K-S p-value < 0.05, but distributed like the Normal distribution model, K-S p-value > 0.05. By these considerations, the Normal distribution model of the standard deviations of corrosion potential and of corrosion current find suitability for the estimation of corrosion noise resistance through use of Eq. (6). This is notwithstanding the fact that all corrosion current data in the study actually scattered like both the Normal and the Weibull distributions. In contrast to these, Fig. 2 showed that the corrosion rate datasets of six steel-reinforced concrete specimens were not scattered like the Normal distribution while all the steel-reinforced concrete specimens in this study exhibited datasets of corrosion rate that scattered like the Weibull distribution. Based on these considerations, also, the Weibull distribution model finds suitability as the descriptive statistics for detailing corrosion rate in this study just as the Normal distribution model constitute the descriptive statistics for detailing corrosion potential and corrosion current.

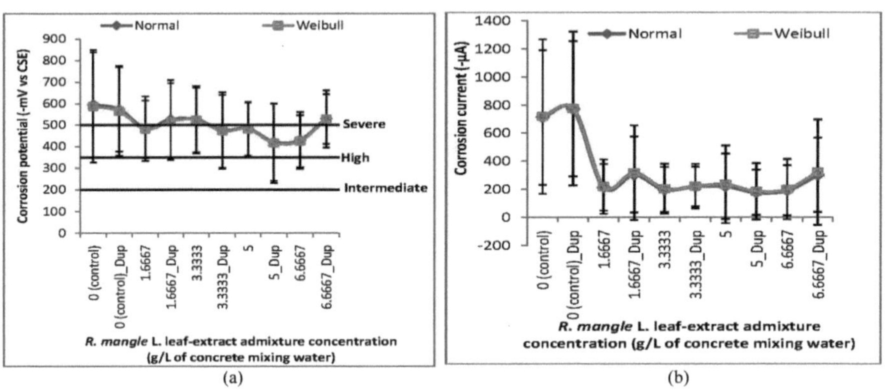

Fig. 1 Plots of modelled results of statistical distribution analyses for **a** corrosion potential, **b** corrosion current

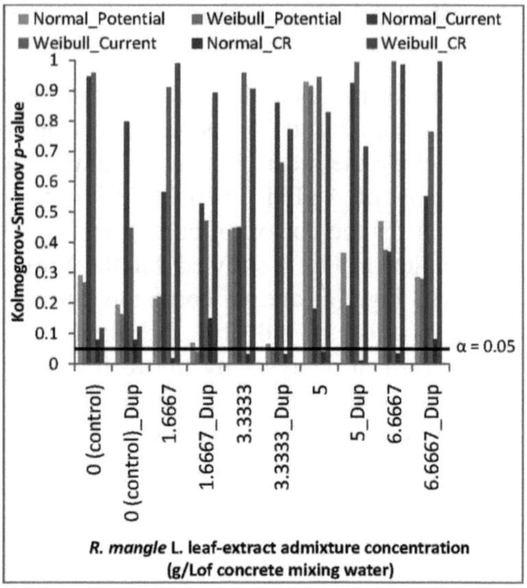

Fig. 2 Plots of Kolmogorov-Smirnov goodness-of-fit analyses of corrosion test-data

Figure 3 presents the results obtained from the corrosion noise resistance by the Normal distribution analyses super-imposed with the results of mean corrosion rate obtained through the Weibull distribution analyses. From this figure, it could be noted that the steel-reinforced concrete specimens that were modelled with high noise resistance tend to exhibit low corrosion rate and vice versa. These agreed with findings from the literature [14, 15, 38]. This bare indication of the existence of some forms of relationship between the corrosion rate and the corrosion noise resistance for the concentrations of *R. mangle* L. admixtures in the steel-reinforced concrete specimens.

The subjecting, therefore, of the noise resistance, R_n, and the *R. mangle* L. admixture concentration, ρ, as independent variables, to correlation fitting modeling

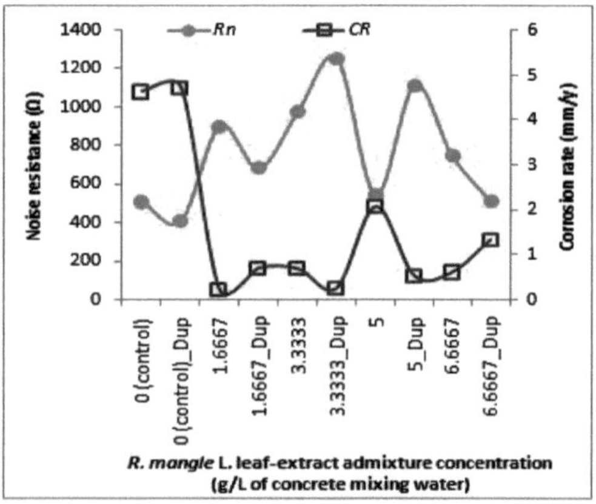

Fig. 3 Plots of corrosion noise resistance and corrosion rate for the steel-reinforced concrete specimens

Table 1 Analysis of variance of the correlation fitting model in Eq. (9)

Source of variations	*df*	SS	MS	*F*	*p*-value
Regression	3	22.6688	7.5563	11.4339	0.0068
Residual	6	3.9652	0.6609		
Total	9	26.6340			

with corrosion rate, *CR*, as the dependent variable gave the relationship which includes:

$$CR = 1.0295 - 0.2104\rho - 8.9939 \times 10^2(1/R_n) + 1.0414 \times 10^6(1/R_n)^2 \quad (9)$$

For this correlation fitting model, the correlation coefficient $R = 92.26\%$ and the Nash-Sutcliffe Efficiency = 85.11%. The Nash-Sutcliffe Efficiency of the correlation fitting model interprets to "very good" model efficiency as per the classification from [39]. Table 1 presents the results of analysis of variance application to the correlation fitting equation, and this showed that the ANOVA p-value = 0.0068. That this p-value < 0.05 implies that that it could not be rejected that there is a significant relationship between the corrosion rate, *CR*, the *R. mangle* L. concentration, ρ, and the noise resistance, R_n.

Applications of the experimental and correlation fitting models/prediction of corrosion rate, averaged over each of the duplicated specimens, to Eq. (7) gave inhibition efficiency, η. The results of these are presented in Fig. 4, in ranking order of *R. mangle* L. admixture performance on steel-rebar corrosion in the 3.5% NaCl-immersed concrete specimens. This showed *R. mangle* L. leaf-extract admixture exhibited inhibition efficiency that ranged from 72.21 ± 19.06 to

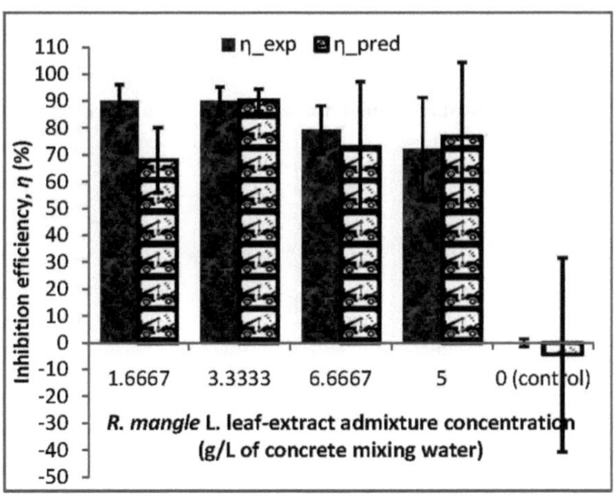

Fig. 4 Experimental and correlated modelled-results of inhibition efficiency performance of *R. mangle* L. leaf-extract admixture on steel-rebar corrosion in 3.5% NaCl immersed-concretes

90.14 ± 5.87% by the experimental model and from 68.01 ± 12.14 to 90.33 ± 4.09% by the correlation prediction model on steel-reinforcement corrosion in the test-environment. By this, both the experimental model and the predictions from correlation fitting analyses exhibited agreements in interpreting to model efficiencies that ranged from "good" to "excellent" according to the model efficiency classification from [39]. The discrepancies, however, are in the inhibition efficiency estimations by the experimental model and the correlation model. For instance, while the 1.6667 g/L *R. mangle* L. leaf-extract was modelled as the optimally inhibiting admixture, η = 90.14 ± 5.87%, on steel-rebar corrosion by the experimental model, the 3.3333 g/L *R. mangle* L. leaf-extract was modelled with the optimal inhibiting performance, η = 90.33 ± 4.09%, by the correlation model. In spite of the discrepancies, it is still worth noting that the 3.3333 g/L *R. mangle* L. leaf-extract, indicated as optimal by the correlation fitting model, exhibited the best agreement in the study with the results from the experimental model from which the modelled inhibition efficiency, η = 89.96 ± 5.35%.

By subjecting the experimental and correlation prediction models of corrosion rate to the Langmuir adsorption isotherm fittings [14], through use of the surface coverage Eq. (8), the plots presented in Fig. 5 were obtained. Also, the fitting parameters of the Langmuir adsorption isotherm modeling and the requisite interpretation of the modelled results [14] are presented in Table 2. These fitting parameters include: the Langmuir equilibrium constant, K_{ads}, the Langmuir adsorption-desorption process; the separation factor, R_L, for indicating adsorption nature [14, 37]; the fitting's coefficient of determination, R, and the Gibbs free energy of adsorption. The figure, and the table, showed that both the experimental and predicted model followed the Langmuir adsorption isotherm. The coefficient of determination R^2 = 98.10% and the Gibbs free energy estimations, $\Delta G_{\text{ads}} = -26.97$ (for the experimental model) while R^2 = 97.84% and $\Delta G_{\text{ads}} = -30.79$ (for the correlated model). Both of these suggest physisorption mechanism as the prevailing

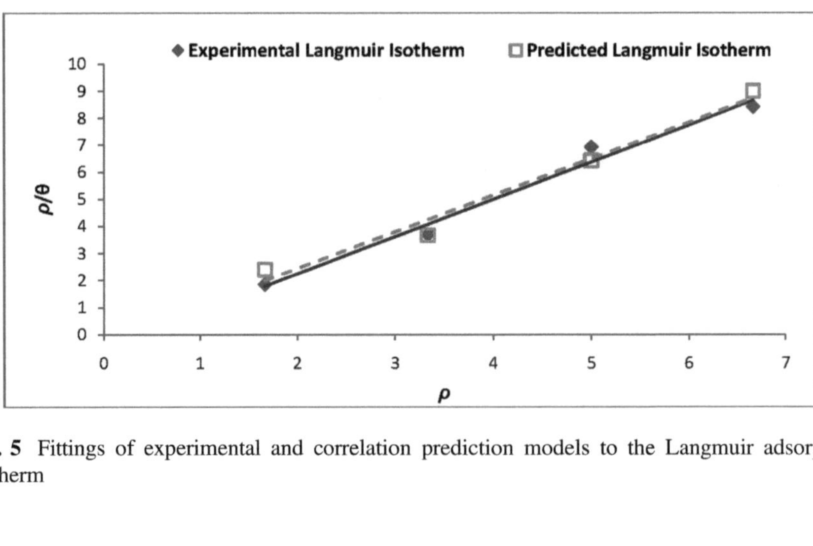

Fig. 5 Fittings of experimental and correlation prediction models to the Langmuir adsorption isotherm

Table 2 Estimated parameters for models of adsorption isotherm

Isotherm parameter	Experimental Langmuir isotherm		Predicted Langmuir isotherm	
	Value	Interpretation	Value	Interpretation
K_{ads}	1.9728		3.8337	
R^2 (%)	98.10	Excellent model-fitting	97.84	Excellent model-fitting
R_L	0.09791	Favourable adsorption	0.05863	Favourable adsorption
ΔG_{ads} (kJ/mol)	−26.9736	Prevalent physisorption	−30.7898	Prevalent physisorption

anticorrosive adsorption mechanism of *R. mangle* L. leaf-extract on concrete steel-reinforcement in the 3.5% NaCl test-solution, simulating saline/marine environment. Also the separation factor R_L was in the range $0 < R_L < 1$ by both the experimental and correlation models, which indicates favourable adsorption by the *R. mangle* L. leaf-extract admixture on steel-reinforcement metal in the 3.5% NaCl-immersed concretes.

Conclusions

From the analyses of electrochemical test-results from steel-reinforced concrete admixed with different concentrations of *R. mangle* L. in this study, it could be concluded that:

- Corrosion rate as independent variable exhibited excellent correlation, $R = 92.26\%$, Nash-Sutcliffe Efficiency = 85.11% and ANOVA p-value = 0.0068, with the independent variables of noise resistance model and the concentration of *R. mangle* L. admixture;
- *R. mangle* L. leaf-extract admixture effectively inhibited steel-reinforcement corrosion in concrete immersed in 3.5% NaCl, simulating saline/marine environment, with inhibition efficiency models ranging from good up to excellent efficiencies on steel-rebar corrosion inhibition performance in the tested medium;
- Both the experimental and correlation fitting models followed the Langmuir adsorption isotherm with physisorption mechanism identified as the prevalent corrosion protection mechanism by *R. mangle* L. leaf-extract on steel-reinforcement metal embedded in concrete immersed in the 3.5% NaCl medium, simulating saline/marine environment.

References

1. Etteyeb, N., Dhouibi, L., Takenouti, H., & Triki, E. (2016). Protection of reinforcement steel corrosion by phenylphosphonic acid pre-treatment PART II: Tests in mortar medium. *Cement & Concrete Composites, 65,* 94–100.
2. Okeniyi, J. O., Popoola, A. P. I., & Loto, C. A. (2015). Corrosion test-data modeling for $C_{10}H_{18}N_2Na_2O_{10}$ performance on steel-rebar in NaCl-immersed concrete. *Corrosion 2015* (NACE International, 2015) Paper No. 5590.
3. Okeniyi, J. O., Omotosho, O. A., Ogunlana, O. O., Okeniyi, E. T., Owoeye, T. F., Ogbiye, A. S., et al. (2015). Investigating prospects of *Phyllanthus muellerianus* as eco-friendly/sustainable material for reducing concrete steel-reinforcement corrosion in industrial/microbial environment. *Energy Procedia, 74,* 1274–1281.
4. Fei, F. L., Hu, J., Wei, J. X., Yu, Q. J., & Chen, Z. S. (2014). Corrosion performance of steel reinforcement in simulated concrete pore solutions in the presence of imidazoline quaternary ammonium salt corrosion inhibitor. *Construction and Building Materials, 70,* 43–53.
5. Birbilis, N., & Cherry, B. W. (2005). Monitoring the corrosion and remediation of reinforced concrete on-site: An alternative approach. *Materials and Corrosion, 56,* 237–243.
6. ASTM G16-95 R04. (2005). *Standard guide for applying statistics to analysis of corrosion data.* West Conshohocken PA: ASTM International.
7. Roberge, P. R. (2003). Statistical interpretation of corrosion test results. In S. D. Cramer & B. S. Covino, Jr. (Eds.), *ASM handbook Vol 13A—Corrosion: Fundamentals, testing, and protection* (pp. 425–429). Materials Park, OH: ASM International.
8. Song, H.-W., & Saraswathy, V. (2007). Corrosion monitoring of reinforced concrete structures: A review. *International Journal of Electrochemical Science, 2,* 1–28.
9. Okeniyi, J. O., Ambrose, I. J., Okpala, S. O., Omoniyi, O. M., Oladele, I. O., Loto, C. A., et al. (2014). Probability density fittings of corrosion test-data: Implications on $C_6H_{15}NO_3$ effectiveness on concrete steel-rebar corrosion. *Sadhana, 39,* 731–764.
10. ASTM C876-91 R99. (2005). *Standard test method for half-cell potentials of uncoated reinforcing steel in concrete.* West Conshohocken PA: ASTM International.
11. Okeniyi, J. O., Ambrose, I. J., Oladele, I. O., Loto, C. A., & Popoola, A. P. I. (2013). Electrochemical performance of sodium dichromate partial replacement models by tri-ethanolamine admixtures on steel-rebar corrosion in concretes. *International Journal of Electrochemical Science, 8,* 10758–10771.
12. Okeniyi, J. O., Oladele, I. O., Ambrose, I. J., Okpala, S. O., Omoniyi, O. M., Loto, C. A., et al. (2013). Analysis of inhibition of concrete steel-rebar corrosion by $Na_2Cr_2O_7$ concentrations: Implications for conflicting reports on inhibitor effectiveness. *Journal of Central South University, 20,* 3697–3714.
13. Ismail, M., Raja, P. B., & Salawu, A. A. (2015). Developing deeper understanding of green inhibitors for corrosion of reinforcing steel in concrete. In H. L. Lim (Ed.), *Handbook of research on recent developments in materials science and corrosion engineering education* (pp. 118–146). Hershey, PA: IGI Global.
14. Okeniyi, J. O., Loto, C. A., & Popoola, A. P. I. (2014). Electrochemical performance of *Anthocleista djalonensis* on steel-reinforcement corrosion in concrete immersed in saline/marine simulating-environment. *Transactions of the Indian Institute of Metals, 67,* 959–969.
15. Okeniyi, J. O., Loto, C. A., & Popoola, A. P. I. (2014). Electrochemical performance of *Phyllanthus muellerianus* on the corrosion of concrete steel-reinforcement in industrial/microbial simulating-environment. *Portugaliae Electrochimica Acta, 32,* 199–211.
16. Perera, L. M. S., Escobar, A., Souccar, C., Ma Remigio, A., & Mancebo, B. (2010). Pharmacological and toxicological evaluation of *Rhizophora mangle* L., as a potential antiulcerogenic drug: Chemical composition of active extract. *Journal of Pharmacognosy and Phytotherapy, 2,* 56–63.

17. Okeniyi, J. O., Loto, C. A., & Popoola, A. P. I. (2014). *Rhizophora mangle* L. effects on steel-reinforced concrete in 0.5 M H_2SO_4: Implications for corrosion-degradation of wind-energy structures in industrial environments. *Energy Procedia, 50,* 429–436.
18. Okeniyi, J. O., Loto, C. A., Popoola, A. P. I., & Omotosho, O. A. (2015). Performance of *Rhizophora mangle* L. leaf-extract and sodium dichromate synergies on steel-reinforcement corrosion in 0.5 M H_2SO_4-immersed concrete. *CORROSION 2015* (NACE International, 2015) Paper No. 5636.
19. ASTM C192/192M-02. (2005). *Standard practice for making and curing concrete test specimens in the laboratory*. West Conshohocken, PA: ASTM International.
20. ASTM G109-99a. (2005). *Standard test method for determining the effects of chemical admixtures on the corrosion of embedded steel reinforcement in concrete exposed to chloride environments*. West Conshohocken, PA: ASTM International.
21. Corbett, R. A. (2005). Immersion testing. In R. Baboian (Ed.), *Corrosion tests and standards: Application and interpretation* (2nd ed., pp. 139–146). West Conshohocken, PA: ASTM International.
22. Okeniyi, J. O., Oladele, I. O., Omoniyi, O. M., Loto, C. A., & Popoola, A. P. (2015). Inhibition and compressive-strength performance of $Na_2Cr_2O_7$ and $C_{10}H_{14}N_2Na_2O_8{\cdot}2H_2O$ in steel-reinforced concrete in corrosive environments. *Canadian Journal of Civil Engineering, 42,* 408–416.
23. ASTM C876-91 R99. (2005). *Standard test method for half-cell potentials of uncoated reinforcing steel in concrete*. West Conshohocken, PA: ASTM International.
24. Okeniyi, J. O., Omotosho, O. A., Ajayi, O. O., & Loto, C. A. (2014). Effect of potassium-chromate and sodium-nitrite on concrete steel-rebar degradation in sulphate and saline media. *Construction and Building Materials, 50,* 448–456.
25. Omotosho, O. A., Loto, C. A., Ajayi, O. O., Okeniyi, J. O., & Popoola, A. P. I. (2014). Investigating potassium chromate and aniline effect on concrete steel rebar degradation in saline and sulphate media. *International Journal of Electrochemical Science, 9,* 2171–2185.
26. Elsener, B., Andrade, C., Gulikers, J., Polder, R., & Raupach, M. (2003). Hall-cell potential measurements—Potential mapping on reinforced concrete structures. *Materials and Structures, 36,* 461–471.
27. Abdelaziz, G. E., Abdelalim, A. M. K., & Fawzy, Y. A. (2009). Evaluation of the short and long-term efficiencies of electro-chemical chloride extraction. *Cement and Concrete Research, 39,* 727–732.
28. McCarter, W. J., & Vennesland, Ø. (2004). Sensor systems for use in reinforced concrete structures. *Construction and Building Materials, 18,* 351–358.
29. Okeniyi, J. O., Popoola, A. P. I., Loto, C. A., Omotosho, O. A., Okpala, S. O., & Ambrose, I. J. (2015). Effect of $NaNO_2$ and $C_6H_{15}NO_3$ synergistic admixtures on steel-rebar corrosion in concrete immersed in aggressive environments. *Advances in Materials Science and Engineering, 2015*, 11.
30. Sastri, V. S. (2011). *Green corrosion inhibitors: Theory and practice*. New Jersey: Wiley.
31. Okeniyi, J. O., Okeniyi, E. T., & Atayero, A. A. (2015). Programming development of Kolmogorov-Smirnov goodness-of-fit testing of data normality as a Microsoft Excel® library function. *Journal of Software & Systems Development, 2015,* 1–15.
32. Okeniyi, J. O., Moses, I. F., & Okeniyi, E. T. (2015). Wind characteristics and energy potential assessment in Akure, South West Nigeria: Econometrics and policy implications. *International Journal of Ambient Energy, 36,* 282–300.
33. Okeniyi, J. O., Obiajulu, U. E., Ogunsanwo, A. O., Odiase, N. W., & Okeniyi, E. T. (2013). CH_4 emission model from the waste of *Sus Domesticus* and *Gallus Domesticus* in Nigerian local farms: Environmental implications and prospects. *Mitigation and Adaptation Strategies for Global Change, 18,* 325–335.
34. Okeniyi, J. O., Omotosho, O. A., Loto, C. A., & Popoola, A. P. I. (2015). Corrosion rate and noise resistance correlation from $NaNO_2$-admixed steel-reinforced concrete. *Asian Journal of Scientific Research, 8,* 454–465.

35. Okeniyi, J. O., Ohunakin, O. S., & Okeniyi, E. T. (2015). Assessments of wind-energy potential in selected sites from three geopolitical zones in Nigeria: Implications for renewable/sustainable rural electrification. *The Scientific World Journal, 2015,* 1–13.
36. Eden, D. A. (2000). Electrochemical noise. In R. W. Revie (Ed.), *Uhlig's corrosion handbook* (2nd ed., pp. 1227–1238). New York: Wiley.
37. Foo, K. Y., & Hameed, B. H. (2010). Insights into the modeling of adsorption isotherm systems. *Chemical Engineering Journal, 156,* 2–10.
38. Kelly, R. G., Inman, M. E., & Hudson, J. L. (1996). Analysis of electrochemical noise for Type 410 stainless steel in chloride solutions. In J. R. Kearns, J. R. Scully, P. R. Roberge, D. L. Reichert, & J. L. Dawson (Eds.), *Electrochemical noise measurement for corrosion applications, ASTM STP 1277* (pp. 101–113). American Society for Testing and Materials.
39. Coffey, R., Dorai-Raj, S., O'Flaherty, V., Cormican, M., & Cummins, E. (2013). Modeling of pathogen indicator organisms in a small-scale agricultural catchment using SWAT. *Human and Ecological Risk Assessment: An International Journal, 19,* 232–253.

Cassia fistula Leaf-Extract Effect on Corrosion-Inhibition of Stainless-Steel in 0.5 M HCl

Olugbenga Adeshola Omotosho, Joshua Olusegun Okeniyi, Cleophas Akintoye Loto, Abimbola Patricia Idowu Popoola, Omokolade Babatunde Ajibola and Adebanji Samuel Ogbiye

Abstract This paper investigates *Cassia fistula* leaf-extract effects on the inhibition of stainless-steel corrosion in 0.5 M HCl. Measurements of corrosion rate were obtained through linear sweep voltammetry (LSV) technique, at the ambient temperature of 28 °C from stainless-steel specimens immersed in the acidic medium, containing different *Cassia fistula* leaf-extract concentrations. Results showed that inhibition effectiveness on stainless-steel corrosion increases with increasing concentration of the leaf-extract. The 10 g/L *Cassia fistula* leaf-extract, the highest concentration of the leaf-extract employed in the study, exhibited optimal inhibition efficiency $\eta = 88.46\%$ on the corrosion of the stainless-steel metal. Adsorption isotherm modelling shows that the experimental data followed the Flory-Huggins isotherm with excellent model efficiency, $r^2 = 90.27\%$, and the Langmuir model with very good model efficiency, $r^2 = 78.83\%$. Other isotherm parameters indicate favourable adsorption and suggest physisorption as the prevalent mechanism of corrosion protection by the leaf-extract on stainless-steel in the acidic chloride environment.

Keywords Linear sweep voltammetry · Corrosion rate · *Cassia fistula* leaf-extract · Stainless steel · Inhibition efficiency · Acidic chloride medium · Adsorption isotherm modelling

O.A. Omotosho (✉) · J.O. Okeniyi · C.A. Loto · O.B. Ajibola
Mechanical Engineering Department, Covenant University, Ota 112001, Nigeria
e-mail: olugbenga.omotosho@covenantuniversity.edu.ng

C.A. Loto · A.P.I. Popoola
Chemical, Metallurgical and Materials Engineering Department,
Tshwane University of Technology, Pretoria 001, South Africa

A.S. Ogbiye
Civil Engineering Department, Covenant University, Ota 12001, Nigeria

© The Minerals, Metals & Materials Society 2017
M.A. Meyers et al. (eds.), *Proceedings of the 3rd Pan American Materials Congress*,
The Minerals, Metals & Materials Series, DOI 10.1007/978-3-319-52132-9_18

Introduction

Hydrochloric acid (HCl) is an acidic solution that is very useful in many industrial applications including surface treatment operation, oil well acidization, pickling and petrol refining operations [1–3]. These widespread applications and usages of HCl generate the attendant problem that acidic chloride exhibits corrosive effects on metal substrates. Although, these corrosive effects are usually addressed using chemical inhibitors, further problems persist from the deleterious effect of many of the known corrosion inhibiting chemicals on the environment [4–6]. Some of these problems include environmental contaminations that are toxic to plant and animal bio-systems, disposal burdens, litigations and, even the expensive nature of many of the chemical inhibitors. For tackling these problems, researchers are shifting focus away from the use of chemical inhibitors [7–15] towards the use of green inhibitors, especially those that could be obtained from plant extracts [3, 12, 16–21]. Other important advantages constituting reasons for the preference of inhibitors from green sources include the fact that they are cheap, easily available, sustainable, even as they are also free of disposal and litigation problems. While several research works had been conducted using different inhibitors on stainless steel corrosion in acidic media [22–29], not many studies have involved the use of green inhibitors on stainless steel in these media. No study has been deliberated on the use of extract from *Cassia fistula* on stainless steel corrosion in HCl.

Usually, stainless steel is known to exhibit self healing properties by which it could immediately form a thin passive layer that is capable of protecting the metal from corrosion attack. However, this protective capability may be insufficient, especially, in the presence of certain aggressive aqueous acids like HCl, at preventing metallic surface attacks that could destroy the protective film layer and render the metal surface susceptible to insidious/localised pitting corrosion [30, 31]. Studies have shown that green inhibitors from organic sources contain heteroatoms that have potentials to donate lone electron pair that could inhibit the metallic corrosion processes [1, 18, 27]. *Cassia fistula* leaf-extract contains phytochemicals like phenolics, flavonoid and proanthocyanidins [32] which are invariably made up of alcohol and carboxylic groups of biochemical heteroatoms. This engenders interest on the fact that extract from this plant could exhibit corrosion inhibiting abilities, especially on stainless steel material. Therefore, the objective of this paper was to investigate the inhibition of stainless steel corrosion in 0.5 M HCl medium in the presence of *Cassia fistula* leaf-extract.

Experimental

Chemical composition of the stainless steel material utilized for the study include: 15.12% Cr, 11.88% Ni and 67.51% Fe. The epoxy-resin mounted metallic specimens of stainless steel metal for the study were cut by use of a guillotine into dimensions of 1 cm × 1 cm. To each of these specimens, pre-experimental surface treatments were applied as per the prescriptions from ASTM D2688-94 R99 [33] for treating stainless steel metal before corrosion experiments, before storing of the specimens in a desiccator prior to the experimental usage.

Leaves of *Cassia fistula* (*C. fistula*) were obtained in fresh form from Ota, Ogun State, Nigeria. The leaves were identified at the Forestry Herbarium, Ibadan, Nigeria, with voucher FHI No. 110261 assigned to the sample deposited at the Herbarium. The *C. fistula* leaves were dried under conditioned atmosphere at 21 °C before being pulverized into powder through the use of grinding machine. From the obtained powdery of the grounded leaves of *C. fistula*, methanolic extract was obtained using procedure that had been detailed in studies [34–36].

The Model K47 corrosion cell system electrochemical cell kit, obtained from Princeton Applied Research®, which has 3-electrode outlets was used for this experimental study. The 3-electrode outlets used for the cell kit system include reference electrode, EDT Direct-Ion double junction [6, 37, 38], graphite counter electrode and the stainless steel sample as the working electrode. Linear sweep voltammetry (LSV) test-technique were then conducted at 28 °C (ambient temperature) using DY2312 Digi-Ivy potentiostat connected to the 3-electrode outlet of the cell kit, in which the 1 cm^2 bare surface area of the epoxy mounted steel sample exhibited contact with the test-solution. Test solution employed in the study include 0.5 M HCl in which different concentrations of *C. fistula* leaf extract were dissolved ranging from 0 g/L (the control sample) up to 10 g/L in increments of 2 g/L. The LSV experiment, executed at a scan rate of 0.1 V/s from an anodic potential of +0.5 V to a cathodic potential of −1.0 V versus corrosion potential, gave readout of corrosion rate (*CR*) using the formula [39]:

$$CR = \frac{0.00327 \times i_{corr} \times eq.wt}{\rho_{ms}} \tag{1}$$

where i_{corr} = corrosion current density (in μA/cm^2), ρ_{ms} = density of metallic sample (g/cm^3) and *eq.wt* = equivalent weight (g). From *CR*, inhibition efficiency, η (%), was estimated for each *C. fistula* leaf-extract concentration through use of the equation [40–42]:

$$\eta = \left(1 - \frac{CR_{\text{sample with } C.\,fistula}}{CR_{\text{sample without } C.\,fistula}}\right) \times 100 \tag{2}$$

In similar manner, *CR* was used for estimation of the surface coverage model, θ, of *C. fistula* leaf-extract on stainless steel metal through the equation [43–45]:

$$\theta = \left(1 - \frac{CR_{\text{sample with } C.\,fistula}}{CR_{\text{sample without } C.\,fistula}}\right) \tag{3}$$

This surface coverage was used for fitting experimental data to the Langmuir and Flory-Huggins adsorption isotherm models, using the respective equations [34, 44, 46, 47]:

$$\frac{\rho}{\theta} = \frac{1}{K_{Lang}} + \rho \tag{4}$$

$$\log \frac{\theta}{CR_0} = \log K_{FH} + n_{FH} \log(1 - \theta) \tag{5}$$

where ρ is *C. fistula* concentration in g/L, K_{Lang} and K_{FH} are, respectively, the Langmuir and Flory-Huggins absorption-desorption constants, CR_0 is the corrosion rate from sample without *C. fistula* and n_{FH} is the Flory-Huggins exponent. The Langmuir adsorption isotherm was used for modelling separation factor, R_L, to indicate adsorption nature and both isotherm fittings were used for modelling the Gibbs free energy of adsorption, ΔG°_{ads}, to indicate adsorption spontaneity and mechanism, through the respective equations [34, 46, 48]:

$$R_L = \frac{1}{1 + K_{Lang} CR_0} \tag{6}$$

$$\Delta G^{\circ}_{ads} = -2.303RT \log_{10}(55.5K_{ads}) \tag{7}$$

where R represents the molar gas constant, 8.314 J/mol K, T represents the absolute temperature (300 K) and the value 55.5 is the concentration of water in solution expressed in molar.

Results and Discussion

Figure 1 shows the results of potentiodynamic polarization obtained from the LSV experiments, for stainless steel in 0.5 M HCl in the absence and presence of varying concentrations of *C. fistula*. These include the potentiodynamic polarization curves in Fig. 1a and the anodic and cathodic Tafel slopes superimposed with the E_{corr} in Fig. 1b, all of which are plotted against the concentrations of the plant extract employed in the study. It is observable from the figure that the cathodic and anodic curves were both influenced by the plant extract concentrations in the experiments but that the cathodic curve exhibited more extensive influence. As could be seen from the E_{corr} plot in Fig. 1b, the polarization curves in Fig. 1a drifted in the cathodic direction relative to the control sample as *C. fistula* leaf-extract concentration increased from 0 to 6 g/L before the curves initiated drifts towards the E_{corr} of the control sample as the extract concentration increased further to 10 g/L. It is worth noting that cathodic potential, in the form of more negative E_{corr} relative to the E_{corr} of the control sample, was maintained from 2 up to 8 g/L *C. fistula*,

Fig. 1 Plots of *C. fistula* effects on stainless steel corrosion in 0.5 M HCl medium **a** potentiodynamic polarization **b** Tafel slopes superimposed with E_{corr}

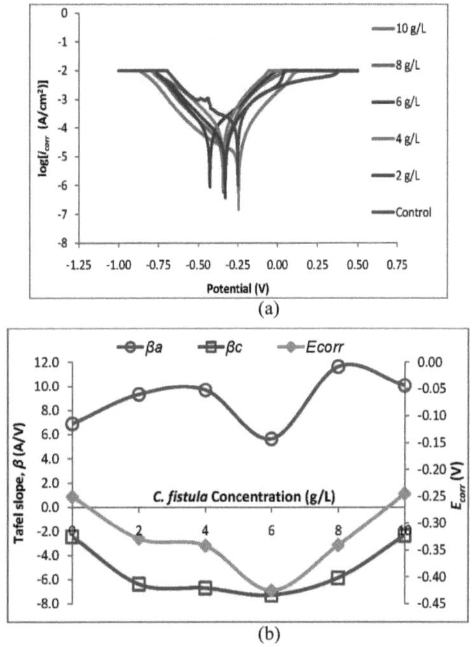

such that only the 10 g/L *C. fistula* exhibited slightly anodic E_{corr} of +0.005 V relative to the control.

Also from Fig. 1b, it could be observed that trend of the plot of cathodic Tafel slope, β_c, exhibited better pattern with the plot of the E_{corr} than the trend of the plot of anodic Tafel slope, β_a, along the changing trends of *C. fistula* leaf-extract concentrations. More specifically, β_c decreases as the E_{corr} decreases and increases as the E_{corr} increases for increasing concentrations of *C. fistula* leaf-extract in the studied test-solution. In comparison, β_a only decreased as the E_{corr} decreased in value from the 4 g/L *C. fistula* to the 6 g/L test-systems and only increased as the E_{corr} increased in value from the 6 to the 8 g/L test-systems while it patterned otherwise relative to E_{corr} values for the other test-systems of *C. fistula* leaf-extract concentrations. These further corroborate the inference that the behaviour of stainless steel in the HCl medium, being studied, exhibited more of cathodic control in the presence of *C. fistula*, and from this it could be implied that cathodic hydrogen evolution mechanisms are under influence of the *C. fistula* leaf-extract. The shifts in corrosion potentials of stainless steel in the HCl medium exhibited highest displacement in E_{corr} value of −175 mV, i.e. in the cathodic direction, thus indicating that *C. fistula* leaf-extract is a cathodic type inhibitor [49, 50].

Figure 2 shows plot of the corrosion rate, estimated by the LSV instrument as per Eq. (1), superimposed with the plot of corrosion inhibition efficiency, as per Eq. (2), for stainless steel in 0.5 M HCl in the absence and presence of different *C. fistula* leaf-extract concentrations. It could be observed from this figure,

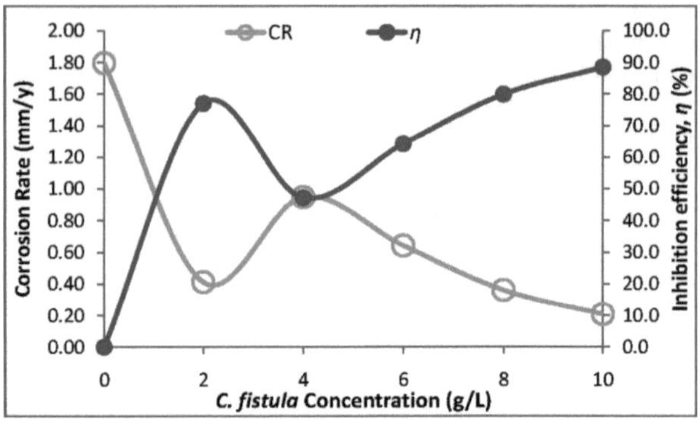

Fig. 2 Plots of corrosion rate and inhibition efficiency on stainless steel in 0.5 M HCl by *C. fistula* leaf-extract

that corrosion rate values exhibited drastic reductions in the presence of the plant extract compared to corrosion rate from the control sample. Consequently, all the concentrations of *C. fistula* leaf-extract employed in the study inhibited stainless steel corrosion in the HCl medium.

With a slight exception, it could be deduced from Fig. 2 that the corrosion rate tends to decrease with increasing *C. fistula* concentration and/or inhibition efficiency tends to increase with *C. fistula* leaf-extract concentration. The exception to this was the performance from the 2 g/L *C. fistula* leaf-extract system, on stainless steel corrosion in 0.5 M HCl, which exhibited lower corrosion rate, and therefore higher corrosion inhibition efficiency effect, than what obtained from the 4 and the 6 g/L *C. fistula* leaf-extract systems. In spite of this slight exception, the 10 g/L *C. fistula* concentration exhibited optimal inhibition efficiency, $\eta = 88.46\%$, in the study, followed by the 8 g/L having $\eta = 79.89\%$, with these trends being as a result of continuous increase in inhibition efficiency, η, from the 4 g/L to that of the 10 g/L *C. fistula* leaf-extract test-systems. Both of these inhibition efficiency performances interpret to very good efficiency models, as the efficiency model classifications from literature [51, 52].

Figure 3 shows the adsorption isotherm modelling of experimental data obtained from the *C. fistula* leaf-extract effect on stainless steel corrosion in 0.5 M HCl by the Langmuir isotherm model, Fig. 3a, and by the Flory-Huggins isotherm model, Fig. 3b. The isotherm parameters evaluated from these fitting models are presented in Table 1.

From the figure, it could be noted that the data-points exhibited closer proximity to the line of the Flory-Huggins model, Fig. 3b, than what obtained from the dispersions of the data-points that could be observed from the line of the Langmuir isotherm fitting, Fig. 3a. These were corroborated by the correlation coefficient estimations, $r^2 = 78.83\%$ by the Langmuir adsorption model, and that of $r^2 = 90.27\%$ by the Flory-Huggins model. The Langmuir fitting correlation coefficient classifies to "very good" model efficiency, while the Flory-Huggins fitting correlation coefficient classifies to "excellent" model efficiency, according to the model efficiency classification detailed in [51]. By requisite substitutions of $K_{Lang} = 0.3761$ and CR_0 into Eq. (6) the obtained separation factor, $R_L = 0.5968$,

Fig. 3 Adsorption isotherm modelling of experimental data from *C. fistula* leaf-extract effect on stainless steel corrosion in 0.5 M HCl **a** Langmuir isotherm model **b** Flory-Huggins isotherm model

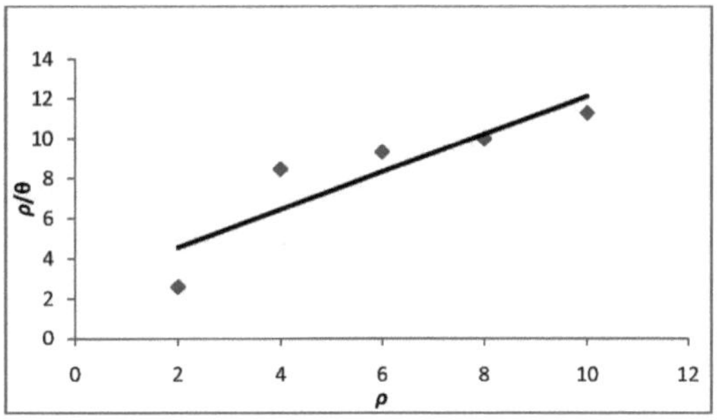

(a)

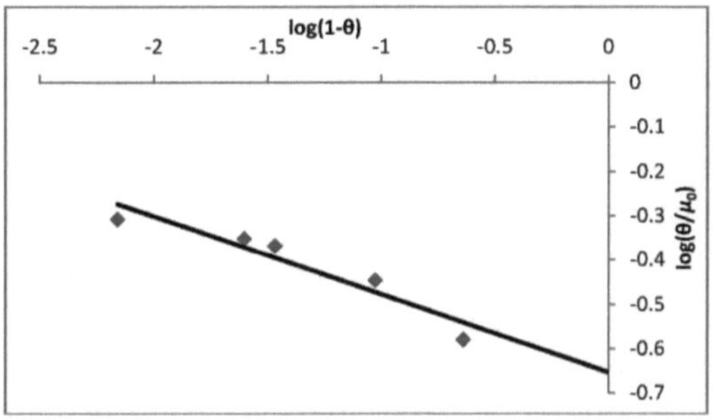

(b)

Table 1 Parameter estimations and interpretations from the adsorption isotherm models

Isotherm parameter	Langmuir isotherm model		Flory-Huggins isotherm model	
	Value	Interpretation	Value	Interpretation
Adsorption constant, K_{ads}	0.3761		0.5201	
Correlation coefficient, r^2 (%)	78.82	Very good fitting model	90.27	Excellent fitting model
Separation factor, R_L	0.5968	Favourable adsorption	–	
Gibbs free energy, ΔG°_{ads} (kJ/mol)	−7.5786	Spontaneous adsorption/physisorption mechanism	−8.3871	Spontaneous adsorption/physisorption mechanism

is such that $0 < R_L < 1$. This, according to separation factor interpretation detailed in reported studies [34, 46], indicates favourable adsorption by the *C. fistula* leaf-extract on the stainless steel in the tested medium. Similarly, substituting requisite values of adsorption constant K_{ads} into Eq. (7) gives the free Gibbs energy as $\Delta G^{\circ}_{ads} = -7.5786$ kJ/mol for the Langmuir model and $\Delta G^{\circ}_{ads} = -8.3871$ kJ/mol for the Flory-Huggins model. The negative values of the free Gibbs energy from both of the adsorption isotherm models indicate spontaneity of the *C. fistula* leaf-extract on the stainless steel. Also, that these values of free Gibbs energy are lower than −20 kJ/mol [34, 44] suggest the prevalence of physisorption mechanism as the corrosion protection mechanism by the *C. fistula* leaf-extract on stainless steel in the acidic chloride medium.

Conclusions

In this paper, inhibition of stainless steel corrosion in 0.5 M HCl in the presence of *C. fistula* leaf-extract has been studied through the use and analyses of test-results from linear sweep voltametry technique. From the study, it could be established that:

- All the different concentrations of *C. fistula* leaf-extract employed in the study inhibited stainless steel corrosion in 0.5 M HCl medium;
- With slight exception, observed in the 2 g/L *C. fistula* leaf-extract system, corrosion rate tends to reduce while inhibition efficiency tends to increase with increasing *C. fistula* concentration;
- Based on these results, 10 g/L *C. fistula* leaf-extract exhibited optimal effectiveness on the inhibition of stainless steel corrosion in the 0.5 M HCl medium through inhibition efficiency $\eta = 88.46\%$, which is followed in effectiveness by the 8 g/L *C. fistula* leaf-extract that exhibited $\eta = 79.89\%$;
- Adsorption isotherm modelling showed that the experimental data fits the Flory-Huggins isotherm model, $r^2 = 90.27\%$, better than the Langmuir isotherm model, $r^2 = 78.83\%$;
- The adsorption isotherm models indicate favourable adsorption, $R_L = 0.5968$ (by the Langmuir model), and prevalent physisorption, $\Delta G^{\circ}_{ads} = -7.5786$ kJ/mol (by the Langmuir model) or $\Delta G^{\circ}_{ads} = -8.3871$ kJ/mol (by the Flory-Huggins model), as the corrosion protection mechanism of the *C. fistula* leaf-extract on stainless steel in the 0.5 M HCl medium.

References

1. Finsgar, M., & Jackson, J. (2014). Application of corrosion inhibitors for steels in acidic media for the oil and gas industry: A review. *Corrosion Science, 86,* 17–41.
2. Rajeev, P., Surendranathan, A. O., & Murthy, C. S. N. (2012). Corrosion mitigation of the oil well steels using organic inhibitors-A review. *Journal of Material and Environmental Science, 3,* 856–869.

3. Omotosho, O. A., Ajayi, O. O., Fayomi, O. S., & Ifepe, V. O. (2011). Assessing the deterioration behavior of mild steel in 2 M sulphuric acid using *Bambusa glauscescens*. *International Journal of Applied Engineering Research Dindigul, 2,* 85–97.
4. Okeniyi, J. O., Popoola, A. P. I., Loto, C. A., Omotosho, O. A., Okpala, S. O., & Ambrose, I. J. (2015). Effect of $NaNO_2$ and $C_6H_{15}NO_3$ synergistic admixtures on steel-rebar corrosion in concrete immersed in aggressive environments. *Advances in Materials Science and Engineering, 2015*, 11 p.
5. Okeniyi, J. O., Ambrose, I. J., Oladele, I. O., Loto, C. A., & Popoola, A. P. I. (2013). Electrochemical performance of sodium dichromate partial replacement models by triethanolamine admixtures on steel-rebar corrosion in concretes. *International Journal of Electrochemical Science, 8,* 10758–10771.
6. Okeniyi, J. O., Omoniyi, O. M., Okpala, S. O., Loto, C. A., & Popoola, A. P. I. (2013). Effect of ethylenediaminetetraacetic disodium dihydrate and sodium nitrite admixtures on steel-rebar corrosion in concrete. *European Journal of Environmental and Civil Engineering, 17,* 398–416.
7. Scendo, M., & Trela, J. (2013). Adenine as an effective corrosion inhibitor for stainless steel in chloride solution. *International Journal of Electrochemical Science, 8,* 9201–9221.
8. Omotosho, O. A., & Ajayi, O. O. (2012). Investigating the acid failure of aluminium alloy in 2 M hydrochloric acid using *Vernonia amygdalina. ITB Journal of Engineering Science, 44,* 77–92.
9. Omotosho, O. A., Ajayi, O. O., Ajanaku, K. O., & Ifepe, V. O. (2012). Environment induced failure of mild steel in 2 M sulphuric acid using *Chromolaena odorata. Journal of Materials and Environmental Science, 3,* 66–75.
10. Omotosho, O. A., Ajayi, O. O., Fayomi, O., & Ifepe, V. O. (2010). Evaluating the deterioration behaviour of mild steel in 2 M sulphuric acid in the presence of *Butyrospermum parkii. Asian Journal of Applied Science, 5,* 74–84.
11. Obot, I. B., & Obi-Egbedi, N. O. (2010). An interesting and efficient green corrosion inhibitor for aluminium from extracts of *Chromolaena odorata* L. in acidic solution. *Journal of Electrochemistry, 40,* 1977–1984.
12. Lebrini, M., Robert, F., & Roos, C. (2010). Inhibition effect of alkaloids extract from *annona squamosa* plant on the corrosion of C38 steel in normal hydrochloric acid medium. *International Journal of Electrochemical Science, 5,* 1678–1712.
13. Raja, P. B., Rahim, A. A., Osman, H., & Awang, K. (2010). Inhibitory effect of *Kopsia singapurensis* extract on the corrosion behavior of mild steel in acid media. *Wuli Huaxue Xuebao/Acta Physico-Chimica Sinica, 26,* 2171–2176.
14. Kasthuri, P. K., & Arulanantham, A. (2010). Eco-friendly extract of Euphorbia hirta as corrosion inhibitor on mild steel in sulphuric acid medium. *Asian Journal of Chemistry, 22,* 430–434.
15. Rajalakshmi, R., Subhashini, S., Leelavathi, S., & Mary, R. F. (2008). Efficacy of sprouted seed extracts of *Phaseolus aureus* on the corrosion inhibition of mild steel in 1 M HCl. *Oriental Journal of Chemistry, 24,* 1085–1090.
16. Okeniyi, J. O., Loto, C. A., Popoola, A. P. I., & Omotosho, O. A. (2015). Performance of *Rhizophora mangle* L. leaf-extract and sodium dichromate synergies on steel-reinforcement corrosion in 0.5 M H_2SO_4-immersed concrete. In: *Corrosion 2015 Conference & Expo*. Houston, TX: NACE International, Paper No. 5636.
17. Omotosho, O. A., Ajayi, O. O., Fayomi, O., & Yussuff, O. (2012). Degradation evaluation of zinc in 2 M hydrochloric acid in the presence of *Bambusa bambos*. *Singapore Journal of Scientific Research, 2,* 14–24.
18. Al-Turkustani, A. M. (2010). Effect of *Ajowan* seeds as safe inhibitor on the corrosion of steel in 2.0 M sulfuric acid. *Modern Applied Science, 4,* 52–61.
19. Ating, E. I., Umoren, S. A., Udousoro, I. I., Ebenso, E. E., & Udoh, A. P. (2010). Leaves extract of *Ananas sativum* as green corrosion inhibitor for aluminium in hydrochloric acid solutions. *Green Chemistry Letters and Reviews, 3,* 61–68.

20. Gunasekaran, G., & Chauhan, L. R. (2007). Corrosion inhibition of mild steel by plant extract in dilute HCl medium. *Corrosion Science, 49,* 1143–1161.
21. Jain, T., Chowdhary, R., Arora, P., & Mathur, S. P. (2005). Corrosion inhibition of aluminum in hydrochloric acid solutions by peepal (*Ficus religiosa*) extracts. *Bulletin of Electrochemistry, 21,* 23–27.
22. Omotosho, O. A., Okeniyi, J. O., Obi, E. I., Sonoiki, O. O., Oladipupo, S. I., & Oshin, T. M. (2016). Inhibition of Stainless Steel Corrosion in 0.5 M H_2SO_4 in the presence of $C_6H_5NH_2$. In: *TMS2016: 145 annual meeting & exhibition: Supplemental proceedings* (pp. 465–472). Cham, Switzerland: Springer International Publishing AG. doi:10.1007/978-3-319-48254-5_56.
23. Herle, R., Shetty, P., Shetty, S. D., & Kini, U. A. (2011). Corrosion inhibition of 304SS in hydrochloric acid solution by N-Furfuryl N'-Phenyl Thiourea. *Portugaliae Electrochimica Acta, 29*, 69–78.
24. Satpati, A. K., & Ravindran, P. V. (2008). Electrochemical study of the inhibition of corrosion of stainless steel by 1,2,3-benzotriazole in acidic media. *Materials Chemistry and Physics,* 109, 352–359.
25. Silva, A. B., Agostinho, S. M. L., Barcia, O. E., Cordeiro, G. G. O., & D'Elia, E. (2006). The effect of cysteine on the corrosion of 304L stainless steel in sulphuric acid. *Journal of Corrosion Science, 48,* 3668–3674.
26. Galal, A., Atta, N. F., & Al-Hassan, M. H. S. (2005). Effect of some thiophene derivatives on the electrochemical behavior of AISI 316 austenitic stainless steel in acidic solutions containing chloride ions II, effect of temperature and surface studies. *Materials Chemistry and Physics, 89,* 28–37.
27. Refaey, S. A. M., Taha, F., & Abd El-Malak, A. M. (2004). Inhibition of stainless steel pitting corrosion in acidic medium by 2-mercaptobenzoxazole. *Applied Surface Science, 236,* 175–185.
28. Abdallah, M. (2003). Corrosion behavior of 304 stainless steel in sulphuric acid solutions and its inhibition by some substituted pyrazolones. *Materials Chemistry and Physics, 82,* 786–792.
29. Abdallah, M. (2002). Rhodanine azosulpha drugs as corrosion inhibitors for corrosion of 304 stainless steel in hydrochloric acid solution. *Journal of Corrosion Science, 44,* 717–728.
30. Olsson, C. O. A., & Landolt, D. (2003). Passive films on stainless steels—chemistry, structure and growth. *Electrochimica Acta, 48,* 1093–1104.
31. Bera, S., Rangarajan, S., & Narasimhan, S. V. (2002). Electrochemical passivation of iron alloys and the film characterization by XPS. *Corrosion Science, 42,* 1709–1724.
32. Bahorun, T., Neergheen, V. S., & Aruoma, O. I. (2005). Phytochemical constituents of *Cassia fistula. African Journal of Biotechnology, 4,* 1530–1540.
33. ASTM D2688-94 R99. (2005). *Standard test methods for corrosivity of water in the absence of heat transfer (weight loss methods).* West Conshohocken: ASTM International, PA.
34. Okeniyi, J. O., Loto, C. A., & Popoola, A. P. I. (2014). Electrochemical performance of *Anthocleista djalonensis* on steel-reinforcement corrosion in concrete immersed in saline/marine simulating-environment. *Transactions of the Indian Institute of Metals, 67,* 959–969.
35. Okeniyi, J. O., Loto, C. A., & Popoola, A. P. I. (2014). Electrochemical performance of *Phyllanthus muellerianus* on the corrosion of concrete steel-reinforcement in industrial/microbial simulating-environment. *Portugaliae Electrochimica Acta, 32,* 199–211.
36. Hameurlaine, S., Gherraf, N., Benmnine, A., & Zellagui, A. (2010). Inhibition effect of methanolic extract of *Atractylis serratuloides* on the corrosion of mild steel in H_2SO_4 medium. *Journal of Chemical and Pharmaceutical Research, 2,* 819–825.
37. Okeniyi, J. O., Oladele, I. O., Omoniyi, O. M., Loto, C. A., & Popoola, A. P. I. (2015). Inhibition and compressive-strength performance of $Na_2Cr_2O_7$ and $C_{10}H_{14}N_2Na_2O_8{\cdot}2H_2O$ in steel-reinforced concrete in corrosive environments. *Canadian Journal of Civil Engineering, 42,* 408–416.

38. Okeniyi, J. O., Oladele, I. O., Ambrose, I. J., Okpala, S. O., Omoniyi, O. M., Loto, C. A., et al. (2013). Analysis of inhibition of concrete steel-rebar corrosion by $Na_2Cr_2O_7$ concentrations: Implications for conflicting reports on inhibitor effectiveness. *Journal of Central South University, 20,* 3697–3714.
39. Canmet, S. P. (2008). Electrochemical polarization techniques for corrosion monitoring. In: L. Yang (Ed.), *Techniques for corrosion monitoring* (pp. 49–85). Cambridge: Woodhead Publishing Limited.
40. Okeniyi, J. O., Loto, C. A., & Popoola, A. P. I. (2015). Corrosion inhibition of concrete steel-reinforcement in saline/marine simulating-environment by *Rhizophora mangle* L. *Solid State Phenomena, 227,* 185–189.
41. Okeniyi, J. O., Loto, C. A., & Popoola, A. P. I. (2014). *Morinda lucida* effects on steel-reinforced concrete in 3.5% NaCl: Implications for corrosion-protection of wind-energy structures in saline/marine environments. *Energy Procedia, 50,* 421–428.
42. Kumar, S., Ladha, D. G., Jha, P. C., & Shah, N. K. (2013). Theoretical study of Chloro-N-(4-methoxybenzylidene)aniline derivatives as corrosion inhibitors for zinc in hydrochloric acid. *International Journal of Corrosion, 2013,* 10 p.
43. Okeniyi, J. O., Loto, C. A., & Popoola, A. P. I. (2015). Evaluation and analyses of *Rhizophora mangle* L. leaf-extract corrosion-mechanism on reinforcing steel in concrete immersed in industrial/microbial simulating-environment. *Journal of Applied Sciences, 15,* 1083–1092.
44. Anejjar, A., Salghi, R., Zarrouk, A., Benali, O., Zarrok, H., Hammouti, B., et al. (2014). Inhibition of carbon steel corrosion in 1 M HCl medium by potassium thiocyanate. *Journal of the Association of Arab Universities for Basic and Applied Sciences, 15,* 21–27.
45. Okeniyi, J. O. (2016). $C_{10}H_{18}N_2Na_2O_{10}$ inhibition and adsorption mechanism on concrete steel-reinforcement corrosion in corrosive environments. *Journal of the Association of Arab Universities for Basic and Applied Sciences, 20,* 39–48.
46. Foo, K. Y., & Hameed, B. H. (2010). Insights into the modeling of adsorption isotherm systems. *Chemical Engineering Journal, 156,* 2–10.
47. Vijayaraghavan, K., Padmesh, T. V. N., Palanivelu, K., & Velan, M. (2006). Biosorption of nickel(II) ions onto *Sargassum wightii*: Application of two-parameter and three-parameter isotherm models. *Journal of Hazardous Materials, B133,* 304–308.
48. Okeniyi, J. O., Loto, C. A., & Popoola, A. P. I. (2015). Investigating the corrosion mechanism of *Morinda lucida* leaf extract admixtures on concrete steel rebar in saline/marine simulating environment. *International Journal of Electrochemical Science, 10,* 9893–9906.
49. Tao, Z. H., Zhang, S. T., Li, W. H., & Hou, B. R. (2009). Corrosion inhibition of mild steel in acidic solution by some oxo-triazole derivatives. *Corrosion Science, 51,* 2588–2595.
50. Ferreira, E. S., Giacomelli, C. F., Gicomelli, F. C., & Spinelli, A. (2004). Evaluation of the inhibitor effect of l-ascorbic acid on the corrosion of mild steel. *Materials Chemistry and Physics, 83,* 129–134.
51. Okeniyi, J. O., Ambrose, I. J., Okpala, S. O., Omoniyi, O. M., Oladele, I. O., Loto, C. A., et al. (2014). Probability density fittings of corrosion test-data: Implications on $C_6H_{15}NO_3$ effectiveness on concrete steel-rebar corrosion. *Sadhana, 39,* 731–764.
52. Coffey, R., Dorai-Raj, S., O'Flaherty, V., Cormican, M., & Cummins, E. (2013). Modelling of pathogen indicator organisms in a small-scale agricultural catchment using SWAT. *Human and Ecological Risk Assessment: An International Journal, 19,* 232–253.

Effect of $C_5H_{11}NO_2S$ on Reinforcing-Steel Corrosion in Concrete Immersed in Industrial/Microbial Simulating-Environment

Joshua Olusegun Okeniyi, Abiodun Oyekola Abioye, Zechariah Chiwonsoko Adikpewun, Adeola Abigail Otesanya, Michael Damilola Eleshin, Olanrewaju Oyewale Gabriel and Oluyori Adeoye

Abstract In this paper, $C_5H_{11}NO_2S$ (Methionine) effect on reinforcing-steel corrosion in concrete immersed in 0.5 M H_2SO_4, simulating industrial/microbial environment, was studied by electrochemical techniques of open circuit potential and corrosion rate. The corrosion test-data were subjected to statistical distribution and tests of significance analyses prescribed by ASTM G16-95 R04. From this, analyzed results showed that the corrosion rate test-data followed the Weibull more than the Normal while the corrosion potential test-data followed both distributions. In spite of these, both the corrosion potential and the corrosion rate models still find agreements in corrosion criteria classification for the tested samples. Samples with $C_5H_{11}NO_2S$ admixture exhibited corrosion rate reductions compared to the control samples. By this, 0.25% $C_5H_{11}NO_2S$ admixture (i.e. wt% cement) exhibited optimal inhibition efficiency, $\eta = 87.95 \pm 7.64\%$, on steel-rebar corrosion in the 0.5 M H_2SO_4-immersed concrete. Experimental data fitted Flory-Huggins adsorption isotherm that indicated physisorption as the prevailing mechanism of $C_5H_{11}NO_2S$ corrosion-protection on the reinforcing-steel in the industrial/microbial simulating-environment.

Keywords Methionine admixture · Steel-rebar corrosion · Industrial/microbial simulating-environment · Statistical tests of significance · Corrosion inhibition efficiency

J.O. Okeniyi (✉) · A.O. Abioye · Z.C. Adikpewun · A.A. Otesanya
M.D. Eleshin · O.O. Gabriel · O. Adeoye
Mechanical Engineering Department, Covenant University, Ota 112001, Nigeria
e-mail: joshua.okeniyi@covenantuniversity.edu.ng

© The Minerals, Metals & Materials Society 2017

M.A. Meyers et al. (eds.), *Proceedings of the 3rd Pan American Materials Congress*, The Minerals, Metals & Materials Series, DOI 10.1007/978-3-319-52132-9_19

Introduction

Corrosion of steel-reinforcement in concrete is a major deterioration problem affecting structural strength and service-life durability of steel-reinforced concrete material. The corrosion-induced premature deterioration of steel-reinforced concrete members generates safety risks to life and loss of property, while the induced repairs and maintenance for averting in-service failure of structural members generate cost-intensive economic and social impacts globally [1–5]. Among other methods, the use of corrosion inhibitor admixture in concrete has been identified as an easily applicable, economical and highly effective approach for protecting concrete steel-rebar from corrosion degradation and for attaining durability of steel-reinforced concrete [1–3, 6–8]. However, it is necessary to conduct research for evaluation and ascertaining corrosion inhibition performance of materials or substances that could be used for inhibiting reinforcing steel corrosion in concrete developed for aggressive service-environments. Among such corrosion-inducing environments to concrete steel-rebar is the acidic sulphate medium usually prevalent in industrial (acid rain) or the microbial (biogenic sulphate oxidizing) environments, usually prevalent in sewage or underground environments [1, 9, 10].

Ascertaining performance of an inhibiting substance through use of corrosion test-results portends the problems that corrosion test-data exhibited more stochasticity that deviates from the prevailing condition in the test-system than many other types of tests [9, 11–13]. Such deviations from the prevailing corrosion conditions in the corrosive test-system make interpretation of corrosion test-results difficult, a fact that could have been responsible for conflicting/contradictory reports on corrosion inhibitor effectiveness [9, 14, 15]. Recommendations from the literature for tackling this problem include the use of statistical analyses, such as probability distribution functions, goodness-of-fit testing and statistical tests of significance, for assisting investigators interpret results from corrosion tests [9, 11–14].

It has been identified in studies [16–18] that organic compounds with π-electron bonds and N-, O-, and S-containing heteroatoms exhibit good effectiveness at inhibiting metallic corrosion in aggressive media. Among such organic chemicals is $C_5H_{11}NO_2S$ (Methionine), which has been found to exhibit good effectiveness at inhibiting corrosion of iron, the highest elemental composition in reinforcing steel, in aggressive environments [16, 18, 19]. However, there is paucity of study in which $C_5H_{11}NO_2S$ has been employed as admixture for inhibiting steel-rebar corrosion in physically cast concrete immersed in acidic sulphate medium, i.e. industrial/microbial environment.

The foregoing considerations constitute motivations for this study on the analyses and evaluation of $C_5H_{11}NO_2S$ performance on steel-reinforcement corrosion in the acidic sulphate medium. Adequate analyses for appropriate performance detailing on corrosion inhibition effectiveness will assist in the transformation of investigative test-results into usable real-life field applications [13]. In this paper, therefore, analyses of test-data from corrosion test-experiments are employed for the evaluation of $C_5H_{11}NO_2S$ (Methionine) performance on the inhibition of concrete steel-rebar corrosion in 0.5 M H_2SO_4 medium. This medium, it could be noted, has been employed as test-environment for simulating industrial/microbial environment in studies [7, 10, 20, 21].

Experimental

The 12 mm diameter reinforcing steel employed in this study was cut into 190 mm rods of specimen for each of which surface preparation was done as prescribed by ASTM G109-99a [22] and described in [4, 10]. Steel-reinforced concrete samples used for the study were cast, cured and prepared according to standard procedure prescribed in ASTM C192/192 M-02 [23] and that had been described in reported works [4, 9].

$C_5H_{11}NO_2S$ (L-(–)-Methionine), which is of chemical structure shown in Fig. 1, was obtained from Oxford Laboratory Chemicals® for use in the present study. As prescribed in [23], the different concentrations of $C_5H_{11}NO_2S$ employed were admixed in duplicates [24, 25] to the samples, during casting. By this, samples with the duplicated $C_5H_{11}NO_2S$ admixture concentration are tagged "Dup" in the study. The $C_5H_{11}NO_2S$ admixed concentrations in the samples were varied from 0% (for the blank samples) in increment of 0.083% up to 0.417% i.e. wt% of the cement employed for casting each concrete paste. This totalled 12 steel-reinforced concrete samples for the study. 150 mm length of the reinforcing steel was centrally placed in each freshly cast concrete sample with the remaining 40 mm covered with self-adhesive tape [26] and bolted at one end for electrochemical connection.

Each 100 mm × 100 mm × 200 mm reinforced concrete sample was immersed lengthwise in bowls containing the 0.5 M H_2SO_4 test-medium, the industrial/microbial simulating-environment. From this, electrochemical measurements of

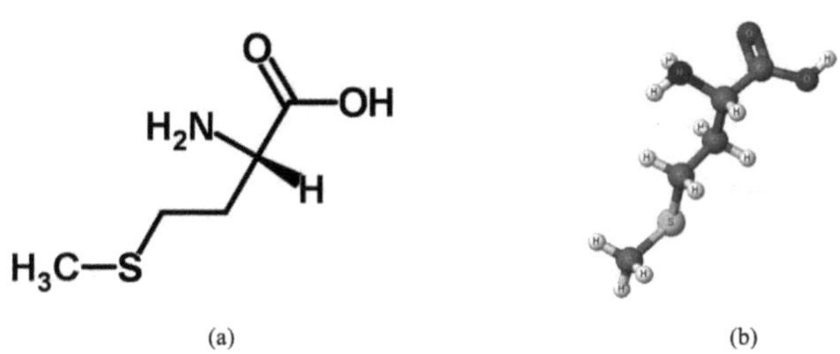

Fig. 1 Structure of $C_5H_{11}NO_2S$ (L-(–)-methionine) **a** molecule **b** optimized ball and stick model

open circuit potential and of corrosion rate, *CR*, were obtained. The *OCP* measurements were taken versus Cu/$CuSO_4$ (CSE), obtained from Tinker and Rasor® [27–29], using high impedance multimeter instrument, as prescribed in ASTM C876-91 R99 [30]. The *CR* measurements were obtained through linear polarization technique that employs the 3-electrode, Model MS1500, LPR Data Logger gotten from Metal Samples® [11, 31, 32]. These electrochemical monitoring of rebar corrosion were taken from the concrete samples in seven days interval for twelve weeks, which along with the 0th day measurements constitute 13 data points of each variable of corrosion test-data measurement for each sample.

As prescribed by ASTM G16-95 R04 [12], the measured test-data were subjected to the statistical analyses of the Normal and Weibull distribution fittings using Kolmogorov-Smirnov goodness-of-fit (K-S GoF) statistics for ascertaining the distribution function that describes the electrochemical test-data better than the other [13, 31, 33–37]. For n number of data points of corrosion test-variable, fitting of the measured test-data, x_i, $i = 1, 2, \ldots, n$, to the Normal distribution model followed usage of this probability distribution for the estimations of the Normal location (mean, μ_N) and Normal scale (standard deviation σ_N) parameters using the maximum likelihood equations [31, 33, 37, 38]:

$$\mu_N = \frac{1}{n}\sum_{i=1}^{n} x_i \tag{1}$$

$$\sigma_N = \sqrt{\frac{1}{n-1}\sum_{i=1}^{n}(x_i - \mu)^2} \tag{2}$$

However, evaluation of the Weibull mean, μ_W, and Weibull standard deviation, σ_N, employs the respective equations [11, 31, 36, 39]:

$$\mu_W = c\Gamma\left(1 + \frac{1}{k}\right) \tag{3}$$

$$\sigma_w = \sqrt{c^2\left\{\Gamma\left(1+\frac{2}{k}\right) - \left[\Gamma\left(1+\frac{1}{k}\right)\right]^2\right\}} \tag{4}$$

From these equations, $\Gamma(\zeta)$ is the gamma function of (ζ).

Equations (3) and (4) require unbiased estimations of the Weibull shape, k, and the Weibull scale c, parameters from the corrosion test test-data, x_i, $i = 1, 2, \ldots, n$ through solution of the simultaneous maximum likelihood equations [31, 39, 40]:

$$\frac{n}{\hat{k}} - n\ \ln(\hat{c}) + \sum_{i=1}^{n}\ln x_i - \sum_{i=1}^{n}\left(\frac{x_i}{\hat{c}}\right)^{\hat{k}}\ln\left(\frac{x_i}{\hat{c}}\right) = 0 \tag{6}$$

$$\hat{c} - \left\{ \frac{1}{n} \sum_{i=1}^{n} x_i^{\hat{k}} \right\}^{\frac{1}{k}} = 0 \tag{7}$$

The K-S GoF tests then measures the difference between the empirical $F^*(x)$ and the theoretical $F(x)$ distribution functions for each n-sized corrosion test-data through use of the statistics [37, 40, 41]:

$$D_n = D(x_1, \ldots, x_n) = \sup_{-\infty < x < \infty} |F^*(x) - F(x)| \tag{8}$$

Also, significance of corrosion rate difference between duplicates of concrete samples (j = 1, 2) having same $C_5H_{11}NO_2S$ admixture concentration was tested by the student's t-test statistics both at equal variance, homoscedastic, and at unequal variance, heteroscedastic, assumptions [9, 13, 24, 42, 43] through the respective formula:

$$t = \frac{|\mu_{N_1} - \mu_{N_2}|}{\sigma_{N_p} \left(\sum_{j=1}^{2} \frac{1}{n_j} \right)^{\frac{1}{2}}} \tag{9}$$

$$t = \frac{|\mu_{N_1} - \mu_{N_2}|}{\left(\sum_{j=1}^{2} \frac{\sigma_{N_j}^2}{n_j} \right)^{\frac{1}{2}}} \tag{10}$$

where σ_{N_j} for each duplicate is estimated from Eq. (2), and, according to ASTM G16-95 R04 [12], σ_{N_p} is the pooled estimate of standard deviation given, for the homoscedastic t-test analyses, by:

$$\sigma_{N_p} = \left(\frac{\sum_{j=1}^{2} (n_j - 1) \sigma_{N_j}^2}{\sum_{j=1}^{2} (n_j - 1)} \right)^{\frac{1}{2}} \tag{11}$$

The probability distribution function, pdf, model that describes the scatter of corrosion rate test-data better finds usefulness for detailing performance by the different concentrations of $C_5H_{11}NO_2S$ employed in the study. This followed usage of the pdf model for corrosion inhibition efficiency, η, and surface coverage, θ, (for adsorption isotherm modelling) evaluations using the respective equations [31, 40]:

$$\eta = \frac{\mu_{pdfCR_{\text{blank sample}}} - \mu_{pdfCR_{\text{admixed sample}}}}{\mu_{pdfCR_{\text{blank sample}}}} \times 100 \tag{12}$$

$$\theta = \frac{\mu_{pdfCR_{\text{blank sample}}} - \mu_{pdfCR_{\text{admixed sample}}}}{\mu_{pdfCR_{\text{blank sample}}}} \tag{13}$$

Results and Discussion

Results of the statistical models of mean corrosion potential and mean corrosion rate by the Normal and the Weibull distributions are presented in Fig. 2, for each sample of steel-reinforced concrete. The plots of mean corrosion potential in Fig. 2a include linear plot of corrosion risk as per ASTM C876-91 R99 [30], while the plots of mean corrosion rate include linear plot of corrosion rate classification according to literature [44, 45]. It could be noted from the figure that both corrosion risk, as per ASTM C876-91 R99 [30] criteria, and corrosion rate reduced, relative to the blank sample, in the presence of $C_5H_{11}NO_2S$ admixture in steel-reinforced concrete. The only exception to this reduction in corrosion condition was observed with the steel-reinforced concrete sample admixed with 0.333% $C_5H_{11}NO_2S$, which exhibited spike of increased corrosion potential that was corroborated with spike of increased corrosion rate in Fig. 2. The increased corrosion condition exhibited by the concrete sample admixed with 0.333% $C_5H_{11}NO_2S$ was not the same as what was obtained from its duplicate sample. This discrepancy of corrosion test-result necessitates needs for the student's *t*-test of significance of differences between the duplicated steel-reinforced concrete samples with similar $C_5H_{11}NO_2S$ admixture concentration. It could also be observed that the Normal mean model patterned like the Weibull model for the corrosion potential analyses while the corrosion rate analyses exhibited over-predictions between the Normal and Weibull mean models. These over-predictions are potent with controversy on which distribution model to employ for requisite detailing of $C_5H_{11}NO_2S$ admixture in the corrosive test-systems. This requires use of the Kolmogorov-Smirnov goodness-of-fit test-statistics [33–35] for identifying the probability distribution model that describes the corrosion test-data better than the other [13].

Figure 3 shows the plots of Kolmogorov-Smirnov goodness-of-fit test-statistics of corrosion test-data scatter like the Normal and the Weibull distribution models. The plots, for corrosion potential datasets in Fig. 3a and for corrosion rate datasets

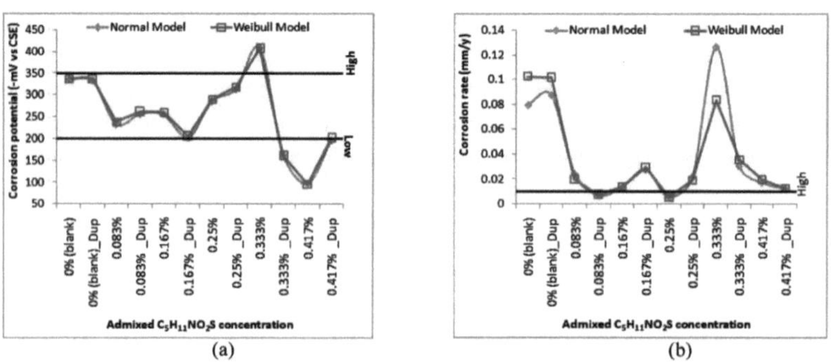

Fig. 2 Statistical distribution analyses of corrosion test-data by **a** corrosion potential **b** corrosion rate

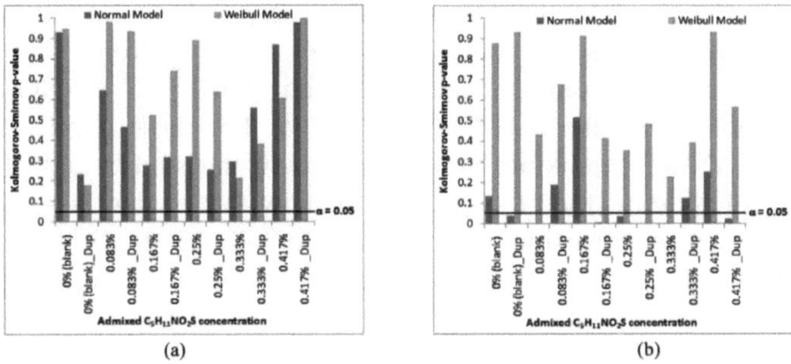

Fig. 3 Kolmogorov-Smirnov goodness-of-fit test-analyses of corrosion test-data distribution like the Normal and the Weibull distribution models **a** corrosion potential analyses **b** corrosion rate analyses

in Fig. 3b, include linear plot of $\alpha = 0.05$ for indicating threshold of significance of the K-S GoF test-criteria.

By this, it could be deduced from Fig. 3a that the datasets of corrosion potential followed both the Normal and the Weibull distributions, K-S *p*-values > 0.05, for all the steel-reinforced concrete samples studied. In contrast to this, corrosion rate datasets from seven steel-reinforced concrete samples were not distributed like the Normal distribution, i.e. K-S *p*-value < 0.05 for these samples, whereas corrosion rate datasets scattered like the Weibull distribution, K-S *p*-values > 0.05, for all the studied steel-reinforced concrete samples. This implies that the Weibull distribution model describes the corrosion rate test-data better than the Normal distribution does, in this study. These results support the use of the Weibull distribution model as the descriptive statistics for detailing $C_5H_{11}NO_2S$ performance on concrete steel-reinforcement corrosion in the test-environment.

Results of the student's *t*-test of significance of differences between each of the duplicate steel-reinforced concrete samples, having similar concentration of $C_5H_{11}NO_2S$ admixture, are shown in Fig. 4, for the homoscedastic and for the heteroscedastic assumptions. The figure also includes linear plot for indicating threshold of significance for the corrosion potential analyses in Fig. 4a and for the corrosion rate analyses in Fig. 4b.

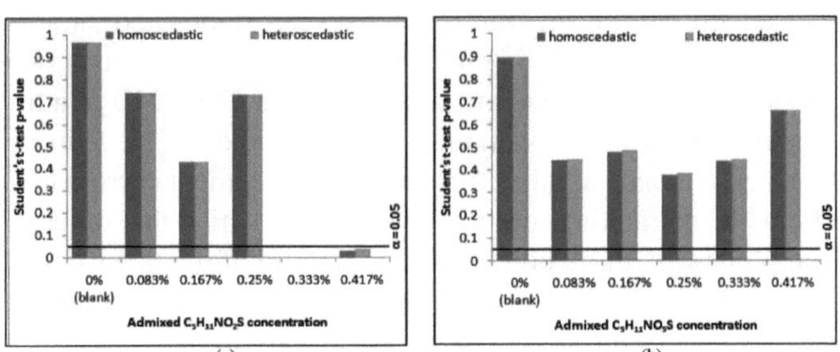

Fig. 4 Student's *t*-tests of significance of differences in the corrosion test-data between duplicate steel-reinforced concrete samples **a** corrosion potential analyses **b** corrosion rate analyses

The plots in Fig. 4a show that the student's t-test p-value > 0.05 for analyses of corrosion potential datasets, except for the datasets from the duplicated steel-reinforced concrete samples admixed with 0.333% $C_5H_{11}NO_2S$ and the duplicated samples with the 0.417% $C_5H_{11}NO_2S$. For these duplicated samples with high $C_5H_{11}NO_2S$ admixture concentrations in the study, student's t-test p-value < 0.05. In contrast, the plots in Fig. 4b show that student's t-test p-value > 0.05 for analyses of corrosion rate datasets for all the duplicated samples of steel-reinforced concretes studied. These result models bare implications that it could not be rejected that the discrepancies of corrosion rate test-results encountered between the duplicates of steel-reinforced concrete samples are only due to chance but they are not significant.

However, for the duplicated samples admixed with 0.333% $C_5H_{11}NO_2S$ and those with the 0.417% $C_5H_{11}NO_2S$, the differences encountered in the corrosion potential measurements are not due to chance but they are significant. Reason for this significance of difference encountered in the corrosion potential from the high concentrations of $C_5H_{11}NO_2S$ admixture in the study is difficult to explain, especially from the consideration that all electrochemical test-measurements were taken under similar experimental test-conditions. However, this exemplifies instance of bias that could make interpretation difficult and lead to contradictory reports in corrosion study. Actually, this form of bias and its attendant difficulty appears not to be unusual in corrosion studies, especially from the fact that ASTM G16-95 R04 [12] indicated it (i.e. bias) to be constituted of not only systematic error component but that it could also combine this with random component. Nevertheless, that this instance of bias is encountered with the corrosion potential analyses, instead of the corrosion rate, bare implication that it is of qualitative rather than of quantitative form of bias [12]. The practical significance of this form of bias could be understood from reported studies [46, 47] where the need for complimenting corrosion potential with other types of corrosion test-technique had been recommended due to the fact that the corrosion potential may not be used for indicating absolute corrosion activity.

Due to the scattering of corrosion rate test-data in the study like the Weibull model, the Weibull distribution analyses of corrosion rate was used for evaluating inhibition efficiency performance of $C_5H_{11}NO_2S$ on steel-rebar corrosion in the H_2SO_4-immersed concrete. Also, that the student's t-test analyses of corrosion rate datasets indicated no significant difference between duplicated samples, the corrosion rate models for the inhibition efficiency evaluations were averaged over each duplicates of steel-reinforced concrete with the same $C_5H_{11}NO_2S$ admixture concentration. The results for these evaluations are presented in Fig. 5, in ranking order of $C_5H_{11}NO_2S$ performance at inhibiting steel-reinforcement corrosion in the 0.5 M H_2SO_4 medium. This ranking order showed that while the other $C_5H_{11}NO_2S$ admixtures employed in the study inhibited steel-reinforcement corrosion in the medium, the 0.25% $C_5H_{11}NO_2S$ exhibited optimal inhibition effect of $\eta = 87.95 \pm 7.64$ on reinforcing steel corrosion in the industrial/microbial

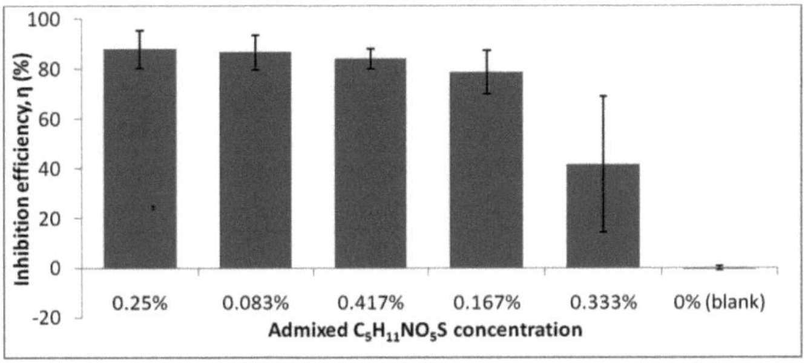

Fig. 5 Plots of evaluated performance, by inhibition efficiency modelling, of $C_5H_{11}NO_2S$ admixture on steel-rebar corrosion in 0.5 M H_2SO_4-immersed concretes

simulating environment studied. This inhibition efficiency model finds comparisons with results obtained in reported study by other researchers in acid rain (industrial simulating medium containing sulphuric acid) environment [48] in which inhibition efficiency as high as 80% was obtained through use of sodium molybdate and sodium nitrite as inhibitors. Also, application of the corrosion inhibition efficiency obtained in this study to the efficiency model [4, 11] of Coffey et al. [49] indicates that the 0.25% $C_5H_{11}NO_2S$ exhibited performance that interprets to "very good" model efficiency on the inhibition of steel-reinforcement corrosion in the studied medium. These results engender support for the use of $C_5H_{11}NO_2S$ (Methionine) as inhibitor of steel-reinforcement corrosion in concrete designed for the industrial/microbial environment.

$C_5H_{11}NO_2S$ surface coverage model on concrete steel-rebar, evaluated from Weibull pdf model of corrosion rate as detailed for the inhibition efficiency modelling, was subjected to different fittings of adsorption isotherms. Results of these showed that the experimental corrosion test-data followed fitting of the Flory-Huggins isotherm model given by [50, 51]:

$$\log\left(\frac{\theta}{\mu_0}\right) = \log K_{FH} + n_{FH}\log(1-\theta) \tag{14}$$

For this Flory-Huggins modelling of corrosion test-data, plot of the adsorption isotherm fitting is presented in Fig. 6 and the estimated parameters from the fitting analysis are presented in Table 1.

From these parameters, the coefficient of determination R^2 = 95.54%, which indicates excellent fitting model as per model efficiency classification from [49]. Also, the negative value of the Gibbs free energy of adsorption ΔG_{ads} indicates spontaneity of the adsorption process by $C_5H_{11}NO_2S$. This value of the ΔG°_{ads} that is also less negative than −20 kJ/mol indicates prevailing physisorption as the mechanism of $C_5H_{11}NO_2S$ on steel-rebar corrosion-protection in concrete immersed in 0.5 M H_2SO_4, simulating industrial/microbial environment.

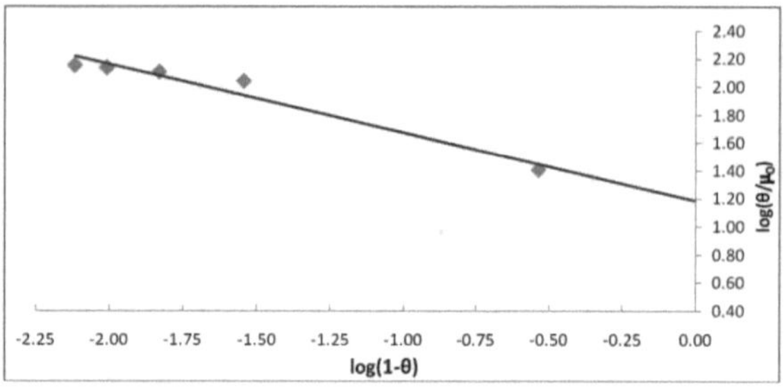

Fig. 6 Flory-Huggins adsorption isotherm fitting of experimental data from $C_5H_{11}NO_2S$ admixed steel-reinforced concrete immersed in 0.5 M H_2SO_4 test-medium

Table 1 Estimated parameters from the Flory-Huggins isotherm modelling of experimental data

Isotherm parameter	Modelled value
K_{ads}	3.2662
R^2 (%)	95.54
ΔG_{ads} (kJ/mol)	−12.9699

Conclusions

From this study, it could be concluded that:

- Concentrations of $C_5H_{11}NO_2S$ admixture reduces both the corrosion risk and corrosion rate, from results of corrosion test-data analyses, in steel-reinforced concrete immersed in 0.5 M H_2SO_4 medium, simulating industrial/microbial environment;
- Corrosion potential finds agreements with corrosion rate in classifications of corrosion risk and corrosion condition, for the $C_5H_{11}NO_2S$ admixed steel-reinforced concrete studied;
- Datasets of corrosion potential scattered like both the Normal and the Weibull distributions while dataset of corrosion rate distributed more like the Weibull than the Normal distribution for all steel-reinforced concrete samples studied, according to the Kolmogorov-Smirnov goodness-of-fit criteria at $\alpha = 0.05$ significant level;
- Student's *t*-test statistics showed that the difference encountered in the corrosion rate test-data between duplicates of steel-reinforced concrete samples, with similar admixture, are not significant;
- Optimal inhibition efficiency performance of $\eta = 87.95 \pm 7.64$ on concrete steel-rebar corrosion was obtained from samples having 0.25% $C_5H_{11}NO_2S$ (per weight of concrete mixing cement) admixture among the 0.5 M H_2SO_4-immersed steel-reinforced concrete samples employed for the study;
- The corrosion test-data from the study followed fitting of Flory-Huggins adsorption isotherm, which indicated physisorption as the prevailing mechanism of the $C_5H_{11}NO_2S$ corrosion-protection on steel-rebar in concrete immersed in 0.5 M H_2SO_4 medium, simulating industrial/microbial environment.

References

1. Okeniyi, J. O., Omotosho, O. A., Popoola, A. P. I., & Loto, C. A. (2016). *Phyllanthus muellerianus* and $C_6H_{15}NO_3$ synergistic effects on 0.5 M H_2SO_4-immersed steel-reinforced concrete: implication for clean corrosion-protection of wind energy structures in industrial environment. In: *AIP Conference Proceedings* (Vol. 1758, No. 030031, pp. 1–8). AIP Publishing. doi:10.1063/1.4959427.
2. Ismail, M., Raja, P. B., & Salawu, A. A. (2015). Developing deeper understanding of green inhibitors for corrosion of reinforcing steel in concrete. In: H. L. Lim (Ed.), *Handbook of research on recent developments in materials science and corrosion engineering education* (pp. 118–146). Hershey, PA: IGI Global.
3. Fei, F. L., Hu, J., Wei, J. X., Yu, Q. J., Chen, Z. S. (2014). Corrosion performance of steel reinforcement in simulated concrete pore solutions in the presence of imidazoline quaternary ammonium salt corrosion inhibitor. *Construction and Building Materials, 70,* 43–53.
4. Okeniyi, J. O., Ambrose, I. J., Oladele, I. O., Loto, C. A., & Popoola, A. P. I. (2013). Electrochemical performance of sodium dichromate partial replacement models by tri-ethanolamine admixtures on steel-rebar corrosion in concretes. *International Journal of Electrochemical Science, 8,* 10758–10771.
5. Mesquita, T. J., Chauveau, E., Mantel, M., Kinsman, N., Roche, V., & Nogueira, R. P. (2012). Lean duplex stainless steels—The role of molybdenum in pitting corrosion of concrete reinforcement studied with industrial and laboratory castings. *Materials Chemistry and Physics, 132,* 967–972.
6. Okeniyi, J. O., Popoola, A. P. I., Loto, C. A., Omotosho, O. A., Okpala, S. O., & Ambrose, I. J. (2015). Effect of $NaNO_2$ and $C_6H_{15}NO_3$ synergistic admixtures on steel-rebar corrosion in concrete immersed in aggressive environments. *Advances in Materials Science and Engineering, 2015,* 11 p. doi:10.1155/2015/540395.
7. Okeniyi, J. O., Oladele, I. O., Omoniyi, O. M., Loto, C. A., & Popoola, A. P. I. (2015). Inhibition and compressive-strength performance of $Na_2Cr_2O_7$ and $C_{10}H_{14}N_2Na_2O_8{\cdot}2H_2O$ in steel-reinforced concrete in corrosive environments. *Canadian Journal of Civil Engineering, 42,* 408–416.
8. Tang, Y., Zhang, G., & Zuo, Y. (2012). The inhibition effects of several inhibitors on rebar in acidified concrete pore solution. *Construction and Building Materials, 28,* 327–332.
9. Okeniyi, J. O., Oladele, I. O., Ambrose, I. J., Okpala, S. O., Omoniyi, O. M., Loto, C. A., et al. (2013). Analysis of inhibition of concrete steel-rebar corrosion by $Na_2Cr_2O_7$ concentrations: Implications for conflicting reports on inhibitor effectiveness. *Journal of Central South University, 20,* 3697–3714.
10. Okeniyi, J. O., Omoniyi, O. M., Okpala, S. O., Loto, C. A., & Popoola, A. P. I. (2013). Effect of ethylenediaminetetraacetic disodium dihydrate and sodium nitrite admixtures on steel-rebar corrosion in concrete. *European Journal of Environmental and Civil Engineering, 17,* 398–416.
11. Okeniyi, J. O., Ambrose, I. J., Okpala, S. O., Omoniyi, O. M., Oladele, I. O., Loto, C. A., et al. (2014). Probability density fittings of corrosion test-data: Implications on $C_6H_{15}NO_3$ effectiveness on concrete steel-rebar corrosion. *Sadhana, 39,* 731–764.
12. ASTM G16-95 R04. (2005). *Standard guide for applying statistics to analysis of corrosion data.* West Conshohocken PA: ASTM International.
13. Roberge, P. R. (2003). Statistical interpretation of corrosion test results. In: S. D. Cramer & B. S. Covino Jr. (Eds.), *ASM Handbook: Vol 13A. Corrosion: Fundamentals, testing, and protection* (pp. 425–429). Materials Park, OH: ASM International.
14. Okeniyi, J. O., Loto, C. A., & Popoola, A. P. I. (2016). Anticorrosion performance of *Anthocleista djalonensis* on steel-reinforced concrete in a sulphuric-acid medium. *HKIE Transactions, 23,* 138–149.

15. Jamil, H. E., Shriri, A., Boulif, R., Bastos, C., Montemor, M. F., & Ferreira, M. G. S. (2004). Electrochemical behaviour of amino alcohol-based inhibitors used to control corrosion of reinforcing steel. *Electrochimica Acta, 49,* 2753–2760.
16. Aouniti, A., Khaled, K. F., & Hammouti, B. (2013). Correlation between inhibition efficiency and chemical structure of some amino acids on the corrosion of Armco Iron in molar HCl. *International Journal of Electrochemical Science, 8,* 5295–5943.
17. Raphael, V. P., Kakkassery, J. T., Shanmughan, S. K., & Paul, A. (2013). Study of synergistic effect of iodide on the corrosion antagonistic behaviour of a heterocyclic phenylhydrazone in sulphuric acid medium on carbon steel. *ISRN Corrosion, 2013,* 1–7.
18. Morad, M. S. (2008). Corrosion inhibition of mild steel in sulfamic acid solution by S-containing amino acids. *Journal of Applied Electrochemistry, 38,* 1509–1518.
19. Oguzie, E. E., Li, Y., Wang, S. G., & Wang, F. (2011). Understanding corrosion inhibition mechanisms—experimental and theoretical approach. *RSC Advances, 1,* 866–873.
20. Okeniyi, J. O., Loto, C. A., Popoola, & A. P. I. (2015). Inhibition of steel-rebar corrosion in industrial/microbial simulating-environment by *Morinda lucida*. *Solid State Phenomena, 227,* 281–285.
21. Gerengi, H., Kocak, Y., Jazdzewska, A., Kurtay, M., & Durgun, H. (2013). Electrochemical investigations on the corrosion behaviour of reinforcing steel in diatomite- and zeolite-containing concrete exposed to sulphuric acid. *Construction and Building Materials, 49,* 471–477.
22. ASTM G109-99a. (2005). *Standard test method for determining the effects of chemical admixtures on the corrosion of embedded steel reinforcement in concrete exposed to chloride environments.* West Conshohocken, PA: ASTM International.
23. ASTM C192/192M-02. (2005). *Standard practice for making and curing concrete test specimens in the laboratory.* West Conshohocken, PA: ASTM International.
24. Okeniyi, J. O., Loto, C. A., & Popoola, A. P. I. (2014). *Rhizophora mangle* L. effects on steel-reinforced concrete in 0.5 M H_2SO_4: Implications for corrosion-degradation of wind-energy structures in industrial environments. *Energy Procedia, 50,* 429–436.
25. Corbett, R. A. (2005). Immersion testing. In: R. Baboian (Ed.), *Corrosion tests and standards: Application and interpretation* (2nd ed., pp. 139–146). West Conshohocken, PA: ASTM International.
26. Ormellese, M., Lazzari, L., Goidanich, S., Fumagalli, G., & Brenna, A. (2009). A study of organic substances as inhibitors for chloride-induced corrosion in concrete. *Corrosion Science, 51,* 2959–2968.
27. Okeniyi, J. O., Loto, C. A., & Popoola, A. P. I. (2015). Evaluation and analyses of *Rhizophora mangle* L. leaf-extract corrosion-mechanism on reinforcing steel in concrete immersed in industrial/microbial simulating-environment. *Journal of Applied Sciences, 15,* 1083–1092.
28. Okeniyi, J. O., Omotosho, O. A., Loto, C. A., & Popoola, A. P. I. (2015). Corrosion rate and noise resistance correlation from $NaNO_2$-admixed steel-reinforced concrete. *Asian Journal of Scientific Research, 8,* 454–465.
29. Omotosho, O. A., Okeniyi, J. O., Ajayi, O. O., & Loto, C. A. (2012). Effect of synergies of $K_2Cr_2O_7$, K_2CrO_4, $NaNO_2$ and aniline inhibitors on the corrosion potential response of steel reinforced concrete in saline medium. *International Journal of Environmental Sciences, 2,* 2346–2359.
30. ASTM C876–91 R99. (2005). *Standard test method for half-cell potentials of uncoated reinforcing steel in concrete.* West Conshohocken, PA: ASTM International.
31. Okeniyi, J. O., Loto, C. A., & Popoola, A. P. I. (2014). Electrochemical performance of *Anthocleista djalonensis* on steel-reinforcement corrosion in concrete immersed in saline/marine simulating-environment. *Transactions of the Indian Institute of Metals, 67,* 959–969.
32. Sastri, V. S. (2011). *Green corrosion inhibitors: Theory and practice*. New York: Wiley.

33. Okeniyi, J. O., Okeniyi, E. T., & Atayero, A. A. (2015). Programming development of Kolmogorov-Smirnov goodness-of-fit testing of data normality as a Microsoft Excel® library function. *Journal of Software & Systems Development, 2015,* 1–15.
34. Frey, J. (2012). An exact Kolmogorov-Smirnov test for the Poisson distribution with unknown mean. *Journal of Statistical Computation and Simulation, 82,* 1023–1033.
35. Okeniyi, J. O., & Okeniyi, E. T. (2012). Implementation of Kolmogorov-Smirnov P-value computation in Visual Basic®: Implication for Microsoft Excel® library function. *Journal of Statistical Computation and Simulation, 82,* 1727–1741.
36. Okeniyi, J. O., Moses, I. F., & Okeniyi, E. T. (2015). Wind characteristics and energy potential assessment in Akure, South West Nigeria: Econometrics and policy implications. *International Journal of Ambient Energy, 36,* 282–300.
37. Okeniyi, J. O., Obiajulu, U. E., Ogunsanwo, A. O., Odiase, N. W., & Okeniyi, E. T. (2013). CH_4 emission model from the waste of *Sus Domesticus* and *Gallus Domesticus* in Nigerian local farms: Environmental implications and prospects. *Mitigation and Adaptation Strategies for Global Change, 18,* 325–335.
38. Krishnamoorthy, K. (2006). *Handbook of statistical distributions with applications*. Florida: Chapman & Hall/CRC, Taylor & Francis Group, LLC.
39. Lai, C.-D., Murthy, D. N. P., & Xie, M. (2006). Weibull distributions and their applications. In: H. Pham (Ed.), *Springer handbook of engineering statistics* (pp. 63–78). Stürtz GmbH, Würzburg Germany: Springer-Verlag London Limited.
40. Okeniyi, J. O. (2016). $C_{10}H_{18}N_2Na_2O_{10}$ inhibition and adsorption mechanism on concrete steel-reinforcement corrosion in corrosive environments. *Journal of the Association of Arab Universities for Basic and Applied Sciences, 20,* 39–48.
41. Kvam, P. H., & Vidakovic, B. (2007). *Nonparametric statistics with applications to science and engineering*. Hoboken, New Jersey: Wiley.
42. Okeniyi, J. O., Loto, C. A., & Popoola, A. P. I. (2014). *Morinda lucida* effects on steel-reinforced concrete in 3.5% NaCl: Implications for corrosion-protection of wind-energy structures in saline/marine environments. *Energy Procedia, 50*, 421–428.
43. Yu, W., Wu, B., Huang, T., Li, X., Williams, K., & Zhao, H. (2006). Statistical methods in proteomics. In: H. Pham (Ed.), *Springer handbook of engineering statistics* (pp. 623–638). Stürtz GmbH, Würzburg Germany: Springer-Verlag London Limited.
44. Söylev, T. A., & Richardson, M. G. (2008). Corrosion inhibitors for steel in concrete: State-of-the-art report. *Construction and Building Materials, 22,* 609–622.
45. Söylev, T. A., McNally, C., & Richardson, M. (2007). Effectiveness of amino alcohol-based surface-applied corrosion inhibitors in chloride-contaminated concrete. *Cement and Concrete Research, 37,* 972–977.
46. Song, H.-W., & Saraswathy, V. (2007). Corrosion monitoring of reinforced concrete structures: A review. *International Journal of Electrochemical Science, 2*, 1–28.
47. Berke, N. S., & Hicks, M. C. (2004). Predicting long-term durability of steel reinforced concrete with calcium nitrite corrosion inhibitor. *Cement & Concrete Composites, 26,* 191–198.
48. Tommaselli, M. A. G., Mariano, N. A., & Kuri, S. E. (2009). Effectiveness of corrosion inhibitors in saturated calcium hydroxide solutions acidified by acid rain components. *Construction and Building Materials, 23,* 328–333.
49. Coffey, R., Dorai-Raj, S., O'Flaherty, V., Cormican, M., & Cummins, E. (2013). Modelling of pathogen indicator organisms in a small-scale agricultural catchment using SWAT. *Human and Ecological Risk Assessment: An International Journal, 19,* 232–253.
50. Okeniyi, J. O., Loto, C. A., & Popoola, A. P. I. (2014). Electrochemical performance of *Phyllanthus muellerianus* on the corrosion of concrete steel-reinforcement in industrial/microbial simulating-environment. *Portugaliae Electrochimica Acta, 32,* 199–211.
51. Eddy, N. O., & Mamza, P. A. P. (2009). Inhibitive and adsorption properties of ethanol extract of seeds and leaves of *Azadirachta indica* on the corrosion of mild steel in H_2SO_4. *Portugaliae Electrochimica Acta, 27,* 443–456.

Fiber Reinforced Concrete Manufactured with Electric Arc Furnace Slag

Vanesa Ortega-López, José A. Fuente-Alonso, Amaia Santamaría, Marta Skaf and Juan M. Manso

Abstract The use of electric arc furnace slag (EAFS), a by-product of the steelmaking industry, as an aggregate in concrete has been demonstrated to be a good practice in its manufacture. Furthermore, the incorporation of fibers in concrete provides a more ductile behavior, increasing their tenacity and load capacity, improving the flexotraction strength and controlling shrinkage cracking. The purpose of this research was to study the performance improvement by reinforcing steel-slag concrete with metallic or synthetic fibers added in different amounts. Some of the properties evaluated were: consistency of freshly mixed concrete by Abrams cone, compressive strength, flexotraction strength and indirect tensile strength. The results show a substantial improvement of the performance of the steel-slag concrete when it is reinforced with fibers. It also fulfils the requirements of "depth of water penetration under pressure" test, even in the worst environmental exposure case.

Keywords Electric arc furnace slag · Reinforced concrete · Fibers

Introduction

Electric Arc Furnace Slag (EAFS) and Ladle Furnace basic Slag (LFS) are by-products of the steel electric industry, produced following the melting of scrap steel and the secondary or basic refining of steel, respectively. The former is usually presented as gravel sized aggregate and the latter, in the form of a grey powder.

V. Ortega-López (✉) · J.A. Fuente-Alonso · M. Skaf · J.M. Manso
University of Burgos (EPS), Calle Villadiego s/n, 09001 Burgos, Spain
e-mail: vortega@ubu.es

A. Santamaría
Department of Mining and Metallurgical Engineering,
UPV/EHU, Alameda Urquijo s/n, 48013 Bilbao, Spain

© The Minerals, Metals & Materials Society 2017
M.A. Meyers et al. (eds.), *Proceedings of the 3rd Pan American Materials Congress*,
The Minerals, Metals & Materials Series, DOI 10.1007/978-3-319-52132-9_20

In Spain, approximately 70% of total steel production (14 MT-millions of tons) is from steel electric furnaces, together with 1MT of EAFS and 0.35 MT of LFS by-products [1].

The steelmaking industry is mainly concentrated in the north of Spain (around 60%). Therefore, it is so important for us to reuse this by-product in order to reduce slags landfilling and contribute to global sustainability.

In the last years, many research works have used EAF and LF slag, in order to reduce the landfilling of the steelmaking waste. Important works on slag characterization have been published [2–5]; It has been demonstrated the suitability of these slags in different applications, such as bituminous mixtures, granular layers, making concrete and mortars, making clinker and others [6–17]. Despite their different applications, approximately 23% of the slag is accumulated in landfill sites.

On the other hand, the construction sector is an important consumer of natural aggregates. Total consumption of natural aggregates in Spain is around 90 MT per year, resulting in a consumption of natural resources and exploitation of quarries.

With this work, we tried to convert the steel slag in steel aggregate for manufacturing steel-slag concrete.

The aim of the investigation carried out in this paper, was based on the reuse of EAF slags in the construction of concrete pavements and concrete slabs reinforced with fibers, in order to slow the spread of the microcracks during the retraction of the concrete and to improve its tenacity and flexotraction.

For that, the slag was used as coarse and medium aggregate in the concrete mixture, and the sizes of sand were completed with 50% of slag and 50% of siliceous sand; which is mean approximately 75% of EAFS and 25% of natural aggregate in the finale concrete mixture. The fine siliceous sand, with rounded morphology, counteracted the effect of the surface irregularity of the EAFS, improving the fluidity throughout the concrete mass [18]. Furthermore, different dosages and types of fibers (metallic and synthetic) were studied in order to determine which are the most suitable for the aim proposed.

Materials

The following materials were used in this research:

- Ordinary Portland Cement (OPC) CEM I 42,5R: The main components determined by X-ray fluorescence analysis (FRX) were: CaO (60.4%), SiO_2 (21.3%), Al_2O_3 (6.1%), Fe_2O_3 (4.0%) and others as Mg and Na. Composition: 95% of Clinker and 5% of limestone filler. Density: 3.15 g/cm^3 and Specific Blaine Surface: 3400 cm^2/g.
- Water from an urban water supply.
- Natural siliceous aggregates: It has been provided in three granulometric sizes: sand 0/4, 4/12 and 12/25 mm; the size limestone aggregate employed in the

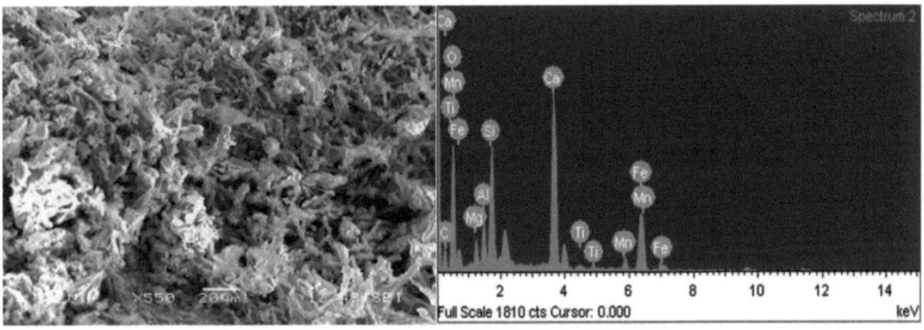

Fig. 1 SEM and microanalysis by X-ray of EAFS

EAFS concrete was the sand 0/4 mm, with fine content according with UNE-EN-933-1 of [19–21] 1.58% and sand equivalent according with UNE-EN 933-8 of 89.

- Super-plasticizing additives/superplasticizers: Policarboxilato modified with water. Density of 1.08 g/cm^3, pH of 5 and solid content of 36%.
- EAFS aggregates: They have been provided in three granulometric sizes: 0/4, 4/10 and 10/20 mm. The lack of fines is appreciated on the sand slag. EAFS is heavy, with density over 3500 kg/m^3 and a very resistant aggregate (loss Angeles Loss: 24%, Flakiness index: 3%), without plasticity. Its main chemical components are iron oxides silicates and aluminates of calcium and magnesium. The components that could provoke expansive processes, as the periclase and free lime were minor than 0.1 and 0.5% respectively, so the volume stability in the concrete would not expected due to its transformation into brucita and portlandite. Figure 1 shows the Scanning Electron Microscope (SEM) and the microanalysis by X-ray of the EAFS.
- Metallic Fibers (Dramix RL-45/50-BN): hard drawn steel filaments with formed ends to improve adhesion. Length: 55 mm, diameter: 1.05 mm, slenderness: 48, density: 7900 kg/m^3; tensile strength > 1000 MPa.
- Synthetic Fibers (SikaFiber M-48): polyolefins curly monofilament. Length: 48 mm, diameter: 0.93 mm, slenderness: 60, density: 910 kg/m^3; tensile strength > 400 MPa.

Experimental Procedure

Some dosages were used for the reference mixtures, which had previously been endorsed by other authors in bibliography:

- **Reference mixture P**, was a mixture with natural aggregates and 30 kg of metallic fibers/m^3 of concrete [21]. The amount of each component per cubic meter was: 804 kg of siliceous sand 0/4 mm, 574 kg of siliceous aggregate

4/12 mm, 464 kg of siliceous aggregate 12/20 mm, 3.63 kg of superplasticizer (1% of the cement), 363 kg of cement, 181 L of water and 30 kg of metallic fibers.
- **Reference mixture E**, was a mixture with 75% of EAFS aggregates and 25% of siliceous sand without fibers [22]. The amount of each component per cubic meter was: 498 kg of siliceous sand 0/4 mm, 514 kg of EAFS sand 0/4 mm, 669 kg of EAFS 4/10 mm, 550 kg of EAFS 10/20 mm, 5.44 kg of superplasticizer (1.5% of the cement), 363 kg of cement and 181 L of water.

The four following studied mixtures had the same dosage that reference mixture E, but they were reinforced with two types of fibers (metallic and synthetic) in different quantities:

- **Mixture EM1**: with 30 kg of metallic fibers per m^3 of concrete.
- **Mixture EM2**: with 45 kg of metallic fibers per m^3 of concrete.
- **Mixture ES1**: with 3.5 kg of synthetic fibers per m^3 of concrete.
- **Mixture ES2**: with 5 kg of synthetic fibers per m^3 of concrete.

For the six mixtures, the following parameters were constant: amount of cement: 363 kg/m^3; relation water/cement (w/c): 0.5; relation coarse aggregate/fine aggregate/cement: 3/3/1.

The performance of all the mixtures was studied as follows with different test:

- Consistency of freshly mixed concrete by Abrams Cone test.
- Mechanical behavior in cured mixtures with Compressive Strength test, Flex-traction Strength test and Indirect Tensile Strength test.
- Resistance to Water Penetration under pressure.

Results and Discussion

Consistency of Freshly Mixed Concrete by Abrams Cone Test According to UNE 12350-2

The consistency of freshly mixed concrete was measured with Abrams Cone before and after including the fibers. The results show that the mixture with natural aggregates and fibers had soft/plastic consistency (58 mm of slump) and the rest of the mixtures manufactured with EAFS or EAFS plus metallic/synthetic fibers had dry consistency (Fig. 2), with value for mixtures E, EM1 and EM2 around 5 mm of slump, and for mixtures ES1 and ES2 around 7 mm of slump.

Fig. 2 Abrams cone of the mixture P-concrete with natural aggregates and the mixture EM1-concrete with EAFS and fibers

Table 1 Compressive strength of the mixtures

Compressive strength (MPa)	P	E	EM1	EM2	ES1	ES2
7 days	39.59	61.43	46.95	68.11	63.53	60.63
28 days	46.30	66.05	54.00	72.62	74.04	66.50
90 days	59.99	68.40	68.35	86.62	82.51	75.33

Compressive Strength According to UNE 83507

The compressive strength was measured on cylindrical concrete specimen, with 15 mm of diameter and 300 mm of height, cured during 7 days, 28 days and 90 days in moist chamber at 20 ± 2 °C and 95% of moisture. The assay was performed in triplicate. According to UNE 83507 the results obtained are shown in Table 1.

Results show higher compressive strength in the concretes with steel-slag aggregates than the concretes with natural aggregates, even when those contained reinforcing fibers. The mixture E (with EAFS without fibers) had an increase of 43% of compressive strength respect to the mixtures P (with natural aggregates and fibers) at 28 days of curing.

EM2 and ES1 are the mixtures with better results in this test, with an increase of 10–12% of compressive strength with respect the same concrete without fibers. These results are according with other authors who said that fiber volumes added to concrete mixes at 0.5, 1.0 and 1.5% by volume of concrete improve the compressive strength between 4 and 19% [23].

Flexotraction Strength According to UNE 83507

The flexotraction strength was measured on concrete specimen with dimensions 150 × 150 × 600 mm and 100 × 100 × 400 cured in moist chamber during

Table 2 Flexotraction strength and Indirect tensile strength of the mixtures

Mixture	Flexotraction strength 28 days (MPa)	Indirect tensile strength 28 days (MPa)
P	5.15	4.33
E	6.80	4.20
EM1	5.95	5.01
EM2	7.00	5.45
ES1	7.13	5.23
ES2	6.88	4.59

28 days. According to the UNE 83509, the load was applied in two spaced points between them 1/3 of the length of the specimen. The results obtained are shown in Table 2.

All the steel-slag concretes had better results than the conventional one, even with fibers. The average of the flexotraction strength for steel-slag concretes with fiber had an increase upper than 30% respect to the reference mixture P.

It was well known that the steel-slag concrete improves the flexotraction strength [24] and that the fibers provide better deformation performance by flexion [25] according to the results here shown for concrete manufactured with EAFS and fibers.

The results were similar to the previous ones of compressive strength, reaching the best results the mixtures EM2 and ES1.

Indirect Tensile Strength According to UNE 12390-6

The indirect tensile strength was measured on cylindrical concrete specimen with diameter of 15 mm and height of 30 mm, cured in moist chamber during 28 days. According to the UNE 12390-6, the load was applied over two opposite lines until failure. The results provide the maximum load that the pavement is able to support. The results obtained were also shown in Table 2.

All the mixtures with fibers had better performance at tensile strength than the mixture E (without fibers), especially the mixtures with EAFS and fibers, EM2 and ES1. It is possible that higher amount of synthetic fibers in mixture ES2 than in mixture ES1 damages the strength characteristics of the hardened concrete, well by its problematic mix process or by the low hydration of the concrete components.

Resistance to Water Penetration Under Pressure According to UNE EN 12390-8

The resistance to water penetration under pressure was measured on cylindrical concrete specimen with diameter of 15 mm and height of 30 mm, cured in moist

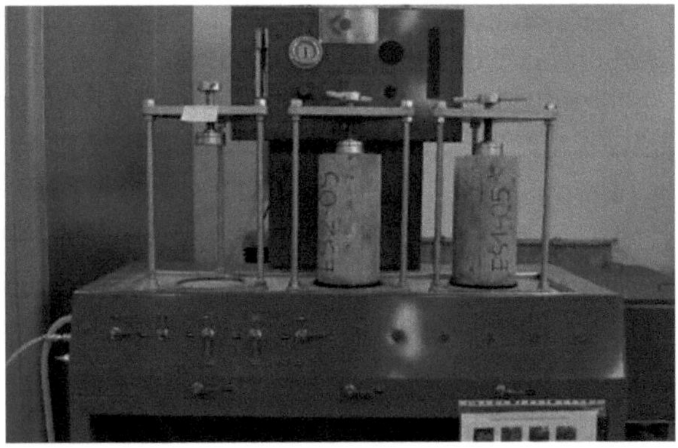

Fig. 3 Testing equipment for water penetration under pressure

chamber during 28 days. According to the UNE EN 12390-8, the water was applied under pressure of 500 ± 50 kPa during 72 h in one of the specimen faces (Fig. 3). After that, the specimen is broken and the depth of water penetration was measured. The results of the maximum height of water penetration were:

- The value for the mixture P was 24 mm.
- The value for the mixture E was 13 mm.
- The value for the mixture EM1 was 21 mm and for the mixture EM2 was 13.5 mm.
- The value for the mixture ES1 was 20.5 mm and for the mixture ES2 was 16.5 mm.

The water penetration surface and the average height of water penetration were also determined.

All the mixtures with EAFS had better resistance to water penetration than the conventional concrete with natural aggregates, which is a more porous concrete. However, in this test, the fibers in the concretes manufactured with EAFS did not improve the results respect to concrete without them.

Even so, all the mixtures fulfilled the requirements of the EHE standard [26], which specifies for the worst environmental exposure cases, the following values: average height of water penetration under 20 mm and maximum height of water penetration under 30 mm. Therefore, these concretes were considered impermeable enough for its use in pavements.

Tenacity According to UNE 83508

The tenacity is the energy required for a prespecified total deterioration or breakage of the material. It is one of the most important characteristics that the fibers provide to the material.

The tenacity by compression was measured on cylindrical concrete specimen with diameter of 15 mm and height of 30 mm. According to the UNE 83508, the results provide the area bounded by the load-deformation curve from the origin 0, and the ordinate corresponding to a deformation of 1.125 mm.

- The value for the mixture EM2 was 1949300 Nmm.
- The value for the mixture ES1 was 1887900 Nmm.

The mixture EM2 had better results than the mixture ES1, which means that metallic fibers provided greater post-cracking strength and higher increase of concrete ductility.

Conclusions

These conclusions can be derived from this research work:

- The concretes here shown, manufactured with EAFS and reinforced with fibers, provide very good results in all tests. Even, the results were close to the requirements for high-performance concretes.
- Volume fibers at around 0.4–0.6% by volume of steel-slag concrete provided concretes with suitable strengths: compression strength, flexotraction strength and indirect tensile strength.
- These steel-slag concretes reinforced with fibers fulfill the requirements of the standards about water penetration for its use in pavements and improve its tenacity behavior.
- The mixtures with better results in all the tests were EM2 (45 kg of metallic fiber per m^3 of concrete) and ES1 (3.5 kg of synthetic fiber per m^3 of concrete). So, increasing the fiber amount does not improve the results in every case; high amounts of synthetic fibers produce problems in the mix process and could hinder the hydration of the components in the mixture.
- Due to the poor workability and docility of these concretes, a water/cement ratio under 0.5 is not recommended, as well as the use of suitable super-plasticizing additives in the mixture is necessary, in order to obtain a proper placement in pavements.

Acknowledgements We express our gratitude to the Spanish Ministry (MINECO) BIA2014-55576-C2-1-R for financing this research work.

References

1. Steel Statistical Yearbook. (2014). World Steel Association.
2. Luxán, M. P., et al. (2000). Characteristics of the slags produced in the fusion of scrap steel by EAF. *Cement and Concrete Research, 30*(4), 517–519.

3. Frías Rojas, M., Sánchez, M. I., & Uria, A. (2002). Study of the instability of black slags from EAF steel industry. *Materiales de Construcción, 52*(267), 79–83.
4. Setién, J., Hernández, D., & González, J. J. (2009). Characterization of ladle furnace basic slag for use as a construction material. *Construction and Building Materials, 23*(5), 1788–1794.
5. Shi, C. (2002). Characteristics and cementitious properties of ladle slag fines from steel production. *Cement and Concrete Research, 32*(3), 459–462.
6. Arribas, I., et al. (2010). *Application of steel slag concrete in the foundation slab and basement wall of the Tecnalia kubik building*. In: *6th European Slag Conference Proceedings*. Madrid, Euroslag.
7. Geiseler, J., & Scholosser, R. (1988). *Investigation concerning the structure and properties of steel slags*. In: *Proceedings of the 3rd International Conference on Molten Slags and Fluxes*. Glasgow, Scotland.
8. Mroueh, U. M., Eskola, P., & Laine-Ylijoki, J. (2001). Life-cycle impacts of the use of industrial by-products in road and earth construction. *Waste Management, 21*(3), 271–277.
9. Manso, J. M., et al. (2005). Ladle furnace slag in construction. *The Journal of Materials in Civil Engineering, 17*, 513–518.
10. Bosela, P., et al. (2009). Fresh and hardened properties of paving concrete with steel slag aggregate. Propiedades para firmes del hormigón fabricado con áridos siderúrgicos. *Carreteras: Revista técnica de la Asociación Española de la Carretera, 4*(166), 55–66.
11. Abu-Eishah, S. I., El-Dieb, A. S., & Bedir, M. S. (2012). Performance of concrete mixtures made with electric arc furnace (EAF) steel slag aggregate produced in the Arabian Gulf region. *Construction and Building Materials, 34*, 249–256.
12. Kim, S. W., Lee, Y. J., & Kim, K. H. (2012). Bond behavior of RC beams with electric arc furnace oxidizing slag aggregates. *Journal of Asian Architecture and Building Engineering, 11*(2), 359–366.
13. Mäkelä, M., et al. (2012). Influence of fly ash and ground granulated blast furnace slag on the mechanical properties and reduction behavior of cold-agglomerated blast furnace briquettes. *ISIJ International, 52*(6), 1101–1108.
14. Nadeem, M., & Pofale, A. (2012). Utilization of industrial waste slag as aggregate in concrete applications by adopting Taguchi's approach for optimizatio. *The Open Civil Engineering, 2*, 96–105.
15. Manso, J. M., et al. (2013). The use of ladle furnace slag in soil stabilization. *Construction and Building Materials, 40*, 126–134.
16. Pellegrino, C., & Faleschini, F. (2013). Experimental behavior of reinforced concrete beams with electric arc furnace slag as recycled aggregate. *ACI Materials Journal, 110*(2), 197–205.
17. Wang, Q., Yang, J., & Yan, P. (2013). Cementitious properties of super-fine steel slag. *Powder Technology, 245*, 35–39.
18. Pellegrino, C., & Gaddo, V. (2009). Mechanical and durability characteristics of concrete containing EAF slag as aggregate. *Cement & Concrete Composites, 31*(9), 663–671.
19. EN Euronorm. *European Committee for Standardization*. Brussels.
20. CEN. *European Committee for Standardization*. Brussels.
21. Turmo, J., et al. (2008). Study of the shear behaviour of fiber reinforced concrete beams. *Materiales de Construcción, 58*(292), 5–13.
22. Manso, J. M., et al. (2011). Design and elaboration of concrete mixtures using steelmaking slags. *ACI Materials Journal, 108*(6), 673–681.
23. Yazici, S., Inan, G., & Tabak, V. (2007). Effect of aspect ratio and volume fraction of steel fiber on the mechanical properties of SFRC. *Construction and Building Materials, 21*(6), 1250–1253.
24. Manso, J. M., Gonzalez, J. J., & Polanco, J. A. (2004). Electric arc furnace slag in concrete. *Journal of Materials in Civil Engineering, 16*(6), 639–645.
25. Bemal, S., et al. (2009). Mechanical behaviour of steel fibre-reinforced alkali activated slag concrete. *Materiales de Construccion, 59*(293), 53–62.
26. Fomento, M. (Ed.). (2008). *Instrucción para realización de obras de hormigón estructural*. Madrid: Comisión Permanente de Hormigón.

Jigs, Hydrocyclones and Sensor-Based Sorting to Value Recycled Aggregate

Régis Sebben Paranhos, Carlos Hoffmann Sampaio, Bogdan Grigore Cazacliu, Raul Oliveira Neto and Maria Alejandra Liendo

Abstract The proposition of this paper is to introduce a sorting platform aiming to increase quality on recycled aggregates by supplementary use of mineral processing techniques and more sorting. The difficulty of liberation is discussed and the methodology currently used in mineral processing is proposed. Jigs, hydrocyclones and sensor-based sorting are equipments considered as having good performances to sort adequately recycled aggregates. On the other hand, new perspectives of sorting and liberation for recycling aggregates are discussed. Based on current process in recycling platforms, the new process with supplementary sorting of the concrete recycled is presented. The gain in density and the reduction in water absorption were studied. The relation between the water content and the density of aggregates is analysed for three quality levels of recycled aggregates. Finally, the gain in density and the reduction in water absorption were linked with rates of replaced aggregates. In our study, replacing a lower quality aggregate by another with medium quality leads an expected gain in density about 4%. On the other hand, if replaced by superior quality, the expected gain will be 8.4%. As consequence, 34% on reduction in water absorption could be obtained too.

Keywords Recycling aggregates · Jigs · Hydrocyclones · Sensor-based sorting

R.S. Paranhos (✉) · R.O. Neto · M.A. Liendo
Federal University of Pampa—UNIPAMPA, Bagé, Brazil
e-mail: regis.paranhos@unipampa.edu.br

R.O. Neto
e-mail: raulneto@unipampa.edu.br

M.A. Liendo
e-mail: alejandra.liendo@unipampa.edu.br

C.H. Sampaio
Federal University of Rio Grande do Sul—UFRGS, Porto Alegre, Brazil
e-mail: sampaio@ufrgs.br

B.G. Cazacliu
GPEM Laboratory, University of Nantes (LUNAM)/IFSTTAR, Nantes, France
e-mail: bogdan.cazacliu@ifsttar.fr

© The Minerals, Metals & Materials Society 2017

M.A. Meyers et al. (eds.), *Proceedings of the 3rd Pan American Materials Congress*,
The Minerals, Metals & Materials Series, DOI 10.1007/978-3-319-52132-9_21

Introduction

The construction industry is a major solid waste generator around the world [21]. Construction and demolition waste (CDW) is generated through the construction, renovation, and demolition processes of residential or commercial buildings, roads, bridges, etc. The CDW material usually is not (re)used and becomes an important environmental problem [17]. They can be stock in landfields, or the inert part is traditionally in roads, dams, etc.

In order to facilitate recycling, automatic optical sorting were developed in mining industry to process different ores. Gülsoy [10] studied optical sorting with CCD cameras in coal beneficiation. The technique was used to improve coal beneficiation for coals with high near-gravity materials. Optical sorting was tested with positive results for materials having significant colour differences. Ergün et al. [8] used optical sorting to treat iron and chromium ores. For iron ores, samples were taken from a magnetite processing plant. The purpose of the study was to concentrate hematite that was mixed with silicates. Optical sorting depends of many factors, such as, particle size, surface conditions, light source, feed rate, etc. Dehler and Robben [7] conducted two tests with borates treatment, by the use of near-infrared sorting. CCD cameras were used as sorting sensors, which have opened new possibilities for sorting in the mining industry. The difference is based on the wavelength of response spectrum.

In urban and metropolitan areas, the building materials recycling industry has used stationary plants to process concrete and masonry demolition waste. A relatively pure, high quality granulates are produced with recycling requirements. Considering these environmental context, the aim of this paper is to make more attractive the sorting systems, i.e. less energy consumption and less emissions. They can produce different flows with more homogeneous composition and provide a specific reuse or a recycling. The process detailed in this paper is based on the use of gravity concentration and sensor-based sorting.

Mineral Processing Applied in CDW

Difficulty of Liberation

Several studies about recycled concrete have been published in the last years [1, 9, 12, 15]. Recycled concrete is produced by reducing the size of concrete debris through multiple crushing stages. According Akbarnezhad et al. [1], depending on the size, the crushed concrete particles can present middlings, which contains particles of natural aggregates mixed with mortars (Fig. 1).

After comminution, a mineral can be totally or partially liberated. The particles liberation depends on the mineralogical origin, the quality of the concrete, the angularity of the original aggregates, the comminution process, size, etc. In order to

Fig. 1 Recycled aggregate (6–10 mm)

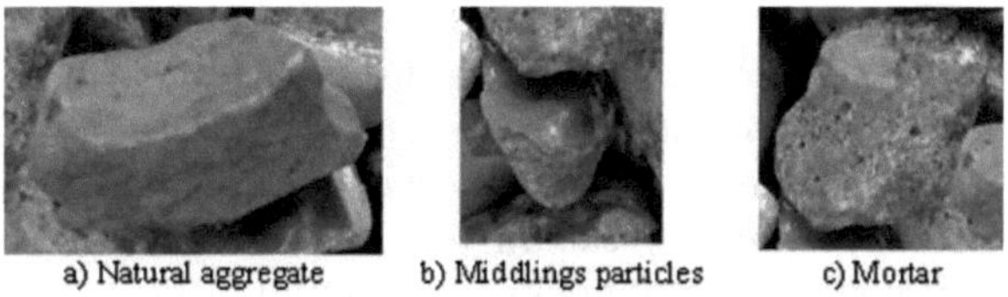

increase liberation, a preparation circuit should be used, to allow more intensive use of the recycled materials. With the beneficiation of recycled concrete material, more liberated particles can be achieved. In this sense, the use of innovative separation techniques in quarries, as sensor-based sorting or/and gravity concentration, will increase the quality of the products generated, as well as the price that they can be sold. These mining techniques, which are common in mining industry, could ensure a more efficient sorting in quarries.

Use of Mineral Processing Techniques with Recycled Concrete

Gravity concentration is defined as the process whereby particles of different sizes, shapes and densities are separated from each other by gravity or by centrifugal forces. It is called gravity concentration once the separation is performed mainly based on the density (specific gravity) [19]. Gravity concentration processes show high mass throughput and low investment and operational costs. Moreover, there is no limit for the maximum particle size and can be used for particles with wide size ranges. Furthermore, reagents are not used and this contributes significantly for low operational costs as well as for low environmental impacts.

Jigging is a separation process which consists of repeated expansion (dilatation) and contraction (compression) of a vertical bed of particles by a pulsating movement of water or air [2, 11]. The result is the stratification of the bed. The particles are deposited in layers with increasing densities from the top to the bottom. Jigs were and still are widely used mainly due to their costs. They present low operational costs, are robust, have high capacity, are easy to operate and can treat ores with a wide particle size range. They are also capable of working with large fluctuations of ore contents, feed rate and solids amount. For this reason, they are largely used in the treatment of alluvial ores.

Water-Only Cyclone (WOC) is a hydrocyclone in which the density is the major separation property. The shape of the hydrocyclone was changed to improve the influence of the particle density and to diminish the influence of the particle size and shape [14, 23]. There are several different names used in the industry to name WOC, such as wide-angle cyclone, circulating bad-concentrator (CBC), autogenous hydrocyclone, etc.

The use of optical sorting devices is increasing each year in the mining industry [13, 18]. There are several reasons for the ever increasing use of this kind of sorting: ability to automate the sorting process, possibility of separation (quantity and accuracy), add an economic value to the final products in relation to the traditional separation process, etc. In recycling, Angulo [3, 4] performed experiments on optical separation of bricks from CDW in Brazil. The main difficulty faced was the non-homogenous Brazilian waste. Angulo identified significant variations in their characteristics such as density and water absorption. The origin of these fluctuations, such as porosity, is due to red tiles, bricks and mortars. Density and water absorption are related to the porosity. The higher is absorption of water, the higher is porosity. That is the main problem of mixed particles in aggregates; porosity has an important impact on the quality of concrete.

New Perspectives of Sorting and Liberation for Recycling Aggregates

Despite techniques for sensor-based sorting, as well as for gravity concentration processes, are widely used in material recycling and in ore treatment [19, 24], the choice of concentration equipment to be installed depends, besides the desired cut precision, on the physical properties of the material to be beneficiated. Therefore, an extensive characterization before the mineral processing is indispensable. The main characteristics to be studied are porosity, water absorption and density distribution because they are associated with liberation. Many different techniques of separation or concentration may be used, alone or associated with others. Gravity concentration processes, as for instance jigs and cyclones, has many positive characteristics suggesting that they can be used to treat CDW.

Current Advanced Process

Many studies show that RCA with different replacement rate provides concrete building of quality. The recent standards provides for the possibility of introducing recycled aggregates in structural concrete, limiting the rate of substitution. In the quarries, the material can be processed by stages: a first stage of sorting by many techniques as manual sorting and/or screening; in a second stage, a sensor-based sorting and/or a gravity concentration is used. In addition, sand fraction had a relative importance in this context. If quarries produce a significant amount of fine particles during the process, on the other hand, recycling plants need to produce all size fractions from recycled aggregates, including sands.

Nowadays, there are the pre-sieving and segregation of ferrous scrap by overhead magnetic separators. The use of picking belts enables separation of large disturbing substances, before material with a particle size of >45 mm is transformed into granulates, mainly by impact crushers. After repeated segregation of separated ferrous scrap, follow a fractional sieving and separation of light substances by air classifiers. This technique allows the production of well qualified, close size fraction granulate mixtures. Products being processed in this way are of high quality and can be assigned as recycling materials (RCA1). In many countries, this material is currently used on roads and terracing. They rarely are used on concrete structures. In general, the aggregate for these applications are obtained from well mastered sources (specific demolition sites, producing a large amount of not contaminated concrete).

Process with Supplementary Sorting of the Concrete Recycled (First Improved Process)

Aiming to increase the recycling rate on concrete formulation, the use of high quality recycled aggregates during recycling off-site or on-site is suggested. In this stage of recycling, gravity concentration by jigs and sensor-based sorting are used to separate different particles (2/20 mm) as bricks, tiles, gypsum and glass, previously concentrated. The aim is to have high quality particles and more liberated particles. These particles have similar colours and could not be separated by conventional optical sorting by colour. Impurities here are can represent around 14% of total waste according Coelho and de Brito [6] and Ulsen et al. [21].

These recycled aggregates, before concentrating, still contain a significant amount of adhered mortar. Firstly, the jig will treat the coarser particles of recycled aggregates in two different density fractions. The fraction with density >2.1 g/cm^3, that represent 90% according Sampaio and Tavares [19], is sent to an optical sorting to generate a concentrate called Recycled Concrete Aggregate (RCA2). The impurities from sensor-based sorting that represent at least 11% considering jig efficiency, and the fraction with density <2.1 g/cm^3 from jig are called Light Recycled Aggregate lower quality and they sent to next stage. The feasibility was showed in Cazacliu et al. [5].

Process with Supplementary Production with High Quality Recycled (Second Improved Process)

This process aims at further increase recycling rates, perhaps beyond 100% substitution of gravel. A priori, this process is developed for recycling or reuses the on-site materials to minimize the production of waste leaving the site. In addition, the lower density recycled sand may have applications in the manufacture of new cements or on public works applications with better binding properties, for instance, their use as an addition (after re-grinding) during production concrete on site. We limit ourselves in this work to study how to sort recycled aggregates and the advantage of increasing the degree of substitution of aggregates at the site.

This stage of sorting receives Light Recycled Aggregate with low quality (density <2.1 g/cm^3) and impurities of jig and sensor-based sorting. All granulates are crushed at size <4 mm in order to liberate mortar that is adhered to the sand fraction (0/4 mm). Gyratory crusher has high efficiency to release aggregates but another type of crusher can be used. Then WOC beneficiate the fines particles of RCA lower quality in two different density fractions. The first fraction with density >2.1 g/cm^3 is sent to Concrete Aggregate High Quality (RCA3). The second fraction with density <2.1 g/cm^3 from WOC is sent to improve terracing (Fig. 2).

This last stage is yet dedicated to increase the quality of the products considered inferior, aiming at the production of top-sands for high quality concrete and earthworks. The advantages are a best concrete, or more extended use of recycled material or less cement consumption, or maybe better road aggregates. The reduction of sorting in the current process can save costs. It means reduction of gypsum still present in the mixture, by the replacement of picking belts. The disadvantages are the costs that will be discussed later.

The efficiency of air jigs are lower than the same type of process using water (water jigs). Air jigs have larger feeding rate as a function of the equipment size, due to the fact that the particles settle in air faster than in water [19]. It makes the investment and operational costs lower. The jig can be used primarily for the removal of part of the light particles (density <2.1 g/cm^3), as a rougher stage. The pre-concentrate (density >2.1 g/cm^3) will be treated in the sensor-based sorting. In this way, many of the low density particles will be removed in the jig and the final concentrate will be quite pure. A major disadvantage of using air jigs for fine particles treatment is its low efficiency. Particles below 2 or 3 mm are treated with very low efficiency in air jigs. In these sizes, wet processes are always recommended.

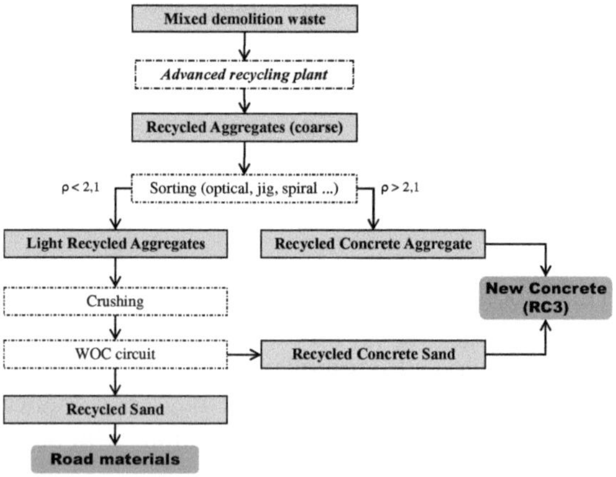

Fig. 2 Scheme of recycling production of high quality recycled concrete production [16]

All the particles will be comminuted under 4 mm (size of sands). In this last stage, a circuit of water-only cyclones was chosen, due to, besides the size of the particles to be treated, its versatility in gravity concentration. Investment costs of a WOC circuit are very low. When they are used in series of 2 or 3 equipment, they present quite good cut efficiencies.

Method of Comparison

Method to Fix the Substitution Rate for a Given Concrete Application

A standard formula is proposed (for example: Aggregates = 1000 kg, Sand = 900 kg, Cement = 280 kg and Water = 180 kg), without recycled aggregate, to allow an identical result at the product RCA3. The products are based on a typical concrete formulation and the density of a mix concrete can be calculated according its proportions and densities in the mixture:

$$D = \frac{FA}{100} \times \left(\frac{subst_{FRA} \times d_{FRA} + (100 - subst_{FRA}) \times d_{FNA}}{100} \right) + \frac{CA}{100} \times \left(\frac{subst_{CRA} \times d_{CRA} + (100 - subst_{CRA}) \times d_{CNA}}{100} \right)$$

where: D—Weighed density of concrete mixture; FA—Percentage or fine aggregates used in the mixture; CA—Percentage or coarse aggregates used in the mixture; substFRA—Replacement level (%) of fine natural aggregates with fine recycled aggregates; substCRA—Replacement level (%) of coarse natural aggregates with coarse recycled aggregates; dFRA—density of fine recycled aggregates; dFNA—density of fine natural aggregates; dCRA—density of coarse recycled aggregates; dCNA—density of coarse natural aggregates.

To solve the above equation, it is necessary to know the water absorption of gravel and sand recycled. These values are determined taking into account the average density of the aggregates.

Price of Transportation and Taxes

It is important to determine transportation costs of aggregates or recycled materials mainly of the construction and demolition works. The price of transportation Pr can be calculated in function of distance, as indicated:

$$\Pr = \left(VC \times d + FC \times \frac{(\frac{d}{s} + t)}{10} \right) \times (1 + m)/w$$

where: Pr = price per tonnes (€/tons); VC = variable costs (€/tons); FC = fixed costs (€/day); d = distance (km); w = weight per truck (tons) = 20 tons fixed; t = time to charge and discharge (h); s = speed (km/h); m = profit margin.

The transport costs by trucks can be related with the transport of virgin aggregates to construction sites or with the transport of CDW to recycling plants. They always have linearity with distance travelled. These results were obtained considering truck speed of 63 km/h, time to charge and discharge of 2 h, fixed costs of 156.01 €/day, variable costs of 0.43 €/km and business's margin of 0.33.

Results and Discussions

Optimal Substitution Rate

The estimation of water absorption of the recycled aggregates from sorting plants, according to the rate of replaced aggregates is presented in Fig. 3. The average absorption and density of natural aggregates were considered as 0.8% and 2650 kg/m^3, respectively. This figure shows clearly that the treatment improves a lot of quality on recycled aggregates changing water absorption (and densities) of the products easily. According to the plant described (Fig. 2), the water absorption of RCA1 was 11.7, 7.0% of RCA2 and 3.6% of RCA3. Densities were 1900, 2085, 2210 and 2650 kg/m^3, respectively.

Gain in Density and Reduction in Water Absorption

Figure 4 shows the possibility of quality increasing of recycled aggregates by the use of a recycling plant. The probable gain on density or reduction on water absorption is linked with the percentages of replaced aggregates. If 50% of RCA1

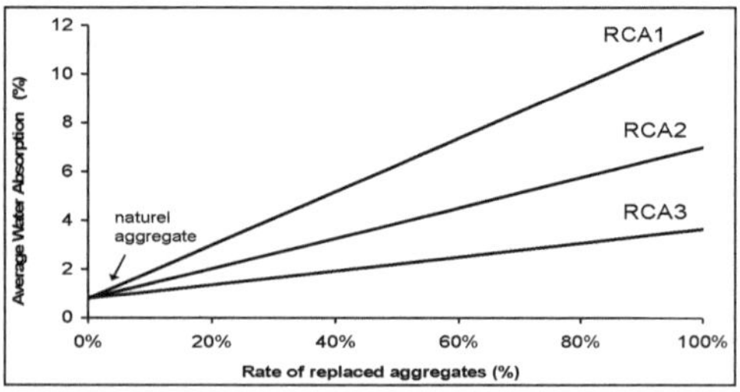

Fig. 3 Estimation of water absorption of recycled aggregates RCA1, RCA2 and RCA3 obtained from treatment according the rate of replaced aggregates

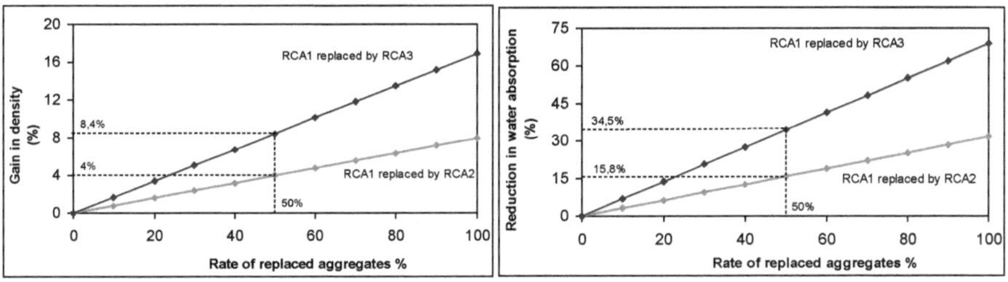

a) Gain in density b) Reduction in water absorption

Fig. 4 The potential gain in density and the reduction in water absorption, linked with rates of replaced aggregates

were replaced by RCA2, the expected gain in density will be 4% (Fig. 4a). On the other hand, if RCA1 were replaced by RCA3, the expected gain will be 8.4%. The same situation concerning reduction on water absorption is showed in Fig. 4b. The levels of quality described in this analysis (i.e. levels of density and water absorption) can be assured by fine adjustment of gravity concentration equipment mainly changing the cut-off grade. The mass balance and the metallurgical balance should be carried out after any adjustments.

By the methodology proposed, the real possibility of gain in quality can be measured by the economic advantages obtained using the processing plant. In this context, the choice of improve or not improve sorting is related with the costs balance including transport, distances and indirect costs.

Conclusions

The improvement of recycled concrete is the main subject of this paper. As an alternative to the heat treatment, the proposition is increase quality on recycled aggregates by supplementary use of mineral processing techniques.

A sorting plant is proposed. Fine and coarse aggregates are treated by gravity concentration processes, which are widely used in mineral processing. During all stages of sorting process, it is possible to adjust the quality of recycled aggregates, i.e. density and water absorption. An approximate graphic method has been used to estimate mixed aggregates proportion. The main advantages of recycling plants are the superior quality of the materials. It is possible to increase the rate of recycling in concrete and save costs. The main disadvantages are the investments in sorting equipment. Their maintenance was not evaluated in this work. However, the gain in density and the reduction in water absorption were studied. In the study, replacing a lower quality aggregate by another with medium quality, for instance RCA1 replaced by RCA2, leads an expected gain in density about 4%. On the other hand, if RCA1 is replaced by RCA3 (superior quality), the expected gain will reach 8.4%. As consequence, there are 34% reductions in water absorption.

Acknowledgements The Coordination for the Improvement of Higher Education Personnel (CAPES) and the CNPq from Brazil, Federal University of Pampa (UNIPAMPA), Federal University of Rio Grande do Sul State (UFRGS) and the French Institute of Science and Technology for Transport, Development and Networks (IFSTTAR) are acknowledged.

References

1. Akbarnezhad, A., Ong, K. C. G., Zhang, M. H., Tam, C. T., & Foo, T. W. J. (2010). Microwave-assisted beneficiation of recycled concrete aggregates. *Construction and Building Materials, 25,* 3469–3479.
2. Agricola, G. (1556). *De Re Metallica* (H. C. Hoover & L. H. Hoover, trans.). New York: Dover Publications (1950).
3. Ângulo, S. C., John, V. M., Carrijo, P. M., Figueiredo, A. D., & Chaves, A. P. (2010). On the classification of mixed construction and demolition waste aggregate by porosity and its impact on the mechanical performance of concrete. *Materials and Structures, 43,* 519–528. doi:10.1617/s11527-009-9508-9
4. Ângulo, S. C., John, V. M., Ulsen, C., Kahn, H., & Mueller, A. (2013). Separação óptica do material cerâmico dos agregados mistos de residuos de construção e demolição. Revista Ambiente construido, Porto Alegre. ISSN 1678-8621.
5. Cazacliu, B., Sampaio, C. H., Miltzarek, G., Petter, C., Le Guen, L., Paranhos, R., et al. (2013). The potential of using air jigging to sort recycled aggregates. *Journal of Cleaner Production, 66,* 46–53.
6. Coelho, A., & de Brito, J. (2013). Economic viability analysis of a construction and demolition recycling plant in Portugal—Part 1: Location, materials, technology and economic analysis. *Journal of Cleaner Production, 39,* 338–352.
7. Dehler, M., & Robben, M. (2012). *NIR versus color sorting of industrial minerals. Sensor based sorting*. Aachen, Germany.
8. Ergün, L. Ş., Gülsoy, Ö. Y., & Gülcan, E. (2012). *Optical sorting of iron and chromite ores. Sensor based sorting*. Aachen, Germany.
9. Gomez-Soberon, J. M. V. (2002). Porosity of recycled concrete with substitution of recycled concrete aggregate. An experimental study. *Cement and Concrete Research, 32,* 1301–1311.
10. Gülsoy, Ö. Y., Ergün, L. Ş., & Gülcan, E. (2012). *Optical sorting of low rank coals—A subsidiary study. Sensor based sorting*. Aachen, Germany.
11. Lyman, G. J. (1992). Review of jigging principles and control. *Coal Preparation, 11,* 145–165.
12. Marie, I., & Quiasrawi, H. (2012). Closed-loop recycling of recycled concrete aggregates. *Journal of Cleaner Production, 37,* 243–248.
13. Nienhaus, K., Pretz, T., & Wotruba, H. (2014). *Sensor technologies: Impulses for the raw materials Industry*. Aachen: Shaker Verlag GmbH. ISBN 978-3-8440-2563-7.
14. O'Brien, E. J. (1976). Water-only cyclones: Their functions and performance. In *Coal age* (pp. 110–114), January.
15. Oikonomou, N. D. (2005). Recycled concrete aggregates. *Cement & Concrete Composites, 27,* 315–318.
16. Paranhos, R. S., Cazacliu, B. G., Sampaio, C. H., Petter, C. O., Neto, R. O., & Huchet, F. (2015). A new methodology to value recycled concrete. *Journal of Cleaner Production.*
17. Raoa, A., Jhab, K., & Misraa, S. (2007). Use of aggregates from recycled construction and demolition waste in concrete. *Resources, Conservation and Recycling, 50,* 71–81.
18. Raulf, K., Pretz, T., & Wotruba, H. (2012). *Potential of sensor technologies in the raw materials of industry. Sensor based sorting*. Aachen, Germany.

19. Sampaio, C. H., & Tavares, L. M. M. (2005). Beneficiamento gravimétrico. Uma introdução aos processos de concentração mineral e reciclagem de materiais por densidade. Editora da Ufrgs.
20. Silva, R. V., Brito, J., & Dhir, R. K. (2014). Properties and composition of recycled aggregates from construction and demolition waste suitable for concrete production. *Construction and Building Materials, 65,* 201–217.
21. Ulsen, C., Kahn, H., Hawlitschek, G., Masini, E., & Angulo, S. (2013). Separability studies of construction and demolition waste recycled sand. *Waste Management, 33,* 656–662.
22. Visman, J. (1962). *Die Sortierung abriebempfindlicher Kohle im Hydrozyklon, proceedings* (pp. 161–170). Harrogate, UK: IV International Coal Preparation Congress.
23. Weyher, L. H. E., & Lovell, H. L. (1969). Hydrocyclone washing of fine coal. *Transactions AIME, 244,* 191–203.
24. Wills, B. A., & Napier Munn, T. (2006). *Mineral processing technology. An introduction to the practical aspects of ore treatment and mineral.* Amsterdam: Elsevier Science & Technology Books.

Performance of Hydraulic Mixes Manufactured with Electric Arc Furnace Slag Aggregates

Amaia Santamaría, Vanesa Ortega-López, Marta Skaf, Ignacio Marcos, José-Tomás San-José and Javier J. González

Abstract Electric arc furnace slag (EAFS) has for many years simply been dumped in landfill sites; over the past few decades many researchers have investigated its reuse in cement mortar and concrete. By doing so, a waste product may be converted into a useful material with added value as a substitute for natural resources, the consumption of which is also minimized. Hydraulic mixes manufactured with EAFS normally have similar or even better hardened properties than mixes manufactured with natural aggregates. One disadvantage in the use of EAFS has been the poorer workability of the mixes, due to its higher density, porosity and water absorption levels. In this research, different EAFS mixes are manufactured and their properties in the fresh and the hardened state are closely analyzed; the results were very promising. The aim of this research is to demonstrate that EAFS concrete can achieve an acceptable workability at the correct dosages.

Keywords Electric arc furnace slag · Self-compacting mixes · Workability · Strength

---

A. Santamaría (✉) · J.-T. San-José · J.J. González
Department of Mining and Metallurgical Engineering,
UPV/EHU, Alameda Urquijo S/N, 48013 Bilbao, Spain
e-mail: amaia.santamaria@ehu.eus

V. Ortega-López
Department of Civil Engineering, University of Burgos,
Calle Villadiego S/N, 09001 Burgos, Spain

M. Skaf
Department of Construction, University of Burgos,
Calle Villadiego S/N, 09001 Burgos, Spain

I. Marcos
Department of Mechanical Engineering, UPV/EHU,
Calle Rafael Moreno "Pitxitxi" N°2, 48013 Bilbao, Spain

© The Minerals, Metals & Materials Society 2017
M.A. Meyers et al. (eds.), *Proceedings of the 3rd Pan American Materials Congress*,
The Minerals, Metals & Materials Series, DOI 10.1007/978-3-319-52132-9_22

Introduction

The steelmaking industry has traditionally represented one of the most important economic activities in the Basque Country (a small region to the north of Spain, covering 1.5% of Spanish territory). Its annual levels of steel production currently stand at around 4 million tons (27% of all steel manufactured in Spain). Since 1986, in the Basque Country, all steel manufacturing is done in electric arc furnaces, generating voluminous amounts of electric arc furnace slag (EAFS). In the past, these slags have been disposed of in dumping sites, although over recent decades, different global research groups have been studying ways of using them as raw materials in the construction sector. Their stable chemical composition based on calcium silicate and iron oxides and their good resistance to abrasion and fragmentation make them particularly suitable for use in hydraulic mixes.

It has been demonstrated [1–4] that concrete manufactured with this type of slag as an aggregate has at least the same mechanical and durability properties as concrete made with natural aggregates, and in some cases it improves them. The main problems for the use of this type of slag are its higher density, which could be an advantage in some applications, and the poorer workability of the mixes, caused by its irregular shapes, higher porosity and outstanding water absorption. This drawback is certainly one of various reasons that explain why this type of slag has yet to be adopted all over the world and is still found in many places at dumping sites.

One of the objectives of this research is to demonstrate that suitable mixes with the required workability may be obtained with the correct mix design. In the early 90s, Okamura presented a self-compacting or self-consolidating [5] concrete, a mix that can fill every corner of formwork spreading purely under its own weight, with no need for vibrating compaction, and with no segregation of coarse aggregates. A similar performance for the mixes in this study is the ambitious challenge set for this research project. The first step was to manufacture self-compacting mortars to evaluate the fines content, water-cement ratio and effects of the admixtures. Following the encouraging results of this first campaign, self-compacting concrete mixes were successfully manufactured, which complied with the initial objectives of this research.

Materials

Two types of cement were used in the present research. In some of the mixtures, a Portland cement type IV/B-V 32.5-N, in others, a Portland cement type I 42.5 R, in accordance with UNE-EN 197-1 standard. Water that contains no harmful compounds that might affect the quality of the hydraulic mixes was taken from the urban mains supply of the city of Bilbao.

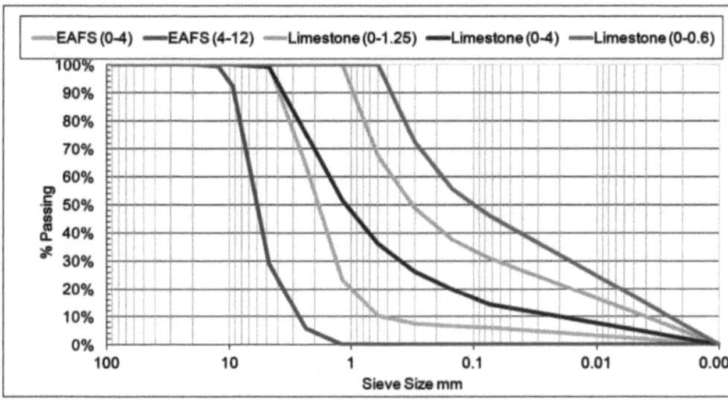

Fig. 1 Grading of aggregates

A commercial crushed natural limestone aggregate, of different sizes, whose grading are shown in Fig. 1, was employed. Its main mineral was calcite (95 wt%) the specific gravity of which was 2.67 Mg/m^3.

Two different sizes of crushed and aged EAFS were employed, as shown is Fig. 1. The main chemical compounds of the EAFS used in this study were Fe_2O_3, CaO, SiO_2 Al_2O_3, with small proportions of MgO MnO, SO_3, Cr_2O_3 P_2O_5 and TiO_2. Its main crystalline components were Wüstite, Ghelenite and Kirsteinite. With regard to its physical properties, it has a water absorption rate of 1.2%, and a specific gravity of 3.42 Mg/m^3.

Two different admixtures (superplasticizer and air entrainment) were added to enhance workability and to decrease the density of the mixtures incorporating heavy EAFS aggregate.

Mortars

Mix Design

Four different mortar mixes were designed to study their performance with regard to workability and hardened properties (Table 1). M1 represents the reference mix with no EAFS. M2, M3 and M4 are mixes manufactured with EAFS. An air-entrainment admixture was added to M3, and M4 was manufactured with CEMIV instead CEMI.

A superplasticizer admixture was used to achieve the self-compacting properties in all the mixtures. Furthermore, a limestone fraction, with particle sizes of under 0.6 mm, was used in the mixes manufactured with EAFS, which has to be added to avoid the segregation of the slag. A good paste (water + particles <1 mm) is essential in the manufacture of self-compacting mixes.

Table 1 Mix design

Mix design (kg/m^3)		SM1	SM2	SM3	SM4
CEM I 42.5 R		551	551	551	
CEM IV/B-V 32.5 N					551
Limestone	Small-size aggregate <5 mm	1584			
	Fine fraction <0.6 mm		474	474	474
EAFS 0–4.5 mm			1259	1259	1259
Water/binder ratio		0.4	0.4	0.4	0.4
Superplasticizer (%)		1.5	1.5	1.6	1.5
Air entraining admixture				0.2%	

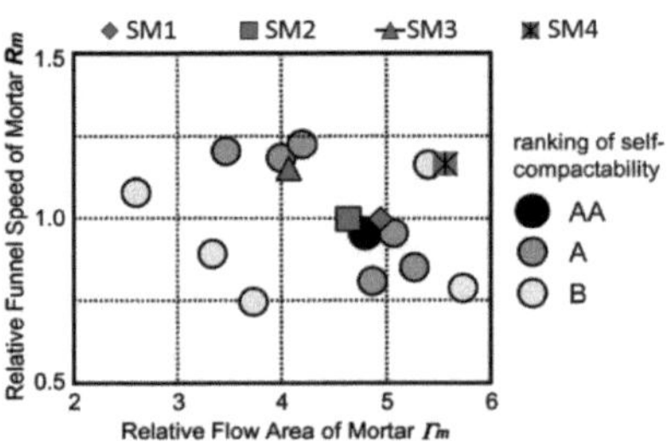

Fig. 2 Ranking of self-compactability

Fresh Properties

The fresh properties were evaluated using the ranking of self-compactability proposed by Okamura [6], as shown in Fig. 2. The required values, Γ_m (Eq. 1) and R_m (Eq. 2), were obtained from the results of the mini-slump cone test and mini-V funnel test [7]:

$$\Gamma_m = \left(d_1 d_2 - d_0^2\right)/d_0^2 \quad (1)$$

where d_1, d_2 represent the after-flow diameter in two orthogonal directions; and d_0 is the diameter of the cone:

$$R_m = 10/t \quad (2)$$

where t (s) is the measured flow-time of the mortar through the funnel.

The workability of SM1 and SM2 was almost perfect. In spite of the air entrainment admixture causing a loss of workability in SM3 the results were really very good. The mix with the worse workability was SM4. Is recommendable to use less Superplasticizer in mixes manufactured with this type of cement, nevertheless the results were also good.

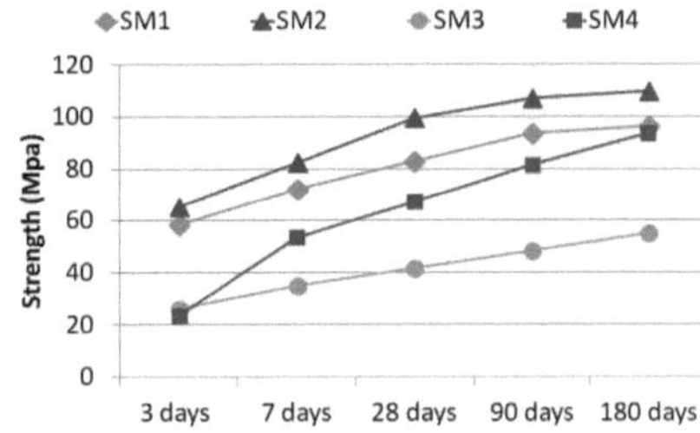

Fig. 3 Compressive strength

Table 2 MIP and CAT results

Property	SM1	SM2	SM3	SM4
Dry density (Mg/m^3)	2.37	2.68	2.24	2.65
MIP bulk density (Mg/m^3)	2.38	2.70	2.37	2.69
MIP app density (Mg/m^3)	2.52	2.91	2.90	2.91
MIP porosity (% vol)	5.3	6.9	18.3	7.2
CAT porosity (% vol)	0.5	1.6	6.2	2
CAT matrix (% vol)	99.5	54.4	53.2	57.6
CAT EAFS (% vol)	–	43.4	40.2	40.1
CAT metallic iron (% vol)	–	0.5	0.3	0.3

Strength

The result of compressive strength at 3, 7, 28, 90 and 180 days are given in Fig. 3. The results showed strength improvements with the use of EAFS as an aggregate. The effect of the air entrainment admixture is detrimental at any age, causing severe loss of strength that cannot compensate the gain in weight. The good effect of fly ash was as expected in the long term results. After 180 days some strength levels reached the significant value of 100 Mpa.

Density, MIP and CAT Analysis

The results of density, MIP (mercury intrusion porosimetry) and CAT (computerized axial tomography) analysis are given in Table 2. It may be seen that the average density increased by 11% in the specimens with EAFS as an aggregate. The air entrainment admixture reduced this value to 20%.

The capillarity porosity increased with the use of EAFS due to the porosity of the aggregates. The use of an air-entrainment admixture increased the occluded air and also changed the pore-distribution size. The use of X-ray mapping CAT distinguished between the different kinds of materials and particles that compose the sample according to their density; dark grey in the regions of lower density and light grey in the regions of higher density. In the case of our mixes, following image

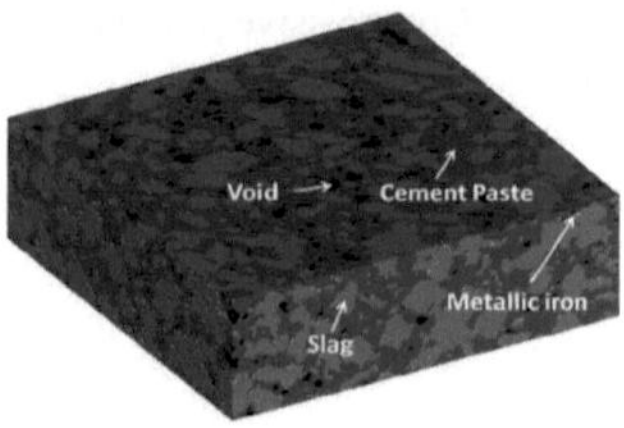

Fig. 4 CAT 3D image

analysis and treatment, it was possible to observe (Fig. 4) the metallic iron (almost white), the EAFS (clear), the cementitious matrix (dark), and the air (black). The results show the higher porosity of SM3 and the metallic iron content in the range 0.3–0.5% in volume, a little bit high for structural applications.

Durability

A durability test based on Spanish standard NLT-361was performed on the mortar mixes to verify their volumetric stability. In this case, the specimens were placed in an autoclave for 48 h, at 0.2 MPa and 130 °C. The results were satisfactory; the samples showed no signs of damage after the test. As expected, the possibility of aggregate expansion, due to their eventual content of expansive compounds, was remote after the suitable treatment of spontaneous aging of EAFS.

Concrete

Following the successful results with the mortars, several attempts were made to manufacture self-compacting concretes, using the method proposed by other authors for natural aggregates, with no success. Instead of the fine limestone used in the mortars, the manufacture of the self-compacting concrete needed fine aggregate with a maximum size of 1.25 mm (Fig. 1). Using this grading, two different self-compacting mixes were prepared with a granulometry shown in Fig. 5 and the mix design shown in Table 3. The evaluation of self-compactability was done with the Abrams cone test and the L-box test. The results are also shown in Table 3.

The self-compacting concrete that was manufactured had a slump flow class SF2, in accordance with the EFNARC [7] and a passing ability of PA1. A visual inspection during the slump flow test showed no segregation of the aggregates.

In the hardened state, the compressive strength of the concrete was evaluated at 7, 28 and 90 days (Fig. 6). As expected, the results of SCCI were higher, reaching a value of 66 MPa at 90 days. The SCCIV reached a value of 36 MPa at the same age. Both of these were good results, when taking into account that the first mix used cement I 42.5 R and the second used cement IV 32.5 N. All of these tests results with the self-compacting concrete are very encouraging.

Fig. 5 Grading of the mixtures

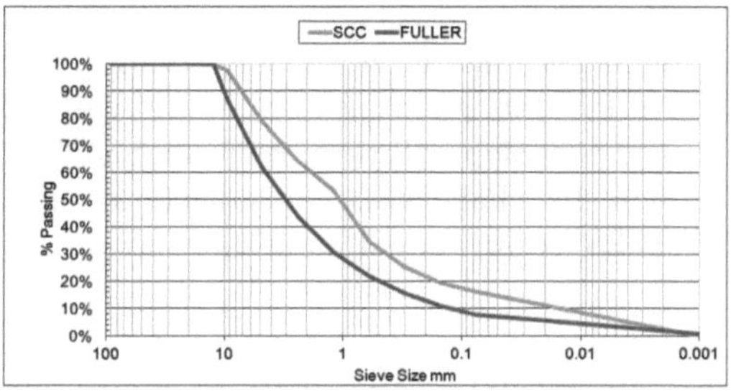

Table 3 Concrete mixes and properties

Mix design (kg/m^3)	SCCI	SCCIV
CEM I 42.5 R	350	
CEM IV/B-V 32.5 N		350
Limestone 0–1.25 mm	950	950
EAFS 0–4.5	575	575
EAFS 4.5–12.5 mm	725	725
Water/binder ratio	0.5	0.5
Superplasticizer (%)	2	2
Slump (mm)	680	700
L-box	0.9	0.9
Density (Mg/m^3)	2.66	2.60

Fig. 6 Compressive strength

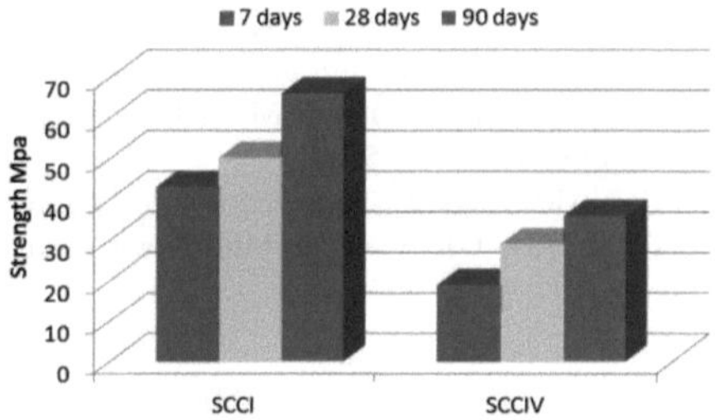

Conclusions

- Self-compacting structural mortar mixes have successfully been designed and manufactured using EAFS as coarse aggregate; careful control of the fine fraction is strongly recommended.
- The global results in terms of mechanical strength are promising in all the mixes. This effect is a consequence of a suitable mortar microstructure, as revealed in the MIP and CAT analyses.
- In addition to the EAFS aggregate, the simultaneous use of fly ash, yielded acceptable results in general, showing good compatibility, and economizing on total consumption of Portland clinker.

- The durability tests performed on the mortars were successful, showing that the aggregates had no non-stable volumetric compounds.
- All the tests performed on the self-compacting concrete are very encouraging. The next step will be to perform durability tests and if the results are positive, then the challenge of manufacturing reliable self-compacting concrete with EAFS aggregate will in all likelihood have been reached.

Acknowledgements The authors wish to express their gratitude to the Basque Regional Government (IT781-13 Research Group), to the Vice-Rectorate of Investigation of the University of the Basque Country (UPV/EHU) for grant PIF 2013, and to the Spanish Ministry (MINECO) for FEDER Funds through project BIA2014-55576-C2-2-R, which all contributed to financing this research work.

References

1. Manso, J. M., Gonzalez, J. J., & Polanco, J. A. (2004). Electric arc furnace slag in concrete. *Journal of Materials in Civil Engineering, 16*(6), 639–645.
2. San-José, J. T., Vegas, I., Arribas, I., & Marcos, I. (2014). The performance of steel-making slag concretes in the hardened state. *Materials and Design, 60,* 612–619.
3. Arribas, I., Vegas, I., San-José, J. T., & Manso, J. M. (2014). Durability studies on steelmaking slag concretes. *Materials and Design, 63,* 168–176.
4. Pellegrino, C., Cavagnis, P., Faleschini, F., & Brunelli, K. (2013). Properties of concretes with black/oxidizing electric arc furnace slag aggregate. *Cement & Concrete Composites, 37,* 232–240.
5. Okamura, H., & Ouchi, M. (1998). Self-compacting high performance concrete. *Progress in Structural Engineering and Materials, 1*(4), 378–383.
6. Okamura, H., & Ouchi, M. (2003). Self-compacting concrete. *Journal of advanced concrete technology., 1*(1), 5–15.
7. EFNARC. (2002). *Specification and guidelines for self-compacting concrete*. Farnham, Surrey GU9 7EN, UK, website: http://www.efnarc.org/. ISBN 953973344.

Porous Asphalt Mixtures with 100% Siderurgic Aggregates

Marta Skaf, Vanesa Ortega-López, Ángel Aragón, José T. San-José and Javier J. González

Abstract In this research, the possibility of making a porous asphalt mixture manufactured completely with recycled aggregates from carbon steel production was explored. Electric arc furnace slag (EAFS) was used as coarse aggregate and ladle furnace slag (LFS) as fine aggregate and filler. Initially, the properties of both slags and their suitability to be used in the manufacture of porous mixtures were analyzed. Then, a series of asphalt mixtures were developed incorporating these slags and they were compared with a reference mixture, made with conventional components. A series of tests were performed, including concepts such as mechanical behavior, durability, moisture susceptibility, rutting resistance, permeability or skid resistance. The results show that it is possible to make a suitable porous asphalt mixture with 100% of steel slag aggregates, complying with the standard requirements and obtaining a durable and environmentally sustainable mixture.

Keywords Steelmaking slag · Ladle furnace slag · Electric arc furnace slag · Porous asphalt · Waste management

---

M. Skaf (✉)
Department of Construction, EPS, University of Burgos,
Calle Villadiego S/N, 09001 Burgos, Spain
e-mail: mskaf@ubu.es

V. Ortega-López · Á. Aragón
Department of Civil Engineering, EPS, University of Burgos, Burgos, Spain
e-mail: vortega@ubu.es

Á. Aragón
e-mail: aragont@ubu.es

J.T. San-José · J.J. González
Department of Engineering of Materials, UPV/EHU, Leioa, Spain
e-mail: josetomas.sanjose@ehu.eus

J.J. González
e-mail: javierjesus.gonzalez@ehu.eus

© The Minerals, Metals & Materials Society 2017

M.A. Meyers et al. (eds.), *Proceedings of the 3rd Pan American Materials Congress*,
The Minerals, Metals & Materials Series, DOI 10.1007/978-3-319-52132-9_23

Introduction

Steel slag is one of the main by-products of the metallurgical industry. Currently, steel production is basically sub-divided into two processes: integral siderurgy and the electric cycle. The electric cycle is the most prevalent in Spain, in carbon steel production, and mainly involves melting recycled scrap into an Electric Arc Furnace (EAF), and then refining the steel through a Ladle Furnace (LF). This process generates over one million tons of slag per year in Spain and over 10 million tons in Europe [1, 2].

Some **uses** have already been found for the slags generated in the electric siderurgy:

- **EAF slag** (oxidizing slag, black slag) has a longer history of research and execution of construction materials and other industrial activities, as quality aggregate:
  - It has traditionally been used as **unbound** material in preparing subgrades, subbases and bases for beds, embankments of roads and railways [3–6]. This slag has good features as coarse aggregate for granular layers due to its wear resistance, hardness and angularity. The main precautions relate to the leaching and swelling of the material.
  - In roads, after an appropriate prior treatment, EAF slag provides a quality aggregate for **bituminous mixtures** manufacturing [6–9]. Its high Polished Stone Value (PSV) and reduced Los Angeles coefficient (LA) makes it a valuable material even for wearing courses.
  - There is a very prolific line of research on the use of EAFS in the manufacture of hydraulic **concrete** as coarse and fine aggregate [10–13]. These concretes show good compressive strength, tensile strength and durability with similar values to conventional materials.
- The reuse of the LFS (basic slag, reducing slag, white slag), the by-product from secondary metallurgy processes, is less widespread:
  - Varying amounts of LFS are usually reintroduced into the steel production process, which is reported to produce beneficial effects on the steels produced and in the black slag that is generated, as well as a reduction in production costs [14].
  - One of the main properties of LFS is its hydraulicity, resulting from its chemical composition, which provides it with cementitious properties [15, 16]. Based on this, one modern-day application for LFS would be as an active or inert addition in the preparation of Portland cement clinker [17, 18]. In fact, cement production is the only use of LFS that is currently approved in Spanish regulations [2].

Although there are several applications of these materials, there is still an important excess of both slags, and a significant amount of them is dumped at landfill sites close to production center, with its consequent environmental and visual impacts. This prompts a search for new alternatives to reduce this volume of waste and undesired landfilling.

Furthermore, road construction requires various different materials; among these materials, bituminous mixes are mainly composed of **aggregates**, traditionally extracted from quarries and gravel pits. Along with the exploitation of limited natural resources, mining, crushing, sieving, washing and transporting natural aggregates, expend significant amounts of energy. Global consumption of natural aggregates is estimated to exceed 30,000 million tons/year.

Porous Asphalt Mixes (PA), also known as Permeable Friction Courses (PFC) are special types of hot bituminous mixtures that are usually used as a thin surface pavement courses. Porous asphalt has a coarse granular skeleton that develops stone-on-stone contact, and a high content of connected air voids, meaning that these mixtures have good drainage properties [19].

These properties provide a mixture with superior performance in terms of safety in wet weather driving, owing to the reduction of splash and spray, the risk of hydroplaning and wet skidding and they also improve the visibility of pavement markings in wet weather [19]. In addition to this, they contribute to noise abatement, reportedly between 4 and 6 dB(A) when compared to a concrete pavement or dense-graded asphalt concrete [20, 21], providing a more comfortable driving.

This research explored the potential use of the electric steelmaking slags to replace the natural aggregates in porous asphalt mixes. EAFS was used as coarse aggregate, and LFS as sand and filler.

Materials

Asphalt mixes are composed of a combination of coarse aggregates (16/2 mm), fine aggregates (2/0.063 mm), filler (<0.063 mm), and binder. The following materials were used in this research:

- CONTROL MIXES: A natural **siliceous aggregate** from a nearby quarry was used as coarse and fine aggregate. Ordinary Portland **cement**, CEM I/42.5 R was used as filler.
- SLAG MIXES: Electric Arc Furnace Slag (**EAFS**) was used as coarse aggregate and Ladle furnace slag (**LFS**) was used as fines and filler. As can be observed in Fig. 1, the LFS used in this investigation is a grayish-white powdery material, with a particle size of 0/2 mm. On the other side, the EAFS is a described as a coarse blackish gray aggregate, with small inclusions of metallic particles. Their physical properties and chemical composition are detailed in Tables 1 and 2, respectively.
- Every specimen was manufactured using a Polymer Modified **Bitumen**, PMB 45/80–60 according to EN 14023 [22].

Fig. 1 EAFS (*left*), LFS (*right*)

Table 1 Physical properties of the siliceous aggregate and the slags

Feature	Standard	Siliceous aggregate	EAFS	LFS
Bulk density	EN 1097-6	2.74 g/cm^3	3.60 g/cm^3	2.83 g/cm^3
Fineness modulus	EN 933-1	2.9	–	4.2
Blaine specific surface	EN 196-6	–	–	2654–3091 cm^2/g
Sand equivalent	EN 933-8	78%	98%	50%
Water absorption	EN 1097-6	1.5%	2.1%	–
Los Angeles coefficient	EN 1097-2	20%	23%	–
Polished stone value (PSV)	EN 1097-8	52%	56%	–
Flakiness index	EN 933-3	18%	3%	–
Crushability index	EN 933-5	100%	100%	–
Plasticity	UNE 103103/ UNE 103104	Non plastic	Non plastic	Non plastic

Table 2 Main chemical composition of the EAFS and LFS used

Component	CaO	SiO_2	MgO	Al_2O_3	Fe_2O_3	MnO
EAFS wt%	27.7	19.1	2.5	13.7	26.8	5.3
LFS wt%	56.7	17.7	9.6	6.6	2.2	0.3

Mix Design

As previously described, two types of mixes were designed: the control mix, with standard components, and the slag containing mix. The particle size distribution of both mixtures was chosen for the grading envelope named PA-11 in the Spanish Standard PG-3 [23]. It is a porous asphalt mix, with a nominal maximum size of 11 mm and a thick mineral skeleton, with a large void ratio (>20%).

In a preliminary phase of the research, a series of initial tests were established to choose the optimum bitumen content (OBC). The choice of OBC was taken on the basis of the results of two tests: the Cantabro test, which provides information on minimum bitumen content, and the binder drainage test, which limits the maximum content.

Samples containing 5% of bitumen proved to be balanced in durability, strength and potential drain down. The final particle size distribution consisted in an 86.5% of coarse aggregates, an 8.2% of fine aggregates and a 5.3% of filler, to maintain the ratio filler/asphalt = 1.

Testing and Results

Volumetric properties: Air void content of the specimen was determined according to EN 12697-8 [22]. The average AVC of the slag specimens was 24.3% whereas the control mixes had an average AVC of 21.1%. This agrees with several references to poor workability and compactibility of mixtures with high percentages of steel slag aggregate, due to its angularity, which results in mixtures with higher air void content [6]. On the other hand, despite being more porous, the slag mixes were heavier, due to the elevated density of the EAF slag.

Permeability is one of the most important characteristics of the porous asphalt mixtures, since it is one of their main advantages over other mixtures. Permeability coefficients (K) of the porous mixtures were assessed using the constant head permeameter in a vertical permeability test described in EN 12697-19 [22]. Slag mixes were found to be more permeable than the control mixes, as the permeability is closely related to the void content [24]. They showed an optimal draining performance, over 10^{-1} cm/s [25].

Abrasion loss was tested by the Cantabro test, in which each Marshall specimen is placed inside the Los Angeles abrasion machine without steel balls and then, the drum is operated for 300 revolutions (Fig. 2). This evaluates the most characteristic property of porous asphalt, which is wear resistance, as the main sing of a correct design, strong cohesion and appropriate mechanical behavior.

Abrasion resistance of the slag mixtures showed to be worse than the control samples. This could be due to their higher air void ratio [26]. Anyway, both complied with the standard requirements for the most demanding uses (<20% PL) [23].

Fig. 2 Two porous asphalt specimen, before and after the Cantabro test

Durability tests consisted in comparing the mechanical behavior of some fresh samples with others after an aging process. The chosen procedure was described in ASTM D-7064 [27], in which aging was achieved by conditioning the specimens in a forced draft oven at 60 °C for 7 days.

Long term performance of the slag samples resulted a bit worse than those made with conventional components, but both exceeded by far the standard requirements, ASTM D-7064 <30% Particle Loss.

Moisture susceptibility: The resistance to moisture damage of the mixes was evaluated by the retained tensile strength or tensile strength ratio (TSR) as specified by EN 12697-12 [22]. Six Marshall specimen were divided into two groups: the control subset, which remains dry at room temperature, and the conditioned subset, which is saturated and submerged in hot water (40 °C) for about 72 h.

The performance of both mixes was similar. Slag mixes performed slightly better than control mixes, which is consistent with other investigations that have studied water susceptibility in bituminous mixtures with steelmaking slags [28, 29]. The rougher texture of the EAF slag and the better affinity with the binder make a more cohesive mixture.

Pavement skid resistance was measured by the British pendulum (TRRL) according to EN 13036-4 [22] on slab specimens. This quality depends on the pavement microtexture and its ability to resist the effect of polishing made by traffic and it is directly related to the amount of accidents in rainy weather.

Skid resistance was enhanced on pavements with slags. The results, both fresh and after an aging treatment, exceeded significantly the control mixture. High Polishing Stone Value (PSV) of steel aggregates ensures optimum performance in this field, and excellent temporal evolution [30].

Resistance to permanent deformations was assessed using the wheel tracking device, according to EN 12697-22 [22]. The repeated loading is performed at 60 °C, by a steel wheel with a solid rubber tire, under a pressure of 70 N, at 53 passes per minute (Fig. 3).

The rutting resistance, was much better in the slag mixtures, in every parameter analyzed. This was attributed to the properties of the coarse aggregate: the high angularity, hardness, shear strength, resistance to wear and polishing of the EAFS particles. These results are consistent with those shown by other researchers, who have studied the effects of the incorporation of slags to porous mixtures [31, 32].

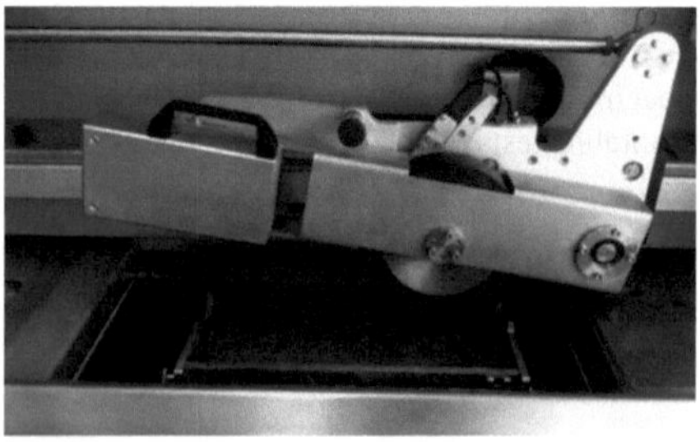

Fig. 3 Wheel-tracking machine used in this research

Conclusions

- The void content in mixes composed only by siderurgic aggregates was higher than the control mixtures. This is due to the superior angularity of the slag, which difficults compaction, resulting in a more porous mixture.
- Mechanical behavior in terms of abrasion loss met the requirements even for heaviest loads. However, the introduction of the slags significantly worsened the conventional results, which is attributable to the higher void content of the slag mixtures.
- Moisture susceptibility improved with the slag aggregates. This is attributed to a better affinity of the binder with the slags, than with the siliceous aggregate. In addition, the rougher texture of slag enhances adhesion.
- Aging produced similar effects on both types of mixtures, far exceeding the standard requirements.
- Resistance to permanent deformation was much better in slag mixtures. This was attributed to the excellent properties of the EAF slag as coarse aggregate: its high angularity, hardness, shear strength, resistance to wear and polishing.
- Skid resistance was optimal for the slag pavements, showing an excellent performance against slipping and skidding. Their rougher texture and higher permeability make them perfectly suitable to rainy regions. Moreover, excellent PSV of EAF slag aggregates ensures excellent long term skidding and slipping performance.

Acknowledgements Our gratitude to the Spanish Ministry of Economy and Competitiveness (MINECO) and FEDER Funds for their financial support through Project BlueCons: BIA2014-55576-C2-1-R.

References

1. EUROSLAG. (2013). *The European slag association.* Position Paper on the Status of Ferrous Slag.
2. CEDEX. (2015). *Catálogo de Residuos utilizables en Construcción.* Available from: http://www.cedexmateriales.vsf.es/view/catalogo.aspx
3. Motz, H., & Geiseler, J. (2001). Products of steel slags an opportunity to save natural resources. *Waste Management, 21*(3), 285–293.
4. Rohde, L., Núñez, W. P., & Ceratti, J. A. P. (2003). *Electric arc furnace steel slag: Base material for low-volume roads* (pp. 201–207). Reno, NV.
5. Behiry, A. E. A. E. M. (2013). Evaluation of steel slag and crushed limestone mixtures as subbase material in flexible pavement. *Ain Shams Engineering Journal, 4*(1), 43–53.
6. FHWA. (1997). *User guidelines for waste and byproduct materials in pavement construction.* US: Federal Highway Administration.
7. Jones, N. C. (2001). The successful use of EAF slag in Asphalt. In *Proceedings of 2nd European Slag Conference.* 2001. Düsselforf: EUROSLAG Eds. Publication n°1. ISSN 1617–5867.

8. Emery, J. J. (1984). Steel slag utilization in asphalt mixes. In *Canadian Technical Asphalt Association Proceedings*.
9. Ali, N. A., Chan, J. S. S., Papagiannakis, T., Theriault, E. G., & Bergan, A. T. (1992). *Use of steel slag in asphaltic concrete*. San Diego, CA, USA: Publ by ASTM.
10. Manso, J. M., Gonzalez, J. J., & Polanco, J. A. (2004). Electric arc furnace slag in concrete. *Journal of Materials in Civil Engineering, 16*(6), 639–645.
11. Papayianni, I., & Anastasiou, E. (2011). Concrete incorporating high calcium fly ash and EAF slag aggregates. *Magazine of Concrete Research, 63*(8), 597–604.
12. Pellegrino, C., & Gaddo, V. (2009). Mechanical and durability characteristics of concrete containing EAF slag as aggregate. *Cement & Concrete Composites, 31*(9), 663–671.
13. Faleschini, F., Alejandro Fernández-Ruíz, M., Zanini, M. A., Brunelli, K., Pellegrino, C., & Hernández-Montes, E. (2015). High performance concrete with electric arc furnace slag as aggregate: Mechanical and durability properties. *Construction and Building Materials, 101*, 113–121.
14. Memoli, F., Mapelli, C., & Guzzon, M. (2007). Recycling of ladle slag in the EAF: A way to improve environmental conditions and reduce variable costs in steel plants. *Iron and Steel Technology, 4*(2), 68–76.
15. Adolfsson, D., Engström, F., Robinson, R., & Björkman, B. (2011). Cementitious phases in ladle slag. *Steel Research International, 82*(4), 398–403.
16. Shi, C. (2002). Characteristics and cementitious properties of ladle slag fines from steel production. *Cement and Concrete Research, 32*(3), 459–462.
17. Richardson, I. G., & Cabrera, J. G. (2000). The nature of C-S-H in model slag cements. *Cement & Concrete Composites, 22*(4), 259–266.
18. Akin Altun, I., & Yilmaz, I. (2002). Study on steel furnace slags with high MgO as additive in Portland cement. *Cement and Concrete Research, 32*(8), 1247–1249.
19. Alvarez, A. E., Martin, A. E., & Estakhri, C. (2011). A review of mix design and evaluation research for permeable friction course mixtures. *Construction and Building Materials, 25*(3), 1159–1166.
20. Ongel, A., Kohler, E., & Harvey, J. (2008). Principal components regression of onboard sound intensity levels. *Journal of Transportation Engineering, 134*(11), 459–466.
21. Kowalski, K. J., McDaniel, R. S., Shah, A., & Olek, J. (2009). Long-term monitoring of noise and frictional properties of three pavements: Dense-graded asphalt, stone matrix asphalt, and porous friction course. *Transportation Research Record,* 12–19.
22. EN Euronorm. European Committee for Standardization: Rue de Stassart, 36. Belgium–1050 Brussels.
23. PG-3 Pliego de Prescripciones Técnicas Generales para Obras de Carreteras y Puentes, PG-3 (General Technical Specifications in Road Construction) Spanish Ministry of Public Works: Madrid.
24. Alvarez, A., Martin, A., & Estakhri, C. (2011). Optimizing the design of permeable friction course mixtures. *Transportation Research Record: Journal of the Transportation Research Board, 2209,* 26–33.
25. Alvarez, A. E., Martin, A. E., & Estakhri, C. (2010). Drainability of permeable friction course mixtures. *Journal of Materials in Civil Engineering, 22*(6), 556–564.
26. Mansour, T. N., & Putman, B. J. (2013). Influence of aggregate gradation on the performance properties of porous asphalt mixtures. *Journal of Materials in Civil Engineering, 25*(2), 281–288.
27. ASTM D 7064. (2004). *Standard practice for open graded friction course (OGFC) mix design*. American Society for Testing and Materials (ASTM). Annual Book of ASTM Standards: West Conshohocken, P.A.
28. Ahmedzade, P., & Sengoz, B. (2009). Evaluation of steel slag coarse aggregate in hot mix asphalt concrete. *Journal of Hazardous Materials, 165*(1–3), 300–305.
29. Kanitpong, K., & Pummarin, K. (2010). Investigation of industrial wastes in hot mix asphalt for moisture damage resistance. *Journal of Solid Waste Technology and Management, 36*(2), 81–90.

30. Li, S., Zhu, K., & Noureldin, S. (2007). Evaluation of friction performance of coarse aggregates and hot-mix asphalt pavements. *Journal of Testing and Evaluation, 35*(6), 571–577.
31. Hainin, M. R., Rusbintardjo, G., Hameed, M. A. S., Hassan, N. A., & Yusoff, N. I. M. (2014). Utilisation of steel slag as an aggregate replacement in porous asphalt mixtures. *Jurnal Teknologi (Sciences and Engineering), 69*(1), 67–73.
32. Wang, Y., & Wang, G. (2011). *Improvement of porous pavement, in Final Report to US Green Building Council*. Greenville, NC: East Carolina University.

Part V
Materials for Oil and Gas Industry

Blends of PVDF with Its Processing Waste: Study of the Mechanical Properties of the Blends Thermally Aged

L.C.M. Cirilo and M.F. Costa

Abstract Offshore oil production is known to demand high performance materials used in equipments and machinery due to severe environmental conditions. Polyvinylidene fluoride (PVDF) has been used as the internal pressure sheath layer in unbonded flexible pipe for oil and gas exploitation when HT/HP conditions are found since it is resistance to most chemicals encountered in such operations, to moisture and high thermal resistant. An increase in the use of PVDF in such installations has been observed and hence an increase in the waste generated. A possible solution to this environmental issue would be the re-use of the recycled PVDF by reprocessing primary PVDF waste together with the neat one. Therefore, $PVDF_{neat}/PVDF_{waste}$ blends loss and storage modulus were evaluated as well as the influence of aging time period and waste composition in these properties, before and after thermal aging.

Keywords PVDF · Thermal aging · PVDF waste · PVDF recycling

Introduction

The polyvinylidene fluoride (PVDF) is a semi-crystalline polymer that presents polymorphism, where the crystalline phase chain can be organized in four different ways: α phase or form II (TGTG′), β phase or form I (TTTT), γ phase or form III (TTTGTTTG′), δ phase or form IV (TGTG′ polar) [1]. The α phase is nonpolar and it is formed from the melt at crystallization temperatures above 110 °C [2]. The β polar phase can be obtained from temperatures below 70 °C during the processing in the presence of suitable solvent. At the range between 70 and 100 °C might occur the generation of a mix of α–β phases [1]. The γ phase is achieved in the

L.C.M. Cirilo (✉) · M.F. Costa
Programa de Engenharia Metalúrgica e de Materiais—PEMM/COPPE,
Universidade Federal do Rio de Janeiro—UFRJ, Av. Horácio Macedo 2.030,
Rio de Janeiro, RJ CEP 21941-972, Brazil
e-mail: marysilvia@metalmat.ufrj.br

© The Minerals, Metals & Materials Society 2017
M.A. Meyers et al. (eds.), *Proceedings of the 3rd Pan American Materials Congress*,
The Minerals, Metals & Materials Series, DOI 10.1007/978-3-319-52132-9_24

crystallization of the α phase near to T_m or by annealing the material near to T_m of the α phase [1]. According to Dohany and Humphrey, the δ phase can be formed by the distortion of the phases α, β or γ when they are exposed to high electric field.

Among the crystalline phases showed by PVDF, the α one had a great industrial interest due to a special set of characteristics, such as resistance to moisture and to most of the chemical species, high thermal stability, low permeability to gases and liquids, resistance to flames and to weathering, high mechanical resistance, resistance to fatigue, resistance to ionizing radiation and to be able to maintain performance for more than 30 years of use in severe conditions, in addition to presenting the lowest cost when compared to others fluoropolymers [2, 3]. Such characteristics enables to be applied in several industrial sectors such as, chemical process, food, pharmaceutical and oil and gas [4]. In the oil and gas exploitation PVDF is used as internal pressure sheath of flexible pipes and umbilical hoses, which are exposed to severe operational conditions, as high temperature and high pressure, chemical products, very aggressive environments (presence of CO_2, H_2S and other gases that sometimes reach supercritical conditions). Such environments are becoming more frequent with the oil exploration in the pre-salt region for example, and the demand for the utilization of these fluoropolymers has increased. As a consequence, a substantial increasing in the generation of its waste originated during the processing of the flexible pipes has been observed.

Nowadays, the processing waste of PVDF as well as other engineering polymers have no disposal protocols and have no reuse stablished since they are materials that have high cost and high density compared to commodity polymers. Disposal of this material in landfills is not a correct environmental practice, once it is not biodegradable, resulting in high environmental impact, because since it reduces the useful area of the landfill that could be occupied by other wastes such as biodegradable polymers.

A possible way to remediate the problem caused by the PVDF waste could be its reinsertion in the production chain, for instance, of the internal pressure sheath of flexible pipes by primary recycling. That is, to reprocess the processing waste together with the neat polymer.

So, the present work aims to produce blends of neat PVDF (PVDFneat) with different proportions of processing waste of PVDF (PVDFwaste) and to assess its structural characteristics and viscoelastic properties, before and after being exposed to thermal aging.

Experimental

Materials and Methods

PVDF of flexible pipes grade (PVDFneat "as-received") in pellets and PVDF processing waste (PVDFwaste "as-received") in an irregular grain size were investigated. Neat PVDF (PVDFneat), processing waste PVDF (PVDFwaste) and blends of $PVDF_{neat}/PVDF_{waste}$ (with 95/5, 80/20 and 50/50 wt% of $PVDF_{waste}$) were prepared by extrusion ($T_{zone\ 4}$ = 230 °C, $T_{zone\ 3}$ = 220 °C, $T_{zone\ 2}$ = 210 °C, $T_{zone\ 1}$ = 190 °C and $T_{cooling}$ = 80 °C) and machined samples of these extruded polymers were exposed to 30 days of thermal aging at temperature of 130 °C to evaluate the influence of waste composition and the exposure time period in the viscoelastic properties of the PVDF blends.

At the end of the ageing period, samples were kept aside, in ambient atmosphere, to stabilize. After 24 h, the samples were analyzed.

FTIR–ATR

To evaluate the crystalline content of different samples of PVDF aged as well as unaged (PVDFwaste "as-received", PVDFneat "as-received" and PVDF blends with different amounts of his waste) FTIR spectroscopy was performed using a Nicolet 6700 (Thermo Scientific Co.) spectrophotometer operating with an attenuated total reflectance (ATR) cell in transmittance mode in the range of 4000–650 cm^{-1} and resolution of 4 cm^{-1}.

XRD Analysis

XRD analysis was performed to identify the crystalline structure and to calculate the degree of crystallinity (%X_c) of the samples. Analysis was performed in a Bruker D8 Discover diffractometer operating with CuKα of wavelength 0.154 nm and 2θ values from 5° to 70°, using a step scan of 0.02°/step and held for 1 s/step.

DSC

DSC analysis was carried out for crystallinity evaluation (% Crystallinity) of the samples before and after thermal aging. DSC analysis were performed in a DSC 8000 calorimeter (Perkin Elmer Co.) under nitrogen atmosphere applying two heating and cooling cycles, 25–210 °C at a rate of 10 °C/min.

Dynamic Mechanical Analysis (DMA)

The DMA analysis was performed using a NETZCH-242C instrument in three-point-bending mode at a frequency of 1 Hz under nitrogen atmosphere. The measures were taken in a temperature range of −100 to 150 °C with heating rate of

2 K/min and dynamic load and amplitude of 4 N and 50 μm respectively. The DMA analysis was applied to investigate the effect of PVDF processing waste (PVDFwaste) on the loss (E″) and storage (E′) modulus curves as well as the loss factor (tanδ).

Results and Discussion

FTIR and XRD Characterization

The FTIR spectra for the PVDFneat "as-received", PVDFwaste "as-received", neat PVDF (PVDFneat), PVDF waste (PVDFwaste) and PVDFneat/PVDFwaste blends (95/5, 80/20 and 50/50 wt%) are shown in Figs. 1 and 2. In Fig. 1, the FTIR spectra of "as-received" samples as well as the PVDFneat and PVDFwaste presents the typical absorption peaks (indicated in the figure) attributed to α and β phase. According to later studies [1, 5, 6] the peaks at 762 cm^{-1} (CF_2 bending and skeletal bending), 795 cm^{-1} (CH_2 rocking), 975 cm^{-1} (CH out-of-plane deformation), 1383 cm^{-1} and at 1423 cm^{-1} are attributed to PVDF α-phase. The peaks associated to β-phase are: 840 cm^{-1} [5], 1277 cm^{-1} (CF out-of-plane deformation) [7], 1430 cm^{-1} [5]. Analyzing the spectra in Fig. 1 it was possible to notice that samples "as-received" have well defined α-phase peaks (at 762, 795, 975 cm^{-1}) but after processing by extrusion, it was possible to observe a decrease in intensity of these peaks and the increase in intensity of the β-phase peaks at 840 and 1430 cm^{-1}. Such reduction is probably associated to conversion of α into β-phase due to drawing. As reported in literature the stretching of α-phase at 80 °C causes $\alpha \rightarrow \beta$ transformation [8, 9] and during extrusion, such process can occur. The presence of β-phase as a result of extrusion also was observed in the unaged XRD patterns presented in Fig. 3a as well as the presence of α-phase in all extruded samples. The α-phase was identified by the presence of peaks at $2\theta = 17.8°$ (100), 18.4° (020), 26.7° (021) and 38.7° (002) [10, 11] and the β-phase by the peaks at $2\theta = 20.06°$ (110) (200) [1]. Some authors attribute this peak to γ-phase (110) crystalline plane, but in this work we attribute this peak to β-phase, once

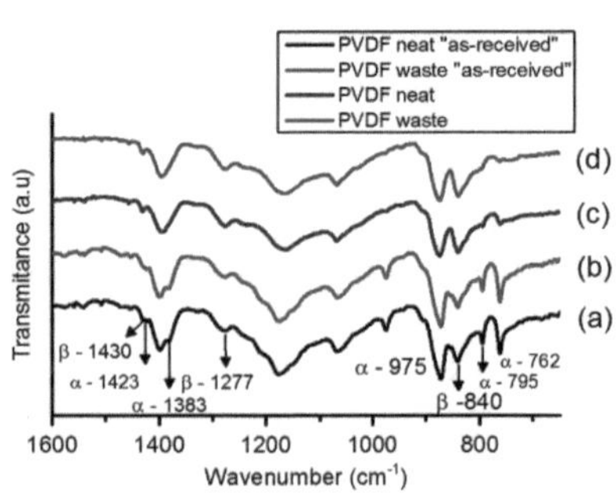

Fig. 1 FTIR-ATR spectra of PVDF samples: (*a*) $PVDF_{neat}$ "as-received", (*b*) $PVDF_{waste}$ "as-received", (*c*) $PVDF_{neat}$ and (*d*) $PVDF_{waste}$

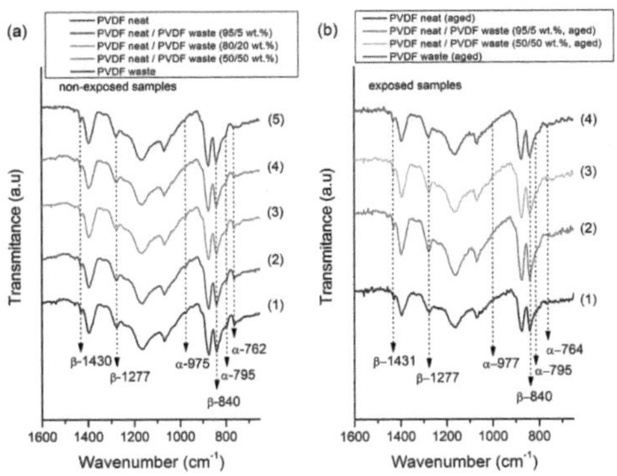

Fig. 2 FTIR-ATR spectra of unaged samples (**a**) and 30 days aged samples at 130 °C (**b**). (*a.1*) $PVDF_{neat}$, (*a.2*) $PVDF_{neat}/PVDF_{waste}$ (95/5 wt%), (*a.3*) $PVDF_{neat}/PVDF_{waste}$ (80/20 wt%), (*a.4*) $PVDF_{neat}/PVDF_{waste}$ (50/50 wt%), (*a.5*) $PVDF_{waste}$, (*b.1*) $PVDF_{neat}$ (aged), (*b.2*) $PVDF_{neat}/PVDF_{waste}$ (95/5 wt%, aged), (*b.3*) $PVDF_{neat}/PVDF_{waste}$ (50/50 wt%, aged) and (*b.4*) $PVDF_{waste}$ (aged)

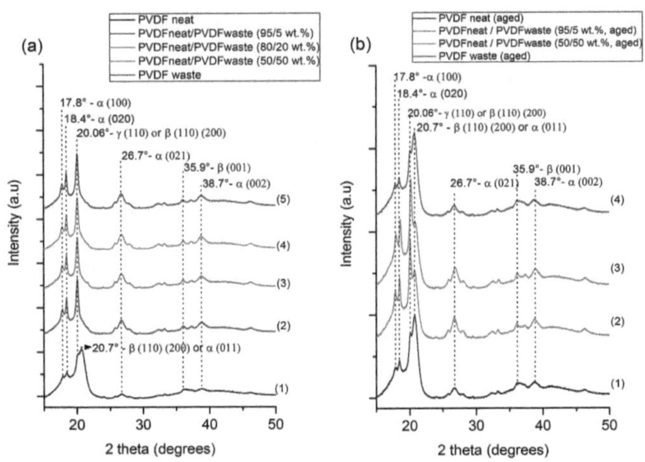

Fig. 3 XRD patterns of unaged (**a**) and 30 days aged at 130 °C (**b**) PVDF samples. (*a.1*) $PVDF_{neat}$, (*a.2*) $PVDF_{neat}/PVDF_{waste}$ (95/5 wt%), (*a.3*) $PVDF_{neat}/PVDF_{waste}$ (80/20 wt%), (*a.4*) $PVDF_{neat}/PVDF_{waste}$ (50/50 wt%), (*a.5*) $PVDF_{waste}$, (*b.1*) $PVDF_{neat}$ (aged), (*b.2*) $PVDF_{neat}/PVDF_{waste}$ (95/5 wt%, aged), (*b.3*) $PVDF_{neat}/PVDF_{waste}$ (50/50 wt%, aged) and (*b.4*) $PVDF_{waste}$ (aged)

that β-phase peak at $2\theta = 35.9°$ (001) [11] is present in the XRD patterns of all samples as well as in the FTIR.

When analyzing the effect of thermal aging on the crystalline phase of the samples, according to Fig. 3b it can be noted a small intensity peak (at $2\theta = 20.7°$ [7]), like a shoulder, of the β-phase peak at $2\theta = 20.06°$. Its presence indicates that after thermal exposition, during 30 days at 130 °C, occurs the ordering of the

Table 1 Storage and loss modulus at 23 °C and degree of crystallinity of neat PVDF, processing waste of PVDF and PVDFneat/PVDFwaste blends (95/5, 50/50 wt%)

Sample	Aging time (days)	E′ (MPa)	E″ (MPa)	Tg (°C) (tanδ)	%Crystallinity (DRX)	% Crystallinity (DSC)
PVDFneat	0	1316.69	83.37	−36.9	43.11	30.08
PVDFneat/PVDFwaste (95/5 wt%)	0	1443.71	88.13	−37.8	43.49	29.40
PVDFneat/PVDFwaste (50/50 wt%)	0	1219.99	77.66	−37.8	41.84	28.65
PVDFwaste	0	1464.05	96.23	−36.9	43.88	29.77
PVDFneat	30	1374.27	107.98	−37.86	43.93	31.25
PVDFneat/PVDFwaste (95/5 wt%)	30	1321.14	104.78	−37.63	44.35	39.29
PVDFneat/PVDFwaste (50/50 wt%)	30	1329.25	106.17	−37.63	46.53	42.11
PVDFwaste	30	1235.5	95.01	−37.63	41.62	30.49

chains of amorphous phase into α-phase [7], due to a secondary crystallization process that will be explained later in this work.

Also, according to DRX results presented in Fig. 3a and Table 1, the presence of PVDF waste on the crystalline phase of unaged PVDF blends decreases the degree of crystallinity when compared with the neat PVDF and PVDF waste. When this samples were exposed to thermal aging at high temperatures (in this works it was used T = 130 °C), an increase on the crystallinity degree is observed (Fig. 3b).

In Fig. 2, it can be noted that the increase of PVDF waste in the PVDF blends do not promote significant changes in the crystalline phase of all samples. However, when these samples were exposed to thermal aging, at 130 °C for 30 days, it can be noticed a decrease in the intensity of the main characteristic band (762–764 cm^{-1} [1, 6, 12]) of α-phase on the FTIR spectra of aged samples, as can be seen in Fig. 3.

DMA

Figure 4 shows the dynamic storage modulus (E'), loss modulus (E'') and the loss factor (tanδ, also known as damping coefficient) curve versus temperature for unaged neat PVDF ($PVDF_{neat}$) at frequency of 1 Hz. It was observed the presence of three thermal transitions (β, γ, α_c), as shown in Fig. 4. The β-relaxation occurs around −40 °C and is associated as the glass-transition temperature [13] (in this work the β-relaxation occurred at tanδ = −37 °C). A 52.3% decrease in the storage modulus was observed at the same temperature (−37 °C). At tanδ = −87 °C it was identify a small transition attributed to chain rotation in the amorphous phase [13] and at tanδ = 95 °C the last transition observed, α_c-relaxation, associated to the chain movement in the crystalline region. These movements are related to shear of

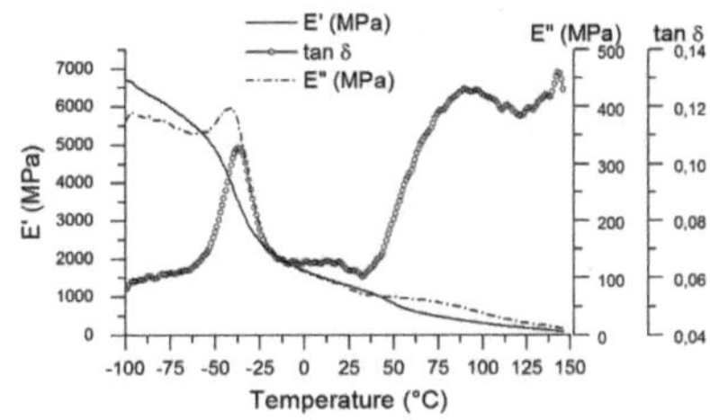

Fig. 4 DMA spectra obtained at a frequency of 1 Hz for unaged neat PVDF

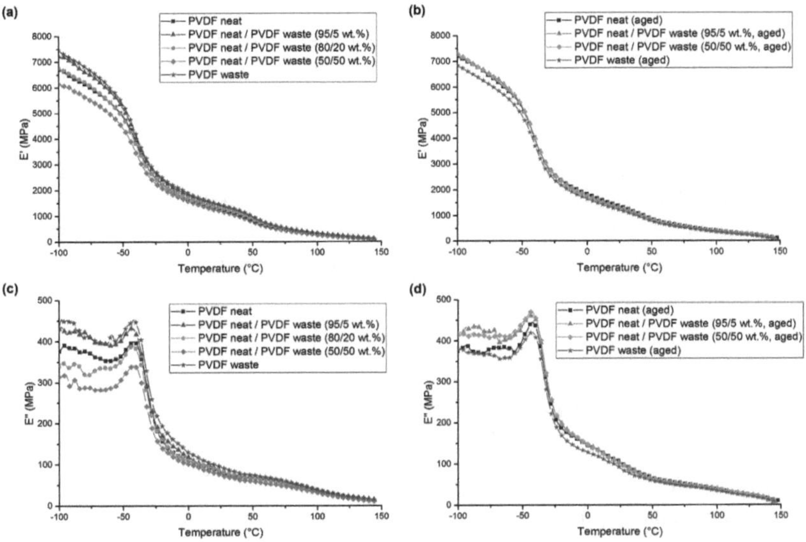

Fig. 5 Dynamic storage modulus and loss modulus curves (at 1 Hz) as function of temperature for neat and waste PVDF as well as PVDF blends (95/5, 80/20 and 50/50 wt%) samples: **a**. unaged E', **b**. 30 days aged at 130 °C E', **c**. unaged E'' and **d**. 30 days aged at 130 °C E''

amorphous layers allowed by elongation of the tie molecules due to the translation of chains within the crystalline lamellae, resulting in an activation of defects mobility in crystals [13, 14].

As neat PVDF, waste PVDF and PVDF blends (thermally aged at 130 °C and unaged) had their E', E'' and loss factor (tanδ) evaluated as function of temperature and the results are shown in Fig. 5. According to Fig. 5a, c for the non-exposed samples to heat, it was possible to notice, in general, that the addition of processing waste PVDF in the PVDF blends led to a decrease in both, storage and loss modulus, and this can be noted at the very beginning of both E' and E'' curves (−100 to −60 °C) and at 23 °C. In Table 1 are shown the values for E' and E'' at 23 °C as well as the calculated degree of crystallinity. This decrease in E' and E'' could be associated with the decrease in the ordering state of chains in the crystalline phase of the analysed polymers. The maximum peak in the E'' versus temperature curve is attributed to the maximum dissipation of mechanical energy as heat that is related to the micro-Brownian motion of segments of the polymer chain that happens when the polymer passes from the glassy to viscoelastic state [12, 15].

However, when the samples were exposed to thermal aging at 130 °C during 30 days, it was possible to notice an increase in the *E*′ and *E*″ (Fig. 5b, d) for PVDF samples of all composition studied in this work. According to Castagnet et al. [14] after hight temperature aging (above 120 °C) secondary crystallization can occur, decreasing the mobility of the amorphous chains present in the crystalline phase, as can be seen by the decrease of α_c-relaxation peaks around tanδ = 95 °C in Fig. 6b. The DSC thermogram (Fig. 7c) of the first heating cycle for aged samples corroborate this observation. The α_c-relaxation temperature varies according to the crystalline morphology and with lamellae thickness [16]. About the effect of PVDF waste in the tanδ versus temperature curves, the PVDF waste has different behavior before and after ageing. Before aging, PVDF blends (95/5, 80/20 and 50/50 wt%) has a lower loss factor (at tanδ ~ 37.8 °C) than neat PVDF and PVDF waste (Fig. 6a). On the other hand, after thermal aging it was noticed the opposite: the blend of neat PVDF with PVDF waste promoted the increase of the loss factor (at tanδ ~ 37.8 °C). According to %Crystallinity results presented in Table 1 and with tanδ curves (Fig. 6), the blend between PVDF waste and neat PVDF led to a slight increase in chains of amorphous phase.

DSC Characterization

DSC analysis was performed on the samples before and after aging. Figure 7 and Table 1 presents the thermograms results for the first heating scan and the degree of crystallinity. Before and after aging it was observed, in all samples, the presence of an endothermic peak at ~125.5 °C.

This same peak was also observed by de Oliveira et al. [17] and Khonakdar et al. [18] and it is associated with *Tm* of HDPE, which is probably present in the grades

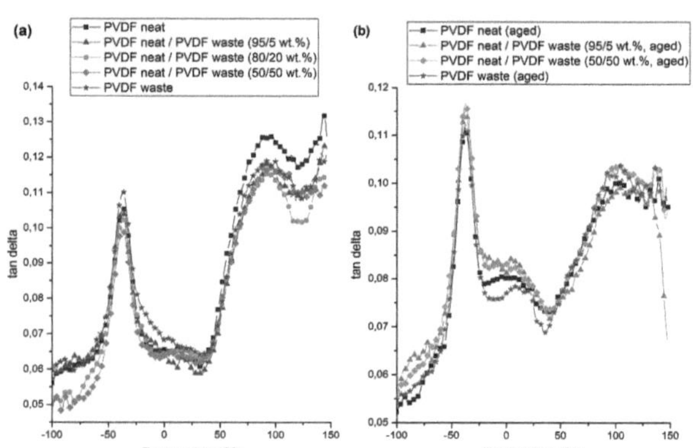

Fig. 6 Loss factor, at 1 Hz, as function of temperature for neat PVDF, processing waste PVDF and PVDFneat/PVDFwaste blends (95/5, 80/20 and 50/50 wt%). **a** unaged and **b** 30 days aged (T = 130 °C) samples

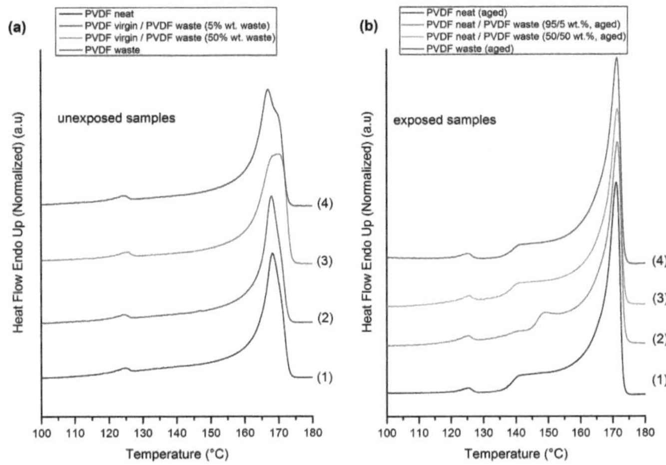

Fig. 7 DSC thermograms of unaged (**a**) and aged (30 days, T = 30 °C) (**b**) sample. (*a.1*) PVDFneat, (*a.2*) PVDFneat/PVDFwaste (95/5 wt%), (*a.3*) PVDFneat/PVDFwaste (50/50 wt%,), (*a.4*) PVDFwaste, (*b.1*) PVDF neat, (*b.2*) PVDFneat/PVDFwaste (95/5 wt%), (*b.3*) PVDFneat/PVDFwaste (50/50 wt%,) and (*b.4*) PVDF waste. (first heating scan; 10 °C/min)

of neat PVDF and PVDF process waste used in this work. After thermal aging at high temperatures (>120 °C), an endothermic peak (~ 141–147 °C) appeared like a shoulder of the main melting peak (~ 171 °C), as can be seen in Fig. 7b. This same trend was observed by Castagnet et al. [14] after annealing PVDF at 95, 120 and 140 °C. This small endothermic peak is an evidence of secondary crystallization process where small-organized volumes are created within amorphous phase at high annealing temperature [14].

Conclusion

For thermally unaged samples, the addition of PVDF waste to produce the PVDF blends (PVDFneat/PVDFwaste) results in a slight decrease of the crystallinity degree when compared to the neat PVDF and PVDF process waste, leading to a slight decrease in the E', E'' and tanδ. The thermal aging induced secondary crystallization of amorphous phase to probably α-phase, leading to a slight increase in the E', E'' and tanδ values. In general, when compared the results of neat PVDF in this work with the result obtained by de Oliveira et al. [17], it was clearly observed that the extrusion conditions applied in this work was responsible for the higher degree of crystallinity (de Oliveira et al. [17] processed the PVDF samples through compression molding). The extrusion conditions were responsible for the $\alpha \rightarrow \beta$-phase transformation that might occurs at T = 80 °C [8, 9]. All in all, as was shown in this work, the presence of PVDF waste in PVDF blends shows a slight influence in the structural properties and dynamic-mechanical properties of studied materials. This is a strong evidence that the PVDF processing waste can be reinsert in the manufacture chain of PVDF based parts, since the chosen processing conditions are suitable for the production of a material that has mechanicals properties very similar to that one who was manufactured applying the neat raw-material (e.g. neat PVDF).

Acknowledgements The authors are grateful for the financial support given by CAPES and thankful to the Mechanical Properties Laboratory (PROPMEC/COPPE/UFRJ) for XRD analysis.

References

1. Gregorio, R. (2006). Determination of the α, β and γ crystalline phases of poly(vinylidene fluoride) films prepared at different conditions. *Journal of Applied Polymer Science, 100,* 3272–3279.
2. Paul, D. R., & Barlow, J. W. (1980). Polymer blends (or alloys). *Journal of Macromolecular Science Part C, 18,* 109–168.
3. Herman, Mark F. (2005). *Encyclopedia of Polymer Science and Technology* (Vol. 1). New York, NY: Willey.
4. Drobny, J. G. (2005). *Rapra Review Reports—Fluoroplastics* (Vol. 16). Shropshire: Rapra Technology.
5. Benz, M., Euler, W. B., & Gregory, O. J. (2002). The role of solution phase water on the deposition of thin films of poly(vinylidene fluoride). *Macromolecules, 35,* 2682–2688.
6. Zheng, J., He, A., Li, J., & Han, C. C. (2007). Polymorphism control of poly(vinylidene fluoride) through electrospinning. *Macromolecular Rapid Communications, 28,* 2159–2162.
7. Dillon, D. R., Tenneti, K. K., Li, C. Y., et al. (2006). On the structure and morphology of polyvinylidene fluoride-nanoclay nanocomposites. *Polymer, 47,* 1678–1688.
8. Gregorio, R., & Ueno, E. M. (1999). Effect of crystalline phase, orientation and temperature on the dielectric properties of poly (vinylidene fluoride) (PVDF). *Journal of Material Science, 34,* 4489–4500.
9. Sajkiewicz, P. (1999). Crystallization behaviour of poly(vinylidene fluoride). *European Polymer Journal, 35,* 1581–1590.
10. Martins, P., Lopes, A. C., & Lanceros-Mendez, S. (2014). Electroactive phases of poly (vinylidene fluoride): Determination, processing and applications. *Progress in Polymer Science, 39,* 683–706.
11. Sun, J., Yao, L., Zhao, Q. L., et al. (2011). Modification on crystallization of poly(vinylidene fluoride) (PVDF) by solvent extraction of poly(methyl methacrylate) (PMMA) in PVDF/PMMA blends. *Frontiers of Materials Science, 5,* 388–400.
12. de J. Silva, A. J., Nascimento, C. R., & da Costa, M. F. (2016). Thermomechanical properties and long-term behavior evaluation of poly(vinylidene fluoride) (PVDF) exposed to bioethanol fuel under heating. *Journal of Materials Science, 51,* 9074–9094.
13. Mekhilef, N. (2001). Viscoelastic and pressure-volume-temperature properties of poly (vinylidene fluoride) and poly(vinylidene fluoride)-hexafluoropropylene copolymers. *Journal of Applied Polymer Science, 80,* 230–241.
14. Castagnet, S., & Girard, D. (2007). Sensitivity of damage to microstructure evolution occurring during long-term high-temperature annealing in a semi-crystalline polymer. *Journal of Materials Science, 42,* 7850–7860.
15. Menczel, J. D., & Bruce Prime, R. (2008). *Thermal Analysis of Polymers: Fundamentals and Applications*. New Jersey, NY: Willey.

16. Mano, J. F., Lopes, J. L., Silva, R. A., & Brostow, W. (2003). Creep of PVDF monofilament sutures: Service performance prediction from short-term tests. *Polymer, 44,* 4293–4300.
17. de Oliveira, G. L., Costa, C., Teixeira, S., et al. (2014). The use of nano- and micro-instrumented indentation tests to evaluate viscoelastic behavior of poly(vinylidene fluoride) (PVDF). *Polymer Testing, 34,* 10–16.
18. Khonakdar, H. A., Jafari, S. H., Wagenknecht, U., & Jehnichen, D. (2006). Effect of electron-irradiation on cross-link density and crystalline structure of low- and high-density polyethylene. *Radiation Physics and Chemistry, 75,* 78–86.

Dynamic Transformation and Retransformation During the Simulated Plate Rolling of an X70 Pipeline Steel

Samuel F. Rodrigues, Clodualdo Aranas Jr, Fulvio Siciliano and John J. Jonas

Abstract The controlled rolling of pipeline steels involves pancaking the austenite and then subjecting it to accelerated cooling. However, the formation of ferrite during rolling decreases the amount of austenite available for microstructure control. Here the formation of ferrite during rolling is simulated using a five-pass rolling schedule applied by means of torsion testing. The first and last pass temperatures were 920 and 860 °C with 15° of cooling between passes. All of the rolling was carried out above the Ae_3 temperature of 845 °C that applies to this steel. Interpass times of 10 and 30 s were employed, which corresponded to cooling rates of 1.5 and 0.5 °C/s, respectively. Samples were quenched before and after the first, third, and fifth passes in order to determine the amount of dynamic ferrite produced in a given pass. The amounts of dynamic ferrite formed and retained increased with pass number. The amounts of ferrite that retransformed increased with pass number. The simulations indicate that ferrite is unavoidably produced during plate rolling and that the microstructures present at the initiation of accelerated cooling do not consist solely of austenite.

Keywords Dynamic transformation · Plate rolling · X70 pipeline steel

S.F. Rodrigues (✉) · C. Aranas Jr · J.J. Jonas
Department of Materials Engineering, McGill University,
3610 University Street, Montreal, QC H3A 0C5, Canada
e-mail: samuel.rodrigues@mail.mcgill.ca; samuel.filgueiras@ifma.edu.br

S.F. Rodrigues
Federal Institute of Education, Science and Technology of Maranhao,
Av. Marechal Castelo Branco 789, São Francisco, São Luís, MA 65076-091, Brazil

F. Siciliano
Dynamic Systems Inc., 323 NY 355, Poestenkill, NY 12140, USA

© The Minerals, Metals & Materials Society 2017

M.A. Meyers et al. (eds.), *Proceedings of the 3rd Pan American Materials Congress*,
The Minerals, Metals & Materials Series, DOI 10.1007/978-3-319-52132-9_25

Introduction

The plate rolling of pipeline steels is carried out primarily within the austenite phase field. One of the aims of thermomechanical processing is to pancake the austenite. Later, on accelerated cooling, transformation to ferrite and bainite takes place and a desirable fine-grained microstructure is formed. Such fine-grained microstructures have the enhanced fracture toughness properties required in pipeline steels. However, if part of the microstructure has already transformed to ferrite during rolling, this component of the structure cannot be subjected to transformation control. Previous work has shown that rolling produces phase transformation of the austenite in the roll bite [1] and that ferrite can be formed everywhere within the austenite phase field [2]. This has been referred to as dynamic transformation (DT) and has been reported by several authors. This phenomenon has been studied and has received considerable attention since the work of Yada and co-workers in the 1980s [3, 4]. They described the results of various compression tests and strip rolling simulations carried out under isothermal conditions and above the Ae_3. They reported that the volume fraction of ferrite increased with strain and decreased with temperature as well as with the holding time after the last pass. Later, they carried out torsion tests in combination with in situ X-diffraction, which permitted direct observation of the phases that were being formed during deformation [5].

In the work of Basabe et al. [6], torsion tests were conducted on a 0.036% Nb microalloyed steel in order to study the mechanisms of nucleation, as well as the effects of strain, strain rate and temperature on the formation of ferrite by DT. It was concluded that dynamically formed ferrite is first nucleated in the form of Widmanstätten plates and that these later coalesce into polygonal grains at strains of 0.5–1.5. They also showed that the reverse static transformation was retarded to a considerable degree by the addition of niobium. They attributed this phenomenon as resulting from dislocation pinning by niobium carbonitride precipitates and also by solute drag due to the presence of Nb in solution. Thus this element can play an important role in steel processing by retarding the reverse transformation during strip and plate rolling [7, 8].

Aranas et al. [1, 9] proposed a thermodynamic explanation for dynamic transformation according to which the driving force for ferrite formation is the softening that takes place during deformation. In this model, the driving force is opposed by the chemical free energy difference between austenite and Widmanstätten ferrite as well as the lattice dilatation work and shear accommodation work that accompany the transformation.

To date, this phenomenon has only been investigated under strip rolling and isothermal plate rolling conditions [1, 8]. The aim of this work was to investigate the role of dynamic transformation when cooling takes place during rolling. Knowledge of the amount of ferrite present before the start of controlled cooling can then be used to improve the cooling/transformation model.

Materials and Methods

The present steel was supplied in form of hot rolled plates with a thickness of 12.5 mm. Its chemical composition is presented in Table 1, together with its corresponding Ae_3 temperatures. These were calculated using the FactSage software. The plates were cut into cylindrical torsion specimens with diameters of 6.35 mm and lengths of 22.0 mm.

The samples were heated to 1200 °C at 1 °C/s and held for 20 min for dissolution of the carbonitrides and austenitization. They were then cooled at 1 °C/s to 1100 °C and maintained at this temperature for 60 s after which the first roughing pass was applied. A second roughing pass was employed after 120 s at this temperature. A strain of 0.4 was applied in each case at a strain rate of 1.0 s^{-1}. After roughing, the samples were held for 60 s to allow complete recrystallization before being cooled to 920 °C. The samples remained at this temperature for 60 s before beginning the finish rolling simulations. For this purposes pass strains of 0.2 were applied at a strain rate of 1.0 s^{-1}.

The samples were deformed while being cooled at 1.5 and 0.5 °C/s. The last pass of each simulation was applied at 860 °C (i.e. 15 °C above the Ae_3) for all these interpass times. Both before and after the 1st, 3rd, and 5th passes of these tests, the samples were quenched to permit quantification of the volume fractions of ferrite formed and retransformed. The thermomechanical schedules for these tests are illustrated in Fig. 1.

For microstructural analysis, the torsion samples were cut perpendicular to the longitudinal axis and also to the rolling direction. The samples were hot mounted and polished using silicon carbide paper grits from 400 to 1200. For final polishing, a diamond paste (3 and 1 μm) suspension was used. The polished samples were then etched with 2% nital for approximately 15 s and then treated with a 10%

Table 1 Chemical composition (mass%) and equilibrium transformation temperatures (°C)

C	Mn	Si	Cr	Nb	N	Orthoequilibrium Ae_3	Paraequilibrium Ae_3
0.047	1.56	0.25	0.21	0.092	0.008	845 °C	810 °C

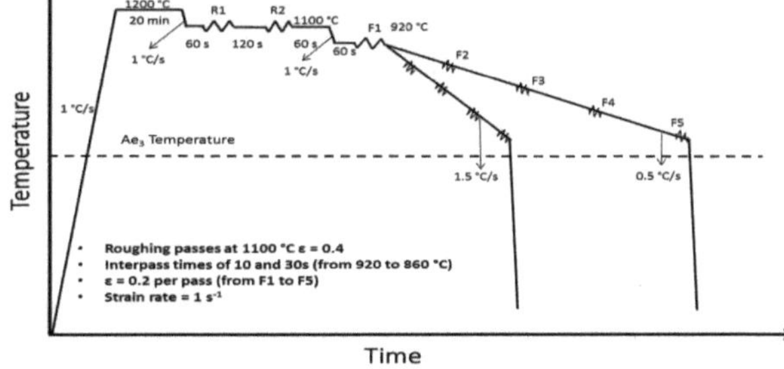

Fig. 1 Torsion testing schedule employed in simulations. The deformation temperatures were 920, 905, 890, 875 and 860 °C and interpass times of 10 and 30 s were employed

aqueous sodium metabisulfite ($Na_2S_2O_5$) solution in order to improve the contrast between ferrite and martensite. The initial microstructural analysis was done using optical microscopy. Later, SEM and EBSD techniques were employed to provide more detailed microstructural information.

Results and Discussions

Stress-Strain Curves

The flow curves determined in this way are displayed in Fig. 2a for the 10 s and Fig. 2b for the 30 s interpass times. The levels of the curves increase progressively as the temperature is decreased but not as sharply as expected from the temperature dependence of the flow stress, as expressed by Eq. 1. This equation represents the slope of a linear fit to the present type of flow stress/temperature data that was deduced from a set of first pass flow curves at a number of decreasing temperatures (not shown here). Strains of 0.04 were employed for this purpose as this value was found to be the critical strain for the onset of DT [8].

Here, $\sigma_{0.04}$, $\sigma_{0.1}$ and σ_p are the flow stresses applicable to applied strains of 0.04, 0.1 and the peak strain, respectively. The slopes ψ depended on the simulation interpass time and took values of 118 and 164 MPa K for the 10 and 30 s intervals, respectively. The broken lines represent the stress levels expected to be achieved in the 3rd, 4th and 5th passes based on the second pass peak and the temperature dependence.

It is evident from the figure that the stresses increase less quickly at the 10 s than at the 30 s interpass time. The rather low rate of increase in flow stress can be attributed to the progressive formation of ferrite. The higher level at the longer interpass time can be attributed to the increased amount of ferrite that retransforms into austenite.

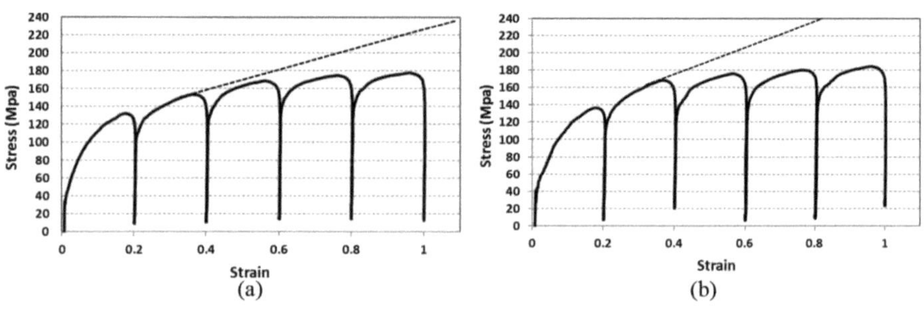

Fig. 2 Stress-strain curves determined according to the schedules of Fig. 1 using pass strains of 0.2 applied at 1 s^{-1}. Interpass times of **a** 10 s and **b** 30 s

$$\frac{d\,\sigma_{0.04}}{d\,(1/T)} = \frac{d\,\sigma_{0.1}}{d\,(1/T)} = \frac{d\,\sigma_{p}}{d\,(1/T)} = \psi \tag{1}$$

Mean Flow Stress

In hot rolling, the evolution of the mean flow stress (MFS) provides insight into the progress of metallurgical phenomena such as work hardening, static and dynamic softening, precipitation, as well as dynamic transformation. These are dependent on the temperature of deformation, as well as the strain, strain rate, interpass time and chemical composition. Here, the MFS's were calculated according to Eq. 2, which specifies the area under the flow curve associated with each rolling pass in Fig. 2.

$$MFS = \frac{1}{\varepsilon_b - \varepsilon_a}\int_{\varepsilon_a}^{\varepsilon_b} (\sigma_{eq} d\varepsilon) \tag{2}$$

In Eq. 2, σ_{eq} is the equivalent stress and $(\varepsilon_b - \varepsilon_a)$ is the equivalent strain applied in a particular pass. The results obtained in this way are presented in Fig. 3 for interpass times of 10 and 30 s.

It can be seen that there is a sharp increase in the MFS from the first to the second pass in both schedules. This is a consequence of strain accumulation. This increase amounts to an average of about 22% for these two interpass times. In the successive passes, the increases amount to averages of 9.5, 3 and 1.8% for the second to the third, third to the fourth and fourth to the fifth passes, respectively. The low MFS increases in the late passes indicate that ferrite (which is softer than austenite) is being formed in each pass. This interpretation is supported by the optical microstructures that will be shown later. It is also important to note that the MFS values increase the least with falling temperature when the length of the interpass time is the shortest. This is because the time available for the back transformation of ferrite into austenite is the least under these conditions.

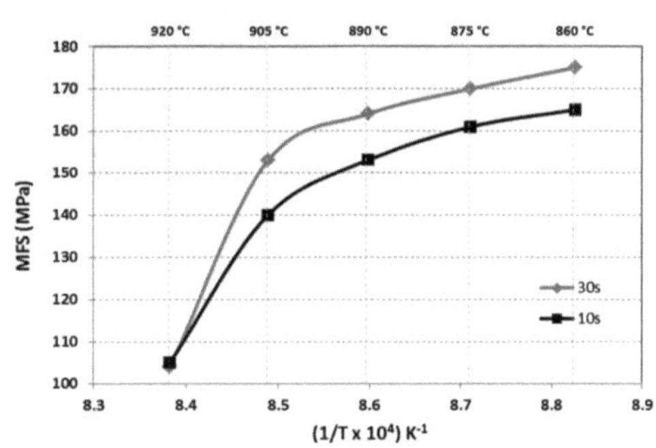

Fig. 3 MFS's curves derived from the stress-strain curves of Fig. 2. The relatively low rate of increase with decreasing temperature is a result of the formation of DT ferrite during each pass

Optical Microscopy

The ferrite volume fractions prior to straining were measured on microstructures of the transverse samples while the fractions present immediately after straining were based on longitudinal microstructures. In both cases, the measurements were carried out about 300 μm below the surface of the sample. This was so as to avoid the oxidized outer layer. The results that correspond to the 10 s interpass time are presented in Fig. 4. Here martensite (prior austenite) appears dark while ferrite is light.

It can be seen that ferrite is already present before the first pass. This small amount was inherited from the roughing passes applied at 1100 °C. Application of the first finishing pass leads to an increase in the quantity of ferrite. The passes that follow produce similar results, with the ferrite fraction increasing each time. The ferrite is mostly present in the form of polygonal grains before the passes. It is also possible to see grain elongation in the micrographs, where the major axis of the grain gradually approaches the horizontal axis. This is particularly visible after the 3th and 5th passes.

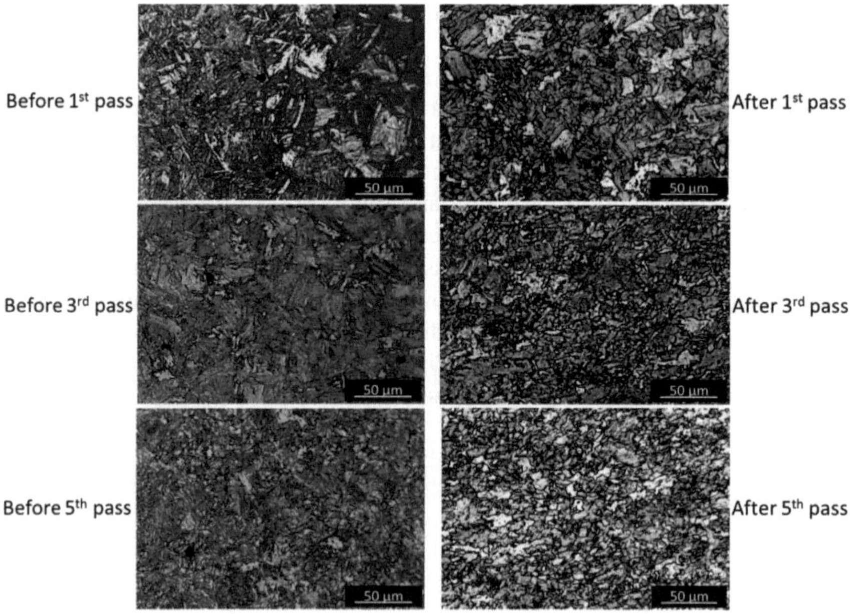

Fig. 4 Optical microstructures of the present material subjected to the 5-pass simulation with interpass times of 10 s. Light regions are ferrite while the dark regions are martensite (prior austenite)

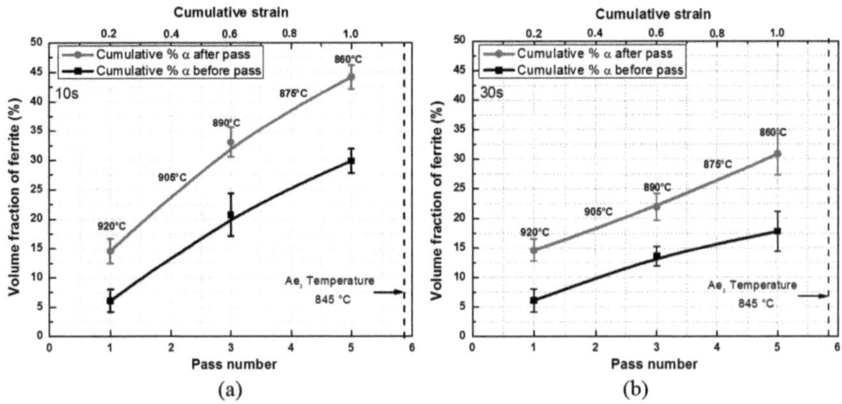

Fig. 5 Dependence of the volume fraction of ferrite formed and retransformed on pass number: **a** 10 s and **b** 30 s. As the pass number increases, the amounts of ferrite formed and retransformed increase

Volume Fraction of Transformed and Retransformed Ferrite Per Pass

The cumulative volume fractions of retained and retransformed ferrite produced using interpass times of 10 and 30 s are displayed in Fig. 5a, b, respectively. Each figure also shows the cumulative strain and deformation temperature. It is clear that the amount of ferrite produced and retained increases with pass number. Note that the volume fractions of both transformed and retransformed ferrite increase sharply as the interpass time is decreased.

The small amount of ferrite (6%) present before the first pass, was produced by double roughing. This increased to 15% after the first finishing pass and then to 33 and 22% after the 3rd pass and 44 and 31% after the 5th pass for the 10 and 30 s interpass times, respectively. Before the third pass, the volume fractions of ferrite present were 20 and 13.5% and before the fifth 30 and 17.5% for the 10 and 30 s schedules, respectively. Despite the presence of Nb, which contributes to solute drag [1, 7], the amount of reverse transformation increased perceptibly with the length of the interpass time.

Production and Retention of Ferrite

The amount of ferrite formed in a given pass is displayed in Fig. 6. This was calculated by subtracting the cumulative amount of ferrite present before each pass from the ferrite present after, as given by Fig. 5. Note that for the second and fourth passes, the amounts were obtained by interpolation. About 8.5% of ferrite was formed during the first pass. When 10 s interpass times were employed, 10.4, 12.3,

Fig. 6 The volume fraction of ferrite formed per pass. The amounts of ferrite increase with pass number but decrease as the length of the interpass time is increased (due to the loss in retained strain)

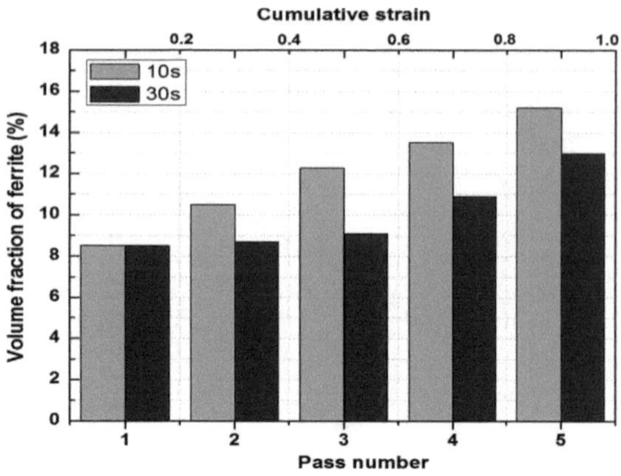

Fig. 7 The volume fraction of ferrite retransformed between passes. The amount of ferrite retransformed increases with pass number and with interpass time

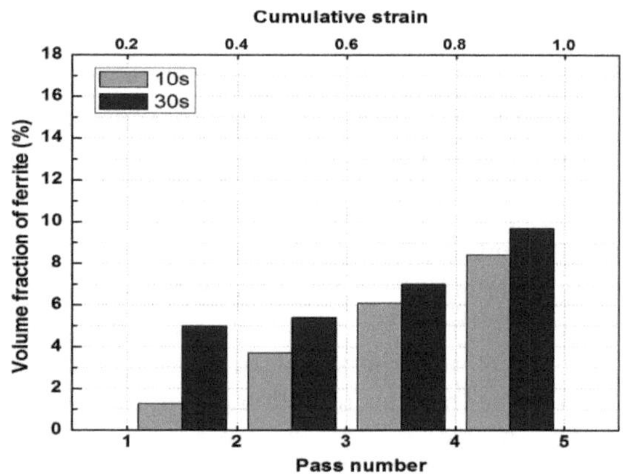

13.6 and 15.2% increases were registered in the second, third, fourth and fifth passes, respectively. Less ferrite was formed during the 30 s schedule: 8.7, 9.1, 10.8 and 13%. The amount of ferrite formed and retained was reduced by about 2.5% per pass on average when the interpass time was increased by 20 s. This is because longer interpass times favor the retransformation of ferrite back into austenite.

Retransformation of Ferrite

The percentage of ferrite retransformed into austenite during a given interpass is illustrated in Fig. 7. These values were obtained from the difference between the cumulative quantity retained and formed in a particular pass and the quantity of ferrite remaining after the interpass interval (i.e. before the next pass). The differences between the quantities of retransformed ferrite formed during the four interpass intervals are greatest during the interval between the first and second passes. This may be because the first pass is applied at the highest temperature. As expected, the longer interpass time favors more retransformation because of the diffusional nature of this mechanism. For the four intervals shown in Fig. 7, the percentages of retransformed ferrite fell in the ranges 1.3, 3.6, 6.1 and 8.4% for the 10 s schedule; and 5.1, 5.5, 7.0 and 9.7% for the 30 s interpass times.

SEM and EBSD Analysis

A scanning electron micrograph of a sample subjected to three passes in the 30 s schedule is illustrated in Fig. 8a. Here two phases are present, ferrite and martensite (prior austenite), where the ferrite is mostly polygonal in form. The microstructure associated with reduced pass strains of 0.05 at a location near the axis of the sample is illustrated in Fig. 8b. (It should be borne in mind that the strain in a torsion sample increases from zero along the axis to a maximum at the surface.) Here, some freshly formed DT ferrite can be seen, which is predominantly plate-like in form. Some typical straight interfaces characteristic of these microstructures are highlighted with white arrows. These interfaces indicate that the newly-formed plates are Widmanstätten in nature and have formed displacively.

The presence of DT ferrite in the present samples was confirmed using EBSD techniques of phase identification on a sample subjected to five passes in the 30 s schedule. This is illustrated here in Fig. 9a, where inverse pole figure (IPF) plots related to the orientations of the ferrite formed are shown and Fig. 9b displays the phase distributions. The red regions represent the ferrite and the unindexed black ones identify the martensite (prior austenite).

The present results show that the lack of load increase cannot be attributed solely to DRX, as suggested by earlier workers [3–5, 10–12].

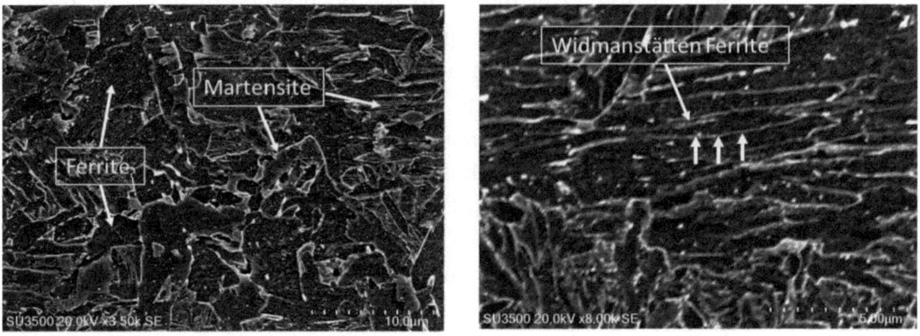

(a) (b)

Fig. 8 Scanning electron micrographs of samples deformed using interpass times of 30 s and quenched: **a** after the 3rd pass (the ferrite is mostly of polygonal form); **b** after the 3rd pass in a region of the sample subjected to pass strains of 0.05; plates of Widmanstätten ferrite can be seen that have formed displacively

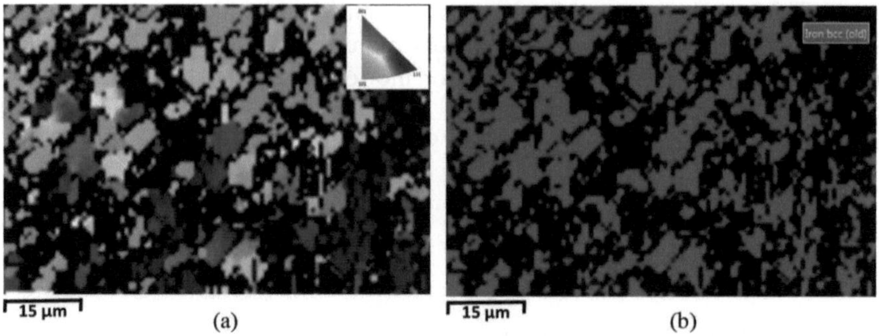

(a) (b)

Fig. 9 EBSD micrographs of the transverse cross-section of a sample subjected to the 30 s schedule. The sample was quenched immediately after straining to the 5th pass; **a** inverse pole figure and **b** phase identification. The ferrite phase is *red* while the prior austenite is *black* (Color figure online)

The significant amounts of ferrite formed during rolling are not available for transformation on the cooling bed. For this reason, it is important to know the volume fraction of ferrite produced during rolling so that improved cooling/transformation models can be developed applicable to plate rolling.

Conclusions

The occurrence of dynamic transformation under plate rolling conditions was examined on an X70 pipeline steel. Torsion tests were carried out under cooling conditions. Analysis of the results led to the following conclusions.

(1) The levels of the flow curves increased only slightly for both interpass times. This behavior is attributed primarily to the occurrence of dynamic transformation during straining, which is responsible for the low rate of increase in the MFS after the second pass. Strain accumulation is responsible for the large increase from the first to the second pass.
(2) The optical micrographs indicate that some ferrite is produced during roughing and that a small amount of this phase is already present before the first finishing pass is applied. The amount of ferrite formed during straining then increases with pass number (i.e. with the retained strain and with proximity to the Ae_3). The amount of ferrite retransformed decreases with the length of the subsequent interpass interval (i.e. with the loss of retained strain due to recovery and recrystallization).
(3) Shorter interpass times lead to the accumulation of more ferrite than longer ones. This is because more of the metastable DT ferrite retransforms back into the more stable austenite during the longer intervals.

(4) The DT ferrite is first produced displacively in the form of Widmanstätten plates; these coalesce into polygonal grains during further straining and holding.
(5) Knowledge of the amount of ferrite present at the beginning of cooling can be used to design improved models for the microstructures formed during transformation on cooling beds.

Acknowledgements The authors acknowledge with gratitude funding received from the Brazilian National Council for Scientific and Technological Development (SFR), the McGill Engineering Doctoral Award (MEDA) program (CAJr), and the Natural Sciences and Engineering Research Council of Canada (JJJ).

References

1. Ghosh, C., Aranas, C., & Jonas, J. J. (2016). Dynamic transformation of deformed austenite at temperatures above the Ae_3. *Progress in Materials Science, 82,* 151–233.
2. Grewal, R., Aranas, C., Jr., Chadha, K., Shahriari, D., Jahazi, M., & Jonas, J. J. (2016). Formation of Widmanstätten ferrite at very high temperatures in the austenite phase field. *Acta Materialia, 109,* 23–31.
3. Matsumura, Y., & Yada, H. (1987). Evolution deformation of ultrafine-grained ferrite in hot successive deformation. *Transactions ISIJ, 27,* 492–498.
4. Yada, H. H., Matsumura, T., & Senuma, T. (1988). *Proceedings of International Conference on Physical Metallurgy of Themomechanical Processing of Steels and Other Metals, ISIJ, THERMEC* (vol. 88, pp 200–207).
5. Yada, H., Li, C. M., & Yamagata, H. (2000). Dynamic $\gamma \rightarrow \alpha$ transformation during hot deformation in iron-nickel-carbon alloys. *ISIJ International, 40,* 200–206.
6. Basabe, V. V., Jonas, J. J., & Ghosh, C. (2014). Formation of Widmanstätten ferrite in a 0.036% Nb low carbon steel at temperatures above the Ae_3. *Steel Research International, 85,* 8–15.
7. Aranas, C., Jr., Rodrigues, S., Grewal, R., & Jonas, J. J. (2015). Ferrite formation above the Ae_3 temperature during the torsion simulation of strip rolling. *ISIJ International, 55,* 2426–2434.
8. Rodrigues, S. F., Aranas, C. Jr., Wang, T., & Jonas, J. J. (2017, In press) Dynamic transformation of an X70 steel under plate rolling conditions. *ISIJ International, 57*, 162–169.
9. Aranas, C. Jr., Jung, I. H., Yue, S., Rodrigues, S. F., & Jonas, J. J. (2016). A metastable phase diagram for the dynamic transformation of austenite at temperatures above the Ae_3. *International Journal of Materials Research, 107*, 881–886.
10. Tong, M., Li, D., Li, Y., Ni, J., & Zang, Y. (2004). Monte Carlo-method simulation of the deformation-induced ferrite transformation in the Fe-C system. *Metallurgical and Materials Transactions A, 35,* 1565–1577.
11. Hanlon, D. N., Sietsma, J., & Zwaag, S. (2001). The effect of plastic deformation of austenite on the kinetics of subsequent ferrite formation. *ISIJ International, 41,* 1028–1036.
12. Sun, X., Luo, H., Dong, H., Liu, Q., & Weng, Y. (2008). Microstructural evolution and kinetics for post-dynamic transformation in a plain low carbon steel. *ISIJ International, 48,* 994–1000.

Effect of Heat Input on the Microstructure and Toughness of Welded API Pipelines

Fernando Guzmán, Moisés Hinojosa, Eduardo Frias and Elisa Schaeffer

Abstract We seek to explain the weld toe-crack failure in terms of crack nucleation and propagation based on the observed phases and their size. Three welded joins were studied, varying the heat inputs in order to modify the weld microstructure. We work with an API X70 PSL2 pipeline steel, varying the ferritic size phases. We performed Charpy impact and Vickers hardness tests; we also determined the chemical composition (especially the oxygen percentage). Based on the relationship between the test results and the microstructure, we achieve a modification of the welded join microstructure and toughness, observing the heat-input effect in the weld oxygen diffusion as well as its effect in the type and size of the phase observed.

Keywords Heat Input · Toughness · Microstructure

Introduction

The demand of products such as natural gas, petroleum derivatives, and oil is rising; it is estimated that the demand will double from 2016 to 2035 [1]. The transport of such materials requires pipelines that withstand a corrosive environment with good welding properties, as well as high toughness [1]. API steels are used in the pipeline manufacture; the available degrees range from X52 to X120, X70 and X80 being

F. Guzmán (✉) · M. Hinojosa · E. Schaeffer
UANL, FIME, San Nicolás de los Garza, Nuevo León, Mexico
e-mail: yoshua_003@hotmail.com

M. Hinojosa
e-mail: hinojosamoises@yahoo.fr

E. Schaeffer
e-mail: elisa.schaeffer@uanl.edu.mx

E. Frias
Tubacero, Monterrey, Nuevo León, Mexico

© The Minerals, Metals & Materials Society 2017

M.A. Meyers et al. (eds.), *Proceedings of the 3rd Pan American Materials Congress*,
The Minerals, Metals & Materials Series, DOI 10.1007/978-3-319-52132-9_26

the most common due to properties such as toughness and ductility. The origin of these desirable properties is that these grades are in the phase frontier combining the polygonal ferritic and pearlite microstructures of the lower degrees and the ferritic-bainitic microstructures (ferrite, bainite, martensite, and retained austenite combinations) of the higher degrees such as X100 and X120 [2].

In this work, we explain the fracture mechanism in the *Weld-Toe crack failure* in the *submerged arc welding* (SAW) process, localized in the *heat affected zone* (HAZ) close to the *fusion line* (FL). The defect is explained in handbooks as a cold fracture for the welding cooling rates, producing a brittle HAZ making the weld-toe crack along the FL; as a solution, handbooks recommend decreasing the cooling rates, applying a preheating before welding. Unfortunately, doing so introduces an extra step in the manufacture process, increasing production time.

In this work, we vary the *Heat Inputs* (HI) that have a direct effect in the cooling rate [3] (high HI is linked with low cooling rates and low HI with high cooling rates). Our hypothesis is that the changes that HI produces in the microstructure alter phase sizes and densities. The *Heat Input* (HI) is the energy (heat) introduced in the material to make the join. The variables governing the HI are the voltage (V), the amperage (A), and the speed (S) [3]:

$$HI = \frac{60VA}{1000S}. \tag{1}$$

Based on the HI, it is possible to estimate the cooling rate in the join (CR) and in the weld center as (CRC)

$$CR \propto \frac{1}{P_t HI}, \tag{2}$$

where P_t is the preheating temperature. Furthermore [4]:

$$CRC = \frac{2\pi k (T_s - T_f)^2}{HI}, \tag{3}$$

where k is the thermal conductivity coefficient, T_s is the initial temperature, and T_f is the final temperature.

Proposed Mechanism

It is known that in the SAW process API X70 PSL2 steels generates a microstructure of epitaxial austenitic grains. Using the unmelted grains as a substrate, this grows in the $\langle 100 \rangle$ preferential direction [5]. When the A3 temperature is surpassed during cooling, different kinds of ferrites start to grow: first, *Alotriomorphic ferrite* (AlF) appears in the austenitic grain boundaries [6]. Controlling its evolution can be achieved through manipulating the cooling rates and the chemical composition. From this phase onward, the *Widmanstatten ferrite* (WF) grows, and finally the *Acicular ferrite* (AF) appears in the sites between the AlF and WF. The generation of AF is influenced by multiple factors as it is generated intergranularly in nonmetallic inclusions such as oxides [7].

Once the weld microstructure is known for the AF, WF and AlF mixture, we propose a stress concentration in the WF tip caused by the phase geometry having a large intensity factor generating a crack in this place. Once a crack is generated, it can propagate through AF or AlF.

Tian et al. [8] propose a crack propagation mechanism across the AF. Byoungchul et al. report that the full AF microstructure have better toughness against the coarse grains of polygonal ferrite, presenting evidence in tests such as Charpy impact tests, DWTT (Drop Weight Tear Test). The last test presents a defect called inverse fracture [9, 10], reported to be present only in coarse grains polygonal ferrite and that the AF avoid this type of failure, as the shape and distribution of this phase do not allow for brittle crack propagation, guiding it in a chaotic manner across of AF grains: if a crack propagates across the AlF there, is a large probability that it propagates in a brittle manner. As stated earlier, the preferential growth direction is $\langle 100 \rangle$ and it is known that $\{100\}$ is the cleavage plane for BCC structures [11]. Also Joo et al. [12] work with similar steels, reporting results of CVN tests showing an effect of the crystallographic plane in the crack propagation mode. Therefore there is a lesser toughness of the polygonal ferrite against the AF.

Based on the above, we propose to decrease the AlF size and consequently the WF, with the goal of increasing the AF density with the purpose of guiding the crack in a chaotic manner, resulting in higher toughness. In order to change the microstructure, we manipulate the HI and consequently the cooling rates, obtaining different phase sizes and densities.

Experimental Procedure

An API X70 PSL2 steel was used with the chemical composition shown in Table 1. Using the HI shown in Table 2, once the were joins welded, we obtained samples for microstructural analysis, CVN, oxygen percent and hardness tests. Three double welded joins of approximately 1.5 m of length were performed (inner and outer). The samples were obtained from 1 m in the middle leaving 0.25 m in each side to stabilize the weld, the cutting samples starts where the join finish because that is the part of the weld where the parameters are most stabilized, starting with the oxide composition sample then the metallographic, harness and finally the CVN. The metallographic preparation was performed under the ASTM E-3 [13] with sandpaper from 220 to 2400, then the polish with cloth and aluminium oxide of 1 μm, then a 3% nital etching during 10–15 s to reveal the microstructure.

Charpy V-notch impact tests were performed under the ASTM E-23 [14] and API 5L 45 ed. [15] standards, both in the external and internal weld as well as in the

Table 1 Chemical composition of API X70 PSL2 steel used in this work

	C	Mn	P	S	Cu	Ni	Si	Cr	Al	Nb	Mo	Ti	V
API X70 PSL2	0.043	1.629	0.012	0.002	0.016	0.175	0.276	0.187	0.033	0.047	0.007	0.016	0.045

Table 2 Oxygen composition in the three samples

Sample	HI	Oxygen (ppm)
1	192	456
2	219	424
3	248	438

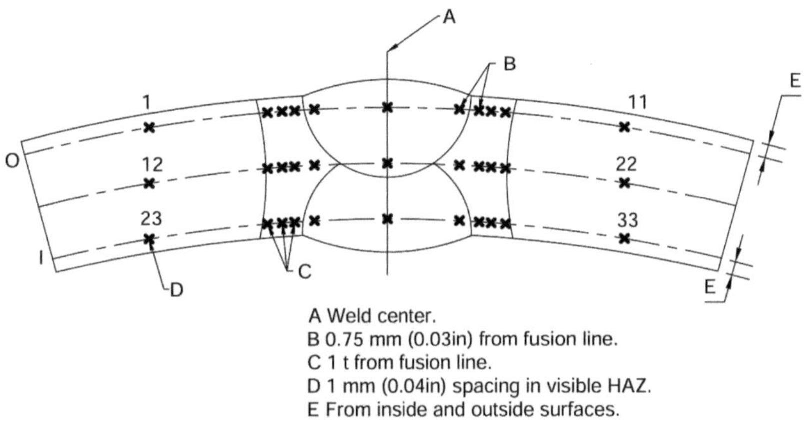

Fig. 1 Indentation pattern based on API 5L 45 ed. [15]

HAZ in full size. The oxygen composition was determined under the ASTM E-1019 [16] standard. The hardness test was performed under the ASTM E-384 [17] and API 5L [15] 45 ed. following the pattern of the latter as shown in Fig. 1.

Results and Discussion

The results of the proposed experiments are presented and discussed in the following sections.

Oxygen Composition

The results from the oxygen-composition analysis are shown in Table 2, showing the heat inputs and the oxygen content in parts per million (PPM). The lowest oxygen content corresponds to the intermediate HI, for which we expect a lower density of AF as there are fewer nucleation sites (oxides). The highest oxygen content corresponds to the lowest HI, implying higher AF content and also eliminating the presence of WF and possibly a thin AlF, the AF preventing the growth of the other phases.

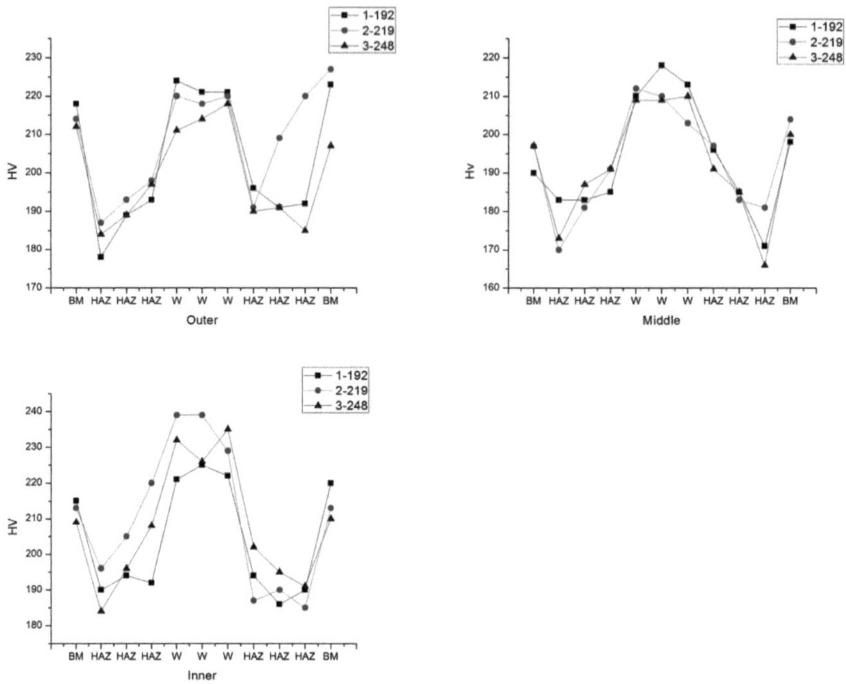

Fig. 2 Hardness Vickers in outer, middle, and inner side, following the pattern shown in Fig. 1

Vickers Hardness

The indentation pattern is shown in Fig. 1 and the test results are shown in Fig. 2 that shows the hardness dispersion in the different weld zones. The higher hardness observed in the weld is due to the phase transformation that results in a mixture of different kind of ferrites; the location is also where fastest cooling rates occur. The HAZ shows the lowest hardness as it is not affected completely: heat only elevates the temperature and causes grain growth without phase transformation. It also exhibits low hardness, which is good for the weld properties, as this is the transition from weld and base metal, and a high stress concentration is present.

Charpy "V" Notch (CVN) Tests

The results of the test CVN are shown in Fig. 3. Starting with the overall results we can see a fairly ductile HAZ, this is expected in view hardnesses obtained, for the

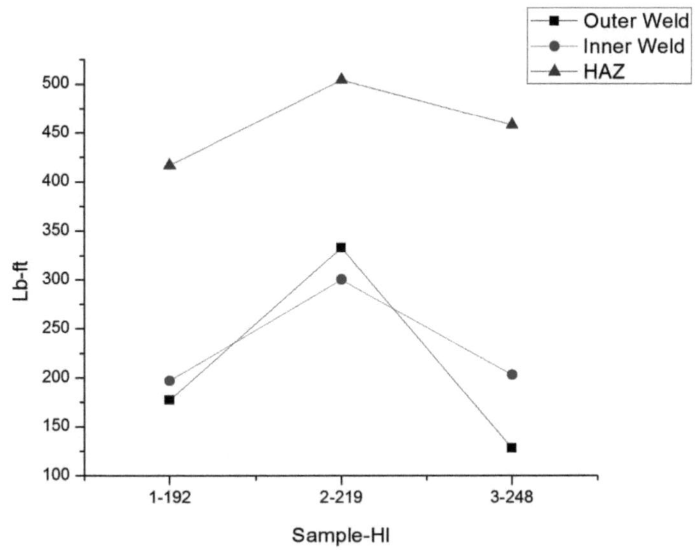

Fig. 3 CVN results in HAZ, outer and inner weld

joins note that the inner weld has a slightly better toughness than the outside, this may be because in the process of welding the inner weld is first done then the outside, in samples 1 and 3 the heat generated in the outside welding promotes a stress relieving in the inner weld changing the microstructure and the toughness a bit, comparing the properties in each join we can observed in join 2 a clear advantage compared to 1 and 3 join properties, keeping in mind the oxygen content in the union 2 is the lowest.

Microstructure

Starting the analysis with the microstructure of the top weld it is observed a clear difference in the sizes of AlF as well as in the presence of WF, the microstructure almost completely of acicular ferrite in Sample 1, with a very small quantity of AlF and no WF, whereas Sample 2 exhibits a common microstructure of a combination of the three ferrites a larger size AF unlike Sample 1. In Sample 3 we observed absence of WF with thicker AlF and an intermediate size of AF. In the bottom weld observe an overheated microstructure without defined morphologies, this result for the generated heat by the external weld affects the inner join changing the microstructure thus losing the defined ferrite morphologies (Fig. 4).

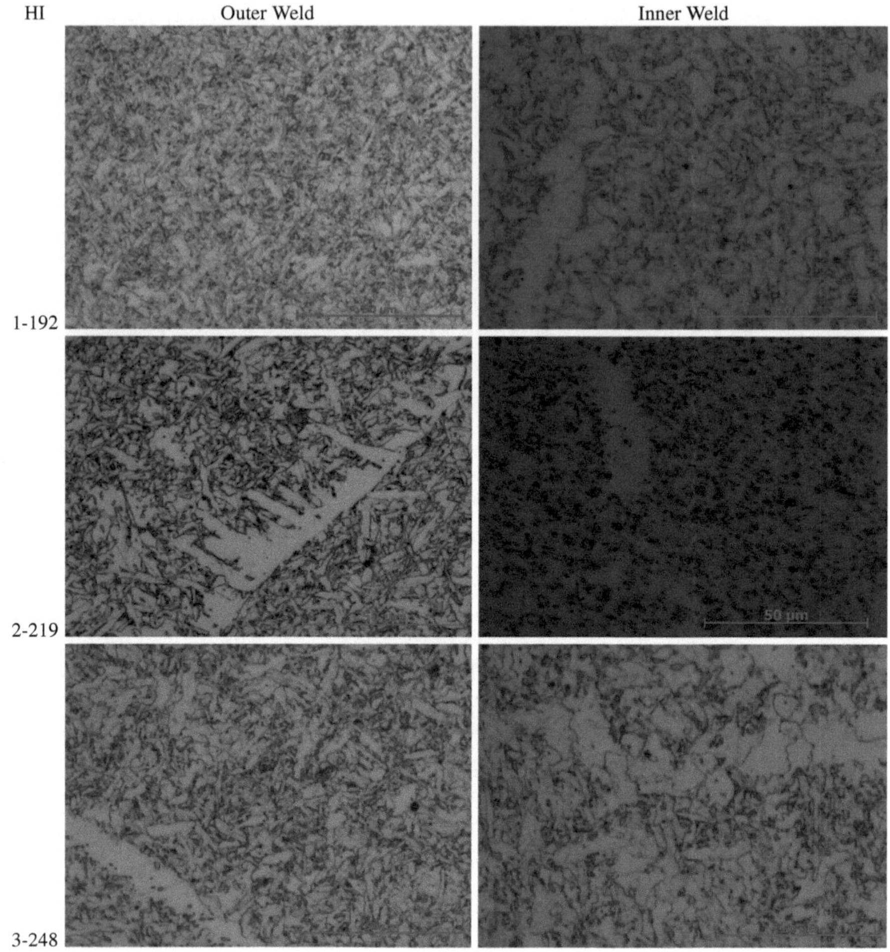

Fig. 4 Microstructure of different HI outer and inner weld

Conclusions

Our results allow to draw the following conclusions:

- It is possible to modify both the microstructure and toughness by manipulating the HI.
- The join of Sample 2 exhibited the highest toughness, the lowest oxygen content, the largest AF, and the presence of WF.
- The oxygen content has a direct effect in the size and density of AF, changing the thickness of AlF and the presence of WF, changing considerably the welding toughness.

- In the proposed mechanism we believe that the crack starts in WF and then it can propagate in two ways: either by ductile manner across the AF, or in a brittle manner across the AlF. In Samples 1 and 3, however, where there is no WF and toughness is the lowest, the size and density of AF is an important factor; this phase can be controlled by the oxygen content which can in turn be controlled by the heat input.
- It appears that the size of AF may affects the dislocation motion: with small sizes the dislocations cannot move freely, producing strain hardening and thus decreasing the toughness of the weld, as observed in the CVN tests.

References

1. Winston Revie, R. (2015). *Oil and Gas Pipelines, Integrity and Safety Handbook* (p.160). New Jersey: Wiley.
2. Malcom Gray, J., & Siciliano, F. *High Strength Microalloyed Linepipe: Half a Century of Evolution*.
3. Scott Funderburk, R. (1999). Key concepts, a look at HEAT input, key concepts in welding engineering. *Welding Innovation, XVI*(1).
4. *Welding Technology, Welding Handbook* (8th ed., Vol. 1). American Welding Society.
5. Kou, S. *Welding Metallurgy* (2nd ed.) NY: Wiley-Interscience.
6. Bhadeshia, H. K. D. H. *Steels, Microstructure and Properties* (3rd ed). Amsterdam: Elsevier Ltd.
7. Kim, Y. M., et al. (2008). Transformation behavior and microstructural characteristics of acicular ferrite in linepipe steels. *Material Science and Engineering A, 478.*
8. Tian, Y., et al. (2015). Effect of ultra fast cooling on microstructure and mechanical properties of pipeline steels. *Journal of Materials Engineering and Performance, 24*(9).
9. Hwang, B., et al. (2005). Effects of microstructure on inverse fracture occurring during drop-weight tear testing of high-toughness X70 pipeline steels. *Metallurgical and Material Transaction A, 36A.*
10. Hwang, B., et al. (2005). Correlation of rolling conditions, microstructure and low-temperature toughness of X70 pipeline steels. *Metallurgical and Material Transaction A, 36A.*
11. Pineau, A., et al. (2016) Failure of metals I: Brittle and ductile fracture. *Acta Materialia, 107.*
12. Joo, M. S., et al. (2014). Toughness anisotropy in X70 and X80 linepipe steels. *Material Science and Technology, 30.*
13. ASTM E-3, Standard Practice for Preparation on Metallographic Specimens.
14. ASTM E-23, Standard test Methods for Notched Bar Impact Testin of Metallic Material.
15. API 5L 45 ed.
16. ASTM E-1019, Standard Test Methods for Determination of Carbon, Sulfur, Nitrogen, and Oxygen in Steel and in Iron, Nickel and Cobalt Alloys.
17. ASTM E-384, Standard Test Method for Microindentation Hardness of Materials.

Evaluation of Non-destructive Techniques (Thermography, Ultrasound and Eddy Current) for Detection of Failures in Metallic Substrates with Composite Anticorrosive Coatings

Marcella Grosso, Priscila Duarte de Almeida, Clara Johanna Pacheco, Iane Soares, João Marcos Rebello, Sergio Damasceno Soares, Isabel Cristina Margarit-Mattos and Gabriela Ribeiro Pereira

Abstract The aim of this study is to evaluate the ability of three nondestructive techniques—ultrasound, thermography and eddy current—for the detection of defects that can occur in metallic substrate with anticorrosive coating and compare the results regarding the advantages and disadvantages of each one of these inspection techniques. For this study, samples were made from a metallic substrate in which a composite anticorrosive coating and two types of defects were introduced: localized corrosion (defects of different geometries and depths inserted into the metallic substrate) and adhesion failure between the coating and the substrate. According to the results, all three techniques showed to be able to detect defects simulating localized corrosion on the substrate. However, the adhesion failure was only effectively detected by thermography and ultrasound and a high correlation between these two techniques was observed.

Keywords Anticorrosive coatings · Thermography · Ultrasound · Eddy current

M. Grosso · P.D. de Almeida · C.J. Pacheco · I. Soares · G.R. Pereira (✉)
Laboratory of Nondestructive Testing, Corrosion and Welding (LNDC/COPPE/UFRJ), Av. Pedro Calmon s/n, LNDC, Ilha Do Fundão, Rio de Janeiro, RJ CEP: 21941-596, Brazil
e-mail: gpereira@metalmat.ufrj.br

J.M. Rebello · I.C. Margarit-Mattos · G.R. Pereira
Department of Metallurgical and Materials Engineering, Federal University of Rio de Janeiro (PEMM/COPPE/UFRJ), Rio de Janeiro, RJ CEP: 21940-915, Brazil

S.D. Soares
PETROBRAS/CENPES, Avenida Horácio Macedo, 950-Ilha Do Fundão, Rio de Janeiro, RJ CEP: 21941-972, Brazil

© The Minerals, Metals & Materials Society 2017
M.A. Meyers et al. (eds.), *Proceedings of the 3rd Pan American Materials Congress*, The Minerals, Metals & Materials Series, DOI 10.1007/978-3-319-52132-9_27

Introduction

Damages caused by corrosion may lead to structural failures, which cause not only high financial losses for industries, but can also promote dramatic consequences for the environment [1]. The use of anticorrosive coatings is one of the several existing methods to control corrosion, which has gained extensive application in various industries. Particularly in oil and gas industry, these coatings can be found in storage tanks, pipelines, heat exchangers and structural components.

The presence of defects, which may arise either during the application of the coating or in service, compromises the effectiveness of this method on the corrosion protection. Although the coatings technology have been presenting significant progress in recent years, the presence of failures is commonly found in the field [1]. Therefore, it becomes extremely important to use non-destructive techniques (NDTs) to inspect this material, both after application of the coating, to ensure the efficiency of its application, as well as periodically to monitor the performance of the coating and the evolution of its integrity in service. However, despite this great importance, there are few studies analyzing the performance of non-destructive techniques for inspection of these materials. This is the fact that motivated the development of this study.

Over the past decades, the infrared thermography has been recognized as a powerful NDT for inspecting and detecting surface and subsurface defects in different kinds of materials [2]. Pulsed thermography is the most common thermographic technique used. This technique consists on briefly heating the sample surface, using an external pulsed heat source, and then recording the surface temperature decay by an infrared camera. The presence of defects reduces the heat flow from the surface to the substrate, so, in the resulting thermal images, defects appear as areas of different temperatures compared to the surrounding sample area. This is the main mechanism to detect defective regions on thermal images [3].

Ultrasonic testing (UT) is a traditional NDT often used for microstructural characterization or defect detection in innumerous types of materials and structures. It is based on the transmission of short pulsed and high frequency ultrasonic waves in the material to be evaluated; the waves travel through the material and whenever they encounter an interface between two media (e.g. defects, back wall etc.), they are reflected back to the source and converted to an electrical signal that can be analyzed for the presence or absence of defects [4].

Eddy current testing (ECT) is commonly used in non-destructive inspections. It is based on the interaction between the primary magnetic field and the tested material. This interaction induces eddy currents within the test piece [5]. The presence of discontinuities, such as corrosion and material losses, is detected by monitoring the changes in the sensed magnetic field caused by perturbation in the eddy currents due to the interaction with defects.

The aim of this study was to evaluate the ability of three traditional NDTs used in the industry—UT, thermography and ECT—in the detection of typical defects presented by metallic substrate with anticorrosive coating. The studied samples consisted on a carbon steel plate on which a composite anticorrosive coating was applied. Two types of defects were artificially inserted in the samples: localized corrosion at the substrate with different geometries and depths, and adhesion failure between the coating and the substrate.

Materials

Sample

The sample consisted on a carbon steel plate of 150 mm × 100 mm × 4.7 mm dimensions (length × wideness × thickness) covered by two layers of a composite coating (epoxy + glass fiber particles). The thickness of the coating on each side of the sample was of approximately 600 μm. The steel plate had six flat bottom holes (FBHs) of several diameters and depths, as presented in Table 1. Before the application of the coating layers, Fe_3O_4 particles were added to the holes in order to reproduce the presence of corrosion product.

During the coating application, some areas also had the adhesion failure defect inserted—an air layer between coating and steel plate—with an expected thickness up to 500 μm.

Non-destructive Methods

Thermography Testing

For the thermographic inspection, an infrared camera (Flir SC5600) was used to collect the thermograms and a set of 5 kW halogen lamps were used to generate a 10 s heat pulse on the sample surface. The tests were carried out with the heat source positioned on the same side of the specimen with respect to the infrared camera (reflection mode). The infrared camera was controlled by a computer which allowed the capturing of thermal images with the same frame of the camera, 30 Hz, as well as the storage of these images. The result of the inspection is shown in a sequence of thermographic images, in which each frame shows the temperature distribution at the surface of the material at a certain time.

Table 1 Description of the defects

FBH ID	Diameter (mm)	Depth (mm)
1	5.15	1.96
2	5.20	2.21
3	5.15	2.86
4	11.04	1.88
5	10.96	2.14
6	10.85	2.73

Ultrasonic Testing

A 5 MHz spherically focused single probe was used in the UT inspections. The application of a focused probe, which converges the acoustic energy to a predetermined focal spot, was necessary to achieve a satisfactory resolution for the detection of defects in this material and the evaluated thickness range [6, 7]. In order to maintain the focal spot on the desired depth, all scanning were performed in an immersion environment. A semi-automated scanning system was used for the probe manipulation and an Olympus Omniscan MX2 was responsible for data recording. Scanning was conducted to both sides of the specimen, simulating both internal and external inspection of the part.

Eddy Current Testing

For the eddy current tests, a commercial ECT system (ZETEC) and an oscilloscope were used to obtain the variation in the signal amplitude during the scan. The eddy current probe for inspection was tested in the frequency range of 5–50 kHz. The scanning of the samples was performed using a mechanical system to move the probe in an automated mode that allows to perform the scan in the x and y direction of the samples. The acquired data was processed in order to visualize the C-scan maps, and used the image processing to eliminate the regions with lift-off variation. These regions represent the variation in the coating thickness [8].

Results

General Results

Figure 1 shows the best thermographic image obtained during the cooling of the sample without any image processing. The criterion for choosing the best image was based on the image that produced the highest thermal contrast between the

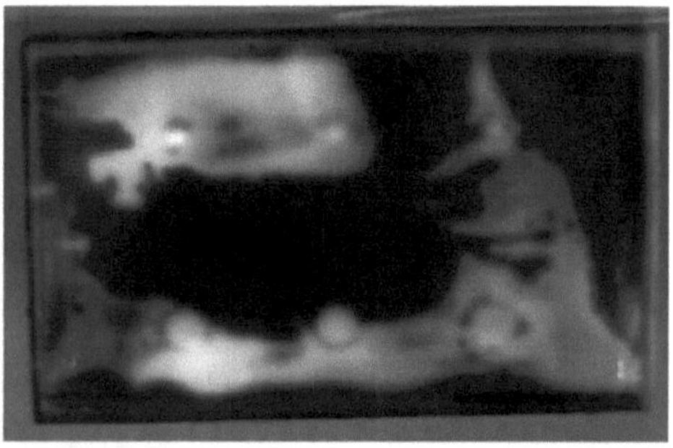

Fig. 1 Thermographic image at the time of the best contrast through the thermographic inspection performed on the same face of insertion of the FBHs

defective region and non-defective region. Analyzing this figure is possible to see extensive lighter areas, i.e., areas with temperatures above the rest of the sample, which can be associated to an adhesion failure defect. Despite the presence of these areas, FBHs can also be observed in the image, so, the thermography can detect this type of defect. Although the thermographic inspection was also done on the opposite surface of the sample, it was not possible to detect any of the defects. This behavior is probably related to the depth of the defects in relation to the surface being inspected by the thermography.

Figure 2 presents ultrasonic C-scan images from scanning made on each side of the sample. In these images, bright colors represent defective areas and dark colors represent non defective areas. Figure 2a shows the inspection on the face in which the FBHs were inserted. None of the holes could be detected through this side due to the existence of a large area with an air layer between the coating and the steel plate. Figure 2b presents the C-scan of the inspection made on the opposite side of the sample. All six holes were successfully detected and no indication of an air layer was observed.

Figure 3 shows the processed C-scan image obtained by the conventional eddy current testing on the face in which the FBHs were inserted. It was possible to

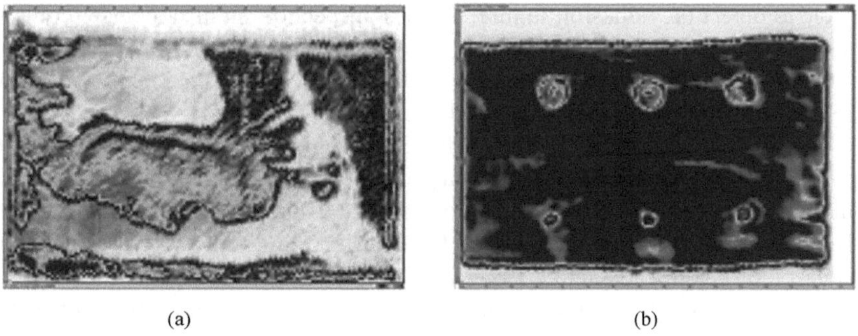

Fig. 2 Ultrasonic C-Scan: **a** scanning on the same face of insertion of the FBHs, **b** scanning on the opposite face

Fig. 3 Processing C-Scan image obtained by conventional eddy current technique. Scanning on the same face of insertion of the FBHs

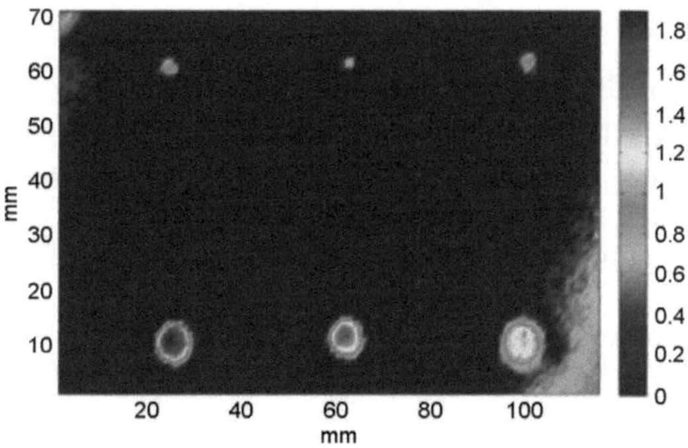

identify six holes on the sample. For larger diameter defects, the amplitude shows values around 1.9 V, and for the defect with smaller diameter the amplitude is about 0.9 V. It was obtained in some regions amplitudes with the similar values in relation to amplitude of small diameter defects. In this case, the results can be produced either by changing the thickness of the coating or boundary effect or defects inserted in the preparation of the sample, but it is not evident the detection of adhesion failure. On the opposite side to FBHs holes, no variation was detected in the amplitude signal, due to the thickness of the sample (lift-off).

Methods Comparison

A comparison among all three techniques was carried out considering the defects, the surface in which the inspection was made and the speed in obtaining and interpreting the results. Thermography and ECT were able to detect all six FBHs in the inspection on the same surface in which they were machined. In the thermographic image, this detection was partially compromised by the coexistence of the adhesion failure, which contributed to the reduction of the thermal contrast and subsequent visualization of the FBHs. ECT detected all six FBHs, however, it was not able to detect the adhesion failure. The UT inspection, in turn, detected only the adhesion failure on this same surface. It is important to highlight that the comparison between the UT and thermographic images shows that there is a strong correlation of the morphology of this defect for both techniques.

UT was the only technique able to detect all six FBHs on the inspection on the opposite surface. This indicates that among the three evaluated techniques, UT would be the best suited for the inspection through external access of the equipment.

Depending on the available system for data recording, all three techniques are capable of producing instantaneous results with real-time assessment. The generated images are of simple interpretation, and all three techniques are equivalent in this aspect.

Conclusions

For this study, the ability of three nondestructive techniques for the detection of defects that can occur in metallic substrate with anticorrosive coating was evaluated. The FBHs were successfully detected using the ECT, UT and thermography, but only the UT and the thermography could detect the adhesion failure between the coating and the substrate.

Acknowledgements The authors would like to thank Petrobras for the technical and financial support, and CAPES and CNPq for the financial support.

References

1. Sørensen, P. A., Kill, S., Dam-Johansen, K., & Weinell, C. E. (2009). Anticorrosive coatings: A review. *Journal of Coatings Technology and Research, 6,* 135–176.
2. Montanini, R. (2010). Quantitative determination of subsurface defects in a reference specimen made of Plexiglas by means of lock-in and pulse phase infrared thermography. *Infrared Physics & Technology, 53,* 363–371.
3. Maldague, X. P. V., & Moore, P. O. (2001). Infrared and thermal testing. *American Society for NonDestructive Testing, 3.*
4. Krautkrämer, J., & Krautkrämer, H. (1990). *Ultrasonic testing of materials* (4th ed.). New York: Springer.
5. Martin, J. G., & Sánchez, E. V. (2011). Non-destructive techniques based on eddy current testing. Javier García-Martín. *Sensors, 11,* 2525–2565.
6. Prevorovsky, M., Landa, M., Blahacek, D., Varchon, J., Rousseau, L., & Ferry, D. (1998). Ultrasonic scanning and acoustic emission of composite tubes subjected to multiaxial loading. *Ultrasonics, 36,* 531–537.
7. Aymerich, F., & Meili, S. (2000). Ultrasonic evaluation of matrix damage in impacted composite laminates. *Composites: Part B, 31,* 1–6.
8. Khan A. N., Khan S. H., Ali F., & Iqbal M. A. (2009). Evaluation of ZrO_2–24MgO Ceramic Coating by Eddy Current Method. *Computational Materials Science, 44.*

High Temperature In-Situ X-Ray Analysis of a Lean Duplex Stainless Steel

Adriana da Cunha Rocha, Andrea Pedroza da Rocha Santos and Gabriela Ribeiro Pereira

Abstract Duplex Steels have been used for different applications in the oil and gas industry due to their high strength and corrosion resistance. In this work, a Lean Duplex Steel (UNS S32304) has been investigated by High Temperature In-Situ X-Ray Diffraction, in order to address phases transformations occurring during high-temperature real time exposure and the formation of deleterious phases such as sigma phase. Samples were subjected to a temperature of 800 °C for different time intervals (30 and 60 min) inside a high temperature furnace attached to an x-ray diffractometer. Formation of sigma phase was observed, both in the ferrite grain boundaries and in the ferrite/austenite interface. The volume of sigma phase was measured by quantitative phase analysis using the Rietveld method and results varied from 2.80 to 15.92% in volume.

Keywords Duplex stainless steel · Sigma phase · In-Situ X-ray diffraction · Rietveld method

Introduction

Duplex stainless steels (DSS) are composed of approximately equal volume fraction of ferrite and austenite, featuring high mechanical resistance and corrosion [1]. They are used in a wide range of applications, such as chemical processing, in

A. da Cunha Rocha (✉) · A.P. da Rocha Santos · G.R. Pereira
Department of Materials Sciences and Metallurgy,
Federal University of Rio de Janeiro (PEMM/COPPE/UFRJ),
CX Postal: 68505 Rio de Janeiro, R.J, Brazil
e-mail: adrirocha@metalmat.ufrj.br

A.P. da Rocha Santos
e-mail: andrea@metalmat.ufrj.br

G.R. Pereira
e-mail: gpereira@metalmat.ufrj.br

© The Minerals, Metals & Materials Society 2017
M.A. Meyers et al. (eds.), *Proceedings of the 3rd Pan American Materials Congress*,
The Minerals, Metals & Materials Series, DOI 10.1007/978-3-319-52132-9_28

maritime environments and in oil and gas industries. In most cases, DSS are chosen based on their strength and State of passivation in various environments.

Recent developments in stainless steels are driven, among other factors, by the low cost of the final material. This fact led to the development of duplex stainless steel subgroup known as "Lean Duplex", which has lower content of alloying elements in relation to standard duplex stainless steels [2]. However, their mechanical resistance is approximately twice that of austenitic steels, as well as having better corrosion resistance and yield superior [3].

When exposed to temperatures above 600 °C though, the balance of alloying elements can be modified due to precipitation of various secondary phases. The most common are precipitated austenite, nitrides, carbides (mostly $M_{23}C_6$ type) and intermetallic elements phases, such as sigma (σ) and chi (χ). The sigma phase is typically enriched with Cr and Mo so, its formation can cause a drastic deterioration in toughness, corrosion resistance and weldability of duplex stainless steels [4].

In this work, S32304 steel samples were subjected to a temperature of 800 °C for 30 and 60 min in an oven coupled to an x-ray diffractometer. Several in-situ scans were performed during the different heating intervals, in order to analyze real-time transformation occurring in the constituent phases. Formation of sigma phase in steel, both on the contours of ferrite grain as ferrite/austenite interfaces was observed. A quantitative analysis, using the Rietveld method, found a variation of 2.80–15.92% in volume of sigma phase. It was then observed that the formation of sigma phase in the temperature of 800 °C occurs significantly after 30 min of heating in amounts that can deeply compromise the properties of the steel.

Materials and Methods

Specimens with dimensions of 20 mm × 20 mm × 1.5 mm were cut and milled from a SSD plate and subjected to metallographic preparation prior to heating. Table 1 present the composition of the Lean Duplex steel (UNS S32304).

Diffraction was performed in a D8 Discover (Bruker AXS), using cobalt Co Kα radiation (λ = 1.789 Å), equipped with a Lynx Eye PS Detector. The primary optics was mounted using a Co Göbel Mirror followed by two slits of 1 and 6 mm and a soller slit with an aperture of 2 cm × 1 cm, that assure that an area of 2 cm^2 was being scanned. The secondary optics consisted of a K_β filter followed by an 8 mm slit and a axial Soller slit with maximum divergence of 2.5°, all mounted in the PSD detector. The scanning data was obtained in the 2θ range of 45°–60°. The step-size applied was 0.001° and the scanning velocity was 0.5 s/step. The Rietveld [5, 6] analysis of each scan was carried out using Diffrac PlusTOPAS (ver 4.2)

Table 1 Lean duplex steel (UNS S32304) composition

Element	Fe	C	Cr	Mo	Mn	Ni	N	Si	V	S, P
wt%	bal	0.04	22.67	0.12	1.29	4.64	0.11	0.39	0.09	<0.01

software, which is a based general non-linear least squares system driven by a scripting language, which focus is in crystallography, solid state chemistry and optimization and, as consequence, has also been applied for quantification through the adoption of the relative weight fraction equation.

For the x ray diffraction analysis, the technique consisted of controlled heating followed by x-ray scans performed during the thermal interval, i.e., a first scan at room temperature (30 °C) was performed and, when the sample hit the target temperature, several scans happened during the thermal cycle. The heating took place at an average rate of 1 °C/s. Subsequently, samples were air cooled to room temperature at an average rate of 2 °C/s. The main phases present were identified by indexation of the peaks revealed in difratogramas, followed by Rietveld calculations.

In addition, samples they were also analyzed by scanning electron microscopy (SEM Zeiss-MA) using a voltage of 30 kV with backscattered electrons (BSE), followed by energy dispersive spectroscopy (EDS) analysis.

Results

Figures 1 and 2 present the diffractograms of sample 1 (S1) and sample 2 (S2) on 30 and 60 min intervals, at 800 °C. It is possible to observe the presence of ferrite (α-Fe), austenite (γ-Fe) and sigma phase (σ), although this latter phase reflections are mostly overleaped by the main peaks of ferrite and austenite.

Phase volume percentage was calculated by Rietveld analysis and values are presented in Table 2.

Observations on the SEM revealed the presence of a darker phase both on the contours of ferrite grain as ferrite/austenite interfaces, as depicted in Fig. 3. EDS

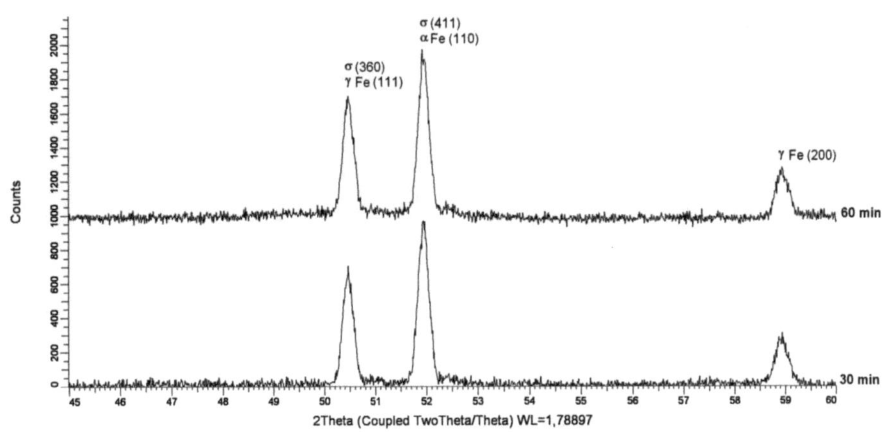

Fig. 1 Diffractograms of sample S1 at 800 °C on a 30 and 60-min thermal interval

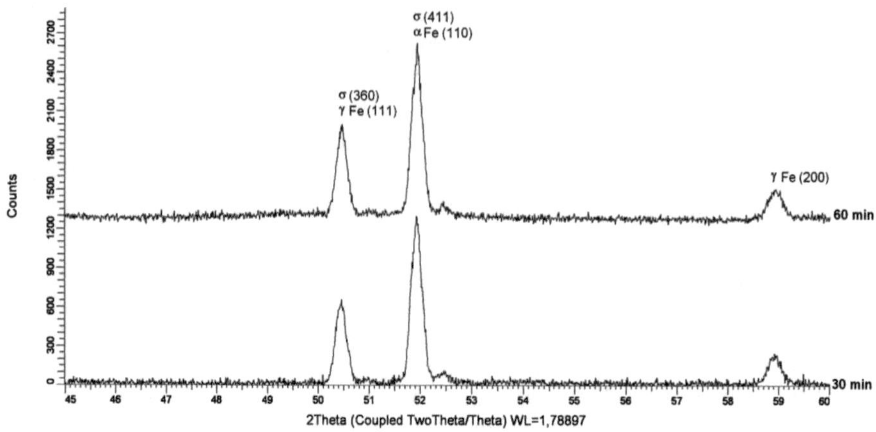

Fig. 2 Diffractograms of sample S2 at 800 °C on a 30 and 60-min thermal interval

Table 2 Calculated values of ferrite, austenite and sigma after different intervals

Samples	Time interval	% volume		
		Ferrite	Austenite	Sigma
S1	Initial scan	63.14	36.86	0
	30 min	47.11	43.53	9.36
	60 min	42.45	41.63	15.92
S2	Initial scan	72.88	27.12	0
	30 min	58.29	38.91	2.80
	60 min	50.12	34.16	15.72

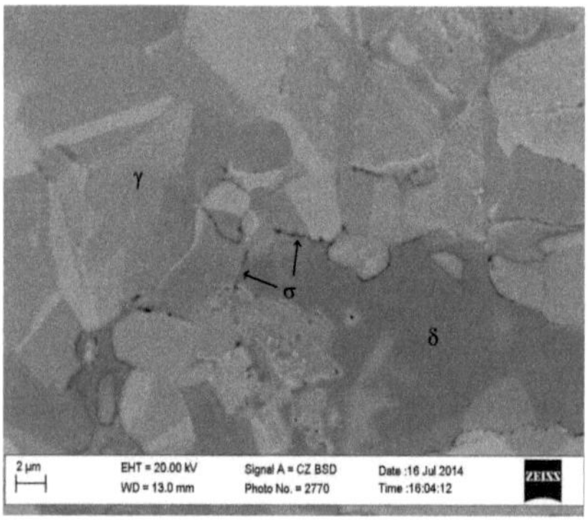

(a) S1 sample

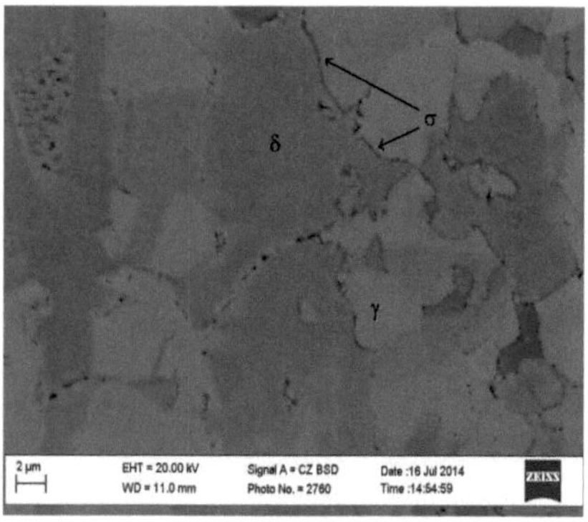

(b) S2 sample

Fig. 3 SEM BSE micrographs of sample S1 (**a**) and S2 (**b**) indicating the areas of sigma phase precipitation in the Lean Duplex Steel matrix. The *lighter grains* correspond to austenite phase, the *grayish grains* to ferrite and the *dark ones* to sigma

Table 3 EDS values obtained from sigma phase precipitates

	Si	V	Cr	Mn	Fe	Ni	Total
S1	0.45	0.65	**30.32**	1.68	**65.10**	1.80	100
S2	0.48	0	**26.38**	1.71	**69.45**	1.99	100

analysis confirmed the high chromium and iron content (Fe–Cr), which are typical sigma phase constituents [4], as presented in Table 3.

Discussion

Samples subjected to a temperature of 800 °C revealed the presence of a precipitated phase (besides the primary ferrite and austenite phases) that was characterized by XRD and SEM as sigma phase.

XRD analysis identified this precipitated as sigma Fe (Cr, Mo) type, with orthorhombic crystal structure.

It was also observed that sigma precipitated in the contours of ferrite grain and in ferrite/austenite interfaces. In fact, many authors have reported this type of precipitation [1, 4, 7–9]. Quantitative phase analysis, using the Rietveld method, indicated that the evolution of this phase differs over time. According to data presented in Table 2, it is noticeable that the ferrite phase volume decreases over time while the sigma phase volume increases. This is in line with the theory that the sigma phase arises from the ferrite grains and grows at the expenses of the same [1, 4, 9].

Analyzing the isothermal sigma phase evolution at the temperature of 800 °C (Fig. 3), continued growth in volume is observed until the phase precipitation stabilizes. Samples S1 and S2 do not reach any common value during the 30-min interval (S1 = 9.36 %vol and S2 = 2.80 %vol), as observed by the plots, but converge to a similar amount when reaching the 60-min interval (S1 = 15.92 %vol and S2 = 15.72 %vol). The extension to which this behavior remains constant was not addressed in this work, however. Scientific studies though, show that the formation of sigma tend to stabilize only after 100 min in high temperature [9].

A simple regression analysis showed that both isothermal curves (S1 and S2) could be fitted according to a 4th degree polynomial expression, indicating that a kinetic behavior for the sigma phase evolution could be addressed by this type of equation. It would be necessary though that further experiments and investigations be performed with more specimens at longer intervals.

Anyhow, it has been observed the formation of a significant amount of sigma phase in both samples (around 15 %vol), what could indeed compromise both the mechanical and corrosion resistance of the steel (Fig. 4).

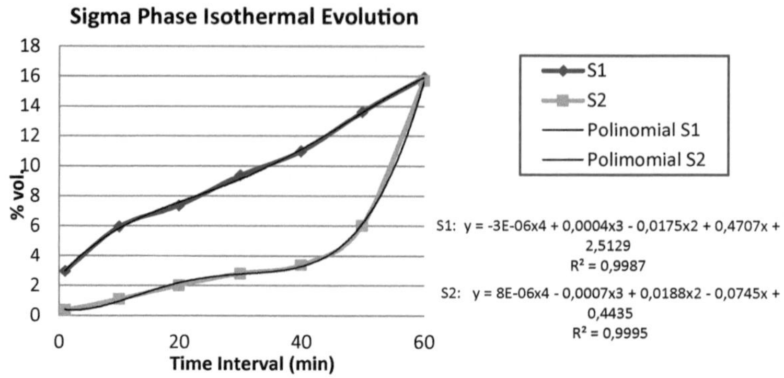

Fig. 4 Plotted data of calculated sigma phase volume percentage versus time, indicating the isothermal evolution of the phase at 800 °C

Conclusions

- UNS32304 steel specimens subjected to a temperature of 800 °C for 30 and 60 min presented significant amounts of sigma phase in their microstructures;
- Sigma phase formation occurred both in the ferrite grain as in the ferrite/austenite interfaces;
- The maximum values of sigma phase calculated were around 15% in volume for both samples, after the maximum exposure time (60 min);
- According to data presented it was observed that ferrite phase volume decreases over time while the sigma phase volume increases;
- It was observed that the precipitation of sigma phase after 60 min of exposure at 800 °C reaches amounts that can compromise the properties of the steel analyzed.

Acknowledgements The authors thank the LNDC for the aid granted.

References

1. Alvarez-Armas, I., & Degallaix-Moreui, S. (2009). *Duplex stainless steels* (1st ed.). London: ISTE.
2. Brytan, Z., & Niagaj, J. (2013). The lean duplex stainless steel welded joint after isothermal aging heat treatment. *Archives of Materials Science and Engineering, 60*(1), 24–31.
3. Hsieh, C., & Wu, W. (2012). Overview of intermetallic sigma (σ) phase precipitation in stainless steels. *ISRN Metallurgy, 2012,* 1–16.
4. Jackson, E. M. L. E. M., VISSER, P. E., & CORNISH, L. A. (1993). Distinguishing between chi and sigma phases in duplex stainless steels using potentiostatic etching. *Materials Characterization, 31*(4), 185–190.
5. Young, R. A. (1995). *The Rietveld method* (1st ed.). New York: Oxford University Press.

6. Rietveld, H. M. (1969). A profile refinement method for nuclear and magnetic structures. *Journal of applied Crystallography, 2*(2), 65–71.
7. Calliari, I., Brunelli, K., Dabala, M., et al. (2009). Measuring secondary phases in duplex stainless steels. *JOM Journal of the Minerals Metals and Materials Society, 61*(1), 80–83.
8. Elmer, J. W., Palmer, T. A., & Specht, E. D. (2007). In situ observations of sigma phase dissolution in 2205 duplex stainless steel using synchrotron x-ray diffraction. *Materials Science and Engineering A, 459*(1–2), 151–155.
9. Fang, Y. L., Liu, Z. Y., Xue, W. Y., et al. (2010). Precipitation of secondary phases in lean duplex stainless steel 2101 during isothermal ageing. *ISIJ International, 50*(2), 286–293.

Structural Integrity of Pipelines

Aníbal C. Di Luch and Nicolás Oyarzábal

Abstract The acceptance standards given in API 1104 Section 9 are based on empirical criteria for workmanship and place primary importance on imperfection length. Such criteria have provided an excellent record of reliability in pipelines service for many years. Alternative acceptance standards for girth welds are presented in Annex A. The use of fracture mechanics analysis and fitness for service criteria for determining acceptance criteria incorporates the evaluation of both imperfection height and imperfection length. This criteria usually provide more generous allowable imperfection length. This approach is applied to a new gas pipeline, API 5L X70 material, 24 inches diameter and 11.9 mm thickness. The applied methodology is presented to perform failure assessment diagram (FAD) and a sensitivity analysis and safety margins or factors are determined.

Keywords Pipelines · Integrity · Welding · Failure

Introduction

This work is part of a comprehensive structural integrity program of pipelines in Argentina, in particular it is applied to a 1810 km length gas pipeline in the North-East region. It is composed mostly of pipes API 5L grade X70 PSL 2, 24 inches nominal diameter and 11.9 mm wall thickness, manufactured by roll bending expand RBE and submerged arc welded SAW inside and outside. Successive sections of approximately 12 m in length that compose the pipeline are

A.C. Di Luch (✉)
Welding Department, CNEA, Buenos Aires, Argentina
e-mail: anibalcd@yahoo.com.ar

N. Oyarzábal
Material Testing Laboratory, ITBA, Buenos Aires, Argentina

© The Minerals, Metals & Materials Society 2017
M.A. Meyers et al. (eds.), *Proceedings of the 3rd Pan American Materials Congress*,
The Minerals, Metals & Materials Series, DOI 10.1007/978-3-319-52132-9_29

welded in situ by qualified welders with Shielded Metal Arc Welding SMAW process downward progression. For the evaluation of the weld joint API Standard 1104 [1] is used. It is common practice to apply Section 9 to assess the welded joint, which is based on empirical criteria and have provided an excellent record of reliability. In this work it was analyzed the implementation of the alternative approach presented in Annex A of this Standard, based on engineering critical assessment (ECA) and fitness for service (FFS). It is needed to perform more complex tests and calculations like CTOD testing to assess both the weld metal and the heat affected zone. Generally, permissible imperfections are larger than those allowed under Section 9 and also takes into account the height and length in its assessment.

The failure assessment diagram FAD is shown in Fig. 1 and the failure assessment curve FAC is given as:

$$K_r = f(L_r) = \left(1 - 0.14L_r^2\right)\left(0.3 + 0.7\exp\left(-0.65L_r^6\right)\right) \tag{1}$$

K_r is the toughness ratio and L_r is the load ratio.

Alternatively, it was also applied the option 2 of BS 7910 [2] failure assessment curve, it requires the mean uniaxial tensile true stress-strain curve.

$$K_r = f(L_r) = \left(\frac{E\varepsilon_{ref}}{L_r\sigma_Y} + \frac{L_r^3\sigma_Y}{2E\varepsilon_{ref}}\right)^{-1/2} \tag{2}$$

E is the elastic modulus, σ_Y is the yield strength and ε_{ref} is the reference strain.

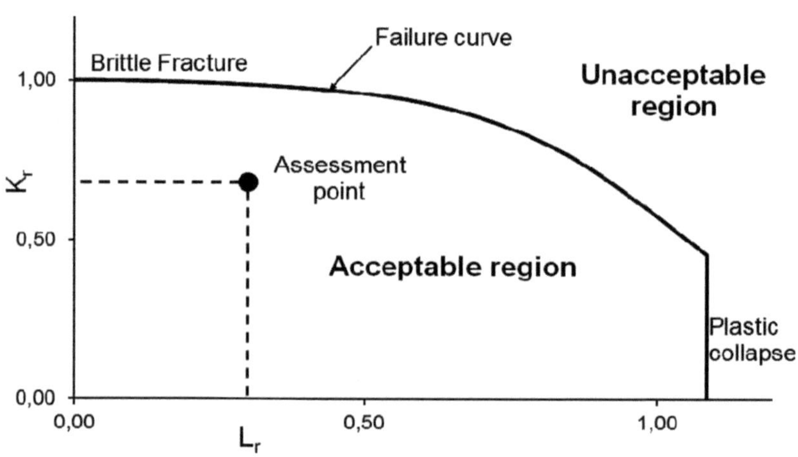

Fig. 1 Failure assessment diagram FAD

Experimental Procedures and Results

The chemical composition of the steel is shown in Table 1 and mechanical properties in Table 2. This is a HSLA steel with good weldability as indicated by the low carbon equivalent values.

The welded joint geometry can be seen in Fig. 2.

The use of Annex A is restricted to the following conditions:

- Circumferential welds between pipes of equal specified wall thickness;
- Nondestructive inspection performed for essentially all welds;
- No gross weld strength undermatching;
- Maximum axial design stress no greater than the SMYS;
- Maximum axial design strain no greater than 0.5%;
- Welds in pump and compressor stations, repair welds, fittings, and valves in the main line are excluded.

A stress analysis was performed to determine the maximum axial design stresses to which the girth welds are subjected to during construction and operation. The highest axial stress occurs during the pipe lowering-in process.

Table 1 Chemical composition of API X70 PSL 2 (wt%)

C	Mn	Si	Ni + Cu	Cr	Mo	Al	Ti + Nb + V
0.06	1.45	0.28	<0.5	<0.4	0.10	0.035	<0.15

Table 2 Mechanical properties of API X70 PSL 2

Grade	YS				UTS			
	Min		Max		Min		Max	
	ksi	MPa	ksi	MPa	ksi	MPa	ksi	MPa
X70	70	483	90	620	82	565	110	758

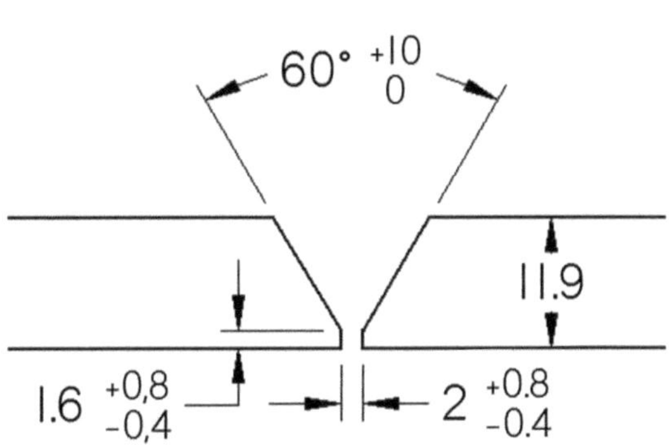

Fig. 2 Weld joint detail

Toughness tests CTOD were performed according to the guidelines of ISO 15653 [3] and ISO 12135 [4] of the weld metal and heat affected zone in 3 different zones of the welded joint, namely: 12, 3 or 9 and 6 as indicated by Fig. 3, giving a total of 6 tests per union. The amount of coupons welded are 5, giving a total of 30 tests. The dimensions of the specimen can be seen in Fig. 4.

The total imperfection length shall be no greater than 12.5% of the pipe circumference (240 mm). The maximum imperfection height shall be no greater than 50% of the pipe wall thickness (6 mm), as shown in Fig. 5.

Fig. 3 Welding zones

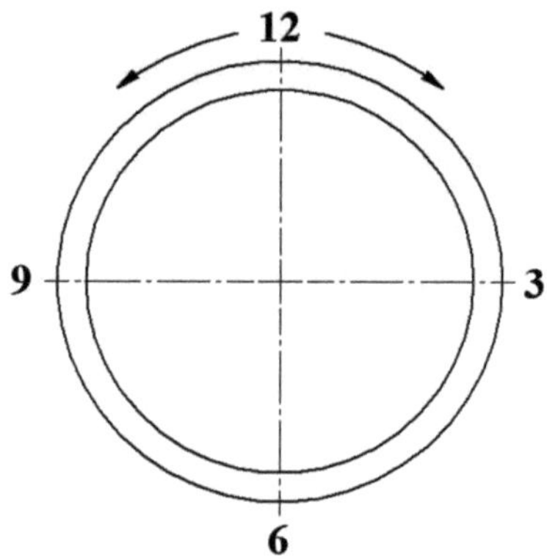

Fig. 4 2B × B specimen

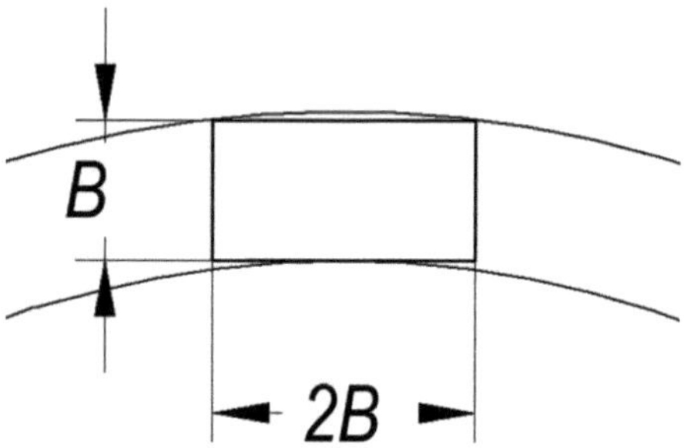

Fig. 5 Imperfection geometry

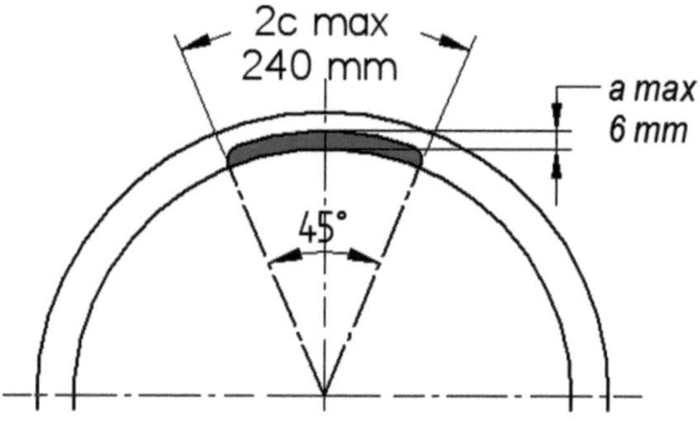

A minimum CTOD value of 0.05 mm is required. Results can be seen in Fig. 6.

Failure assessment diagram FAD provides a quantitative way to assess the acceptability of a planar type imperfection combining two criteria, brittle fracture and plastic collapse. Failure assessment curve FAC, the load ratio L_r and toughness ratio K_r are shown in Fig. 7 and are calculated following the annex directions.

The axial stress range was suggested to be between 50 and 70 Ksi, so three different levels within this range was assessed. It was also used four different CTOD values within a range from 0.05 to 0.30 mm. This combination of values leads to a sensitivity analysis as regards imperfection length and height.

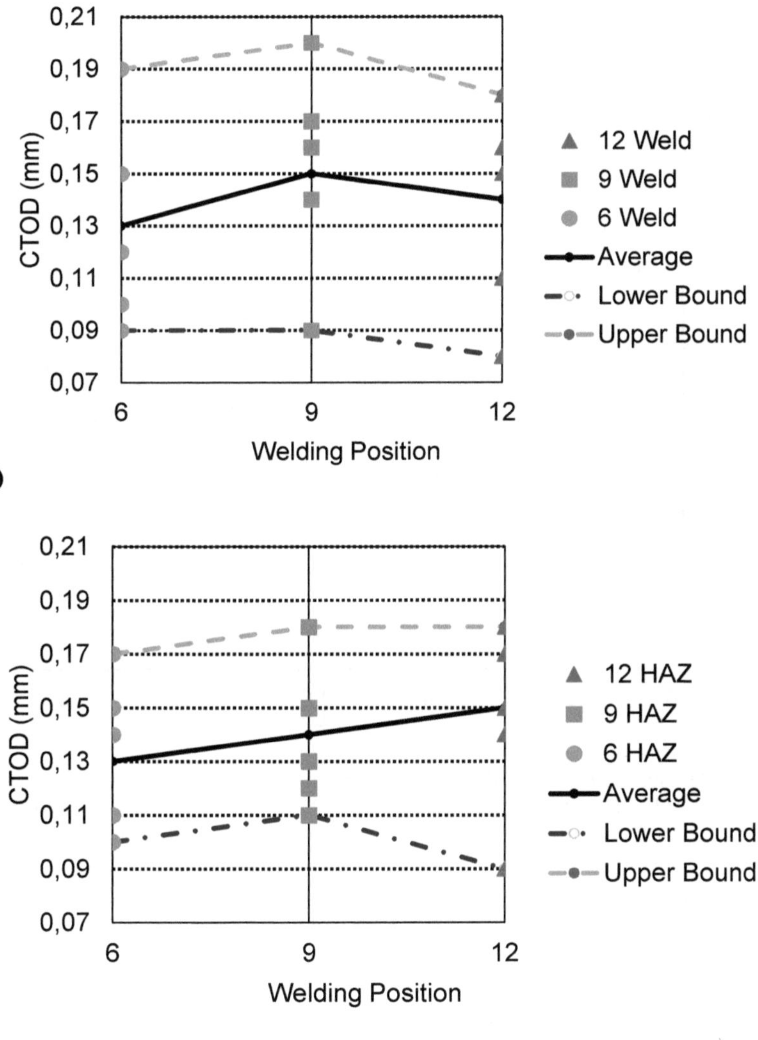

(a)

(b)

Fig. 6 CTOD results for **a** WELD and **b** HAZ, showing the mean value, upper and lower bounds

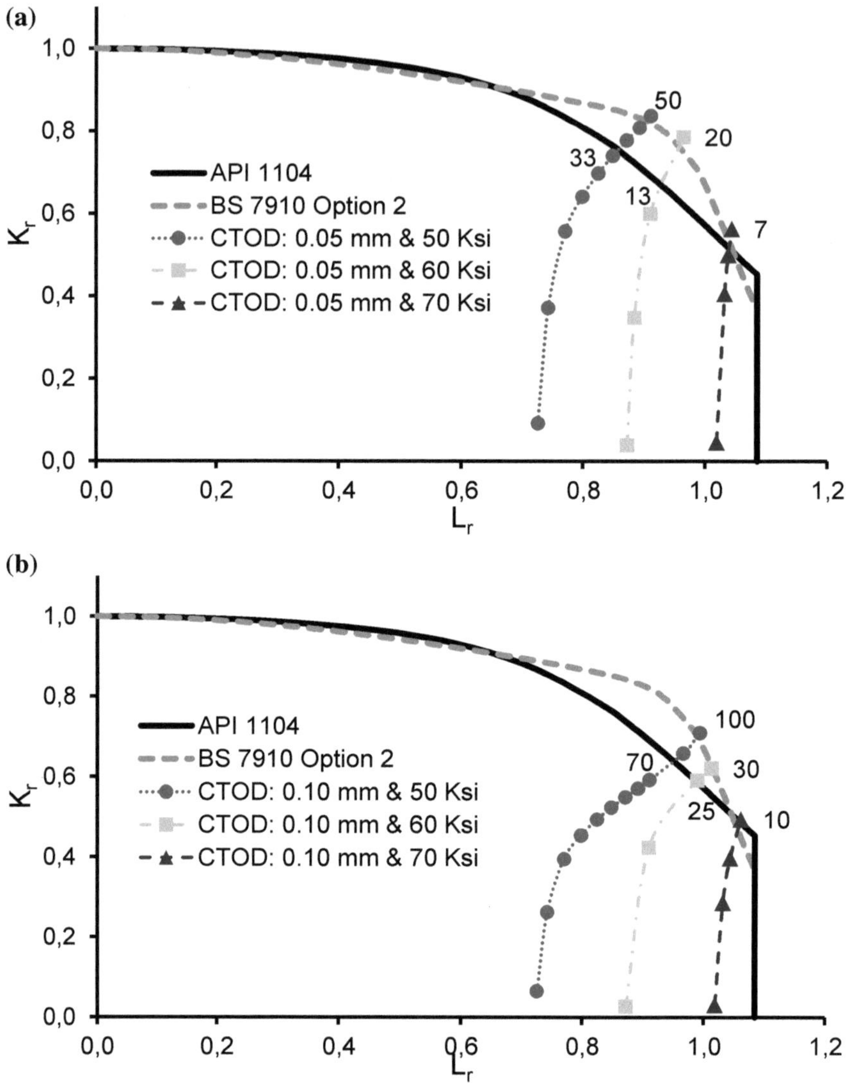

Fig. 7 FAD with the FAC for API 1104 (*black*) and for BS 7910 (*blue*). Three levels of axial stress are considered: 50 (*green*), 60 (*yellow*) and 70 (*red*) ksi. and four different CTOD values ranging from 0.05 mm (**a**), 0.10 mm (**b**), 0.20 mm (**c**), 0.30 mm (**d**). Numbers over the curves represent the imperfection half-length in mm (Color figure online)

(c)

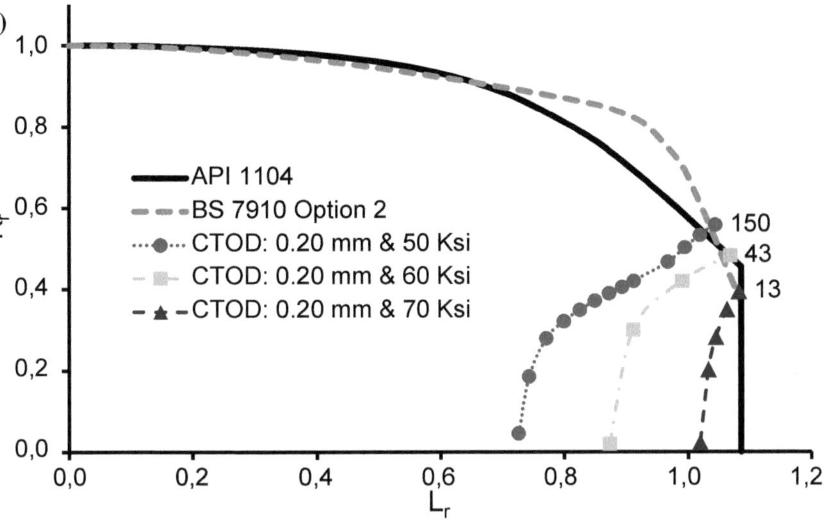

(d)

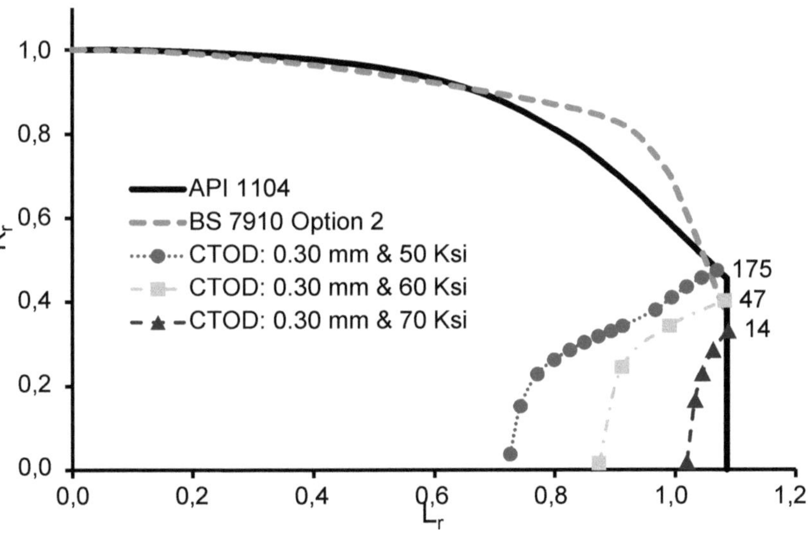

Fig. 7 (continued)

Conclusions

In the present work it was shown that the ECA is a powerful tool to assess the structural integrity of pipelines. Two different approaches were used to define the FAC being the API 1104 more conservative and the BS 7910 allows more generous imperfection dimensions. Finally, a sensitivity analysis limits the imperfection length taking into account the CTOD and the axial stress values.

References

1. American Petroleum Institute. (2013). API Standard 1104. Welding of Pipelines and Related Facilities.
2. The British Standards Institution. (2015). BS 7910:2013 + A1. Guide to methods for assessing the acceptability of flaws in metallic structures.
3. International Standard. (2010). ISO 15653. Method of test for the determination of quasistatic fracture toughness of welds.
4. International Standard. (2002). ISO 12135. Unified method of test for the determination of quasistatic fracture toughness.

Part VI
Materials for Transportation and Lightweighting

Advantages of Hot Compression in the Manufacture of Al-B_4C Composites

Lucio Vázquez, Dulce Y. Medina, Ángel D. Villarreal, David A. López, Gilberto Rangel, Elizabeth Garfias and Manuel Vite

Abstract The aim of this work was to prove that preparation of Al-B_4C by hot compression at 350 °C (HC) followed by sintering at 550 °C would improve mechanical properties with respect to manufacturing by cold compression (CC) followed by sintering at the same temperature, the later experiments were carried out in a previous work. Samples with aluminum matrix adding 0, 3, 5 and 7% were prepared by powder metallurgy technique and tested. The mechanical properties were better using HC. Remarkably, resistance to compression was 3.5 times larger for HC. Vickers hardness, resistance to wearing and impact, as well as density were higher for HC. Scanning electron microscopy of CC specimens exhibit a cellular microstructure while HC ones display a smooth appearance. The better properties of the HC samples are explained due to an improved flow of material at higher temperatures of compression.

Keywords Aluminum-boron carbide (Al-B_4C) · Composites · Hot compression · Cold compression · Mechanical properties

Introduction

Regulations in many countries are pushing to improve fuel efficiency and control of emissions in transportation vehicles: bicycles, buses, planes, and spaceships. Therefore, the necessity of reducing the ratio weight/strength in structural materials

L. Vázquez (✉) · D.Y. Medina · Á.D. Villarreal · D.A. López ·
G. Rangel · E. Garfias
Departamento de Materiales, Universidad Autónoma Metropolitana,
Av. San Pablo 180, Col. Reynosa, C.P. 02200 Tamaulipas,
Mexico City, Mexico
e-mail: vbl@correo.azc.uam.mx

M. Vite
Sección de Posgrado e Investigación, ESIME, Instituto Politécnico Nacional,
Unidad Profesional Adolfo López Mateos, Av. Instituto Politécnico Nacional S/N,
C.P 07338 Mexico City, Mexico

© The Minerals, Metals & Materials Society 2017
M.A. Meyers et al. (eds.), *Proceedings of the 3rd Pan American Materials Congress*,
The Minerals, Metals & Materials Series, DOI 10.1007/978-3-319-52132-9_30

is astringent. Scientific laboratories all around the world are working to develop materials with this characteristic, composite metal matrix composites (MMC) occupy an important place within these materials [1].

The production of metal and ceramic powders by fast cooling has allowed the fabrication of metal matrix composites (MMC) by compacting and sintering, a technique which decreases segregation, feature which is observed in the process of adding ceramic particles into the matrix liquid and shaking up. Composites made of boron carbide powders within an aluminum matrix are used for applications which require high specific strength, hardness, wear resistance and corrosion resistance [2, 3]. Literature on composites reports that additions of about 50% Al_2O_3 to Al allows to yield modulus values of 180 GPa with good ductility [4]. Besides superior mechanical properties, composites must have a good fluidity at the compacting temperature which allows filling of all the cavities within a matrix, feature which allows to make parts of complex geometry.

Materials, Experimental Procedure and Testing

Aluminum powder, purity 99.99%, particle size: 50 meshes. Boron carbide powder, purity >99.0%, particle size: 100 meshes.

Specimen Preparation

Five tests were carried out for each experiment, but the average value of the results are presented in the graphs.

An OHAUS balance was employed to weight the powders. A LABMILL 800 with steel rotating rollers and ceramic wall, 6 mm diameter zirconia spheres were the tools used for milling.

100 g of Al, B_4C powders were weighted in the relations: 100% Al, 97% Al/3% B_4C, 95% Al/5% B_4C, Al 93%/7% B_4C and putted in the mill together with 900 g of ceramic balls to have a relationship balls: experimental material 9:1. The mill rotated at 340 rpm and milling time was 16 h. These experimental parameters showed in previous work that does not produce coalescence [2].

A H13 bar alloy, for hot compression, was selected to make compacting dies. Dies were fabricated to make composite specimens with circular cross section for compression, density, and hardness testing, on the one hand, and square and rectangular transverse section, on the other hand, for impact and wear testing, respectively.

A thin film of oil SAE 240 was applied to the die wall before placing the powder into the die cavity. Preliminary, experiments at 300 and 400 °C and pressures of 250 and 350 MPa were carried out, to define conditions which produced the best properties. A temperature of 300 °C and pressure of 250 MPa were selected as the best conditions.

The powder was compacted, first, at room temperature during 10 min and 250 MPa, to avoid ignition of the aluminum powder, after that, temperature was increases slowly to reach 300 °C, keeping the pressure constant, conditions, which remained during 10 min. Afterwards, the specimen was removed from the die.

The specimens were sintered during 3 h at 550 °C within a box furnace using a nitrogen atmosphere to avoid oxidation. Conditions which were found satisfactory in a previous work [2].

Mechanical and Physical Characterization

The compression tests for each composition were carried out in an Instron Machine, The specimen dimensions were 2 cm height, 2 cm diameter, the same sample geometry was used for micro hardness and density. It was obtained an engineering stress against strain curve.

Micro hardness was carried out in a Buehler micro hardness apparatus. Before testing, samples were prepared by polishing the transversal faces with increasing fineness sandpaper until the 600 grade and polished with alumina of 0.3 μm particle size. It was applied a 50 g load. Ten hardness determinations were carried out for each specimen to avoid errors due to the presence of pores or hard particles.

Density determinations were carried on using an Ohaus balance with capacity to weigh 310 g. Archimedes principle was applied, samples were weighted without and with immersion in water.

Wearing test. Samples of 1.0 cm × 2.75 cm cross section and 6 cm height were used for this test. The specimen is hold by an arm as a rotatory wheel (200 rpm) covered with rubber applies pressure, while a stream of silica sand particles, 70 meshes size, falls from a chute, at a rate of 340 g/min. Three runs of 5 min each were carried out for each sample. The testing procedure is detailed in the ASTM G65 Standard, Test Method for Measuring Abrasion Using the dry sand/rubber Wheel Apparatus. The specimen is weighed before and after the 3 runs to measure the wear which experienced the sample.

Impact test. Cylindrical specimens, 6 cm length and a 2 cm diameter, without notch (trials were made to machine a notch, but samples crumbled), therefore the tests were carried out on square bars 6 cm long, without notch for each composition were tested in an impact machine Tinius-Olsen.

Electron Microscopy

An electron microscope Karl Zeiss was used to examine the microstructure of the composites. A cross section of a cylindrical sample was polish until a finish mirror, washed with acetone with ultrasonic agitation.

Results

Mechanical and Physical Tests

Figure 1 Compression strength (maximum stress) versus composition plot obtained from compression testing.

Figure 2 Vickers hardness versus composition curves.

Figure 3 Presents the curve density versus composition.

Figure 4 Loss of weight percent after wearing test.

Figure 5 Exhibits the graph absorbed energy versus composition after impact test.

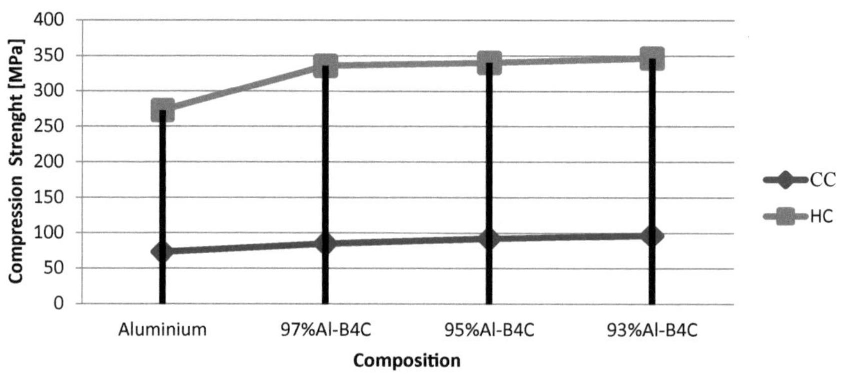

Fig. 1 Compression strength versus composition for CC and HC compression

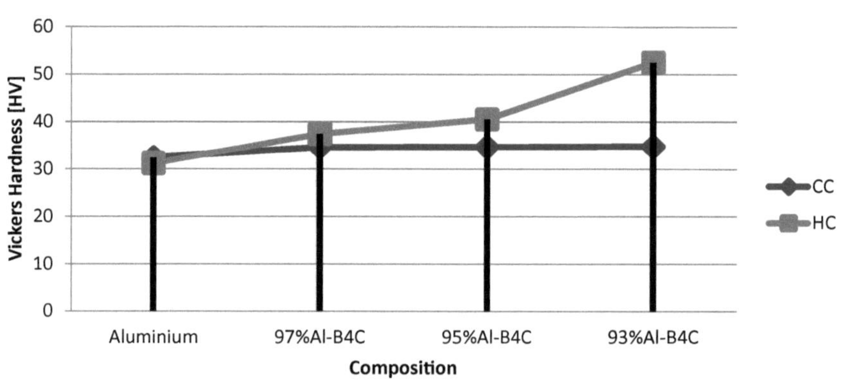

Fig. 2 Vickers hardness versus composition for CC and HC compression

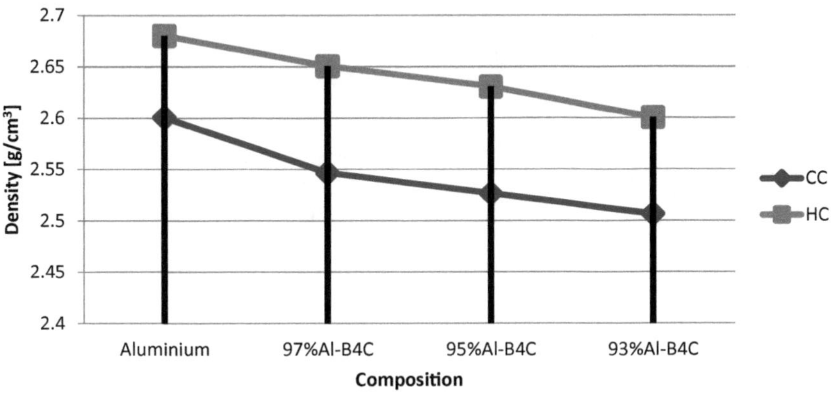

Fig. 3 Density versus composition for CC and HC compression

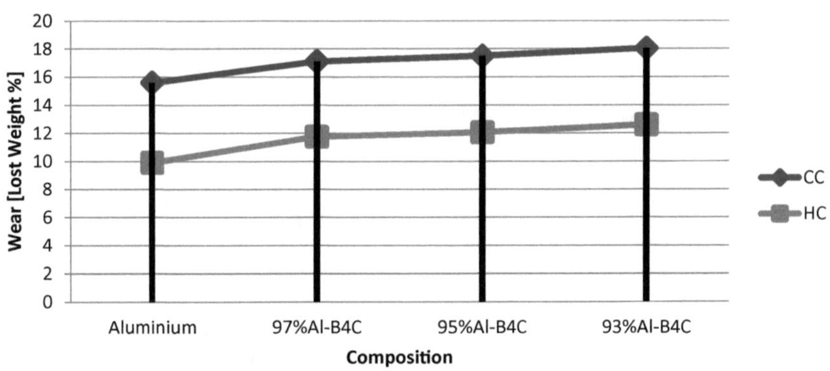

Fig. 4 Loss of weight percent for CC and HC compression

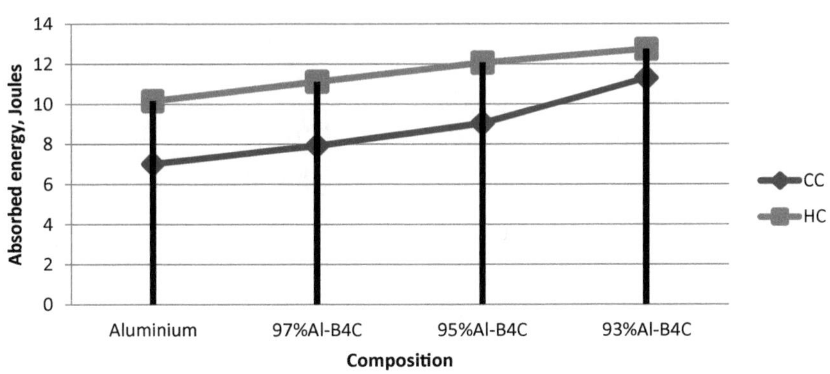

Fig. 5 Absorbed energy versus composition for CC and HC compression

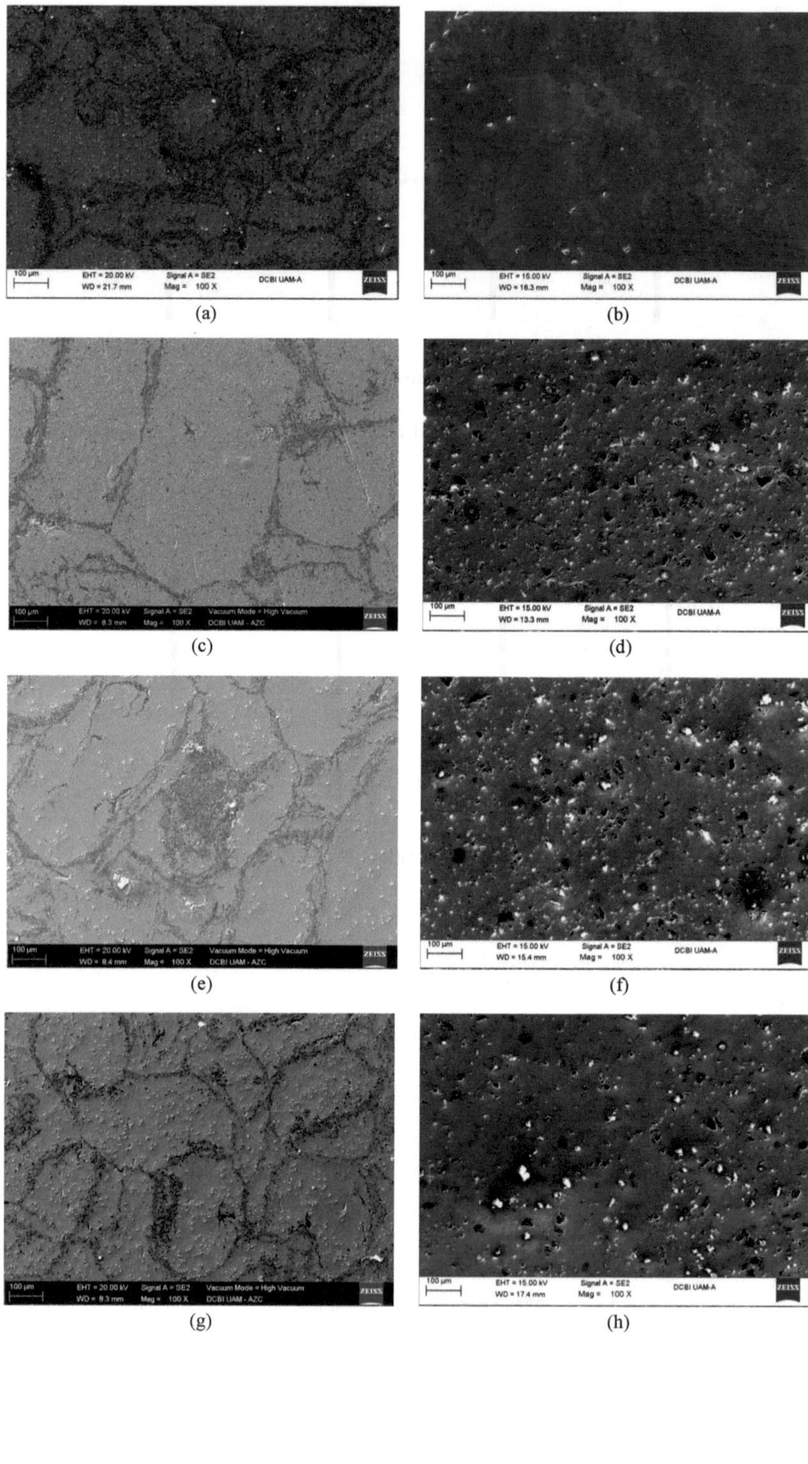
(a)
(b)
(c)
(d)
(e)
(f)
(g)
(h)

◄ **Fig. 6** **a** Aluminum CC, cellular microstructure; **b** Aluminum HC 100×, smooth Surface. *Black particles* are boron carbides, aluminum oxides are the *white particles* in both samples. Composites 93%Al–7%B_4C, **c** CC, cellular microstructure 100×, fine and well distributed carbides; **d** HC, 100×. B_4C, *black particles*, are rather well distributed within the matrix. *White particles* were identified as Al_2O_3. 95%Al–5%B_4C, **e** CC, cellular microstructure, 100×; **f** HC, 100×, smooth surface, boron carbides are uniformly distributed within the aluminum matrix, *white particles* are Al_2O_3. 97%Al–3%B_4C, **g** CC, cellular microstructure, 100×; **h** HC, smooth surface, B_4C particles well distributed, 100×

Electron Microscopy

Figure 6a, c, e, and g electron microscopies at 100× of CC samples. Figure 6b, d, f and h electron microscopies of HC samples.

Discussion

Composite specimens, B_4C-Al, prepared by cold and hot compression showed higher compression strength than aluminum powder compressed under the same conditions, strength increases with content of reinforcement. Findings that coincides with a report by Morensen and Lorca [4].

Specimens prepared at 300 °C and 250 MPa rendered specimens with a compression strength about 3.5 times higher than samples made by cold compression, Fig. 1. Authors of this work assume that fluidity of aluminum improves with temperature, therefore interface matrix-particles is more coherent, interface becomes stronger and the constraint imposed by the ceramic on the matrix increases [4], thus, compression strength increases [5], since the load is transferred across the interface.

Composite specimens exhibit higher hardness in cold and hot compression than compressed aluminum powder under the same conditions, hardness increases with content of reinforcement, Fig. 2. Causes of this behavior are the same as for compression strength: improved fluidity at the higher temperature leads to a more coherent interface matrix-particles. Smallman [5] describes that hardness increases as the interface particle-matrix is more coherent. Experiments with aluminum carbide particles in an aluminum matrix pressed at 600 °C, displayed an increase of micro-hardness from 750 MPa in cold compression to 1550 MPa in hot compression [6].

Density is lower in composites prepared by both cold and hot compression than in compressed aluminum powder, Fig. 3, because B_4C has lower density than aluminum (2.51 g/cm^3 against 2.7 g/cm^3). Therefore, density decreases as B_4C content increases, one reason for application of these composites in the aerospace industry [7]. Hot compressed specimens are denser than the HC compression ones, Fig. 3. At higher temperature, enhanced fluidity decreases pores and cavities, thus density is about 0.08 g/cm^3 higher for HC specimens in relation to CC ones.

The loss of weight by wearing was higher for CC samples than for HC ones, Fig. 4. The span decreases with the content of B_4C, it is 57% for specimens with the lowest content of reinforcement, it decreases to about 35% for compositions in the middle of the curve and it decreases to about 10% for the composites with the highest content of B_4C, Fig. 4. The resistance to wearing improves as the reinforcement increases which is an expected result. These results agree with those obtained by [7–9].

Hot compressed samples presented more toughness as shown by the higher values of absorbed energy in relation to cold compressed ones, Fig. 5, Jung and Kang, report [10] promising indentation toughness values for B_4C-Al composites.

Electron microscopic at 100×, for all the compositions, show that the cold compressed samples exhibit a cellular microstructure, Fig. 6a, c, e and g, on the other hand, a rather smooth surface with B_4C particles arising from the surface is a characteristic presented by the hot compressed samples, which look well compacted, Fig. 6b, d, f and h. Black particles are particle carbides, white particles are aluminum oxides, identified by the electron microscopy spectrum cold compressed composites present smaller carbide particles than hot compressed specimens (2–5 and 15–20 μm, respectively). This size difference is attributed to the initial B_4C particle size of the starting materials.

Conclusions

Preparation of specimens prepared by hot compression, 300 °C and 250 MPa, rendered better mechanical properties than those prepared by cold compression: higher values of compression strength, Vickers hardness, wear resistance, and toughness.

Electron microscopy observations show that these improved properties correspond to more tightly compacted specimens obtained by hot compression.

High values of compression resistance, hardness, wearing resistance, and toughness are related with tightly compacted particles. Manufacture of specimens by hot compression exhibit higher density.

High compression leads to a better properties than cold compression.

References

1. Sillekens, W. H., Jarvis, D. J., Vorozhtsov, A., Bojarevics, V., Badini, C.F., Pavese, M., et al. (2014). *Metallurgical and Meterials Transactions A, 45A*, 3349–3361.
2. Vazquez, L., Hernández, E., Altamirano, A., Cortés, V., Garfias, E., Refugio, E., et al. (2014). *Proceedings of Advanced Composites for Aerospace, Marine, and Land Applications, TMS 2014, 143 Annual Meeting and Exhibition* (pp. 23–33).
3. Dowson, G. *Introduction to Powder Metallurgy the Process and Its Products*. European Powder Metallurgy Association, Education and Training.

4. Mortensen, A., & Lorca, J. (2010). *Annual Review of Materials Research, 40,* 243–270.
5. Smallman, R. E. (1980). *Modern Physical Metallurgy, Butterworth* (3rd ed., p. 411), UK.
6. Vorozhtsov, S., Kolarik, V., Promakhov, V., Zhukov, I., Vorozhtsov, A., & Kuchenreuther-Hummel, V. (2016, May). *JOM,* (68).
7. Callister, W.D. Jr., & Rethwish, D. G. (2014). *Materials Science and Engineering, An Introduction* (9th ed., p. 660).
8. Moghadam, A. D., Shultz, B. F., Ferguson, J. B., Omrani, E., Rohatgi, P. K., & Gupta, N. (2014). *JOM Journal of the Minerals Metals and Materials Society, 66,* 872–881.
9. Christmas, T., Needleman, A., & Suresh, S. (1989). *Acta Metallurgica, 37,* 3029–3050.
10. Jung, J., & Kang, S. (2004). *Journal of the American Ceramic Society, 87*(1), 47–57.

An Improved Silicon Carbide Monofilament for the Reinforcement of Metal Matrix Composites

Michael V. Rix, Mark Baker, Mark J. Whiting, Ray P. Durman and Robert A. Shatwell

Abstract As part of ongoing research in the UK, TISICS have developed an improved 140 μm carbon coated silicon carbide monofilament for the reinforcement of metal matrix composites. The monofilament is fabricated in a single reactor using a high speed chemical vapor deposition process at a rate of 8 m/min (26 ft/min). Statistical analysis of monofilament properties over two years of production has demonstrated excellent reproducibility of the process. The monofilaments have an average tensile strength of 4.0 ± 0.2 GPa with a Weibull modulus of 50 ± 10. Composites incorporating the monofilaments show similar low variability in yield and tensile strength with the latter exhibiting a mean value above 90% of the maximum theoretical strength predicted by the rule of mixtures. By varying the volume fraction and orientation of the monofilament reinforcement, composite properties can be tailored to fit design requirements. Examples are given of demonstrator components made for the European aerospace sector.

Keywords Silicon carbide · Monofilament · Fibre · SiC monofilament

Introduction

Silicon carbide (SiC) monofilaments are continuous ceramic fibers. They have extremely high specific strength and stiffness as well as good creep resistance and can maintain these properties at elevated temperatures [1]. This material is an excellent candidate for the reinforcement of metal matrix composites (MMC) where the inherent brittleness of the SiC can be mitigated. The monofilaments are typically produced through chemical vapor deposition (CVD) of silicon carbide on to a

M.V. Rix (✉) · M. Baker · M.J. Whiting
Department of Mechanical Engineering Sciences,
University of Surrey, Guildford, Surrey GU2 7XH, UK
e-mail: mrix@tisics.co.uk

M.V. Rix · R.P. Durman · R.A. Shatwell
TISICS Ltd., Farnborough, Hampshire GU14 7QU, UK

© The Minerals, Metals & Materials Society 2017
M.A. Meyers et al. (eds.), *Proceedings of the 3rd Pan American Materials Congress*,
The Minerals, Metals & Materials Series, DOI 10.1007/978-3-319-52132-9_31

carbon or tungsten filament substrate [2], although recent progress has been made on producing a lower strength monofilament through a polymer route [3]. Whilst the properties of SiC reinforced MMC are attractive the manufacturing economics are poor at low production volumes. A significant factor in the expense is the need to provide an appropriate interface coating between the fiber and the matrix. In both SCS-6 and previous TISICS monofilaments, this has required two consecutive passes through CVD reactors. To overcome this obstacle, TISICS have developed a single stage deposition process capable of depositing the bulk silicon carbide and protective coating in one pass through a reactor. This reduces the overall cost of the process by taking out a major step in the production line. The new process is also 60% faster than the coating stage it has replaced.

Silicon Carbide Monofilament

SM3256 is the latest monofilament developed by TISICS and is the successor to SM1140+ [4] and SM2156 [1]. Stoichiometric silicon carbide is deposited by resistively heating the tungsten core filament in an atmosphere of dichloromethylsilane (CH_3SiHCl_2), propene, argon and hydrogen. Following this, layers of silicon carbide and carbon are deposited. The entire process takes place within a single CVD reactor and the multi-layered monofilament as shown in Fig. 1 is produced at 8 m/min.

The ultimate tensile strength (UTS) of each batch of monofilament is measured and the results analyzed using a Weibull distribution. Tensile testing is performed with an Instron Model 1011 tensile tester using a 100 N load cell. 30 samples with a gauge length of 25 mm are taken at random from a 10 m length of monofilament.

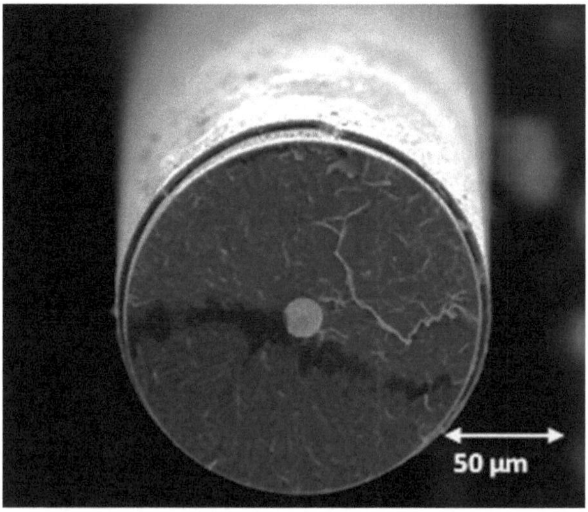

Fig. 1 Secondary electron image of the SM3256 cross-section. The tungsten core, bulk silicon carbide and outer coating are clearly visible

The diameter of each sample is measured to the nearest micron using a laser gauge. The UTS is calculated from the measured pull force and the sample diameter. A set of measurements and the Weibull plot from one production run is shown in Fig. 2a, b. A plot of mean UTS for one hundred consecutive production runs is shown in Fig. 2c. The small variation of UTS of 4.0 ± 0.2 GPa demonstrates the

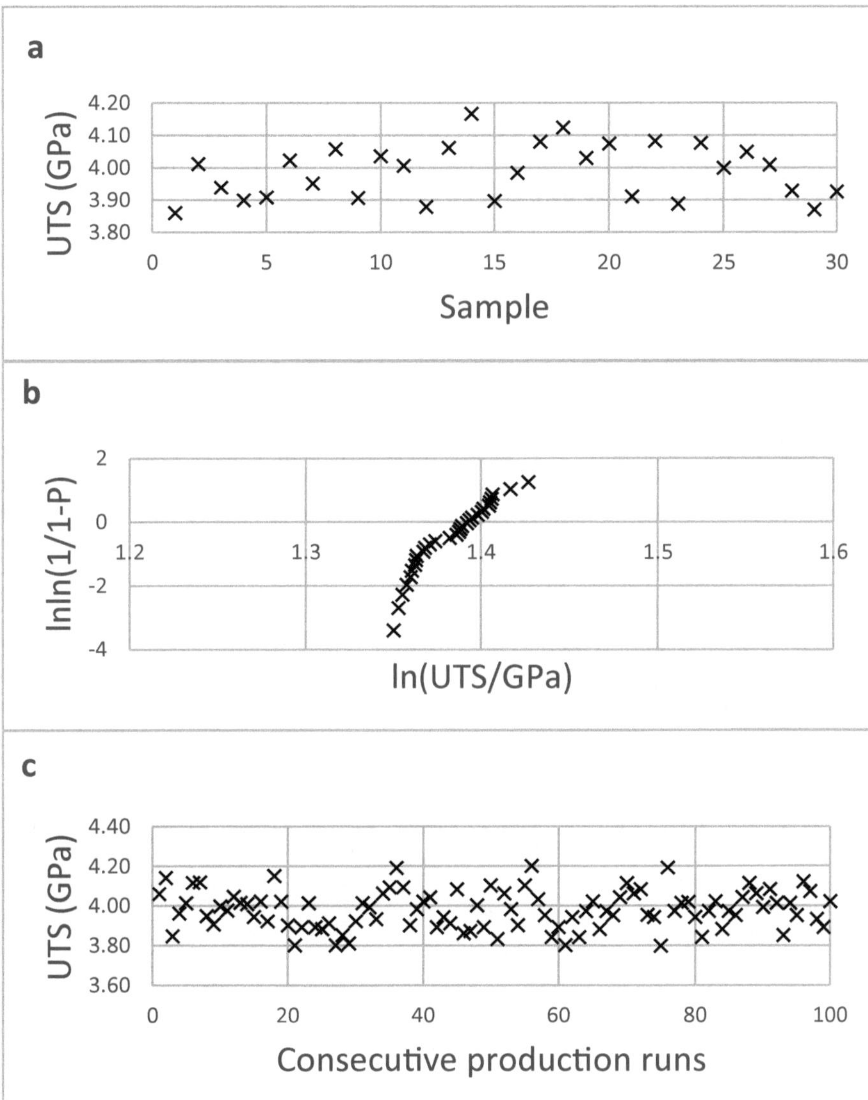

Fig. 2 **a** UTS in GPa of 30 monofilament samples from one production run, **b** Weibull plot of the UTS from 2.2a with a modulus of 51, **c** Mean UTS in GPa for one hundred consecutive production runs

reliability of the process. The Weibull modulus of 51 is exceptionally high compared to equivalent ceramic monofilaments [5].

Metal Matrix Composite Performance

SM3256 has been optimized for use in titanium matrices but can also be incorporated successfully into aluminum and potentially magnesium matrices. The metal matrix composite is most commonly fabricated through a foil/fiber lay-up technique, but also as matrix coated fibers (MCF) [6] before consolidation through a proprietary hot isostatic pressing (HIPing) cycle. The foil/fiber layers are encapsulated in a steel can before de-gassing to remove contaminants. The HIP chamber is brought up to a pressure of 100 MPa and a temperature above 900 °C with a single temperature ramp at a heating and cooling rate of under 10 °C/min. The HIP cycle lasts for several hours.

The current manual lay-up process allows for precise control of the distribution and volume fraction of the monofilaments. Automated processes are under development for high production rates. A cross-section of a 33% volume fraction uniaxially reinforced titanium MMC panel is shown in Fig. 3.

The uniaxial reinforcement results in the composite having anisotropic mechanical properties with vastly increased longitudinal tensile strength and slightly reduced transverse tensile strength compare to the matrix material. Tensile testing of the composite is performed using a Phoenix Calibration 190 kN tensile tester. MMC panels are sectioned by wire erosion into 10 mm wide, 1.475 mm thick 'dogbone' specimens with a gauge length of 25 mm. The UTS is calculated from the measured pull force and the cross-sectional area of the sample. This is compared to the theoretical maximum longitudinal UTS predicted by the rule of mixtures (ROM). The longitudinal UTS and corresponding Weibull plot of a set of

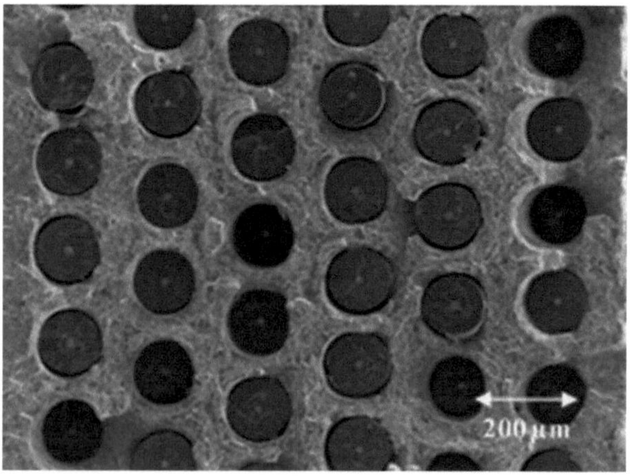

Fig. 3 Secondary electron image of tensile test specimen cross-section from a MMC panel with 6-ply uniaxial monofilament reinforcement using the foil/fiber process

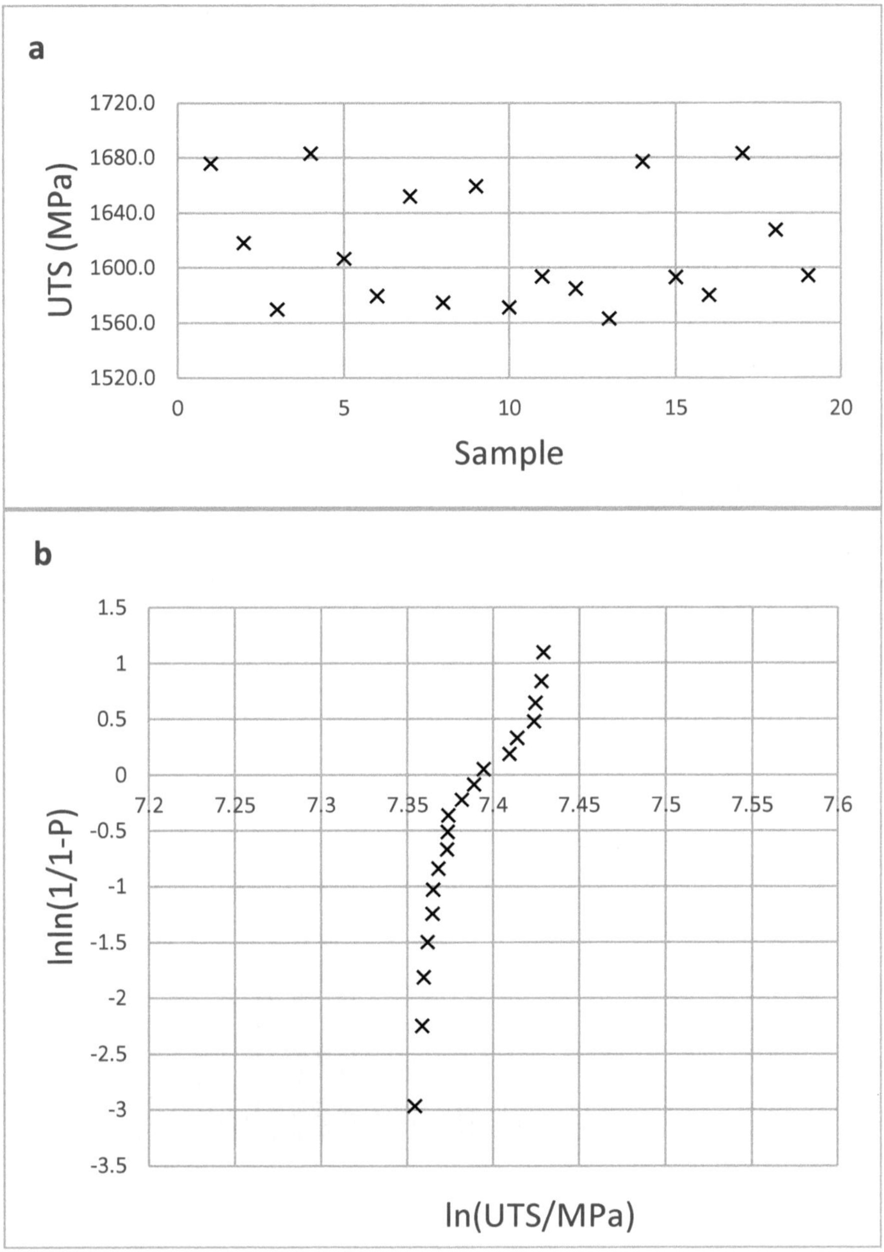

Fig. 4 **a** Longitudinal UTS in MPa of 18 dogbone Ti-3Al-2.5V MMC samples, **b** Weibull plot of the UTS from 3.2a with a modulus of 37

18 ‘dogbone’ samples taken from 5 separately manufactured composite panels are shown in Fig. 4a, b.

Discussion

The development of a single pass process for the manufacture of silicon carbide monofilaments removes the need for expensive additional CVD reactors in the production line. The overall decrease in capital and operating cost in combination with the increased production speed and the prevention of handling damage to uncoated monofilament allows for the economically competitive mass-production of the material. This is aided by the use of very long lengths of tungsten core which are readily available. This has been one of the major obstacles to the application of MMCs in industry [7]. Continued development in production capacity will allow for more widespread use.

In addition to the improved economic efficiency of production SM3256 has an extremely narrow strength distribution when compared to an equivalent monofilament as shown in Fig. 5. The tensile strengths of 1750 SM3256 samples are overlaid on 6500 SCS-6 samples.

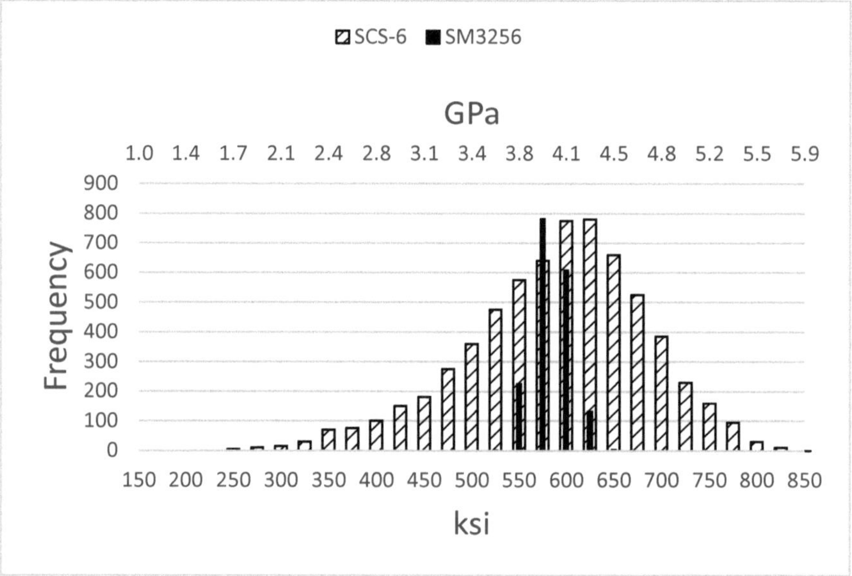

Fig. 5 Comparison of UTS in ksi and GPa of SCS-6 manufactured by Speciality Materials Inc. and SM3256 manufactured by TISICS Ltd. (SCS-6 tensile test data reproduced from [8])

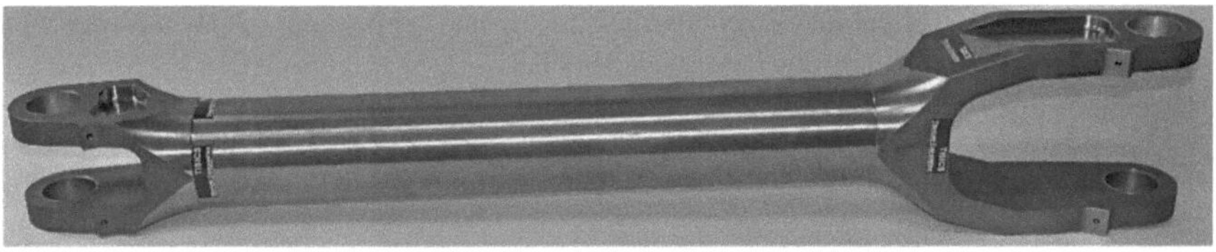

Fig. 6 SiC monofilament reinforced titanium MMC 1.16 m long landing gear demonstration component. Designed for the single-aisle Airbus A320

The greater control over monofilament properties means that the reinforcing monofilament can be treated as a material with near uniform strength, this aids in the design of MMC components.

The monofilaments have a very high Weibull modulus and therefore a very narrow strength distribution, however the resulting MMC test specimens have a slightly wider distribution. Additionally the MMC specimens achieve between 90 and 97% of the maximum theoretical strength predicted by the ROM. While this performance is very good there is a clear discrepancy in the material properties. Investigation of the tensile test specimens suggests that minor faults in monofilament distribution and matrix movement during consolidation are responsible. This highlights the importance of precise control over the composite manufacturing process in order to fully exploit the monofilament properties. Application of this technology allows for significant weight savings when compared to traditional materials. A full-size titanium MMC landing gear demonstration component is shown in Fig. 6, this achieved a 30% reduction in weight from the original forged aluminum model.

References

1. Chollon, G., Naslain, R., Prentice, C., Shatwell, R., & May, P. (2005). High temperature properties of SiC and diamonc CVD-monofilaments. *Journal of the European Ceramic Society, 25*, 1929–1942.
2. Zhang, R., Yang, Y., Shen, W., Wang, C., & Luo, X. (2010). Microstructure of SiC fiber fabricated by two-stage chemical vapor deposition on tungsten filament. *Journal of Crystal Growth, 313*, 56–61.
3. Andreas, N. (2014). Fabrication of large diameter SiC monofilaments by polymer route. *Journal of the European Ceramic Society, 34*, 1487–1492.
4. Cheng, T. T., Jones, I. P., Shatwell, R. A., & Doorbar, P. (1999). The microstructure of sigma 1140 + SiC fibres. *Materials Science and Engineering A, 260*, 139–145.
5. Le Petitcorps, Y., Lahaye, M., Pailler, R., & Naslain, R. (1988). Modern boron an SiC CVD filaments: A comparative study. *Composites Science and Technology, 32*, 31–55.
6. Ward-Close, C. M., Chandrasekaran, L., Robertson, J. G., Godfrey, S. P., & Murgatroyde, D. P. (1999). Advances in the fabrication of titanium metal matrix composite. *Materials Science and Engineering A, 263*, 314–318.

7. Rawal, S. (2001). Metal-matrix composites for space applications. *JOM Journal of the Minerals Metals and Materials Society, 53*(4), 14–17.
8. Hutson, A. L., & Kleek, J. J. (2009). *Quick evaluation of materials and processes* (Report AFRL-RX-WP-TR-2010-4175 approved for public release, Air Force Research Laboratory Materials and Manufacturing Directorate Wright-Patterson Air Force Base).

Analysis of Coir Fiber Porosity

Fernanda Santos da Luz and Sergio Neves Monteiro

Abstract Lignocellulosic natural fibers exhibit high variation in mechanical properties values due to their heterogeneity. Recent studies have shown that the dispersion of these properties depends on the diameter of the fiber. A possible explanation for the tendency of low mechanical properties with large diameter is the high probability of defects into the fiber. However, no study has yet investigated the relation between the diameter and defects of natural fibers. Hence, in the present work the total porosity of coir fiber (*Cocos Nucifera L.*) was estimated. A statistical analysis was carried out on a batch of about 100 fibers and the geometric density was measured by using a stereomicroscope. The closed and open porosity as well as the density of the fiber were quantified by helium pycnometry method. The values obtained were correlated with the diameter of each analyzed fiber. Results made it possible a more detailed knowledge of the porous structure of the fiber.

Keywords Coir fibers · Lignocellulosic fibers · Porosity · Diameter

Introduction

In the last decade, several studies have been focused on lignocellulosic fibers (LCFs) due to specific features such as, their renewable character, low environmental impact, low density, low cost, high availability and good mechanical properties, which allow the development of many composite materials [1, 2]. Thus, the application of LCF as a replacement for the synthetic fibers in polymer composites has already succeeded in industrial scale products. An example of this is the

F.S. da Luz · S.N. Monteiro (✉)
Department of Mechanical and Materials Engineering,
Military Institute of Engineering, Praça General Tibúrcio, 80,
Rio de Janeiro 22290-270, RJ, Brazil
e-mail: snevesmonteiro@gmail.com

F.S. da Luz
e-mail: fsl.santos@gmail.com

© The Minerals, Metals & Materials Society 2017

M.A. Meyers et al. (eds.), *Proceedings of the 3rd Pan American Materials Congress*,
The Minerals, Metals & Materials Series, DOI 10.1007/978-3-319-52132-9_32

use of composites reinforced with a variety of LCFs in the automobile industry to manufacture internal components such as panels, cushions and seat back linings [3–6]. As a side note the coir and sugarcane bagasse fibers, that are considered waste, could be used as reinforcements of polymer composites with lower commercial values [7–10].

However, these fibers present a high dispersion in their dimensional values, which is not a desirable feature. Therefore, several papers have reported the correlation between the mechanical properties and LCFs dimensions [11–15]. Bledski and Gassan [16] observed a trend of increase strength to smaller diameters of pineapple fiber. The same behavior was observed for elastic modulus of coir fibers which are extracted from the mesocarp of the coconut fruit (*Cocos Nucifera L.*) [5]. Recent works [17–25] have shown an inverse correlation between strength and diameter for various species of LCFs. This seems to be a common characteristic for all fibers due to their heterogeneous nature [19–23].

The coir fibers, object of this study, are found abundantly in Brazil, with 30% of their total production concentrated in the Northeast, being the Bahia state the largest producer [26]. About 85% by weight of coconut fruit corresponds to the fibers. These are considered as waste. Research for new applications for these fibers is increasing due to their advantages such as, high availability, low cost, high degradation resistance because of their high content of lignin and low density [27, 28]. Therefore, the objective of the present work (PW) was to estimate the quantity of porosity of coir fibers. The relation between the diameter and defects for these fibers was investigated allowing a more detailed knowledge of the porous structure of the coir fiber.

Materials and Methods

Concerning the results achieved in the present work, it should be mentioned that the geometric densities were determined by precise diameter measurements that was conducted in a stereo microscope Zeiss, Stemi 508 model, at different points along the fiber axis. The equivalent diameter was obtained by the average of these values. The association between the high dispersion values of diameter for different species of LCFs and the heterogeneous characteristics was already discussed [1]. Diameter intervals were defined according to the natural dispersion found in the experimental lot. The statistical distribution of diameters was based on 100 randomly picked fibers from the lot.

Figure 1 shows the histogram corresponding to the frequency of diameter distribution obtained for the coir fibers. These fibers were supplied by the Brazilian firm "Coco Verde Reciclado". Another aspect in Fig. 1 is that the range of intervals is limited for the lot size. Therefore, if the lot with a larger number of fibers was investigated, thinner and thicker fibers possibly would be found, extending the range of the histograms.

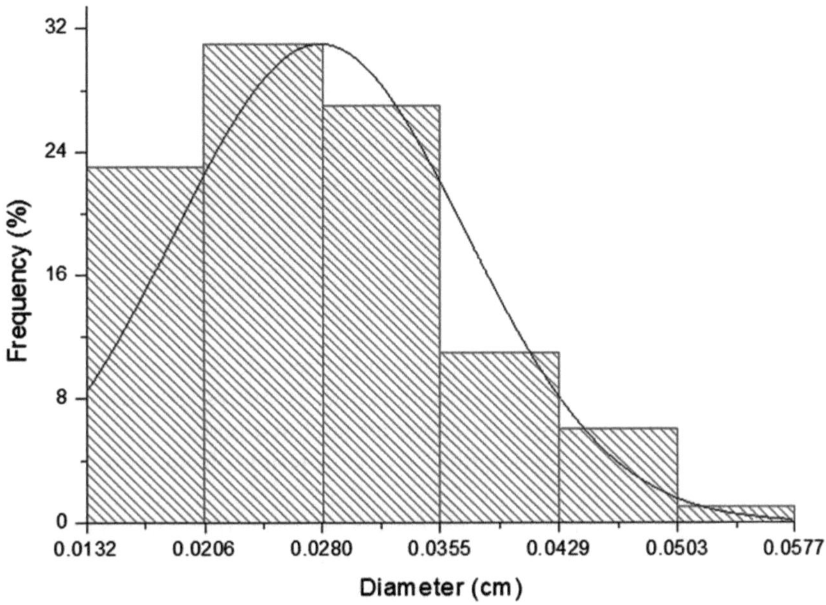

Fig. 1 Histogram for the mean equivalent diameters of coir fiber

The geometric density (ρ_g) was determined from the measurements of length, weight and diameter for the experimental lot. Due to the irregular shape of cross section fiber, measurements were taken at five positions along the fiber, equally spaced, and this process was repeated by rotating the fiber in 90°.

Helium pycnometer model AccuPyc 1330 was used to measure the apparent density of fibers and an analytical balance was used to weigh the fibers. Weights were measured with a precision of 0.001 g. The sample was placed into the pycnometer sample holder and weighed. The sample and container were then placed into the cell chamber. The helium gas was admitted into the cell chamber filling the fibers open lumen and pores. This procedure was repeated ten times for each diameter interval shown in Fig. 1. The absolute density considered was 1540 kg/m^3 same as cellulose. This slight drop of density in LCFs is due to the presence of porosity including the lumen. The total porosity (P_T) of fibers was calculated using the apparent (ρ_{ap}) and absolute density (ρ_{abs}) of the fibers (Eq. 1).

$$P_T = \left(1 - \frac{\rho_{ap}}{\rho_{abs}}\right) * 100 \quad (1)$$

Equation 2 was used to determine the open porosity (P_o).

$$P_o = \left(1 - \frac{\rho_g}{\rho_{ap}}\right) * 100 \tag{2}$$

The close porosity was obtained from the difference between total and open porosity.

Results and Discussion

The absolute density excludes all the pores and lumen and is therefore a measure of the solid matter of the fibers. The apparent density is always less than that because the buoyancy effect caused by the trapped air in closed pores.

According to the procedure described above the results obtained are presented in Table 1.

Experimentally, it was noted that the density of defects (porosity) in LCFs varies with the size of the cross section, i.e., equivalent diameter. Greater equivalent diameters present comparatively higher density of porosity, both in the surface and inside the volume of the fiber. Hence, thicker fibers tend to be more heterogeneous than thinner ones. For all LCFs, it was determined earlier [2, 16] that the thicker fibers display a rupture associated with more heterogeneous microstructure and more participations of microfibrils. By contrast, the thinner fibers reveal a comparatively more homogeneous rupture associated with relatively less microfibrils.

It can be concluded that the combination of the geometric density and the pycnometry method for measuring apparent density allow the determination of fiber open and close porosity. Therefore, the results of the present work support the trend observed in previous studies [2, 16].

Table 1 Density and porosity of coir fibers

Equivalent diameter (μm)	Geometric density (kg/m^3)	Apparent density (kg/m^3)	Total porosity (%)	Open porosity (%)	Closed porosity (%)
173	874	1410	43.24	38.01	5.23
244	815	1422	47.08	42.69	4.39
316	734	1380	52.33	46.80	5.53
402	656	1312	57.40	49.99	7.40
466	652	1324	57.67	50.77	6.91
577	620	1261	59.77	50.87	8.90

Summary and Conclusions

The next few decades should see a growing interest in the application of composites reinforced with natural fibers. These fibers, especially the lignocellulosic, LCFs, are abundant, biodegradable, renewable, and neutral with respect to CO_2 emissions that cause global warming. Each lignocellulosic fiber has a specific surface micro-morphology that affects the mechanical behavior of the composites in distinct ways and can significantly improve their strength and toughness. The incorporation of these fibers is promising for a future of less expensive, stronger, and environmentally friendly composites.

The results obtained in this work indicate that the amount of defects is one of the reasons for the inverse correlation between tensile strength and LCF diameter noted in previous studies.

Acknowledgements The authors acknowledge the support to this investigation by the Brazilian agencies CNPq, CAPES, and FAPERJ.

References

1. Monteiro, S. N., et al. (2009). Natural-fiber polymer-matrix composites: Cheaper, tougher, and environmentally friendly—An overview. *JOM Journal of the Minerals Metals and Materials Society, 61*(1), 17–22.
2. Monteiro, S. N., et al. (2011). Natural lignocellulosic fibers as engineering materials—An overview. *Metallurgical and Materials Transactions A, 42A*, 2963–2974.
3. Suddell, B. C., et al. (2002). A survey into the application of natural fibre composites in the automobile industry. In *Proceedings of the 4th International Symposium on Natural Polymers and Composites—ISNAPol* (pp. 455–461). São Paulo, Brazil: ABPol.
4. Marsh, G. (2003). Next step for automotive materials. *Materials Today, 6*(4), 36–43.
5. Hill, S. (1997). Cars that grow on trees. *New Scientists, 153*(2067), 36–39.
6. Zah, R., et al. (2007). Curaua fibers in automobile industry—A sustainability assessment. *Journal of Cleaner Production, 15*, 1032–1040.
7. Hill, C. A. S., & Khalil, H. P. S. A. (2000). The effect of environmental exposure upon the mechanical properties of coir or oil palm fiber reinforced composites. *Journal of Applied Polymer Science, 77*, 1322–1330.
8. Rout, J., et al. (2001). The influence of fibre treatment on the performance of coir-polyester composites. *Composites Science and Technology, 61*, 1303–1310.
9. Monteiro, S. N., Lopes, F. P. D., & d'Almeida, J. R. M. (2005). Mechanical strength of polyester matrix composites reinforced with coconut fiber wastes. *Revista Materia, 10*(4), 571–576.
10. Monteiro, S. N., et al. (1998). Sugar cane bagasse waste as reinforcement in low cost composites. *Advanced Performance Materials, 5*(3), 183–191.
11. Kulkarni, A. G., Satyanarayana, K. G., Sukumaran, K., & Rohatgi, P. K. (1981). Mechanical behavior of coir fibers under tensile load. *Journal of Materials Science, 16*, 905–914.
12. Kulkarni, A. G., Satyanarayana, K. G., Rohatgi, P. K., & Vijayan, K. (1983). Mechanical properties of banana fibers. *Journal Materials Science, 18*, 2290–2296.
13. Murkherjee, P. S., & Satyanarayana, K. G. (1984). Structure and properties of some vegetable fibers—Part 1: Sisal fiber. *Journal of Materials Science, 19*, 3925–3934.

14. Mukherjee, P. S., & Satyanarayana, K. G. (1986). Structure and properties of some vegetable fibers—Part 2: Pineapple fiber. *Journal of Materials Science, 21*, 51–56.
15. Murkherjee, P. S., & Satyanarayana, K. G. (1986). Structure and properties of some vegetable fibers—Part 3: Talipot and palmyrah fibres. *Journal of Materials Science, 21*, 57–63.
16. Bledzki, A. K., & Gassan, J. (1999). Composites reinforced with cellulose-based fibers. *Progress in Polymer Science, 4*, 221–274.
17. Monteiro, S. N., Satyanarayana, K. G., & Lopes, F. P. D. (2010). High strength natural fibers for improved polymer matrix composites. *Materials Science Forum, 638–642*, 961–966.
18. Inacio, W. P., Lopes, F. P. D., & Monteiro, S. N. (2010). Tensile strength as a function of sisal fiber diameter through a Weibull analysis. In *Proceedings of Biomaterials Symposium, 1st TMS-ABM International Materials Congress*, Rio de Janeiro, Brazil (pp. 1–10).
19. Margem, F. M., Bravo Neto, J., & Monteiro, S. N. (2010). Ramie fibers mechanical properties evaluation by the Weibull analysis. In *Proceedings of 19th Brazilian Congress on Materials Science and Engineering,* Campos do Jordão, Brazil (pp. 1–10) (in Portuguese).
20. Ferreira, A. S., Monteiro, S. N., & Lopes, F. P. D. (2009). Curaua fiber mechanical properties evaluation by the Weibull analysis. In *Proceedings of 64th International Congress of the Brazilian Association for Metallurgy and Materials*, Belo Horizonte, Brazil (in Portuguese) (pp. 1–12).
21. Bevitori, A. B., Silva, I. L. A., & Monteiro, S. N. (2010). Weibull analysis of the tensile strength variation with diameter for jute fibers. In *Proceedings of Biomaterials Symposium, 1st TMS-ABM International Materials Congress*, Rio de Janeiro, Brazil (pp. 1–10).
22. Costa, L. L., Loiola, R. L., & Monteiro, S. N. (2010). Tensile strength of bamboo fibers: Weibull analysis to characterize the diameter dependence. In *Proceedings of Biomaterials Symposium, 1st TMS-ABM International Materials Congress*, Rio de Janeiro, Brazil (pp. 1–10).
23. Santafe, H. P. G., Jr., Monteiro, S. N., & Costa, L. L. (2009). Weibull distribution as an instrument of statistical analysis for coir fiber tensile tests. In *Proceedings of 64th International Congress of the Brazilian Association for Metallurgy and Materials*, Belo Horizonte, Brazil (pp. 1–12) (in Portuguese).
24. Nascimento, D. C. O., Motta, L. C., & Monteiro, S. N. (2010). Weibull analysis of tensile tested piassava fibers with different diameters. In *Proceedings of Characterization of Minerals, Metals and Materials Symposium—TMS Conference*, Seattle, WA (pp. 1–8).
25. Portela, T. G. R., Costa, L. L., Lopes, F. P. D., & Monteiro, S. N. (2010). Characterization of fibers from different parts of the buriti palm tree. In *Proceedings of Characterization of Minerals, Metals and Materials Symposium—TMS Conference*, Seattle, WA (pp. 1–7).
26. IBGE. (2016). *Systematic survey of agricultural production* (Vol. 29, no. 7, p. 5). Rio de Janeiro, RJ: IBGE.
27. Jayavani, S., Deka, H., Varghese, T. O., & Nayak, S. K. (2015). Recent development and future trends in coir fiber reinforced green polymer composites: Review and evaluation. *Polymer Composites*.
28. Holbery, J., & Houston, D. (2006). Natural-fiber-reinforced polymer composites in automotive applications. *JOM Journal of the Minerals Metals and Materials Society, 58* (11), 80–86.

Ballistic Performance in Multilayer Armor with Epoxy Composite Reinforced with Malva Fibers

Lucio Fabio Cassiano Nascimento, Luis Henrique Leme Louro, Sérgio Neves Monteiro, Alaelson Vieira Gomes, Édio Pereira Lima Júnior and Rubens Lincoln Santana Blazutti Marçal

Abstract A multilayer armoring system (MAS) is commonly formed by three layers. The initial layer is normally composed by a ceramic with high compressive strength, which absorbs most of the kinetic projectile energy. The subsequent composite layer was formed by epoxy matrix reinforced with natural malva fibers (*Urena lobata*, *Linn*), in the form of pure or hybrid fabric with jute fibers, in order to absorb part of the kinetic energy, and to retain ceramic and projectile shrapnel. A third layer formed by aluminum alloy, was included as a penetration restrictor for bullet and fragments by plastic deformation. The ballistic efficiency was evaluated by penetration of the 7.62 × 51 mm ammunition into a clay witness backing the armor. The results showed a great potential by epoxy composites reinforced with malva fabric as compared to other natural fibers and materials traditionally used in personal protection, such as Kevlar™ aramid.

Keywords Ballistic test · Multilayered armor · Malva fabric · Composites

Introduction

In this global scenario of major developments and uncertainties, the development of ballistic armor systems, more efficient, it is essential to ensure personal safety and vehicle, both civil and military level. A balance formed by mobility, penetration resistance and ability to absorb the energy of impact is critical to making an effective ballistic armor. However, the improvement of these factors often

---

L.F.C. Nascimento (✉) · L.H.L. Louro · S.N. Monteiro
A.V. Gomes · É.P.L. Júnior · R.L.S.B. Marçal
Militar Institute of Engineering—IME, Rio de Janeiro, Brazil
e-mail: lucio_coppe@yahoo.com.br

S.N. Monteiro
e-mail: snevesmonteiro@gmail.com

© The Minerals, Metals & Materials Society 2017
M.A. Meyers et al. (eds.), *Proceedings of the 3rd Pan American Materials Congress*,
The Minerals, Metals & Materials Series, DOI 10.1007/978-3-319-52132-9_33

negatively influence each other, for example, an increase in penetration resistance may cause a reduction in mobility due to the necessary increase in thickness and therefore weight, in the case of shielding monolithic made of steel. In this context, the armor with ceramic components combined with other materials such as polymer composites, are used when you want a good weight/ballistic protection. These are called multilayer armor systems (MAS). The MAS consist of the synergistic use of specific materials, to promote adequate protection against various threats, without prejudice to the mobility of the structure, i.e., allowing achieve the same or lower densities to those achieved in monolithic armor. The goal is not only to absorb the impact, but also prevent the penetration of fragments. In MAS typically the initial layer consists of a hard and brittle ceramic with high compressive strength, which has the function to absorb most of the kinetic energy of the projectile, breaking and destroying its tip. The subsequent layer is usually formed by a lighter ceramic material, in order to absorb part of the kinetic energy produced by shrapnel and projectile ceramic. This layer composite materials are used, for example, composites of glass fibers, carbon fibers, aramid Kevlar™ and Twaron™ fabric and ultra high molecular weight polyethylene fibers such as Dyneema™ and Spectra™. In these MAS may be added a third layer, typically of a ductile metal, which has the function of restricting the penetration of the bullet and fragments generated by plastic deformation mechanisms. However, synthetic materials that usually make up the composite of the intermediate layer have drawbacks, mainly linked to the high cost of acquisition and processing, as well as possible environmental damage, after the end of its useful life. In this context, natural lignocellulosic fibers (NFLs) may play this role because they have many advantages, among which: good specific properties (low density), low toxicity, biodegradability, low cost, etc. In addition to enabling the economic and technological development of less favored regions in Brazil, which are major producers of natural fibers, such as the northern and northeastern regions.

In this work we were used NFLs of malva (Urena lobata, Linn), in the form of fabrics, pure and hybrids with jute fibers (Corchorus capsularis) for final application of epoxy matrix composites within MAS. It evaluated the efficiency of these ballistics MAS by measuring its penetration into a clay witness backing the armor after 7.62×51 mm military ammunition impact. The NIJ standards specify that body armor should stop a projectile and prevent its penetration into a clay witness backing the armor to a depth not exceeding 1.73 in. (44 mm).

Materials and Methods

Figure 1 illustrates schematically the side view of the multilayered armor system (MAS) arrangement used in this investigation. The front layer (A), first to be hit by the projectile, was a 10 mm thick hexagonal tile with 31 mm of side dimension and made of 4 wt% Nb_2O_5 doped Al_2O_3 impact resistant ceramic [1, 2]. Ceramic tiles were fabricated by sintering Al_2O_3 powder (0.3 μm of particle size) supplied by

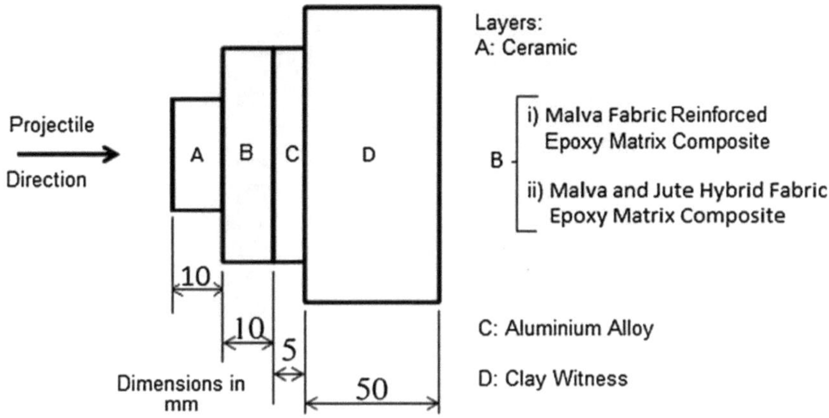

Fig. 1 Schematic diagram of the multilayered armor

Treibacher Schleifmittel as commercial purity mixed with Nb_2O_5 powder (0.69 μm of particle size) supplied by the Brazilian firm CBMM as 99% pro-analysis. Sintering was carried out as the route proposed by Trindade et al. [3].

The intermediate layer (B) with dimensions 150 mm × 120 mm and 10 mm in thickness: (i) 30% vol. of malva fabric, or (ii) 30% vol. of malva and jute hybrid fabric (70% malva and 30% jute). The malva fabrics and hybrids were supplied by the Castanhal Textile Company. Fabrics were cut to dimensions 150 mm × 120 mm, dried at 70 °C in a laboratory stove for 24 h and placed inside a steel mold. An initially fluid diglycidyl ether of the bisphenol-A (DGEBA) epoxy resin, mixed with a phr 13 stoichiometric fraction of triethylenetetramine (TETA) as hardener, was poured onto the mold. A pressure of 5 MPa was applied and the composite plate cured for 24 h [4–6]. For malva and jute fibers was used as an initial reference density of 1.40 g/cm^3 [7, 8] and the epoxy resin (DGEBA-TETA) value of 1.11 g/cm^3 [9]. The back layer (C) was a 150 × 120 mm 5052-H34 aluminum alloy sheet with 5 mm in thickness. These layers were bonded in the composite with commercial Sikaflex™ glue from SikaCo. Figure 2 shows a MAS mounted before tests.

In direct contact with this metallic back layer, a block of clay witness simulated a human body protected by the MAS. Modeling clay, compressed to avoid air bubbles, was commercially supplied by American Express. The trauma in the clay duplicates the plastic deformation imposed by the projectile impact on the aluminum back layer. Figure 3a–c respectively shows the clay witness, multilayer armor system mounted on it and measurement penetration after ballistic impact.

The ballistic tests were conducted at the Brazilian Army shooting range facility, CAEX, in the Marambaia peninsula, Rio de Janeiro. All tests, 5 for each type of MAS, were carried out according to the NIJ 0101.04 standards using 7.62 × 51 mm military ammunition. Fractured samples of each MAS component after the ballistic test were analyzed by scanning electron microscopy (SEM) in a

Fig. 2 Multilayer armor system mounted before tests

Fig. 3 **a** Clay witness; **b** Multilayer armor system mounted on the it; **c** Measurement penetration after ballistic impact

model QUANTA FEG250 FEI microscopes operating with secondary electrons at 20 kV.

Results and Discussions

Ballistic tests were conducted in MAS with intermediate layer malva fabric (MF) reinforced epoxy matrix composite and malva and jute hybrid fabric (HF) (70% malva/30% jute) reinforced epoxy matrix composite too. The results are shown in Table 1.

Table 1 Results of ballistic tests for MAS

Intermediate layer	Specification	MAS number tested	Average value of depth in clay witness (mm)
Epoxy—reinforced with 30% vol. malva fabric	E—30%MF (100% malva)	5	21.48 ± 1.640
Epoxy—reinforced with 30% vol. hybrid fabric (70/30)	E—30%HF (70/30)	5	23.16 ± 3.117

In both groups tested penetration in the clay witness was less than 44 mm, which is the value provided as lethal to humans by NIJ 0101.04. Thus, there is evidence that the MAS with intermediate layer of epoxy composite with 100% malva fabric or hybrid malva and jute (70/30) in a percentage of 30% vol. of fibers is effective against the threat of ammunition 7.62 × 51 mm. In none of the 10 shots were drilling the aluminum plate that was the last layer of MAS. In Figs. 4 and 5 are shown the armor arrangements before and after the ballistic impact.

For visual analysis, it is seen that the principal failure mechanism of the composite of both groups was delamination of layers, showing great absorption of impact energy. However, there is simultaneously the occurrence of other mechanisms that can dissipate energy and greatly influence the ballistic performance, such as the fracture strain of the fibers [10]. Figure 6a, b is scanning electronic microscopy showing the breaking of fibers and the retention of ceramics particles.

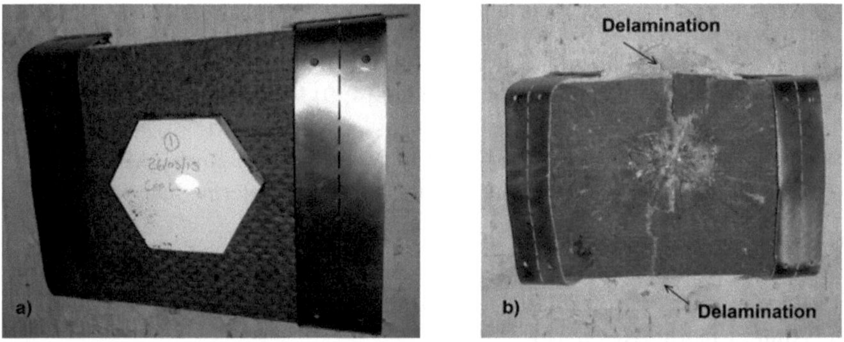

Fig. 4 Multilayer armor composite reinforced with malva fabric (30% vol. fibers). **a** Before the impact; **b** After the impact

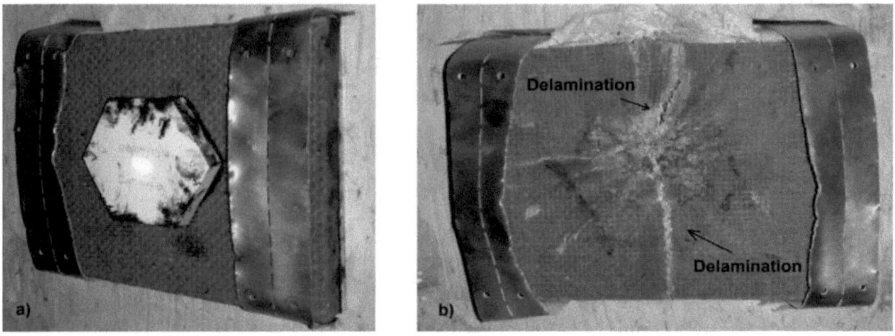

Fig. 5 Multilayer armor composite reinforced with hybrid fabric malva and jute (70/30)–(30% vol. fibers). **a** Before the impact; **b** After the impact

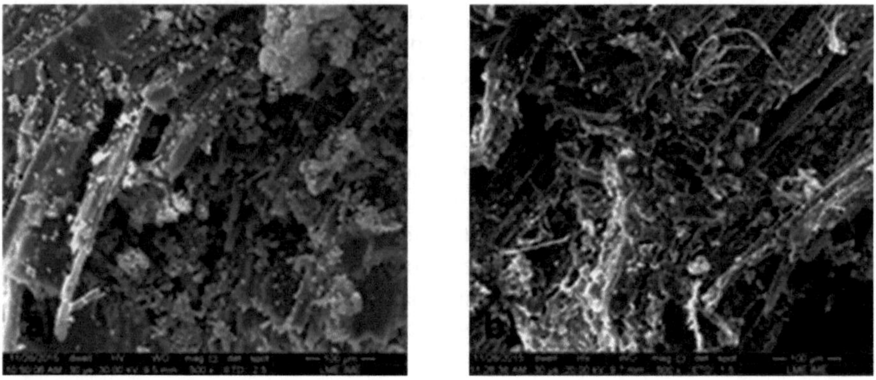

Fig. 6 Scanning electronic microscopy after ballistic impact (500×). **a** Malva fabric; **b** Malva and jute hybrid fabric (70/30)

Analyzing the results of other authors who worked with composites reinforced with natural fibers, with applications in multilayer armor, you can see the great potential of malva in this area of knowledge [4, 9], even for traditional aramid. In Table 2 is shown some of indentation values of works cited.

The statistical analysis was applied for analysis of variance (ANOVA) of the means of indentations in MAS that had the following intermediate layers:

(a) Epoxy—reinforced with 30% vol. 100% malva fabric;
(b) Epoxy—reinforced with 30% vol. hybrid fabric (70% malva/30% jute);
(c) Epoxy—reinforced with 30% vol. 100% jute fabric;
(d) Aramid fabric (Kevlar™).

In Table 3 are shown the results of the statistical analysis.

Table 2 Average values of depth in clay witness after ballistic tests

Specification	Average values of depth in clay witness (mm)	References
Epoxy—reinforced with 30% vol. Jute fabric	20.67 ± 3.109	[4]
Aramid fabric	22.67 ± 2.787	[9]
Epoxy—reinforced with 30% vol. Malva fabric	21.48 ± 1.640	This work

Table 3 Analysis of variance of average depth obtained for MAS indicated

Variation causes	DF	Sum of squares	Mean square	**F calculated**	**F critical (tabulated)**
Treatments	3	20.07	6.69	**1.46**	**3.24**
Residual	16	73.21	4.58		
Overall	19	93.27			

From the results obtained in the analysis of variance (Table 3) accepts the hypothesis that the means are equal with 95% confidence level, because the statistical "F", we have: F calculated < F critical (tabulated).

Conclusions

The MAS having epoxy intermediate layer reinforced with 30% vol. malva fabric or 30% vol. hybrid fabric malva and jute (70/30) meet the requirements of the standard N.I.J. 0101.04 because the average depth obtained on the clay witness, after impact of 7.62 mm ammunition has been less than 1.73 in. (44 mm). Therefore, one can consider the material suitable for use in ballistic armor in MAS.

The main failure mechanism observed by visual inspection after the ballistic impact was delamination layer, which enables the absorption of large part of the kinetic energy from the projectile and the ceramic front layer shrapnel.

Through analysis of variance it was found that there was no significant difference between the mean values of depth in MAS epoxy reinforced. with 30% vol., 30% vol. hybrid fabric malva and jute (70/30), 30% vol. jute fabric and aramid (Kevlar™). With that, notes the similarity of the malva and jute in applications such as ballistic composites for reinforcing purposes. And promising use of malva, compared to aramid fabric, which is the material traditionally used in vests protection personnel.

Acknowledgements The authors acknowledge the support to this investigation by the Brazilian agencies CNPq, CAPES and FAPERJ, Castanhal Textile Company for supplying the hybrid fabrics malva/jute and UENF for supplying the malva fibers.

References

1. Gomes, A. V. (2004). *Behavior ballistic alumina with added nióbia and variation of the target geometry*. Thesis (Doctorate in Materials Science). Military Engineering Institute, Rio de Janeiro (in Portuguese).
2. Trindade, W. (2012). *Influence of geometry and microstructure on dynamic behavior of the additive alumina niobia*. Thesis (Doctorate in Materials Science), Military Engineering Institute, Rio de Janeiro (in Portuguese).
3. Trindade, W., Gomes. A. V., & Louro, L. H. L. (2013). Preparation of a new ligand elimination route alumina ceramics. *Journal of Military Science & Technology/4th Quarter*, 71–79 (in Portuguese).
4. Luz, F. S. (2014). *Behavior ballistic multilayer armor with composite epoxy reinforced with jute fiber*. Master's thesis of the Graduate Program in Materials Science at the Military Institute of Engineering (in Portuguese).
5. Milanezi, T. L. (2015). *Synergistic Behavior ballistic armor multilayer with rami fiber*. Master's thesis of the Graduate Program in Materials Science at the Military Institute of Engineering (in Portuguese).

6. Braga, F. O. (2015). *Ballistic behavior of a multilayered armoring using composite polyester-Curauá as middle layer*. Master's thesis of the Graduate Program in Materials Science at the Military Institute of Engineering (in Portuguese).
7. Agopyan, V., Savastano, H., Jr. (1997). Use of alternative materials based on plant fibers in construction: Brazilian experience. In *Iberoamerican Seminar* (in Portuguese).
8. Oliveira, J. T. S. (1998). *Characterization of eucalyptus wood for construction*. 429f. Thesis (Doctorate in Civil Engineering)—Polytechnic School of the University of São Paulo, São Paulo (in Portuguese).
9. Silva, L. C. (2014). *Ballistic behavior of Epoxy-Curauá composite in multilayer armor*. Thesis (Doctorate in Materials Science). Military Engineering Institute, Rio de Janeiro (in Portuguese).
10. Morye, S. S., Hine, P. J., Duckett, R. A., Carr, D. J., & Ward, I. M. (2000). Modelling of the energy absorption by polymer composites upon ballistic impact. *Composites Science and Technology, 60*, 2631–2642.

Curaua Non-woven Fabric Composite for Ceramic Multilayered Armors: A Lightweight, Natural, and Low Cost Alternative for Kevlar^{TM}

Fábio de Oliveira Braga, Augusto Corrêa Cabral, Édio Pereira Lima Jr., Sergio Neves Monteiro and Foluke Salgado de Assis

Abstract Advanced ceramics have been extensively applied for ballistic protection, when high levels of protection and low weight are demanded. However, their spalling characteristics require the suitable backing materials to collect the fragments generated on the impact. Synthetic fiber laminates, as Kevlar^{TM} and Dyneema^{TM}, are currently the most used solutions despite of their high cost and non-sustainable conditions. Therefore, several materials are being studied to replace the synthetic fiber laminates as second layer, including natural fiber composites, which are light, low cost and sustainable materials. Among these, Amazon curaua fiber composites are promising, due to its known high strength and high modulus. In the present work, novel curaua non-woven fabrics polymer composites have been investigated as part of the ceramic armor system. Ballistic tests were performed following NIJ 0101.06 armor standard. The results showed that these composites are promising alternatives to the synthetic fiber laminates as ceramic backing materials.

Keywords Composites · Curaua fiber · Non-woven fabric · Multilayered armor · Ballistic test

Introduction

Both local and world class armed conflicts are matter of personal concern, especially to law-enforcement officers and military personnel. Additionally, as weapon and ammunition technology become highly developed, it becomes a greater challenge to researchers and materials engineers to develop personal armor systems for ballistic protection. For heavy ammunition, such as 7.62 × 51 mm caliber bullet,

F. de Oliveira Braga (✉) · A.C. Cabral · ÉdioP. Lima Jr.
S.N. Monteiro · F.S. de Assis
Militar Institute of Engineering—IME, Rio de Janeiro, Brazil
e-mail: fabio_obraga@yahoo.com.br

© The Minerals, Metals & Materials Society 2017

M.A. Meyers et al. (eds.), *Proceedings of the 3rd Pan American Materials Congress*,
The Minerals, Metals & Materials Series, DOI 10.1007/978-3-319-52132-9_34

for example, a monolithic layer (e.g. steel plate) cannot provide a satisfying protection unless relatively thick pieces are employed, which compromises lightness and portability [1, 2]. For these types of threats, Multilayered Armor Systems (MAS) usually performs better because they combine synergistically the ballistic properties of different light materials [3, 4]. The MAS are typically composed of a ceramic frontal layer, a composite or fiber laminate as second layer, and a ductile metal back layer. According to Monteiro et al. [4], the ceramic front erodes and deforms the projectile's tip, and also absorbs most of the incident energy ($\sim$56.6%); The second layer collects the fragments of the ceramic and projectile's fragments, by mechanisms as mechanical incrustation, and Van der Waals or electrostatic forces, and absorbs part of the incident energy (36.7%); The back layer absorbs the rest of the energy (6.7%). The system is designed to prevent lethal trauma of the wearer.

Currently, synthetic fiber laminates as aramid (Kevlar^{TM} or Twaron^{TM}) and ultra-high molecular weight polyethylene (Dyneema^{TM} or Spectra^{TM}) have been the preferred materials to protect against the light ammunition and to integrate the protective systems against heavier caliber bullets, such as 7.62 mm. In those systems, the synthetic fiber laminates are commonly applied as ceramic backing, in the second layer of the MAS [1, 2].

However, due to the economic and environmental issues, other materials are being considered for the second layer of the MAS, including natural fiber reinforced composites, which have demonstrated good mechanical and ballistic performance [2, 5, 6]. Among the natural fibers, the fiber extracted from Amazon plant *Ananas erectifolius* named "curaua" have interesting mechanical properties such as high strength and high modulus, which is making their polymer composite a promising one for ballistic applications [7, 8].

Monteiro et al. [2] have already studied the ballistic behavior of curaua unidirectional fiber composites subjected to ballistic impact, as a stand-alone target and as part of a MAS. In the present work, the main objective is to study the ballistic performance of a novel curaua non-woven fabric composite integrating a 3-layered MAS, comparing its performance and mechanisms of fracture to aramid (Kevlar^{TM}) laminates.

Materials and Methods

Ballistic tests were conducted with MAS targets composed of a front 10 mm thick Al_2O_3 doped with 4 wt% Nb_2O_5 ceramic, backed by a 10 mm thick plate of either 30 vol.% of curaua fabric reinforced composite or Kevlar^{TM} laminates. Each MAS had a third layer, a 5 mm thick 5052 H34 aluminum alloy. The layers were bonded together with a polyurethane based rubber.

The curaua non-woven fiber fabric was acquired from the Brazilian company Pematec Triangel (Fig. 1a). The areal density is approximately 0.830 kg/m^2. An electronic image (Fig. 1b) shows that it is composed of randomly oriented fibers.

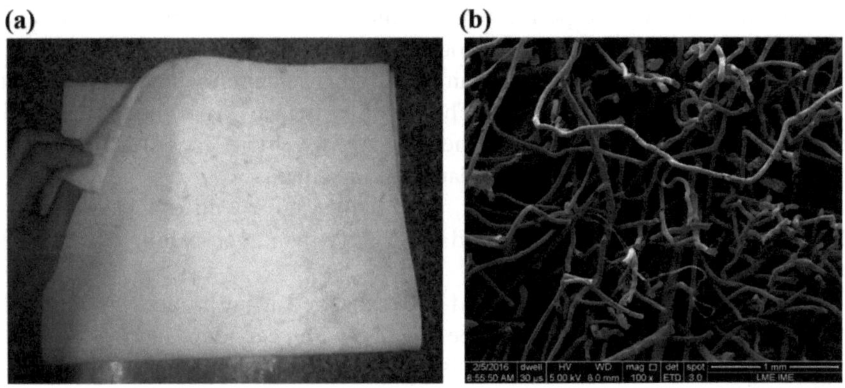

Fig. 1 Curaua fiber non-woven fabric: **a** general macroscopic aspect; **b** electronic micrograph, 100×

Fig. 2 Curaua fabric MAS positioned in front of the clay witness, before the ballistic test

The fabrics were cut and dried (60 °C for 24 h) for the production of composites. They were prepared by compression molding at room temperature (25 °C), in the volumetric fraction of 30%. The fiber fabrics were carefully positioned in the mold, in layers, intercalating with the resin-hardener mixture, being kept under the pressure of 5 MPa for 24 h. The polymeric matrix was epoxy diglycidyl ether of bisphenol-A (DGEBA) resin, hardened with triethylene tetramine (TETA, 13 phr), both produced by Dow Chemical and acquired from Brazilian company Resinpoxy.

The Du Pont Kevlar™ laminate was supplied by LFJ Blindagem (Brazil). It is composed of 18 plies of aramid woven fabric with areal density of 0.460 kg/m^2, bonded with polychloroprene, producing a laminate. The 5052 H34 aluminum alloy sheet was acquired from Matalak Co. (Brazil).

Before the tests, a block of Roma type clay witness, which is supplied by Corfix Ltd. (Brazil), was placed behind the target in direct contact with the aluminum alloy sheet, as shown in Fig. 2.

The ballistic tests were performed by following the NIJ 0101.06 international standard [9], in the Brazilian Army shooting range facility (CAEX, Rio de Janeiro), using 7.62 × 51 mm (9.7 g) NATO ammunition. The experimental arrangement is schematically represented in Fig. 3. The shooting device was a model B290 High Pressure Instrumentation (HPI), and the projectile's velocity measuring device was a model SL-520P Weibel doppler radar. Five specimens were tested.

The clay witness indentation (Fig. 4) was measured with a special Mitutoyo caliper of 0.01 mm accuracy, and used to quantify the target's ability to absorb the projectile's energy.

The fracture aspects were observed and registered using a model Quanta FEG 250 FEI scanning electron microscope (SEM), with secondary electrons contrast. An energy-dispersive chemical analysis (EDS) was performed.

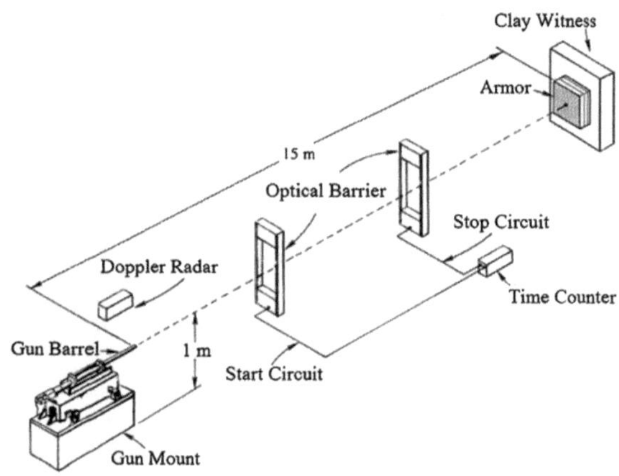

Fig. 3 Experimental arrangement for the ballistic tests

Fig. 4 Indentation in the clay witness, after the ballistic test

Results and Discussions

Table 1 shows the depth of indentation in the clay witness for the non-woven fiber fabric curaua composite, in comparison with similar multilayered armor systems of the same thickness reported by the literature [2, 5]. The previous indentation results are slightly better than the present one, however, some other aspects should be considered.

First, the international standard NIJ 0101.06 specifies that the target should not be perforated, and also prevent penetration into the clay witness backing the armor to a depth not exceeding 1.73 in. (44 mm). Thus, according to these criteria, the curaua fabric composite provides a satisfactory protection against 7.62 mm bullet.

Besides that, it is important for practical purposes that the second layer keeps its integrity after the impact, and this can be clearly observed in the Fig. 5.

A sample within the impact zone of the projectile was analyzed by SEM, using several magnifications, and some results can be seen in Figs. 6, 7 and 8. In Fig. 6, two special details stand out: Very thin particles were deposited over the whole surface (over the fibers), and also, relatively thick particles in some isolated spots. Monteiro et al. [4] have already discussed the role of the second layer in the multilayered armor, and this includes capturing the ceramic and projectile's fragments, besides absorbing part of the kinetic energy. Thus, it is reasonable to suppose that thick particles could be the projectile's fragments eroded by the ceramic

Table 1 Average depth of indentation in the clay witness backing different multilayered armors

Intermediate material layer	Indentation (mm)	Reference
Epoxy reinforced with 30 vol.% curaua non-woven fiber fabric	28 ± 3	PW[a]
Epoxy reinforced with 30 vol.% aligned curaua fiber	22 ± 3	[2]
Plain epoxy plate	21 ± 2	[2]
Aramid fiber fabric laminate	21 ± 3	[5]

[a]Present work

Fig. 5 Curaua fabric MAS positioned in front of the clay witness, after the ballistic test

Fig. 6 SEM micrograph showing the surface of the non-woven fiber fabric composite subjected to ballistic test (1000×)

Fig. 7 Thin particles observed on a curaua fiber, in the composite surface (5000×)

Fig. 8 Thick particles observed on curaua fabric composite surface (2000×)

layer, and the thin particles are the ceramic fragments itself. Figure 7 is a close image showing the surface of one fiber, and its ability to collect the fragments. Figure 8 is a close image of thick particles of approximately 50 μm.

Table 2 is the chemical composition of the two points in Fig. 8. The high percentage of Pb in the point 2 confirms that the thick particles are fragments of the projectile, which is eroded by the ceramic front. Point 1 shows high percentage of Al, Nb and O, confirming that the fine particles are fragments of the ceramic tile. The Pt observed was due to the conductive layer for SEM analysis. Monteiro et al. [4] performed a similar analysis for a Kevlar™ MAS, and observed the same fracture mechanisms. Figure 9a, b shows the macroscopic and microscopic aspects of a similar Kevlar™ MAS subjected to ballistic test, performed in the present work. The results show the macroscopic integrity and the microscopic capture of the fragments by the second layer. These similarities, in conjunction with the indentation results, suggest that curaua non-woven fabric composites are promising for Kevlar™ replacing in the present application.

Table 2 Chemical composition of the particle in two points, obtained by EDS/SEM

Element	Composition (%)	
	Point 1	Point 2
Al	33.75	0.33
Nb	21.60	19.30
O	31.02	2.30
Pt	12.78	5.45
Pb	–	72.62

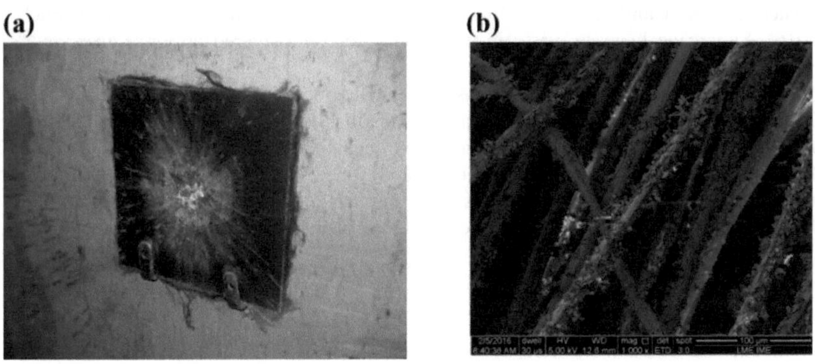

Fig. 9 Aramid MAS after the ballistic test: **a** macroscopic aspect; **b** microscopic aspect (1000×)

Conclusions

Ballistic tests following NIJ standard were performed and curaua non-woven fabric epoxy composites were evaluated as the second layer of a Multilayered armor system. The results showed that these composites are promising for this application, since the targets were not perforated and penetration of the clay witness did not reach 44 mm. Microscopic analysis showed that the same fracture mechanisms can be observed when comparing the curaua MAS with Kevlar™.

Acknowledgements The authors thank the Brazilian agencies CAPES and CNPq for the financial support, and CAEX for performing the ballistic tests.

References

1. Medvedovski, E. (2006). Lightweight ceramic composite armour system. *Advances in Applied Ceramics, 105*(5), 241–245.
2. Monteiro, S. N., Louro, L. H. L., Trindade, W., Elias, C. N., Ferreira, C. L., Lima, E. S., et al. (2015). Natural curaua fiber-reinforced composites in multilayered ballistic armor. *Metallurgical and Materials Transactions A, 46*(10), 4567–4577.
3. Akella, K., & Naik, N. K. (2015). Composite armour—A review. *Journal of the Indian Institute of Science, 95*(3), 297–312.
4. Monteiro, S. N., Lima, E. P., Jr., Louro, L. H. L., Silva, L. C., & Drelich, J. W. (2014). Unlocking function of aramid fibers in multilayered ballistic armor. *Metallurgical and Materials Transactions A: Physical Metallurgy and Materials Science, 46*, 37–40.
5. Monteiro, S. N., Milanezi, T. L., Louro, L. H. L., Lima, E. P., Jr., Braga, F. O., Gomes, A. V., et al. (2016). Novel ballistic ramie fabric composite competing with Kevlar™ fabric in multilayered armor. *Materials & Design, 96*, 263–269.
6. Monteiro, S. N., Candido, V. S., Braga, F. O., Bolzan, L. T., Weber, R. P., & Drelich, J. W. (2016). Sugarcane bagasse waste in composites for multilayered armor. *European Polymer Journal, 78*, 173–185.
7. Monteiro, S. N., Lopes, F. P. D., Barbosa, A. P., Bevitori, A. B., Silva, I. L. A., & Costa, L. L. (2011). Natural lignocellulosic fibers as engineering materials—An overview. *Metallurgical and Materials Transactions A: Physical Metallurgy and Materials Science, 42A*, 2963–2974.
8. Oliveira, F. H., Helfer, A. L., & Amico, S. C. (2012). Mechanical behavior of unidirectional curaua fiber and glass fiber composites. *Macromolecular Symposium, 319*, 83–92.
9. NIJ Standard 0101.06. (2008). Ballistic resistance of body armor. US Depart. of Justice.

Effect of Forging on Microstructure, Texture and Compression Behavior of Extruded AZ31B

D. Toscano, S.K. Shaha, B. Behravesh, H. Jahed, M. Wells, B. Williams and J. McKinley

Abstract Forging is a common method employed in the fabrication of automotive components. In this study, extruded AZ31B magnesium alloy was semi-close die forged at a temperature of 500 °C with a ram rate of 40 mm/s. Microstructural study indicated a bimodal grain structure with weaker texture in the forged material compared to the as-extruded material. Uniaxial compression tests indicated a remarkable improvement of fracture strain from 36 to 61% with a reduction of ultimate compressive strength between 4 and 22% in the forged samples compared to the as-extruded samples. It is attributed to the modification of microstructure and texture decreases twinning and increases the slipping activity resulting the improvement of ductility and reduced strength at room temperature.

Keywords Forging · Extrusion · Microstructure · Texture · Compression

Introduction

Concerns on climate change have led to the Environmental Protection Agency (EPA) and National Highway Traffic Safety Administration (NHTSA) in the United States of America to legislate a fleet-wide automotive efficiency target of 54.5 miles per gallon (mpg) by 2025 [1]. One method of reducing vehicular emissions is to

© Her Majesty the Queen in Right of Canada, as represented by the Minister of Natural Resources, 2016.

D. Toscano (✉) · S.K. Shaha · B. Behravesh · H. Jahed · M. Wells
Department of Mechanical and Mechatronics Engineering, University of Waterloo,
200 University Ave W, Waterloo, ON N2L 3G1, Canada
e-mail: dtoscano@uwaterloo.ca

B. Williams · J. McKinley
CanmetMATERIALS, Natural Resources Canada,
183 Longwood Road South, Hamilton, ON L8P 0A1, Canada

© The Minerals, Metals & Materials Society 2017

M.A. Meyers et al. (eds.), *Proceedings of the 3rd Pan American Materials Congress*,
The Minerals, Metals & Materials Series, DOI 10.1007/978-3-319-52132-9_35

reduce the curb weight of the vehicle [2]. Consequently, automotive manufacturers are investing considerable resources in the implementation of light-weight materials for the construction of future vehicles. Magnesium alloys due to their high specific strength and excellent machinability are one such family of materials that show potential for use in structural automotive applications [3–5]. A common method employed in the fabrication of automotive components is hot forging which involves severe plastic deformation at elevated temperatures. Such a thermo-mechanical process is expected alter the microstructure and texture of the material in question thereby altering the properties of the final product [6–8]. To that end this paper aims to examine the effect of forging on the microstructure, texture and mechanical properties of extruded AZ31B Mg-alloy.

Experimental Procedure

Cylinders of 20 mm diameter and 45 mm length were extracted from a 63.5 mm diameter extruded AZ31B billet. These cylinders were heated to a temperature of 500 °C and placed between the isothermal dies of a hydraulic press. The lower die was flat whereas the upper die was profiled to include a step and a central rib. The forging geometry was chosen to simulate a closed die forging process by constraining material flow. A schematic representation of the final forged shape is displayed in Fig. 1. Graphite was used as lubricant and the forging was performed at a displacement rate of 40 mm/s to simulate a manufacturing process. Due to multiple flow paths, a range of strain rates from about 0.1 to 10 s^{-1} could be

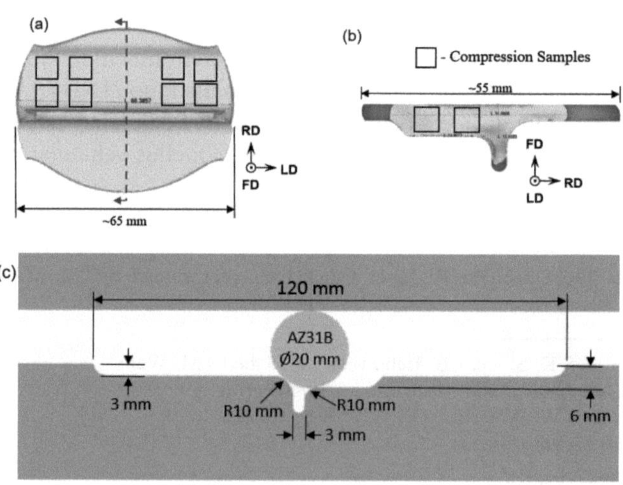

Fig. 1 A schematic of the forged geometry **a** view parallel to forging direction (FD), with *squares* showing compression sample extraction locations **b** view parallel to the longitudinal direction (LD) showing compression sample locations and **c** die geometry (dimensions in mm)

achieved during forging of this geometry at 40 mm/s. After forging, the components were air cooled to room temperature.

Superficial hardness measurements of the extruded and forged material were performed using a Rockwell 15T scale on a United Instruments Tru-Blue II digital hardness tester equipped with a 1/16 in. steel ball indenter. Microstructure was captured using an optical microscope whereas texture analysis was performed using a Bruker D8-Discover equipped with an advanced 2D detector. The microstructure and texture analysis followed the procedures detailed in [9, 10].

Cubic compression specimens of 6 mm edge length were machined from the as-extruded material and from the stepped portion of the forgings (Fig. 1). These cuboid specimens were tested in compression using a MTS uniaxial load frame with a capacity of 50 kN and at a displacement rate of 1 mm/min. Strain was measured using ARAMIS 3D 5MP Digital Image Correlation (DIC) system.

Results and Discussion

Microstructure and Texture

Optical micrographs of the as-received material are included in Fig. 2; the as-extruded alloy exhibited a mild bimodal grain distribution with an average grain size of 32.5 μm. Texture analysis of the as-received material revealed that the c-axis of the hexagonal close packed (hcp) unit cell in the grains is aligned perpendicular to the extrusion direction (ED). Similar type of texture and microstructure were observed by others indicating c-axis oriented perpendicular to the extrusion direction [11].

Microstructure analysis of the forged material indicated a strong bimodal grain structure in the interior of the forging with the presence of larger grains near position 3 and position 4 (Fig. 3). The overall average grain size of the forged material was calculated to be 39.9 μm. The grain size decreased to 26.1 μm at the joint or angle of the forged materials (position 2). However, the microstructure of position 1 of the forged materials shows equiaxed grains structure. This is due to the severe plastic deformation and recrystallization at higher temperature. The increase

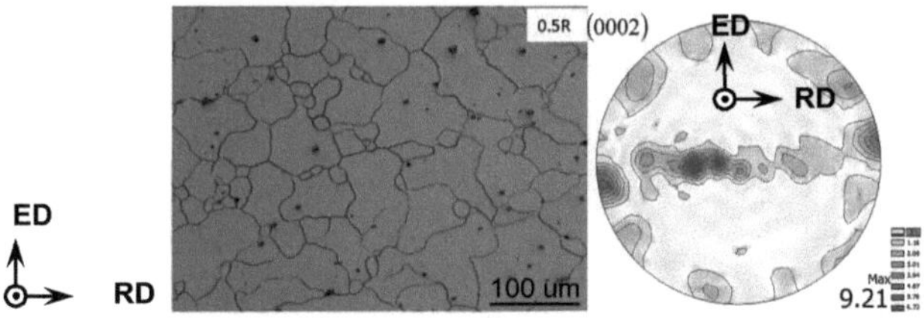

Fig. 2 The initial microstructure and (0002) pole figure of extruded AZ31B Mg-alloy at the half radius (0.5R) of the 63.5 mm billet

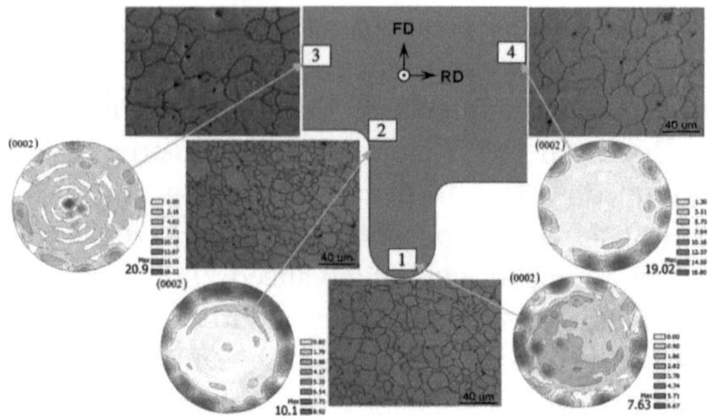

Fig. 3 Optical micrographs and (0002) pole figures of the forged AZ31B Mg-alloy in different locations. Note that *directions* indicated apply to all micrographs and pole figures

in bimodality may be due to the high forging temperature causing grain growth following recrystallization. Texture analysis of the forged material indicated a change from extruded texture such that c-axis orientations were nearly parallel to the localized deformation directions. The (0002) pole intensity increased to ~20.9 MRD (Multiples of Random Distributions) at the position 3 (Thin section) while at the tip of the forged component, the intensity reached 7.63 MRD. The weaker texture in points (1) and (2) than in points (3) and (4) is due to the fact that severe plastic deformation in points (1) and (2) and thermo-mechanical processes lead to dynamic recrystallization which modified the microstructure and texture [12].

Superficial Hardness

An intensity map of the measured superficial hardness values for the extruded and forged material is included in Fig. 4. Measurements revealed no statistically significant difference in hardness values in the forged material compared to the extruded AZ31B. The hardness value of the forged sample in the in-plane direction was ~53 HRB and through thickness was ~57 HRB. The hardness distribution was relatively homogenous considering the sensitivity of the scale used for

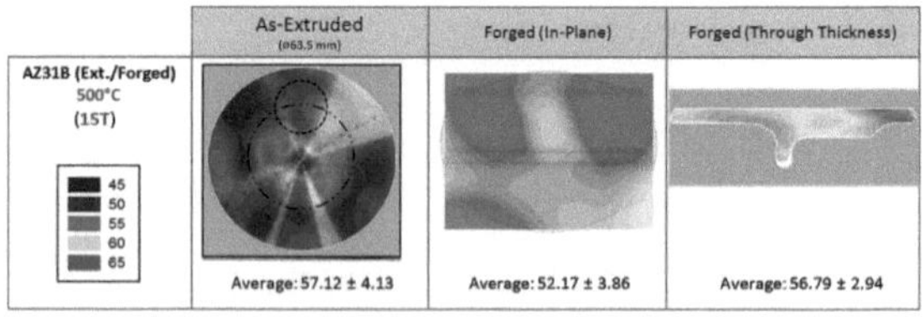

Fig. 4 A comparison of superficial Rockwell hardness map between the extruded and forged AZ31B Mg-alloy. The smaller *dashed circle* in the extrusion map represents the extraction location of the forged billet

presentation; this should manifest in the material displaying relatively homogenous material properties throughout the forging. It is seen that through thickness hardness values appeared to positively correspond to the texture intensity (Figs. 3 and 4), which means that the area showing strong texture exhibits higher hardness along the ED. Also, the large surface grains could also account for the lower in-plane hardness values illustrated in Fig. 4 [13].

Compression Properties

Compression at room temperature along the ED of the as-extruded material resulted in severe twinning apparent by the strong sigmoidal shape of the corresponding stress-strain curves (Fig. 5). In contrast, compression along RD of the extruded material displayed primarily monotonic hardening behavior suggesting primarily slip dominated plasticity. The deformation behavior of the as-received material is consistent with the measured texture and has been documented in other studies [14, 15]. The forged LD and RD specimens exhibited moderate twinning in compression due to c-axis orientations being near perpendicular to the loading direction, Fig. 3 (the LD/RD/FD directions in forged samples are shown in Fig. 1). By contrast, compression in FD displayed monotonic hardening due to the unfavorable orientation for twinning. Considering the c-axis orientation at the stepped portion of the forging (Fig. 3, position 4), basal slip is likely the dominant mechanism for plasticity in FD.

The ultimate compressive strength and failure strain of the tested specimens are summarized in Fig. 6. It is evident that the forging process resulted in a substantial increase in overall failure strain with only a little decrease in ultimate compressive strength of the material. Specifically LD specimen of the forging exhibited a 22% decrease in strength and 36% increase in failure strain compared to the extruded ED specimen. Similarly, RD and FD forged specimens demonstrated 11 and 4%

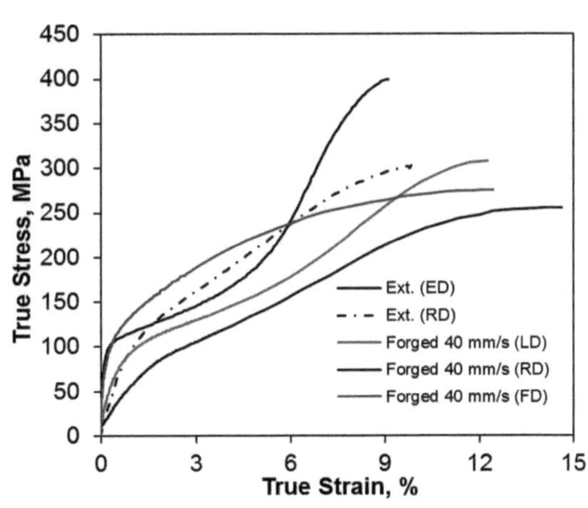

Fig. 5 Typical True stress-strain compression curves for as-received and forged AZ31B Mg-alloy in different directions

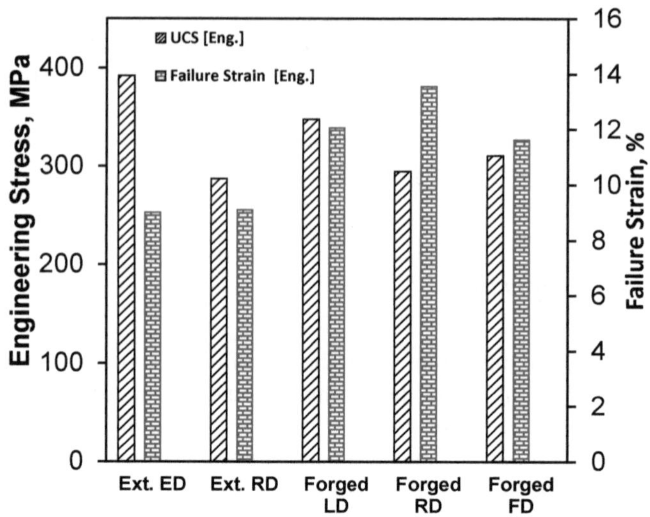

Fig. 6 A comparison of compression properties between as-received and forged AZ31B Mg-alloy

decrease in strength, and 61 and 36% increase in failure strength, respectively, compared to the RD specimens of the extruded material. Higher strength in the ED compared to LD directions is a result of impeded dislocation motion due to high twin activation and twin boundary density in extruded material during compression. Similarly, basal texture in the RD and FD directions of the forged material results in lower twin activation and twin boundary density compared to the ED direction therefore exhibiting lower overall strength due to easier dislocation movement. In all directions for the forged material, increase in failure strain may be partially attributed to a change in texture favoring higher strain accommodation of slip based plasticity over low strain accommodation of twinning. A slightly refined grain structure at the sample extraction location (Fig. 3) may also contribute to the marked increase in failure strain [16].

Conclusion

The effect of semi-closed die forging on the microstructure, texture and mechanical properties of extruded AZ31B was examined.

- The forging process resulted in a modification of grain distribution with refined bimodal grains to equiaxed grains depending on the location and flow of the material during forging.
- Strong basal texture was modified by forging and the local texture of the forged product demonstrated c-axis alignment parallel to localized deformation direction.

- Reduction in the density of twin bands during deformation resulted in the forged material displaying lower strength compared to the extruded material.
- Increase in failure strain compared to the as-received material is partially attributable to a shift from twin based plasticity to slip based plasticity complemented by a refined interior microstructure.

Acknowledgements The financial support of the Natural Sciences and Engineering Research Council of Canada (NSERC) through the Automotive Partnership Canada (APC) program under APCPJ 459269–13 grant with contributions from Multimatic Technical Centre, Ford Motor Company, and Centerline Windsor are acknowledged. The authors would also like to thank L. Blaga of CanmetMATERIALS, Hamilton for forging trials and, G. Yu for design of the forging die and Dr. Y. Ding of the University of Waterloo for assistance with SEM and OM.

References

1. EPA Office of Transportation and Air Quality. (2012). *EPA and NHTSA set standards to reduce greenhouse gases and improve fuel economy for model years 2017–2025 cars and light trucks*. Washington: United States Environmental Protection Agency.
2. Kulekci, M. K. (2008). Magnesium and its alloys applications in automotive industry. *The International Journal of Advanced Manufacturing Technology*, 851–865.
3. Huang, G., Han, T., Pan, F., Zhang, H., & Li, J. (2014). Improving low-cycle fatigue properties of rolled AZ31 magnesium alloys by pre-compression. *Material & Design, 58,* 439–444.
4. Albinmousa, J., Jahed, H., & Lambert, S. (2011). Cyclic behaviour of wrought magnesium alloy under mutliaxial loading. *International Journal of Fatigue*, 1403–1416.
5. Albinmousa, J., Jahed, H., & Lambert, S. (2011). Cyclic axial and cyclic torsional behaviour of extruded AZ31B magnesium alloy. *International Journal of Fatigue, 33*(8), 1127–1139.
6. Martin, E., & Jonas, J. J. (2010). Evolution of microstructure and microtexture during the hot deformation of Mg–3% Al. *Acta Materialia, 58*(12), 4253–4266.
7. Gryguc, A., Jahed, H., Williams, B., & McKinley, J. (2015). MagForge—Mechanical behaviour of forged AZ31B extruded magnesium in monotonic compression. *Materials Science Forum, 828–829,* 291–297.
8. Gryguc, A., Shaha, S., Jahed, H., Wells, M., Williams, B., & McKinley, J. (2016). Tensile and fatigue behaviour of as-forged AZ31B extrusion. *Fracture and Structural Integrity, 38,* 251–258.
9. Sarker, D., Friedman, J., & Chen, D. L. (2015). De-twinning and texture change in an extruded AM30 magnesium alloy during compression along normal direction. *Journal of Materials Science and Technology, 31*(3), 264–268.
10. Roostaei, A. A., & Jahed, H. (2016). Role of loading direction on cyclic behaviour characteristics of AM30 extrusion and its fatigue damage modelling. *Materials Science and Engineering A, 670,* 26–40.
11. Zheng, S., Yu, Q., & Jiang, Y. (2013). An experimental study of fatigue crack propagation in extruded AZ31B magnesium alloy. *International Journal of Fatigue, 47,* 174–183.
12. Yang, X., Yi, J., Ni, S., Du, Y., & Song, M. (2016). Microstructural evolution and structure-hardness relationship in an Al-4wt.%Mg alloy processed by high-pressure torsion. *Journal of Materials Engineering and Performance, 25,* 1909–1915.
13. Choi, S. H., Shin, E. J., & Seong, B. S. (2007). Simulation of deformation twins and deformation texture in an AZ31 Mg alloy under uniaxial compression. *Acta Materialia, 55* (12), 4181–4192.

14. Barnett, M. R., Keshavarz, Z., Beer, A. G., & Atwell, D. (2004). Influence of grain size on the compressive deformation of wrought Mg–3Al–1Zn. *Acta Materialia, 52,* 5093–5103.
15. Jahadi, R., Sedighi, M., & Jahed, H. (2014). ECAP effect on the microstructure and mechanical properties of AM30 magnesium alloy. *Materials Science and Engineering A, 593,* 178–184.
16. Koike, J., Kobayashi, T., Mukai, T., Watanabe, H., Suzuki, M., Maruyama, K., et al. (2003). The activity of non-basal slip systems and dynamic recovery at room temperature in fine-grained AZ31B magnesium alloys. *Acta Materialia, 51*(7), 2055–2065.

Effect of Hypoeutectic Sc Additions to Al-4.5 wt% Cu Under Different Cooling Rates

A.-A. Bogno, J. Valloton, H. Henein, M. Gallerneault and D. Herlach

Abstract Lightweight materials are the best response to improving performance and efficiency of sport and transportation industrial products. Aluminum is one of the most attractive lightweight materials due to its low density and a high strength to weight ratio achievable through cold working and/or heat treatment. Age hardenable Al–Cu alloys are one of the strongest aluminium alloys available. Sc, though very expensive, is renowned for yielding the highest strength increase per atomic percent of any alloying addition in Al through grain refinement of hyper-eutectic Sc compositions and precipitation hardening. This paper studies the solidification of Al-4.5 wt% Cu with minor Sc additions (hypo-eutectic compositions) over a wide range of cooling rates. The objective is to determine the minimum Sc addition for a maximum strengthening effect while reducing typical processing steps. Based on the microstructures and mechanical properties analyses, a cost and time effective processing route is proposed for the 2000 series aluminum alloys.

Keywords Solidification · Aluminum–copper–scandium · Cooling rate

Introduction

Forming about eight percent of the earth's crust, aluminum has been identified as the most common metal on earth. It is the third most plentiful element known to man after oxygen and silicon (sand). Lightness is the outstanding and best known

A.-A. Bogno (✉) · J. Valloton · H. Henein
Department of Chemical and Materials Engineering,
University of Alberta, Edmonton, AB T6G 2G6, Canada
e-mail: bogno@ualberta.ca

M. Gallerneault
Alcereco Inc., Kingston, ON K7L3N6, Canada

D. Herlach
DLR, Institute of Materials Physics in Space, Linder Höhe,
51147 Köln, Germany

© The Minerals, Metals & Materials Society 2017
M.A. Meyers et al. (eds.), *Proceedings of the 3rd Pan American Materials Congress*,
The Minerals, Metals & Materials Series, DOI 10.1007/978-3-319-52132-9_36

characteristic of aluminum. A material's weight is an important aspect taken into consideration during selection for any industrial applications especially when motion is involved. Minimizing weight not only results in greater economy of operation but also saves energy and offers lower shipping, handling and erection costs. Thanks to the low weight and the possibility of high strength when properly alloyed, aluminum is considered as the backbone of the aerospace industry. It is also used in cooking and packaging to name a few of its applications.

The term aluminum refers to a family of alloys and therefore the properties of a particular aluminum product depend on the chosen alloy. Addition of Transition Metals (TM) such as Cu and Sc to aluminum results in the formation of finely dispersed precipitates upon heat treatment. Al–Cu (2000 series) is one of the most widely used Al-base alloys due to the relevant high age hardening effect of Cu characterized by precipitation of finely dispersed Guinier–Preston (GP) zones, θ', followed by the stable θ phase through heat treatment.

It has been found that, hyper-eutectic compositions (>0.55 wt% Sc) in binary Al–Sc can yield good grain refining when slowly cooled [1]. It is also reported that by fast cooling during solidification (100 °C/s) and quenching, a high supersaturation of Sc was achieved which subsequently promoted age hardening upon heat treatment, through the precipitation of finely dispersed Al_3Sc particles [1]. The Sc grain refining effect results from its ability to promote small equiaxed grains formation instead of long dendrites resulting in porosity and hot-cracking. It was shown by Norman et al. [2] that hypereutectic Al–Sc (>0.55 wt% Sc) yields an effective grain refining of as-cast microstructures at a cooling rate of 1000 °C/s. They also showed that the change from large dendritic grains to fine spherical grains is realized by the addition of Sc to Al. When combined with other elements such as Zr, Norman and coworkers found that this grain refining limit shifted to a lower Sc level. A lower Sc level would be economically very interesting because of its high cost. Also, it is still a challenge to obtain full precipitation hardening from Al_3Sc in the common heat treatable alloys such as the 2000 series aluminum alloys.

The goal of this paper is to investigate the effects of hypo-eutectic Sc compositions (<0.55 wt% Sc) on age-hardenable Al–Cu at various solidification histories in order to find a solidification/processing route that yields maximum benefit for a minimum Sc addition. In order to achieve this goal, Al-4.5 wt% Cu containing respectively 0.1, 0.2 and 0.4 wt% Sc are solidified under cooling rates varying from 1 to 10^4 °C/s by Differential Scanning Calorimetry (DSC), Electro-Magnetic Levitation (EML) and Impulse Atomization (IA). Heat treatment of the samples is then carried out and their mechanical properties are measured and compared.

Experimental and Methods

The experimental work includes a series of solidification of Al-4.5 wt% Cu with different Sc level (0.0, 0.1, 0.2 and 0.4 wt%) by Impulse Atomization (IA), Electromagnetic Levitation (EML) and Differential Scanning Calorimetry (DSC).

The raw materials in the form of pellets were prepared by alloying 99.99% pure Al with commercial purity Cu and Sc.

DSC

DSC experiments were performed in a Setaram Labsys Evo 1600 DSC using two alumina crucibles (sample and reference crucibles) and platinum-rhodium DSC rods. A furnace, regulated by means of an S-type thermocouple (Pt/Pt Rh 10%), was used to heat the samples after 20 min purging with argon. A scanning rate of 2 °C/min up to the temperature of 850 °C was applied during melting and the samples were solidified at different cooling rates varying from 0.5 up to 50 °C/min. Temperature measurement was achieved by a thermocouple placed between the two alumina crucibles. Prior to the experiments, the calorimeter had been calibrated for temperature and heat measurements for a wide range of temperatures using standard samples of Al, Ag, Zn, Sn, and Au.

EML

EML is a containerless solidification technique for the processing of electrically conducting samples. An alternating current flowing through a water-cooled levitation coil produces an electromagnetic field. A conducting sample placed within this field is levitated by the Lorentz force which compensates for the gravitational force. Meanwhile, Eddy currents are induced in the sample and it is heated and melted by ohmic losses. To solidify the sample, cooling jets of inert gas are used, with the temperature of the sample being monitored continuously with a two-color pyrometer.

Using an alumina holder, each pre-alloyed sample ($\sim$0.45 g) was inserted into the levitation coil in an ultra-high vacuum chamber. The chamber was evacuated to a pressure of 10^{-7} mbar and backfilled with high purity inert gas (helium 6.0) in order to limit contamination of the melt. The levitating droplet was melted and overheated to a temperature 100–200 °C above its liquidus temperature to remove oxides and contaminants. To cool the sample below its liquidus temperature and induce solidification, jets of helium 6.0 were then used. The solidified sample, a quasi-sphere of 6–7 mm in diameter (depending on the weight), was retrieved for subsequent characterization. A detail description of the process is given in [3].

IA

A quantity of 300–400 g of investigated samples were melted in a dense graphite crucible by induction heating up to 850 °C (~200 °C above the liquidus temperature) under an inert atmosphere of choice. Atomization was achieved by the principle described in [4]. The atomized droplets rapidly solidified during their fall by losing heat to the stagnant inert atmosphere contained in an almost oxygen free (10 ppm) chamber. The solidified droplets were then collected in an oil filled beaker, washed, dried and sieved into different size ranges. Droplets of sizes varying from 212 to 1000 μm have been investigated.

Analysis Techniques

In order to study resulting solidification microstructures of the generated samples, X-ray diffraction analysis was performed using a wide range of angles (2θ) from 5° to 90° with a step of 0.02° and a holding time of 0.60 s at each step. The current and the voltage of the X-ray tube during the analysis were 38 mA and 38 kV respectively. The radiation was the CoKα1 with a wavelength to calculate d-spacing = 1.78899 Å. SEM-Back Scattered Electrons (BSE), Transmission Electron Microscopy (TEM) and Energy Dispersive X-ray Spectroscopy (EDX) were also carried out on the ground and polished samples in order to reveal microstructures/scale of microstructures and then identify the precipitated phases.

The scale of the microstructure was determined by Cell Spacing measurement. Indeed, Secondary Dendrite Arms Spacing are approximated by the dendrite cell intervals (center-to-center distance between two cells). The cell intervals are approximately equal to the size of the cells so that using line intercepts method accordingly with ASTM E112-13, measurements of cell intervals were performed on the micrographs obtained by SEM (BSE). Hardness of as-solidified as well as heat treated samples were carried out using a Buhler VH 3100 microhardness machine. The device was calibrated using a manufacturer provided steel block. Five indentations were randomly applied on each sample with a load of 100 gf held for 10 s.

Results and Discussions

Scale of Microstructures

Figure 1 shows the solidification microstructures of Al-4.5 wt% Cu with different Sc content obtained at the cooling rate of 0.8 °C/s using DSC. As can be observed, the morphology of the structures changes from long dendrites to equiaxed cells as Sc contained is increased from 0.0 to 0.4 wt%.

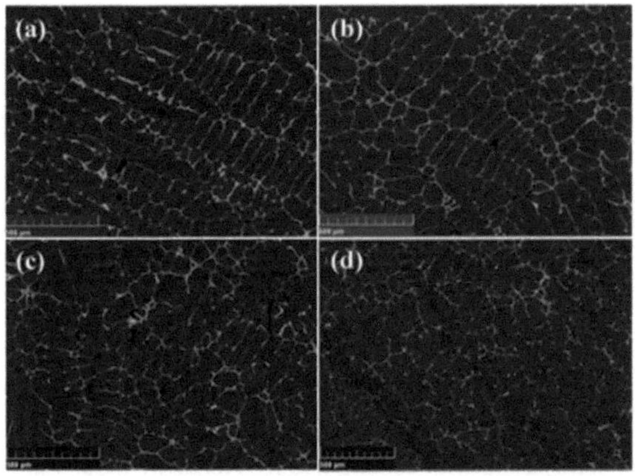

Fig. 1 Representative solidification microstructures of investigated Al-4.5 wt% Cu with different Sc addition cooled at 0.8 °C/s using DSC; **a** 0.0 wt% Sc **b** 0.1 wt% Sc **c** 0.2 wt% Sc **d** 0.4 wt% Sc

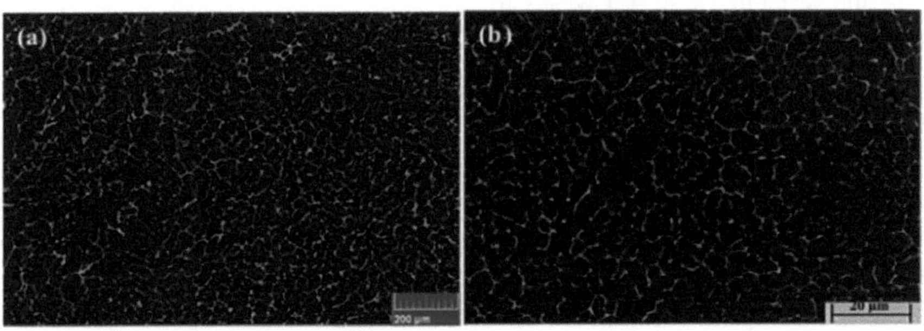

Fig. 2 Representative solidification microstructures of investigated Al-4.5 wt% Cu with 0.4 wt% Sc obtained by **a** EML (15 °C/s) and **b** IA (3500 °C/s)

At higher cooling rates obtained respectively by EML (5–15 °C/s) and IA (10^2–10^4 °C/s), the microstructures consist essentially of equiaxed grains as can be seen in Fig. 2.

Sc effect on microstructures scale is found to be negligible; instead grain refinement is very dependent on the cooling rate. Cell spacing variation with solidification cooling rate is plotted for EML and IA samples (Fig. 3). While cooling curve can be obtained directly by pyrometry during EML [3, 5], a thermal model for atomization [6–8] is used to estimate the cooling rates of IA samples.

As shown in Fig. 3, cell spacing variation with solidification cooling rate for the investigated alloys follows the equation $\lambda_2 = A\dot{T}^{-n}$, where λ_2 represents the cell spacing, $\dot{T}$ the solidification cooling rate and A and n are constants, as described by Eskin et al. [10]. The values of A and n found in this study are in the range of values published in [11] by Mullis and co-workers in the estimation of cooling rates during close-coupled gas atomization using secondary dendrite arm spacing measurement.

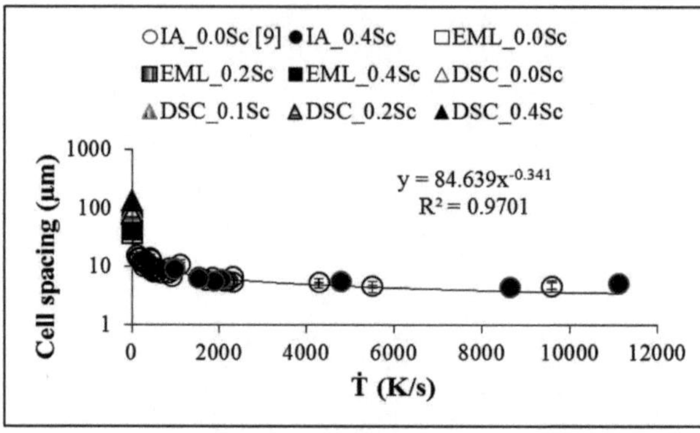

Fig. 3 Variation of cell spacing with cooling rate for Al-4.5 wt% Cu with up to 0.4 wt% Sc of different thermal histories. Published data by Wiskel et al. [9] are also incorporated

Clearly the addition of Sc in the hypoeutectic range of compositions does not grain refine the structure of Al-4.5 wt% Cu.

Mechanical Properties

Vickers micro-hardness variation with Sc content is plotted for various cooling rates. Figure 4 show the hardness variation with Sc for a slowly cooled sample (0.8 °C/s) by DSC in as-solidified as well as heat treated conditions.

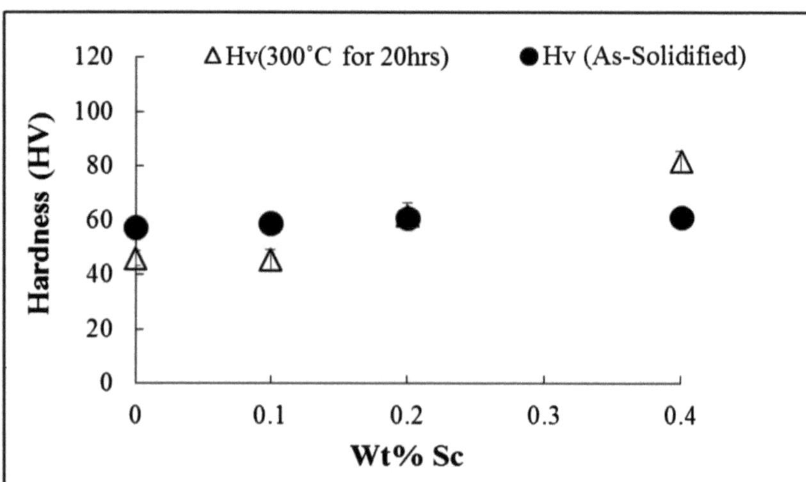

Fig. 4 Variation of Vickers micro-hardness with Sc level in Al-4.5 wt% Cu (Sc) solidified at 0.8 °C/s before and after aging at 300 °C for 20 h

As can be seen in Fig. 4 (at the cooling rate of 0.8 °C/s), hardness does not vary with Sc level; it is constant at 60 HV. When aged at 300 °C for 20 h, it is found to increase from 60 to 80 HV at the Sc level of 0.4 wt%. Figure 5 shows the variation of Vickers micro-hardness with cooling rate at all level of Sc. It can be seen that the hardness does not significantly vary for all levels of Sc. Thus, it can safely be said that cooling rate and Sc addition has no effects on hardness within the limits of this investigation.

Samples with the lowest (solidified by DSC) and the highest (atomized) cooling rates have been aged and their hardness after aging have been measured and compared with the as-solidified and as-atomized samples. The results presented in Fig. 6 show that upon heat treatment, there is negligible increase of hardness for Sc level less than 0.4 wt%. At 0.4 wt% Sc, however, hardness is found to increase by a factor of 2 after aging at 300 °C for 20 h.

Samples with 0.4 wt% Sc content solidified at 3500 °C/s were then subsequently aged at the temperatures of 350 and 365 °C respectively for 20 h. Figure 6 shows the variation of hardness with these aging temperatures. As can be seen hardness decreases as the aging temperature increases. This result suggests that 300 °C might be the optimum aging temperature for hypo-eutectic Al–Cu with hypo-eutectic Sc composition. This aging temperature is in the annealing temperature range

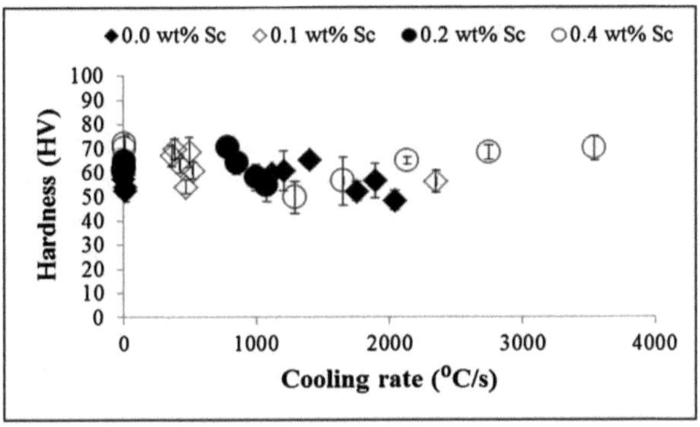

Fig. 5 Variation of Vickers micro-hardness with cooling rate for Al-4.5 wt% Cu with different Sc levels

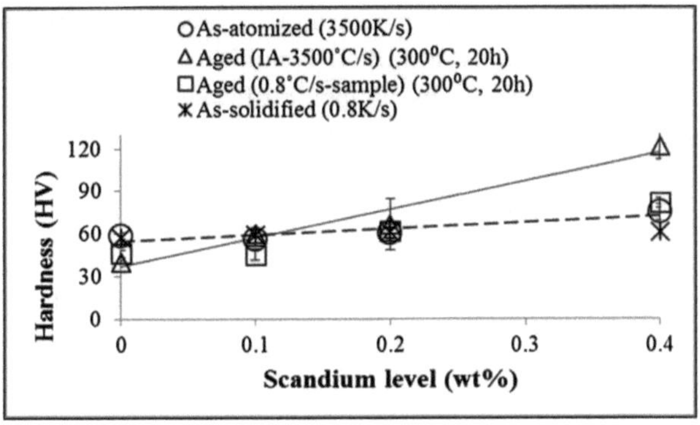

Fig. 6 Variation of Vickers micro-hardness with Sc level in Al-4.5 wt% Cu (Sc) under various thermal histories

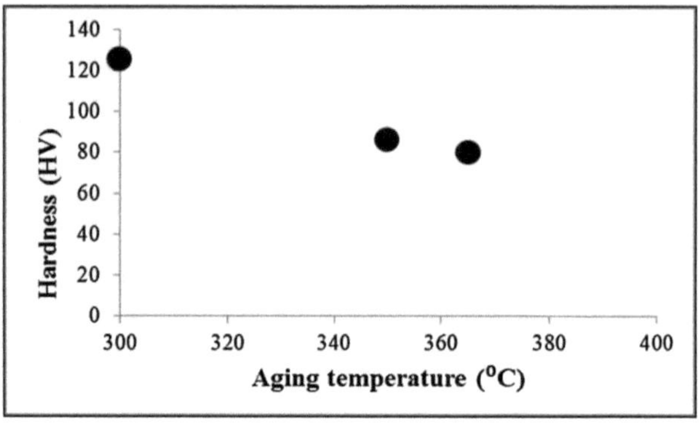

Fig. 7 Variation of Vickers micro-hardness with aging/holding time of IA-Al-4.5 wt% Cu (cooling rate: 3500 °C/s)

250–350 °C within which Al–Sc present a great precipitation hardening potential from Al_3Sc [12] (Fig. 7).

Conclusions

Al-4.5 wt% Cu with different level of Sc addition (up to 0.4 wt%) were generated under low, medium and high cooling rate conditions respectively by DSC, EML and IA.

At low cooling rate (0.8 °C/s) no grain refining effect of Sc is observed. Instead, the microstructure morphology changes from long dendrites to equiaxed cells with increasing level of Sc. At medium and high cooling rates, the Sc effect on structure morphology was overshadowed by the combined effects of high cooling rate and high undercooling induced by container-less solidification resulting in equiaxed grains.

Cell spacing variation with cooling rate for the investigated alloys is found to follow an empirical law of secondary dendrite arms spacing commonly found in literature.

The mechanical properties of the investigated samples in both the as-solidified and heat treated conditions have been evaluated through vickers micro-hardness measurements. Cooling rate and Sc level are found to have negligible effects on hardness in the as solidified condition. However at the level of 0.4 wt% Sc the hardness increased by a factor of 2 after aging of atomized samples. This is attributed to the precipitation of Al_3Sc and Al_2Cu.

Acknowledgements The authors are grateful to the Natural Sciences and Engineering Research Council of Canada (NSERC), Novelis, the Canadian Space Agency (CSA), and the European Space Agency (ESA) for their financial support. Also the staff at the Institute of Materials Physics in Space (DLR) is acknowledged for their cooperation and help during EML experiments.

References

1. Royset, J. (2007). Scandium in aluminium alloys overview: Physical metallurgy, properties and applications. *Metallurgical Science and Technology, 25*(2), 11–21.
2. Norman, A. F., Prangnell, P., & McEwen, R. (1998). The solidification behaviour of dilute aluminium-scandium alloys. *Acta Materialia, 46*(16), 5715–5732.
3. Herlach, D. M. (2012). Containerless undercooling of drops and droplets. In *Solidification of containerless undercooled melts* (pp. 1–30).
4. Henein, H. (2002). Single fluid atomization through the application of impulses to a melt. *Materials Science and Engineering A, 326*(1), 92–100.
5. Bogno, A.-A., Valloton, J., Natzke, P., Yin, S., Herlach, D. M., & Henein, H. (2015). Scandium effects on Nucleation undercooling in Al–Cu droplets generated by impulse atomization and electro-magnetic levitation. In *Materials Science and Technology Conference and Exhibition 2015, MS and T 2015* (Vol. 2).
6. Wiskel, J. B., Henein, H., & Maire, E. (2002). Solidification study of aluminum alloys using impulse atomization: Part I—Heat transfer analysis of an atomized droplet. *Canadian Metallurgical Quarterly, 41*(1), 97–110.
7. Prasad, A., Mosbah, S., Henein, H., & Gandin, C.-A. (2009). A solidification model for atomization. *ISIJ International, 49*(7), 992–999.
8. Prasad, A., & Henein, H. (2008). Droplet cooling in atomization sprays. *Journal Materials Science, 43*(17), 5930–5941.
9. Wiskel, J. B., Navel, K., Henein, H., & Maire, E. (2002). Solidification study of aluminum alloys using impulse atomization: Part II. Effect of cooling rate on microstructure. *Canadian Metallurgical Quarterly, 41*(2), 193–204.
10. Eskin, D., Du, Q., Ruvalcaba, D., & Katgerman, L. (2005). Experimental study of structure formation in binary Al–Cu alloys at different cooling rates. *Materials Science and Engineering A, 405*(1–2), 1–10.
11. Cochrane, R. F., Adkins, N. J., Mullis, A. M., & Farrell, L. (2013). Estimation of cooling rates during close-coupled gas atomization using secondary dendrite arm spacing measurement. *Metallurgical and Materials Transactions B, 44*(4), 992–999.
12. Røyset, J., Ryum, N., Bettella, D., Tocco, A., Jia, Z., Solberg, J. K., et al. (2008). On the addition of precipitation- and work-hardening in an Al–Sc alloy. *Materials Science and Engineering A, 483,* 175–178.

References

Izod Impact Tests in Polyester Matrix Composites Reinforced with Fique Fabric

Artur C. Pereira, Sergio N. Monteiro, Foluke S. Assis and Henry A. Colorado

Abstract The fique fibers are studied worldwide as an alternative of synthetic fibers in composites. This study evaluated the impact resistance of this type of composite. Specimens were made with up to 30% in volume of fique fabric in an Izod normalized mold. The fique fabric was embedded with polyester resin and cured at room temperature for 24 h. The specimens were tested in Izod impact pendulum and the fracture surfaces were examined by scanning electron microscopy (SEM). The impact resistance of composites increased linearly with the relative amount of fique fabric reinforcing the composite. This performance was associated with the difficulty of rupture imposed by the fique fabric as well as the type of cracks resulting from the interaction jute fiber/polyester matrix that corroborate the energy absorption at the impact test.

Keywords Fique fabric · Composite · Polyester matrix · Izod impact tests

Introduction

Natural fibers, especially those lignocellulosic obtained from plants, offer economical environmental and technical advantages in comparison to synthetic fibers for application as the reinforcement of polymeric composites [1–3]. Some of these lignocellulosic fibers like jute, sisal, cotton, flax, hemp, coconut and sawdust are, since long time, being added to polymers to enhance the properties [4]. This historical fact dates back to the first industrial polymer, the Bakelite, which was more than one hundred years ago incorporated with sawdust to improve its impact resistance and reduce cost [5]. For the industry, low cost is certainly an important

A.C. Pereira · S.N. Monteiro · F.S. Assis (✉)
Military Institute of Engineering, IME, Praça Gen., Tibúrcio,
n°80 Urca, Rio de Janeiro, RJ 22290-270, Brazil
e-mail: foluke.assis@hotmail.com

H.A. Colorado
Universidad de Antioquia, Calle 67 #53-108, Medellín, Antioquia, Colombia

© The Minerals, Metals & Materials Society 2017
M.A. Meyers et al. (eds.), *Proceedings of the 3rd Pan American Materials Congress*,
The Minerals, Metals & Materials Series, DOI 10.1007/978-3-319-52132-9_37

incentive associated with the use of lignocellulosic fibers that usually have a commercial price around five times lower than that of glass fiber, the cheapest among the synthetic fibers.

Environmental issues are, additionally gaining attention owing to worldwide problems related to climate changes and pollution. This is nowadays a major advantage for the natural fibers that are renewable, biodegradable and recyclable. By contrast to glass fiber composites that cannot be recovered, natural fiber composites can be completely burnt to produce energy [6]. Moreover, lignocellulosic fibers are neutral with respect to CO_2 emission, the main responsible for global warming [7].

Technical advantages are also associated with the use of lignocellulosic fibers in polymeric composites. According to Zah et al. [5], the application of natural fiber composites is rapidly increasing in the automobile industry with annual growth rates above 20%. Both interior and exterior components are already on the market and a major reason is the technical advantage of a higher impact resistance. This is of great importance in case of a crash event and applies equally for an automobile head rest or a cyclist helmet.

Earlier works [8, 9] on impact resistance of polymeric composites were conducted with different short-cut randomly oriented lignocellulosic fiber reinforcement. In these works, Izod impact tests with notched and fixed specimens resulted in absorbed energy values lower than 60 J/m for all fibers investigated as polypropylene composite reinforcement. Recent works on the impact resistance of thermoset polymer composites reinforced with long and aligned lignocellulosic fibers [10–15], revealed a much higher value for the impact energy. In particular, a polyester composite reinforced with 40% of fique fabric reached 170 J/m [11], which is more than three times the maximum obtained by any short-cut and randomly oriented lignocellulosic fiber composites [9]. This remarkable result served as motivation for a work to confirm it with a different impact method, the Charpy test.

According to the ASTM D 256 norm, there are significant differences between both tests, Charpy and Izod [15] that could lead to distinct results. These differences are shown in Fig. 1.

This figure reveals that the Charpy specimen, with a minimum length of 124 mm, is free-standing on the support during the impact, Fig. 1a, with a hammer, which strikes exactly at the opposite side of the notch. By contrast, the Izod specimen has a maximum length of 63 mm and is fixed to the support during the impact, Fig. 1b, which strikes at a point 22 mm away from the notch. In practice, the Izod test simulates better the actual situation of a component fixed into a system, which is hit at a point away from a stress raiser like a grove or a flange.

Base on these considerations, the objective of the present work was to assess the impact resistance, by means of Izod tests, of polymeric composites reinforced with different amounts of fique fabric.

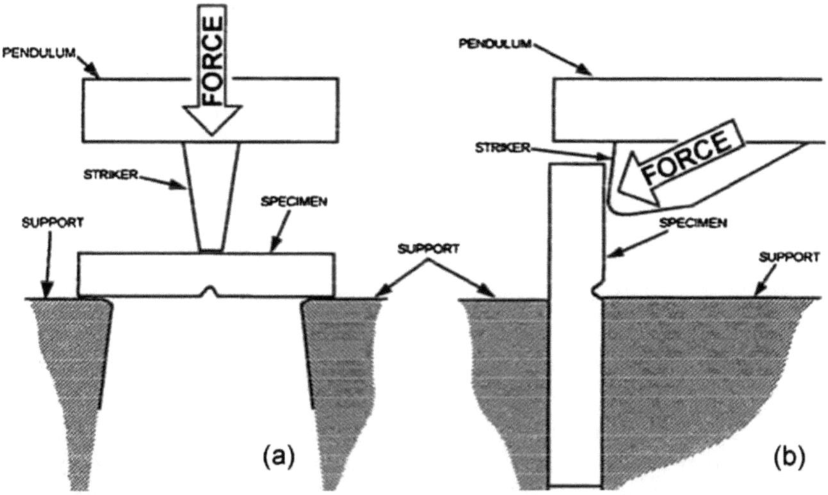

Fig. 1 The Charpy **a** and the Izod **b** impact test methods

Fig. 2 Fique plant **a** and its fibers **b**

Experimental Procedure

The fique fabric was obtained from the Colombia. The typical aspect of a fique plantation and a bundle of soft fibers are shown in Fig. 2.

From the as-received lot of fique, one hundred fibers were separated for a statistic dimensional analysis. The distribution of diameter is presented in Fig. 3. This figure reveals that dimensions of the fique fabric as any other lignocellulosic fiber [1–3], are heterogeneous with a significant dispersion in values. This is considered a condition for long fiber in terms of composite reinforcement and assures an effective strengthening of the matrix [17].

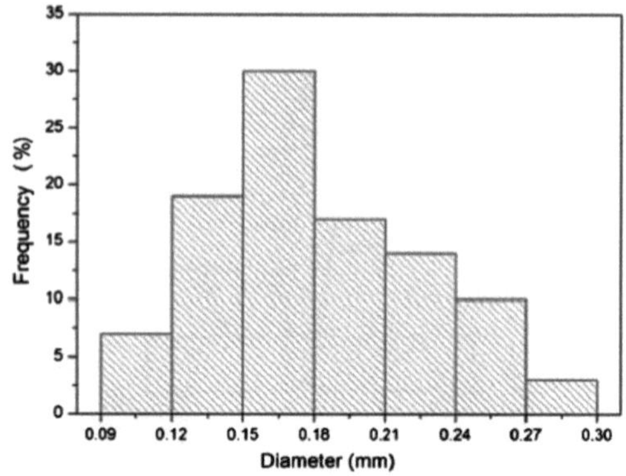

Fig. 3 Histogram of the statistical distribution of diameter of fique fabric

After cleaning and room temperature (RT) drying, aligned fique fabric were mixed in amounts of 0, 10, 20, and 30% in volume with a commercial orthophthalic polyester resin added with 0.5% methyl-ethyl-ketone hardener. Plates of these composites were press molded and allowed to cure at RT for 24 h. Standard notched specimens, 63 × 12.7 × 10 mm for Izod impact testing, according to the ASTM D256 norm, were cut from the plate along the direction of alignment of the fibers. The notch with 2.54 mm in depth, angle of 45° and a tip curvature radius of 0.25 was machined according to DIN 847 norm. For each condition, 10 specimens were tested to assure a statistical validation. The specimens were impact tested with a EMIC hammer pendulum.

The impact fracture surface of the specimens was analyzed by scanning electron microscopy, SEM in a model JSM-6460 LV Jeol microscope. Gold sputtered SEM samples were observed with secondary electrons imaging at 15 kV.

Results and Discussion

The variation of the Izod impact energy with the amount of fique fabric in the polyester composite is shown in Fig. 4.

In this figure it should be noticed that the fique fabric incorporation into the polyester matrix significantly improves the impact toughness of the composite. Within the standard deviation, this improvement can be considered as an exponential function with respect to the amount of fique fabric up to 30%. The relatively high dispersion of values, given by the error bars associated with the higher fiber percentage points in Fig. 4 is a well known non-uniform characteristic of the lignocellulosic fibers [2]. The values shown in this figure are consistent with results reported in the literature. The reinforcement of a polymeric matrix with both synthetic [18] and natural [8, 10] fibers increases the impact toughness of the composite. Table 1 compares values of impact toughness of polymeric composites with different natural fibers.

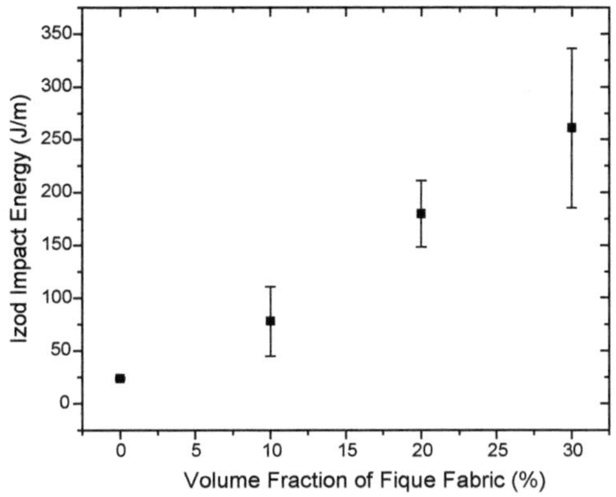

Fig. 4 Izod impact energy as a function of the amount of fique fabric

Table 1 Impact toughness of polymeric composites reinforced with natural fibers

Composite	Amount of fiber (%)	Fiber condition in the composite	Izod impact toughness (J/m)	Reference
Jute/polypropylene	50	Short-cut randomly oriented	39	[8]
Sisal/polypropylene	50	Short-cut randomly oriented	51	[8]
Flax/polypropylene	50	Short-cut randomly oriented	38	[8]
Wood/polypropylene	50	Short-cut randomly oriented	28	[8]
Fique/polypropylene	50	Short-cut randomly oriented	54	[8]
Coir/polypropylene	50	Short-cut randomly oriented	46	[8]
Coir/polyester	40	Long and aligned	121	[19]
Fique/polyester	30	Long and aligned	260	This work

In this work, using long and aligned fique fabric, the impact toughness is significantly higher than the values reported for polypropylene composites reinforced with 50% of short cut and randomly oriented lignocellulosic fibers. The greater impact resistance of the polyester in comparison with the polypropylene matrix could be one reason for the superior performance of the present result. However, there are other important factors related to the impact fracture characteristic of polymeric reinforced with long and aligned natural fibers.

The relatively low interface strength between a hydrophilic natural fiber and a hydrophobic polymeric matrix contributes to an ineffective load transfer from the matrix to a longer fiber. This results in relatively greater fracture surface and higher impact energy needed for the rupture [20]. Another factor is the flexural compliance of a long fiber during the impact test, which will be further discussed.

The incorporation of long and aligned fique fabric results in a marked change with respect to pure polyester (0% fiber) in which a totally transversal rupture occurs. Even with 10% of fiber, the rupture is no longer completely transversal. This indicates that the cracks nucleated at the notch will initially propagate transversally through the polyester matrix, as expected in a monolithic polymer. However, when the crack front reaches a fiber, the rupture will proceed through the interface. As a consequence, after the Izod hammer hit the specimen, some long fibers will be pulled out from the matrix but, owing to their compliance, will not break but simply bend. In fact, for amounts of fiber above 10%, the specimens are not separated at all. For these amounts of long fique fabric, part of the specimen was bent enough to allow the hammer to continue its trajectory without carrying away the top part of the specimen, as expected in a Izod test. The value of the impact toughness in this case cannot be compared with others in which the specimen is totally split apart. Anyway, the fact that a specimen is not completely separated in two parts underestimates the impact toughness. In other words, had all the fibers been broken, the adsorbed impact energy would be higher.

The SEM analysis of the Izod impact fracture permitted to have a better comprehension of the mechanism responsible for the higher toughness of polyester composites reinforced with long fique fabric. Figure 5 shows the aspect of the fracture surface of a pure polyester (0% fiber) specimen. With lower magnification, the upper darker layer in the fractograph, Fig. 5a, corresponds to the specimen notch, revealing the machining horizontal marks. The smoother and light gray layer underneath corresponds to the transversal fracture surface. The fracture in Fig. 5 suggests that a single crack was responsible for the rupture with the roughness in Fig. 5b, being associated with voids and imperfections during the processing.

Figure 6 presents details of the impact fracture surface of a polyester composite specimen with 30% of fique fabric. This fractograph shows an effective adhesion between the fibers and the polyester matrix, where cracks preferentially propagate. Some of the fibers were pulled out from the matrix and others were broken during the impact. By contrast, the part of the specimen in which the rupture preferentially occurred longitudinally through the fiber/matrix interface reveal that most of the fracture area is associated with the fiber surface.

This behavior corroborates the rupture mechanism of cracks that propagate preferentially in between the fique fabric surface and the polyester matrix due to the

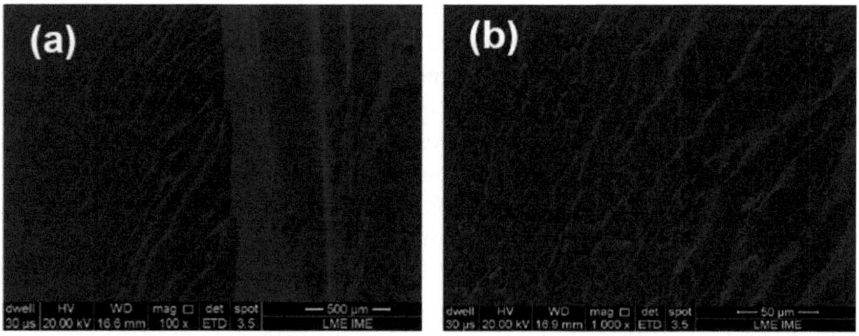

Fig. 5 Izod impact fracture surface of pure polyester specimen (0% fiber): **a** general view; **b** detail of the polyester transversal fracture

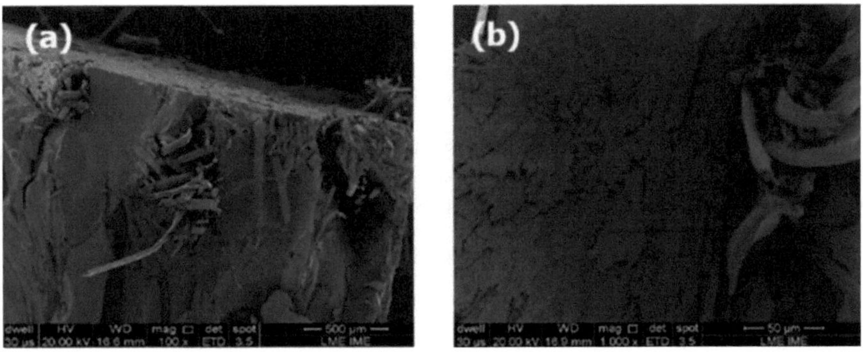

Fig. 6 Impact fracture surface of a polyester composite reinforced with 30% fique fabric: **a** 30× and **b** 500×

low interfacial strength [20]. The greater fracture area, Fig. 6, associate with the long and aligned fibers acting as reinforcement for the composite, justify the higher absorbed impact energy, Fig. 4, with increasing amount of fique fabric.

Conclusions

Composites of aligned fique fabric reinforcing a polyester matrix display a significant increase in the toughness, measures by the Izod impact test, as a function of the amount of the fiber.

Most of this increase in toughness is apparently due to the low fiber/polyester matrix interfacial shear stress. This results in a higher absorbed energy as a consequence of a longitudinal propagation of the cracks throughout the interface, which generates larger rupture areas, as compared to a transversal fracture.

Amounts of fique fabric above 10% are associated with incomplete rupture of the specimen owing to the bend flexibility, i.e., flexural compliance, of the fique fabric.

Acknowledgements The authors thank the support to this investigation by the Brazilian agencies: CNPq, CAPES, FAPERJ and TECNORTE/FENORTE. It is also acknowledged the permission to the use of the impact equipment of IQ/UFRJ and the SEM microscope by the PEMM from COPPE/UFRJ.

References

1. Nabi Sahed, D., & Jog, J. P. (1999). Natural fiber polymer composites: A review. *Advances in Polymer Technology, 18*, 221–274.
2. Bledzki, A. K., & Gassan, J. (1999). Composites reinforced with cellulose-based fibers. *Progress in Polymer Science, 4*, 221–274.

3. Mohanty, A. K., Misra, M., & Hinrichsen, G. (2000). Biofibers, biodegradable polymers and biocomposites: An overview. *Macromolecular Materials and Engineering, 276*(277), 1–24.
4. George, J., Sreekala, M. S., & Thomas, S. (2001). A review on interface modification and characterization of natural fibre plastic composites. *Polymer Engineering and Science, 41*(9), 1471–1485.
5. Zah, R., Hischier, R., Leão, A. L., & Braun, I. (2007). Curaua fibers in the automobile industry—A sustainability assessment. *Journal of Cleaner Production, 15,* 1032–1040.
6. Corbière-Nicollier, T., Gfeller, L. B., Lundquist, L., Leterrier, Y., Manson, J. A. E., & Jolliet, O. (2001). Life cycle assessment of biofibres replacing glass fibres as reinforcement in plastics. *Resource, Conservation and Recycling, 33,* 267–287.
7. Mohanty, A. K., Misra, M., & Drzal, L. T. (2002). Sustainable biocomposites from renewable resources: Opportunities and challenges in the green material world. *Journal of Polymers and the Environment, 10,* 19–26.
8. Leão, A. L., Tan, I. H., & Caraschi, J. C. (1998). Curaua fiber—A tropical natural fibre from Amazon—Potential and applications in composites. In *Proceedings of the International Conference on Advanced Composites* (pp. 557–564). Hurghada, Egypt.
9. Leão, A. L., Caraschi, J. C., & Tan, I. H. (2000). *Natural polymers and agrofiber composites.* In E. Frollini, A. L. Leão & H. C. Matoso (Eds.) (pp. 257–271). São Carlos: UNESP.
10. Monteiro, S. N., Aquino, R. C. M. P., Lopes, F. P. D., Carvalho, E. A., & d'Almeida, J. R. M. (2006). Charpy impact notch toughness of piassava fibers reinforced polyester matrix composites (in Portuguese). *Matéria (Rio de Janeiro), 11*(3), 204–210.
11. Monteiro, S. N., & Lopes, F. P. D. (2007). Impact tests in polymeric composites reinforced with curua fibers (in Portuguese). In *Proceedings of the 62nd International Congress of the Brazilian Association for Metallurgy and Materials* (pp. 871–879). Vitoria, Brazil.
12. Monteiro, S. N., Margem, F. M., & Santos, L. F. L., Jr. (2008). Impact tests in polyester matrix composites reinforced with ramie fibers (in Portuguese). In *Proceedings of the 63rd International Congress of the Brazilian Association for Metallurgy and Materials* (pp. 563–571). Santos, Brazil.
13. Ferreira, A. S., Monteiro, S. N., Lopes, F. P. D., & Aquino, R. C. M. P. (2008). Analysis of the impact strength of epoxy composites reinforced with aligned curaua fibers (in Portuguese). In *Proceedings of the 63rd International Congress of the Brazilian Association for Metallurgy and Materials* (pp. 805–813). Santos, Brazil.
14. Costa, L. L., Monteiro, S. N., Santafe, H. P., Jr., Lopes, F. P. D., & Santos, L. F. L., Jr. (2008). A characterization of the impact resistance of coir fiber reinforced polyester matrix composites (in Portuguese). In *Proceedings of the 63rd International Congress of the Brazilian Association for Metallurgy and Materials* (pp. 572–580). Santos, Brazil.
15. ASTM D 256–84. (1984). *Standard test method for impact resistance of plastic and electrical insulating materials* (pp. 81–102). USA: ASTM.
16. Monteiro, S. N., & d'Almeida, J. R. M. (2006). Pullout tests in lignocellulosic fibers—A methodology of analysis (in Portuguese). *Matéria (Rio de Janeiro), 11*(3), 189–196.
17. Callister, W. D., Jr. (2000). *Materials science and engineering—An introduction* (5th ed.). New York, NY: Wiley.
18. Fu, S. Y., Lauke, B., Mäder, E., Hu, X., & Yue, C. Y. (1999). Fracture resistance of short-glass-fiber-reinforced and short-carbon-fiber-reinforced poly-propylene under Charpy impact load and dependence on processing. *Journal of Materials Processing Technology, 89* (90), 501–507.
19. Monteiro, S. N., Costa, L. L., Lopes, F. P. D., & Terrones, L. A. H. (2008). Characterization of the impact resistance of coir fiber reinforced polyester composites. In *Proceedings of the TMS Conference* (pp. 1–6). New Orleans, LA, USA.
20. Yue, C. Y., Looi, H. C., & Quek, M. Y. (1995). Assessment of fibre-matrix adhesion and interfacial properties using the pullout test. *International Journal of Adhesion and Adhesives, 15,* 73–80.

Izod Impact Tests in Polyester Matrix Composites Reinforced with Jute Fabric

Foluke S. de Assis, Artur C. Pereira, Fábio O. Braga and Sergio N. Monteiro

Abstract Jute fibers are among the lignocellulosic fibers with greater potential for use as fabric reinforcing polymer composites. This study evaluated the impact resistance of this type of composite. Specimens were made with up to 30% in volume of jute fabric in an Izod normalized mold. The jute fabric was embedded with polyester resin and cured at room temperature for 24 h. The specimens were tested in Izod impact pendulum and the fracture surfaces were examined by scanning electron microscopy (SEM). The impact resistance of composites increased linearly with the relative amount of jute fabric reinforcing the composite. This performance was associated with the difficulty of rupture imposed by the jute fabric as well as the type of cracks resulting from the interaction jute fiber/polyester matrix that corroborate the energy absorption at the impact test.

Keywords Jute fabric · Composite · Polyester matrix · Izod impact tests

Introduction

A significant change in the transportation sectors mainly the aerospace and automotive, occurred last century with the introduction of stronger and lighter materials. The most relevant example was the polymer composite reinforced with synthetic fiber. They are nowadays replacing conventional materials such as the aluminum alloys in airplane body owing to superior specific strength. In recent years, however, environmental problems in a global scale are imposing a reversion in the use of synthetic fiber specially the glass fiber. In spite of its higher specific strength, this fiber and its polymer composites can neither be recycled nor incinerated in a thermal plant to generate electricity [1]. Moreover, the fabrication, processing and

F.S. de Assis · A.C. Pereira · F.O. Braga (✉) · S.N. Monteiro
Military Institute of Engineering, IME, Praça Gen. Tibúrcio, nº80 Urca, Rio de Janeiro, RJ 22290-270, Brazil
e-mail: fabio_obraga@yahoo.com.br

© The Minerals, Metals & Materials Society 2017
M.A. Meyers et al. (eds.), *Proceedings of the 3rd Pan American Materials Congress*, The Minerals, Metals & Materials Series, DOI 10.1007/978-3-319-52132-9_38

transportation of a glass fiber composite are associated with CO_2 emission, the major responsible for global warming [2].

A possible substitute for glass fiber without the above-mentioned environmental problems, could be a natural fiber. In fact, natural fibers obtained from cellulose-rich plants, also called lignocellulosic fiber, are already being used in several engineering fields, particularly the automobile industry [3–6].

In recent years a growing number of publications [7–12] have been dedicated to lignocellulosic fibers and their polymer composites. The results justify the current application as new engineering materials with advantages not only in terms of environmental issues but also economical, societal and even technical [10–12]. Moreover, recent studies [13, 14] show that the polymeric composites reinforced with lignocellulosic fibers can be applied in ballistic armor as intermediate layer in multilayered armor.

The major challenge faced by the lignocellulosic fiber and their composites is to shorten the distance to the performance of the corresponding synthetic ones. Here, drawbacks such as non-uniform properties and limited dimension as well as weak adhesion to polymeric matrix, are still impeding technical factors to the expansion on the use of lignocellulosic fibers [10].

The jute fiber (*Corchorus capsularis*) has been investigated in many studies, principally polymeric composites reinforced with jute fiber. These composites can replace the products of wood, and also they can be utilized as materials of construction or auto parts [10]. Then the jute fiber present very interesting potential improvement as seen before in literature. This work evaluated the Izod impact resistance of polyester composites reinforced with jute fabric.

Experimental Procedure

The jute fabric (Fig. 1) investigated in this work was commercially supplied by Companhia Têxtil Castanhal do Pará.

Fig. 1 Jute fabric

After cleaning at room temperature, the jute fabric was mixed in amounts of 0, 10, 20 and 30% in volume with unsaturated polyester resin to prepare the composites. Plates of the composites with 13 mm thickness were fabricated in a rectangular steel mold with dimensions of 152 × 125 mm.

The still liquid polyester resin, together with 0.5% catalyst based on methyl ethyl ketone, was poured into the jute fabric inside the mold. The composite thus formed was allowed to cure for 24 h at room temperature. The plate of each different composite was then cut in bars measuring 10 × 125 × 12.7 mm. These bars were used for preparation of samples for Izod impact test, according to ASTM D256 [16].

The samples were impact tested in a PANTEC pendulum with Izod configuration. The impact energy was obtained using an 11 J power hammer for composites with 0, 10, 20 and 30% of jute fabric. For each condition, relative to a certain volume fraction of fibers, 10 specimens were used and the results were statistically interpreted.

The fracture surface of specimens after impact test was analyzed by scanning electron microscopy, SEM, in a SSX-550 model Shimadzu equipment, operating with secondary electrons at an accelerating voltage of 15 kV.

Results and Discussion

Table 1 shows the results of Izod impact energy values with their respective standard deviations for pure polyester and for composites with different volume fractions of jute fabric.

Based on the results of Table 1, the change in Izod impact energy with the fraction of jute fabric is shown in Fig. 2.

In this figure it should be noted that the incorporation of fabric in the polyester matrix significantly increases the Izod notch toughness of the composite. The values shown in Fig. 2 are consistent with results reported in the literature for synthetic fibers [15] as well as for lignocellulosic fibers [16–18], in which the incorporation of fibers substantially increases the tenacity of polymer matrix composites.

Figure 3 shows that, macroscopically, the incorporation of jute fabric causes a strikingly different behavior as compared to the pure polyester (0% fiber). For fibers fractions of 20 and 30%, the specimens were not even completely broken upon impact. This is due to the relatively high tensile strength characteristic of the jute

Table 1 Energy impact Izod for polyester matrix reinforced with jute fabric

Jute fabric (vol.%)	Energy (J/m)
0	21.8 ± 2.5
10	30.4 ± 7.8
20	50.5 ± 8.7
30	83.1 ± 10.5

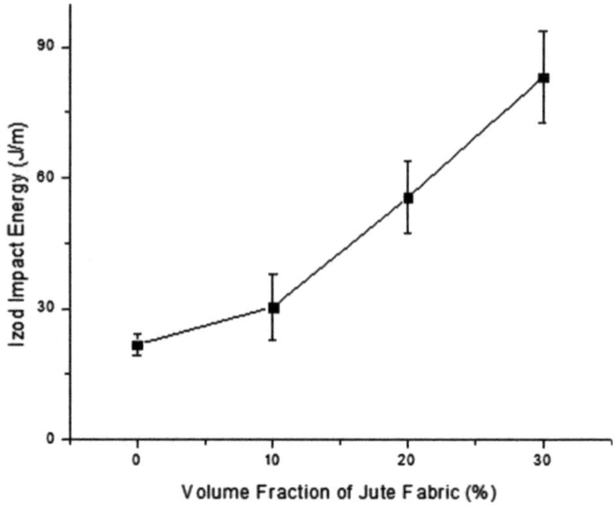

Fig. 2 Izod impact energy as a function of the amount of jute fabric

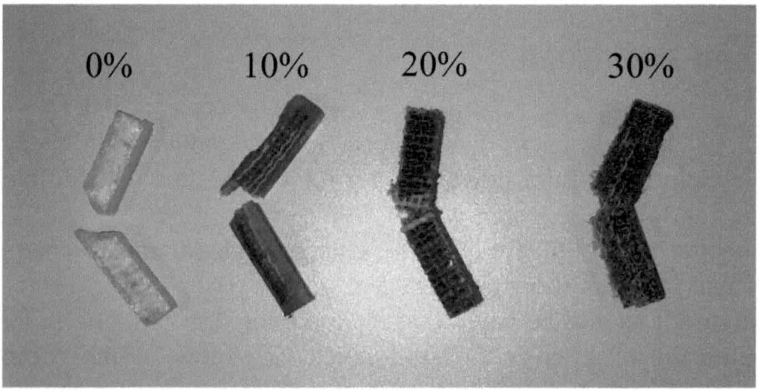

Fig. 3 Typical macrostructural aspects of Izod specimens of polyester composites reinforced with different volume fractions of jute fabric

fiber [12], which avoids a total collapse. For these larger fiber fractions, the impacted specimen is bent enough to allow the hammer to continue its trajectory, while the top part of the folded specimen remained attached to the other part without separating. The non-occurrence of rupture upon impact indicates a high toughness of the composite.

The general SEM (Fig. 4) aspect of the surface ruptured by Izod impact of a specimen with 30% by volume of reinforcing jute fibers shows sticking out of the polyester matrix. Some fibers were pulled out from the matrix by crack propagation through the surface due to the relatively low interfacial tension. A longitudinal crack propagation results in two aspects favorable to increase the energy absorbed

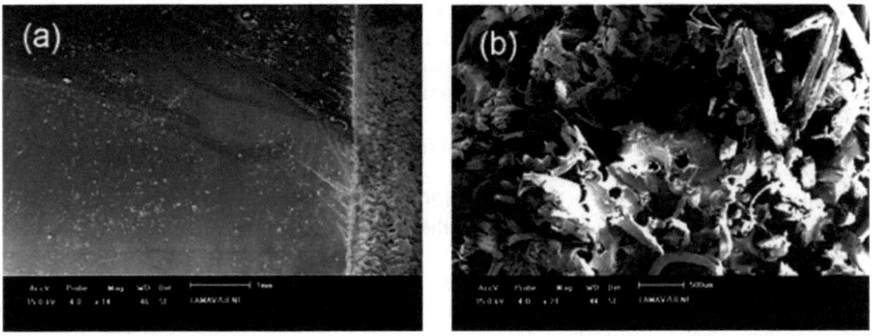

Fig. 4 Impact fractured surface of **a** the specimen of pure polyester resin and **b** the composite with 30% in volume of jute fabric

during impact. First, the longitudinally separated area is larger than the section through the matrix. This is associated with a higher energy and, consequently, higher impact resistance [17, 18]. Second, jute fabric, being kept intact, hinders the separation of the specimen in two parts.

Conclusions

Composites made of jute fabric used as reinforcement of polyester matrix showed an almost linear increase in notch toughness, measured by Izod impact tests, as compared to the pure polyester resin.

The incorporation of 30% jute fabric in the polyester matrix significantly improves the toughness of the composite, 83 J/m, as compared to pure polyester resin, with 22 J/m.

Most of this increase in toughness is apparently due to the low jute fabric/polyester matrix interfacial shear stress. This results in a higher absorbed energy as a consequence of the longitudinal propagation of the cracks throughout the interface, which generates larger rupture areas.

References

1. Wambua, P., Ivens, I., & Verpoest, I. (2003). Natural fibers: Can they replace glass and fibre reinforced plastics? *Composites Science and Technology, 63,* 1259–1264.
2. Gore, A. (2006). *An inconvenient truth. The planetary emergency of global warming and what we can do about it*. Emmaus, Pennsylvania, USA: Rodale Press.
3. Hill, S. (1997). Cars that grow on trees. *New Scientists, 153*(2067), 36–39.
4. Larbig, H., Scherzer, H., Dahlke, B., & Poltrock, R. (1998). Natural fiber reinforced foams based on renewable resources for automotive interior applications. *Journal of Cellular Plastics, 34,* 361–379.

5. Marsh, G. (2003). Next step for automotive materials. *Materials Today, 6*(4), 36–43.
6. Zah, R., Hischier, R., Leão, A. L., & Brown, I. (2007). Curaua fibers in automobile industry—A sustainability assessment. *Journal of Cleaner Production, 15,* 1032–1040.
7. Bledzki, A. K., & Gassan, J. (1999). Composites reinforced with cellulose-based fibers. *Progress in Polymer Science, 4,* 221–274.
8. Nabi Sahed, D., & Jog, J. P. (1999). Natural fiber polymer composites: A review. *Advances in Polymer Technol*ogy, *18*, 221–274.
9. Mohanty, A. K., Misra, M., & Drzal, L. T. (2002). Sustainable biocomposites from renewable resources: Opportunities and challenges in the green material world. *Journal of Polymers and the Environment, 10,* 19–26.
10. Satyanarayana, K. G., Guimarães, J. L., & Wypych, F. (2007). Studies on lignocellulosic fibers of Brazil. Part I: Source, production, morphology, properties and applications. *Composites: Part A, 38,* 1694–1709.
11. Crocker, J. (2008). Natural materials innovative natural composites. *Materials Technology, 2–3,* 174–178.
12. Monteiro, S. N., Lopes, F. P. D., Ferreira, A. S., & Nascimento, D. C. O. (2009). Natural fiber polymer matrix composites: Cheaper, tougher and environmentally friendly. *JOM Journal of the Minerals Metals and Materials Society, 61,* 17–22.
13. Monteiro, S. N., Lima, E. P., Louro, L. H. L., Silva, L. C., & Drelich, J. W. (2014, November). Unlocking function of aramida fibers in multilayered ballistic armor. *Accept for publication in Metallurgical and Materials Transactions A.*
14. Monteiro, S. N., Candido, V. S., Braga, F. O., Bolzan, L. T., Weber, R. P., & Drelich, J. W. (2016, March). Sugarcane bagasse waste in composites for multilayered armor. *European Polymer Journal.*
15. Leão, A. L., Tan, I. H., & Caraschi, J. C. (1998). Curaua fiber—A tropical natural fiber from Amazon—Potential and applications in composites. In *International Conference on Advanced Composites, Hurghada, Egito, Maio*, pp. 557–564.
16. Monteiro, S. N., Margem, F. M., & Santos, L. F. L., Jr. (2008, November). Ensaios de impacto Izod em compósitos poliméricos reforçados com fibras de rami. In *18° Congresso Brasileiro de Engenharia e Ciência dos Materiais, CBECIMAT 2008, Porto de Galinhas, PE* (pp. 1–12).
17. Monteiro, S. N., Ferreira, A. S., & Lopes, F. P. D. (2009, March). Izod impact energy of polyester matrix composites reinforced with aligned curaua fibers. In *Mineral, Metals & Materials Characterization Symposium—TMS Conference, San Francisco, EUA* (pp. 1–8).
18. Monteiro, S. N., Costa, L. L., Lopes, F. P. D., & Terrones, L. A. H. (2008, March). Characterization of the impact resistance of coir fiber reinforced polyester composites. In *Mineral, Metals & Materials Characterization Symposium—TMS Conference, New Orleans, LA, USA* (pp. 1–6).

Microstructure and Hardness of Subzero Quenched and Heat Treated Ti-6Al-4V Alloy

Abdelrahman Abbas, Andrew Seif, Iman El-Mahallawi and Waleed Khalifa

Abstract Titanium is one of the most important materials nowadays with promising lightweight demanding applications. However, despite its high strength-to-weight ratio, high temperature stability and high corrosion resistance, it has relatively low hardness. It is shown in this work that enhanced hardness values could be obtained for Ti-6Al-4V Alloy after heat treatment consisting of subzero quenching in a medium made up of dry ice and alcohol, followed by an aging treatment. The proposed heat treatment resulted an increase of 25% in the hardness of the alloy, compared to 5% reported in literature.

Keywords Ti-6Al-4V · Heat treatment · Hardness

Introduction

Titanium alloys have gained significance in medical, aerospace, and automotive industrial applications, recently. This increasing usage is due to its properties, specifically its high strength to weight ratio. Titanium and its alloys also enjoy other important properties like high corrosion resistance, and high strength under elevated temperature, depending on the presence of other alloying elements [1]. The allotropic behavior of titanium allows its existence in two phases, which makes it possible tailor heat treatment processes to enhance its final properties by changing

A. Abbas · A. Seif
Faculty of Engineering, Mechanical Engineering Department,
The British University in Egypt, El Sherouk, Egypt

I. El-Mahallawi (✉) · W. Khalifa
Faculty of Engineering, Metallurgical Engineering Department,
Cairo University, Giza, Egypt
e-mail: ielmahallawi@bue.edu.eg

© The Minerals, Metals & Materials Society 2017
M.A. Meyers et al. (eds.), *Proceedings of the 3rd Pan American Materials Congress*,
The Minerals, Metals & Materials Series, DOI 10.1007/978-3-319-52132-9_39

its microstructure. Wear behavior is also affected by heat treatment and microstructure [2]. Titanium and its alloys may exist in two phases; namely α phase which has an HCP structure and is usually present below the transus temperature and β phase which has a BCC structure and is present above the transus temperature. With the addition of a group of elements (Al, O, N, Ga and C) known as α stabilizers, the α phase is stabilized. Similarly, with the addition of another group of elements (V, Mo, Nb, Cr, Ni) known as stabilizers, the β phase is retained.

Ti-6Al-4V alloy is considered one of the most important and common titanium alloys, and belongs to the ($\alpha + \beta$) type alloys. This category is designed for high strength-to-weight ratio and is distinguished by its ability to be heat treated, resulting higher strength than the alpha phase, but with a less formability. This type has a huge range of different composition, where Sn, Mo, Zr and Nb may be added to give additional properties. The microstructure of the alloy at room temperature consists of alpha and beta. With the presence of a sufficient amount of beta stabilizing elements, the beta phase is retained at room temperature by fast cooling from the $\alpha + \beta$ range above the transus temperature. The Beta phase (β) transforms by the fast cooling in water or oil to martensite (α') which has an HCP crystal structure or (α'') which has an orthorhombic crystal structure. The presence of more amounts of β phase stabilizers favors the formation of more α' than the α''. However, by slow cooling the β phase transforms to α Widmanstätten lath type structure.

Generally, titanium alloys are heat treated for a variety of reasons. Solution treating and aging treatments are well known among titanium heat treatment regimes and are imposed to achieve selected mechanical properties. Previous authors [2–5] have studied the effect of different heat treatment cycles consisting of cooling by air and water from above the transus temperature on the microstructure and properties of the Ti-Al-4V alloy, and the effect of subsequent aging was investigated. The recommended industrial heat treatment for Ti-6Al-4V alloy is solution treatment at 955–970 °C, followed by water quenching and aging at 480–595 °C for about 4–8 h [6, 7]. However, one challenge still remains which is represented by the gap between the hardness and strength of the Ti-6Al-4V. Usually, the increase in its ultimate tensile strength after solution treatment and aging is accompanied by drop in the ductility of the alloy.

The aim of this work is to study the influence of heat treatment on the microstructure, hardness and the tensile strength of forged Ti-6Al-4V Alloy, with special emphasis on hardness. Therefore, the heat treatment cycles will be tailored in this work towards improving the hardness of the alloy Ti-6Al-4V. In this research, the effect of cooling media and quenching temperature on the final alloy properties will be studied. A faster cooling media composed of dry ice and alcohol is used for the first step in heat treatment (quenching) from different temperatures, followed by aging at 550 °C for 12 h.

Table 1 Chemical composition of the as-received bars

Element	Al	V	Nb	Fe	C	O	N	H
wt%	6.2	4	–	0.1	0.02	0.03	0.01	0.002

Experimental Program

The material used for this work in as-received condition was Ti-6Al-4V bars with 10 mm diameter. The composition of this alloy is shown in Table 1. The β transus temperature of this alloy was determined to be 953 °C, according to the provided data sheet.

Three temperatures were chosen for solution treatment: 1000, 850 and 650 °C. The samples were held in the furnace for 30 min at each treatment temperature. The 1000 °C represents the above β transus temperature. The 850 °C temperature is taken lower than the β transus from the (α + β) region, and the 650 °C is taken below the Ms temperature, which is 850 °C [1, 6]. The samples were cooled from all temperatures in different cooling media: air, water and dry ice/alcohol with a ratio 2:1 (−45 °C). Four samples cooled from different temperatures in dry ice/alcohol medium were aged at 550 °C for 12 h (Table 2).

The microstructure of all cases was studied and the micro-hardness (HV500g) of all samples was determined before aging and after aging for selected samples. In view of the hardness test results, the tensile strength of the two conditions of solution treatment and quenching in dry ice/alcohol before and after aging was determined. The tensile test was carried out using a universal testing machine at a rate of 20 mm/min, and the specimens were prepared according to ASTM E8.

In order to investigate the microstructure details and the distribution of V and other elements in the different phases, SEM study and EDX analysis were made. The fracture surfaces of the samples used for the tensile test were also studied.

Results and Discussion

Microstructure Results

A schematic illustration of the different developed microstructures (using etchant consisting of 5% HF-10% HNO_3 and 85% distilled water) after solution treatment of a cast Ti-6Al-4V model titanium is shown in Fig. 1 [8]. The obtained microstructures in this work were compared and identified according to this illustration. Figure 2 shows the microstructure obtained after cooling from a temperature above the transus temperature (1000 °C) and using different cooling media. It is clear from Fig. 2a, d that the rapid cooling in dry ice medium caused the formation of martensite structure α′ and retained β. This phase transformation process does not go through the (α + β) region but goes directly to below the Ms which is

Table 2 Conditions of samples selected for different tests

Sample No.	11	1	2	3	4	5	6	7	8	9	10
Solution treated for 30 min and cooled ST	As forged	1000 °C/ dry ice	1000 °C/ water	1000 °C/ air	980 °C/ dry ice	850 °C/ dry ice	850 °C/ water	850 °C/ air	650 °C/ dry ice	650 °C/ water	650 °C/ air
Aging: ST dry ice/alcohol + aged at 550 °C 12 h		x			x	x			x		
Micro-hardness	x	x	x	x	x	x	x	x	x	x	x
Tensile test[a]	x	x									

[a]In addition to the two shown conditions, the tensile test was made on a third sample (solution treated in dry ice/alcohol and aged at 650 °C for 24 h)

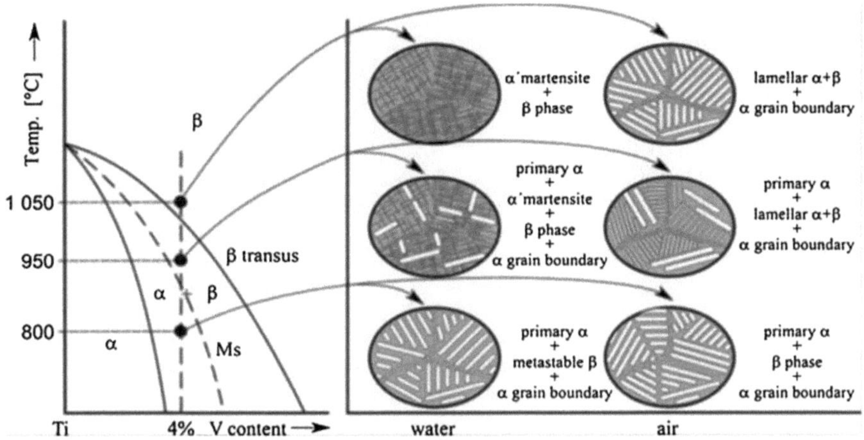

Fig. 1 Schematic illustration of the different developed microstructures (using etchant consisting of 5% HF-10% HNO3 and 85% distilled water) after solution treatment of a cast Ti-6Al-4V model titanium [8]

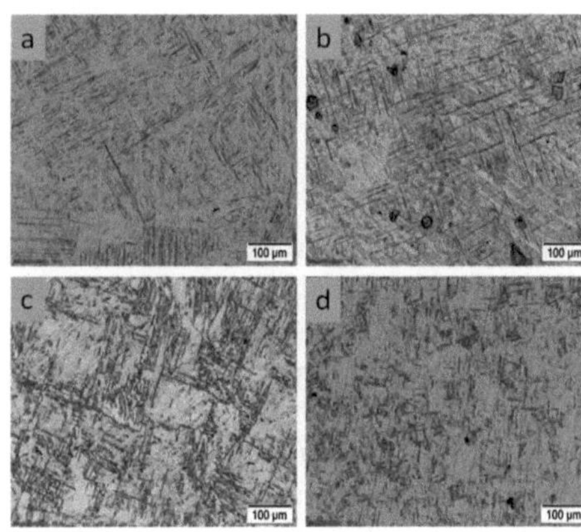

Fig. 2 Microstructure obtained after cooling: **a**, **d** dry ice martensite structure α′ and retained β, **b** water quenched coarser martensite lathes, **c** air cooled Widmanstätten structure

considered the point of martensite formation. According to literature [1, 8], the α′ martensite + retained β phase coexist in the microstructure after rapid cooling. This meta-stable structure gives potential for property and structural changes by aging treatment. Cooling in water produces similar microstructure, with coarser martensite lathes (Fig. 2b). Cooling in air produces Widmanstätten structure (primary α + lamellar (α + β) + α grains boundaries), see Fig. 2c. Heating in furnace up to the solution treatment and holding for 30 min causes coarse grain structure (Fig. 2a–c), while putting specimen in furnace and holding for 30 min produced much finer microstructure (Fig. 2d).

Figure 3 shows the microstructure of the specimens which were cooled from below the β transus temperature; solution treated from above the Ms temperature (850 °C), and below the Ms temperature (650 °C). The micrographs of Fig. 3 show

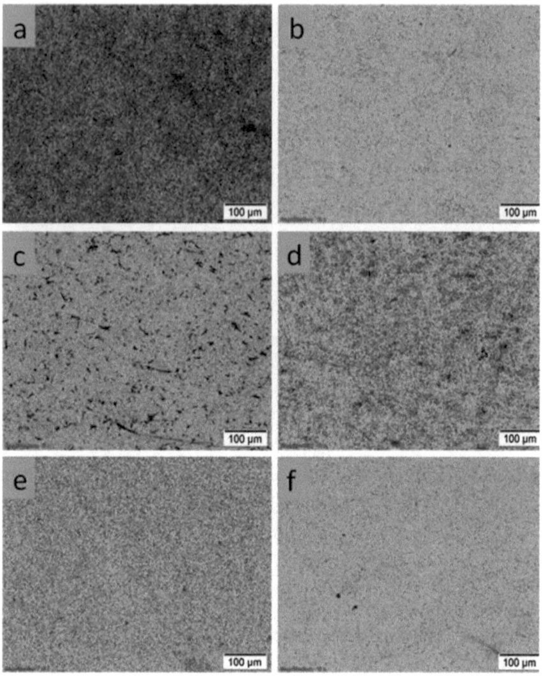

Fig. 3 Microstructure of specimens cooled from below the transus in different media: cooling in dry ice medium: **a** solution at 850 °C and **b** solution at 650 °C; cooling in water: **c** solution at 850 °C and **d** solution at 650 °C; and cooling in air: **e** solution at 850 °C and **f** solution at 650 °C

almost identical microstructures consisting of primary α and β phases, meaning that the different cooling media has no effect upon cooling from below the transus temperature. The microstructure consists mainly of primary α and β phase. Other phases may exist depending on the solution treatment temperature and cooling rate [1, 9].

Figure 4 shows the microstructure of the specimens which were selected for aging, before and after aging. As seen from Fig. 4, apart from the darkening of the aged structure [9], there is no significant difference between the microstructure of the samples, which were solution treated below the Ms Temperature, after and before aging treatment (Fig. 4e–h). The samples which were cooled in dry ice medium showed retained beta phase before aging and this phase was transformed to a fine alpha phase very fine needle-like microstructure α′ martensite (Fig. 4a, b), which induced higher hardness to these samples, as will be shown. The samples which were cooled from 980 °C showed an equiaxed (primary) α structure in a lamellar α and β matrix (Fig. 4c, d).

The mechanical properties of the Ti-6Al-4V alloy is largely determined by the size and arrangement of the two phases α and β. The two extreme cases of phase arrangements are the lamellar microstructure, which is generated upon cooling from the β phase field, and the equiaxed microstructure, which is a result of the recrystallization process. Once the temperature falls below the transus temperature α nucleates at grain boundaries and then grows as lamellae into the (prior) β grain

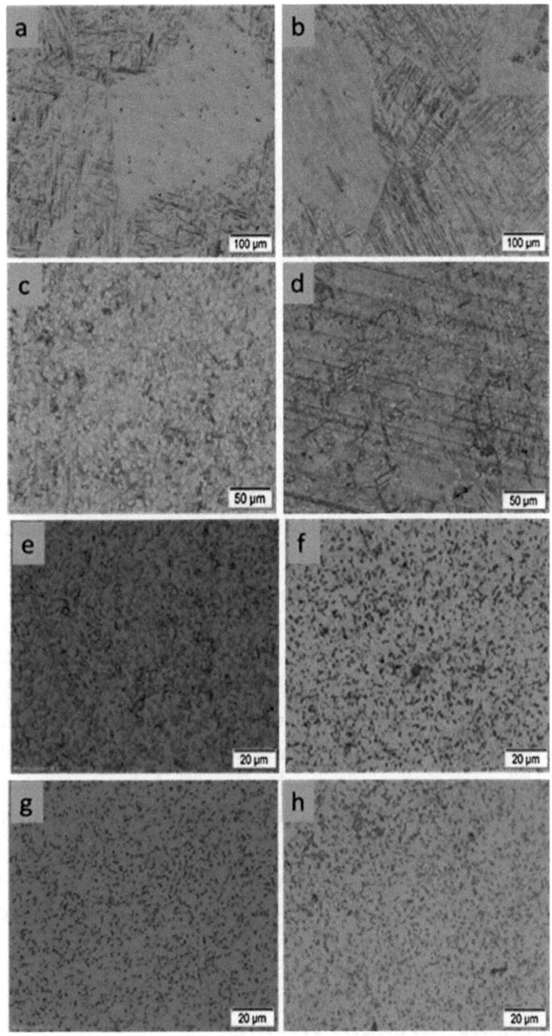

Fig. 4 Microstructure of specimens before and after aging, sample 1: **a** before aging, **b** after aging, sample 4: **c** before aging, **d** after aging; sample 5: **e** before aging, and **f** after aging; and sample 8 **g** before aging, and **h** after aging

[1]. While rapid cooling of Ti alloys can produce some retained β in case the β has sufficient stabilizer, retained β depends on the alloy composition. Rapid quenching also leads to a martensitic transformation of β, resulting in a very fine needle-like microstructure which is different from the martensite known in steels, as it only brings about a moderate hardening effect [1]. Martensite and ω (thermal) phases can also form by a not fully understood mechanism. On the other hand, slow cooling from the β range will provide Widmanstätten α phase by nucleation and growth. The morphology of this Widmanstätten α changes from aligned α lathes to basket wave's arrangements [6, 9, 10].

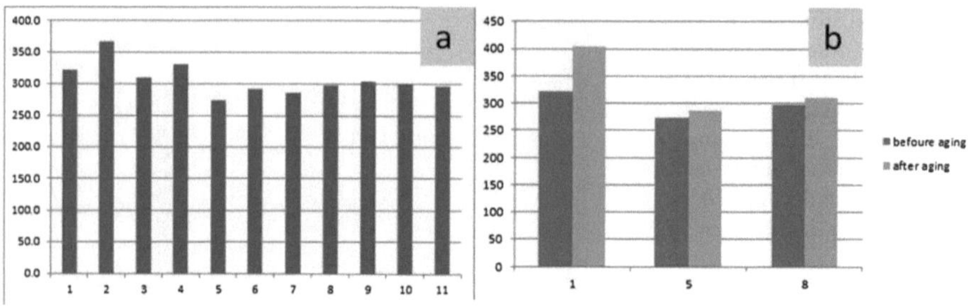

Fig. 5 **a** VHN micro-hardness results of the samples before aging; **b** HV hardness results before and after aging

Microhardness Results

Figure 5a shows the micro-hardness results for all studied samples, from which it is shown that the highest hardness values were obtained for the sample which was solution treated at 1000 °C followed by water cooling. The second highest hardness was obtained for the sample which was cooled in the dry ice/alcohol medium (samples 2 and 1, respectively). Figure 5b shows the hardness changes after aging for samples 1, 5 and 8. It is shown from Fig. 5b that the average hardness of sample 2 increased significantly after aging from 320 to 404 HV, resulting an increase of 26 and 36% compared to the solution treated and as received conditions, respectively. The increased hardness for the rapid cooled samples is attributed to the presence of the retained β phase and the martensite, while the increase in hardness after aging is attributed to the decomposition of the retained β phase into fine α phase and precipitates, and the decomposition of martensite structure α′ to α + β (α′ → α + β) [5]. However, at this stage of work those precipitates were not fully identified and further characterization by SEM and TEM is needed.

Tensile Strength Results

The tensile test results for three selected conditions are shown in Table 3. The results show that after solution treatment at 1000 °C and cooling in dry ice/alcohol the tensile strength dropped compared to the as-forged condition from 934 to 917 MPa (2%). After aging at 650 °C for 24 h, the tensile strength was significantly increased to 960 MPa (5%). The 0.2% yield strength showed an increase from 870 MPa in the as received condition to 925 MPa in the solution treated and aged condition (about 6%). The increase in yield strength in the solution treated and aged condition is attributed to the decomposition of the retained β phase into fine α phase and other unidentified precipitates.

Table 3 Tensile strength of quenched and tempered samples

Condition	Sample ID	0.2% yield stress (MPa)	Tensile strength (MPa)	Elongation (%)
As forged as received	11	870	934	10
Solution treated for 30 min at 1000 °C and cooled in dry ice/alcohol	1	824	917	5.01
Solution treated for 30 min at 1000 °C and cooled in dry ice/alcohol and aged at 650 °C for 24 h	1 + aged	925	960	5.3

SEM Study

Figure 6a shows the SEM image of the fracture surface of the solution treated at 1000 °C and quenched in dry ice sample, the appearance of facets indicates the brittle nature of the fracture. The SEM image also shows the presence of dark agglomerates, which are identified by EDX analysis to be α′ martensite enriched with Ti oxides and carbides. Figure 6b shows a higher magnification image, which shows the presence of light spheres in the size range from 10 to 50 μm. EDX analysis of these spheres indicate that they are retained β phase with enriched V content. EDX analysis in the areas outside the spheres show less V content. These observations support the previous explanation given for the reason for observed increase in hardness after solution treatment at 1000 °C and quenching to be attributed to the presence of the retained β phase and the martensite.

Figure 7a shows the SEM image obtained for the sample No. 1 which was solution treated, quenched in dry ice and aged for 24 h. The microstructure reveals very fine needle-like microstructure α′ martensite embedded in retained β matrix. The fracture surface of the solution treated and aged sample is shown in Fig. 7b, c, which reveals the appearance of ductile fracture features and even some ridges and dimples. It is also shown that some fine new grains appear in the matrix, and that the dark agglomerates have decreased in intensity. These observations support the previous explanation given for the reason for the increase in hardness after aging to be attributed to the decomposition of the retained β phase into fine α phase and precipitates, and the decomposition of martensite structure α′ to α + β ($\alpha' \rightarrow \alpha + \beta$).

Discussion

This work was done to study the effect of sub-zero quenching on the microstructure and hardness of the Ti-6Al-4V alloy. As far as we are aware, quenching in dry ice/alcohol media was not investigated before. Using this cooling media was adopted to investigate options for increasing the hardness of the Ti-6Al-4V alloy

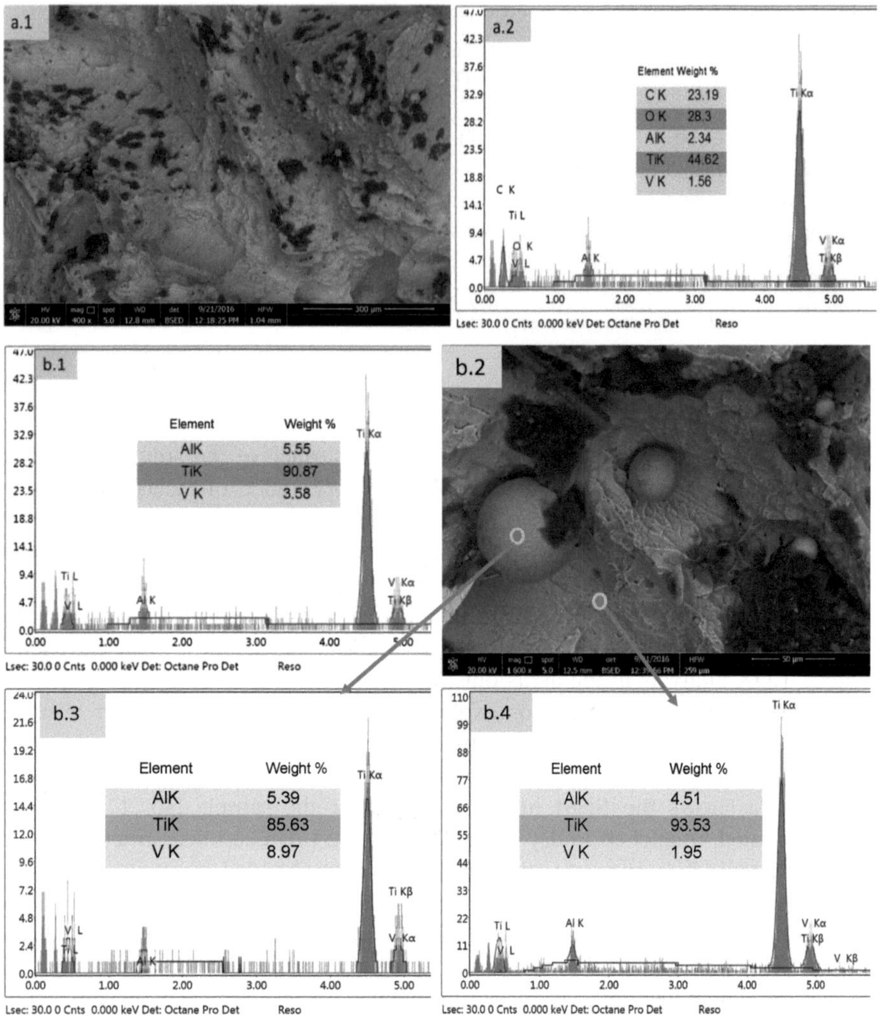

Fig. 6 SEM fracture surface of the sample solution treated at 1000 °C and quenched in dry ice revealing (**a.1**, **a.2**) presence of martensite, (**b.1**, **b.2**, **b.3**, **b.4**) EDX of overall surface and showing the retained β

which possesses good strength and low surface hardness. In this work it was possible to induce an increase in hardness of 26% after aging from the solution treated condition, for the first time in literature. The main microstructural changes brought about after fast cooling in dry ice/alcohol media was the increased area of the retained β phase, which appeared as an unetched area. This super saturated structure is believed to transform after aging too finer α platelets as shown in Figs. 4b and 7b, c, however advanced microscopic investigation and phase identification is still needed.

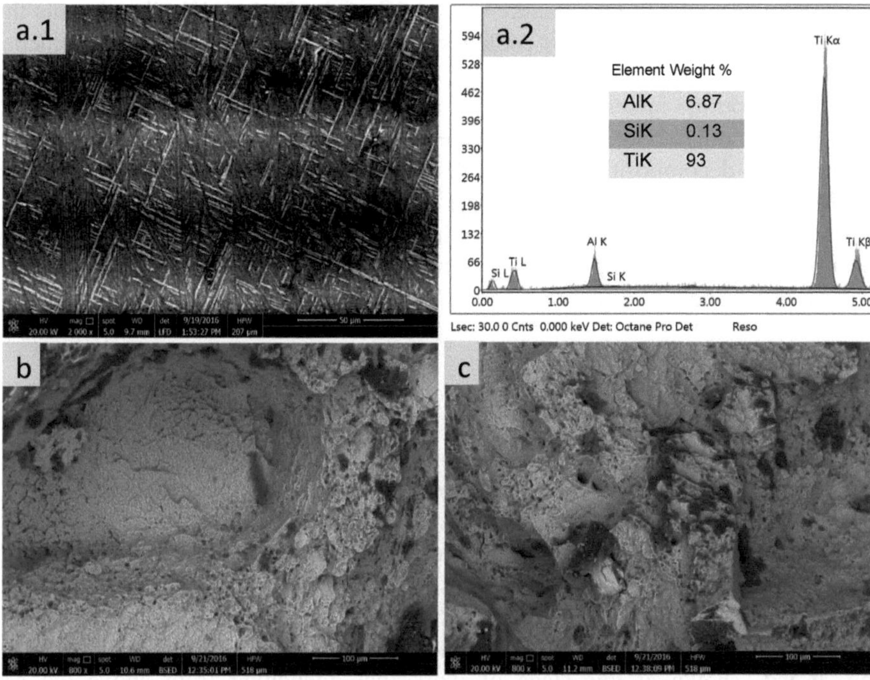

Fig. 7 SEM of the sample solution treated at 1000 °C and quenched in dry ice and aged revealing (**a.1**, **a.2**) presence of fine needle-like α′ martensite embedded in retained β matrix (**b**) fracture surface showing new born α grains on the expense of retained β (**c**) fracture surface showing dimples and ridges ductile regions

Tatsuro et al. [3] showed that solution treatment at 930 °C for 60 s, followed by water quenching resulted an increase in the hardness of the alloy from 310 to 360 HV on the expense of ductility. Subsequent aging at 480 °C resulted an increase in hardness to 380 HV mounting to about 6% increase from the solution treated condition, as well as enhancement of yield strength by 25% and retain of initial ductility. The increase in hardness after solution treatment was attributed to the formation of α′ martensite and retained β phase, and no change in microstructure was reported after aging. However, this work has revealed that changes in microstructure appear after aging and they can be revealed by SEM.

Also, a previous study [5] has investigated the mechanical properties and the microstructure of cast Ti-6Al-4V alloy after solution treatment at 950 °C for 0.5 h followed by water quenching, and aging at 450–650 °C for 1–24 h. The microstructure of the solution treated samples was found to consist of acicular α′ martensite transformed from β phase with α and retained β phases. After aging, the α′ martensite transformed to α + β phases. The hardness was shown to slightly increase with aging time, but the increase in the hardness was no more than about 3% compared to the solution treated condition. Similarly, another study [8] showed that solution treatment at temperatures of 950 and 1050 °C, followed by water quenching and aging at temperature of 500 °C of the Ti-6Al-4V alloy, results the formation of α′ martensite. Air-cooled samples had a microstructure of lamellar α + β. Similar to this work, no change of the microstructure was revealed after aging. The reported increase in hardness after aging relative to the solution treated condition was only about 6% [8]. Pinke et al. [8] concluded from his work that the creation and decomposition of the α′ martensite is more significant than the transformation of the retained β on the effect on hardness. However, this work shows that this argument needs further investigation as the opposite is shown by this work and the retained β and its decomposition after aging resulted in significant increase in hardness after aging.

Conclusions

This work focuses on enhancing the hardness of Ti-6Al-4V alloy by heat treatment consisting of solution treatment and aging. The results showed that the solution treatment above the beta transus temperature followed by quenching either in water or dry ice medium, results higher hardness values of the alloy. Subsequent aging at 550 °C for 12 and 24 h results further enhancement in hardness, yield strength and tensile strength by 36, 6 and 5%, respectively compared to the as received condition. It was shown at this stage of work that quenching from above the Ms Temperature results microstructures consisting of α′ martensite and retained β, while quenching from below Ms shows microstructures consisting of primary α + less retained β + α on the grain boundaries. Air cooling from above beta transus results equiaxed primary α phase in β matrix, while quenching in dry ice results more retained beta, which decomposes while aging into fine α phase increasing the strength. The current results open further investigations using additional techniques (such as XRD) that can differentiate the solid phases, the % weight of the phases and reveal the microstructure of these phases with crystal parameters of each.

References

1. Leyens, C., & Peters, M. (2003). *Titanium and titanium alloys: Fundamentals and applications* (pp. 1–36). Weinheim: Wiley-VCH.
2. Swalla, D. R., Neu, R. W., & Mcdowell, D. L. (2004). Microstructural characterization of Ti-6Al-4V subjected to fretting. *Journal of Tribology, 126*(4), 809–816.
3. Morita, T., Hatsuoka, K., Iizuka, T., & Kawasaki, K. (2005). Strengthening of Ti-6Al-4V alloy by short-time duplex heat treatment. *Materials Transactions, 46*(7), 1681–1686.
4. Reda, R., Nofal, A., & Hussein, A. (2013). Effect of single and duplex stage heat treatment on the microstructure and mechanical properties of cast Ti-6Al-4V alloy. *Metallography, Microstructure, and Analysis, 2*(6), 388–393. doi:10.1007/s13632-013-0103-7

5. Oh, S.-T., Woo, K.-D., Lee, T., & Lee, H.-C. (2015). Effects of heat treatment on mechanical properties of VAR-Cast Ti-6Al-4V alloy. In *Proceedings of the World Congress on Mechanical, Chemical, and Material Engineering (MCM 2015)*, Barcelona, Spain, July 20–21, 2015.
6. Donachie, M. J., Jr. (2001). Heat treating titanium and its alloys. *Heat Treating Progress, ASME.*
7. Technical Data Sheet. (2012). ATI Ti-6Al-4V, Grade 5 titanium alloy (UNS R56400). Pittsburgh, PA 15222-5479 U.S.A.: Allegheny Technologies Incorporated.
8. Pinke, P., Caplovic, L., & Kovacs, T. (2011). The influence of heat treatment on the microstructure of the casted Ti-6Al-4V titanium alloy. *Slovak University of Technology Bratislava. Web*, 11.
9. Gammon, L. M., Briggs, R. D., Packard, J. M., Batson, K. W., Boyer, R., & Domby, C. W. (2004). Metallography and microstructures of titanium and its alloys. In G. F. Vander Voort (Ed.), *ASM handbook: Metallography and microstructures* (Vol. 9, pp. 899–917).
10. Sridhar, G., Gopalan, R., & Sarma, D. (1987). A microstructural characterization of solution-treated titanium alloy Ti-6Al-4V. *Metallography, 20*(3), 291–310.

Nano-additive Reinforcement of Thermoplastic Microballoon Epoxy Syntactic Foams

Kerrick R. Dando and David R. Salem

Abstract Syntactic foams comprised of glass microballoons have gained considerable attention over the past several years due to mechanical and thermal properties that are advantageous for use as a core material in naval and aerospace applications. Recently, advancements in the production of thermoplastic microballoon syntactic foams have allowed for an increase in microballoon volume fraction (up to 90 volume fraction), with corresponding lower densities but reduced mechanical properties. In this work, carbon nanofibers and halloysite nanotubes were incorporated in thermoplastic microballoon-based syntactic foam to enhances its mechanical properties, and the effects of these two nanoscale reinforcements are compared. X-Ray micro-computed tomography (MCT) was employed to analyze the microstructure of the materials produced, and scanning electron microscopy was used to assess the dispersion of nano-additives within the resin. Through characterization of the tensile and compressive strength properties of these materials, it was observed that dramatic mechanical property enhancements can be engineered through additions of either nano-additive at specific loading levels.

Keywords Syntactic foam · Carbon nanofiber · Nano-reinforcement

Introduction

Syntactic foams consisting of 0.3–0.74 volume fraction of glass microballoons in epoxy resin matrices have gained considerable attention in recent years due to their low density combined with high compressive strength [1], low moisture absorption [2], and high-energy absorption [3]. Consequently, there has been extensive evaluation of the tensile [4, 5], compressive [1, 6, 7] and hygrothermal [2] properties of

---

K.R. Dando (✉) · D.R. Salem
Composite and Polymer Engineering (CAPE) Laboratory, South Dakota School of Mines and Technology, 501 E. St. Joseph St., Rapid City, SD 57701, USA
e-mail: kerrick.dando@mines.sdsmt.edu

© The Minerals, Metals & Materials Society 2017
M.A. Meyers et al. (eds.), *Proceedings of the 3rd Pan American Materials Congress*,
The Minerals, Metals & Materials Series, DOI 10.1007/978-3-319-52132-9_40

these materials. The uniquely tunable properties of syntactic foams have led to their increased commercial use, primarily in aerospace [8] and marine [9] applications.

A significant limitation of hollow glass epoxy syntactic foams is the density ceiling ($\sim$0.35 g/cc) due to packing limitations of the rigid, polydisperse glass spheres [1, 3]. To overcome this limitation, expandable hollow thermoplastic microballoons have been used to enable the production of syntactic foams with microballoon volume fractions of up to 0.95 and densities as low as 0.07 g/cc [10]. However, while density is reduced with the incorporation of large volume fractions of thermoplastic microballoons, so too are mechanical properties. This is due to the large amount of purposely placed voids within the composite.

One alternative to remedy this mechanical property deficiency is to incorporate nano-additives, lending mechanical property enhancements with negligible impact on composite density [11–14]. Previous work has shown that small additions of carbon nanofibers in glass microballoon epoxy syntactic foams led to improvement in tensile strength properties [5]. Halloysite hollow nanotubes have been incorporated for strength/toughness enhancements of resin systems, showing promise as a potential nano-additive in syntactic foam materials [15–17].

The present work compares the structure and mechanical properties of thermoplastic microballoon epoxy syntactic foams loaded with varying weight % of either carbon nanofibers or halloysite nanotubes.

Experimental

Materials

SC-15 resin, manufactured by Applied Poleramic Inc. was selected as the epoxy resin system. SC-15 is a low-viscosity (550 $\pm$ 45 cP), two-phase toughened epoxy cured with a cycloaliphatic amine [9]. Acetone was used as the solvent, reducing the viscosity of the epoxy resin to 57 $\pm$ 12 cP, for improved wetting and increased working time. D15 hollow-thermoplastic spheres ("microballoons") produced by Akzo Nobel, and K11 hollow-glass spheres produced by 3 M were used to produce the highly-loaded syntactic foams. The D15 and K11 microballoons have a true particle density of 0.015 $\pm$ 0.001 and 0.11 g/cm^3 and a diameter range of 60–90 and 30–120 μm, respectively. The carbon nanofibers (PR-19-XT-PS) were produced by Pyrograf Products, having a density of 1.95 g/cm^3 with diameter and length of 100–200 and 30–100 μm, respectively. The halloysite nanotubes (NN-HNT200) were produced by Naturalnano, with a density of 2.3 g/cm^3 and diameter and length of 40–200 and 0.5–1.2 μm, respectively.

Sample Fabrication

Preparation of Resin

Dispersion of the carbon nanofiber (CNF) and halloysite nanotube (HNT) additives in epoxy resin followed identical preparation methods. Master batches containing 1 wt% of either additive were prepared to allow for production of the subsequent 5 syntactic foam sample types. First, part A of the epoxy resin was weighed out, followed by the weighing and addition of either additive. This modified part A, containing the additive, was then speed mixed using a speed mixing unit (Hauschild Speedmixer DAC 1100 FVZ) to distribute the nanomaterial into the resin. The resin was mixed for 10 min at 1500 rpm followed by 15 min at 2500 rpm and finally 15 min at 4500 rpm using a mechanical mixer with a high-shear impeller. This modified resin was then sonicated for 1 h at an amplitude of 4, followed by 30 min at an amplitude of 20.

Syntactic Foam Fabrication

For epoxy syntactic foam sample formation, SC-15 resin and curing agent were speed mixed (130:100, resin: curing agent by weight ratio) to allow for adequate dispersion of the curing agent and ensuring even cure of the material. Varying amounts of the epoxy part A master batch (described above) were added depending on the desired wt% of nano-additive targeted for that material. Acetone was then added to the epoxy and speed mixed. Next, D15 hollow thermoplastic spheres were added to the epoxy solution and speed mixed. This mixture was then transferred into a 114.3 mm × 114.3 mm × 12.7 mm mold and packed down using a tongue depressor. The mold, containing the mixture, but without a lid, was placed in an oven at 30 °C for 30 min to facilitate evaporation of the acetone diluent. After 30 min, the mold-lid was clamped in place for the remainder of the cure period. Excess resin exited through the clearance space between mold parts during the 24 h cure period at 60 °C. Once cured, the syntactic foam plaque was de-molded, and the density of the syntactic foam was evaluated (gravimetrically) and compared to the theoretical density and material mass-input values to verify that the intended loadings were achieved.

Property Characterization

Microstructure Analysis

Syntactic foam specimens were imaged using x-ray MCT (Xradia MicroXCT-400), a non-destructive method allowing 3D morphological characterization. X-ray MCT utilizes x-rays to capture sample image "slices" that can be analyzed in 2-dimensional form, or compiled together to produce a 3-dimensional image or video. Specimen dimensions employed for analyses were: 6.35 mm × 6.35 mm 12.7 mm. Multiple cross-sectional image slices, acquired on each sample using x-ray MCT, were analyzed using ImageJ© software to determine microballoon morphology and volume fraction. The volume fraction calculated from image analysis was compared to the empirically measured sample density as a cross-reference.

Epoxy specimens containing 1 wt% of either halloysite nanotubes or carbon nanofibers were analyzed to determine if the nano-additives were adequately dispersed during the dispersion method. The fracture surfaces were analyzed using scanning electron microscopy (Zeiss Supra 40 Variable-Pressure Field Emission SEM), where all images were acquired using a secondary electron detector.

Compressive Strength

Specimens (25.4 mm × 25.4 mm × 19.05 mm) of epoxy syntactic foams were prepared for compression analysis, and tested using a MTS Q-Test 10 Elite Controller (5 kN Load Cell) in accordance with ASTM D 695 [18]. Samples were compressed to 80% strain at a rate of 0.5 mm/min. to analyze initial compression, yield point, and densification region. Load-displacement data obtained from these tests were then employed for the calculation of compressive yield strength and modulus. Compressive yield strength is denoted as the first point on the stress-strain diagram at which an increase in strain occurs without an increase in stress. For each sample type, fifteen specimens were tested (5 each, cut from 3 molded plaques) to determine the mean compressive strength values.

Tensile Strength

Tensile analyses were performed on 25.4 mm × 25.4 mm × 12.7 mm specimens using a MTS Q-Test 10 Elite Controller (5 kN Load Cell) in accordance with ASTM D 1623 [19]. Sample tests were carried out until yield, denoted as the point where failure occurs, where strain rates were adjusted to produce failure in 2–5 min. Load-displacement data obtained from these machines were then used for the calculation of Young's Modulus and yield strength. For each sample type, fifteen specimens were tested (cut from 3 molded plaques) and the mean yield strength and strength value determined.

Results

Structure and Morphology

Once the master batches were produced, epoxy specimens containing 1 wt% of either halloysite or carbon nanofibers were cast, fractured and analyzed using scanning electron microscopy (Fig. 1). The fracture surface was analyzed to determine if the mixing method performed adequately in dispersing nano-additives and preventing or minimizing agglomerations that are detrimental to composite performance properties.

From Fig. 1, we can see images of the fracture surfaces for both epoxy samples. The images are representative of what was seen throughout the analysis of the fracture surfaces of both types of materials. Prominent protrusion of carbon nanofibers or halloysite nanotubes was noted across the material fracture surface, with little to no agglomerations noted.

Syntactic foam samples containing different weight percentages of either nano-additive were then produced and imaged using x-ray MCT to comparatively analyze the foam microstructures (Fig. 2). This analysis method allows for artifact-free characterization of thermoplastic microballoon morphology, the

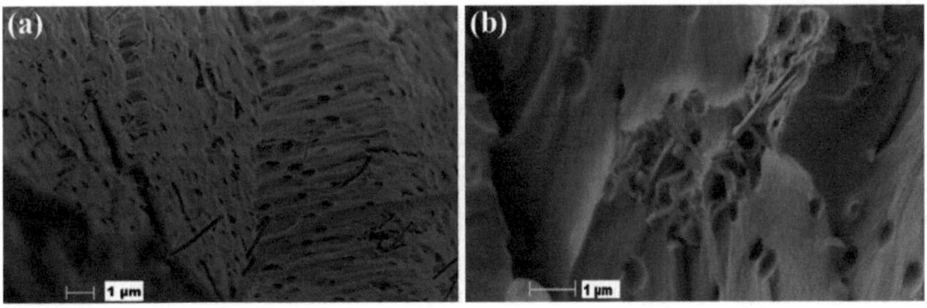

Fig. 1 Scanning electron microscope images of **a** 1 wt% CNF epoxy and **b** 1 wt% HNT epoxy samples

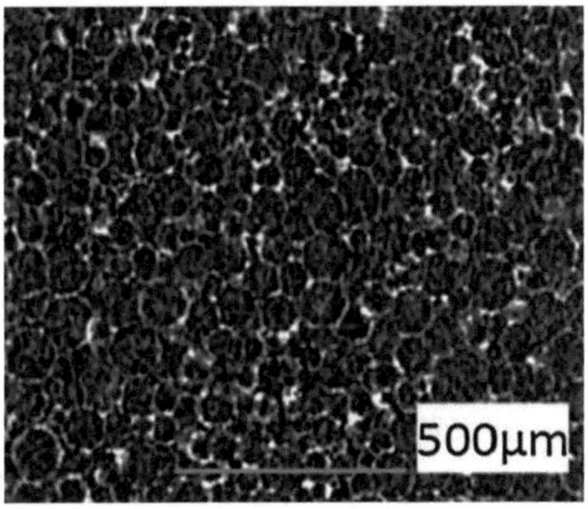

Fig. 2 X-ray micro CT cross sectional image of a 0.9 volume fraction thermoplastic microballoon epoxy syntactic foam containing 1 wt% halloysite nanotubes

microballoon dispersion in single-filler and mixed-filler systems and the volume fraction microballoon loading (by 2D image analysis) for comparison with values obtained by density measurements.

Due to instrument resolution limitations, the nano-additives could not be directly visualized using this analysis method, however the internal microstructure of the syntactic foam structure was well defined. Utilizing this tool, we were able to get an understanding of how these high microballoon loadings could be achieved. During the curing cycle, the heat produced from the exothermic reaction (~ 190 °C) of the epoxy matrix causes the thermoplastic microballoon shells to soften and expand. In so doing, excess resin is expelled and the microballoon shells deform to contours that efficiently fill the mold volume, and increase the microballoon volume fraction in the process. ImageJ© software was used to analyze the epoxy syntactic foams which had nominal 0.90 microballoon volume fractions based on gravimetric density. Syntactic foam microballoon volume fractions were calculated from the analyses of 90 individual image slices (of the type shown in Fig. 2). The measured syntactic foam microballoon volume fractions were virtually identical to values calculated from gravimetric analyses thereby validating that the correct densities and loadings were achieved.

Compressive Properties

Following microstructural analysis of the syntactic foam specimens, samples were prepared and tested for their compressive properties. Figure 3 shows the compressive strength (a) and modulus (b) properties of thermoplastic microballoon syntactic foams loaded with varying wt% of carbon nanofibers, where error bars shown in all graphs represent the 95% confidence interval. A dramatic increase in both compressive strength and modulus (150 and 200%, respectively) are noted with the addition of 0.125 wt% carbon nanofibers. There is an additional increase (15 and 20%, respectively) in strength and modulus properties as CNF loading is

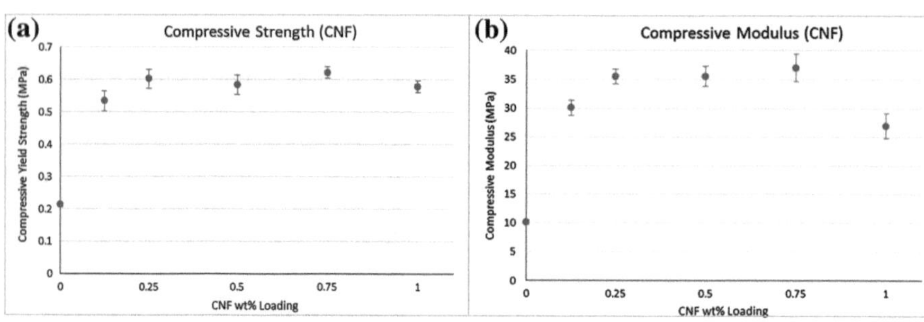

Fig. 3 **a** Compressive strength and **b** compressive modulus of thermoplastic microballoon syntactic foams reinforced with varying wt% of carbon nanofibers

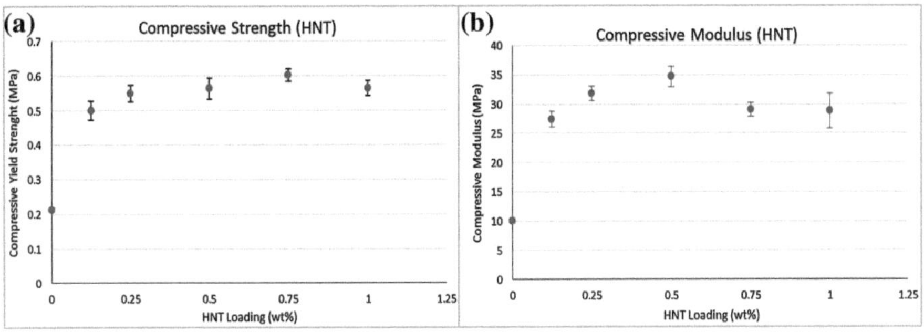

Fig. 4 **a** Compressive strength and **b** compressive modulus of thermoplastic microballoon syntactic foams reinforced with varying wt% of halloysite nanotubes

further increased to 0.25 wt%. Thereafter, the compressive strength and modulus of these materials remained essentially unchanged up to CNF loadings of 0.75 wt%. At CNF loadings of about 1 wt% the strength and modulus properties appear to drop off. Overall, strength and modulus enhancements as large as 180 and 250% respectively can be achieved with a 0.25 wt% addition of CNF.

Figure 4 shows the compressive strength (a) and modulus (b) properties of thermoplastic microballoon syntactic foams loaded with varying wt% of halloysite nanotubes. Dramatic increases in compressive strength and modulus (135 and 170% respectively) were noted with the addition of 0.125 wt% halloysite nanotubes. A further overall increasing trend in strength properties with increasing HNT wt% up until 0.75 wt% is apparent, with a subsequent drop off at 1 wt%. A similar trend was noted for modulus, where the reduction in modulus enhancement was observed at the 0.75 wt% HNT loading. Overall, strength enhancement as large as 183% was achieved with a 0.75 wt% addition of HNT and a modulus enhancement as large as 244% was achieved with a 0.5 wt% addition of HNT.

Tensile Properties

Figure 5 shows the tensile strength (a) and modulus (b) properties of thermoplastic microballoons loaded with varying wt% of carbon nanofibers. As with compression analysis, a dramatic increase in both tensile strength and modulus (110 and 165% respectively) was noted with the initial addition of 0.125 wt% carbon nanofibers. These material properties remained essentially invariant with increasing CNF loading up to 0.75 wt%, where a statistically significant drop in tensile strength is observed. Overall, tensile strength and modulus enhancements as large as 119 and 168% respectively were achieved with a 0.5 wt% addition of CNF.

Figure 6 shows the tensile strength (a) and modulus (b) properties of thermoplastic microballoons loaded with varying wt% of halloysite nanotubes. A dramatic increase in tensile strength and modulus (133 and 173% respectively) was noted

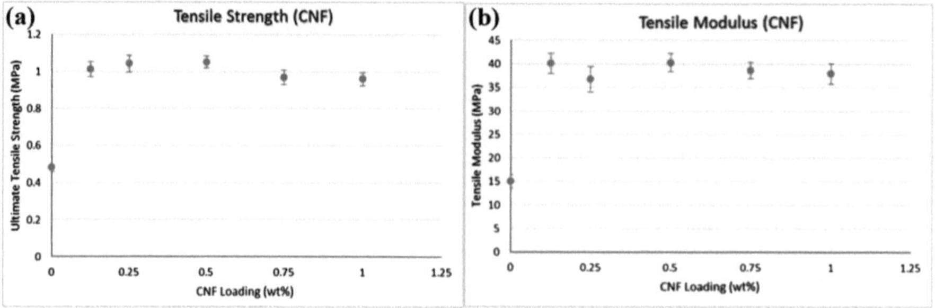

Fig. 5 **a** Tensile strength and **b** tensile modulus of thermoplastic microballoon syntactic foams reinforced with varying wt% of carbon nanofibers

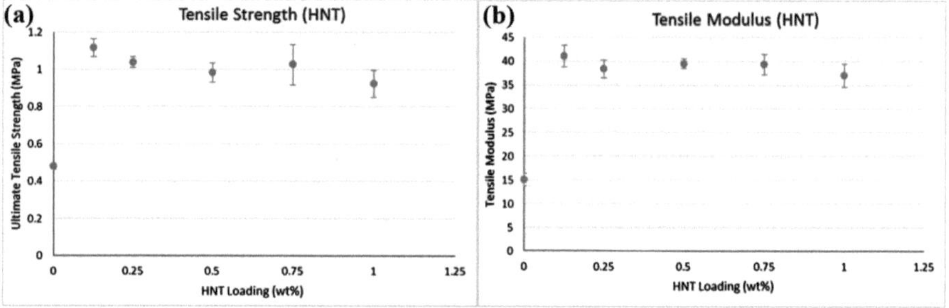

Fig. 6 **a** Tensile strength and **b** tensile modulus of thermoplastic microballoon syntactic foams reinforced with varying wt% of halloysite nanotubes

with the initial addition of 0.125 wt% halloysite nanotubes. Tensile modulus properties remained essentially unchanged with increasing HNT loadings, while there is a statistically significant drop in tensile strength after 0.125 wt% (0.75 wt% discounted). Tensile strength and modulus enhancements as large as 133 and 173% respectively were achieved with a 0.125 wt% addition of HNT.

Conclusions

This work highlights a consistent and repeatable method for the production of highly loaded thermoplastic microballoon syntactic foams utilizing carbon nanofibers or halloysite nanotubes as a reinforcement. A consistent method for mixing of the nano-additives in epoxy resin was determined, showing adequate dispersion and aggregation of each respective nanomaterial with little to no agglomerations. From mechanical analysis, it was observed that small wt% inclusions of each nano-additive into the syntactic foam matrix can lead to large enhancements in strength and modulus with negligible impacts on composite foam density.

Compressive strength and modulus enhancements as large as 180 and 250% respectively can be achieved with a 0.25 wt% addition of CNF and increases of 165 and 244% respectively can be achieved with a 0.5 wt% addition of HNT. Tensile strength and modulus enhancements as large as 110 and 165% respectively can be achieved with a 0.125 wt% addition of CNF and increases of 133 and 173% respectively can be achieved with a 0.125 wt% addition of HNT.

Acknowledgements The authors acknowledge the National Aeronautics & Space Administration's Experimental Program to Stimulate Competitive Research (EPSCoR) program for the financial support of our research through grant (NASA Proposal # 11-EPSCoR-0049). Acknowledgements are also due to the Composite and Polymer Engineering (CAPE) Laboratory and staff for equipment usage, guidance and technical assistance with experimentation.

References

1. Gupta, N., Woldesenbet, E., & Sankaran, S. (2001). Studies on compressive failure features in syntactic foam material. *Journal of Materials Science, 36*(18), 4485–4491.
2. Gupta, N., & Woldesenbet, E. (2003). Hygrothermal studies on syntactic foams and compressive strength determination. *Composite Structures, 61*(4), 311–320.
3. Gupta, N. (2007). A functionally graded syntactic foam material for high energy absorption under compression. *Materials Letters, 61*(4), 979–982.
4. Gupta, N., & Nagorny, R. (2006). Tensile properties of glass microballoon-epoxy resin syntactic foams. *Journal of Applied Polymer Science, 102*(2), 1254–1261.
5. Colloca, M., Gupta, N., & Porfiri, M. (2013). Tensile properties of carbon nanofiber reinforced multiscale syntactic foams. *Composites Part B Engineering, 44*(1), 584–591.
6. Karthikeyan, C. S., & Sankaran, S. (2004). Elastic behaviour of plain and fibre-reinforced syntactic foams under compression. *Materials Letters, 58*(6), 995–999.
7. d'Almeida, J. R. M. (1999). An analysis of the effect of the diameters of glass microspheres on the mechanical behavior of glass-microsphere/epoxy-matrix composites. *Composites Science and Technology, 59*(14), 2087–2091.
8. *Mechanical property allowables generated for the solid rocket booster composite nose cap.* (2000). United States. NASA. *NASA Technical Report No. NASA/TM 2000-201252.* Marshall Space Flight Center, AL.
9. Watkins, L. (1998). Syntactic foam buoyancy for production risers. In *Proceedings of the Seventh International Conference on Offshore Mechanical and Artic Engineering.* Houston, TX.
10. Dando, K., & Salem, D. (2016). Production and characterization of epoxy syntactic foams highly loaded with thermoplastic microballoons. Manuscript Submitted for Publication.
11. Gupta, N., & Maharsia, R. (2005). Enhancement of energy absorption in syntactic foams by nanoclay incorporation for sandwich core applications. *Applied Composite Materials, 12*(3–4), 247–261.
12. Maharsia, R., Gupta, N., & Jerro, H. D. (2006). Investigation of flexural strength properties of rubber and nanoclay reinforced hybrid syntactic foams. *Materials Science and Engineering A, 417*(1), 249–258.
13. Maharsia, R. R., & Jerro, H. D. (2007). Enhancing tensile strength and toughness in syntactic foams through nanoclay reinforcement. *Materials Science and Engineering A, 454*, 416–422.
14. Wouterson, E. M., Boey, F. Y., Wong, S. C., Chen, L., & Hu, X. (2007). Nano-toughening versus micro-toughening of polymer syntactic foams. *Composites Science and Technology, 67*(14), 2924–2933.

15. Deng, S., Zhang, J., Ye, L., & Wu, J. (2008). Toughening epoxies with halloysite nanotubes. *Polymer, 49*(23), 5119–5127.
16. Tang, Y., Ye, L., Deng, S., Yang, C., & Yuan, W. (2012). Influences of processing methods and chemical treatments on fracture toughness of halloysite–epoxy composites. *Materials and Design, 42*, 471–477.
17. Tang, Y., Deng, S., Ye, L., Yang, C., Yuan, Q., Zhang, J., et al. (2011). Effects of unfolded and intercalated halloysites on mechanical properties of halloysite–epoxy nanocomposites. *Composites Part A Applied Science and Manufacturing, 42*(4), 345–354.
18. ASTM Standard D695-15. (2015). *Standard test method for compressive properties of rigid plastics*. West Conshohocken, PA: ASTM International.
19. ASTM Standard D1623-09. (2009). Standard test method for tensile and tensile adhesion properties of rigid cellular plastics. West Conshohocken, PA: ASTM International.

Nanocomposites Mechanical and Tribological Properties Using Graphene-Coated-SiC Nanoparticles ($GCSiC_{NP}$) for Light Weight Applications

A. El Ghazaly, M. Shokeir, S.N. El Moghazi, A. Fathy, M.M. Emara and H.G. Salem

Abstract In the current work, Aluminum Alloy 2124-SiC/Graphene nanocomposite is fabricated via high energy milling followed by uniaxial cold compaction at 525 MPa, sintered at 450 °C, and followed by hot extrusion at 4:1 extrusion ratio. SiC nanoparticles (SiC_{NP}) powders the G-micron-clusters forming G-coated-SiC_{NP} ($GCSiC_{NP}$) reinforcement filler, used for the reinforcement of AA2124 matrices via milling. The processed nanocomposite combines the properties suitable for dry wear resistant and self-lubricating solids. It is anticipated that the formation of $GCSiC_{NP}$ decreases the agglomeration of SiC_{NP} producing uniform dispersion of the $GCSiC_{NP}$ reinforcement within the Aluminum matrices. Mechanical and wear resistance of the processed $GCSiC_{NP}$ nanocomposites were characterized compared to the milled AA2124 and AA2124-SiC_{NP} nanocomposites processed under similar milling conditions. FESEM and XRD are used for the investigation of the milled powders crystallite size, lattice strain, and phases as well as powder morphology.

Keywords Dual function · Self-lubricating nanocomposites · G-coated-SiC_{NP} · Ball milling · Wear resistance

A. El Ghazaly
American University, Cairo, Egypt

M.M. Emara
Manufacturing Engineering Department, Canadian International College, El Tagamoa El Khamis, South of Police Academy, P.O. Box 59, New Cairo 11835, Egypt

M. Shokeir · S.N. El Moghazi · A. Fathy · H.G. Salem (✉)
Mechanical Engineering Department, American University in Cairo, American University in Cairo, AUC Avenue, P.O. Box 74, New Cairo 11835, Egypt
e-mail: hgsalem@aucegypt.edu

© The Minerals, Metals & Materials Society 2017

M.A. Meyers et al. (eds.), *Proceedings of the 3rd Pan American Materials Congress*, The Minerals, Metals & Materials Series, DOI 10.1007/978-3-319-52132-9_41

Introduction

Over the last decades, growing interest of both automotive and aerospace sectors in aluminum-based composites properties has been evident. Those multiple industries are now expecting a lift that takes them to new levels of safety, comfort, and efficiency. High performance tribological parts such as pistons, cylinder heads, cylinder liners, drive shafts, and brake disks represent applications which motivated the research and development of aluminum-based metal matrix composites [1–3]. Their enhanced specific strength, higher thermal conductivity, compared to ceramics and lower coefficient of thermal expansion (CTE) are of some great importance to multiple industrial applications [4–7].

Metal and refractory (ceramic based) carbides including SiC, TiC, Al_2O_3, and $NiAl_3$, among others, are commonly used as hardening fillers in the aluminum matrices to enhance their wear resistance. The addition of SiC (3.18 g/cm^3) to Al-matrices is commonly used as reinforcement fillers in Al-matrices for such applications [8–11]. SiC reinforcements are denser than that of Al-alloys (2.7 g/cm^3); hence the SiC reinforcements increase the overall density of the composite depending on their content [12]. Another drawback to the addition of SiC particles is the increased hardness of composite, resulting in machining difficulty, especially for the cast components. Moreover, incorporation of such hard particles within the Al-matrices is associated with a loss in toughness of the composite; therefore an optimization of the ceramic filler content and size is required to achieve the combined strength/toughness [13]. Identification of the various parameters controlling the composite's final properties was investigated by many researchers. A number of the investigated parameters were the reinforcing particles size, particle distribution within the matrix, particle content, and dispersion method [14, 15]. The reduced size of the reinforcement phase down to the nano-scale is of significant importance. It results in a remarkable improvement of mechanical properties with excellent dispersion. Many applications require materials that combine both temperature and corrosion resistance, especially if the component is under heavy load. However, one of the encountered issues is the agglomeration of the nanoparticles into coarse clusters within the Al-matrices due to their poor wettability and high surface-to-volume ratio. Clustering results in forming stress concentration sites leading to premature failure of the nanocomposite [16–18]. A number of processing techniques were employed by various researchers aiming at enhancing the distribution of the micron and nanoscale ceramic particles within the aluminum matrices. Some of the techniques include stir casting [19], squeeze casting [19], friction stir processing (FSP) [20] and high energy ball milling of mixed powders. High energy ball milling is one of the most popular techniques used for enhancing the distribution of the nanoparticles. However, high energy ball milling results in hardening of the mixed nanoceramic-Al-matrix powders resulting in the formation of incompressible powder that is necessary during the compaction and sintering steps. To overcome this problem, increasing the sintering temperatures or durations becomes inevitable which results in the loss of the strengthening and hardenability gained by prior milling [21–24]. Conventionally, composites based on ceramic particles addition to Al-matrices require external lubrication, which plays an important role in reducing wear between the rubbing surfaces and hence increases wear resistance. Conversely, self-lubricating solids are more desired than those requiring external lubrication, which should be applied periodically, especially when there is difficulty in accessing the rubbing surfaces. In the self-lubricating solids the continuous release of the lubricating element occurs automatically, which decreases the wear rates and lowers the coefficient of friction in the wear components. Graphite is one of the elements used as filler in the self-lubricating solids.

Several efforts were made to improve the strength, hardness, and wear resistance of the self-lubricating solids, to counteract the overall composite loss in strength and hardness associated with the soft graphite (Gr) additions [25–28]. This includes the use of ultrafine graphite particles and carbon nanotubes (CNTs) [29]. Moreover, graphene (G) particles have recently been used to reinforce pure aluminum [30] and 2124 Al-alloy [31] in order to increase both hardness and tensile strength. Work conducted by Salem et al. [9] on AA2124-(G) composite milled and hot extruded powders consolidates showed that VHN was highest for the 3 wt% graphene composites. This was coupled with lowest wear loss, wear rates, and coefficient of friction (COF) via formation of smooth surfaces covered with graphene lubricating layer. To overcome the drawbacks of using ceramic reinforcements in strong/hard composites as well as the drawbacks of using CNTs and Gr in self-lubricating solids, dual function composites emerged and became the focus of many researchers Alizadeh et al. [32] investigated the creep and tribological behavior of Al5083 based hybrid composites reinforced with carbon nanotubes (CNTs) and boron carbide (B_4C). Fallahdoost et al. [33] studied the dual functions of TiC nanoparticles on the tribological behavior of mechanically alloyed Al7075/graphite composites. Kaushik and Rao [34] worked on the tribological behavior of stir cast AA6082 reinforced with SiC and Gr. Li et al. [13] employed chemical vapor deposition for growing CNTs on SiC particles followed by ball milling with Al-matrices. Liu et al. [35] investigated the influence of introducing graphene oxide and graphene nano-sheets on the mechanical and physical properties of pure Al processed via mechanical milling. Mechanical and tribological behavior of Al/SiC/Ni-coated graphite Al6061 composite prepared via semi solid deformation of the Ni coated graphite with the SiC and Al6061 were investigated by Guo and Tsao [36]. Regardless of the processing technique used in producing the dual composites, it was mainly based on the introduction of the ceramic and G/Gr individually in the Aluminum matrices, except for the Ni-coated Gr particles mixed with SiC in Al6061.

Accordingly, in the current research work, ball milled (BM) and hot extruded (HE) AA2124 nanocomposite reinforced with G-coated-SiC_{NP} ($GCSiC_{NP}$) compared with that reinforced with similar content of SiC-nanoparticles (SiC_{NP}) and plain AA2124 are investigated. The processed nanocomposite is designed so that it combines both properties of the cremate-reinforced Al-alloys with their ultrahigh strength and dry wear resistance and the self-lubrication graphene reinforced

Al-alloys, with enhanced load bearing high performance tribological applications. Accordingly, evaluation of the powder morphologies throughout the processing stages as well as the physical and mechanical properties including scratch resistance of the hot extruded nanocomposites compared to the plain alloy are conducted.

Experimental Procedure

Materials and Processing

The composite powders were fabricated from prealloyed Al2124 aluminum micron-powders, graphene micronclusters and SiC_{NP}. The AA2124 powders were supplied by the Aluminum Powder Company limited (APC) with a chemical composition of Al-3.9 Cu-1.5 Mg-0.65 Mn-0.1 Si-0.1 wt%. Al2124 polycrystalline powder morphology and grain size were reported in other publications [22]. The particle size ranged between 5 and 85 μm with an average size of 40 μm, while the internal grain size was 1 μm with an average crystallite size of 87 nm as measured by XRD [21, 24]. SiC_{NP} powders were received from American Elements with an average particle size ranging between 64–05 nm (Fig. 1b), which are agglomerated in clusters about 8 μm in average size as shown in Fig. 1a. G-powders were received in the form of 5–10 nm monolayers, which were clustered into 15 μm particles due to high surface area as shown in Fig. 1c, d at low and high magnification, respectively. Graphene powders were purchased from Skyspring Nanomaterials (SSnano). G-coated SiC_{NP} ($GCSiC_{NP}$) powders were processed via high energy BM of 3 wt% G and 5 wt% SiC_{NP}. The Graphene content of 3 wt% was selected based on the reported highest mechanical and tribological properties produced in the ball milled and extruded AA2124-G composites [31]. $GCSiC_{NP}$ powders were fabricated via mixing under argon atmosphere using a TURBULA Shaker for 1 h at 96 rpm. This was followed by ball milling of the mixed powders under Argon atm in stainless steel jars containing 5 and 10 mm stainless steel hardened balls using the RETSCH 400MA high energy ball mill. The milling speed

Fig. 1 SEM micrographs showing the as received (AR) **a**, **b** SiC_{NP} and **c**, **d** graphene micronclusters, at low and high magnifications, respectively

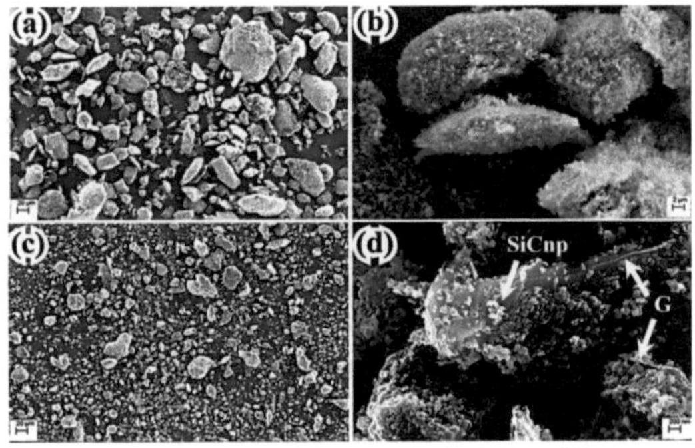

Fig. 2 SEM images for **a**, **b** mixed $GCSiC_{NP}$ and **c**, **d** milled $GCSiC_{NP}$, at low and high magnification, respectively. *Arrows* point at G and SiC_{NP}

was maintained at 400 RPM. The ball-to-powder ratio (BPR) was 15:1, while the milling time was 2 h.

Figure 2a–d shows the mixed and milled powders of G and SiC, respectively; the $GCSiC_{NP}$ powders are represented at low and high magnifications, respectively. It is clear from the images that mixing resulted in the formation of uniform coating of the SiC_{NP} on the G-micron-clusters (Fig. 2a, b). Due to the huge particle size scale ratio between the SiC_{NP} and the G-micron-clusters, the ceramic nanoparticles almost fully covered the surfaces G-clusters forming powdered particles (Fig. 2b). On the other hand, milling for 2 h of the mixed powders resulted in de-clustering of both the SiC_{NP} and graphene. The obvious refinement of the graphene particles is shown in Fig. 2c, which was associated with the uniform distribution of the SiC_{NP} on the G-cluster surfaces (Fig. 2d). The G-mono-layered structure with uniform distribution of SiC_{NP} was evident in the milled powders (pointed at by arrows) as shown in Fig. 2d.

AA 2124 powders were mixed with 0.5 wt% SiC_{NP} and 5 wt% $GCSiC_{NP}$ for 1 h at 96 rpm. Knowing that the 5 wt% $GCSiC_{NP}$ constituted 3/8 wt% G and 5/8 wt% SiC_{NP}. Ball milling (BM) of the mixed powders of the AA2124-wt% SiC_{NP} and $GCSiC_{NP}$ compared to the plain milled AA2124 powders was carried out at 400 rpm and 15:1 BPR for 2 h. Consolidation of the milled powders were green compacted in a die under pressure of 525 MPa at room temperature followed by sintering at 450 °C for 75 min followed by hot extrusion at 4:1 extrusion ratio. The extruded rods were 10 mm in diameter and 89 mm long. The produced rods were then air-cooled down to room temperature.

Characterization

Density was measured using Mettler Toledo densitometer based on Archimedes principles. Cylindrical specimens 10 mm in diameter were cut from the extruded rods and immersed in xylene, the auxiliary fluid that was used. Hardness of the plain milled alloy and composite powders, post hot extrusion, was measured on sections cut perpendicular (transverse) and parallel (longitudinal) to the extrusion direction using Mettler Toledo micro-hardness. The hardness test was conducted using 1 kg over dwell time of 15 s. The remaining parts of the extrudates were machined according to the ASTM E8 standard into tensile specimens and then tested at room temperature at 0.1 mm/min strain rate; their tensile properties were characterized using Instron 100 kN UTM. Wear resistance characterization of the composite was conducted MTS Nanoindenter m/c scratch test. The test was carried out via XP nano-scratch method, where a Berkovich diamond tip was used to scratch the HE nanocomposite ground and polished surfaces (cut perpendicular to the extrusion direction) under single pass wear test at a constant load of 50 mN over a scratch distance of 700 μm. The average of 4 scratches per sample was investigated.

XRD was employed for the investigation of the as-received and composite powders before and after milling and as a function of graphene content post hot extrusion. Cu-α radiation was employed at 30 kV and 40 mA at a scan speed of 0.01°/s. A LEO FESEM (30 kV) and 1 nm resolution was employed for the characterization of the powders' morphologies at the various stages of processing. Investigation of the nanoscratched surfaces morphology (width and features) was facilitated by the FESEM.

Results and Discussion

Nanocomposite Powders Processing

The low energy associated with the mixing process led to a uniform distribution of the SiC_{NP}-nanoclusters on the surfaces of the aluminum powders as pointed out by arrows in Fig. 3a, b for the AA2124-5 wt% SiC_{NP} at low and high magnification, respectively. Figure 3c, d shows the mixed AA2124 with 5 wt% $GCSiC_{NP}$ powders, where clusters and individual NP of SiC are uniformly coating the graphene micron-clusters and the AA2124 particles as well (Fig. 3d). This was an indication

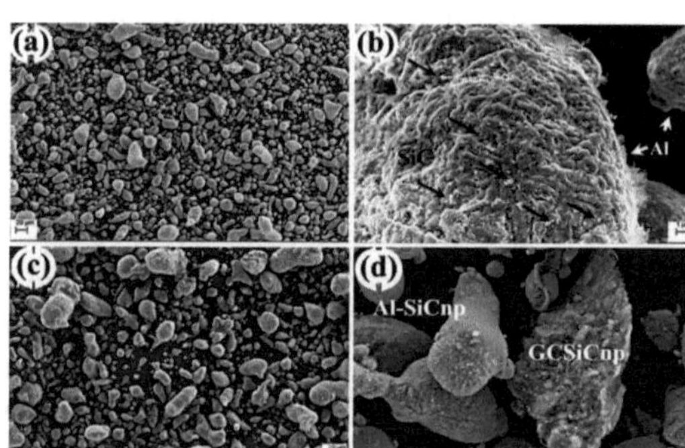

Fig. 3 SEM images for mixed AA2124-with 5 wt% **a**, **b** SiC_{NP} and **c**, **d** $GCSiC_{NP}$, at low and high magnification, respectively. *Black Arrows* point at SiC_{NP}

that the SiC_{NP} were not strongly bonded to the G-clusters, which was evident by the observed SiC_{NP} on the surfaces of the Al-particles.

Figure 4a, b shows SEM images for the AA2124-AR and BM powders, respectively. Evidence for cold welding of the soft AR-particles associated with partial fragmentation and refinement (10 μm average particle size) was observed via 2 h BM [37]. Al_2O_3 was also observed either in the form of fully fragmented or partially delaminated films on the surfaces of the milled Al-particles as pointed out by the arrows in Fig. 4b and the corresponding XPS scans for Oxides (Fig. 4c). Similar observations were evident by Salem et al. [21]. Ball milling of the mixed AA2124 with 5 wt% SiC_{NP} nano-clusters for 2-h was not enough for producing a significant refinement of the Al-particles (20 μm in average size) as evident in Fig. 5a. Cold welding between the individual AA2124 particles and the AA2124 with SiC_{NP} was the dominant mechanism [22, 37]. However, BM resulted in de-clustering of the SiC_{NP}, which was manifested by the impinged individual SiC_{NP} on the surfaces of the polycrystalline Al-softer particles as pointed at by white arrows in Fig. 5b. This agrees with BM model described by Varol [37] for the effect of reinforcement relative size and hardness that of the soft matrix particles.

Ball milling of the mixed AA2124-5 wt% $GCSiC_{NP}$ powders is shown if Fig. 5c, d at low and high magnifications, respectively. Comparing the particle size of the BM AA2124 with SiC_{NP} and $GCSiC_{NP}$ at a low magnification (Fig. 5a, c), respectively show evidence for spherodization and slight reduction in particle size

Fig. 4 SEM images for AA2124 **a** (AR), **b** BM and **c** XPS scans for oxides in the BM powders

Fig. 5 SEM images for milled AA2124-5 wt% **a**, **b** SiC_{NP} and **c**, **d** $GCSiC_{NP}$, at low and high magnification, respectively. *White arrows* point at SiC_{NP} and *red arrows* point at Al_2O_3 delaminated and fragmented films (Color figure online)

compared to the BM nanocomposite powders with SiC_{NP}. The captured images show the disappearance of the G-clusters, while the SiC_{NP} were clearly observed on the surfaces of the cold welded particles. It is suggested that the Al-particles were coated with the $GCSiC_{NP}$ micronclusters during the BM stage. This agrees with the work of Salem et al. [31] who worked on the BM of AA2124-G hot extruded at 300 °C, showing an evidence for the encapsulation of the Al-particles with graphene thin layer elongated along the extrusion direction. It is also suggested that during the cold welding graphene could be entrapped between the flaked Al-particles forming layered microstructure [31, 38]. Figure 5b, d also show evidence for Al_2O_3 films fragmented and segregated within the Al-particles films pointed at by red arrows for both nanocomposites, respectively.

Physical and Mechanical Properties of the Extruded Alloy and Nanocomposite BM-Powders

Table 1, lists the influence of reinforcement type (SiC_{NP} and $GCSiC_{NP}$) compared to the as milled plain AA2124 on the densification degree, and the mechanical properties' variations post the hot extrusion (HE). The calculated percent relative density (RD%) was based on the ratio of the experimental measured density of the hot extrudates to the theoretical density calculated based on rule of mixtures. From the displayed results BM of the mixed powders for 2 h followed by HE at 450 °C resulted in relatively high densification, which can be revealed by the achieved > 99% relative densities. Addition of 5 wt% SiC_{NP} and $GCSiC_{NP}$ resulted in the enhancement of the composite densification compared with the plain BM AA2124 hot extruded powders [26, 38]. As commonly known, 2-h of BM was mainly responsible for the cold welding of the soft Al-particles and was not enough for the formation of refined, hard, and hence incompressible particles [21, 22]. However, the measured slight increase in density with the addition of 5 wt% SiC_{NP} (99.7%) compared to the plain BM powders (99.1%) could be attributed to the oxide layers readily fragmented and induced by the impingement of the SiC_{NP} on the surfaces during milling. This allowed the exposure of oxide free surfaces, which facilitated diffusion during HE. The relatively ultrafine nano-scaled ceramic hard particles of

Table 1 Relative density (RD%), micro (Hv), yield strength, ultimate tensile strength, and elongation % for the hot extruded (HE) BM plain AA2124 and nanocomposites

Hot extruded-BM powders	RD%	(HV)		σ_Y (MPa)	σ_{UTS} (MPa)	Elong. (%)
		T	L			
AA2124	99.1	95	94	150	290	9.8
AA2124-5 wt% SiC_{NP}	99.7	131	120	235	299	6.4
AA2124-5 wt% $GCSiC_{NP}$	99.9	249.5	234.5	290	350	3.4

SiC compared to the soft-coarse AA2124 did not allow for the SiC_{NP} segregation on the boundaries and triple junctions but rather caused their impingement on the surfaces and inside the cold welded soft particles [37], which facilitated consolidation during HE. Addition of 5 wt% $GCSiC_{NP}$ enhanced the nanocomposite densification (99.9%). This could be explained by the formation of $GCSiC_{NP}$ layer around the Al-particles. Graphene forms a layer surrounding the Al-particles, which facilitates the deformation of the composite powders via its lubrication effect and hence provides a soft media for closing pores during extrusion [31, 36, 38]. This disagrees with the reported results by Jafari et al. [39] who reported a significant decrease in density (82%) when 1.5 wt% MWCNT were added to AA2024 BM and hot pressed discs at 450 °C.

Table 1, also displays the variation of hardness on the extrudates' transverse and longitudinal sections of the nanocomposites compared to the plain AA2124. HV-values exhibited by the AA2124 reinforced with 5 wt% $GCSiC_{NP}$ were almost 100 and 162% higher than that of 5 wt% SiC_{NP} and the plain AA2124 extrudates, respectively. The exhibited HV-values increase applies for both the transverse and longitudinal sections. The reported values for the plain BM for 2 h is lightly higher than that reported by Carreno and coworkers for 2024 at 2 h BM powders [40]. The highest achieved hardness (95-vs.-70, respectively). According to Carreno, HV-values were achieved for the AA2024-2.5 and 5 wt% SiC_{NP} (106 and 102 HV), while in this current research the measured hardness for the AA2124-5 wt% SiC was 131 HV. The higher hardness values reported in the current research could be due to the slight difference in composition between AA2124 and 2024 and mainly due to the difference in consolidation method used. The employment of cold compactions followed by hot extrusion (current work) compared to pressure-less sintering (Carreno's work) could explain the observed difference in the results. Moreover, the processed AA2124-$GCSiC_{NP}$ nanocomposite (current research) displayed higher hardness (250-HV) compared to that reported by Varol et al. [37] who investigated the mechanical properties of AA2024-5 wt% B_4C (199-Hv) [40]. The displayed tensile behavior for the nanocomposites compared to the plain alloy BM extrudates, shows agreement with the displayed Hv-results, where the highest yield and ultimate tensile strength and lowest ductility was achieved by the AA2124-5 wt% $GCSiC_{NP}$ compared to that reinforced with SiC_{NP} and the plain aluminum.

Tribological Properties: NanoScratch Test

Figure 6a presents the on-load and residual scratch depth and Fig. 6b represents the scratch width variation for the BM and HE nano-composites compared to the plain AA2124 alloy consolidates. The sliding direction is represented by the arrow shown

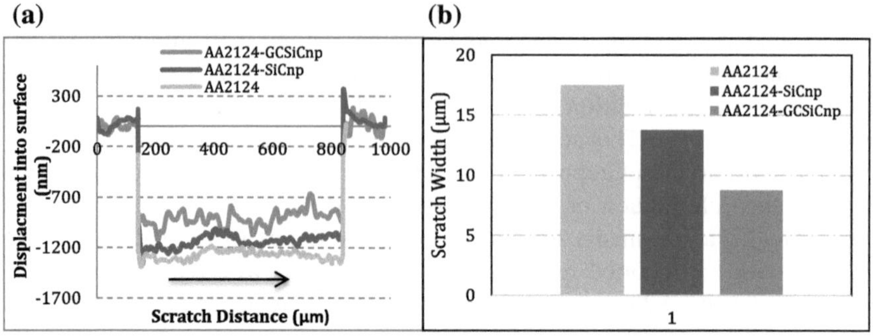

Fig. 6 Nanoscratch test results at constant load of 50 mN over a scratch distance of 700 μm for the BM-HE nanocomposites compared to the plain alloy **a** scratch depth as a function of distance and **b** scratch width

on the diagram. It is clear from the nano-scratch profile that addition of $GCSiC_{NP}$ in the Al-matrices resulted in high resistance to the indenter penetration, which was depicted by the lowest average penetration depth (−873 nm) and lowest scratch width of 8.5 μm in average size. AA2124-reinforced with SiC_{NP} displayed higher average scratch depth (−1155 nm) and wider scratches about 13.7 μm in average size. The highest scratch depth was recorded for the plain AA2124 alloy BM extrudates (−1240 nm), which was associated with 18.5 μm width. It is also observed that presence of graphene as reinforcement in the nanocomposite results in fluctuating scratch depth profile, while the AA2124-SiC_{NP} and the plain one displayed more uniform scratch depth profile. The presence of graphene layers confronts the Berkovich tip and forces it to shift upwards [41].

Figure 7 shows SEM images for the scratches produced on the surfaces of the extruded nanocomposites compared to the plain alloy. Results displayed for the crack depth and width were averaged over 10 scratches per condition. It is clear that deformation by ploughing was significantly dominating the scratching behavior of both the plain AA2124 (Fig. 7a) and the AA2124-5 wt% SiC_{NP} (Fig. 7b), with excessive damage on the scratch edges [42]. Conversely, the AA2124-$GCSiC_{NP}$ (Fig. 7c) showed the least damage manifested by the symmetrical marks on the scratch edges. The $GCSiC_{NP}$ reinforced nanocomposite shows uniform appearance with limited amount of debris compared to the other two conditions. Evidence for smeared appearance on the edges of the plain alloy (Fig. 7a, circled) resulting from plastic deformation associated with indenter tip and surface of the rod [42]. On the other hand, high magnification images for the SiC_{NP} reinforced nanocomposite shows evidence for adhesive wear, which was manifested by the torn out alligator skin-like feature (Fig. 7b, circled).

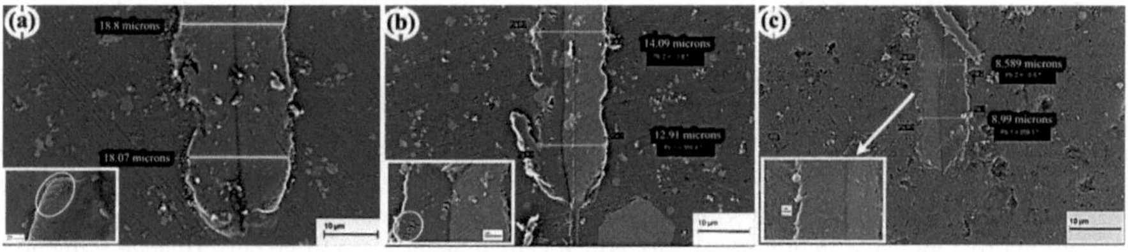

Fig. 7 SEM images for the scratches of the HE **a** plain AA2124 alloy, **b** AA2124-SiC_{NP} and **c** AA2124-$GCSiC_{NP}$ nanocomposites

Conclusions

In the current research, the processing of almost fully consolidated mixed and BM AA2124-$GCSiC_{NP}$ and SIC_{NP} were carried to produce hot extrudates which were conducted at 0.67Tm. BM of G micron-clusters with SiC_{NP} nanoclusters produced G-clusters fully coated (powdered) with de-clustered SiC_{NP}. Addition of 5 wt% $GCSiC_{NP}$ instead of 5 wt% SiC_{NP} produced a nanocomposite with uniform distribution of G-coated SiC_{NP} within the soft aluminum matrix. Nonetheless, cold welded particles of $GCSiC_{NP}$ were present on the AA2124 and this is associated with BM. BM of $GCSIC_{NP}$ with AA2124 induced the highest lattice strain and finest crystallite size compared to SiC_{NP} addition. AA2124-$GCSiC_{NP}$ compared to AA2124-SiC_{NP} and plain AA2124, displayed higher consolidation densities, higher hardness, higher tensile properties and higher resistance to scratching. The enhanced hardness, strength and resistance to scratch provide preliminary indicators for the production of nanocomposites with dual function for ultrahigh wear resistance-self-lubricating component.

Acknowledgements Authors of the work would like to acknowledge the Youssef Jameel Science and Technology Research Center (YJSTRC) for facilitating the characterization of the tested composites. Extended gratitude is given to the effort exerted by Mr. Zakarya Taha and Eng. M. Bakr for their technical assistance.

References

1. Basavarjappa, S., Chandramohan, G., & Paulo Davim, J. (2007). Applications of Taguchi techniques to study dry sliding wear behavior of metal matrix composites. *Materials & Design, 28*(4), 1393–1398.
2. Palanikumar, K., Rajasekaran, T., & Paulo Davim, J. (2010). Modelling and analysis on wear behavior of metal matrix composites. In J. Paulo Davim (Ed.), *Tribology of composite materials* (pp. 157–174). New York, NY: NOVA Publishers.
3. Paulo Davim, J. (Ed.). (2012). *Wear of advanced materials.* London, UK: Wiley-ISTE.
4. Niskanen, P., & Mohn, W. R. (1988). Versatile metal matrix composites. *Advanced Materials and Processes, 133*(3), 39–41.

5. Hunt, M. (1989). Making metal matrix composites stronger and tougher. *Mechanical Engineering, 7,* 43–46.
6. Kempfer, L. (1990). Materials take hypersonic leap into space. *Mechanical Engineering, 8,* 19–22.
7. Hunt, M. (1990). Form and function in metal matrix composites. *Mechanical Engineering, 6,* 27–31.
8. Macke, A. J., Schultz, B., & Rohatgi, P. K. (2012). Metal matrix: Composites offer the automotive industry an opportunity to reduce vehicle weight, improve performance. *Advanced Materials and Processes, 170*(3), 19–23.
9. Christy, T. V., Murugan, N., & Kumar, S. (2010). A comparative study on the microstructures and mechanical properties of Al 6061 alloy and the MMC Al 6061/$TiB_2/12_p$. *Journal of Minerals & Materials Characterization & Engineering, 9*(1), 57–65.
10. Miracle, D. B. (2005). Metal matrix composites—From science to technological significance. *Composites Science and Technology, 65*(15), 2526–2540.
11. El-Garaihy, W. H., Oraby, S. E., Rassoul, E. S. M., & Salem, H. G. (2015). On the effect of SiC content and processing temperature on relative density and hardness of hot compacted aluminum AA6061 composite-mathematical empirical and response surface approach. *Journal of Materials Science Research, 4*(3), 1–14.
12. Prabu, S. B., Karunamoorthy, L., Kathiresan, S., & Mohan, B. (2006). Influence of stirring speed and stirring time on distribution of particles in cast metal matrix composite. *Journal of Materials Processing Technology, 171*(2), 268–273.
13. Li, S., et al. (2016). Enhanced mechanical behavior and fabrication of silicon carbide particles covered by in-situ carbon nanotube reinforced 6061 aluminum matrix composites. *Materials and Design, 107,* 130–138.
14. Ibrahim, I. A., Mohamed, F. A., & Lavernia, E. J. (1991). Particulate reinforced metal matrix composites—A review. *Journal of Materials Science, 26*(5), 1137–1156.
15. Suh, Y. S., Joshi, S. P., & Ramesh, K. T. (2009). An enhanced continuum model for size-dependent strengthening and failure of particle-reinforced composites. *Acta Materialia, 57*(19), 5848–5861.
16. Wang, Z., Song, M., Sun, C., Xiao, D., & He, Y. (2010). Effect of extrusion and particle volume fraction on the mechanical properties of SiC reinforced Al-Cu Alloy. *Materials Science and Engineering A, 527*(24–25), 6537–6542.
17. Tjong, S. C. (2007). Novel nanoparticle-reinforced metal matrix composites with enhanced mechanical properties. *Advanced Engineering Materials, 9*(8), 639–652.
18. Kollo, L., et al. (2011). Nano-silicon carbide reinforced aluminum produced by high-energy milling and hot consolidation. *Materials Science and Engineering A, 528*(21), 6606–6615.
19. Shalaby, E. A. M., Churyumov, A. Y., Solonin, A. N., & Lotfy, A. (2016). Preparation and characterization of hybrid A359/(SiC + Si_3N_4) composites synthesized by stir/squeeze casting techniques. *Materials Science and Engineering A, 674,* 18–24.
20. Jeon, C. H., et al. (2014). Material properties of graphene/aluminum metal matrix composites fabricated by friction stir processing. *International Journal of Precision Engineering and Manufacturing, 15*(6), 1235–1239.
21. Salem, H. G., & Sadek, A. W. (2010). Fabrication of high performance PM nanocrystalline bulk AA2124. *Journal of Materials and Engineering Performances, 19*(3), 356–367.
22. Salem, H. G., El-Eskandarany, S., Kandil, A., & Abdel Fattah, H. (2009). Bulk behavior of ball milled AA2124 nanostructured powders reinforced with TiC. *Journal of Nanomaterials,* 1–12.
23. Sadek, A., Salem, H. G., & Attallah, M. (2009). *Nanocrystalline powder consolidation in AA2124 using uniaxial compaction and severe plastic deformation.* Paper Presented at the San TMS 2009 Annual Meeting and Exhibition, San Francisco, California, February 15–19.
24. Sadek, A., & Salem, H. G. (2009). Controlling the processing parameters for consolidation of nanopowders into bulk nanostructured material. In: T. Hinklin & K. Lu (Eds.), *Processing of nanoparticle structures and composites* (pp. 1–12). Hoboken, NJ: Wiley.

25. Kumar, P. H. G., & Xavior, M. A. (2014). Graphene reinforced metal matrix composite (GRMMC): A review. *Procedia Engineering, 97,* 1033–1040.
26. Sharma, P., Khanduja, D., & Sharma, S. (2015). Production of hybrid composite by a novel process and its physical comparison with single reinforced composites. *Materials Today: Proceedings, 2*, 2698–2707.
27. Flores-Zamora, M. I., Estrada-Guel, I., Gonalez-Hernandez, J., Miki-Yoshida, M., & Martinez-Sanchez, R. (2007). Aluminum-graphite composite produced by mechanical milling and hot extrusion. *Journal of Alloys and Compounds, 434–435*, 518–521.
28. Deaquino-Lara, R., et al. (2014). Structural characterization of aluminum alloy 7075-graphite composites fabricated by mechanical alloying and hot extrusion. *Materials and Design, 53,* 1104–1111.
29. Al-Qutub, A., Khail, A., Saheb, N., & Hakeem, A. (2013). Wear and friction behavior of Al6061 alloy reinforced with carbon nanotubes. *Wear, 297*(1), 752–761.
30. Bartolucci, S. F., et al. (2011). Graphene-aluminum nanocomposites. *Materials Science and Engineering A, 528,* 7933–7937.
31. Ghazaly, A., Seif, B., & Salem, H. G. (2013). *Mechanical and tribological properties of AA2124-graphene self-lubricating nanocomposite.* Paper Presented at the San TMS 2013 Annual Meeting and Exhibition, Antonio, Texas, March 2–7, 2013.
32. Alizadeh, A., Abdollahi, A., & Biukani, H. (2015). Creep behavior and wear resistance of Al 5083 based hybrid composites reinforced with carbon nanotubes (CNTs) and boron carbide (B_4C). *Journal of Alloys and Compounds, 650,* 783–793.
33. Fallahdoost, H., Nouri, A., & Azimi, A. (2016). Dual functions of TiC nanoparticles on tribological performance of Al/graphite composites. *Journal of Physics and Chemistry of Solids, 93,* 137–144.
34. Kaushik, N Ch., & Rao, R. N. (2016). Effect of grit size on two body abrasive wear of Al 6082 hybrid composites produced by stir casting method. *Tribology International, 102,* 52–60.
35. Liu, J., et al. (2016). Graphene oxide and graphene nanosheet reinforced aluminum matrix composites: Powder synthesis and prepared composite characteristics. *Materials and Design, 94,* 87–94.
36. Ted Guo, M. L., & Tsao, C.-Y. A. (2000). Tribological behavior of self-lubricating aluminum/SiC/graphite hybrid composites synthesized by the semi-solid powder-densification method. *Composites Science and Technology, 60,* 65–74.
37. Varol, T., & Canakci, A. (2013). Effect of particle size and ratio of B_4C reinforcement on properties and morphology of nanocrystalline Al2024-B_4C composite powders. *Powder Technology, 246,* 462–472.
38. Tabandeh-Khorshid, M., Omrani, E., Menezes, P. L., & Rohatgi, P. K. (2016). Tribological performances of self-lubricating aluminum matrix nanocomposites: Role of graphene nanoplatelets. *Engineering Science and Technology, and International Journal, 19,* 463–469.
39. Jafari, M., Abbasi, M. H., Enayati, M. H., & Karimzadaeh, F. (2012). Mechanical properties of nanostructured A2024-MECNT composite prepared by optimized mechanical milling and hot pressing methods. *Advanced Powder Technology, 23,* 205–210.
40. Carreno-Gallardo, C., Estrada-Guel, I., Lopez-Melendez, C., & Martinez-Sanchez, R. (2014). Dispersion of silicon carbide nanoparticles in a AA2024 aluminum alloy by a high-energy ball mill. *Journal of Alloys and Compounds, 586,* S68–S72.
41. Shokrieh, M. M., Hosseinkhani, M. R., Naimi-Jamal, M. R., Tourani, H. (2013). Nanoindentation and nanoscratch investigations on graphene-based nanocomposites. *Polymers Testing, 32*(1), 45–51.
42. Hodge, A. M., & Nieh, T. G. (2004). Evaluating abrasive swear of amorphous alloys using nanoscratch technique. *Intermetallics, 12,* 741–748.

Synthesis of Energetic Composites in Ti–Al–B–C System by Adiabatic Explosive Compaction

Mikheil Chikhradze and Fernand D.S. Marquis

Abstract Recent developments in materials science have increased the interest towards the bulk (energetic/energy) materials and the technologies for their production. The unique properties which are typical for the composites fabricated in Ti–Al–B–C systems make them very attractive for aerospace, power engineering, machine and chemical applications. In addition, aluminum matrix composites (AMCs) have great potential as structural materials due to their excellent physical, mechanical and tribological properties. Because of good combinations of thermal conductivity and dimensional stability AMCs are found to be also potential materials for electronic packaging/application. The methodology/technology for the fabrication of bulk materials from ultrafine-grained powders of Ti–Al–B–C system are described in this paper. It includes the results of theoretical and experimental investigation for selection of powder compositions, determination of thermodynamic conditions for blend preparation and optimal technological parameters for mechanical alloying and adiabatic compaction. For the consolidation of mixtures, the explosive compaction technology was applied at room temperatures.

Keywords Synthesis · Explosive compaction · Nano powders · Energetic materials

M. Chikhradze (✉)
G. Tsulukidze Mining Institute, 7 E. Mindeli Str., 0186 Tbilisi, Georgia
e-mail: m.chikhradze@gtu.ge

F.D.S. Marquis
San Diego State University, 5500 Campanile Drive,
San Diego, CA 92182-8010, USA

M. Chikhradze
F. Tavadze Institute of Metallurgy and Materials Science,
10 E. Mindelistr, 0111 Tbilisi, Georgia

M. Chikhradze
Georgian Technical University, 75 Kostava Str., 0175 Tbilisi, Georgia

© The Minerals, Metals & Materials Society 2017

M.A. Meyers et al. (eds.), *Proceedings of the 3rd Pan American Materials Congress*,
The Minerals, Metals & Materials Series, DOI 10.1007/978-3-319-52132-9_42

Introduction

Intermetallics fabricated in Ti–Al–B–C system are characterized by unique physical-mechanical properties. They have high specific strength under tensile and compression conditions, good high temperature corrosion, oxidation and wear resistant properties [1–6].

According to the phase diagrams in binary and ternary system the composites/intermetallics may be obtained with wide spectrum of phase composition, in crystalline and amorphous (brittle and ductile) structures. Depending on the composition and structure, the synthesized composites may have different specific properties. The potential of the system for development of new structural/composite materials in different thermodynamic conditions is very attractive. The increased attention to this system is caused by the development of nanomaterials, because nano sized grained materials exhibit the unique complex of properties in comparison with the large grain materials. It is expected that the ultrafine grained intermetallics obtained in Ti–Al–B–C system should be characterized with significantly improved physical and mechanical properties. It was prior motivation for investigations in the above-mentioned system.

Current State of the Problem

Fine grained composite materials of Ti–Al–B–C system, prepared in the form of micromechanical blends, solid solutions and intermetallic compounds are of great practical interest because of improved mechanical properties in comparison with coarse grain material (>1 μm) [7–11].

Practical implementation of the nanomaterials novel properties depends on the improvement of synthesis methods to manufacture nanocrystalline materials in a large scale. Widespread application of nanocrystalline materials requires the low cost production of the industrial applicable quantity of nanopowder and efficient technology of synthesis/consolidation of nanoparticles for obtaining bulk materials.

The laboratory methods of fine grained powders preparation (condensation from vapor, electrodeposition, chemical, inert gas condensation and etc.) are mainly used for research, because of high cost and low output capacity. Large volume fraction of atoms in the grain/particles boundaries and as a result increased chemical activity of nanoparticles create additional problems for handling and consolidation.

Several conventional methods are known for obtaining bulk ultrafine grained/nanostructured materials, including: hot isostatic pressing (HIP), spark plasma synthesis (SPS), mechanical alloying (MA), laser engineered net shaping (LE), and etc. The HIP normally requires use of high pressure and high temperature for extended period of time. Because of significant coarsening of the ultrafine particles/grains, nanostructure effects are decreased. MA involves repeated cold welding, fracturing, and re-welding of powder particles in a high-energy ball mill.

Due to specific advantages offered by this technique, MA is used to synthesize a variety of ultrafine grained materials and nanocomposites. SPS method has significant limitations for production of large scale nanocomposites. An attractive method is self-propagating high temperature syntheses (SHS). SHS is based on the combustion of powder exothermal systems, running in stable regime. One of the promising technological type in the set of SHS technologies is SHS-compacting, when the red-hot product, right after the end of the reaction, undergoes a deformation effect over the entire volume. By this method, a wide spectrum of materials have been synthesized (Multilayer Functional Gradient Alloys, SIGMA type) on the basis of refractory compounds TiC, TiB_2, Cr_3C_2. The further development of SHS became a synthesis in thermal explosion mode. This method includes preheating, combustion and densification stages.

The disadvantages of described technologies can be classified as the following problems:

The first one is connected with nanopowders preparation that include the control of particles sizes; chemical instability, separation and etc.; storage and handling for consolidation and obtaining bulk samples.

The second part of problems is connected with fabrication of bulk nanostructured samples that include limitations on the sizes and geometry of bulk material; energy consumption; needs for complicated facility/equipment; difficulties to control grains sizes; significant coarsening of structure upon high temperature conditions for extended period of time; practical impossibility of integration of different technological regimes in one cycle is (syntheses-cladding-welding).

Based on the above-mentioned issues, the main objects of the investigations were the following.

I. Ball milling technology for obtaining amorphous and nanopowder materials as the precursors for synthesis of bulk materials.
II. Shock wave compaction technology for fabrication of the bulk nanostructured materials.

Experimental Procedures

Precursors and Selection of Powder Composition

As starting elemental powders there were used: coarse powders of Ti, Al (Fig. 1), C and amorphous Boron. Preliminary selection of blend compositions was made on the basis of theoretical investigations. Different initial compositions of Ti–Al–B–C system consisting of Ti/Al, 5Ti/5Al/20B and 3Ti/2Al/1C elemental molar ratios were prepared for experiments. Phase equilibrium system was determined based on the Gibbs's principle-minimization of total energy.

Fig. 1 SEM images of initial materials

Fig. 2 SEM picture of the Ti–Al–C blend. Processing time—5 h

Ultrafine Powder Fabrication

Precursors were classified by vibratory sieves within the particle size ranges: (−0.063, +0) mm; (−0.16, +0.063) mm; (−0.315, +0.16) mm; and +0.315 mm. The powders were weighed and mixed to produce the blend.

For MA, amorphization and nanopowder production, the high energetic "Fritsch" Planetary premium line ball mill (Fig. 3) was used. The mill was equipped with Zirconium Oxide jars and balls. Ratio ball to powder by mass was 10:1. The time of the processing was varied in range: 1; 2; 5; 15; 20 h. Rotation speed of the jars was 500 rpm.

In all processing conditions, diffraction lines of elementary Al, Ti, B, C, as well of titanium and aluminum oxides were identified by X-ray investigations. During the MA, the syntheses of Titanium alluminades in Ti–Al–B system (TiAl, $TiAl_3$, Ti_3Al) are confirmed when the processing time exceeds 5 h. The SEM pictures of the 3Ti/2Al/1C and 5Ti/5Al/20B nanoblends obtained upon MA in Ball mill are shown in Fig. 2. Tendency of amorphization and nanopowder formation is

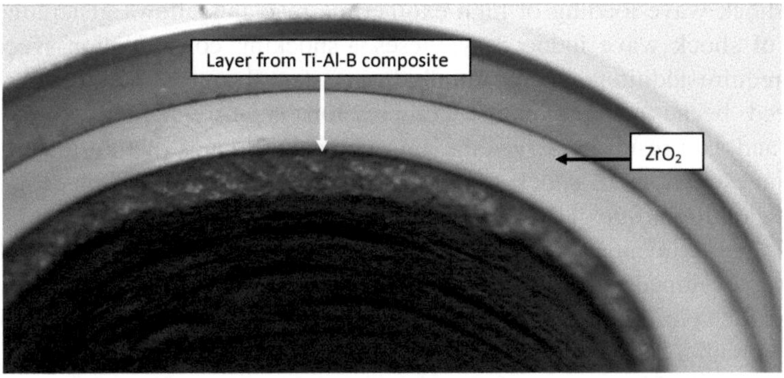

Fig. 3 Result of self-induced reaction in the ball mill jar. The wall of Zirconium oxide jar is a quite coated with the thick layer of synthesized multiphase Ti–Al–B intermetallides. Processing time is 36 h. The adhesion of the composite to the wall is strong

proportional to processing time and confirmed by structural investigations and particles size measurements.

The nanopowder preparation requires continuous control and reliable protection against spark and strong mechanical interaction. The powder becomes aggressive and super reactive and the risk of the combustion self-initiation is very high. The reaction can be self-induced during the long time milling. Therefore, the long time technological regimes should be realized under special media for prevention of non-controlled combustion/reaction. The result of self induced reaction during the ball milling is presented in Fig. 3.

The anti spark precautions in the powder mixture operations are necessary as well as on any further stages of bulk composites fabrication.

Explosive Compaction Technique

The next important stage of the research was selection of efficient/rational technology for fabrication of bulk nanocomposites. As an alternative to conventional compaction technologies, in this work it is proposed the shock wave compaction of nano powders for fabrication of bulk samples.

The motivation was derived from preliminary works showing that the explosive consolidation (EC) of metal-ceramic compositions is not only feasible but can produce materials of almost theoretical densities [12–18]. It was clear that the preliminary ball milling (due to fragmentation, mechanical alloying and critical reduction of particles sizes) should significantly increase the sintering ability of the blend and improve the compacting process and mechanical alloying of selected powder compositions. The major advantages of EC for bulk nanomaterials production are realization of high pressure, short processing time, and super high cooling rate (adiabatic cooling).

The shock wave loading of high exothermic reactants allows generating in situ process of shock wave induced syntheses + shocking consolidation. Technology doesn't require additional energy supply from external sources. High cooling rate is guaranteed because the dynamic compression is accompanied with adiabatic cooling and as a result preserves the amorphous and nanostructure materials.

The investigations of shock induced SHS were the subject of investigations of leading scientific groups [19–22].

The EC experiments were performed at the underground explosive chamber. The base has developed infrastructure (Chamber, Storage, Security service) and is equipped for investigation of explosive and blasting technologies. The chamber is connected with a cable system to observation station, which is equipped by sensors and high speed video camera for registration and observation of detonation and shock wave parameters. General view of experimental base and tunnel is presented on Fig. 4.

For shock wave generation (explosive compaction experiments) the industrial explosives were used in the experiments. The major energetic characteristics of explosives are presented in Table 1.

a b

Fig. 4 General views of experimental base: **a** tunnel to chamber; **b** inside explosive chamber

Table 1 Some energetic characteristics of explosive materials

Name of explosives	Energy of explosion, $E_{м}$ (kJ/kg)	Gravimetric density, ρ (g/cm^3)	Pressure on St. 3, $P \times 10^{-9}$ (N/m^2)	Gas volume of explosion, V (l/kg)	Speed of detonation, D (km/s)
NH_4NO_3	1439	1.0	1.5	980	1.8–2.0
ANFO (AN)	3815	0.8	4–5	980–990	2.8
79%NH_4NO_3 + 21% $C_6H_2(NO_2)_3CH_3$	4300	0.8–1.0	10	895	3.6–4.2
60%AN + 40%, ПП9/7	2920	0.98	3	948–976	5.2
$C_3H_6O_6N_6$	5439	1.1–1.82	20	950	8.6

Consolidation of the samples was performed in two stages. The powder blend was loaded in the carbon steel tube container and at the first stage the pre-densification of the mixtures was performed under static press loading (intensity of loading P = 500–1000 kg/cm^2). Cylindrical container/tube was closed from the both sides.

A card box was filled with the powdered explosive and placed around the cylindrical powder container. The experiments were performed at room temperature. The shock wave pressure (loading intensity) was varied in the range of 3–20 GPa. The explosive was detonated by electrical detonator. The explosive compaction schemes are shown in (Fig. 5).

Results of Explosive Compaction

The fabrication of bulk nanocomposites from nanopowders requires selection of the compaction technological parameters. The three main factors must be considered for optimization of shock wave compaction regime: (1) Selection of explosive, mass and geometry (developed pressure, detonation velocity, positive phase duration, impulse, configuration); (2) Selection and determination of powder container parameters (material, mechanical behavior, geometry, internal diameter, free volume, wall thickness, dimensions); (3) Powder related parameters (composition, charging density, particle sizes and their distribution). Number of investigations was dedicated by several researchers to investigation of shock

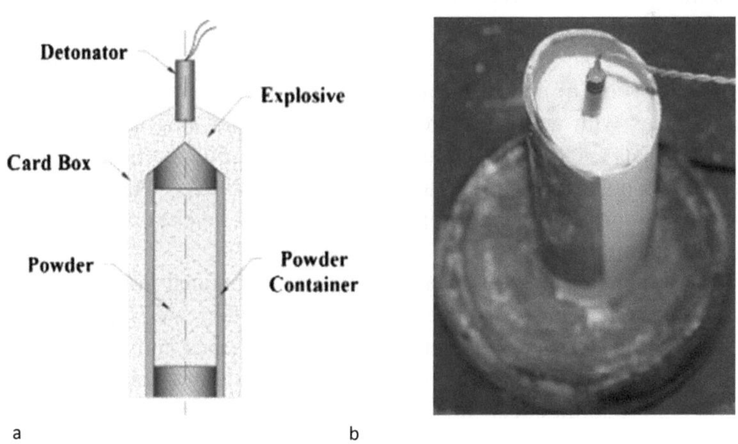

Fig. 5 Scheme of explosive compaction setups: **a** schematic view of assembly for fabrication of bulk rod; **b** general view of assembly

compaction process. Professor Pruemmer [12] performed detailed analysis of explosive compaction of powders. But, for explosive compaction of nanopowders (with dramatically changed free surfaces, reactivity and exothermal rate) the topic under discussion requires additional theoretical and experimental investigations for obtaining accurate data on technological parameters. Selection of the container material for particular cases needs the detailed investigations as well. For selection of container parameters in cylindrical axis symmetric experimental set up the criteria for selection of container (internal diameter d and wall thickness δ) can be expressed in the following form:

$$A < E_p + E_{con.} E_{destr.} = A_{destr.} \times M;\ M = \rho \times \pi h \left[d^2 - (d - 2\delta)^2\right]/4 = \rho \times \pi h \delta (d - \delta).$$

where A—full energy of explosive; E_p—energy consumed on plastic deformation of the container; $E_{con.}$—energy consumed on consolidation of the powder; $A_{destr.}$—energy for full destruction of material's unit mass; M—mass of container; ρ—density of material; h—length of cylindrical tube.

For preliminary selection of explosives and configuration around the sample, computer modeling was used. The calculated results were validated experimentally.

The optimal shock wave loading pressure is varied in the range of (7–10) GPa. In these conditions the configuration of loading/unloading waves in powder and container allows one to initiate the syntheses in the reaction mixture, to simultaneously consolidate it and to fix the phase composition under adiabatic cooling.

If shock wave pressure and developed energy exceed the strength limits of the container the resulting effects are destruction. The characteristic pictures obtained experimentally are presented on Fig. 6.

In particular cases/set ups, the tangential stresses developed on the border of metal/container (Fe/Cu) and blend/intermetallic provide the welding/joining of the metallic surfaces with nanostructured intermetallic and as a result the Functional Gradient Materials (FGM) are obtained. The welding zone represents a metal matrix reinforced by nanoparticles and synthesized intermetallides. As a result the distribution of the hardness and other properties from metal layer to nanocomposite layer are changed smoothly.

The photographs of the container and compacts obtained in optimal regimes, as well as samples cut from the long dimensional cylinders are shown in Fig. 7.

The bulk compacts were recovered in different shapes and prepared for investigations. The density of specimens was determined (cut from different part of samples) by the Archimedean method. The microstructure was studied by SEM.

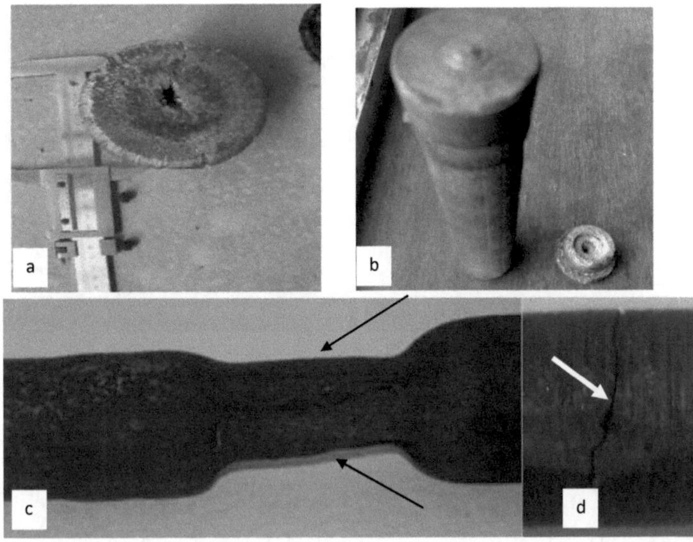

Fig. 6 Explosive loaded samples: **a** cross section of container, P = 20 GPa; **b**, **c** destructed container; **d** destructed sample (P = 20 GPa; A $\geq$ E_p + E_{com})

Fig. 7 Photograph of samples obtained in optimal technological regimes: bulk TiAlB rods before mechanical treatment

The SEM pictures of bulk samples are presented in Fig. 8. By investigations, it was established that the structure of bulk samples is not uniform. Structure is presented by nanosized and coarse grains.

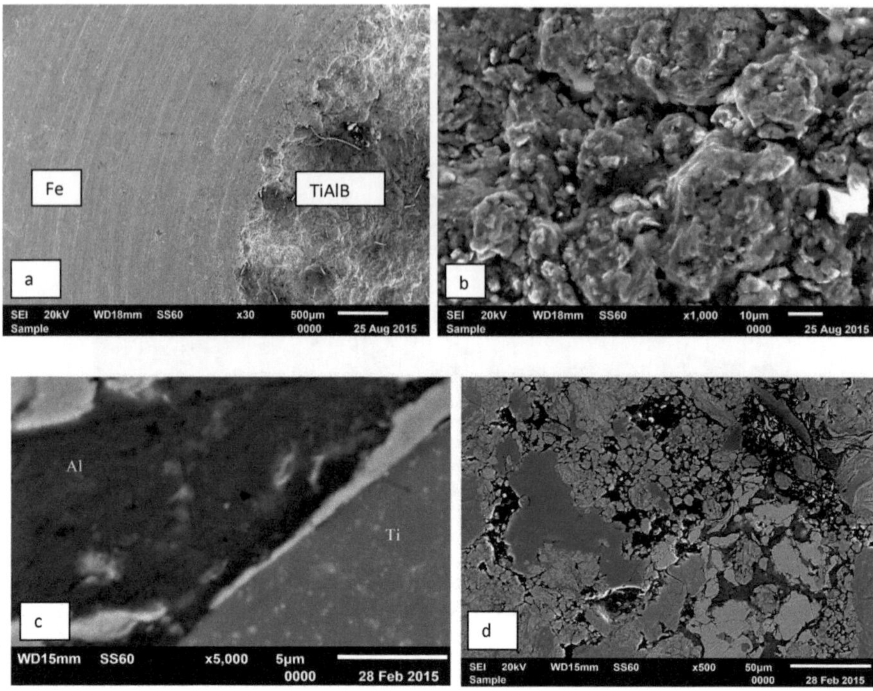

Fig. 8 SEM pictures of bulk composites obtained by shock wave consolidation P = 10 GPa, T = 20 °C; TiAlB (**a**, **b**); TiAlC (**c**, **d**)

Conclusions

- The effective technology/regimes for obtaining nanopowders and nanocomposites in Ti–Al–B–C composition has been elaborated;
- Rational technology for the fabrication of bulk amorphous and nanostructured materials by shock wave induced synthesis has been selected.

Acknowledgements The work is supported by the research grants of Shota Rustaveli National Science Foundation #YS15_2.2.10_84.

References

1. The Second World Space Congress, held October 10–19, 2002 in Houston, TX, USA, p. I-4-03IAF. Abstracts, 34th COSPAR Scientific Assembly.
2. Mania, R., Dabrowski, M., et al. (2003). Some application of TiAl micropowders produced by self-propagating high temperature syntheses. *International Journal of Self-Propagating High-Temperature Synthesis, 12*(3), 159–164.

3. Levashov, E. A., Senatulin, B. R., et al. (2002). Peculiarities of the functionally graded targets in combustion wave of the SHS-system with working layer Ti-Si-B, Ti-Si-C, Ti-B-N, Ti-Al-B, Ti-C, book of abstracts. In *IV International Symposium on SHS, Technion* (p. 35). Haifa, Israel, 17–21 February, 2002.
4. Merzhanov, A. G., Pityulin, A. N. (1995). Self-propagating high-temperature synthesis in production of functionally graded materials. In *Proceedings of 3rd International Symposium on FGM* (pp. 87–94). Lausanne, Switzerland.
5. Pityulin, A. N., Sytschev, A. E., Rogachev, A. S., Merzhanov, A. G. (1995). One-stage production of functionally graded materials of the metal-hard alloy type by SHS compaction. In *Proceedings of 3rd International Symposium on FGM* (pp. 101–108). Lausanne, Switzerland.
6. Tavadze. (2005). A new SHS method for special ferroalloys production. *First Armenian-Israel workshop on SHS (AIW-2005).* Yerevan (p. 33).
7. Das, N., Dey, G. K., et al. (2005). On amorphization and nanocomposite formation in Al-Ni-Ti system by mechanical alloying. *PRAMANA Journal of Physics, Indian Academy of Sciences, 65*(5), 831–840.
8. Zhang, Z. H., Han, B. Q. (2006). *Syntheses of nanocrystalline aluminum matrix composites reinforced with in situ devitrified Al-Ni-La amorphous particles.* University of California Postprints, Paper 39.
9. Hebeisen, J., Tylus, P., Zick, D., Mukhopadhyay, D. K., Brand, K., Suryanarayana, C., et al. (1996). Hot Isostatic pressing of nanostructured γ-TiAl powders. *Metals and Materials, 2*(2), 71–74.
10. Groza, J. R. (1993). Nonconventional pressure-assisted powder consolidation methods. *Journal of Materials Engineering and Performance, 2*(2), 283–290.
11. Suryanarayana, C., Klassen, T., Ivanov, E. (2011). Syntheses of nanocomposites and amorphous alloys by mechanical alloying. *Journal of Materials Science, 46*, 6301–6315.
12. Prummer, R. (1987). *Explosive working of porous materials*. Berlin, Heidelberg, New York: Springer.
13. Thadhani, N. N. (2005). Shock-induced chemical reactions in exotermic intermetallic-forming powder mixture systems. In *Proceeding of ICCES'05* (p. 394), December 1–10, 2005, India.
14. Kecskes, L., Peikrishvili, A., Chikhradze, N., Dgebuadze, A. (2002). Hot explosive fabrication of nano-crystalline W-based powders. In *Advances in Powder Metallurgy & Particulate Materials.* Orlando, USA.
15. Kecskes, L. J., Woodman, R. H., Chikhradze, N., Peikrishvili, A. (2004). Processing of aluminum nickelides by hot explosive consolidation. *International Journal of Self-Propagating High-Temperature Synthesis, 13*, #1.
16. Chikhradze, N., Staudhammer, K., Marquis, F., Chikhradze, M. (2005). Explosive compaction of Me-Boron containing composite powders. In *Proceeding of Powder Metallurgy World Congress & Exhibition, PM2005* (Vol. 3, pp. 163–173). Prague, Czech Republic.
17. Mamniashvili, G., Chikhradze, N., et al. (2006). Shock-wave compaction and investigation of Fe-Ni-Al powder composition. *Physica Mettallov I Metalovedenie.*
18. Chikhradze, N., Politis, C., Henein, H. (2010). Formation of ultrafine grained bulk Si and Si-Ge alloys by shock wave compaction technology. In *Proceeding of PM2010 World Congress—Nanotechnology* (Vol. 1, pp. 321–326).
19. Tavadze, G., Oniashvili, G. (1998). SHS technology—Resource save technology for obtaining materials. *Metsniereba da teqnika*, #6.
20. www.ism.ac.ru/handbook/31fgm.htm
21. Lu, L., Lai, M. O., & Wang, H. Y. (2000). Syntheses of titanium diboride TiB_2 and Ti-Al-B metal matrix composites. *Journal of Materials Science* (Springer, Netherlands), *35*, #1.
22. Oniashvili. Design and SHS of new functionally gradient materials (FGM). In *VII International Symposium on SHS*. Crakow.

Tensile and Impact Properties of Two Fiber Configurations for Curaua Reinforced Composites

Fábio de Oliveira Braga, Noan Tonini Simonassi, Augusto Corrêa Cabral, Sérgio Neves Monteiro and Foluke Salgado de Assis

Abstract Natural fibers have been extensively investigated in the past decades, due to their good properties, lightweight, low cost and renewable nature. From the *ananas erectifolius* plant, high strength and high modulus curaua fibers can be obtained. Their remarkable properties make them adequate to several high performance applications. In the present work, tensile and impact properties of two fiber configurations for curaua reinforced composites were investigated: a non-woven fiber fabric (NWFF) and high percentage continuous and aligned fiber (HPCAF) composites, using epoxy and polyester as polymeric matrix, respectively. The results showed that the fabric configuration does not effectively reinforce the polymer on tensile load, in spite of promoting significant improvement on the impact properties of the composite. The latter configuration results in both high strength and tough composites, however, its ability to resist to impacts depends on the direction of load.

Keywords Natural fibers · Curaua fibers · Composites · Mechanical properties

Introduction

Natural fiber composites have been extensively studied, especially because of their environmental advantages and low cost [1–4]. Among the several existing natural fibers, increasing attention is being given to the curaua fibers, due to its remarkable properties, as high strength and high modulus [4–8]. These fibers are extracted from the leaves of the *Ananas erectifolius*, native plant from the Amazon region. Values as high as 3000 MPa for tensile strength and 80 GPa for Young's modulus were reported [4].

F. de Oliveira Braga (✉) · N.T. Simonassi · A.C. Cabral
S.N. Monteiro · F.S. de Assis
Militar Institute of Engineering—IME, Rio de Janeiro, Brazil
e-mail: fabio_obraga@yahoo.com.br

© The Minerals, Metals & Materials Society 2017

M.A. Meyers et al. (eds.), *Proceedings of the 3rd Pan American Materials Congress*,
The Minerals, Metals & Materials Series, DOI 10.1007/978-3-319-52132-9_43

Curaua fiber reinforced composites are especially promising in the automotive industry, because they could replace glass fiber composites for automobile parts, since they can achieve satisfying properties, with less disposal problems [5–8]. Although curaua fiber composites can reach tensile strength of 160 MPa (untreated fibers) [6], Gomes et al. [5], highlighted that curaua composites with tensile strength around 130 MPa cannot be successfully applied to replace the glass fibers for these applications, whose composites can achieve 288 MPa [6]. To increase the safety factor for curaua composites, different fiber configurations and treatments can be tested [5, 6]. In the present work, two fiber configurations were tested: a non-woven fiber fabric (NWFF) and high percentage continuous and aligned fibers (HPCAP). Both tensile and impact properties were chosen as a measure of the composites performance. Thus, the objective of this work is to study the influence of fiber configuration (NWFF and HPCAP) in the mechanical performance of the composites.

Materials and Methods

The curaua fibers were acquired from the Brazilian company Pematec Triangel, as both bundles (Fig. 1) and non-woven fiber fabrics (NWFF, Fig. 2a). The areal density of the fabric was approximately 0.830 kg/m^2. An electronic image (Fig. 2b) shows that it is composed of randomly oriented fibers.

The fibers were manually cleaned, cut and dried at 60 °C for 24 h for the production of composites. Continuous and approximately aligned fiber composites were prepared by compression molding, at room temperature (25 °C), in the volumetric fractions of 30, 40, 50, 60, 70 and 80%. The fibers were positioned carefully in the mold, the resin with hardener added, and the mixture was kept under the pressure of 30 MPa, for 24 h. The present fractions of fibers are considered relatively high for natural fiber composites, especially 60–80%, and thus these composites were designated as "high percentage continuous and aligned fiber" (HPCAF) composites in the present work. For the NWFF composites, the

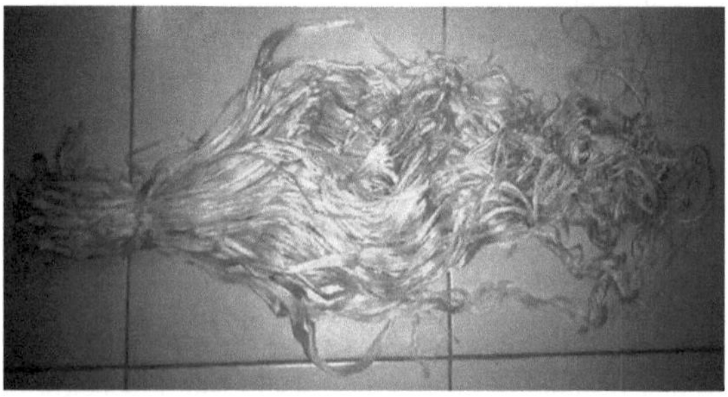

Fig. 1 Curaua fiber bundle

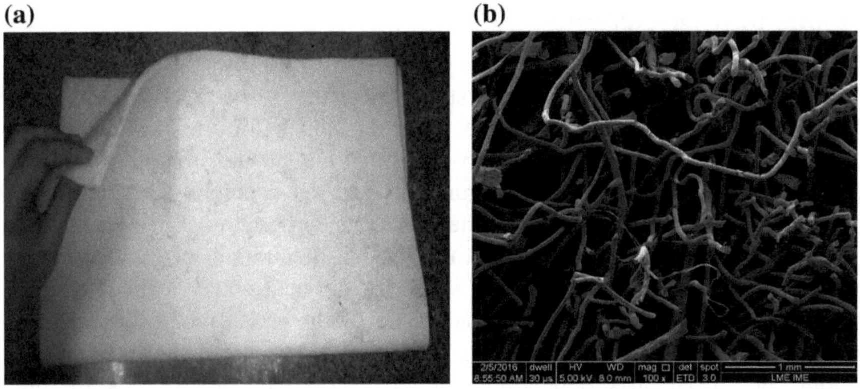

Fig. 2 Curaua non-woven fiber fabric (NWFF): **a** general macroscopic aspect and **b** electronic micrograph of the fabric, 100×

polymeric matrix was an epoxy diglycidyl ether of bisphenol-A (DGEBA) resin, hardened with triethylene tetramine (TETA, 13 phr), both acquired from Brazilian company Resinpoxy. For the HPCAF composites, the polymeric matrix was an orthophthalic polyester resin, hardened with methyl ethyl ketone (0.5%), acquired from the same company.

The same fabrication procedure was performed for the NWFF composite production, although no providence was taken about the alignment of fibers, and just the 30 vol.% composites were produced. An epoxy composite reinforced 30 vol% of continuous and aligned curaua fibers was produced for comparison. The latter composites were designated ECAF (epoxy reinforced with continuous and aligned fibers), which differs from 30 vol.% HPCAF composites by the polymeric matrix. The tensile and impact properties of the composites were evaluated.

The tensile tests were performed until fracture in a model 3565 Instron machine, available in LNDC/UFRJ/Brazil, using a 30 kN load cell. The samples were produced following the dimensions specified by the ASTM D638 standard. Four specimens of each were tested (NWFF, ECAF and HPCAF).

The Izod impact tests were performed in a model CH/IZ-25 Pantec machine, using a 22 J hammer, at room temperature (20 ± 2 °C). The rectangular specimens (65 × 12 × 12 mm) were produced following the dimensions specified by the ASTM D256 standard. For the NWFF composites, the notches were made either parallel or normal to the curaua fabric plane (designation P and N, respectively). For the ECAF composite, the notches were made either parallel (P) or normal (N) to the fiber axis. Seven specimens of each were tested.

The fracture aspects were observed and registered using a model JSM-5800LV Jeol scanning electron microscope (SEM), with secondary electrons contrast.

Results and Discussions

Table 1 shows the results of tensile strength for the 30 vol% NWFF and ECAF composites, in comparison with the reported values for the neat epoxy resin (DGEBA/TETA) [9, 10]. It can be observed that the tensile strength of the curaua continuous and aligned fiber configuration (ECAF) is slightly higher than the strength of the epoxy resin [10]. The NWFF incorporation, however, does not provide a satisfactory reinforcement, and the tensile strength of the NWFF composite is significantly lower than the value for the neat epoxy resin. This can be attributed to the non-alignment characteristic of the fibers in the fabric, in conjunction with the low strength fiber-matrix interface. Figure 3 shows intense fiber pullout and fiber-matrix debonding, which clearly indicate the lack of affinity between the hygroscopic curaua fibers and the hydrophobic epoxy matrix, and thus the latter cannot properly transfer the loads to the former, decreasing the tensile strength of the material [8, 11]. Despite that, when fibers are aligned (as in ECAF, Fig. 4), the compressive loads perpendicular to the load axis might help the matrix to transfer part of the stress to the fibers, slightly increasing the tensile strength.

Table 1 Tensile strength of the non-woven fiber fabric composites (NWFF) in comparison with 30% ECAF composite and neat epoxy resin (DGEBA/TETA)

	NWFF	ECAF	DGEBA/TETA
σ (MPa)	26 ± 1	60 ± 3	50 ± 9 [9]; 53 ± 2 [10]

Fig. 3 SEM micrograph showing the surface of the NWFF composite subjected to tensile test (100×)

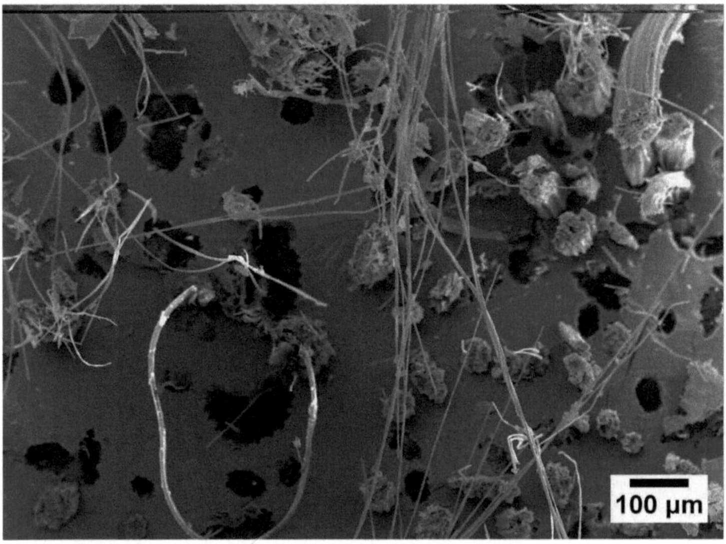

Fig. 4 SEM micrograph showing the surface of the ECAF composite subjected to tensile test (100×)

Table 2 Izod impact energy of the non-woven fiber fabric composites (NWFF) in comparison with 30% ECAF and neat polyester resin (DGEBA/TETA)

Material					
Property	NWFF/N	NWFF/P	ECAF/N	ECAF/P	DGEBA/TETA
E (J/m)	433 ± 36	400 ± 35	410 ± 55	24 ± 1	39 ± 9 [11]
E (10^3 J/m^2)	38 ± 3	35 ± 1	35 ± 5	2.2 ± 0.1	0.18 [10]

Table 2 shows the Izod impact energies of the NWFF and ECAF composites, for the N and P configurations, in comparison with the neat epoxy resin. From the results, it can be seen that the Izod impact energies for the curaua fiber composites are significantly higher (~10 times) than the values reported for the epoxy resin [10, 11]. The higher toughness can be attributed to the low interface strength between the matrix and the fibers that, even though decreases the tensile strength, results in greater fracture surface, and, consequently, better impact properties are obtained [8]. Figures 5 and 6 shows intense fiber pullout indicating the weak fiber-matrix interface for the NWFF and ECAF/N composites, respectively. Despite that, for the ECAF/P composites (if the sharp notch is parallel to the fiber axis), the contrary happens, and the fibers weakens the structure subjected to impact, since it is very easy to separate fibers and the matrix (low interface strength), as can be seen by the fiber-matrix debonding in Fig. 7.

Figure 8 shows the tensile strength of several HPCAF polyester composites, from 30% fibers up to 80%. A growth tendency of the tensile strength can be observed until the fraction 60%, reaching a maximum near 202.7 MPa, and then

Fig. 5 SEM micrographs showing the surface of the NWFF composite subjected to Izod impact test (200×)

Fig. 6 SEM micrographs showing the surface of the ECAF/N composite subjected to Izod impact test (100×)

Fig. 7 SEM micrographs showing the surface of the ECAF/P composite subjected to Izod impact test (500×)

Fig. 8 Tensile strength of the HPCAF composites as a function of the fiber fraction

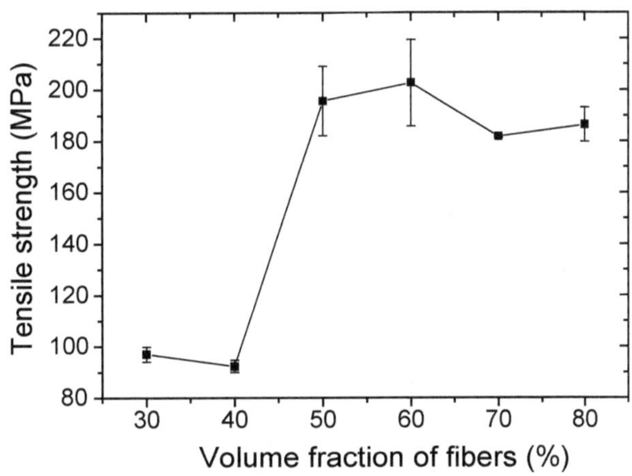

decreasing slightly for higher fractions of fibers. The value for 30 vol% composite (97.0 MPa) was smaller than the previously reported result of 160 MPa [6]. The difference can be attributed to the composite fabrication: While in the present work the composites were compressed at room temperature, Oliveira and co-workers [6] performed the same process at 95 °C, and this could lead to a higher degree of polymerization of the matrix. Increasing the volumetric fraction of curaua fibers, however, the results overcome 160 MPa reaching the value of 202.7 MPa, that is much higher than the 130 MPa that Gomes and co-workers [5] mentioned as a limit value for automobile parts application.

Conclusions

The incorporation of curaua non-woven fiber fabrics as reinforcement for epoxy polymer is not satisfactory when tensile loads are applied, decreasing the strength of the material. However, the impact toughness of these composites is nearly 10 times higher than the values reported for the neat epoxy resin (DGEBA/TETA). Also, these fabrics reinforce the epoxy polymer for impact loads both in parallel and normal-to-fabric plane directions of the composite, differing from the continuous and aligned fibers, whose composites have an impact weak direction (parallel to fiber axis). Besides that, the incorporation of high percentage continuous and aligned curaua fibers, such as 60–80 vol.%, reinforcing polyester, results in high strength composites, with tensile strength reaching 202.7 MPa (60 vol.% composites), making those promising for automobile applications.

Acknowledgements The authors thank the Brazilian agencies CAPES and CNPq for the financial support, and LNDC/UFRJ for performing the tensile tests.

References

1. Faruk, O., Bledzki, A. K., Fink, H., & Sain, M. (2014). Progress report on natural fiber reinforced composites. *Macromolecular Materials and Engineering, 299*, 9–26.
2. Faruk, O., Bledzki, A. K., Fink, H., & Sain, M. (2012). Biocomposites reinforced with natural fibers: 2000-2010. *Progress in Polymer Science, 37*, 1552–1596.
3. Satyanarayana, K. G., Arizaga, G. G. C., & Wypych, F. (2009). Biodegradable composites based on lignocellulosic fibers—An overview. *Progress in Polymer Science, 34*, 982–1021.
4. Monteiro, S. N., Lopes, F. P. D., Barbosa, A. P., Bevitori, A. B., Silva, I. L. A., & Costa, L. L. (2011). Natural lignocellulosic fibers as engineering materials—An overview. *Metallurgical and Materials Transactions A: Physical Metallurgy and Materials Science, 42A*, 2011–2963.
5. Gomes, A., Matsuo, T., Goda, K., & Ohgi, J. (2007). Development and effect of alkali treatment on tensile properties of curaua fiber green composites. *Composites: Part A, 38*, 1811–1820.
6. Oliveira, F. H., Helfer, A. L., & Amico, S. C. (2012). Mechanical behavior of unidirecional curaua fiber and glass fiber composites. *Macromolecular Symposium, 319*, 83–92.
7. Zah, R., Hischier, R., Leão, A. L., & Braun, I. (2007). Curauá fibers in the automobile industry—A sustainability assessment. *Journal of Cleaner Production, 15*, 1032–1040.
8. Monteiro, S. N., Lopes, F. P. D., Nascimento, D. C., Ferreira, A. S., & Satyanarayana, K. G. (2013). Processing and properties of continuous and aligned curaua fibers incorporated polyester composites. *The Journal of Materials Research and Technology, 2*(1), 2–9.
9. D'Almeida, J. R. M., Menezes, G. W., & Monteiro, S. N. (2003). Ageing of the DGEBA/TETA epoxy system with off-stoichiometric compositions. *Journal of Materials Research, 6*(3), 415–420.
10. Garcia, F. G., Soares, B. G., Pita, V. J. R. R., Sánchez, R., & Rieumont, J. (2007). Mechanical properties of epoxy networks based on DGEBA and aliphatic amines. *Journal of Applied Polymer Science, 106*, 2047–2055.
11. Amaral, C. R., Rodriguez, R. J., Garcia, F. G., Junior, L. P. B., & Carvalho, E. A. Impact of aliphatic amine comonomers on DGEBA epoxy properties. *Polymer Engineering & Science, 54*(9), 2132–2138.

Thermo-mechanical Behavior of Nanostructure Polyacrylic Polymer Based on Al_2O_3 and Bentonite Nanoparticles

Rubén Castillo-Pérez, Mireya Lizbeth Hernández-Vargas, Oscar Hernández-Guerrero, Bernardo Fabián Campillo-Illanes and Osvaldo Flores-Cedillo

Abstract Polymer-nanocomposites are being used for numerous applications, ranging from car bumpers to advanced optoelectronic devices. Understanding the impact of nano-fillers on the composite mechanical properties is critical to the success of all of these applications. This work focuses on the influence of inorganic spheres-particles such as Al_2O_3 and platelets, such as Bentonite clay, on the thermo-mechanical properties of the acrylic polymer. The nanocomposites were characterized by Fourier transform infrared spectroscopy (FTIR), thermogravimetric analysis, uniaxial tension, nanoindentation and scanning electron microscopy (SEM). Films drawn from the latex exhibited excellent optical transparency and no evidence of aggregation was detected by SEM and X-ray energy dispersive spectroscopy (EDS). The analysis of the thermal and mechanical properties in the presence of nanoparticles platelets (Bentonite), showed an increase in the glass transition temperature ($T_{g)}$, decomposition temperatures ($T_{dec)}$, the Young's

R. Castillo-Pérez (✉) · M.L. Hernández-Vargas (✉) · B.F. Campillo-Illanes
Facultad de Química, Universidad Nacional Autónoma de México,
Ciudad Universitaria, 04519 Ciudad de México, Mexico
e-mail: kstper@hotmail.com
e-mail:

R. Castillo-Pérez · M.L. Hernández-Vargas
Posgrado de Ingeniería, Facultad de Química,
Universidad Nacional Autónoma de México, Ciudad Universitaria, 04519
Ciudad de México, Mexico

O. Hernández-Guerrero
Facultad de Ciencias Químicas e Ingeniería, Centro de Investigación
En Ingeniería Y Ciencias Aplicadas, Universidad Autónoma del
Estado de Morelos, Av. Universidad 1001, 62209 Cuernavaca, MOR, Mexico

B.F. Campillo-Illanes · O. Flores-Cedillo
Instituto de Ciencias Físicas, Universidad Nacional Autónoma de México,
Av. Universidad s/n, Col. Chamilpa, 62210 Cuernavaca, MOR, Mexico

© The Minerals, Metals & Materials Society 2017
M.A. Meyers et al. (eds.), *Proceedings of the 3rd Pan American Materials Congress*,
The Minerals, Metals & Materials Series, DOI 10.1007/978-3-319-52132-9_44

modulus and their hardness. On the other hand, in the presence of spherical nanoparticles (Al_2O_3), the thermal (T_g and T_{dec}) and mechanical properties decrease, due to the morphology, dimension (e.g., aspect ratio) and degree of dispersion of the nanofillers in the polymer matrix.

Keywords Polymer · Nanocomposite · Layered silicate · Mechanical properties · Al_2O_3

Introduction

Polymer composites have been proposed or are being used for numerous applications, ranging from car bumpers to advance optoelectronic devices. Different types of nanofillers are currently utilized, e.g. carbon nanotubes, electrospun polymeric nanofibers, graphemes, and hybrid organic–inorganic nanoparticles such as polyhedral silsesquioxane (*POSS*), and others. In particular, the spherical Al_2O_3 nanoparticle, are attractive nano-reinforcers of polymer due to enhancement of physical properties. It has been reported that Al_2O_3 nanoparticles increased Young's modulus, retardant flame, among others. On the other hand, laminar nanoparticles and bentonite nanoclays, are attractive nano-reinforcers of polymer due to enhancement of physical properties. It has been reported that nanoclays nanoparticles increased Young's modulus, thermal stability, barrier properties against vapor and water permeation and morphology. The properties of these nanocomposites materials depend of the polymer matrix synergized with the surfaces chemistry, the sizes, degree of dispersion, and the shape of nanoscale filler to define thermo-mechanical properties [1]. The degree of dispersion of the filler in a polymer matrix has a significant influence on the thermo-mechanical properties; however the filler have a strong tendency to agglomerate and makes it difficult the dispersion. There are three common methods to produce polymer nanocomposites: melt compounding being favored by industry as no solvent is required [2, 3], in situ polymerization [4–6], and solvent dissolution of polymer matrix and mixing with the nanoparticles. The latter being the least favored as solvent removal and disposal is required, adding cost to the process.

Alumina oxide is a spherical nanoparticle, was reported that this nanofiller improved mechanical property, retardant flame in polymer matrix [7]. The bentonite nanoclays is a laminar particle, it is being reported that this nanoparticle improved the mechanical properties, thermal stability, barrier properties against vapor and water permeation and morphology. The clays nanocomposites exhibit a three form of composite: exfoliation, intercalation and aggregates composites [8]. In the present work, we synthesized polyacrylic nanocomposites, using nanoparticles as seeds, via emulsion polymerization in batch mode, i.e., in situ polymerization. The thermal and mechanical properties of the nanocomposites are discussed.

Experiment

Materials

Bentonite nanoclay [$Al_2O_3 \cdot SiO_2 \cdot H_2O$, CAS Number: 1302-78-9] and the monomers butyl acrylate, methyl methacrylate, and acrylic acid were purchased from the Aldrich Chemical Co. The monomers were purified using an inhibitor remover purchased from Aldrich prior to the polymerization experiments. Al_2O_3 nanoparticles were purchased from Fermont, Productos Químicos Monterrey S.A de C.V. (Monterrey, N.L. Mexico). The nanoparticles were used as-received.

Chemical Synthesis

Al_2O_3 and bentonite nanocomposites containing 1 wt% of Al_2O_3 and bentonite nanoparticles (based on the total monomer weight) were synthesized by seeded emulsion polymerization in batch. The emulsion polymerization was carried out under reflux conditions at 77 ± 3 °C for 3.5 h using a radical initiator and tensioactive agent. Al_2O_3 or bentonite were dispersed in distilled water. A mixture of butyl acrylate, methyl methacrylate, acrylic acid, tensioactive agent and radical initiator were added to the flask and Al_2O_3 or bentonite was added before initiating the polymerization. The resultant Al_2O_3-nanocomposites or bentonite-nanocomposites latex was filtered to remove un-reacted material. Finally, the reactor was cooled down to room temperature and the pH was set at 8.5–9.5 by ammonia solution. In both cases the metastable nanostructured emulsions with dispersed Al_2O_3 and bentonite nanoparticles thus obtained were stable for at least six months. Films were cast from the emulsions on Teflon™ Petri dishes first drying at room temperature for 3 days and then in vacuum at 90 °C for 4 h.

Thermal Analysis

Thermal decomposition temperatures were determined by high resolution modulated thermogravimetric analysis (MTGA), under nitrogen atmosphere, using a Q500 TGA manufactured by TA Instruments (New Castle DE, USA). The thermal program consisted of: heating rate of 2 °C/min, period of 200 s and temperature modulation of ±2 °C.

Uniaxial Tension

Elastic mechanical modulus was determined by uniaxial tensile deformation using the mini-tensile tester TST-350 manufactured by Linkam Scientific Instruments (Tadworth, Surrey, UK). Tensile deformation was carried out at 5 mm/min, room temperature. Tensile tests were repeated five times, and tests were carried out using a fresh sample each time.

Nanoindentation

The nanoindentation tests were performed with a TI 750 Ubi L indentor system. A Berkovich diamond indenter was used, using standard load. The maximum displacement control was 600 nm, with a load and unload function of 25 s. The grid/test number for each sample was of 10 × 10, 100 in total per sample.

Scanning Electron Microscopy

Elemental analysis and identification of aggregates of the acrylate nanocomposite was further investigated by scanning electron microscopy (SEM). For SEM the JSM 5900 LV manufactured by JEOL Ltd. (Japan) operating at 20 kV equipped with an Oxford Link Isis-energy dispersive X-ray spectrometer (EDS) was utilized.

Results and Discussion

Covalent binding of polyacrylic monomers chains and the incorporation of layered silicate (Bentonite nanoclay) and Al_2O_3 nanoparticles via emulsion polymerization were verified by FTIR spectroscopies using an attenuated total reflectance accessory (ATR-FTIR), spectra are shown in Fig. 1. Polymerization chemical bond from reaction of double bond C=C of acrylate group from the monomers with the initiator molecule was observed at 1640 cm^{-1}. The attenuation of vibration band for the group −C=C− confirmed formation of covalent binding of acrylic chains. Other characteristic vibration bands corresponding to the structure of acrylic monomers ($C–H_2$, $C–H_3$, C=O, C–O–C) and chemical structure of Bentonite nanoparticles (Si–O–Si) were clearly identified.

Moreover the films as-cast were examined using optical microscopy in order to investigate whether the materials remained optically transparent, as expected in well-dispersed nanoparticles or showed evidence of nanoparticle agglomeration. Figure 2 shows the optical micrographs of neat acrylate and the nanocomposites.

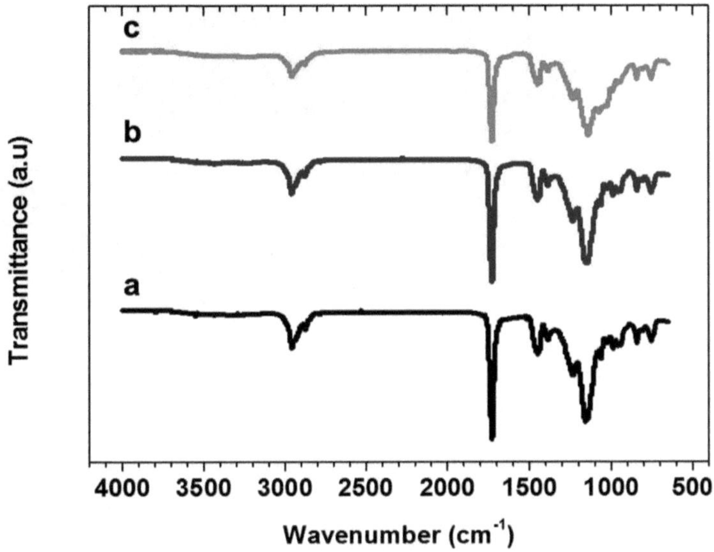

Fig. 1 FTIR spectra of *a* Neat polyacrylic, *b* Ac–Al_2O_3, *c* Ac–Bentonite

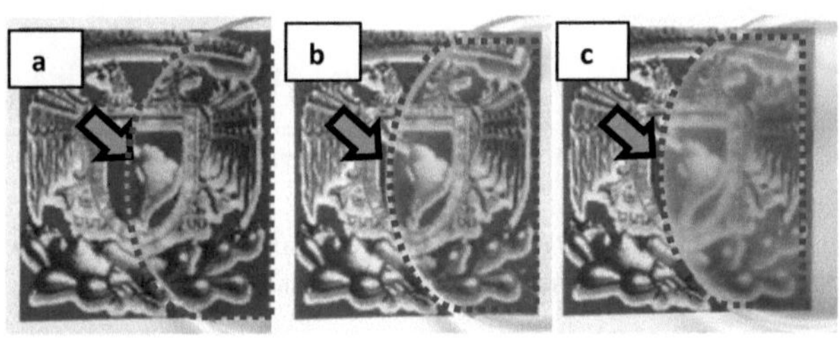

Fig. 2 Pictures obtained of the polyacrylic nanocomposites. **a** Neat acrylic, **b** Ac–Al_2O_3, **c** Ac–Bentonite. White light conditions

The neat acrylate is an optically transparent polymer, as shown in Fig. 2a. Incorporating Al_2O_3 nanoparticle, Fig. 2b, the coatings kept their optical transparency. On the other hand, the nanostructured coatings with bentonite, is slightly opaque, as show in Fig. 2c.

Elemental mapping further confirmed the dispersion of Al_2O_3 and bentonite in the polyacrylic matrix. Figure 3 shows a SEM micrograph and the corresponding elemental maps. As expected, it can be seen that elements such as C and O predominate throughout the sample due to the organic nature of the matrix. Moreover, EDS analysis confirmed the presence of Al and Si elements in the nanocomposites, respectively (see inset in Fig. 3).

The glass transition temperature T_g values were obtained by small-strain oscillatory shear using the stress-controlled CVO rheometer. Dynamic temperature ramps enabled to identify when the material suffered a transition from glass-like to

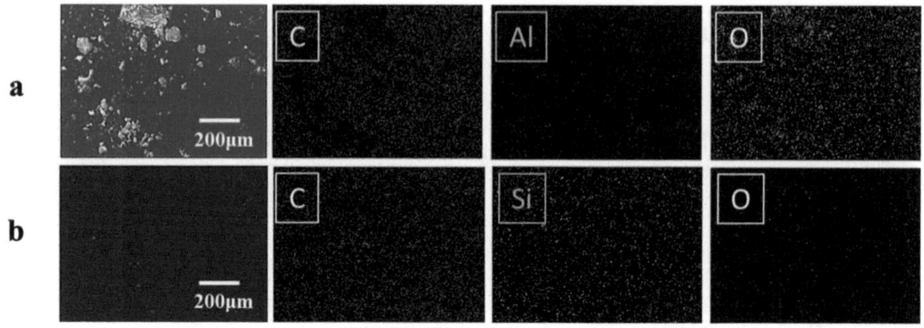

Fig. 3 SEM micrograph and the element mapping of acrylic surface of **a** Al_2O_3 nanocomposite and **b** bentonite nanocomposite, containing 1 wt% nanoparticles

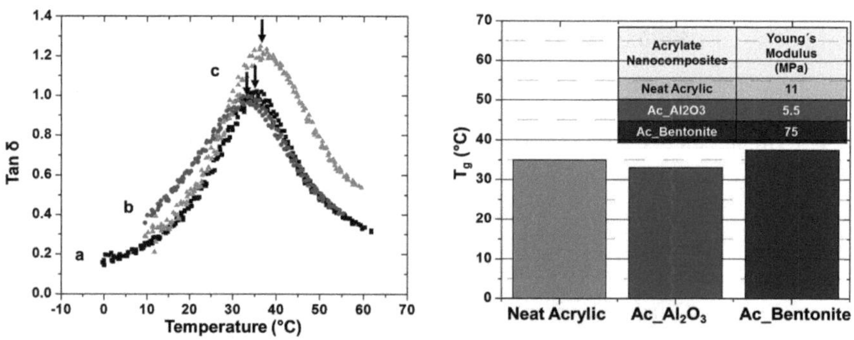

Fig. 4 Glass transition temperature *Tg* (temperature at maximum loss angle δ) as a function of nanoparticles type, determined by Dynamic Mechanical Analysis (DMA)

rubber-like. During this transition there is a temperature where there is maximum dissipation of energy and the mechanical damping, Tan δ, goes through a maximum value. This temperature is associated with the material's T_g, and the results are shown in Fig. 4. The results showed that T_g decreased when adding spherical nanoparticles and increased when adding nanoplatelets.

The mechanical properties were determined by uniaxial tensile tests, and elastic Young's modulus was determined from stress-strain curves, the results are shown in Fig. 5. The neat polyacrylic showed in Fig. 5a an elongation about 150%, typical of these materials. On the other hand, polyacrylic/with Al_2O_3 nanospheres particles as it can be seen in Fig. 5b, were it shows a drop in Young's modulus. However, in the case of bentonite (nanoplatelets), the coating exhibited a significant increase of the Young's modulus of about 600%, see Fig. 5c. However, it is important to note, that the mechanical modulus is increased at the expense of the reduction of strain at fracture, and a reduction of the impact strength.

The morphology of these films was investigated via SEM and Fig. 6a–c shows micrographs of fractured surface of (a) the neat polyacrylate and hybrid

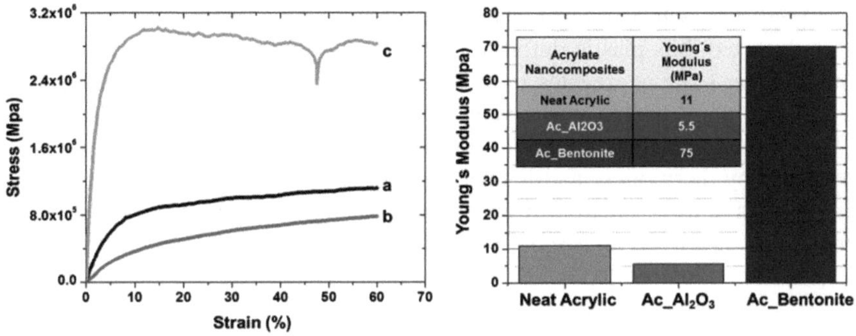

Fig. 5 Stress-strain curves of as-cast nanocomposites. *a* Neat polyacrylic; nanocomposites containing the following nanoparticles: *b* Al_2O_3 and *c* bentonite. Uniaxial tensile tests are straining at 5 mm/min and room temperature

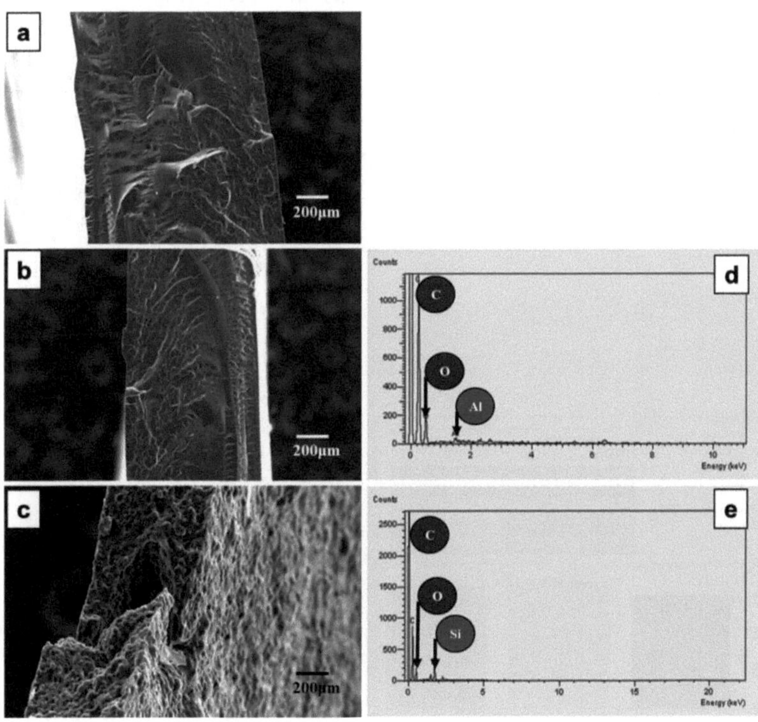

Fig. 6 SEM micrographs of acrylate hybrid nanocomposites fractured in liquid nitrogen, containing **a** neat polyacrylic [0], **b** Al_2O_3 at 1 wt% and **c** bentonite at 1 wt%. EDS plot in inset shows the presence of **d** Al and **e** Si in the nanocomposites

nanocomposites with (b) Al_2O_3 and (c) bentonite nanoparticles. As expected, the morphology exhibits plastic deformation for all materials and confirm the absence of micrometer-scale aggregates in all of them. Moreover, EDS analysis confirmed the presence of Al and Si in the nanocomposites (see inset in Fig. 6b, c).

The thermal decomposition temperatures of acrylate nanocomposites films were determined by thermogravimetric analysis, TGA. Figure 7a, b shows the graphs of mass derivative as a function of temperature and the thermal degradation temperature, T_{dec}, respectively. It can be seen that the neat acrylate is thermally stable with negligible mass loss up to ca. 280 °C, and then by about 365 °C there is the onset of thermal decomposition.

Figure 7b showed the nanocomposites that exhibited significantly higher thermal stability, at only 1 wt% nanoparticles the decomposition temperature T_{dec} has increased up to 347 and 333 °C, respectively; some 36 and 22 °C relative to the neat acrylate.

Surface nanohardness is also related to molecular chain mobility. Therefore, measuring this property may indicate whether the incorporation of the nanofiller (Al_2O_3 and Bentonite) contributed appreciably to a decrease in molecular mobility.

The surface nanohardness was evaluated by nanoindentation tests. The nanoindentation test results are shown in Fig. 8. The nanohardness in presence of Al_2O_3 nanoparticles, decrease, due to the free volume increase that facilitates the chains mobility. On the other hand, this property increases as bentonite is incorporated, due to restriction of chains mobility, by the structure of the nanocomposite synthetized (exfoliated, intercalated).

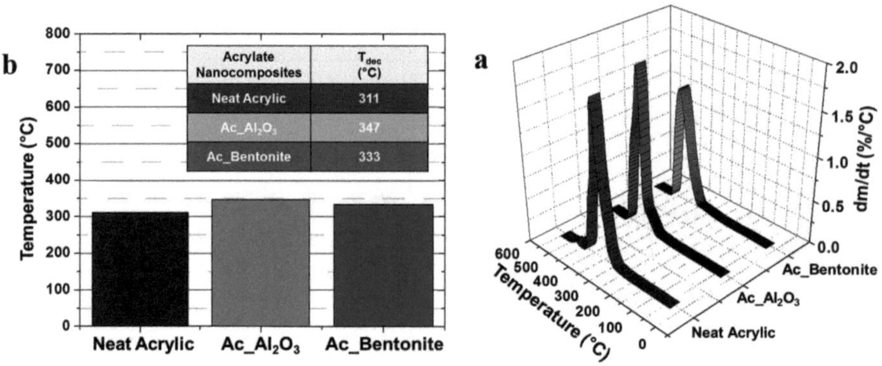

Fig. 7 **a** Graph of temperature derivatives of mass loss and **b** thermal degradation temperature, T_{dec}

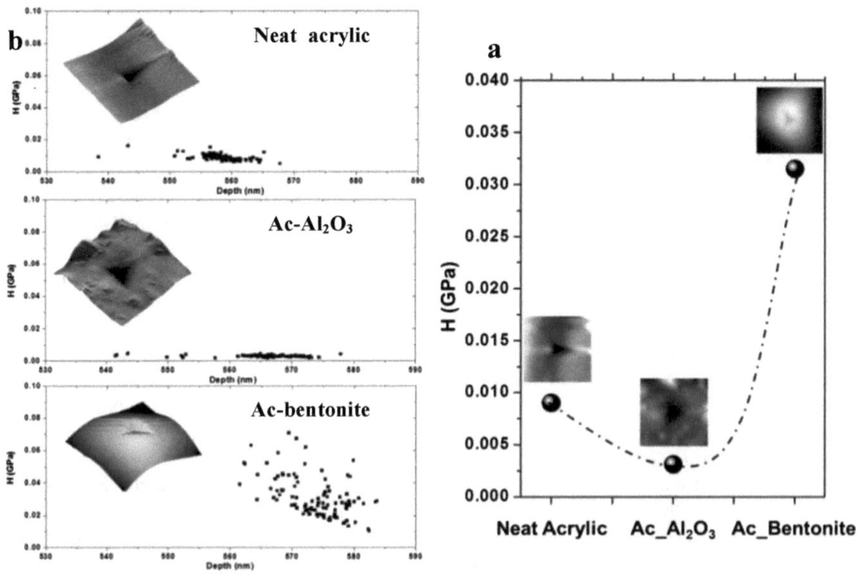

Fig. 8 **a** Effect of nanofiller incorporation on hardness of nanocomposites evaluated up to 100 points. **b** Average harness values obtained

Conclusion

The nanocomposites were synthesized by seeded batch emulsion polymerization using untreated (non-surface modified) Al_2O_3 and bentonite nanoclay up to 1 wt%. It is shown that the geometry and size of the nanoparticles incorporated, impact the properties of the polyacrylic matrix. Al_2O_3 and Bentonite films were optically transparent. When the morphology is spherical (Al_2O_3) the Young's modulus, nanohardness and thermal properties decreased. However, there was a significant increase of these properties in the presence of layered silicates (Bentonite). Mechanical Young's modulus increased nearly 70-fold, and also hardness and glass transition temperature increases.

Acknowledgements R. Castillo-Pérez, M.L. Hernández-Vargas and O. Hernández-Guerrero were supported by graduate scholarships from the Mexican Council for Science and Technology (CONACyT). Thanks to Institute of Physical Sciences at UNAM (Campus Morelos) and engineering postgraduate in the Faculty of Chemistry at UNAM. Also thanks to the DGAPA Papiit project number IN115616, to Mr. Ivan Flores for the nanoindentation facility, I.Q.M. Itzel Reyes Chaparro and Q. Ivan Puente Lee (Faculty of chemistry, UNAM) for the they help in SEM.

References

1. Crosby, A. J., & Lee, J. Y. (2007). Polymer nanocomposites: The "Nano" effect on mechanical properties. *Polymer Reviews, 47*, 217–229. doi:10.1080/15583720701271278
2. Romero-Guzmán, M. E., Romo-Uribe, A., Ovalle-García, E., Olayo, R., & Cruz-Ramos, C. (2008). Microstructure and dynamic mechanical analysis of extruded layered silicate PVC nanocomposites. *Polymer Advance Technology, 19*, 1168–1176.
3. Romero-Guzmán, M. E., Floresa, O., Flores, A., Romo-Uribe, A., Alvarado-Tenorio, B., & Campillo, B. (2011). Cold-drawn induced microstructure in PVC-bentonite nanocomposites. *Polymer Advance Technology, 22*, 836–846.
4. Chorng-Shyan, C. (2008). *Principles and applications of emulsion polymerization*. Hoboken, New Jersey, USA: Wiley.
5. Provder, T., & Baghdachi, J. (Eds.). (2007). Smart coatings. In *ACS Symposium Series 957*. Washington, DC: American Chemical Society.
6. Hernández-Vargas, M. L., Valerio-Cárdenas, C., Romo-Uribe, A. (2011). Nanocomposite coatings incorporating nanosilica particles into polyacrylics. In *Polymer Chemistry Proceedings 2011*. American Chemical Society.
7. Ash, B. J., Rogers, D. F., & Wiegand, C. J. (2002). Mechanical properties of AI2O3/polymethylmethacrylate nanocomposites. *Polymer Composites, 23*(6), 1014–1025.
8. Tong, X., Zhao, H., Tang, T., Feng, Z., & Huang, B. (2002). Preparation and characterization of poly(ethyl acrylate)/bentonite nanocomposites by in situ polymerization. *Journal of Polymer Science Part A: Polymer Chemistry, 40*, 1706–1711.

Thermo-mechanical Properties of Waterborne Acrylate Hybrid Nanocomposites

Mireya Lizbeth Hernández-Vargas, Rubén Castillo-Perez, Oscar Hernández-Guerrero, Osvaldo Flores-Cedillo and Bernardo Fabián Campillo-Illanes

Abstract Waterborne acrylate hybrid nanocomposites containing 1 wt% nanoparticles were synthesized by in situ polymerization using SiO_2 and Fe_2O_3 nanoparticles. The synthesized latex hybrids were characterized by thermogravimetric analysis, uniaxial tension, nano-indentation and scanning electron microscopy (SEM). Films drawn from the latex exhibited excellent optical transparency and no evidence of aggregation was detected by SEM and X-ray energy dispersive spectroscopy (EDS) analysis confirmed the presence of silica and iron throughout the films. The thermal properties showed a decrease of the glass transition temperature in the presence of Fe_2O_3 nanoparticles as demonstrated by the dynamic mechanical analysis. The nano-SiO_2 induced significantly higher thermal stability, as decomposition temperatures of the nanocomposites increases by as much as 40 °C relative to the neat acrylic. On the other hand, the Young's modulus and hardness of the nanocomposites films decrease about 30 and 65%, denoting a modification of macromolecular dynamic by the SiO_2 and Fe_2O_3 nanoparticles, respectively.

Keywords Acrylate · Nanocomposites · SiO_2 · Fe_2O_3 · Mechanical properties

---

M.L. Hernández-Vargas (✉) · R. Castillo-Perez · B.F. Campillo-Illanes
Facultad de Química, Universidad Nacional Autonoma de Mexico,
Ciudad Universitaria, 04510 Ciudad de Mexico, Mexico
e-mail: mireya.l.hdezvargas@gmail.com

M.L. Hernández-Vargas · R. Castillo-Perez
Posgrado de Ingeniería, Facultad de Química, Universidad Nacional
Autonoma de Mexico, Ciudad Universitaria, 04519 Ciudad de Mexico, Mexico

O. Hernández-Guerrero
Facultad de Ciencias Químicas e Ingeniería, Centro de Investigación en Ingeniería y Ciencias
Aplicadas, Universidad Nacional Autonoma de Mexico, Av. Universidad 1001,
62209 Cuernavaca, Morelos, Mexico

M.L. Hernández-Vargas · R. Castillo-Perez · O. Hernández-Guerrero ·
O. Flores-Cedillo · B.F. Campillo-Illanes
Instituto de Ciencias Físicas, Universidad Nacional Autonoma de Mexico,
Av. Universidad s/n, Col. Chamilpa, 62210 Cuernavaca, Morelos, Mexico

© The Minerals, Metals & Materials Society 2017
M.A. Meyers et al. (eds.), *Proceedings of the 3rd Pan American Materials Congress*,
The Minerals, Metals & Materials Series, DOI 10.1007/978-3-319-52132-9_45

Introduction

Nanocomposites are characterized by at least one dimension in the nanometer range [1]. The properties of the nanocomposites are contributed to the properties of the filler particle, like shape, size, volume fraction, and the nature of the interphase that sometimes develop at the interface of the two components [2]. Different types of nanofillers are currently utilized, e.g. nanoclays, inorganic nanoparticles, carbon nanotubes, and hybrid organic–inorganic nanoparticles such as polyhedral silsesquioxane *(POSS)*, to refer some [3]. One of the most attractive nanoparticles is SiO_2 with several favorable advantages such as relatively inexpensive, nontoxic, biocompatible, high thermal resistant and especially its ability to reinforce polymer matrix's mechanical properties [4]. On the other hand, the iron oxide *(*Fe_2O_3*)*, the most common oxide of iron has important magnetically properties too [5, 6]. From the standpoint of basic research, iron (III) oxide is a convenient compound for the general study of polymorphism and the magnetic and structural phase transitions of nanoparticles [5]. The most frequent polymorphs structure "*alpha*" hematite having a rhombohedral-hexagonal and prototype corundum structures [5]. Generally, the semiconductor properties of the hematite are extremely useful in solar energy conversion, photocatalyse, water splitting, etc. [5]. In the present work, we synthesized polyacrylic nanocomposites, using nanoparticles as seeds, via emulsion polymerization in batch mode, i.e., in situ polymerization. The final thermal and mechanical properties of waterborne acrylate hybrid nanocomposites are discussed.

Experiment

Materials

Fumed SiO_2 nanoparticles and the monomers butyl acrylate, methyl methacrylate, and acrylic acid were purchased from the Aldrich Chemical Co. The monomers were purified using an inhibitor remover purchased from Aldrich prior to the polymerization experiments. Fe_2O_3 nanoparticles were purchased from Fermont, Productos Químicos Monterrey S.A de C.V. (Monterrey, N.L. Mexico). The nanoparticles were used as-received.

Chemical Synthesis

SiO_2 and Fe_2O_3 nanocomposites containing 1 wt% of SiO_2 and Fe_2O_3 nanoparticles (based on the total monomer weight) were synthesized by seeded emulsion polymerization in batch. The emulsion polymerization was carried out under reflux conditions at 77 ± 3 °C for 3.5 h using a radical initiator and tensioactive agent. SiO_2 or Fe_2O_3 were dispersed in distilled water. A mixture of butyl acrylate, methyl methacrylate, acrylic acid, tensioactive agent and radical initiator were added to the flask and SiO_2 or Fe_2O_3 was added before initiating the polymerization. The resultant SiO_2-nanocomposites or Fe_2O_3-nanocomposites latex was filtered to remove un-reacted material. Finally, the reactor was cooled down to room temperature and the pH was set at 8.5–9.5 by ammonia solution. In both cases the metastable nanostructured emulsions with dispersed SiO_2 and Fe_2O_3 nanoparticles thus obtained were stable for at least six months. Films were cast from the emulsions on Teflon™ Petri dishes first drying at room temperature for 3 days and then in vacuum at 90 °C for 4 h.

Thermal Analysis

Thermal decomposition temperatures were determined by high resolution modulated thermogravimetric analysis (MTGA), under nitrogen atmosphere, using a Q500 TGA manufactured by TA Instruments (New Castle DE, USA). The thermal program consisted of: heating rate of 2 °C/min, period of 200 s and temperature modulation of ±2 °C.

Uniaxial Tension

Elastic mechanical modulus was determined by uniaxial tensile deformation using the mini-tensile tester TST-350 manufactured by Linkam Scientific Instruments (Tadworth, Surrey, UK). Tensile deformation was carried out at 5 mm/min, room temperature. Tensile tests were repeated five times, and tests were carried out using a fresh sample each time.

Nano-Indentation

The nanoindentation tests were performed with a TI 750 Ubi L indentor system. A Berkovich diamond indenter was used, using standard load. The maximum displacement control was 600 nm, with a load and unload function of 25 s. The grid/test number of each sample was of 10 × 10, 100 in total per sample.

Scanning Electron Microscopy

Elemental analysis and identification of the aggregates of the acrylate nanocomposite was observed by scanning electron microscopy (SEM). For SEM observations, a JSM 5900 LV manufactured by JEOL Ltd. (Japan) operating at 20 kV equipped with an Oxford Link Isis-energy dispersive X-ray spectrometer (EDS) was employed.

Results and Discussion

Films about 1 mm thick were cast from the emulsions, and dried at room temperature in a fume hood and then under vacuum at 90 °C, as described in the experimental section. The films exhibited excellent optical transparency with no evidence of macro-aggregates, as shown in Fig. 1.

Further elemental mapping confirmed the dispersion of SiO_2 and Fe_2O_3 in the polyacrylic matrix. Figure 2 shows a SEM micrograph and their corresponding elemental mapping. As expected, it can be seen that elements such as C and O, predominate throughout the sample due to the organic nature of the matrix. Moreover, EDS analysis confirmed the presence of Si and Fe elements in the nanocomposites, respectively (see inset in Fig. 2).

The glass transition temperature, T_g, values were obtained by small-strain oscillatory shear using the stress-controlled CVO rheometer. Dynamic temperature ramps enabled to identify when the material suffered a transition from glass-like to rubber-like. During this transition there is a temperature where there is maximum dissipation of energy and the mechanical damping, Tan δ, goes through a maximum value, see Fig. 3a. This temperature is associated with the material's T_g, and the results are shown in Fig. 3b. The results showed that T_g decreased when adding Fe_2O_3 nanoparticles and increased when adding SiO_2 nanoparticles.

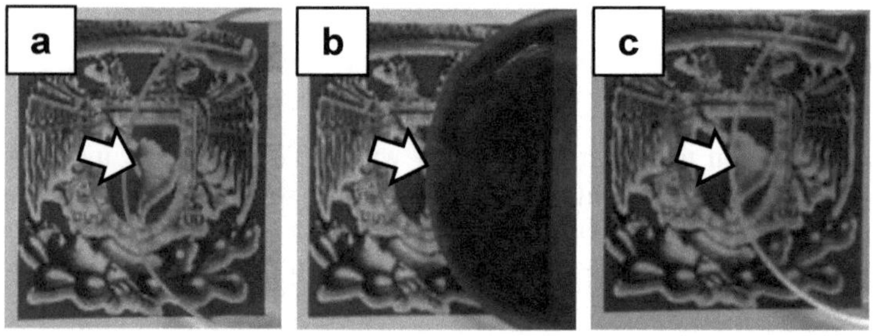

Fig. 1 Pictures obtained of acrylic nanocomposite films cast from respective emulsions: **a** Neat acrylic without nanoparticles, **b** Fe_2O_3 and **c** SiO_2 nanoparticles

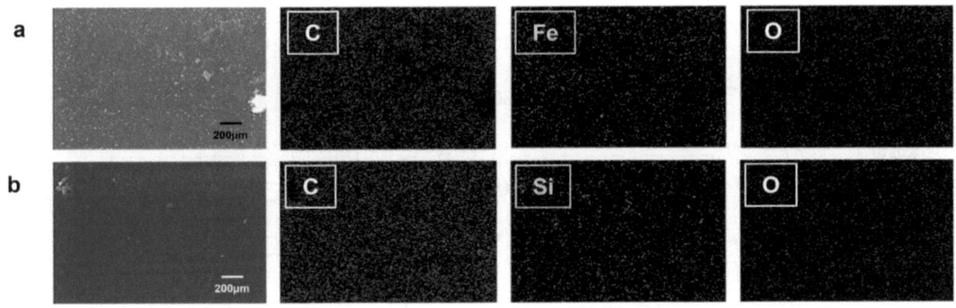

Fig. 2 SEM micrograph and element mapping of the acrylic surface of **a** Fe_2O_3 nanocomposite and **b** SiO_2 nanocomposite, containing 1 wt% nanoparticles

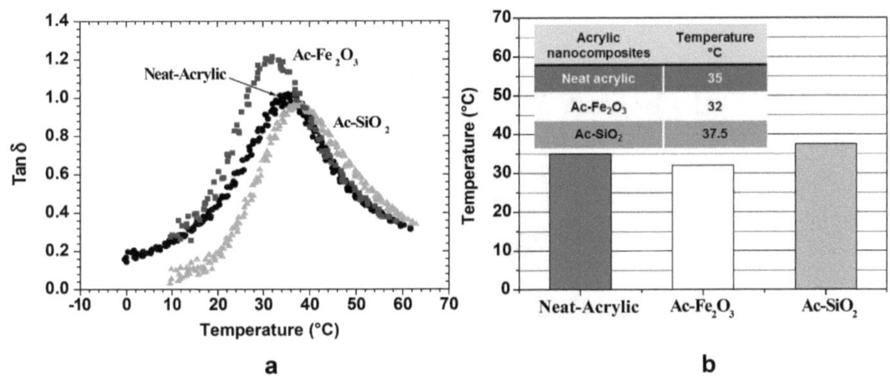

Fig. 3 **a** Maximum dissipation of energy and the mechanical damping, Tan δ and **b** glass transition temperature, T_g, as a function of nanocomposite type, determined by Dynamic Mechanical Analysis (DMA)

The mechanical properties were determined by uniaxial tensile tests at room temperature, and elastic Young's modulus was determined from stress-strain curves, the results are shown in Fig. 4a. The stress traces exhibit elastic deformation with no apparent yield stress, typical of elastomeric materials. The slope of the initial linear region enables the determination of Young's modulus, E, the results are showed in Fig. 4b. It can be seen that Young's modulus of the nanocomposites films decrease about 84 and 30% with Fe_2O_3 and SiO_2 nanoparticles, respectively.

The morphology of these films was investigated via SEM and Fig. 5a–c shows micrographs of fractured surface of (a) the neat polyacrylic and hybrid nanocomposites with (b) SiO_2 and (c) nanoparticles. As expected, the morphology exhibits plastic deformation for all materials and confirm the absence of micrometer-scale aggregates in all of them. Furthermore, EDS analysis confirmed the presence of Si and Fe in the nanocomposites (see inset in Fig. 5d, e). Additionally, the EDS analysis showed the presence of S, which is due to the initiator type utilized in the polymerization, ammonium persulfate (APS).

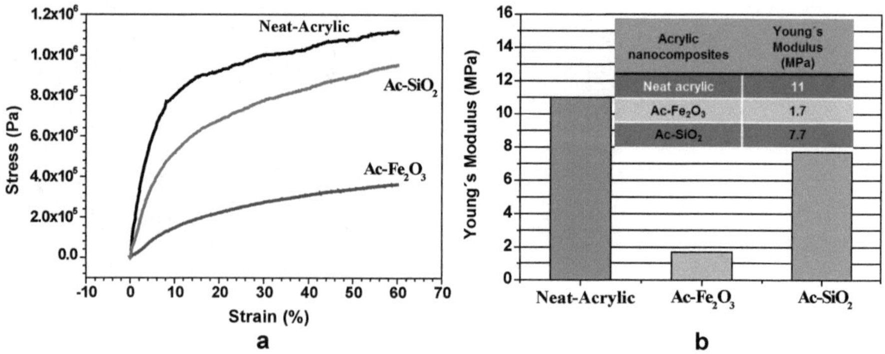

Fig. 4 **a** Stress-strain curves and **b** Young's modulus of acrylic nanocomposite

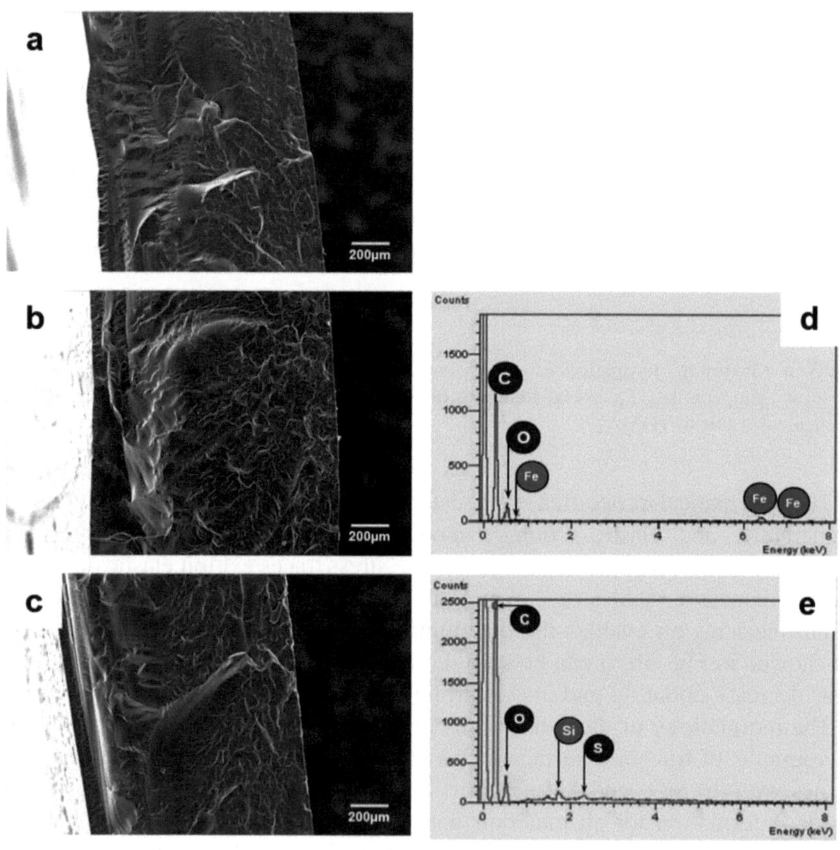

Fig. 5 SEM micrographs of acrylate hybrid nanocomposites fractured in liquid nitrogen, containing **a** neat acrylic, **b** Fe_2O_3 at 1 wt% and **c** SiO_2 at 1 wt%. EDS patterns in inset, shows the presence of **d** Fe and **e** Si in the nanocomposites

Surface nanohardness is also related to molecular chain mobility [7, 8]. Therefore, measuring this property may indicate whether the incorporation of the nanofiller (Fe_2O_3 and SiO_2) contributed appreciably to a decrease in molecular mobility.

Surface hardness was evaluated by nanoindentation tests. The nanoindentation test results are shown in Fig. 6. Hardness in presence of nanoparticles (nanofiller), decrease, due to the free volume increase that facilitates the chains mobility. It can be seen that hardness of the nanocomposites films decrease about 65 and 40% with Fe_2O_3 and SiO_2 nanoparticles, respectively.

Thermal decomposition temperatures of acrylate nanocomposites films were determined by thermogravimetric analysis, TGA. Figure 7a, b shows the graphs of mass derivative as a function of temperature and the thermal degradation temperature, T_{dec}, respectively. It can be seen that the neat acrylate is thermally stable with negligible mass loss up to ca. 280 °C, and then by about 365 °C there is the onset of thermal decomposition.

Figure 7b showed the nanocomposites exhibited significantly higher thermal stability, at only 1 wt% nanoparticles the decomposition temperature, T_{dec}, has increased up to 347 and 366 °C, respectively; some 36 and 55 °C relative to the neat acrylate.

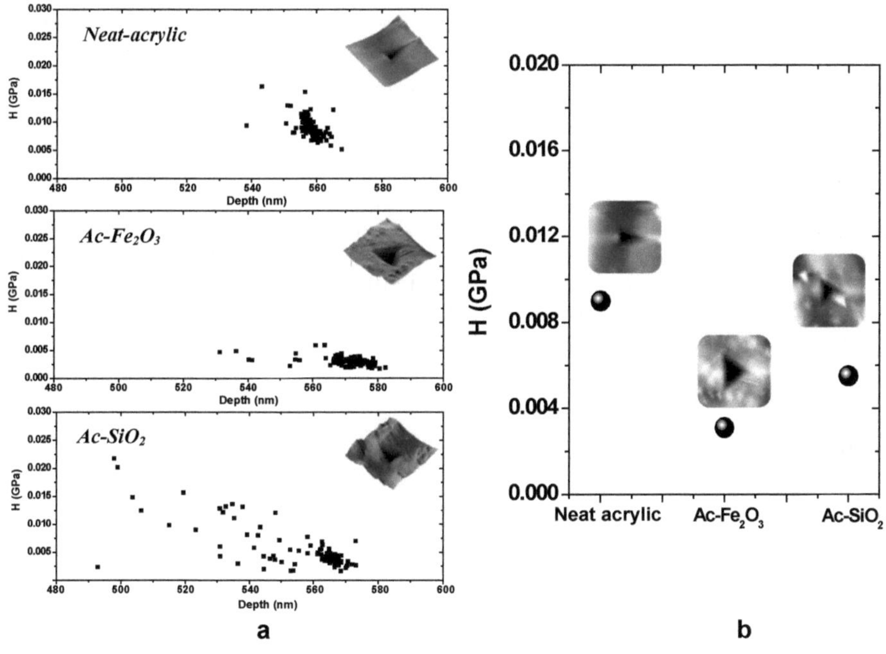

Fig. 6 **a** Effect of nanofiller incorporation on hardness of nanocomposites evaluated up to 100 points. **b** The average hardness values obtained

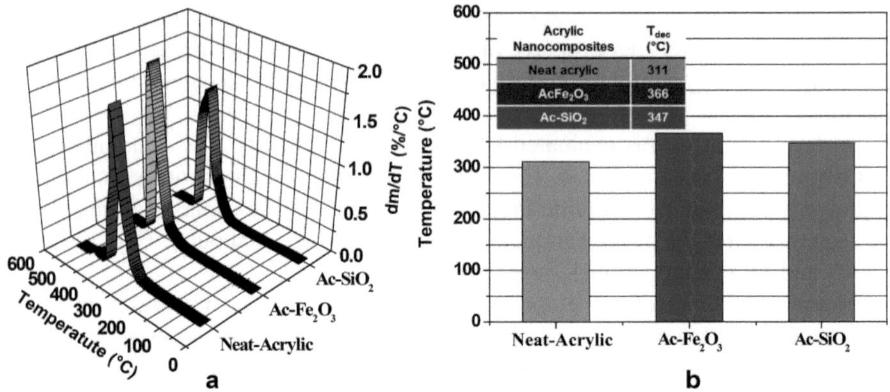

Fig. 7 **a** Graph of temperature derivatives of mass loss and **b** thermal degradation temperature, T_{dec}, for the nanocomposites

Conclusions

These results demonstrated the feasibility of synthesizing stable nanostructured emulsions, incorporating Fe_2O_3 and SiO_2 nanoparticles in concentrations of 1 wt%. Films as-cast were optically transparent. SEM and X-ray energy dispersive spectroscopy analysis confirmed the absence of micrometer-scale nanoparticles aggregates in all nanocomposites as well as the presence of S and Fe throughout the films. Mechanical Young's modulus and hardness decrease with incorporation of these nanoparticles. However, there was a significant increase in thermal decomposition temperatures due to the presence of Fe_2O_3 and SiO_2 nanoparticles.

Acknowledgements M.L. Hernández-Vargas, R. Castillo-Pérez and O. Hernández-Guerrero were supported by graduate scholarships from the Mexican Council for Science and Technology (CONACyT). Thanks to Institute of Physical Sciences at UNAM (Campus Morelos) and engineering postgraduate program in the Faculty of Chemistry at UNAM. Thanks to the DGAPA-UNAM under PAPIIT project number IN115616. Also, the authors thanks to Ivan Flores for the nanoindentation testing, I.Q.M. Itzel Reyes Chaparro and Q. Rafael Iván Puente Lee (Faculty of Chemistry, UNAM) for their help in SEM.

References

1. Mittal, V. (2010). Polymer nanocomposite: Synthesis, microstructure and properties. In V. Mittal (Ed.), *Optimization of polymer nanocomposite properties* (pp. 1–19). Weinheim: Wiley-VCH.
2. Hussain, F., & Hojjati, M. (2006). Polymer-matrix nanocomposites, processing, manufacturing, and application: An overview. *Journal of Composite Materials, 40*(17), 1511–1565.
3. Castillo-Perez, R., Romo-Uribe, A., Baghdachi, J. (2015). Dynamic mechanical analysis and morphology of nanostructured acrylic coatings. In *MRS Symposium Proceedings* (Vol. 1767, pp. 81–86). doi:10.1557/opl.2015.230

4. Wang, X., Wang, L., Su, Q., & Zheng, J. (2013). Use of unmodified SiO_2 as nanofiller to improve mechanical properties of polymer-based nanocomposites. *Composites Science and Technology, 89,* 52–60.
5. Chirita, M., & Grozescu, I. (2009). Fe_2O_3–Nanoparticles, physical properties and their photochemical, and photoelectrochemical applications. *Chemical Bulletin of "POLITEHNICA" University (Timisoara), 54*(68), 1–8.
6. Laurent, S., Forge, D., Port, M., Roch, A., Robic, C., Vander Elst, L., et al. (2008). Magnetic iron oxide nanoparticles: Synthesis, stabilization, vectorization, physicochemical characterizations, and biological applications. *Chemical Reviews, 108,* 2064–2110.
7. Hocine, N. A. (2010). Mechanical property enhancement of polymer nanocomposites. In V. Mittal (Ed.), *Optimization of polymer nanocomposite properties* (pp. 123–138). Weinheim: Wiley-VCH.
8. Innocentini-Mei, L. H. (2010). Thermomechanical properties of nanocomposite. In V. Mittal (Ed.), *Optimization of polymer nanocomposite properties* (pp. 351–368). Weinheim: Wiley-VCH.

Thermo-mechanical Properties of Copolymer/Clay Nanocomposites: A Comparative Study of Production Method by In-situ and Solution Mixture

Oscar Hernandez-Guerrero, Rubén Castillo-Pérez, Mireya Lizbeth Hernández-Vargas and Bernardo Fabián Campillo-Illanes

Abstract The field of polymer/clay nanocomposites is a new route for preparing polymers with enhance properties like higher modulus, thermal stability, better barrier properties, chemical resistance and electrical conductivity. In this work, nanoclay montmorillonite (MMT) was incorporated as reinforcement to a copolymer matrix, containing 0, 5% and 10 wt% of nanoclay. Two different processes were used to prepare the nanocomposites: (a) via in-situ (MMT-IN) and the (b) solution mixing (MMT-B). These nanocomposites show an increase in their glass transition temperature, T_g, studied by differential scanning calorimetry (DSC), and also a rise by an order of magnitude in the Young's modulus for both series. On the other hand, the study of thermogravimetric analysis (TGA) for the two series IN and B with addition of montmorillonite clay nanoparticles showed a decreasing behavior in thermal decomposition temperatures, T_{dec}, relative to the neat copolymer matrix.

Keywords Montmorillonite · Nanocomposites · Young's modulus

O. Hernandez-Guerrero (✉)
Facultad de Ciencias Químicas e Ingeniería, Centro de Investigación en Ingeniería y Ciencias Aplicadas, Universidad Autónoma del Estado de Morelos, Av. Universidad 1001, 62209 Cuernavaca, Morelos, Mexico
e-mail: oscar.hernandez@uaem.mx; zracos@gmail.com

O. Hernandez-Guerrero · R. Castillo-Pérez · M.L. Hernández-Vargas
B.F. Campillo-Illanes
Instituto de Ciencias Físicas, UNAM, Av. Universidad s/n, Col. Chamilpa, 62210 Cuernavaca, Morelos, Mexico

R. Castillo-Pérez · M.L. Hernández-Vargas · B.F. Campillo-Illanes
Facultad de Química, Universidad Nacional Autónoma de México, Ciudad Universitaria D.F, 04510 Mexico city, Mexico

R. Castillo-Pérez · M.L. Hernández-Vargas
Posgrado de Ingeniería, Facultad de Química, Universidad Nacional Autónoma de México, Ciudad Universitaria D.F, 04519 Mexico city, Mexico

© The Minerals, Metals & Materials Society 2017
M.A. Meyers et al. (eds.), *Proceedings of the 3rd Pan American Materials Congress*, The Minerals, Metals & Materials Series, DOI 10.1007/978-3-319-52132-9_46

Introduction

Clay/polymer nanocomposites are a typical example of nanotechnology. This class of material uses smectite-type clays, such as hectorite, montmorillonite, and synthetic mica, as fillers to enhance the properties of polymers. Another advantage of clay/polymer nanocomposites is that the optical properties of the copolymer are not significantly affected. The thickness of individual clay layers is much smaller than the wavelength of visible light so that well exfoliated clay/polymer nanocomposites should be optically clear [1]. Because of the nanometer-size dispersed particles, these nanocomposites exhibit markedly improved mechanical, thermal, optical, and physicochemical properties in comparison with pristine polymers or microscale transition composites, as first demonstrated by Kojima et al. [2].

However, it is very difficult to achieve this structure; as a result, semi exfoliated-intercalated structure is formed. The degree of exfoliation depends on many factors such as the chemical surface treatment of the clay particles, compatibility between layered silicates and polymer matrix [3], and their processing conditions [4], which implies that the method of synthesis has an effect of the nanocomposite structure. Three strategies are mainly used to prepare nanocomposites: in-situ intercalative polymerization [4], melt blending, and the solution blending method [5, 6]. The morphology of composites strongly depends on the preparation method as well as the selection of the organic modifier [7]. In the present work these nanocomposites of clay/polymers are studied by two different processes and inorganic nanofiller montmorillonite (MMT). Subsequently, this evaluation is focused on the comparative study of the mechanical properties and thermal clay nanocomposite/pure polymer with nanoclay amounts of 5% and 10 wt % for both processes [8].

Experiment

The polymer is an acrylic binder with composition 65BA/33.5MMA/1.5STY synthesized via a staged-acid redox emulsion polymerization process at the former Rohm and Haas R&D Laboratories (Spring House PA, USA). The clay is a sodium montmorillonite, PVG grade, manufactured by Nanocor. The nanocomposites were prepared by in-situ polymerization process. In the latter, the nanoparticles are incorporated into a monomer solution and the polymerization is initiated in their presence. An important feature is that the monomer provides dispersion of the nanoparticles during the reaction, causing greater interaction between the generated copolymer and the nanoparticles. Furthermore, an advantage of this procedure is the attainable nanocomposites up to 50% by weight of nanoparticles. By solution blending, the nanoparticles are suspended in a polar organic solvent by mechanical stirring or ultrasound and then is added the dissolved polymer in a second solvent. Once it is achieved, i.e. the polymer chains are intercalated between layers of the nanoparticles, the solvent is evaporated to obtain the nanocomposite [9]. Films of 2 mm in thickness were formed in a 100 mm diameter Petri dishes by drying them at room T. Then, to remove water excess, the samples were baked during 3 h at 90 °C.

Thermal decomposition temperatures were determined by thermogravimetric analysis (TGA), under nitrogen atmosphere, using a Q500 TGA manufactured by TA Instruments (New Castle DE, USA). Temperature scans were carried out at 10 °C/min. The thermal transitions were determined by differential scanning calorimetry (DSC). For DSC experiments, the DSC6000™ calorimeter manufactured by Perkin Elmer (Connecticut, USA) was used. Temperature and enthalpy calibration were carried out using analytical grade indium (Tm = 156.6 °C). The thermal transitions were determined at a heating rate of 20 °C/min under dry nitrogen atmosphere. Elastic mechanical modulus was determined by uniaxial tensile deformation using the mini-tensile tester TST-350 manufactured by Linkam Scientific Instruments (Tad worth, Surrey, UK). Tensile deformation was carried out at 5 mm/min, at room temperature.

Results and Discussion

Figure 1, shows the pictures of the films obtained by the two different processes, the in-situ and mixed solution of 5 and 10% weight concentration of MMT. The films of 2 mm in thickness remain optically transparent, exhibiting a good dispersion of the layered silicates through the copolymer matrix.

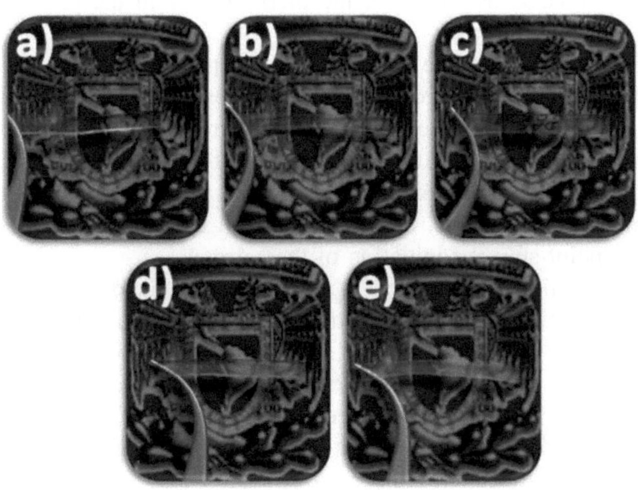

Fig. 1 Pictures of the nanocomposite films obtained of 2 mm in thickness prepared by the in-situ polymerization method: **a** Acrylic coating control, **b** 5 wt%, and **c** 10 wt%, and by solution blending method, **d** 5 wt% and **e** 10 wt%

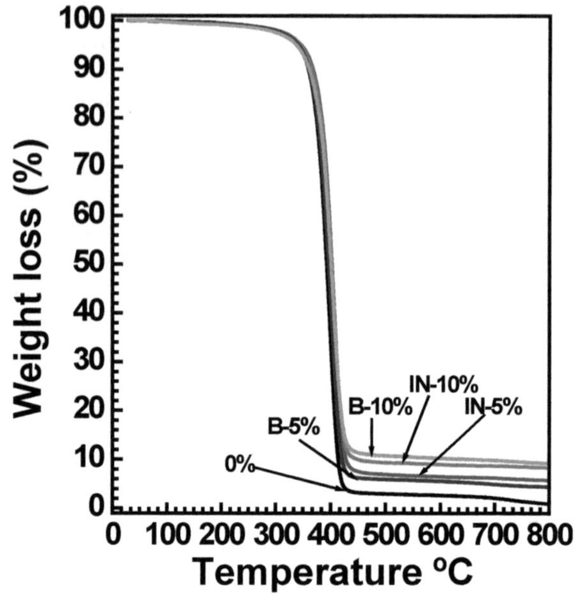

Fig. 2 Thermograms showing the mass loss for the polyacrylic with MMT variation as a function of temperature °C

The thermal stability was studied by thermogravimetric analysis, and the results are shown in Fig. 2, the thermal degradation of nanocomposites in both processes, in-situ and mixed in solution have the same behavior which recognize that they achieve their mass loss at the same temperature, as the final degradation. This may be due to, that even if it is the same material, only MMT-nanoparticle are increased in the polymer matrix, as shown in the thermograms in the residue of the final degradation.

In Fig. 3, can be seen the influence in thermal stability by increasing the MMT-nanoparticle which decrease to a decomposition T of about up to 5% by mass. It can be observed that the degradation T decreases as the concentration of the MMT-nanoparticle increases [10]. Regarding the IN series, it shows that it should increase from 5% to relative higher concentrations, also shows the initiation of the decay at higher concentrations. For the B series, it is stable around 5% MMT, and also from this point start the initiation of their decay.

The glass transition temperature, T_g, of the acrylic nanocomposites was determined by DSC, their results are shown in Fig. 4. The scans were carried at 10 ° C/min under dried N_2 atmosphere; the scans correspond to a second heating. The pristine polyacrylic showed a glass transition temperature T_g, of ≈8.4 °C. The results showed that T_g slightly increased by the addition of nanoclay (MMT). It is important to note, that the process solution mixture produces nanocomposites with slightly higher glass transition temperature.

Since the coatings studied in the present work are polyacrylic and in turn these are classified as amorphous polymers, the thermal properties exhibited by these materials, will be only the temperature associated with the glass transition

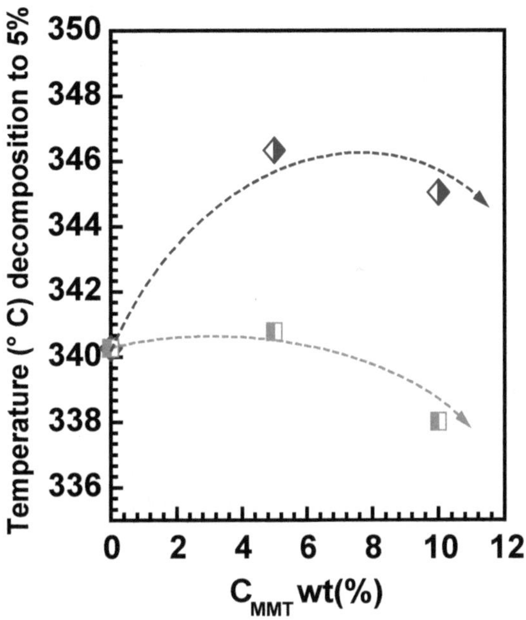

Fig. 3 Decomposition of 5% as a function of the MMT concentration in the two methods in-situ IN (◆) and Solution blending B (◧) series

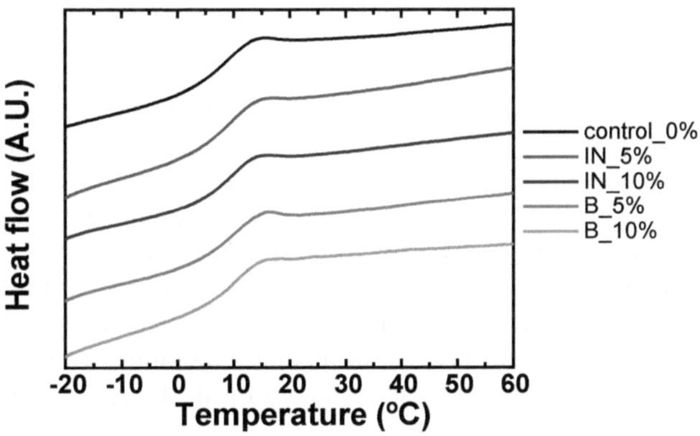

Fig. 4 DSC heating scans obtained for both series IN and B

temperature (T_g). That is because its chains are not arranged in a crystalline structure, but are scattered in any order. In the case of these nanocomposites in Table 1 exhibited an endothermic step associated with the glass transition temperature with an increase of 2 °C above the control [11, 12].

Table 1 Results of the maximum T as the glass transition T_g by TGA and the Young's modulus of both series samples IN and B

Samples	Young's modulus (MPa)	DSC T_g (°C) second heat	T_{max} (°C)
Control	0.73	8.44	398
IN_5%	0.97	8.53	407
IN_10%	2.78	8.93	406
B_5%	1.00	9.61	401
B_10%	1.42	10.41	403

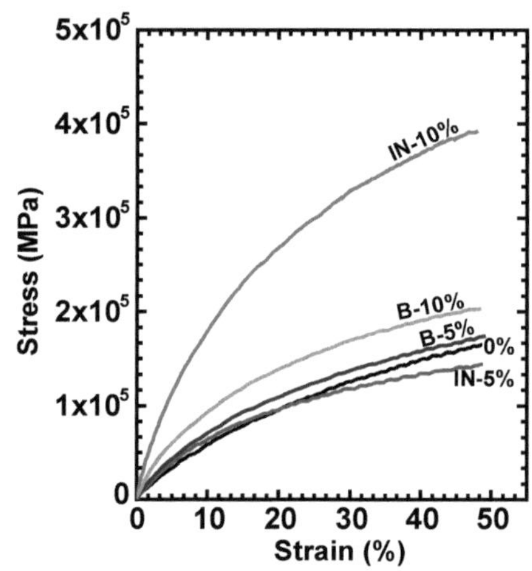

Fig. 5 Stress-stress deformation curves for the materials of the IN and B methods

The mechanical properties of polyacrylic layered-silicate were determined via uniaxial tensile deformation at 5 mm/min and room temperature, as shown in Fig. 5 shows the stress and strain deformation curves for in-situ and solution methods for 5 and 10% by weight. It can be observed for both series an elongation increase of about 50%, typical of acrylic. The IN-10 wt% MMT, predominate higher stress, this behavior is due to a good interaction during the process. In Fig. 6 shows the Young's modulus as function of MMT concentration for both series (IN and B) It can be observed a significant increase in Young's modulus of until 380% for IN process and 195% for B process. The summary of all the results are displayed in Table 1.

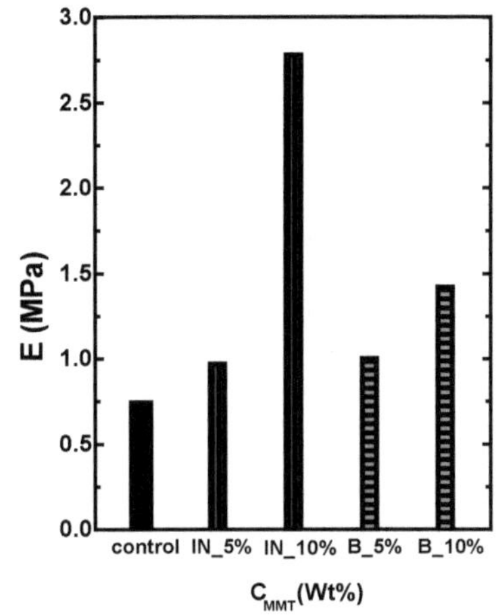

Fig. 6 The Young's modulus as a function of the wt% concentration of MMT in (IN) and (B) methods

Conclusions

These nanocomposites polyacrylates were synthesized by two methods, in-situ (IN) and solution mixture (B). The results showed that increasing the nanoclay in the copolymer matrix in low and high concentrations retains its optical clarity. This determines an interaction in both materials. The thermogravimetric analysis (TGA) the nanoparticle clay produces a negative bias to the copolymer matrix in both materials. The study by DSC in the glass transition shows a small increase for the two processes, and in the other hand these nanocomposites increases an order of magnitude in reinforcement mechanical Young's modulus, as the concentration of nanoparticle increases.

Acknowledgements O. Hernández-Guerrero was supported by graduate scholarship from the Mexican Council for Science and Technology (CONACyT).

References

1. Gao, F. (2004). Clay/polymer composites the history. *Materials Today, 7*(11), 50–55.
2. Tong, X., Zhao, H., Tang, T., Feng, Z., & Huang, B. (2002). Preparation and characterization of poly(ethyl acrylate)/bentonite nanocomposites by in situ emulsion polymerization. *Journal of Polymer Science Part A: Polymer Chemistry, 40*, 1706–1711.
3. Alexandre, B., Langevin, D., Mederic, P., Aubry, T., Counderc, H., Nguyen, Q. T., et al. (2009). "Water barrier properties of polyamide 12/montmorillonite nanocomposite membranes": Structure and volume fraction effects. *Journal of Membrane Science, 328*, 186–204.
4. Zha, W., Han, C. D., Moon, H. C., Han, S. H., Lee, D. H., & Kim, J. K. (2010). "Exfoliation of organoclay nanocomposites based on polystyrene-block-polyisoprene-block-poly(2 vinylpyridine) copolymer": Solution blending versus melt blending. *Polymer, 51*, 936–952.

5. Madaleno, L., Schjødt-Thomsen, J., & Pinto, J. C. (2010). Morphology, thermal and mechanical properties of PVC/MMT nanocomposites prepared by solution blending and solution blending þ melt compounding. *Composite Science and Technology, 70*, 804–814.
6. Lee, J. L., Zeng, C., Cia, X., Han, X., Shen, J., & Xu, G. (2005). Polymer nanocomposite foams. *Composite Science and Technology, 65*, 2344–2363.
7. Zulfiqar, S., Ahmad, Z., Ishaq, M., & Sarwer, M. I. (2009). Aromatic–aliphatic polyamide/montmorillonite clay composite materials; synthesis, nanostructure and properties. *Materials Science and Engineering A, 525*, 30–36.
8. Soundararajah, Q. Y., Karunaratne, B. S. B., & Rajapakse, R. M. G. (2010). Mechanical properties of poly(vinyl alcohol) montmorillonite nanocomposites. *Journal of Composite Materials, 44*, 303.
9. Covarrubias-Gordillo, C. A., Farías-Cepeda, L., Pérez-Aguilar, N. V., & Hernández-Hernández, E. (2013). Nanocompuestos a base de polímeros dispersos y nanofibras de carbón. *Revista Iberoamericana de Polímero, 14*(3).
10. Dlamini, D. S., Mishra, S. B., Mishra, A. K., & Mamba, B. B. (2011). Comparative studies of the morphological and thermal properties of clay/polymer nanocomposites synthesized via melt blending and modified solution blending methods. *Journal of Composite Materials, 45*(21), 2211–2216.
11. Cardoso, J., Romo-Uribe, A., & Flores, A. (2012). Nanostructure and viscoelasticity of layered silicate nanocomposite-electrolyte supports. *Journal of Applied Polymer Science, 123*, 944–955.
12. Romero-Guzmán, M. E., Flores, O., Flores, A., Romo-Uribe, A., Alvarado-Tenorio, B., & Campillo, B. (2011). Cold-drawn induced microstructure in PVC-bentonite nanocomposites. *Polymers for Advanced Technologies, 22*, 836–846.

Part VII
Minerals Extraction and Processing

Biotechnological Recycling of Precious Metals Sourced from Post-consumer Products

Norizoh Saitoh, Toshiyuki Nomura and Yasuhiro Konishi

Abstract Although conventional chemical or thermal recycling techniques are often the most appropriate means of recovering precious and rare metals, biological methods provide an attractive and eco-friendly alternative strategy in which microorganisms are applied to metal extraction. Recently, we have focused on metal ion-reducing microorganisms, *Shewanella* bacteria that are able to separate and concentrate platinum group metals and gold from solution into microbial cells. In a process, *Shewanella* bacteria can then be processed to generate metallic nanoparticles at room temperature, within 60 min. When processing the aqua regia leachate of waste LSI (large scale integration), *Shewanella* bacteria were able to rapidly and selectively collect gold ions from acidic solution. The biotechnological procedure also has the potential to allow the recovery of platinum group metals such as platinum, palladium, and rhodium from the leachate of spent automotive catalysts. We also found that *Shewanella* bacteria are able to absorb indium from dilute solution. This phenomenon can be applied to use microorganisms, an inexpensive biological material, as an adsorbent. We achieved selective adsorption of indium ions in *Shewanella* bacteria from the leachate of waste LCD (liquid crystal display) panels by adjusting its pH, because the pH range necessary for *Shewanella* bacteria to act as an effective adsorbent differs for different metal ions. Our proposed biological method enables the rapid and highly efficient recovery of precious and rare metals by using microbial reactions.

Keywords Recycling · Precious metals · Rare metals · Biotechnology · Biomineralization · Biosorption

---

N. Saitoh · T. Nomura · Y. Konishi (✉)
Department of Chemical Engineering, Osaka Prefecture University,
1-1 Gakuen-Cho, Sakai, Osaka 599-8531, Japan
e-mail: yasuhiro@chemeng.osakafu-u.ac.jp

© The Minerals, Metals & Materials Society 2017

M.A. Meyers et al. (eds.), *Proceedings of the 3rd Pan American Materials Congress*,
The Minerals, Metals & Materials Series, DOI 10.1007/978-3-319-52132-9_47

Introduction

Many post-consumer products, from electronics to automotive catalysts, contain either precious or rare metals, such as gold, platinum group metals (PGMs), indium, gallium, and rare earth metals. Although conventional chemical or thermal recycling techniques are often the most appropriate means of recovering precious and rare metals, biological methods provide an attractive and eco-friendly alternative strategy, in which microorganisms are used to separate and concentrate soluble metals from dilute solutions into microbial cells. Biological systems generally operate at ambient temperature and atmospheric pressure. Consequently, microbial recovery of metals would be an environmentally friendly method for recycling precious and rare metals from post-consumer products.

We have recently developed new recovery technologies that are designed to reduce and deposit the precious metal ions of Au(III) and PGMs (Pt(IV), Pd(II) and Rh(III)) as metallic nanoparticles, a process known as "biomineralization" [1–9]. We have also proposed the separation of rare metal In(III) ions from very dilute solutions and its concentration within microbial cells, a process known as "biosorption" [10, 11]. This paper describes new recycling technologies for a sub-set of precious and rare metals from post-consumer products in an environmentally friendly manner, such as the use of metal ion-reducing microorganisms to recover gold and PGMs from leachates of waste LSI (large scale integration) and spent automotive catalysts, and the use of microorganisms as adsorbents for the selective separation and concentration of indium from leachates of used LCD (liquid crystal display) panels.

Metal Ion-Recovering Microorganisms

For the recovery of precious and rare metals from aqueous solutions, we have focused on the metal ion-reducing bacteria, *Shewanella algae* (ATCC 51181) and *Shewanella oneidensis* (ATCC 700550). The *Shewanella* bacteria (Fig. 1) are a type of ferric iron-reducing microorganism and thus are able to reduce Fe(III) ions to Fe(II) ions by electron transfer. The electrons are obtained from the oxidation of organic acid salts such as lactate and formate.

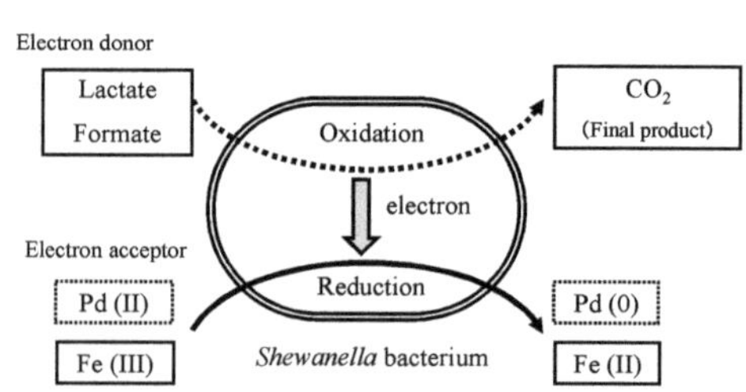

Fig. 1 A conceptual diagram for the microbial reduction of metal ions by *Shewanella* bacteria

The *Shewanella* bacteria are safe non-pathogenic microorganisms that can multiply rapidly in the presence of general inexpensive nutrients. Therefore, it is not difficult to prepare the large quantities of *Shewanella* bacteria necessary for the microbial recovery of precious and rare metals.

Recycling Technology of Precious Metals Based on Biomineralization

Microbial Reduction and Deposition of Soluble Precious Metals Using the Shewanella Bacteria

We recognized that the reduction potential of Fe(III) ions is almost equal to the potential for the reduction of precious metals ions such as Au(III) and PGMs (Pt (IV), Pd(II), and Rh(III)). Therefore, we investigated the use of the ferric ion-reducing bacteria, *S. algae* and *S. oneidensis*, to accumulate precious metal ions within the microbial cells, where the ions can be reduced to metallic nanoparticles.

Figure 2 shows the reduction of aqueous Pd(II) ions by *S. algae* cells at 25 °C and pH 7. We obtained these data simply by mixing *S. algae* cells with an aqueous $PdCl_2$ solution and then measuring the liquid-phase Pd(II) concentration as a function of time. When *S. algae* cells were combined with sodium formate as an electron donor, the aqueous Pd(II) concentration rapidly decreased from 5.0 mol/m^3 to almost zero, within 30 min. In this case, the bacterial cell suspension changed from yellow-orange to black, providing visible evidence of the plasmon absorption of palladium nanoparticles. The marked decrease in aqueous Pd(II) concentration presumably reflected its microbial reduction to metallic nanoparticles by the *Shewanella* bacteria.

Figure 3 shows a flow diagram of the method for concentrating palladium in bacterial cells. By drying wet, palladium-containing cells at 50 °C for 12 h, the palladium content in the dried cells was found to be 10%, which is 200 times

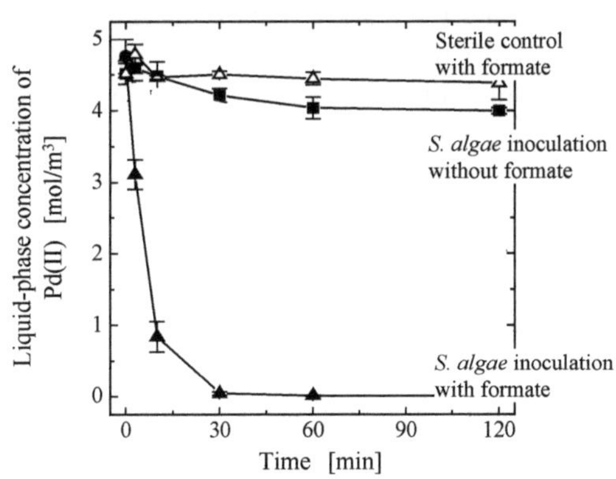

Fig. 2 Microbial reduction and deposition of soluble Pd(II) by *S. algae* cells at 25 °C and pH 7: (*filled triangle*) 8 × 10^{15} cells/m^3 *S. algae* cells in the presence of 50 mol/m^3 formate; (*filled square*) 8 × 10^{15} cells/m^3 *S. algae* cells; (*open triangle*) sterile control containing no *S. algae* cells in the presence of 50 mol/m^3 formate

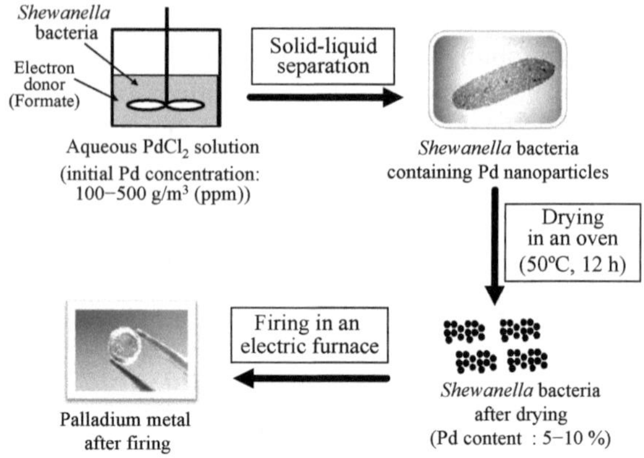

Fig. 3 Flow diagram of method for concentrating and recovering palladium collected in bacterial cells

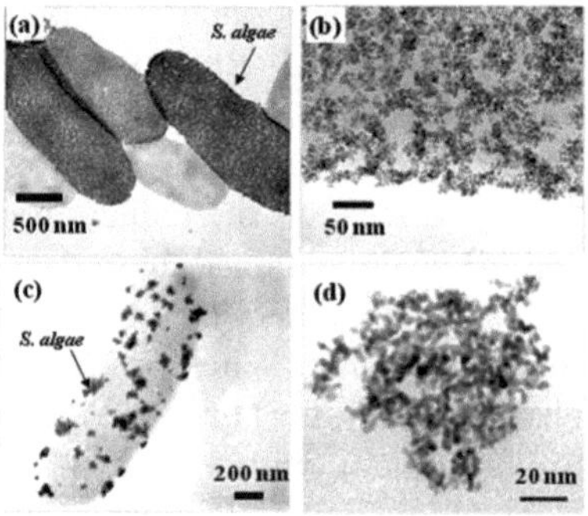

Fig. 4 Low- and high-magnification TEM images for biogenic metal nanoparticles deposited in *S. algae* cells at 25 °C and pH 7. **a** and **b** palladium, **c** and **d** platinum

greater than the 500 g/m^3 concentration of Pd(II) in the starting solution. Furthermore, firing the dried cells in an electric furnace produced metallic palladium.

The *Shewanella* bacteria were able to reduce soluble precious metals (Au(III), Pt (IV), and Rh(III)) to insoluble metal nanoparticles at room temperature within a short operation period of 120 min after cell inoculation. To confirm the reductive deposition of precious metals, *S. algae* cells were observed using transmission electron microscopy (TEM) after 120 min of exposure to aqueous solutions of soluble precious metals. As shown in Fig. 4, *Shewanella* bacteria are rod-shaped cells, approximately 0.5 μm wide and 2 μm long. The TEM images show individual discrete Pd nanoparticles about 10 nm in diameter in the bacterial cells. Thin sections of bacterial cells were also examined to clarify the exact deposition location of the nanoparticles. From this TEM image (Fig. 5), it is evident that the biogenic nanoparticles were localized in the periplasmic space, where the soluble substrate in aqueous solution has ready access to the bacterial cell. Considering the position of the formed nanoparticles, the bio-substances that contribute to the deposition of soluble precious metals is likely to reside in the periplasmic space.

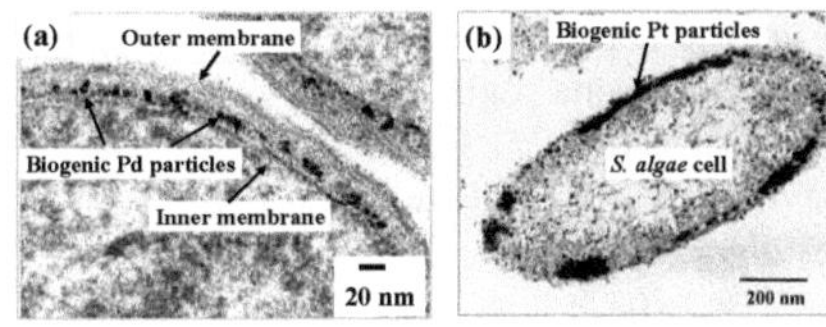

Fig. 5 TEM images for a thin section of biogenic nanoparticles deposited in *S. algae* cells at 25 °C and pH 7. **a** palladium, **b** platinum

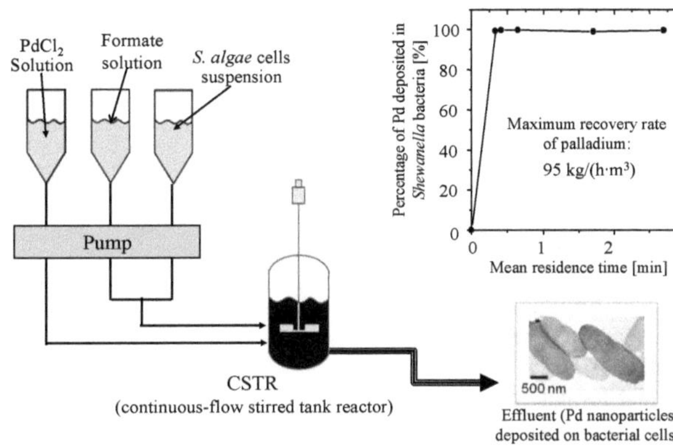

Fig. 6 Microbial recovery of soluble palladium in a continuous-flow stirred tank reactor

To allow the processing of larger amounts of palladium solution, microbial reduction and recovery runs were performed using a stirred tank reactor under continuous operation (Fig. 6). An aqueous $PdCl_2$ solution and *S. algae* suspension were continuously supplied to the reactor, while the resulting product was continuously discharged. At a feed palladium concentration of 5.0 mol/m^3, almost 100% recovery of the palladium was maintained under steady-state operation, even when the mean residence time was decreased to 20 s. Under these operating conditions, we were able to achieve rapid and highly efficient recovery of palladium at a rate of 95 kg/(h m^3). Generally, microbial treatment has the disadvantage of being very slow, but this biotechnological procedure has the capacity to perform rapid and highly efficient recovery of precious metals at room temperature.

Recycling Process of Precious Metals Using the Biomineralization Method

To confirm the validity of our new biological recovery method, we attempted to extract gold sourced from post-consumer products. When processing the aqua regia leachate of waste LSI (large scale integration), *Shewanella* bacteria were able to rapidly and selectively collect gold ions from the solution. Importantly, the *Shewanella* bacteria did not react with other metal ions, such as copper and iron. The ability to selectively recover only gold under strongly acidic conditions, such as at pH 1.3, demonstrates the potential for commercialization of this microbial

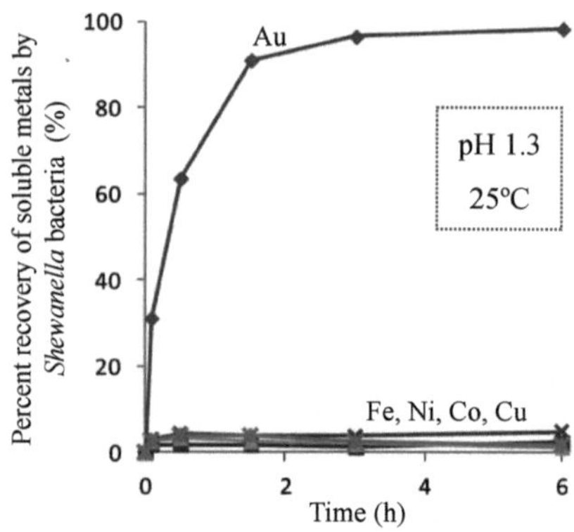

Fig. 7 Microbial recovery of gold from aqua regia leachate of waste LSI at 25 °C, pH 1.3, and 5.0×10^{15} cells/m^3 *S. algae* cells

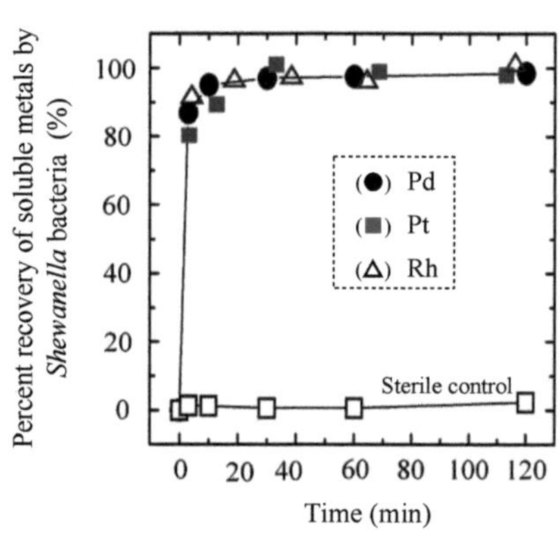

Fig. 8 Microbial recovery of PGMs from aqua regia leachate of spent automotive catalysts at 25 °C, pH 6, an initial formate concentration of 200 mol/m^3, and 8.0×10^{15} cells/m^3 *S. algae* cells

recovery process, comprising a new system for bio-recovery of precious metals from waste electrical and electronic equipment (Fig. 7).

This biotechnological procedure also has the potential to allow the recovery of PGMs from the leachate of spent automotive catalysts. *Shewanella* bacteria were able to successfully perform the complete reduction and deposition of PGM ions such as Pt(IV), Pd(II), and Rh(III) (Fig. 8). The ability to collect PGM ions suggests that this represents a new process for the microbial recovery of PGMs from spent catalysts.

Figure 9 shows flow diagrams comparing the conventional chemical process and the new bioprocess for the recycling of precious metals. In contrast to the conventional process, the new bioprocess is integrated, unifying a multi-step method into a single-step procedure that separates and concentrates precious metals from a dilute solution, enabling the formation of metal nanoparticles. However, the addition of an electron donor and pH adjustment is required for the bioprocess to operate.

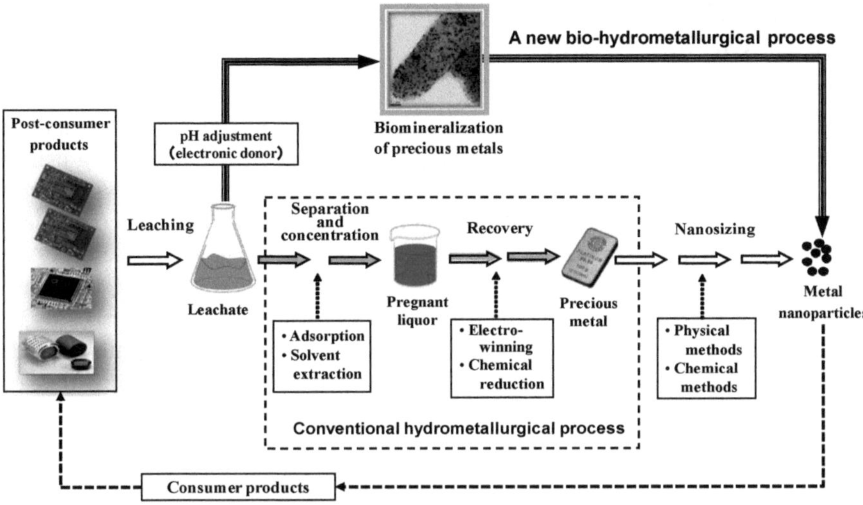

Fig. 9 Recycling flows of precious metals using the new and conventional technologies

Recycling Technology of Rare Metals Based on Biosorption

Microbial Adsorption of Soluble Rare Metals Using Shewanella Bacteria

The *Shewanella* bacteria can also be applied as an inexpensive bio-material for recovering soluble rare metals from aqueous solutions because they have a cell surface consisting of biological materials containing functional groups, which are responsible for the adsorption of rare metal ions. We used *Shewanella* bacteria to separate rare metal ions, In(III), from dilute solutions.

Shewanella bacteria used in the microbial reduction and deposition of precious metals are also applicable to the adsorption of rare metal ions from dilute solutions. Figure 10 shows the adsorption of aqueous In(III) ions by *S. algae* cells at 25 °C and pH 3.0. The concentration of *S. algae* cells was maintained at 8.0×10^{15} cells/m^3. The aqueous concentration of In(III) ions rapidly decreased within a short operation period of 10 min after cell inoculation, and the percent recovery of In(III) ions increased with decreasing initial In(III) concentration. In particular, at an initial In(III) concentration of 0.09 mol/m^3, In(III) ions in the aqueous solution were completely recovered within 5 min by *S. algae* cells.

The amount of indium collected per unit biomass was 0.47 mmol-In(III)/g-dry cells. This value is relatively high compared with other typical heavy metal biosorption values reported elsewhere; e.g., 0.20 mmol-Cr(III)/g-biomass at pH 3.0, 0.43 mmol-Cu(II)/g-biomass at pH 5.0, 0.31 mmol-Mn(II)/g-biomass at pH 5.0, 0.17 mmol-Au(III)/g-biomass at pH 3.5, and 0.15 mmol-Se(VI)/g-biomass at pH 3.0 [12, 13].

Process for Recycling a Rare Metal (Indium) by the Biosorption Method

We attempted to recover indium from waste LCD (liquid crystal display) panels. To leach indium from waste LCD panels, we proposed hydrothermal leaching using a dilute HCl solution at 120 °C. Using this method, 100% of the indium was leached out in just 5 min. However, in addition to indium, aluminum and other components were found to coexist in the leachate of waste LCD panels. The pH range necessary for *Shewanella* bacteria to act as an effective adsorbent differs for different metal ions (Fig. 11). Thus, by adjusting the pH of the LCD panels leachate to pH 2.8–3.0, we predicted that the adsorption of Al(III) ions would be inhibited, allowing the selective adsorption of In(III) ions by the *Shewanella* bacteria. Using this pH

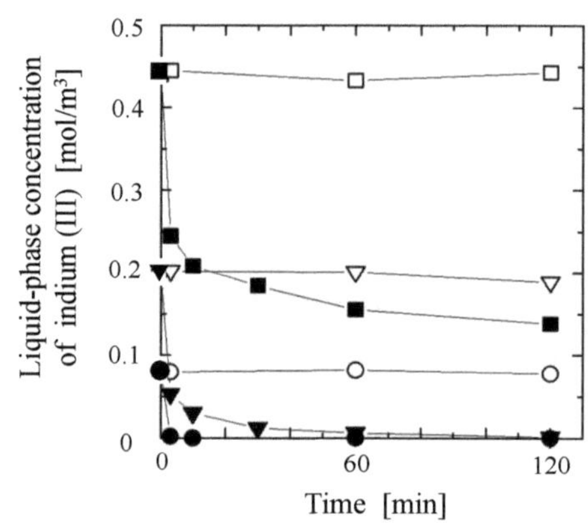

Fig. 10 Microbial adsorption of In(III) ions from dilute solutions by *S. algae* cells at pH 3 and 8.0×10^{15} cells/m^3 *S. algae*. (*filled square*) 0.45 mol/m^3 $InCl_3$ solution, (*filled triangle*) 0.20 mol/m^3 $InCl_3$ solution, (*filled circle*) 0.09 mol/m^3 $InCl_3$ solution, (*open square, open inverted triangle, open circle*) sterile control without *S. algae* cells

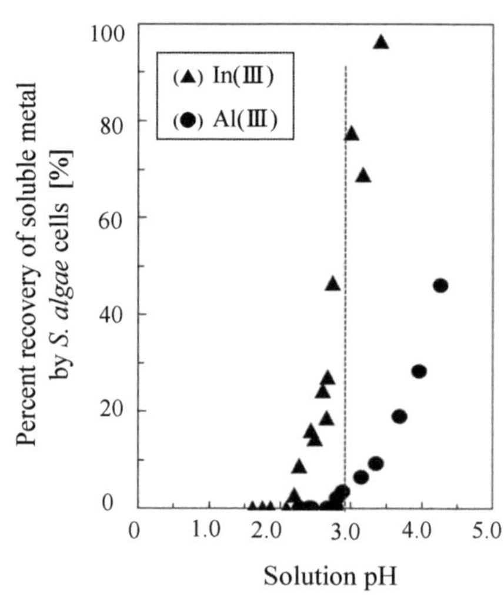

Fig. 11 Effect of solution pH on the percent recovery of In(III) and Al(III) by *S. algae* cells

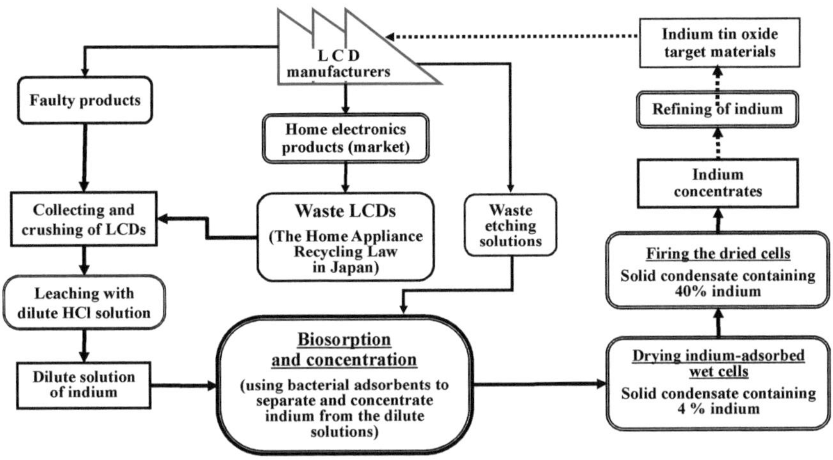

Fig. 12 Recycling flow of indium from waste LCD panels using the new biotechnology process

adjustment, we were successfully able to selectively separate indium from the LCD panels leachate.

Figure 12 shows an indium recycling process based on applying the eco-friendly biotechnology to waste LCD panels and etching solutions. The benefits of this biosorption method are: significant reductions in energy and material consumption, with a small environmental load compared with conventional technology; indium from dilute solution can be selectively separated and concentrated rapidly inside the cell, within 10 min; after combusting indium-containing microbial cells, concentrates of 40% indium content can be easily recovered.

Conclusions

We successfully developed new recycling technologies based upon eco-friendly biotechnology, to extract precious metals (gold and PGMs) and rare metals (indium) from post-consumer products. Our proposed recycling biotechnologies are linked to the development of low-cost and eco-friendly recycling processes that enable the low-energy and rapid recovery of precious and rare metals, using microbial reactions at room temperature and atmospheric pressure. Moreover, because a highly efficient process is achievable in a small compact unit, it can potentially be introduced at localized collecting points for post-consumer products, operating as a regionally distributed technology that will contribute effectually towards constructing recycling systems for precious and rare metals.

Acknowledgements This work was supported by the Japan Society for the Promotion of Science KAKENHI Grant Nos. 20360411 and 23360406. Parts of this research were performed under the Environment Research and Technology Fund Numbers K2135, K22082, and K2308.

References

1. Konishi, Y., Tsukiyama, T., Ohno, K., Saitoh, N., Nomura, T., & Nagamine, S. (2006). Intracellular recovery of gold by microbial reduction of $AuCl_4^-$ ions using the anaerobic bacterium *Shewanella algae*. *Hydrometallurgy, 81,* 24–29.
2. Konishi, Y., Tsukiyama, T., Saitoh, N., Nomura, T., Nagamine, S., Takahashi, Y., et al. (2007). Direct determination of the oxidation state of gold deposits in the metal-reducing bacterium *Shewanella algae* using X-ray absorption near-edge structure spectroscopy (XANES). *Journal of Bioscience and Bioengineering, 103,* 568–571.
3. Konishi, Y., Tsukiyama, T., Tachimi, T., Saitoh, N., Nomura, T., & Nagamine, S. (2007). Microbial deposition of gold nanoparticles by the metal-reducing bacterium *Shewanella algae*. *Electrochimica Acta, 53,* 186–192.
4. Konishi, Y., Ohno, K., Saitoh, N., Nomura, T., Nagamine, S., Hishida, H., et al. (2007). Bioreductive deposition of platinum nanoparticles on the bacterium *Shewanella algae*. *Journal of Biotechnology, 128,* 648–653.
5. Ogi, T., Saitoh, N., Nomura, T., & Konishi, Y. (2010). Room-temperature synthesis of gold nanoparticles and nanoplates using *Shewanella algae* cell extract. *Journal of Nanoparticle Research, 12,* 2531–2539.
6. Tamaoki, K., Saito, N., Ogi, T., Nomura, T., & Konishi, Y. (2010). Microbial reduction and recovery of palladium using metal ion-reducing bacterium *Shewanella algae*. *Kagaku Kogaku Ronbunshu, 36,* 288–292.
7. Ogi, T., Makita, K., Tamaoki, K., Saitoh, N., & Konishi, Y. (2011). Synthesis of gold nanoparticles by the metal ion-reducing bacterium *Shewanella oneidensis*. *Journal of the Society of Powder Technology, 48,* 160–166.
8. Ogi, T., Honda, R., Tamaoki, K., Saitoh, N., & Konishi, Y. (2011). Direct room-temperature synthesis of a highly dispersed Pd nanoparticle catalyst and its electrical properties in a fuel cell. *Powder Technology, 205,* 143–148.
9. Tamaoki, K., Saitoh, N., Nomura, T., & Konishi, Y. (2013). Microbial recovery of rhodium from dilute solutions by the metal ion-reducing bacterium *Shewanella algae*. *Hydrometallurgy, 139,* 26–29.
10. Higashi, A., Saitoh, N., Ogi, T., & Konishi, Y. (2011). Recovery of Indium by biosorption and its application to recycling of waste liquid crystal display panel. *Journal of the Japan Institute of Metals, 75,* 620–625.
11. Ogi, T., Tamaoki, K., Saitoh, N., Higashi, A., & Konishi, Y. (2012). Recovery of indium from aqueous solutions by the Gram-negative bacterium *Shewanella algae*. *Biochemical Engineering Journal, 63,* 129–133.
12. Michalak, I., & Chojnacka, K. (2010). Interactions of metal cations with anionic groups on the cell wall of the macroalga Vaucheria sp. *Engineering in Life Sciences, 10,* 209–217.
13. Niu, H., & Volesky, B. (2003). Characteristics of anionic metal species biosorption with waste crab shells. *Hydrometallurgy, 71,* 209–215.

Biotechnologies for Wastewater Treatment in the Mineral Industry

Natalia R. Barbosa, Sueli M. Bertolino, Renata G.S. Cota and Versiane A. Leão

Abstract Biotechnological processes are important alternatives for water recycling both in hydrometallurgical and mineral processing operations and this paper is focused on two of such technologies: (i) manganese bioremediation and (ii) sulphate reduction. While high concentrations are related to hydrometallurgical operations or AMD generation, manganese is one of the most difficult metals to remove from wastewaters. It is demonstrated herein that a bacterial consortium enriched from a mine water was able to remove 99.7% Mn^{2+} from a solution containing 50 mg/L. Molecular studies revealed *Stenotrophomonas, Bacillus* and *Lysinibacillus* genera in the sample. Cell metabolism resulted in a pH increase and catalysed chemical Mn^{2+} oxidation. Subsequently, sulphate reduction by sulphate reducing bacteria (SRB) was addressed. A fluidized bed reactor (FBR), in which there was immobilized biomass along with fluidization enabled a high bacterial population (>10^9 cells/mL) in the bioreactor and thus a large sulphate reduction efficiency (97%) for a specific sulphate reducing rate of 0.186 ± 0.015 g SO_4^{-2}/gVSS.d.

Keywords Mining water · Sulphate · Manganese · Biotechnology · Bioremediation

Introduction

Acid mine drainage (AMD) can be defined as the resulting drainage produced during natural oxidation of sulphide minerals, mainly pyrite (FeS_2) and pyrrhotite (FeS_1-x) contained in rocks exposed to the action of air and water [1]. AMD may

N.R. Barbosa · R.G.S. Cota · V.A. Leão (✉)
Universidade Federal de Ouro Preto, Campus Morro do Cruzeiro S/N, Bauxita, Ouro Preto, Minas Gerais, Brazil
e-mail: versiane@demet.em.ufop.br

S.M. Bertolino
Universidade Federal de Uberlândia, ICIAG, Uberlândia, Minas Gerais, Brazil

© The Minerals, Metals & Materials Society 2017
M.A. Meyers et al. (eds.), *Proceedings of the 3rd Pan American Materials Congress*, The Minerals, Metals & Materials Series, DOI 10.1007/978-3-319-52132-9_48

have a pH that can reach values as low as two and contain toxic metals in concentrations often exceeding the limits defined by environmental regulations worldwide. It also has irreversible consequences to the environment once started and is considered one of the most pressing environmental issues in mining countries [2].

As the pH is reduced as a result of sulphides oxidation, a plethora of metals along with sulphate ions are produced and the concentrations of these pollutants depend on the geology of the region and drainage chemistry. Traditionally such effluents are treated by increasing the pH to precipitate metals. However, the removal efficiency depends on the type of metal ion. Copper, zinc and cadmium for instance are efficiently removed by a pH increase whereas metals with more complex solution chemistry, such as manganese remain a challenge. In addition, metal precipitation techniques are in a different development stage as compared to sulphate removal. There is no universally accepted or adopted procedure to remove sulphate and although the addition of lime to increase the pH of the drainage induces gypsum precipitation, the sulphate concentration is reduced to concentrations around 1200–2000 mg/L, which does not comply with most environmental regulations. Therefore, a procedure to specifically remove sulphate needs to be included in the drainage treatment flowsheet.

Bioremediation of Mn-bearing mine waters relies on Mn^{2+} oxidation to MnO_2, which precipitates from solution. This is a kinetically controlled process and some microorganisms have been proposed to catalyse such oxidation [3, 4]. Conversely sulphate may be converted to sulphide ions by sulphate reducing bacteria (SRB) and the latter oxidized to elemental sulphur. Thus, the current work sought to investigate both biotechnologies in the context of mine water remediation.

Materials and Methods

Manganese Bioremediation

Enrichment: A mine water with a high manganese content (140 mg/L) was selected for bacterial enrichment. Firstly, a sample of this water was subjected to filtration in a 0.22 μm membrane (Millex Millipore) and then the membrane was inoculated into 10 mL of K medium (0.15 g/L $MnSO_4.H_2O$; 1 mg/L $FeSO_4.7H_2O$; 2.0 g/L peptone; 0.5 g/L yeast extract and 10 mM HEPES as buffer) and incubated at 30 °C, for 24 h under constant stirring (150 min^{-1}). Subsequently, the culture was transferred to 90 mL of K medium and incubated in similar conditions. Bacterial isolates were produced from the CL enrichment after growth in agar using the same K medium and 50 mg/L Mn^{2+} (as $MnSO_4.H_2O$), at 30 °C until a brown color was observed (indicating manganese oxidation). Those colonies with a brown color were selected for further experiments.

Manganese removal tests: These experiments were carried out with ten isolates from the mine water. Specifically, 200 mL of K growth medium were transferred to glass flasks containing 50 mg/L Mn^{2+} as $MnSO_4.H_2O$, at 30 ± 1 °C. Stirring (150 min^{-1}) was provided by a New Brunswick Scientific, C25KC orbital shaker and all experiments were carried out in duplicate during 1 week. Furthermore manganese removal by the enrichment CL was investigated in a bioreactor (Bioflo 110—New Brunswick Scientific) with a final volume of 1L, during 14 days at 30 ± 1 °C under 150 min^{-1} (mechanical) stirring. Aliquots were collected periodically for determination of Mn concentration (ICP-OES Agilent 725), pH and either optical density at 600 nm (isolates) or bacterial counts (sample CL).

Sulphate Reduction

SRB enrichment: The first SRB sample (LG) used in the experiments was enriched from sediment collected from an urban pond located in the city of Ouro Preto in Minas Gerais, Brazil. The second sample (UASB) was an anaerobic sludge of a UASB reactor located at Arrudas Sewage Treatment Plant in the city of Belo Horizonte, Brazil. Aliquots of 10% (v/v) were inoculated into glass flasks (200 mL) containing 100 mL of Postgate C (0.5 g/L KH_2PO_4, 1.0 g/L NH_4Cl, 0.06 g/L $MgSO_4.7H_2O$; 0.1 g/L $FeSO_4.7H_2O$; 0.25 g/L yeast extract; 2.96 g/L Na_2SO_4) [5]. The flasks containing the culture medium were sealed and incubated in a microbiological incubator at 35 °C. SRB growth was detected by the formation of a black precipitate resulting from the production of iron sulphide.

Batchwise sulphate reduction: Bacterial growth and sulphate reduction tests were carried out with 2 different carbon source (ethanol and lactate) as well as at two different pH values −5.5 and 7.0. The COD/Sulphate ratio was set at either 1.0 in the experiments with lactate or 2.0–3.6 with ethanol. The effect of COD/SO_4^{2-} ratio (0.67–2.0) on sulphate reduction was also assessed.

Continuous experiments: The fluidized bed reactor used in the continuous experiments is described elsewhere [6]. It operated inside a fume hood at 23 ± 2 °C and activated carbon (2.0–3.0 cm diameter, 566 m^2/g surface area) was the biomass carrier. A pump accounted for biomass recirculation working at 166 L/h (75 m/h superficial velocity) and 86% bed expansion. That was achieved by recycling the effluent from the outlet reactor port. The reactor was 74-cm long and the internal diameter of the sulphate reduction zone was 5.0 cm corresponding to a nominal volume of 1.2 L. The Postgate C mineral medium was used during the start-up and continuous operation of the reactor. It was fed upwards by a second peristaltic pump so that the HRT was 10 h.

Results and Discussion

Manganese Bioremediation

The CL consortium was enriched from a high manganese mine water using the K growth medium. This enrichment removed the element from solutions containing up to 300 mg/L Mn^{2+} as listed in Table 1, which also shown pH changes during the tests. Among the three manganese concentrations investigated, the experiment with 50 mg/L Mn^{2+} had the highest Mn^{2+} removal (99.7%), while the experiments with 200 and 300 mg/L Mn solutions showed similar removal efficiencies (60%). Either low or no manganese removal was observed in the abiotic controls at the three different concentrations. The solution pH increased in the biotic experiments and reached 7.86 and 7.62 in the 50 and 300-mg/L experiments respectively. Conversely the pH did not change significantly in the abiotic controls, which suggested that the pH increase was promoted by bacterial growth.

Figure 1a details the profiles for manganese concentration, pH and bacterial counts during a 14-day experiment using 50 mg/L Mn^{2+}. The Mn^{2+} concentration decreased mostly after the 12 day whereas the pH steadily increased during the whole experiment. Moreover, the bacterial population leveled off within two days and remained stable until the seventh day. Subsequently a slight decrease in cell counts was observed, reaching 2.5×10^8 cells/mL on the 12th day (when the manganese removal rate was the greatest). Therefore, manganese removal was not related to bacterial growth because it occurred more intensively at the end of the experiment during the stationary growth phase.

Subsequently twenty bacterial strains were isolated from the enriched sample (CL) and ten were tested for their Mn^{2+} removal capabilities. Manganese removal (%), optical density (at 600 nm) and pH values during a seven-day experiment are depicted in Fig. 1b, c. Among the isolates identified as *Bacillus* (B1, B2, B3 and B4), isolate B2 had the lowest (33.5%), while isolated B3 showed the best manganese removal capacity (81.8%) (Fig. 1b). Three isolates identified as *Stenotrophomonas* (S1, S2 and S3) have also been tested and showed basically the same metal removal [70.9, 66.4 and 78.7% (Fig. 1c) removal, respectively]. From the isolates identified as *Lysinibacillus* (L1, L2 and L3), strain L2 revealed the

Table 1 Mn^{2+} removal from synthetic solutions by the CL enrichment and pH change during the experiments

Mn^{2+} conc. (mg/L)	Type of test	Mn^{2+} removal (%)	Initial conc. (mg/L)	Final conc. (mg/L)	Initial pH	Final pH
50	Biotic	99.7	45.8	0.2	7.36	7.86
	Abiotic	0	35.7	37.8	7.37	7.24
200	Biotic	61.2	210.7	81.8	7.27	7.47
	Abiotic	15.5	176.5	151.6	7.51	7.37
300	Biotic	58.1	279.0	117.4	7.14	7.62
	Abiotic	0	256.7	258.9	7.48	7.33

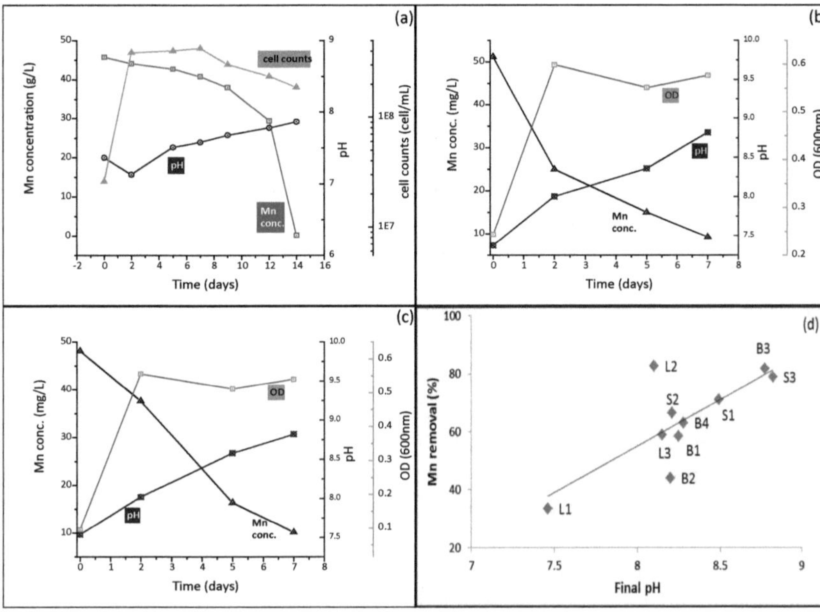

Fig. 1 Main parameters recorded during manganese removal by the bacterial enrichment CL (**a**), *Bacillus* (**b**) and *Stenotrophomonas* (**c**) selected isolates. Correlation between final pH and manganese removal for all isolates from the CL enrichment (**d**)

highest removal (82.7%) of all isolates. A correlation between final pH and manganese removal during the tests by all isolates was demonstrated (Fig. 1d).

The important effect of pH on manganese removal is because Mn^{2+} oxidation by oxygen is thermodynamically favourable but kinetically slow and Hydroxyl ions (OH^-) are important catalysts of chemical manganese oxidation. Particularly the oxidation rate is second order [7] with respect to OH^- and increases 10 times as the pH is changed from pH 8.0–8.8 [8]. Solid Mn oxy-hydroxides are also catalysts of Mn^{2+} oxidation and $Mn(OH)_2$ is particularly important because if its solubility limit is exceed the precipitated Mn^{2+} hydroxide is quickly oxidized to Mn^{3+} and then Mn^{4+}. Su et al. [9] stated that chemical oxidation of Mn^{2+} was negligible for pH values below 8.2, but it increased rapidly to 70%, at pH 8.5. Similarly, Learman et al. [3] investigated biotic and abiotic manganese removal at pH 7.2 and reported 33 and 66% removal by biotic and abiotic contributions, respectively. These results imply that chemical- and biological manganese oxidation usually occurs concomitantly whenever pH is raised to values above the 8.0–8.5 range depending on the microorganism studied. The formation of oxidized manganese compounds was confirmed in the current work [10] as the pH increased to values above 8.5. Summarizing, because a pH increase was not observed in the abiotic tests it was postulated that the microbial metabolism accounted for a pH increase which catalysed manganese oxidation and its removal from solution.

Sulphate Reduction

Firstly, batchwise sulphate reduction by two different microbial enrichments was investigated in the presence of either lactate or ethanol as carbon source at both pH 7.0 and 5.5. Subsequently these two enrichments were utilizing to inoculate a fluidized bed reactor containing activated carbon as biomass carrier.

A previous work demonstrated that the of the cited enrichments (LG01) showed larger growth rates using lactate as the carbon source (6.38×10^6 cell/mL/h) as compared to ethanol (1.61×10^6 cell/mL/h), at pH 7.0. The number of generations for LG01 sample was 5.2–3.5, with lactate and ethanol, respectively at the same pH. When the culture was subjected to more acidic initial conditions (pH 5.5), the outcome in terms of number of generations was equal to 2.9 and 1.7, for lactate and ethanol, respectively. Similarly, for the UASB01 sample grown in lactate there was faster growth in neutral conditions (pH = 7.0) than in slightly acid pH (5.5), as the growth rate values of 2.34×10^6 and 9.04×10^6 cell/mL/h, respectively, were achieved, corresponding to a number of generations of 5.3 and 3.7, respectively [11].

The pH is an important factor in biological sulphate reduction as values below 6.0 and above 9.0 have been reported to inhibit sulphate reduction [12]. It was observed that the samples inoculated at neutral pH (7.0) showed higher sulphate removal efficiencies. For sample LG01, sulphate reduction attained 71% (1000 mg/L SO_4^{2-} remained in solution) with lactate as the carbon source as compared to 50% (1500 mg/L residual SO_4^{2-}) using ethanol. When sample LG01 was inoculated in more acidic pH (5.5), a decrease in SRB activity was verified which was expressed by lower sulphate removal efficiency, i.e., 43% with lactate and 16% with ethanol. This may be explained by the faster growth rate in the presence of lactate. For instance the growth rate was 220×10^4 cell/mL/h with lactate, whereas a figure two orders of magnitude lower was observed with ethanol (5.83×10^4 cell/mL/h). The findings of Cao et al. [13] indicated a high sulphate removal for a COD/SO_4^{2-} ratio of 3.0 and 7.0 as the initial pH, in the presence of lactate, confirming the results of the current work [11].

Subsequently the COD/SO_4^{2-} ratio was modified so that its effect on sulphate reduction could be determined. Figure 2 shows the change of sulphate concentration with time at different COD/SO_4^{2-} ratios using either lactate or ethanol as the carbon source for both enrichments. It is implied from Fig. 2 that even when working with high values of the COD/SO_4^{2-} ratios in the ethanol-containing reactor, the final sulphate concentrations were higher than when lactate was used as a carbon source. For instance, applying ethanol was the carbon source and a COD/SO_4^{2-} ratio of 1.8 produced a final sulphate concentration of 2.3 g/L, whereas a lower COD/SO_4^{2-} ratio (1.0) using lactate resulted in 1.2 g/L residual sulphate with sample USAB 01 (Fig. 1a). Moreover, sulphate removal was more efficient with sample LG01 (Fig. 1b) (67% sulphate removal with a final concentration of 900 mg/L) compared with that achieved by the UASB01 sample (55% sulphate removal—1200 mg/L residual sulphate). Further tests with the LG01 sample

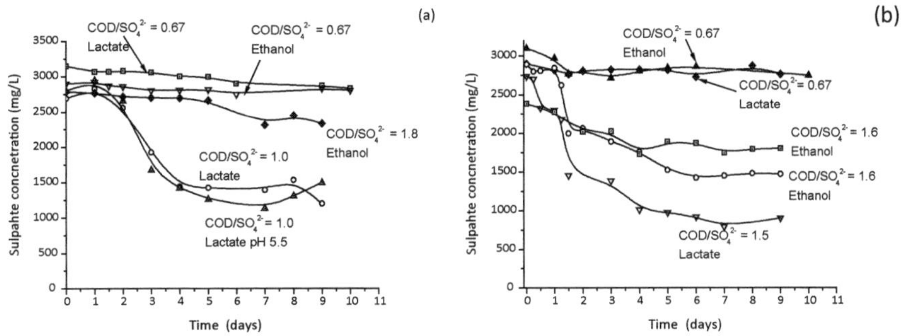

Fig. 2 Effect of carbon source, pH and the COD/SO_4^{2-} ratio on sulphate reduction by the samples UASB01 (**a**) and LG01 (**b**) in batch experiments

revealed that the COD/SO_4^{2-} ratio of 2 resulted in very low residual sulphate concentrations ($\sim$80 mg/L) [14]. This value corresponded precisely to the theoretical COD/SO_4^{2-} ratio when the incomplete oxidation of organic matter is assumed.

$$2CH_3COHCOO^- + SO_4^{2-} \rightarrow 2CH_3COO^- + HS^- + H^+ + 2HCO_3^- \qquad (1)$$

The enrichments UASB01 and LG01 were used to inoculate a fluidized-bed reactor which operated continuously by around 600 days with a COD/SO_4^{2-} ratio above 2 because of the incomplete substrate oxidation (lactate) by the different SRB groups (*Desulfovibrio, Desulfotomaculum, Desulfobulbus, Desulfomona*) detected in the reactor [15]. Overall the FBR operated quite stably with higher than 90% sulphate removal yields. However, only the phase in which the sulphate removal was the highest is discussed in the current study (between day 210 and 311 of reactor operation).

Worldwide discharge limits for sulphate in industrial wastewaters vary between 250 and 500 mg/L [16, 17]. For a target value of 250 mg L^{-1}, it can be seen in Fig. 3a that the FBR consistently produced residual sulphate concentrations below that limit, which corresponded to a sulphate reducing yield of 93.1 $\pm$ 1.49%. The average COD oxidation yield was 58.22 $\pm$ 2.91% for an inlet COD concentration set at 5.0 g/L as shown in Fig. 2b.

Lactate oxidation produces alkalinity (Eq. 1), which buffers the reactor at pH values which facilitates SRB growth. The alkalinity profile is presented in Fig. 3c as bicarbonate alkalinity (BA) and enabled the pH values to remain in the optimum range (above 8) for SRB growth [18], without any external alkalinity requirement. In addition the profiles of both the specific sulphate reduction and COD consumption rates are presented in Fig. 2d and the specific sulphate reduction rate was 0.186 $\pm$ 0.015gSO_4^{-2}/gVSS.d, whereas the specific COD consumption rate was 0.253 $\pm$ 0.009gCOD/gVSS.d. Such low COD removal can be justified by the incomplete oxidation of the organic matter and the consequent acetate production (Eq. 1) as discussed next.

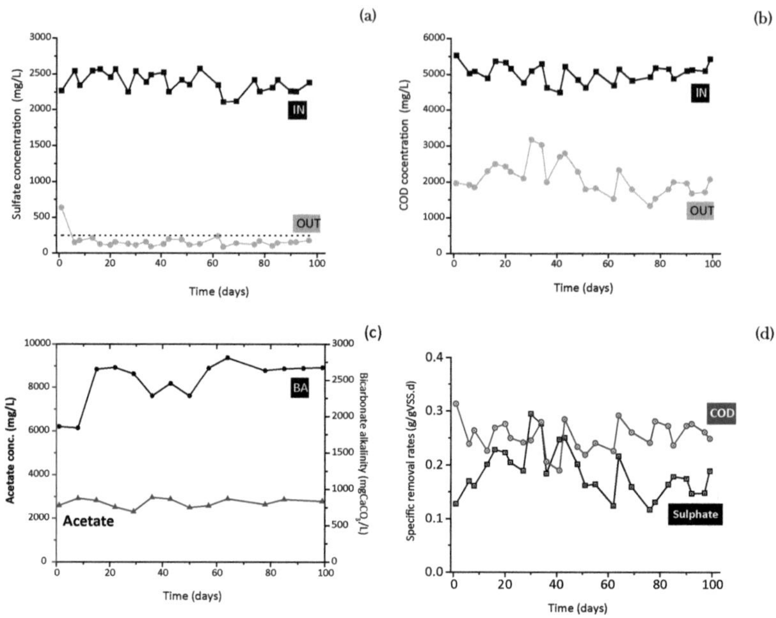

Fig. 3 Parameters monitored during sulphate reduction with lactate: COD (**a**), sulphate (**b**), alkalinity and acetate concentrations (**c**) and specific removal rates (**d**). The data shown represented the period in which the best reactor performance was achieved

From the four different fatty acids analysed in the system (acetic, propionic, valeric, butyric), only acetate (Fig. 3c) was detected, indicating incomplete lactate oxidation during sulphate reduction in the continuous experiment (Eq. 1). Furthermore, the absence of propionate suggested no competition between SRB and fermentative bacteria for the organic substrate during sulphate reduction. This outcome may be explained by the by the high SRB population (above 10^9 free cells/mL) measured in the reactor, a result of both fluidization (which improved mass transfer) and SRB retention by carbon particles. Therefore the FBR consistently produced high sulphate reduction provided that a COD/SO_4^{2-} ratio above 2 is employed.

Conclusions

Removing manganese and sulphate are two challenges for an efficient water management in the mineral industry. Among the plethora of technologies available biotechnological alternatives can be also considered and while sulphate reduction has already been industrially proposed, manganese bioremediation is usually associated with passive mine water treatments and no strict control of the microbial population is carried out in this latter case. The main issue with sulphate reduction is the high consumption of the organic source as COD/SO_4^{2-} ratios above 2 were required for an efficient sulphate reduction as demonstrated herein. Conversely the current work indicates that indirect manganese removal may play a key role because manganese oxidation was catalysed the high pH values resulting from bacterial activity.

Acknowledgements Funding provided by Vale and the agencies FINEP, CAPES, CNPq and FAPEMIG are acknowledged.

References

1. Akcil, A., & Koldas, S. (2006). Acid mine drainage (AMD): Causes, treatment and case studies. *Journal of Cleaner Production, 14*(12–13), 1139–1145.
2. White III, W. W., Lapakko, K. A., & Cox, R. L. (1999). The environmental geochemistry of mineral deposits, part A: Theory and background. In G. S. Plumlee & M. Logsdon (Eds.), *Society of Economic Geologists Reviews in Economic Geology* (Vol. 7A, pp. 325–338).
3. Learman, D. R., Wankel, S. D., Webb, S. M., Martinez, N., Madden, A. S., & Hansel, C. M. (2011). Coupled biotic–abiotic Mn(II) oxidation pathway mediates the formation and structural evolution of biogenic Mn oxides. *Geochimica et Cosmochimica Acta, 75*(20), 6048–6063.
4. Miyata, N., Tani, Y., Iwahori, K., & Soma, M. (2004). Enzymatic formation of manganese oxides by an Acremonium-like hyphomycete fungus, strain KR21-2. *FEMS Microbiology Ecology, 47*(1), 101–109.
5. Postgate, J. R. (1963). Versatile medium for the enumeration of sulfate-reducing bacteria. *Applied Microbiology, 11*(3), 265–267.
6. Bertolino, S. M., Melgaço, L. A., Sá, R. G., & Leão, V. A. (2014). Comparing lactate and glycerol as a single-electron donor for sulfate reduction in fluidized bed reactors. *Biodegradation, 25*(5), 719–733.
7. von Langen, P. J., Johnson, K. S., Coale, K. H., & Elrod, V. A. (1997). Oxidation kinetics of manganese (II) in seawater at nanomolar concentrations. *Geochimica et Cosmochimica Acta, 61*(23), 4945–4954.
8. Morgan, J. J. (2005). Kinetics of reaction between O_2 and Mn(II) species in aqueous solutions. *Geochimica et Cosmochimica Acta, 69*(1), 35–48.
9. Su, J., Bao, P., Bai, T., Deng, L., Wu, H., Liu, F., et al. (2013). CotA, a multicopper oxidase from *Bacillus pumilus* WH4, exhibits manganese-oxidase activity. *PLoS ONE, 8*(4), e60573.
10. Barboza, N. R., Amorim, S. S., Santos, P. A., Reis, F. D., Cordeiro, M. M., Guerra-Sá, R., et al. (2015). Indirect manganese removal by *Stenotrophomonas* sp. and *Lysinibacillus* sp. isolated from Brazilian mine water. *BioMed Research International, 2015*, 14.
11. Barbosa, L. d. P. (2009). *Cultivo de bactérias redutoras de sulfato (BRS) e s ua aplicação na biorremediação de efluentes ácidos contendo metais* (Master thesis). Universidade Federal de Ouro Preto (p. 130).
12. Widdel, F. (1988). Microbiology and ecology of sulfate- and sulfur-reducing bacteria. *Biology of anaerobic microorganisms* (pp. 469–585). New York: Willey.
13. Cao, J., Zhang, G., Mao, Z., Fang, Z., & Yang, C. (2008). Presipitation of valuable metal from bioleaching solution by biogenic sulfides. *Minerals Engineering, 135*(1–3), 40–46.
14. Barbosa, L. P., Costa, P. F., Bertolino, S. M., Silva, J. C. C., Guerra-Sá, R., Leão, V. A., et al. (2014). Nickel, manganese and copper removal by a mixed consortium of sulfate reducing bacteria at a high COD/sulfate ratio. *World Journal of Microbiology & Biotechnology, 30*(8), 2171–2180.

15. Bertolino, S. M., Silva, L. A. M., Aquino, S. F., & Leão, V. A. (2015). Comparison of UASB and fluidized-bed reactors for sulfate reduction. *Brazilian Journal of Chemical Engineering, 32*, 59–71.
16. INAP. (2003). *Treatment of sulphate in mine effluents*. International network for acid prevention.
17. World Health Organization (WHO). (2011). *Guidelines for drinking-water quality*.
18. Barton, L. L. (1995). *Sulfate-reducing bacteria* (Vol. 8, p. 336). New York: Plenum Press.

Dissolution Thermodynamics of Smithsonite in Alkaline Iminodiacetate Aqueous Solution

Aichun Dou, Lei Yu, Mingru Su and Yunjian Liu

Abstract The thermodynamic model of the dissolution of smithsonite ($ZnCO_3$) in alkaline iminodiacetate aqueous solution (Ida^{2-}–H_2O system) was constructed according to the thermodynamic equilibrium equations in the system. Thermodynamic results of total concentration of Zn^{2+}, free concentration of Zn^{2+}, free concentration of Ida^{2-}, free concentration of ${CO_3}^{2-}$, species distribution of Zn^{2+} and species distribution of Ida^{2-} changed with total concentration of Ida^{2-}, and pH value were investigated respectively in the system at 298 K. The total concentration of Zn^{2+} increases with increasing total concentration of Ida^{2-} at constant pH value. The optimal pH area for the dissolution of $ZnCO_3$ in Ida^{2-}–H_2O system is pH 8–11. In this area, the total concentration of Zn^{2+} increases smoothly with increasing of pH, above 90% of Ida^{2-} is used for the formation of $Zn(Ida)_2^{2-}$, and above 99% of Zn^{2+} is formed as $Zn(Ida)_2^{2-}$.

Keywords Thermodynamic · Smithsonite · $ZnCo_3$ · Ida^{2-} · Zinc oxide

Introduction

Zinc sulphide ores are easily upgraded by flotation, and are exhausted with the development of zinc industry. Secondary zinc oxide materials such as zinc oxide ores and zinc containing dusts draw more and more attention as a result. Secondary zinc oxide materials usually contain valuable metals such as Pb, Cd, Cu and Ni, along with impurities such as Ca, Mg, Fe, Si and C. It is a significant concern to separate valuable metals from impurities efficiently in these secondary zinc oxide materials.

A. Dou (✉) · L. Yu · M. Su · Y. Liu
School of Material Science and Engineering, Jiangsu University, 301 Xuefu Road, Zhenjiang, Jiangsu Province, People's Republic of China
e-mail: aichund@163.com

L. Yu
e-mail: 15751003673@163.COM

© The Minerals, Metals & Materials Society 2017
M.A. Meyers et al. (eds.), *Proceedings of the 3rd Pan American Materials Congress*, The Minerals, Metals & Materials Series, DOI 10.1007/978-3-319-52132-9_49

Several hydrometallurgical methods have been developed for dealing with complex secondary zinc oxide materials. Acid leaching [1–5] is a traditional method for zinc extraction. It is also as the leading process in zinc metallurgy for its high efficiency. Acid leaching is more suitable for those secondary materials with high zinc content. Otherwise, the consumption of acid is very high and the impurities in leaching liquor are difficult to separate from valuable metals. The complex-leaching process was introduced to treat these secondary materials. Ammonia leaching and alkaline leaching are both selective methods for Zn extraction by the formation of complexes. In ammonia leaching [6–10], NH_3, serving as the main ligand, can combine with Zn^{2+} and bring it into solution with little Ca, Mg, Fe and Si dissolution. However, ammonia in the system evaporates easily, which causes environment pollution. OH^- is used as a leaching ligand to coordinate with Zn^{2+} for zinc extraction from materials in alkaline leaching [11–16]. It is a challenge for facilities to resist the corrosion by strong basic liquors in these operations. Moreover, SiO_2 will be dissolved in the leaching liquor as silicates under the high concentration of OH^- in the process.

Many works have been done about ammonia leaching and alkaline leaching for dealing with these secondary materials. Good techno-economic indicators have been obtained in operation in these two processes, but ligands are only limited to NH_3, OH^- and Cl^-. In order to introduce some new ligands more appropriate to leach these secondary zinc oxide materials, some works had been done by authors [17–21] about complex-leaching process with selectivity in alkaline iminodiacetate aqueous solution (Ida^{2-}–H_2O system). Ida^{2-}, as a novel investigated ligand, its coordination ability with Zn^{2+} is stronger than ligands of NH_3, OH^- and Cl^-. Moreover, Ida^{2-} is a friendly ligand to environment. There are few researches on treating with secondary zinc oxide materials in Ida^{2-}–H_2O system. Studies on thermodynamics in Zn^{2+}–CO_3^{2-}–Ida^{2-}–H_2O system have not been detailed reported yet, and which may cause doubts in practice. Smithsonite ($ZnCO_3$) is one of the main phases of Zn in zinc oxide ores. It is necessary to get the dissolution characteristics of $ZnCO_3$ in iminodiacetate aqueous solution for zinc exaction. In this paper, the dissolution thermodynamics of $ZnCO_3$ in Ida^{2-}–H_2O system was investigated. In addition, this paper describes the ions and their concentrations. The total concentration and the free concentration of ions are unified as $[M]_T$ and $[M]$ respectively for simplicity.

Thermodynamic

Thermodynamic Model

In Zn^{2+}–CO_3^{2-}–Ida^{2-}–H_2O system, reactions contain the dissociation reaction of insoluble compound ($ZnCO_3$), complexion of Ida^{2-} and OH^- with Zn^{2+}, protonation of Ida^{2-} and CO_3^{2-} with H^+ and the dissociation reaction of H_2O. The thermodynamic model can be constructed with the thermodynamic equilibrium equations from these reactions in the system.

Dissociation Equilibrium of $ZnCO_3$

In aqueous solution, the dissociation reaction of $ZnCO_3$ is as follows

$$ZnCO_3 \rightleftharpoons Zn^{2+} + CO_3^{2-} \tag{1}$$

The solubility product constant of $ZnCO_3$ can be written as

$$K_{sp} = [Zn^{2+}][CO_3^{2-}] \tag{2}$$

It can be seen from Eq. (1), when one mole of Zn^{2+} is released from $ZnCO_3$, one mole of CO_3^{2-} is also released. So an equation can be confirmed as

$$[Zn^{2+}]_T = [CO_3^{2-}]_T \tag{3}$$

The concentration of $CO_{2(g)}$ in the aqueous solution is too low to be considered in the system.

Equilibrium of Zn^{2+}

In the Zn^{2+}–CO_3^{2-}–Ida^{2-}–H_2O system, both Ida^{2-} and OH^- can combine with Zn^{2+}. Reactions and their critical stability constants can be described as

$$Zn^{2+} + rIda^{2-} \rightleftharpoons Zn(Ida)_r^{2-2r} \quad \beta_r = \frac{[Zn(Ida)_r^{2-2r}]}{[Zn^{2+}][Ida^{2-}]^r} \quad (r = 1, 2) \tag{4}$$

$$Zn^{2+} + jOH^- \rightleftharpoons Zn(OH)_j^{2-j} \quad \beta_j = \frac{[Zn(OH)_j^{2-j}]}{[Zn^{2+}][OH^-]^j} \quad (j = 1, 2, 3, 4) \tag{5}$$

Thus, the total concentration of Zn^{2+} can be expressed as

$$[Zn^{2+}]_T = [Zn^{2+}]\left(1 + \sum_{r=1}^{2} \beta_r [Ida^{2-}]^r + \sum_{j=1}^{4} \beta_j [OH^-]^j\right) \tag{6}$$

Equilibrium of Ida^{2-}

Ida^{2-} not only combines with Zn^{2+} for complexes, but also be protonated by H^+ in Zn^{2+}–$CO_3{}^{2-}$–Ida^{2-}–H_2O system, because Ida^{2-} is a weak acid radical. The protonation of Ida^{2-} with H^+ and the reaction constants can be written as

$$\mathrm{Ida}^{2-} + x\mathrm{H}^+ \rightleftharpoons \mathrm{H}_x\mathrm{Ida}^{x-2} \quad \beta_x = \frac{[\mathrm{H}_x\mathrm{Ida}^{x-2}]}{[\mathrm{Ida}^{2-}][\mathrm{H}^+]^x} \quad (x = 1, 2, 3) \tag{7}$$

According to Eqs. (4) and (7), the total concentration of Ida^{2-} can be represented as

$$[\mathrm{Ida}^{2-}]_\mathrm{T} = [\mathrm{Ida}^{2-}]\left(1 + \sum_{x=1}^{3} \beta_x[\mathrm{H}^+]^x\right) + [\mathrm{Zn}^{2+}]r\sum_{r=1}^{2} \beta_r[\mathrm{Ida}^{2-}]^r \tag{8}$$

Equilibrium of $CO_3{}^{2-}$

Because $CO_3{}^{2-}$ is also a weak acid radical, $CO_3{}^{2-}$ can also be protonated by H^+ in Zn^{2+}–$CO_3{}^{2-}$–Ida^{2-}–H_2O system. The protonation of $CO_3{}^{2-}$ with H^+ and its reaction constants can be described as

$$\mathrm{CO}_3^{2-} + y\mathrm{H}^+ \rightleftharpoons \mathrm{H}_y\mathrm{CO}_3^{y-2} \quad \beta_y = \frac{[\mathrm{H}_y\mathrm{CO}_3^{y-2}]}{[\mathrm{CO}_3^{2-}][\mathrm{H}^+]^y} \quad (y = 1, 2) \tag{9}$$

$CO_3{}^{2-}$ cannot be combined with Zn^{2+} for complexes, so the total concentration of $CO_3{}^{2-}$ is expressed as

$$[\mathrm{CO}_3^{2-}]_\mathrm{T} = [\mathrm{CO}_3^{2-}]\left(1 + \sum_{y=1}^{2} \beta_y[\mathrm{H}^+]^y\right) \tag{10}$$

Dissociation Equilibrium of H_2O

The dissociation reaction of H_2O is represented as

$$\mathrm{K_w} = [\mathrm{H}^+][\mathrm{OH}^-] \tag{11}$$

Table 1 Thermodynamic data of related species at 298 K

Species	$\lg\beta_i$	Ref.	Species	$\lg\beta_i$	Ref.	Species	$\lg\beta_i$	Ref.
$^*ZnCO_3$	−10	[22]	$Zn(OH)_{2(aq)}$	11.30	[23]	$HIda^-$	9.34	[24]
$^*H_2O$	−14	[23]	$Zn(OH)_3^-$	14.14	[23]	H_2Ida	11.95	[24]
Zn(Ida)	7.24	[24]	$Zn(OH)_4^{2-}$	17.66	[23]	H_3Ida^+	13.77	[24]
$Zn(Ida)_2^{2-}$	12.52	[24]	HCO_3^-	9.57	[22]			
$Zn(OH)^+$	4.40	[23]	$H_2CO_{3(aq)}$	15.59	[22]			

$^*\lg\beta_i$ of $ZnCO_3$ and H_2O mean log K_{sp} and log K_w respectively

Calculation

There are eight unknowns containing $[Zn^{2+}]$, $[Zn^{2+}]_T$, $[CO_3^{2-}]$, $[CO_3^{2-}]_T$, $[Ida^{2-}]$, $[Ida^{2-}]_T$, $[H^+]$ and $[OH^-]$ in Eqs. (2), (3), (6), (8), (10) and (11). Constants for these equations can be obtained in handbooks of thermodynamic data or by calculation. Thus, if two of the eight unknowns were given, the other six unknowns can be calculated by these six equations. In operation, the pH value ($[H^+]$ or $[OH^-]$) and the total concentration of Ida^{2-} ($[Ida^{2-}]_T$) are under control and are often used for calculating the other six unknowns. The related thermodynamic data at 298 K are listed in Table 1.

Results and Discussion

In this paper, the concentration of H^+ ($[H^+]$) and the total concentration of Ida^{2-} are set as constants for calculation. Calculated results were all represented as 3D graphs which varied with pH value (6–12) and $[Ida^{2-}]_T$ (0.5–2.5 mol L^{-1}).

Total Concentration of Zn^{2+}

The variety of total concentration of Zn^{2+} is shown in Fig. 1. Figure 1 shows that $[Zn^{2+}]_T$ increases with increasing $[Ida^{2-}]_T$, and there are three stages change of $[Zn^{2+}]_T$ versus the pH value. At pH 6–8, $[Zn^{2+}]_T$ increases gradually. It means more and more free Ida^{2-} was released from H_yIda^{y-2} (Eq. 7) with decreasing acidity in the system, and more free Ida^{2-} participates in the complexation with Zn^{2+} (Eq. 4). At pH 8–11, the protonation of Ida^{2-} almost disappears, and the free Ida^{2-} is almost completely used for combining with Zn^{2+}. It causes $[Zn^{2+}]_T$ to increases smoothly in this area. At pH 11–12, OH^- takes part in the complexation with Zn^{2+} (Eq. 5). The complexation of both Ida^{2-} and OH^- with Zn^{2+} causes the rapid increase of $[Zn^{2+}]_T$ in high alkalinity.

Fig. 1 Impact of pH and $[Ida^{2-}]_T$ on $[Zn^{2+}]_T$

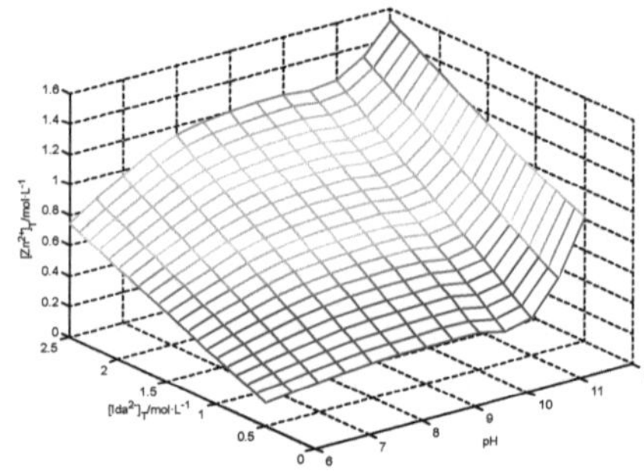

It seems that $ZnCO_3$ can be dissolved efficiently in high alkalinity (pH > 11). But impurities such as As and Si which are undesirable in leaching liquor can also be dissolved, which is bad for the selective leaching process. Thus the proper pH area for complex-leaching of $ZnCO_3$ is from 8–11.

Free Concentration of Zn^{2+}

The result of free concentration of Zn^{2+} is an important index for complexation of Ida^{2-} with Zn^{2+}. It means that the more free Zn^{2+} is present in the Zn^{2+}–$CO_3{}^{2-}$–Ida^{2-}–H_2O system, the less Zn^{2+} is combined with Ida^{2-} and OH^-. The variety of $[Zn^{2+}]$ versus pH and the total concentration of Ida^{2-} is shown in Fig. 2. It can be seen from Fig. 2 that $[Zn^{2+}]$ decreases sharply at pH 6–8, and is very low when pH is above 8. At pH 6–8, part of the Ida^{2-} is also protonated in the system, so there has not enough free Ida^{2-} to combine with Zn^{2+}. Meanwhile, the protonation of Ida^{2-} is weakened with the decrease of the concentration of H^+. H_yIda^{y-2} can release more free Ida^{2-} for complexion with Zn^{2+}, which causes the decrease of $[Zn^{2+}]$ with increasing pH. The decrease of $[Zn^{2+}]$ changes smoothly later which can be seen clearly in Fig. 3. It indicates that Zn^{2+} is almost coordinated by Ida^{2-} and OH^- in this area.

Fig. 2 Impact of pH and $[Ida^{2-}]_T$ on $[Zn^{2+}]$

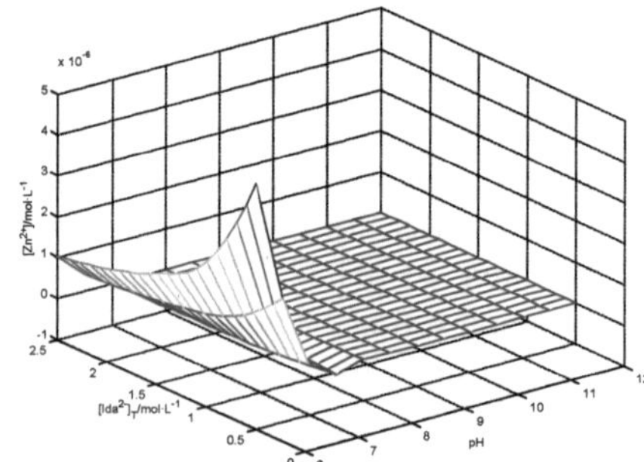

Fig. 3 Impact of pH and $[Ida^{2-}]_T$ on log $[Zn^{2+}]$

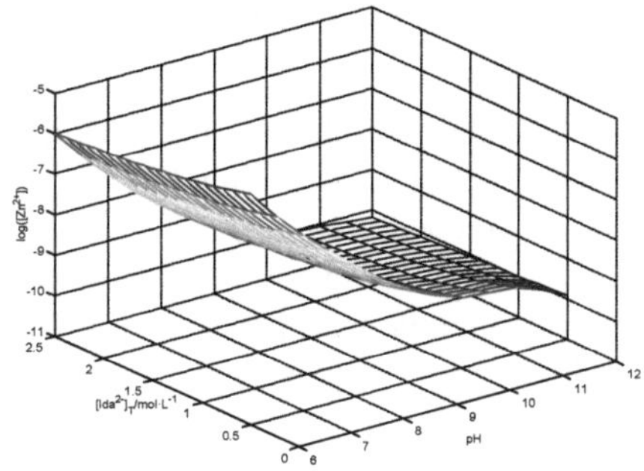

Free Concentration of Ida^{2-}

The result of free concentration of Ida^{2-} ($[Ida^{2-}]$) is an important index for proceedings of protonation of Ida^{2-} and complexion of Ida^{2-} with Zn^{2+}. With the increase of pH, the protonation of Ida^{2-} is weakened which can release free ion of Ida^{2-}. Meanwhile, the complexion of Ida^{2-} with Zn^{2+} can consume the free ion of Ida^{2-}. So $[Ida^{2-}]$ is controlled by the protonation of Ida^{2-} and the complexion of Ida^{2-} with Zn^{2+}. The variety of $[Ida^{2-}]$ versus pH and the total concentration of Ida^{2-} is shown in Fig. 4.

Figure 4 shows that $[Ida^{2-}]$ always increases with increasing pH. It indicates that $[Ida^{2-}]$ which released is always over the consumption. But there are also two stages of the variety in Fig. 4. $[Ida^{2-}]$ increases more quickly at pH 6–10 than at pH 10–12. Protonated of Ida^{2-} disappears when the pH is above 10, so there is no more free ion of Ida^{2-} released from H_yIda^{y-2}. But Zn^{2+} still needs free Ida^{2-} to form higher level complex $Zn(Ida)_2^{2-}$. Thus $[Ida^{2-}]$ increases more slowly at pH 10–12 than at pH 6–10.

Fig. 4 Tendency of $[Ida^{2-}]$ changed with pH and $[Ida^{2-}]_T$

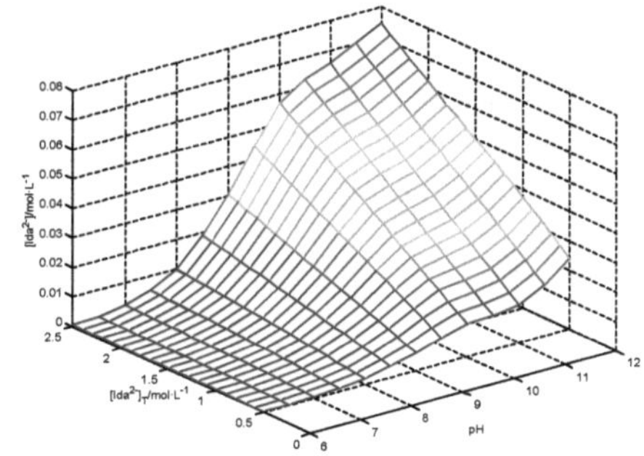

Free Concentration of CO_3^{2-}

CO_3^{2-} cannot combine with Zn^{2+} in the Zn^{2+}–CO_3^{2-}–Ida^{2-}–H_2O system, but the protonation of CO_3^{2-} with H^+. The free concentration of CO_3^{2-} is restricted by the free concentration of Zn^{2+} from the solubility product rule of $ZnCO_3$ (Eq. 2). So the variety of $[CO_3^{2-}]$ versus pH and the total concentration of Ida^{2-} exhibits opposite change with $[Zn^{2+}]$ in Figs. 2 and 3. It can be seen that $[CO_3^{2-}]$ increases gradually with increasing pH in Fig. 5.

Because $[CO_3^{2-}]_T$ equals $[Zn^{2+}]_T$ all the time Zn^{2+}–CO_3^{2-}–Ida^{2-}–H_2O system for ions of all CO_3^{2-} and Zn^{2+} come from the dissolution of $ZnCO_3$ (Eq. 2). It indicated that $[CO_3^{2-}]_T$ increases with the increase of pH according to Fig. 1. Meanwhile, the protonation of CO_3^{2-} is also weakened with the increase of pH. Thus can cause the increasing of $[CO_3^{2-}]$ with the increase of pH in Fig. 5. For there are no reactions for CO_3^{2-} with Zn^{2+} but its protonation with H^+, $[CO_3^{2-}]$ equals $[CO_3^{2-}]_T$ when the protonation of CO_3^{2-} is disappeared in high pH area. Moreover, $[CO_3^{2-}]_T$ equals $[Zn^{2+}]_T$ all the time Zn^{2+}–CO_3^{2-}–Ida^{2-}–H_2O system. Thus, $[CO_3^{2-}]$ should equal $[Zn^{2+}]_T$ when the protonation of CO_3^{2-} is disappeared in high pH area. It can be seen clearly from Figs. 5 and 1 at pH 11–12, which indicates that protonated CO_3^{2-} almost disappears as the pH approaches 11.

Species Distribution of Zn^{2+}

The species of zinc are Zn^{2+}, Zn(Ida), $Zn(Ida)_2^{2-}$, $Zn(OH)^+$, $Zn(OH)_{2(aq)}$, $Zn(OH)_3^-$ and $Zn(OH)_2^{2-}$ in the Zn^{2+}–CO_3^{2-}–Ida^{2-}–H_2O system. The concentrations of Zn^{2+} in these species were all counted to the total concentration of Zn^{2+}. So the distribution of Zn^{2+} means the concentration of Zn^{2+} from one certain specie occupies the percentage of $[Zn^{2+}]_T$. The results of the distribution of Zn^{2+} in the Zn^{2+}–CO_3^{2-}–Ida^{2-}–H_2O system is shown in Fig. 6.

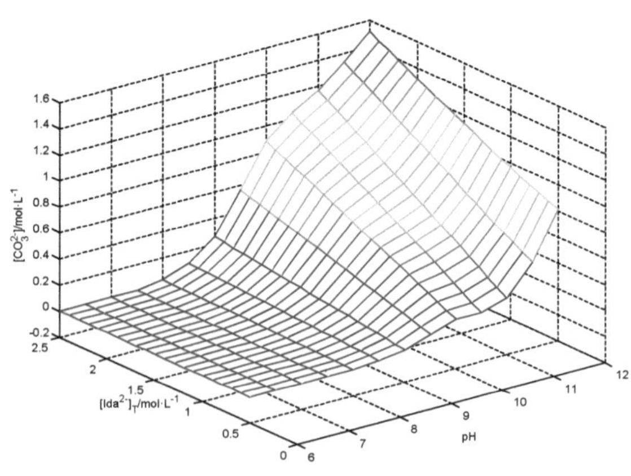

Fig. 5 Tendency of $[CO_3^{2-}]$ changed with pH and $[Ida^{2-}]_T$

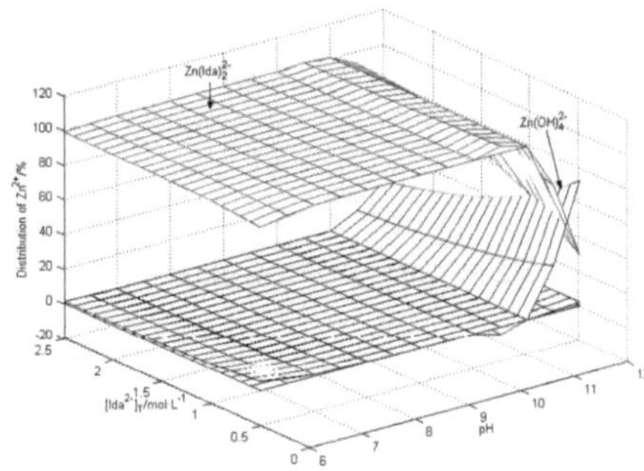

Fig. 6 Species distribution of Zn^{2+} in Zn^{2+}–$CO_3{}^{2-}$–Ida^{2-}–H_2O system

Figure 6 shows that $Zn(Ida)_2^{2-}$ is the main complex of Zn^{2+} at pH 6–11. And the distribution of Zn^+ from $Zn(Ida)_2^{2-}$ is over 95%. When pH is above 11, OH^- takes part in the complexion with Zn^{2+}. It can be seen from Fig. 6 that the distribution of Zn^+ from $Zn(OH)_4^{2-}$ increases strongly, and the distribution of Zn^+ from $Zn(Ida)_2^{2-}$ decreases sharply with increasing pH at pH 11–12. Thus, it can be proved that both Ida^{2-} and OH^- complex with Zn^+ to increase of $[Zn^{2+}]_T$ in Fig. 1 at pH 11–12.

Species Distribution of Ida^{2-}

The species which containing Ida^{2-} are Ida^{2-}, Zn(Ida), $Zn(Ida)_2^{2-}$, $HIda^-$, H_2Ida and H_3Ida^- in the Zn^{2+}–$CO_3{}^{2-}$–Ida^{2-}–H_2O system. The concentrations of Ida^{2-} in these species were all counted to the total concentration of Ida^{2-}. So the distribution of Ida^{2-} means the concentration of Ida^{2-} from one certain specie occupies the percentage of $[Ida^{2-}]_T$. The results of the distribution of Ida^{2-} in Zn^{2+}–$CO_3{}^{2-}$–Ida^{2-}–H_2O system is shown in Fig. 7.

Figure 7 shows that the protonation of Ida^{2-} persists until the pH reaches 10. It can prove the variety of $[Ida^{2-}]$ in Fig. 4. In Fig. 7, from pH 6 to 10, the protonation of Ida^{2-} is weakened as $[H^+]$ decreases; the distribution of Ida^{2-} from $HIda^-$ decreases from 41 to 0.5%. Meanwhile, free Ida^{2-} released from $HIda^-$ is used for

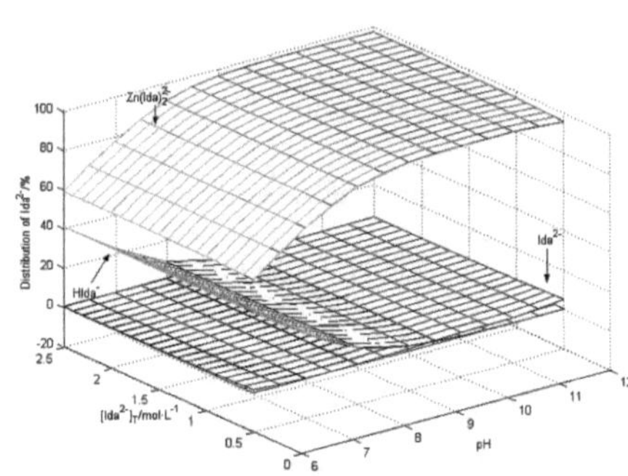

Fig. 7 Species distribution of Ida^{2-} in Zn^{2+}–$CO_3{}^{2-}$–Ida^{2-}–H_2O system

complexion with Zn^{2+} as $Zn(Ida)_2^{2-}$, it causes increased distribution of Ida^{2-} from $Zn(Ida)_2^{2-}$. More than 97% of the $[Ida^{2-}]_T$ is present as $Zn(Ida)_2^{2-}$ when pH approaches 10. The distribution of Ida^{2-} from $Zn(Ida)_2^{2-}$ continues increasing smoothly through pH 11. From pH 11–12, the distribution of Ida^{2-} from $Zn(Ida)_2^{2-}$ decreases slightly. OH^- is as a competitive ligand for complexion with Zn^{2+}, which can replace part of the ligand of Ida^{2-} in $Zn(Ida)_2^{2-}$ in this pH area.

Conclusions

(1) The total concentration of Ida^{2-} has a significant effect on the total concentration of Zn^{2+}. The total concentration of Zn^{2+} increases with increasing total concentration of Ida^{2-} at a given pH value.

(2) The pH value is the most important factor which affects the total concentration of Zn^{2+}. In the area of pH 6–8, because of the protonation of Ida^{2-} declines with decrease of acidity in the system, the total concentration of Zn^{2+} increases gradually with the increase of pH. In the area of pH 8–11, because of the vast majority of the ligand Ida^{2-} is used for the complexion with Zn^{2+} as $Zn(Ida)_2^{2-}$, the total concentration of Zn^{2+} increases smoothly with the increase of pH. While in the area of pH 11–12, because of OH^- takes part in the complexion with Zn^{2+} in Zn^{2+}–$CO_3{}^{2-}$–Ida^{2-}–H_2O system, both Ida^{2-} and OH^- complex with Zn^+ cause the total concentration of Zn^{2+} increases rapidly in high alkalinity.

(3) The optimal pH area for smithsonite dissolution in Ida^{2-}–H_2O system is pH 8–11. In this area, above 90% of Ida^{2-} is used for the formation of $Zn(Ida)_2^{2-}$, and above 99% of Zn^{2+} is formed as $Zn(Ida)_2^{2-}$.

Acknowledgements The author appreciates professor YANG Tianzu for his help in this work. The project was supported by the Natural Science Foundation of China (51604124), the Natural Science Foundation of Jiangsu Province (BK20140558) and the Advanced Foundation of Jiangsu University (13JDG097).

References

1. Souza, A. D., Pina, P. S., Santos, F. M. F., da Silva, C. A., & Leão, V. A. (2009). Effect of iron in zinc silicate concentrate on leaching with sulphuric acid. *Hydrometallurgy, 95*, 207–214.
2. Li, C. X., Xu, H. S, Deng, Z. G., Li, X. B., Li, M. T., & Wei, C. (2010). Pressure leaching of zinc silicate ore in sulfuric acid medium. *Transactions of Nonferrous Metals Society of China, 20*, 918–923.
3. Sun, Y., Sheng, X. Y., & Zhai, Y. C. (2015). Thermodynamics and kinetics of extracting zinc from zinc oxide ore by the ammonium sulfate roasting method. *International Journal of Minerals, Metallurgy and Materials, 22*, 467–474.

4. Kul, M., Oskay, K. O., Simsir, M., Subutay, H., & Kirgezen, H. (2015). Optimization of selective leaching of Zn away electric arc furnace steelmaking dust using response surface methodology. *Transactions of Nonferrous Metals Society of China, 25*, 2753–2762.
5. Kukurugya, F., Vindt, T., & Havlik, T. (2015). Behavior of zinc, iron and calcium from electric arc furnace (EAF) dust in hydrometallurgical processing in sulfuric acid solutions: Thermodynamic and kinetic aspects. *Hydrometallurgy, 154*, 20–32.
6. Yin, Z. L., Ding, Z. Y., Hu, H. P., Liu, K., & Chen, Q. Y. (2010). Dissolution of zinc silicate (hemimorphite) with ammonia-ammonium chloride solution. *Hydrometallurgy, 103*, 215–220.
7. Yin, Z. L., Ding, Z. Y., Hu, H. P., & Chen, Q. Y. (2010). Dissolution kinetics of zinc silicate (hemimorphite) in ammoniacal solution. *Hydrometallurgy, 104*, 201–206.
8. Xia, Z. M., Yang, S. H., Tang, M. T., Yang, T. Z., Liu, Z. H., Tang, C. B., et al. (2013). Cycle leaching of low grade zinc oxide ores in MACA system for preparing zinc. *The Chinese Journal of Nonferrous Metals, 23*, 3455–3461 (in Chinese).
9. Tang, M. T., Zhang, J. L., Wang, B., Yang, S. H., He, J., Tang, C. B., et al. (2011). Cycle leaching of low grade zinc oxide ores in MACA system. *The Chinese Journal of Nonferrous Metals, 21*, 214–219 (in Chinese).
10. Miki, T., Chairaksa-Fujimoto, R., Maruyama, K., & Nagasaka, T. (2016). Hydrometallurgical extraction of zinc from CaO treated EAF dust in ammonium chloride solution. *Journal of Hazardous Materials, 302*, 90–96.
11. Yuan, T. C., Cao, Q. Y., & Li, J. (2010). Effects of mechanical activation on physicochemical properties and alkaline leaching of hemimorphite. *Hydrometallurgy, 104*, 136–141.
12. Zhang, Y. C., Deng, J. X., Chen, J., Yu, R. B., & Xing, X. R. (2013). Leaching of zinc from calcined smithsonite using sodium hydroxide. *Hydrometallurgy, 131–132*, 89–92.
13. Chen, A. L., Zhao, Z. W., Xu, D., Liu, X. H., & Chen, X. Y. (2013). Effect of sodium silicate on the solubility of zinc oxide of the Na_2O-ZnO-H_2O equilibrium system. *Hydrometallurgy, 136*, 46–50.
14. Zhang, Y. C., Deng, J. X., Chen, J., Yu, R. B., & Xing, X. R. (2014). A low-cost and large-scale synthesis of nano-zinc oxide from smithsonite. *Inorganic Chemistry Communication, 43*, 138–141.
15. Chairaksa-Fujimoto, R., Maruyama, K., Miki, T., & Nagasaka, T. (2016). The selective alkaline leaching of zinc oxide from electric arc furnace dust pre-treated with calcium oxide. *Hydrometallurgy, 159*, 120–125.
16. Stefanova, A., Aromaa, J., & Forsen, O. (2013). Alkaline leaching of zinc from argon oxygen decarburization dust from stainless steel production. *Physicochemical Problems of Mineral Processing, 49*, 37–46.
17. Dou, A. C., Yang, T. Z., Yang, J. X., Wu, J. H., & Wang, A. (2011). Leaching of low grade zinc oxide ores in Ida^{2-}-H_2O system. *Transactions of Nonferrous Metals Society of China, 21*, 2548–2553.
18. Dou, A. C., Yang, T. Z., Wu, J. H., Zhang, D. C., Yang, J. X., & Wang, A. (2011). Recovery of zinc and regeneration of lixiviant from Zn^{2+}-Ida^{2-}-CO_3^{2-}-H_2O system. *The Chinese Journal of Nonferrous Metals, 21*, 3218–3225 (in Chinese).
19. Yang, T. Z., Dou, A. C., Lei, C. M., Ren, J., & Liu, Z. Z. (2010). Ligand selection for complex-leaching valuable metals in hydrometallurgy. *Transactions of Nonferrous Metals Society of China, 20*, 1148–1153.
20. Dou, A. C., Yang, T. Z., Liu, Y. J., & Su, M. R. (2015). Cycle leaching of low grade zinc oxide ores in Ida^{2-}-H_2O system. *The Chinese Journal of Nonferrous Metals, 25*, 3229–3236 (in Chinese).
21. Dou, A. C. (2012). *Fundamental and technological study on treating low grade zinc oxide ore with alkaline iminodiacetate aqueous solution (Ph.D. thesis)*, pp. 69–118. Central South University, China (in Chinese).

22. Martell, A. E., & Smith, R. M. (1974). Critical stability constants (Vol. 4. pp. 37–38). New York and London: Plenum Press.
23. Dean, J. A. (2003). *Lange's handbook of chemistry* (15th ed., pp. 6.143–6.144, 6.120, 8.81–8.82, J. F. Wei, Trans.). Beijing: Science Press.
24. Martell, A. E., & Smith, R. M. (1974). Critical stability constants (Vol. 1, pp. 116–118). New York and London: Plenum Press.

Effect of Ethylenediamine on Smithsonite Flotation

Chao Lv, Shuming Wen, Shaojun Bai and Kun Yang

Abstract Smithsonite, a typical zinc oxide mineral, has been developed for many years as an alternative source. However, restricted to inferior ability of floating, zinc oxide is one well-known refractory mineral with poor selectivity and high regent consumption. In this paper, ethylenediamine ($NH_2CH_2CH_2NH_2$) was selected to active flotation of smithsonite using dodecylamine as collector. The effect of ethylenediamine on flotation efficiency was conducted; the results showed that without addition of ethylenediamine, the recovery of smithsonite was only 32.85% when the usage of dodecylamine-hydrochloride as collector was 5×10^{-4} mol/L. The optimum dosage of ethylenediamine was 6×10^{-3} mol/L and flotation recovery could be obviously improved to 92% under the same usage of collector. This finding may promote the recovery of refractory zinc oxide mineral resource in future.

Keywords Smithsonite · Ethylenediamine · Flotation · Activator

C. Lv · S. Wen (✉) · S. Bai (✉) · K. Yang
State Key Laboratory of Complex Nonferrous Metal Resources Clean Utilization, Kunming University of Science and Technology, Kunming 650093, Yunnan, People's Republic of China
e-mail: shmwen@126.com

S. Bai
e-mail: baishaojun830829@126.com

C. Lv
e-mail: lvchao0711@126.com

K. Yang
e-mail: truepsyche@sina.com

C. Lv · S. Wen · S. Bai
Faculty of Land Resource Engineering, Kunming University of Science and Technology, Kunming 650093, Yunnan, People's Republic of China

K. Yang
Faculty of Metallurgical and Energy Engineering, Kunming University of Science and Technology, Kunming 650093, Yunnan, China

© The Minerals, Metals & Materials Society 2017

M.A. Meyers et al. (eds.), *Proceedings of the 3rd Pan American Materials Congress*,
The Minerals, Metals & Materials Series, DOI 10.1007/978-3-319-52132-9_50

Introduction

To ensure future demands of zinc sources, smithsonite has been developed as alternative of zinc metal for many years [1, 2]. Because of high activity to water dipoles on the surfaces of zinc oxide, resulting in low natural floatability of smithsonite compared to sphalerite. So, zinc oxide is one well-known refractory zinc resource mineral compared with sphalerite, due to its poor selectivity and high regent consumption. Specially, when the pulp containing slime minerals, such as iron oxide, calcite, chlorite and kaolinite et al. the process of flotation became worse [3–6].

In the past few years, many techniques, such as flotation, pyrometallurgy and hydrometallurgy, etc., have been attempted to treat zinc oxide ores, among which sulfidization xanthate or aliphatic amine flotation method was the most commonly and commercially used method for concentration and pretreatment of zinc oxide minerals [7–9].

In general, many papers were focused on high-efficiency collectors, mechanism sulfidation of smithsonite and inhibitor of gangues. Activator of smithsonite was mainly focused on Na_2S because of its importance during the flotation of zinc oxide. However, there were few reports about the research on the other activators except for the Na_2S series. It was known that what could be used as the flotation organic activators were those regents that could chelate with the metal ion on the surface of minerals and did not contain else hydrophilic radical [10].

In this work, an organic activator—ethylenediamine ($H_2NCH_2CH_2NH_2$) was selected to perform the flotation behavior of smithsonite. Different addition concentration was investigated for testing the effect of ethylenediamine on the flotation recovery of smithsonite by dodecylamine cationic collector. Meanwhile, its possible mechanism was also investigated.

Experimental

Materials and Methods

The experimental pure smithsonite ore was obtained through manual removal of gangue minerals such as cerussite, calcite, and quartz. The pure smithsonite samples were manually ground in an agate mortar and then sieved using standard test sieves to obtain the desired particle size fractions. Its chemical composition and XRD pattern were shown in Table 1 and Fig. 1. From them, it could be known that the

Table 1 The main chemical composition of smithsonite ore (mass fraction, wt%)

Zn	Fe	Pb	S	SiO_2	Al_2O_3	MgO	CaO
49.66	0.44	0.51	0.07	0.88	0.23	0.85	1.09

Fig. 1 X-ray diffraction pattern of the smithsonite

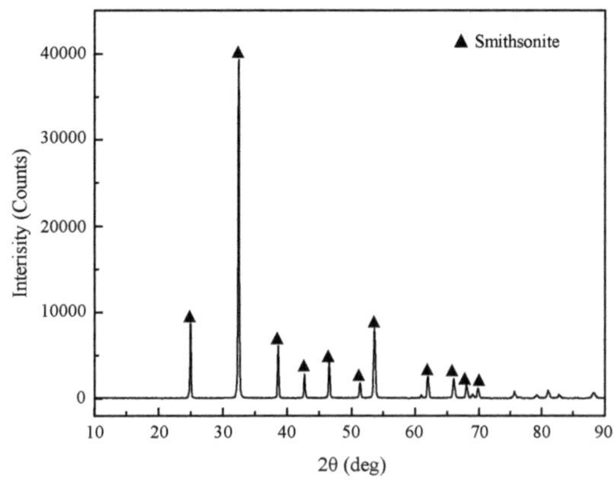

smithsonite was mainly composed of Zn, and a small quantity of impurities, it had a high purity of 95.5%. The main phase in this ore was smithsonite JCPDS card (no. 831765). Thus the smithsonite could be regard as pure mineral for experiments.

The reagents used in the experiment were hydrochloric acid, ethylenediamine and dodecylamine (DDA) with analytical grade. Dodecylamine-hydrochloride ($R_{12}NH_2$–HCl) solution with a concentration of 1×10^{-3} mol/L was prepared, the mole ration of dodecylamine to hydrochloride was 1:1. Ethylenediamine solution with a concentration of 1×10^{-2} mol/L was prepared. High-purity deionized water produced by a Milli Q5O system with a resistivity of 18.25 MΩ was used throughout the experimental work.

Micro-flotation Experiments

The micro-flotation experiments were conducted in a mechanical agitation flotation machine. The mineral suspension was prepared by adding 2.0 g of pure smithsonite (−74 to −45 μm) to 50 mL of solution. Freshly prepared ethylenediamine solutions of desired concentrations was added to the pulp for 5 min, then the pulp was conditioned with dodecylamine-hydrochloride ($R_{12}NH_2$–HCl) 5×10^{-4} mol/L for 3 min and floated for 3 min. Contrast flotation test was also performed without ethylenediamine under the same condition. After the flotation tests, the concentrate and tailing products were filtered, dried and weighed to determine the flotation recovery.

Adsorption of Dodecylamine on the Smithsonite Surface

The adsorption experiment was performed in a 100 ml beaker with magnetic stirring at the room temperature. Typically, 2.0 g of smithsonite samples with 50 ml aqueous solution put in the clean beaker, freshly prepared ethylenediamine solution with required concentration and dodecylamine (5×10^{-4} mol/L) was added to the beaker at natural pH for stirring 10 min, respectively. Subsequently, a centrifuge was used for solid-liquid separation. The liquid obtained from the separation was stored in closed vials and analyzed for the total concentrations of dodecylamine by means of ultraviolet spectrophotometer to determine the adsorption of dodecylamine on the surface of smithsonite.

Results and Discussion

Flotation Study of Smithsonite

The results with addition of ethylenediamine as activator on micro-flotation recovery rate of smithsonite was given in Fig. 2. It showed that the optimum concentration of ethylenediamine was 6×10^{-3} mol/L, and the corresponding recovery rate was 91.07% with dodecylamine-hydrochloride being 5×10^{-4} mol/L. When the concentration of ethylenediamine was lower than 6×10^{-3} mol/L, recovery rate of smithsonite increased with increasing ethylenediamine concentration. While higher ethylenediamine usage would result in decreasing of recovery rate of smithsonite, this was may be due to the competitive adsorption of excrescent ethylenediamine and dodecylamine on the surfaces of smithsonite. From the above analysis, it can be deduced that added amount of

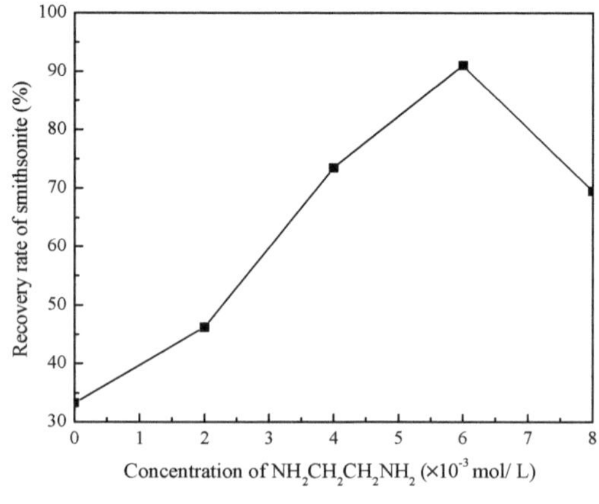

Fig. 2 The effect of addition concentration of ethylenediamine on micro-flotation recovery rate of smithsonite

Fig. 3 The effect of ethylenediamine concentration on dodecylamine adsorption on the surface of smithsonite

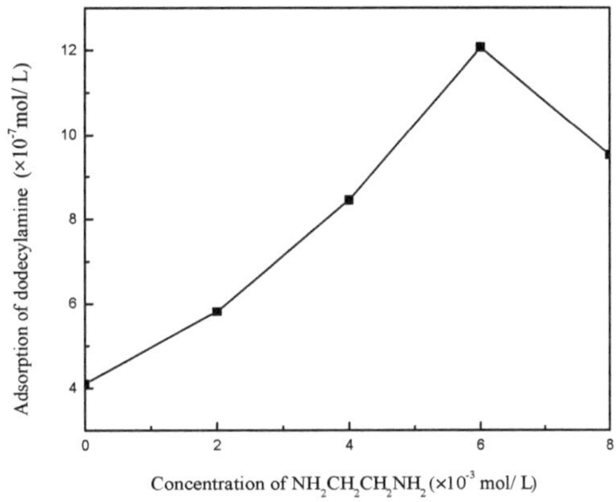

ethylenediamine played a significant role during smithsonite flotation, and the optimum amount of ethylenediamine was 6×10^{-3} mol/L.

Effect of Ethylenediamine on the Adsorption of Dodecylamine on the Smithsonite Surface

Figure 3 demonstrated the adsorption concentrations of dodecyl amine with different initial concentrations of ethylenediamine. Results showed that the adsorption of dodecylamine on the surface of smithsonite increased with increasing concentration of ethylenediamine when the concentration of ethylenediamine below 6×10^{-3} mol/L, which indicated that the activation reaction occurred because ethylenediamine in the pulp solution were transformed onto the mineral surface through chelation function. When the initial concentration of ethylenediamine exceeded 5×10^{-4} mol/L, the adsorption of dodecylamine onto the smithsonite surface would be suppressed by excessive ethylenediamine in pulp solutions, which was in good agreement with the results of micro-flotation tests.

The Mechanism Analysis

1. Ethylenediamine ($H_2NCH_2CH_2NH_2$) hydrolysis in water generated OH^-, making the pulp become alkaline, during the experiments, the usage of ethylenediamine was 10^{-3} mol/L, pH of pulp was about 9–10, heighten the surface electronegativity of smithsonite, which was beneficial to the adsorption of cationic collector $R_{12}NH_2{\cdot}R_{12}NH^{3+}$ on the smithsonite surface.
2. When using the optimal addition of ethylenediamine 6×10^{-3} mol/L, Zn^{2+} on the surfaces of smithsonite or in the solution was prior to bond with

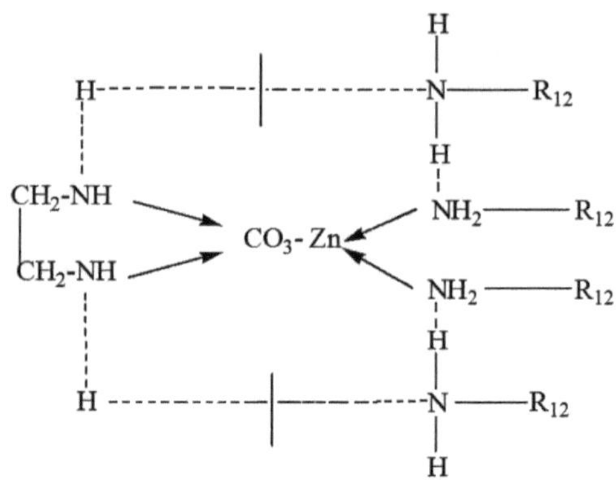

Fig. 4 Model of DDA adsorption on smithsonite after ethylenediamine activization

ethylenediamine, complex of zinc diamine was formed, hindering production of hydrophilic Zn(OH)n. Meanwhile, ethylenediamine would erode $ZnCO_3$ particles generating more Zn^{2+} active sites on surfaces of smithsonite, which benefits to the adsorption of collector and promote floatability of smithsonite.

3. After the chelation of ethylenediamine with Zn on the surfaces of smithsonite, dodecylamine would hydrogen bond with ethylenediamine through (N–H···N), as well as hydrophobic force between non-polar groups of ethylenediamine and dodecylamine, leading to more dodecylamine molecules adsorption on the surfaces of smithsonite.

Based on the above explanation, the activization mechanism of ethylenediamine could be modeled as in Fig. 4.

Conclusion

The present investigation introduced a kind of organic activator ($NH_2CH_2CH_2NH_2$) for flotation of smithsonite when using dodecylamine as collector. Based on the results and discussion, the following conclusions can be drawn:

1. When without addition of ethylenediamine, the recovery of smithsonite was below 40% by dodecylamine directly flotation.
2. Proper addition of ethylenediamine, which would promote flotation of smithsonite remarkably, under the optimum dosage of ethylenediamine 6×10^{-3} mol/L and dodecylamine 5×10^{-4} mol/L, the recovery of smithsonite reached to 91%.
3. Excessive ethylenediamine would suppress the flotation of smithsonite, this may be due to the results of competition adsorption between ethylenediamine and dodecylamine on surfaces of smithsonite.

Acknowledgements The authors would like to express their gratitude for the financial support from the National Science Foundation for Young Scientists of China (Grant No. 51404118).

References

1. Hosseini, S. H., & Forssberg, E. (2006). Adsorption studies of smithsonite flotation using dodecylamine and oleic acid. *Minerals & Metallurgical Processing, 23,* 87–96.
2. Hosseini, S. H., & Forssberg, E. (2007). Physicochemical studies of smithsonite flotation using mixed anionic/cationic collector. *Minerals Engineering, 20,* 621–624.
3. Mehdilo, A., Irannajad, M., & Zarei, H. (2014). Smithsonite flotation from zinc oxide ore using alkyl amine acetate collectors. *Separation Science and Technology, 49,* 445–457.
4. Fa, K. Q., Miller, J. D., Jiang, T., & Li, G. H. (2005). Sulphidization flotation for recovery of lead and zinc from oxide-sulfide ores. *Transactions of Nonferrous Metals Society of China, 15* (5), 1138–1144.
5. Li, Y., Wang, J., Wei, C., Liu, C. X., Jiang, J. B., & Wang, F. (2007). Sulfidation roasting of low grade lead–zinc oxide ore with elemental sulfur. *Minerals Engineering, 20*(6), 621–624.
6. Wu, D., Wen, S., Deng, J., Liu, J., & Mao, Y. (2015). Study on the sulfidation behavior of smithsonite. *Applied Surface Science, 329,* 315–320.
7. Irannajad, M., Ejtemaei, M., & Gharabaghi, M. (2009). The effect of reagents on selective flotation of smithsonite–calcite–quartz. *Minerals Engineering, 22*(9–10), 766–771.
8. Mehdilo, A., Irannajad, M., & Zarei, H. (2014). Smithsonite flotation from zinc oxide ore using alkyl amine acetate collectors. *Separation Science and Technology, 49,* 345–457.
9. Li, M., Liu, F., Wang, X., & Wang, G. (2015). Optimization of flotation process of zinc oxide ore by response surface methodology. In *International Conference on Materials Engineering and Information Technology Applications (MEITA 2015)* (pp. 1000–1004).
10. Li, M. X., Jian, S., & Zhao, W. J. (2013). Effect of Ca^{2+} and Mg^{2+} on Floatability of Smithsonite and Hemimorphite. *Advanced Materials Research, 813,* 298–301. (Trans Tech Publications).

Extraction of Gold from Sands and Slimes Tailings Dump from Mazowe Mine, Zimbabwe

Alain M. Bantshi and Peter Makuvise

Abstract The Mazowe Gold Mine is managed by Metallon Gold (Zimbabwe) and is situated in the west-central part of the Harare greenstone belt. Orebodies here generally comprise shear zones which are in-filled with gold-bearing sulphides and quartz. The mineralised zones are up to 1 m in width, have average grades of 4–5 g/t. Mazowe Mine is one of the oldest mines in Zimbabwe, and exploration and development in this region dates back to 1890, with over 1.4 million ounces of gold produced to date. Ore is processed in a single plant which consists of conventional crushing and milling and carbon-in-leach facility. Baldmin Projects was awarded the contract for the design, construction and commissioning of the 60,000 tonnes per month plant at Mazowe mine in 2014 for the processing of old tailings dump from previous operations at Mazowe mine. The average grade of the tailings is 1.1 g/t. This paper describes the test work, process design of a processing system at the Mazowe Mine that will improve the extraction efficiency from gold tailings material. The recovery is derived from the separation of the gold contained within the sulphides of the tailings material and a dedicated process stream to recover that gold. The sulphides are separated by means of flotation process and that resulting concentrate is subjected to regrind below 25 μm to enable a greater liberation of the gold. The fine material is then subjected to a dedicated leach process to recover the now exposed gold particles. The case study is an example of the challenges of using existing technology and methodology to a well-established process and there by attaining improved efficiencies and tangibles business benefits.

Keywords Extraction · Gold tailings · Gold tailing dump

A.M. Bantshi (✉) · P. Makuvise
Baldmin Projects, PO Box 59892, Kengray 2001, South Africa
e-mail: alainb@baldminprojects.co.za; alainbantshi@gmail.com

P. Makuvise
e-mail: peter@baldminprojects.co.za

© The Minerals, Metals & Materials Society 2017

M.A. Meyers et al. (eds.), *Proceedings of the 3rd Pan American Materials Congress*, The Minerals, Metals & Materials Series, DOI 10.1007/978-3-319-52132-9_51

Introduction

Metallon Gold Zimbabwe embarked on an on-site pilot plant flotation testing campaign on dump ores at its different subsidiary Mines in Zimbabwe in 2013. The pilot plant test works were carried out using flotation conditions in order to generate more data for engineering design and to produce float concentrates for gold dissolution test work. The objective of the pilot scale test plant was to prove that gold can be recovered economically from the tailings with little reprocessing and low mining and handling costs. Metallon Gold Zimbabwe engaged Baldmin Projects to carry out the engineering design of the plant with the data generated from the metallurgical test works and pilot test works on assays, milling, flotation and cyanidation.

The test work completed comprised of:

- Head assays
- Milling curves
- Flotation Test (Laboratory and Pilot Plant)
- Diagnostic leach test work on flotation concentrate samples as received and Intensive cyanidation at a finer grind.

The objective of the test works was to generate data to develop a process design criteria (PDC) for engineering design. This paper provides details of the work, results, conclusions arrived at and recommendations made from the work completed.

Ore Mineralogy

Pyrite is the dominant sulfide ($\geq 95\%$ of the ore minerals). It occurs in the form of massive and layered ores, of which both fine-(grain diameters ≤ 50 µm) and coarse-grained (mm-size) varieties are present. The pyrite grains are compact, homogeneous, and idiomorphic to hypidiomorphic in shape. They usually form aggregates that are locally fractured. Pyrrhotite, sphalerite, chalcopyrite, galena, arsenopyrite and cobaltite are present in minor quantities in the ores. Gold recoveries of $\geq 90\%$ are achieved by a combination of gravity separation and cyanide leaching, indicating that gold is present in a particulate form. Gold occurs as roundish inclusions (5–30 µm in diameter) in pyrite or in rarer instances in gangue, and as thin films along grain contacts or cracks in pyrite. Occasionally, gold also forms large (up to 200 µm in diameter), irregular grains in pyrite or gangue. Silver contents of the gold, analysed by microprobe, range between 7 and 20 wt% Ag in gold grains of different samples collected throughout the mine. The silver grains of different samples collected throughout the mine. The silver contents of gold vary in an unsystematic manner with little correlation with depth or from one reef to the other. The Fig. 1 illustrates the scanningelectron microscopes images of polished samplefrom Mazowe gold Mine.

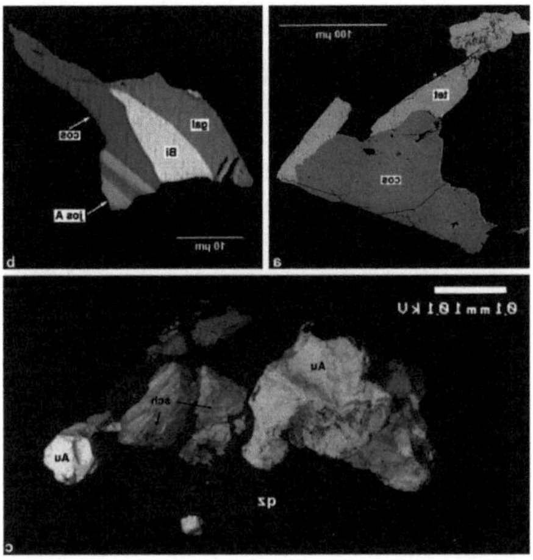

Fig. 1 **a-c** Photomicrographs. **a** Cosalite (*cos*) and tetradymite (*tet*) in quartz (*black*). Bucks Reef, 6-level, Mazowe Mine. Polished section, backscatter electron image. **b** Galena (gal), native bismuth (*Bi*), cosalite (cos) and joseite A (*jos A*) in quartz gangue. Nucleus Reef, 10-level, Mazowe Mine. Polished section, backscatter electron image. **c** Well-crystallised gold (*Au*) and Scheelite (*sch*) on quartz (*qz*). Fragment of quartz vein, Stori's Golden Shaft Mine. Scanning electron microscope image

Table 1 Head assay of tailing samples

Section	North	South	East	West
	2.28	0.76	1.6	0.44
	1.56	0.84	1.86	1.1
	1.12	0.84	0.84	0.88
	1.1	2.14	1	1.54
	1.42	1.14	1.1	1.46
Average	1.5	1.14	1.28	1.08
Composite	1.3	1.04	1.14	0.9

Sample Characteristics

Samples were collected from four sloping sides of the tailing dam. Holes approximatively 1 m^3 were dug on the base, in the middle and near the top samples were scalped from the inside of the holes, on all 5 exposed surfaces. For each face, all samples were mixed into one, which was thoroughly mixed before samples were taken for assaying.

Chemical Composition

The gold assay of the samples as well as a more complete chemical analysis of the composite sample are presented in Tables 1 and 2.

Table 2 Chemical composition

Element	Sample
	Composite
Au g/t	1.173
Ag g/t	0.124
Pb	0.158
Cu	0.21
Bi	1.476
Fe	0.55
Se	0.2
Sb	0.114
S	0.984
Te	0.727
Ca	0.87

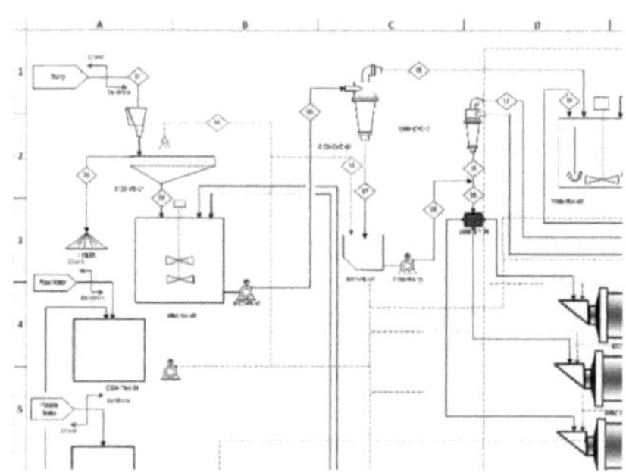

Fig. 2 Recommended flowsheet for processing Mazowe gold tailings

Size Distribution

The particle size and gold distributions of thesample are presented in Table 3.

The table shows that approximately 45% of the oreand 80% of the gold reports in the −75 μm fraction.

The fine size fraction <75 which constitutes 11.6% by mass gave the highest gold value 1.32 g/t Au.

Generally, the coarse material (>150 and >300) are lower grade compared to the finer size fractions.

The coarser size fractions (>150 and >300) constitute 68.9% by mass and contain 64.4% of total gold.

The finer size fractions (<75 and >75) constitute 31.1% by mass and contain 35.6% of total gold.

The results do not support the assumption that the coarse size fraction is higher grade compared to fine size fraction.

Table 3 Particle size distribution

Sieve size	Mass (kg)	Grade (g/t Au)	% mass	Au content %
>300	35	1.02	33.3	31.6
>150	37.4	0.99	35.6	32.8
>75	20.5	1.17	19.5	21.3
<75	12.2	1.32	11.6	14.3
Total	105.1	1.07	100	100

Gold Recovery Tests

Gold occurrence as liberated grains and inclusions in sulphides or silicates indicated that flotation and cyanide would be most adequate methods of concentration for this ore. Gravity concentration was not considered suitable for this material due to the relatively fine distribution of the gold in the tailing sample.

Therefore, test work to recover gold from the ore sample was conducted by flotation, cyanidation and a combination of both.

Flotation

Flotation tests were conducted to determine the best reagents regime and grinding conditions for optimum gold recovery. Selection of the most adequate flotation reagents was based on the metallurgical results of the flotation tests. Several combinations of collectors were tried at natural pH of about 6.5. The results of the rougher and cleanertests are presented in Table 4.

No activating agent was used. MIBC was the frother employed in all tests. These tests were carried out under an arbitrarily chosen grinding of approximately 80% −75 μm.

A summary of the flotation result using various collectors is presented in Tables 5 and 6 presents the results of optimum conditions.

The following combination of reagents was selected:

- Unicol PX2 (100 g/t)
- Aero 3418 Xanthate (50 g/t)
- Oreprep 501 (40 g/t)

Under these conditions a concentrate of approximately 25 g/l Au for a gold recovery in excess of 90% was obtained. Overall recovery is 88% after leaching the concentrate.

This is without cleaning the concentrate. If the concentrate is cleaned, recovery drops slightly to 86%.

Optimum particle size for gold flotation was investigated in a series of tests carried out under various grinding conditions but using the same reagents regime.

Table 4 Rougher and cleaner tests

Test	(%) Weight	Assay Au (g/t)	Dist. (%)
Rougher concentrate			
Test 1–2	12.4	8.95	93
Test 5–6	7.4	12.83	91
Average	9.9	10.89	92
Cleaner concentrate			
Test 1–2	2.3	46.41	87
Test 5–6	1.8	49.19	86
Average	2	47.8	87
Cleaner tails			
Test 1–2	10.2	0.62	5
Test 5–6	5.6	0.91	5
Average	7.9	0.77	5

Table 5 Flotation results using various collectors

Reagent		Flotation concentrate			
Collector and flotation reagents	g/t	% weight	Assay Au (g/t)		Dist., Au
			Rougher	Cleaner	
SNPX	100	12.43	8.95	46.41	92.7
SK49	50				
XP200	40				
XP200	300	7.4	12.83	49.19	91.1
SNPX	100				
SK49	50				
XP200	50				
Unicol PX2	150	24.2	4.77	10.12	93.9
Aero 3418	50				
Aero 3418	40				
SNPX	100	14.68	8.85	35.24	92.6
SK42	50				
XP200	40				

The optimum grinding size for gold flotation was considered to be approximately 80% −75 µm. At this grind a gold recovery of about 90% would be obtained for a concentrate grade in excess of 20 g/t Au. Floating at coarser size will result in a significant drop in gold recovery while a finer grind might be too costly and unjustified by the minor increase in gold extraction.

Gold distribution per size fraction of the flotation concentrate was not examined. However, a screen test indicated that 95% of the flotation concentrate reported in the −75 µm fraction. On the basis of the gold distribution per size fraction of the original sample it is believed that at least 95% of the gold recovered in flotation concentrate reported in the −75 µm fraction.

Table 6 Summary of optimum conditions and results

Item	Unit	Pilot result
Grade	g/t	1.06
Float recovery—roughers	%	93.9
Float recovery—cleaners		91.4
Leach recovery—24 h B/Roll	%	94
Aggregate recovery	%	88.3
Production at 60,000 tpm	Oz per month	1,798
Cash cost—projected	\$/oz.	459.1
Profitability	\$/oz.	790.9
	Reagents suite	Unicol PX2 (100 g/t)
		Aero 3418 (50 g/t)
		Oreprep 501(40 g/t)

Table 7 Head assay of flotation concentrate

Item	Description	Mazowe (g/t)
1	Head assay	24.16
2	Head assay	24.67
3	Head assay	26.83
Average		25.22

Gold Dissolution Test Work

The objective of the test work was to develop dissolution conditions to recover gold from flotation concentrates.

The gold dissolution test wok completed on the Mazowe concentrate samples was mainly developed to establish conditions for the recovery of gold using intensive cyanidation and to determine the deportment of gold in the constituent minerals. The flotation concentrates from Mazowe contained 25.22 g/t of gold; approximately 95.2% was amenable to intensive cyanidation and 0.1% of the gold was preg-robbed. The gold associated with minerals susceptible to a mild oxidative pre-leach such as calcite, dolomite, pyrrhotite and haematite was 2.1% and gold associated with sulphide minerals that require severe oxidative pre-treatment (i.e. pyrite, arsenopyrite etc.) was quantified to be 0.7% of the total gold. Approximately 1.5 and 0.1% of the gold was associated with carbonaceous materials and quartz respectively. Intensive cyanidation recovery on Mazowe increased marginally to 96.5% after grinding the sample finer to 80% −25 μm, these results were consistent with the diagnostic leach results where 4.6% of the total gold was occluded in sulfides, carbonaceous and quartz materials.

Diagnostic leach tests were carried out on the Mazowe flotation concentrate samples “as received”. The head assay is concentrate samplesare presented in Table 7. The objective of the test was to establish the deportment of gold in the minerals constituting the ore.

This was achieved through sequential solubilisation of the least-stable minerals using different pre-treatment processes with subsequent extraction of the associated gold by cyanidation/CIL. The first two steps carried out were of two discrete tests comprising of direct cyanidation and cyanidation in the presence of carbon (CIL); these were undertaken with the objective of determining the preg-robbed gold.

The gold associated with HCl digestible minerals i.e. minerals susceptible to a mild oxidative pre-leach such as calcite, dolomite, pyrrhotite, haematite etc. was quantified by first subjecting the CIL residue to hot HCl pre-treatment followed by CIL dissolution of the acid treated residue. The gold associated with sulfide minerals dissolution of the acid treated residue. The gold associated to sulphide minerals (i.e. pyrite, arsenopyrite etc.) was quantified by subjecting the second CIL residue to severe oxidative pre-treatment using hot HNO_3 follow by CIL dissolution of the acid-treated residue. To quantify the gold associated with carbonaceous material such as kerogen, third CIL residue was subjected to complete oxidation through roasting and followed by CIL dissolution of the calcined product. The gold remaining in the final residue was assumed to be associated with quartz.

Acid treatment was carried out in mechanically-agitated vessels and cyanide dissolution was carried out for 24 h using the bottle roll method. Cyanidation conditions comprised of: 5 kg/t NaCN, pH of 11 adjusted using lime and pulp density of 30% solids. Table 8 presents the diagnostic leach test results on the Mazowe float concentrate samples.

Approximately 95.3% of the total gold found in the Mazowe float concentrate sample was amenable to gold recovery through direct cyanidation (intensive) in which the recovered was 24.017 g/t. the results also indicated that there was minimum preg-robbing of gold in the concentrate during leaching, with approximately 0.12% of the total gold at a grade of 0.03 g/t. the gold associated with HNO_3 digestive minerals accounted for 0.73% of the total gold at a grade of 0.185 g/t. approximately 1.48% of total gold at a grade of 0.374 g/t was associated with carbonaceous material and this gold can only be recovered through total roasting. The remaining gold that was assumed to be associated with quartz was 0.34% of the total gold with a grade of 0.086% g/t.

Table 8 Diagnostic leach test results

Gold association	Tails	
	Gold grade (g/t)	Gold distribution (%)
Available to direct cyanidation	24.017	95.24
Preg-robbed (CIL)	0.03	0.12
HCl digestive minerals	0.525	2.08
HNO3 digestive minerals	0.185	0.73
Carbonaceous matter	0.374	1.48
Silica (balance)	0.086	0.34
Total	25.22	100

Table 9 Intensive cyanidation results

Head Au g/t	Grade Au g/t	25.2
Leach results "as received"	Tails Au g/t	1.17
	Recovery %	95.4
Leach results @ 80%—25 μm	Tails Au g/t	0.88
	Recovery %	96.5

Intensive Cyanidation Results

In order to establish the effect of grinding finer on gold dissolution, each sample was milled to 80% -25 μm prior to intensive cyanidation. The intensive cyanidation conditions comprised of:

- pH of 10.5–11 adjusted using lime
- cyanide addition of 5 kg/t
- pulp density of 30% solids and
- Dissolution time of 24 h.

All the leach residue was analysed for the Au only. Cyanide consumption was determined using the titration method. The cyanidation results obtained are presented in Table 9.

The Mazowe concentrate samples responded well to intensive cyanidation at both grind sizes. Gold dissolution achieved on Mazowe flotation concentrate samples using intensive cyanidation "as received" and on samples milled to 80% −25 μm was 95.4 and 96.5% respectively. These results confirmed the diagnostic leach results which indicated that there was minimum number of refractory minerals of sulphides, carbonaceous material and quartz in the Mazowe sample which added to 4.6% of total gold, the gold that was not solubilised at both grind was therefore concluded to be finely occluded with theses minerals.

Flotation Followed by the Cyanidation of the Concentrate

Since both flotation and direct cyanidation of the tailing sample resulted in high gold recoveries, it was considered appropriate to investigate the possibility of gold extraction by a combination of these two technologies, namely: flotation followed by cyanidation of the flotation concentrate. The combined process presents the following advantages versus the individual technologies.

Cyanidation of the flotation concentrate rather than that of the entire ore will result in a significant cost reduction for the cyanidation process because only a small fraction of the plant feed will undergo the cyanide leaching. Also, cyanide leaching will produce a gold precipitate of much higher concentration in precious metals than that of the flotation product.

The flotation of the tailings sample to produce the concentrate for the cyanidation tests was carried out under previously established conditions.

The cyanidation of flotation concentrate was performed after regrinding the material almost entirely below −25 μm (80% −25 μm) and for a duration of 24 h.

Conclusions

The optimum conditions and results of the gold recovery tests are presented in Table 6.

The dissolution results indicated that the Mazowe flotation concentrate sample responded well to intensive cyanidation and that the gold associated with refractory minerals was not more than 5%.

The response of the ore to all these metallurgical processes was very good. Approximately 90% of the gold contained in the ore was recovered in a concentrate or a cyanide solution. However, the recommended method for treating this gold tailing material was flotation followed by the cyanidation of the flotation concentrate.

The reasons for selecting this metallurgical process against flotation or cyanidation, as single processes, were:

- The flotation alone produced a concentrate too low in gold content (less than 100 g/t) to be accepted by smelters, which prefer to process high-grade gold cyanidation precipitates.
- Cyanidation of the flotation concentrate (10% by weight) rather than that of the entire ore plant feed is likely to present a significant economic advantage, despite the extra cost of flotation and somewhat lower gold extraction.

The regrind of the flotation concentrate was not economically justified because of low increased in the gold recovery from 95.4 to 96.5%. So, the regrind process was not considered in the recommended process flowsheet of the gold tailings retreatment plant.

The recommended flowsheet sketched in Fig. 2 consisted of: feed classification, grinding the tailing to approximately −75 μm (80% −75 μm), bulk flotation, thickening of flotation concentrate followed by 24 h' cyanidation and adsorption of gold by carbon-in-leach (CIL).

The feasibility study indicated that the economics of the proposed process to recover gold from the tailings dump appeared to be favourable. The conceptual flowsheet of the plant consisted of one 60,000 tonnes per month flotation module and a CIL treatment unit of approximately 240 tonnes per day, for leaching and carbon adsorption.

At 60,000 tonnes per month the dump will give over 1900 oz. per month, easily making Mazowe tailing retreatment project a 3000-oz. operation.

In conclusion, the combined efforts of all involved in this project succeeded in demonstrating that the gold values from Mazowe tailings dump could be turned into a profitable operation by a relatively simple process. The process utilizes more or less conventional flotation and cyanidation techniques and although no new or novel technology is involved.

Acknowledgements The authors would like to thank the project team members, staff and management of Baldmin Projects as well as the Metallon gold team that contributed to the successful implementation of the this project.

Gold Recovery from Waste Solutions of PCBs Gold Plating Process Using Hydro Cyclone Reactor for Demonstration Study

Mooki Bae, Sookyung Kim and Jae-chun Lee

Abstract This paper presents a novel methodology to recover gold from waste solutions of printed circuit boards (PCBs) generated during the gold electro-plating. Studies have been carried out using two different sizes of hydrocyclone (357.00 and 381.87 mm height) at various parameters such as current, flow rate, electrolyte concentration, cathode materials, and anodic area to recover the gold. ~96% gold was recovered using short-cone type hydrocyclone (357.00 mm) at following conditions: applied current 51.3 A/m^2, flow rate 7.7 m/s, using stainless steel as cathode material. However, approximately same amount of gold can be also recovered with long-cone type hydrocyclone (381.87 mm). But, the possibility for damage of long-cone type hydrocyclone might be higher than short cone type hydrocyclone because reaction occurs only at small reaction surface area during gold recovery. Therefore, short-cone type hydrocyclone has been considered for further study to recover gold.

Keywords Gold · Recovery · Cyclone reactor · Plating solutions · Demonstration

Introduction

Gold is one of the valuable metal and apart from being used in jewelry; it has a wide range of applications in a fields of electronics, chemical and corrosion resistance materials due to its specific physical and chemical properties. During the etching,

M. Bae · S. Kim (✉) · J. Lee
Resources Recycling, Korea University of Science and Technology,
Daejeon 34132, Korea
e-mail: skkim@kigam.re.kr

M. Bae
e-mail: muki.bae@kigam.re.kr

M. Bae · S. Kim · J. Lee
Mineral Resource Research Division, Korea Institute
of Geoscience and Resources, Daejeon 34132, Korea

© The Minerals, Metals & Materials Society 2017

M.A. Meyers et al. (eds.), *Proceedings of the 3rd Pan American Materials Congress*,
The Minerals, Metals & Materials Series, DOI 10.1007/978-3-319-52132-9_52

rinsing, chemical and mechanical polishing of electronic devices and articles, enormous amount (~50000 tons/year in KOREA) of waste solution generated which contains 30–500 ppm of gold that is much higher than the (0.005 g/t) ore/primary resources [1]. Hence, it is necessary to treat the waste solution to recover the precious metal as well as to minimize the generation of waste solution/effluent. Various researches used the different techniques like co-precipitation [2], solvent extraction with dibutyl carbitol (DBC) [3], adsorption with activated carbon and ion exchange with variety of adsorbent used to recover the gold from waste solution [4–7]. Ok and Jeon et al. used the Dowex 21K XLT resin to recover the gold with due to its excellent selectivity toward to adsorb gold cyanide complex [8]. Strong basic resin like Purolite A-500 [9], Dowex G-55 [10] has advantages to achieve high adsorption capacity due to formation of strong complex with basic resin. Arima et al. and Lee et al. [11, 12] used the zinc, copper and aluminum powder to cement the gold from thiosulfate solution due to their lower reduction potential than gold. ~100% gold precipitate with zinc and aluminum while copper with ~93 gold precipitate at pH ~9.5. Cementation processes has a disadvantage to take long time to recover the gold from waste solution. Recently, Donnan dialysis technique used to recover Au(III) from aqua regia solutions of electronic waste containing low concentrations of Au(III). Almost complete removal of Au(III) was achieved in four transport cycles [13]. Few researchers employed the biological methods as an eco-friendly and cost-effective approach to recover precious metal ions from aqueous solutions [14, 15]. However, biological methods have not been applied in practical recycling processes due to very low concentrations of precious metals, strong acidity, and high concentrations of base metals [16]. The pre-treatment followed by electro-winning employed for a high recovery rate of gold might be difficult while processing a diluted solution due to a higher consumption of electricity. Hence, hydrocyclone used to recover the gold from waste solution, rinse water containing low concentration of gold. Previously, a hydrocyclone process has been used to recover gold from a PCBs waste solution containing ~30 ppm gold. Using a platinum-cathode and stainless steel anode with an applied current 35–45 A and 5.3–7.7 m/s solution flow rate, approximately 99% gold was recovered [17, 18]. Present study focused on the demonstration of process at small scale to optimize the following process parameters such as flow rate, applied current, electrolyte concentration etc., for gold recovery.

Experimental

Electrolytes were prepared by dissolving $Au[CS(NH_2)_2]_2$ (Aldrich, 99.9%) in water and the pH was adjusted with hydrochloric acid. The chemical composition of the synthetic solution used is shown in Table 1. The waste solution generated during gold electro-plating from company of Korea was contained 213, 302, 412, and 502 mg/L of gold.

Table 1 Composition of synthetic solution

Au (mg/L)	Concentration (%)
213–502	5–10

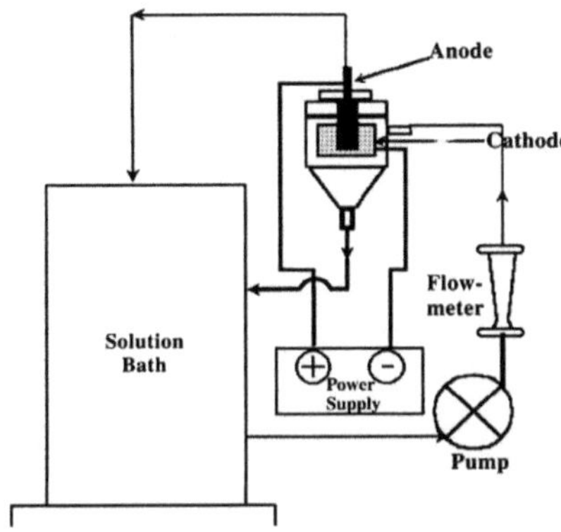

Fig. 1 Schematic diagram of the cyclone electrolytic cell

An electrochemical reactor, utilizing the principal features of the well-known hydrocyclone, was fabricated for gold recovery in dilute solution. Figure 1 shows the schematic diagram of the hydrocyclone [19]. The aqueous in the solution bath containing the low concentration of gold was supplied to the hydrocyclone at a constant flow rate by the rotary pump and discharged into the top and bottom of the hydrocyclone after formation of the turbulent flow along the wall of the cathode then returned to the solution bath. Through such a continuous process, the reduction of gold was happened within electrolytic cell. The stainless steel (SUS 304) and platinum coated cylinder served as cathode, having an inner diameter of 55 mm and a height of 50 mm; where as, the stainless steel (SUS 304) was used as an anode. As shown in Fig. 2, the specific anode electrode was made with small size hole to protect the limiting of reaction speed this being so can secure the sufficient reaction area. These two electrodes were connected to a D.C. power supply. All experiments were performed under potentiostatic conditions, and samples were withdrawn periodically to measure the Au concentration by ICP-OES (inductively coupled plasma-atomic emission spectrometer, Jobin-Yvon Equipment Co., JY-38 plus).

Results and Discussion

Hydrocyclone Design

The recovery of gold using hydrocyclone is affected by specially designed two types of electrolytic cell. Based on previous studies on the recovery of gold from waste solution generated during the gold electro-plating by using small

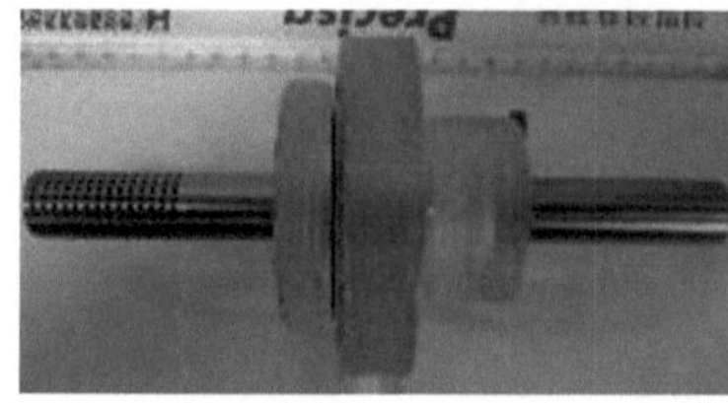

Fig. 2 The Anode electrode

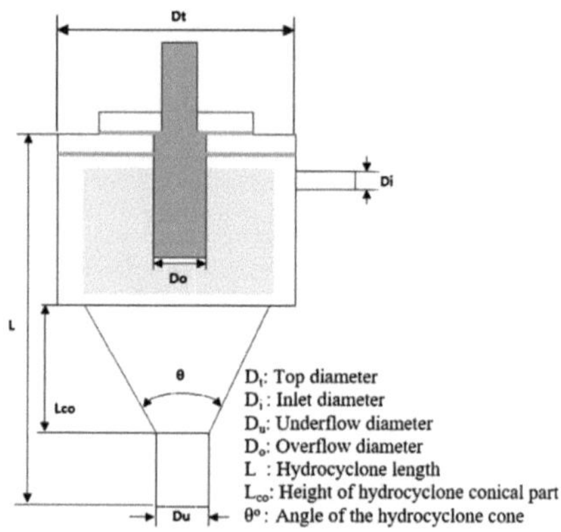

Fig. 3 Cross section of the hydrocyclone

hydrocyclone [17, 18], the following results were considered to design the hydrocyclone for the demonstration studies.

- The longer the length of the cone occurs a tendency that the fine particles are discharged to the upper portion of the hydrocyclone.
- Apex diameter is small, the fluid flow in the lower portion of the hydrocyclone is reduced.
- The greater inlet velocity would increase the separation efficiency, however creating the problems like, noise and vibration.
- The proportion of Do/Du is increased with increasing the critical particle diameter.
- The smaller the ratio of Du/Dc the concentration efficiency is increased.

In present studies, the hydrocyclone reactor used in the electrolytic cell was fabricated in different geometric shapes of the short-cone type and long-cone type. Figure 3 shows the representative cross-section of the hydrocyclone and the dimensions of the two-type hydrocyclone reactor used in this study are listed in Table 2.

In order to investigate the velocity of flow and processing capacity of the designed two-type hydrocyclone reactor the running water was used for the preliminary experiments. The possibility for damage of long-cone type hydrocyclone might be higher than short cone type hydrocyclone because load of rotary pump increase with increasing of flow rate and reaction occurs only at small reaction

Table 2 Dimensions of the 2-type hydro cyclone reactor

	Short type (mm)	Long type (mm)
D_t	71.50	66.50
D_i	26.40	21.20
D_o	30.25	21.20
D_u	37.90	19.06
L	357.00	381.87
L_{co}	189.00	348.00
θ	9.00	20.00

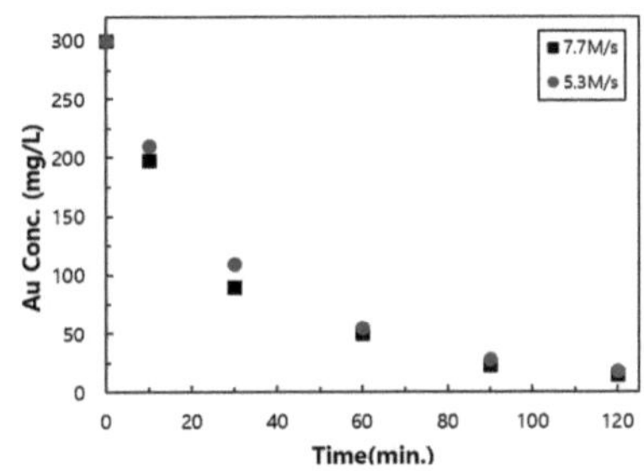

Fig. 4 Effect of flow rate on the recovery of gold (Temperature, 27 °C; electrolyte concentration, 5.0 vol.%; current density, 51.3 A/m^2; initial gold concentration, 302 mg/L)

surface area during gold recovery due to small area of cathode. Therefore, short-cone type hydrocyclone has been considered for further study to recover gold.

Effect of Flow Rate

The effect of flow rate on electrolytic recovery of gold was investigated at a room temperature of 27 °C. Figure 4 presents the results of flow rate experiments obtained with 5.3 and 7.7 m/s. As shown in Fig. 4, the recovery rate of gold was slightly increased with increase in the flow rate at the initial stage. The recovery rate, however, gradually diminished in 60 min. The concentration of remaining gold in the solution was around 15.3 and 18.6 mg/L, respectively. Therefore, gold recovery using hydrocyclone reactor may be available from 100 L solution which is contained around 300 mg/L gold within 6 h. This indicated that the waste solution treatment of 12 tons per month is theoretically possible.

Effect of Current Density

To investigate the influence of current density on the recovery rate of gold, different current density with 38.5, 51.3 A/m^2 were tested at 27 °C and 5.0 vol.% concentration of electrolyte. Figure 5 shows the effect of current density on the recovery

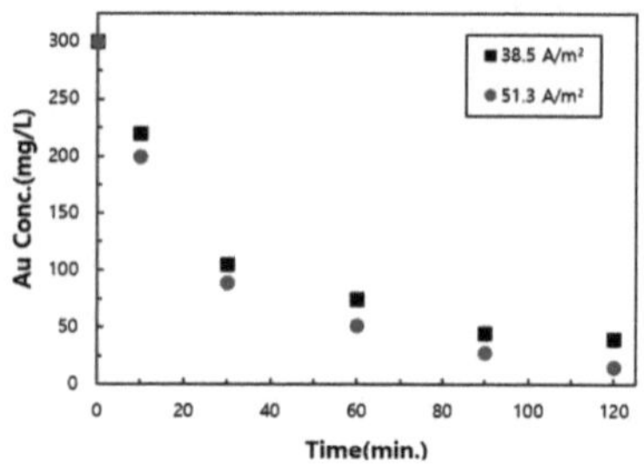

Fig. 5 Effect of applied current on the recovery of gold (Temperature, 27 °C; electrolyte concentration, 5.0 vol.%; flow rate, 7.7 m/s; initial gold concentration, 302 mg/L)

behavior of Au. As the current density increase, the recovery rate of gold slightly increases. The reason for the increase in recovery rate is the effect of current density applied in the constant current mode was shown to increase the amount of quantity of electrodeposition by following reaction:

$$W_T = \frac{A \times I \times t}{n \times F} = ZIt \tag{1}$$

W_T	mass of metal (g)
A	atomic weight of metal (g/mol)
I	current (A)
n	number of electros transferred
F	Faraday's constant
Z	A/nF (electrochemical equivalent)
A/n	chemical equivalent

Concentration of Electrolyte and Initial Gold

The percentage recovery of gold as a function of electrolyte concentration in the range 5.0, 10.0 vol.% at 27 °C and 51.3 A/m^2 current density is given in Fig. 6a. Raising the concentration of electrolyte had no effect on the extent of gold recovery till 120 min. Figure 6b shows the pattern of gold recovery rate from the gold bearing solution, as a function of concentration of gold at a fixed electrolyte concentration (5.0 vol.%), and the applied current density of 51.3 A/m^2. As can be seen, the overall recovery behavior, regardless of the initial gold concentration shows a similar pattern. In particular, the recovery rate has been reached to ~96% after 90 min when the concentration of gold is low or relatively high. The results of recovery rate were relatively decreased compared to the previous study reported by other researcher because making hydrocyclone reactor larger for the demonstration study [18].

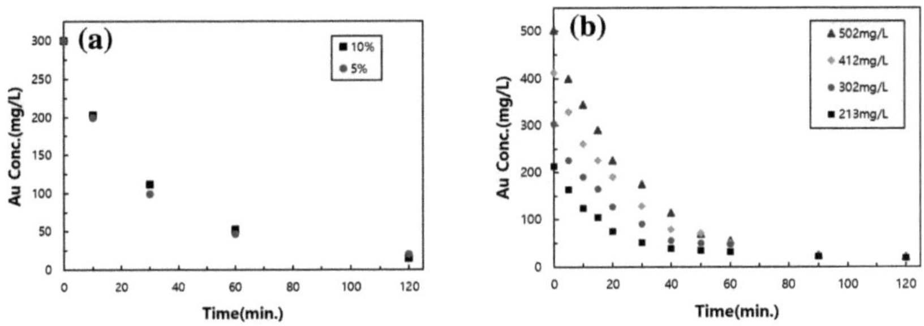

Fig. 6 Recovery behavior with **a** electrolyte concentration, **b** Au concentration in initial stage (Temperature, 27 °C; electrolyte concentration, 5.0–10.0 vol.%; flow rate, 7.7 m/s; initial gold concentration, 213–502 mg/L; current density, 51.3 A/m^2)

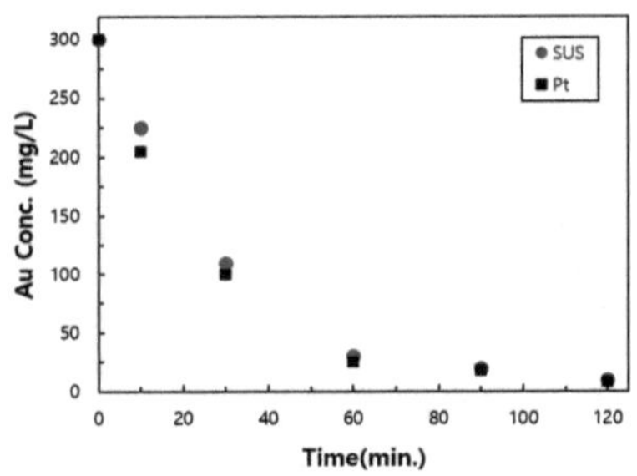

Fig. 7 Recovery behavior depending on cathode electrode materials (Temperature, 27 °C; electrolyte concentration, 5.0 vol.%; flow rate, 7.7 m/s; initial gold concentration, 302 mg/L; current density, 51.3 A/m^2)

Cathode Materials

The recovery behavior of gold was investigated while two kinds of the cathode materials such as stainless steel(SUS304) and platinum coated stainless steel (SUS304) in 5.0 vol.% electrolyte at 27 °C and 120 min operating time at 51.3 A/m^2 current density. Figure 7 shows the gold recovery rate with time in 120 min with accordingly the cathode materials. As can be seen, the increasing time from 10 to 120 min resulted in the gold recovery of around same between SUS304 and platinum coated SUS304 as a cathode materials. Thus, for further experiments SUS304 which is easy to process mechanically was selected to maximize the economic efficiency.

Conclusions

This study demonstrated that recovery of gold from waste solutions of printed circuit boards (PCBs) generated during the gold electro-plating was influenced by effect of designed hydrocyclone reactor. The initial concentration of the gold of the solution, the concentration of electrolyte, the type of the cathode electrode materials was not significantly affect the recovery rate. However, the effect of current density applied in the constant current was shown to increase the recovery rate of gold. And the mass transfer was influenced the deposition rate of gold by increasing of the flow rate.

Acknowledgements This study was supported by the R&D Center for Valuable Recycling (Global-Top R&BD Program) of the Ministry of Environment (Project No. 2016002250004).

References

1. Korea Ministry of Environment. (2015). Industrial wastewater treatment and occurrence (Report No. 10605, Korea National Institute of Environmental Research, 2015).
2. Zhao, Y. Z. (2006). The enrichment and separation of race gold, Pt and Pd from the ores based on co-precipitation. *Gold, 27*, 42–44.
3. Byoung, H. J., et al. (2009). Processing of high purity gold from scraps using diethylene glycol di-N-butyl ether (dibutyl carbitol). *Hydrometallurgy, 95*(3–4), 262–266.
4. Lam, K. F., Fong, C. M., & Yeung, K. L. (2007). Separation of precious metals using selective mesoporous adsorbents. *Gold Bulletin, 40*(3), 192–198.
5. Chang, Y. C., & Chen, D. H. (2006). Recovery of gold(III) ions by a chitosancoated magnetic nano-adsorbent. *Gold Bulletin, 39*(3), 98–102.
6. Kardira, R. M., et al. (2010). Adsorption of the complex ion $Au(CN)_2^-$ onto sulfur-impregnated activated carbon in aqueous solutions. *Journal of Colloid and Interface Science, 349*(2), 602–606.
7. Tasdelen, C., et al. (2009). Gold recovery from dilute gold solution using DEAE-cellulose. *Hydrometallurgy, 96*(3), 253–257.
8. Ok, Y. S., & Jeon, C. (2014). Selective adsorption of the gold-cyanide complex from waste rinse water using Dowex 21K XLT resin. *Journal of Industrial and Engineering Chemistry, 20*(4), 1308–1312.
9. Rajasingam, R., et al. (2006). Selective elution of the gold cyanide complex from anion exchange resin using mixed solvents. *Minerals Engineering, 19*(9), 896–903.
10. Zhang, H., & Dreishinger, D. (2002). The adsorption of gold and copper onto ion-exchange resins from ammoniacal thiosulfate solutions. *Hydrometallurgy, 66*, 67–76.
11. Arima, H., et al. (2002). Gold cementation from ammonium thiosulfate solution by zinc, copper and aluminium powder. *Materials Transactions, 43*(3), 485–493.
12. Lee, H. Y., Kim, S. G., & Oh, J. K. (1997). Cementation behavior of gold and silver onto Zn, Al, and Fe powders from acid thiourea solutions. *Canadian Metallurgical Quarterly, 36*(3), 149–155.
13. Agarwal, C., Cattrall, R. W., & Kolev, S. D. (2016). Donnan dialysis based separation of gold (III) from electronic waste solutions using an anion exchange pore-filled membrane. *Journal of Membrane Science, 514*, 210–216.
14. Das, N. (2010). Recovery of precious metals through biosorption—A review. *Hydrometallurgy, 130*(1–4), 180–189.

15. Ju, X., et al. (2016). Effective and selective recovery of gold and palladium ions from metal wastewater using a sulfothermophilic red alga, Galdieria sulphuraria. *Bioresource Technology, 211*, 759–764.
16. Avci, E. (1988). Electrolytic recovery of gold from aqueous solutions. *Electrochimica Acta, 33*(10), 1263–1266.
17. Lee, C. K., et al. (2012). Electrowinning of precious metals using a modified electrochemical cyclone reactor. *Advances in Materials Science Research, 13*, 257–273.
18. Kim, S. K., & Son, J. S. (2013). Method of gold recovery using cyclone electrolytic cell (Korea Patent, No. 10–1274476, 2013).
19. Kim, S. K., et al. (2004). Electrowinning of platinum using a modified cyclone reactor. *Resources Processing, 51*(1), 48–51.

Improving Quality of Coke Made from Chinese Xinjiang Gas Coal with High Strength Modifier

Qiang Wu, Zizong Zhu, Guojing Shi, Feng Wang, Zilong Wang and Yangyang Xie

Abstract Xinjiang gas coal, one of the Chinese low rank coking coal, was modified by high strength modifier. By measuring final contraction of coke residue index (x), it revealed that Xinjiang gas coal with high x about 38 mm has an adverse effect on coke quality. The coke quality was evaluated by 2 kg coke-oven coking experiment. The results suggested that the coke strength after reaction (CSR) of the coke increased by 21% and the coke reactivity index (CRI) decreased by 18% when 0.5 wt% high strength modifier was added into gas coal. The increase of the diameter of carbon network layer and the decrease of the disordered and unsystematic inherent polycondensation in the Semicoke contraction stage may be responsible for the modified results. The modified mechanism was proposed based on thermo gravimetric (TG) and FTIR spectrometer characterizations of the modified coals, and X-ray diffraction (XRD) analysis of modified cokes.

Keywords Low rank coking coals · High strength modifier · Coke quality · Carbon network layer · Inherent polycondensation

Introduction

China abounds in coal resource but little metallurgical coking coals have been found [1, 2]. Xinjiang coal resource account for 42% of the total coal resource in China [3]. However, it mainly consists of long-flame coal, non-caking coal and slight caking coal, which are of low degree of metamorphism, of low ash, low sulfur, and low phosphorus. Coking coals are 19% of Xinjiang coal, while the coal is the most abundant.

Q. Wu (✉) · Z. Zhu · G. Shi · F. Wang · Z. Wang · Y. Xie
College of Material Science and Engineering,
Chongqing University, Chongqing 400044, China
e-mail: 1015312785@qq.com

Z. Zhu
e-mail: zhuzizong@163.com

© The Minerals, Metals & Materials Society 2017

M.A. Meyers et al. (eds.), *Proceedings of the 3rd Pan American Materials Congress*,
The Minerals, Metals & Materials Series, DOI 10.1007/978-3-319-52132-9_53

The Xinjiang coking coals quality and properties are different from other coking coals in China, which may result from the late coal forming period (Jurassic) and particular geological conditions in Xinjiang [4]. Xinjiang coals are known for having high reactivity. The coke made using Xinjiang coking coal has high CRI and low CSR. For example, the cokes (CRI > 70%, CSR < 25%) produced using Xinjiang Avril ditch coals cannot be directly used in blast furnaces larger than 1000 m^3 [5, 6].

Metallurgical cokes are porous material, whose size and shape mainly depend on the coking coals used in the blend [7, 8]. To improve the coke quality and reduce the dosage of coking coal, some research has considered adding plasticity additives such as crude tar, petroleum coke, plastic, and coal tar pitch to coke. Barranco et al. [9] added refuse-derived fuel (RDF) to blending coal for coking. Adding < 7.5 wt% RDF to coke had little influence on the coke quality, but the cost of coke will be decreased. Qian et al. [10] found that the anthracite and green petroleum coke bonded well to the coal coke matrix, and the most likely mechanism for their incorporation was transitional bonding.

Much attention has been paid to the production of coke by using additives [11–13]. However, few published papers concentrate on improving the quality of coke produced by Xinjiang gas coal. In the present work, a coal powder additive was used in the cokemaking process to improve the quality of coke made by Xinjiang gas coal. This paper reports the changes of structural parameters and transformation of micro-constituents in the coke, and summarizes the impact of modifier on the coke strength and quality.

Experimental

Materials

One gas coal sample and one high strength modifier were used. The gas coal was obtained from Baicheng (Xinjiang, China). The sample used was milled to <3 mm and dried for 2 h at 80 °C before use. The high strength modifier was provided by Chongqing University in China. The coal sample was subjected to proximate analysis using a muffle roaster according to China Standard GB/T 212-2008 [14]. The plastic layer index of the coal sample was measured in accordance with China Standard GB/T 479-2000. The measurement of coal caking index (G_{RI}) was carried out according to the National Standard of China (GB5447-85), which is based on the Roga index.

Kg Coke Oven Experiment

The carbonization experiments were carried out in an electrically heated experimental coking oven at Chongqing University. About 2.5 kg coal sample was loaded into a cylinder-shaped steel can (100 mm in diameter and 500 mm in height). The coal density was 1.05 t m^{-3} and moisture content was adjusted to 10 wt%. Then the filled steel can with a cover was placed in the experimental coking oven and heated at 3–6 °C/min to 400 °C, at 5 °C/min to 1000 °C, and held at 1000 °C for 4 h, then cooled to room temperature. The coke produced was subjected to further evaluation.

Reactivity Test

Coke samples were tested for reactivity by PL-500F coke reaction determiner based on the standard method GB/T4000-2008. The reactivity test denoted CRI was carried out in a high temperature reaction electric furnace. The coke reactivity index (CRI) was calculated as the percentage of weight loss after the reaction. After the coke was weighed into a cylindrical drum (140 mm internal diameter and 700 mm length) and subjected to 600 rotations at a speed of 20 rpm, the weight percent of coke particles (>10 mm in size) was used as the indicator of coke strength after reaction (CSR). The CRI and CSR were reported as the average value of two runs.

FTIR and TG Measurements

FTIR were measured on a Nicolet iS5 FT-IR spectrometer at a resolution of 4 cm^{-1}. Samples for the FTIR measurement were prepared by mixing 1 mg of coal sample with 200 mg of KBr and the mixture was pressed to form a pellet. The FTIR spectrum were simplified by curve-fitting analysis with a commercially data-processing program (PeakFit software). Thermogravimetric (TG) analysis was carried out on a Netzsch STA 449F3 analyzer. About 8 mg of sample was placed in an alumina pan and heated from 25 to 1000 °C at a rate of 10 °C/min under 20 ml/min nitrogen gas flow.

X-Ray Diffraction (XRD)

Powder X-ray diffraction experiments were performed in reflection mode on a Bruker D8 Focus Powder Diffractometer with Cu K_α radiation (40 kV, 40 mA) as the X-ray source. Coke samples were pulverized to less than 0.074 mm. Coke samples were then packed into a silicon holder and scanned over an angular range from 10° to 55° 2θ at a scan speed of 0.6° 2θ per min, using a step size of 0.02° 2θ. The XRD patterns were smoothed and deconvoluted by software facilities [15].

Bragg's Law (Eq. 1) was used to calculate the interlayer spacing between aromatic planes of carbon crystallites, d_{002}. The lateral size (L_a) and stacking height (L_c) of the crystallite were calculated by Eqs. (2) and (3), respectively [16, 17].

$$d_{002} = \frac{\lambda}{2\sin\theta_{(002)}} \tag{1}$$

$$L_a = \frac{1.84\lambda}{\beta_{(100)}\cos\theta_{(100)}} \tag{2}$$

$$L_c = \frac{0.91\lambda}{\beta_{(002)}\cos\theta_{(002)}} \tag{3}$$

where λ ($\lambda = 0.15418$ nm) is the wavelength of the X-ray. $\theta_{(002)}$ and $\theta_{(100)}$ are the peak position of 002 band and 100 band, respectively. $\beta_{(002)}$ and $\beta_{(100)}$ are the half high width of crystal plane 002 and 100, respectively.

Results and Discussion

Effect of Modifier on Coal Characterization Data

The modified experiment of gas coal was designed for improving coke quality. The mass percentages of the high strength modifier was 0.5 wt% in the gas coal. Table 1 gives a summary of proximate and processing analysis results of the raw and modified gas coal.

It can be observed from Table 1 that the changes of C, H, N, and S in modified gas coal were negligible. An slight increase in V_{daf} and A_d content of the modified gas coal compared to that of raw gas coal could be observed in Table 1, whilst modified gas coal, had a rather low FC_{ad} number at 56.23 wt%. This may be

Table 1 Basic characteristics of raw and modified gas coal samples

Sample	Ultimate analysis (wt%, daf)				Proximate analysis (wt%)			Processing analysis		
	C	H	N	S	V_{daf}	A_d	FC_{ad}	$G_{R.I}$	*x* (mm)	*y* (mm)
Raw gas coal	76.93	4.83	1.31	0.51	39.01	6.96	56.41	91	38.3	12.4
Modified gas coal	77.01	4.93	1.32	0.51	39.54	7.12	56.23	90	37.9	12.7

C carbon; *H* hydrogen; *N* nitrogen; *S* sulfur; *daf* dry and ash-free; *A* ash yield; *d* dry basis; *V* volatile matter; *FC* fixed carbon; *ad* air-dry basis; $G_{R.I}$ caking index; *y* maximum thickness of plastic layer; *x* final contraction of coke residue

Table 2 Characterization data of coke samples

Sample	Proximate analysis (wt%)			Reactivity (%)	
	A_d	V_{daf}	FC_{ad}	CRI	CSR
Raw coke	11.53	0.95	87.43	61.3	25.4
Modified coke	11.87	0.98	86.71	43.3	46.4

because the high strength modifier mainly consisted of inorganic compounds, which will form the ash after burning. In addition, the $G_{R.I}$, x, and y numbers of modified gas coal almost keep the same as that of raw gas coal as shown in Table 1, suggesting that high strength modifier might have a slight influence on basic characteristics of the gas coal.

Effect of Modifier on Coke Characterization Data

Table 2 shows the proximate analysis and reactivity data of the raw coke sample and modified coke sample at 0.5 wt% high strength modifier.

Table 2 shows that adding high strength modifier to gas coal slightly increases ash content and reduces fixed carbon and volatilization components. The changes of ash content, volatile matter, and fixed carbon in the modified coke were similar to those in the modified gas coal. The change of the CRI and CSR of modified coke is significant compared to that of the raw coke, as shown in Table 2. The CRI decreased from 61.3% of the raw coke to 43.3% of the modified coke while the CSR increased from 25.4 of the raw coke to 46.4 of the modified coke. This suggests that high strength modifier can improve the hot strength of coke and has little effect on the ash content.

In order to probe the mechanism for the reduction of CRI and enhancement of CSR in the modified coke, FTIR and TG measurements were used to discover the structural change of gas coal after adding high strength modifier. The XRD measurements were also used to reveal the crystallite parameter changes of the modified coke.

Effect of High Strength Modifier on Functional Groups During Pyrolysis

Figure 1 shows the FTIR spectra of raw gas coal and modified gas coal with 0.5 wt% high strength modifier during the pyrolysis process. To explore the influence of high strength modifier on functional groups in the carbonization process, the raw and modified gas coal samples were heated at 450 and 750 °C for an hour, and were then analyzed by FTIR at room temperature.

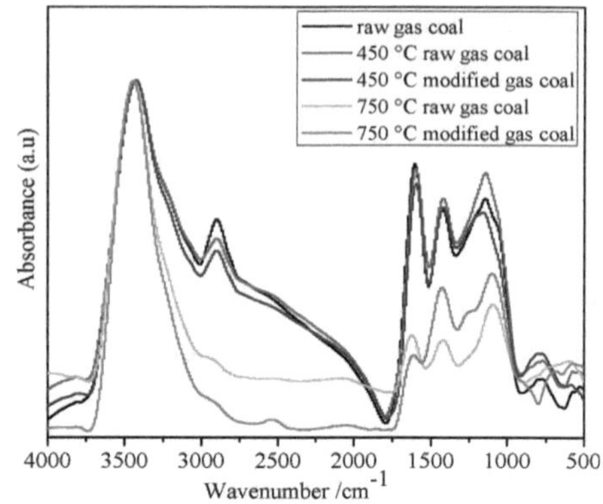

Fig. 1 FTIR spectra of raw gas coal and modified gas coal at different tmperature

With the increasing of pyrolysis temperature, the intensity of bands near 3400 and 1650 cm^{-1} for the raw gas coal decreased, which are assigned to self-associated OH hydrogen bonds and carbonyl (C=O) stretching band [18, 19] in coal respectively. This suggests a decrease of self-associated OH hydrogen bonds and carbonyl bands in the raw gas coal (Fig. 1). Besides, the peaks at around 2920 and 2850 cm^{-1} corresponding to the aliphatic groups disappeared after pyrolysis above 750 °C. This may be because the unstable carbonyl groups and weak aliphatic groups were partially decomposed at high temperature owing to the deep dehydration and polycondensation reactions. However the changes of intensity of the band near 1650 cm^{-1} was mild in the raw coal at 450 °C. This may be because the carbonyl groups in the aromatic linkages have a higher stability below 700 °C [20]. The change of modified gas coal after heat treatment was similar to that of the raw gas coal, as shown in Fig. 1. Simultaneously, the intensity of bands near 3400 and 1650 cm^{-1} in the raw gas coal almost stayed the same as that of the modified gas coal at 750 °C while the intensity of bands near 3400 and 1650 cm^{-1} in the modified gas coal are a relatively high value compared with the raw gas coal at 450 °C (in Fig. 1). The high strength modifier may help to increase the stability of the carbon-hydrogen bond in the macro-molecules at low temperature.

It can also be observed from the spectrum of raw gas coal in comparison with the spectrum of modified gas coal at 750 °C that the modifier seemed to induce an increase of the CH groups in the aliphatic chain and aromatic ring, which was indicated by the increased intensity of the bands near 1370 and 1095 cm^{-1} in the modified gas coal. This may be because high strength modifier can slow the acutely inherent polymerization of the semicoke contraction stage, resulting in a decrease of carbon activity point in the modified coke.

Effect of High Strength Modifier on Weight Variation During Pyrolysis

Figure 2 showed the TG profiles of raw gas coal and modified gas coal with 0.5 wt% high strength modifier. An obvious weight loss of raw gas coal beginning at about 380 °C can be observed in Fig. 2, indicating pyrolytic reactions taking place. However, with the increase of temperature, the TG profiles showed a steady weight loss before 380 °C, which reflected the evaporation of pre-existing fragments. Figure 2 showed that the weight losses of raw gas coal were larger than that of modified gas coal before 380 °C, suggesting more light fragments remained in the modified gas coal. This may be because the decomposition of aggregated structure can be inhibited by adding high strength modifier. A smaller weight loss of the modified gas coal than that of the raw gas coal between 550 and 700 °C is apparent in Fig. 2, probably because the polycondensation reactions were slowed by adding modifier in the semi-coke stage.

Simultaneously, a sharper weight loss of modified gas coal compared to that of raw gas coal can be observed between 700 and 770 °C (in Fig. 2), suggesting much polycondensation reactions of macro-molecular taking place in the modified gas coal in comparison with raw gas coal. However the weight loss of raw coal came gently, indicating the polycondensation reactions having been basically completed between 700 and 770 °C. This may be because the high strength modifier helped to restrain the fleet polycondensation reactions at low temperature, and then to form more macro-molecular substances in the semi-coke stage. The decrease of polycondensation reactions on micro-molecule in the modified gas coal may be responsible for the smaller weight loss of modified gas coal in the pyrolysis process.

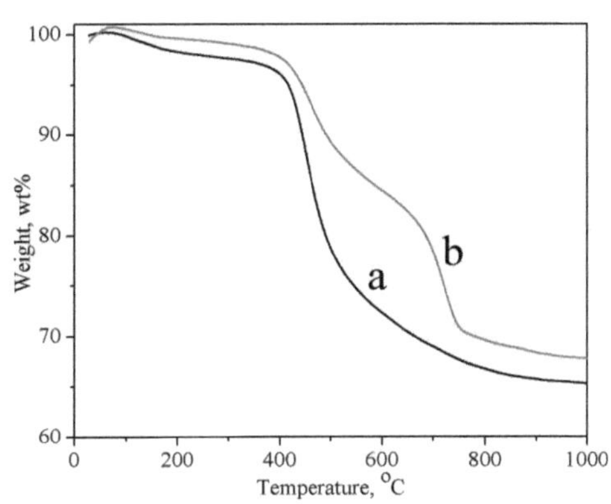

Fig. 2 TG curves of raw gas coal (*a*) and modified gas coal (*b*)

Effect of the High Strength Modifier on the Structural Parameters

The sharp quartz peak (not shown in Fig. 3, removed for deconvolution purposes), was observed at around 26° 2θ in the XRD patterns. Two broad peaks were found in the 2θ range of 20°–30° and 40°–48°, which is similar to the peak structure observed by Sonibare et al. [21] in carbonized bituminous coal.

The stacking in aromatic layers and the distance between points in an aromatic layer were related to the 002 and 100 bands respectively, which were usually observed at about 25° and 45° 2θ [16, 22]. Besides, the aliphatic side chains and irregular packing of buckled aromatic layers, corresponding to γ band, were associated with the packing distance of coal crystallites [23]. The ratio of ordered graphitic carbon to amorphous carbon was usually measured by the ratio of the areas of 002 band and γ band [24]. The ratio in raw coke was about 2.87, whilst for modified coke it was about 2.94, suggesting that the modifier helps to increase the ordered graphitic structures, which improves the quality of the coke. Moreover, the inter-layer spacing between aromatic planes of carbon crystallites of raw coke (d_{002} at 0.357 nm) was greater than that in the modified coke (0.352 nm), indicating a lower degree of graphitization or poorer crystallinity and the presence of some aliphatic carbon in the raw coke [25]. Furthermore, the 002 band of modified coke was narrower than the corresponding band in the raw coke, again suggesting a higher degree of crystalline order than that in the raw coke. Simultaneously, the stacking height, L_c in the modified coke was about 1.73 nm compared to only about 1.62 nm for the raw coke; bigger crystallite size in the modified coke indicates a better degree of graphitization than that of the raw coke.

The bigger proportion of well-ordered graphitic structure, the better crystallinity of the graphitic structure, and the bigger size of the crystallites in the modified coke all help to explain their lower reactivity in comparison with that of raw coke, which was consistent with the higher value of CSR and lower value of CRI in the modified coke mentioned above.

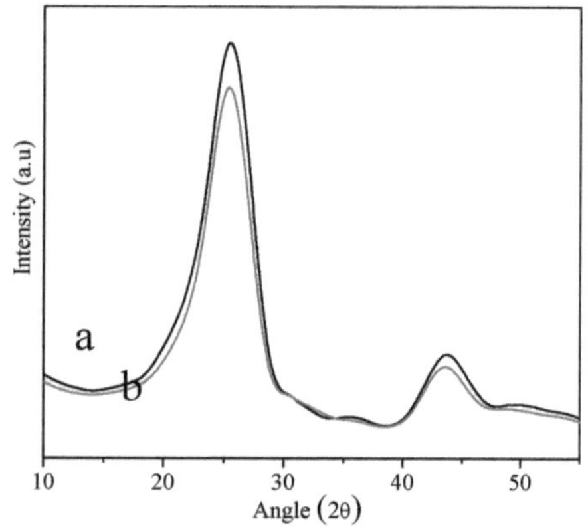

Fig. 3 The XRD patterns of raw coke (*a*) and modified coke (*b*) with 0.5 wt% high strength modifier

Conclusion

Coke prepared from Xinjiang gas coal showed high reactivity about 61.3% and low CSR about 25.4%, owing to the high number x about 38.3 mm in coal and sharp polycondensation reactions taking place in the semi-coke stage. The high strength modifier played an important role in improving the hot strength, which resulted in the CRI decrease of 18% while the CSR increased by 21% in the modified coke.

The high strength modifier slowed the velocity of polycondensation reactions in the plastic stage. This decreases disorder micro-molecule fragments in the macro-molecular network of the coal, thus favoring the decrease of d_{002} and increase of L_a and L_c in the pyrolysis stage. The low reactivity in the modified coke compared to that in the raw coke was probably related to the greater extent and smaller disorder of their graphitic structure.

Acknowledgements The support of The National Natural Science Foundation of China (51044005) and The Key Project of Science and Technology of Chongqing (CSTS.2010AB4084), are greatly acknowledged for the funding of this work.

References

1. He, J. H. (2006). Optimize coal blending to save high-quality coking coal. *ShanXi Coal, 26*(4), 53–55. (in Chinese).
2. Wu, K. H., Chen, Y. F., & Yu, H. B. (2005). Resources and production of coking coal and anthracite in China. *China Metallurgy, 15*(7), 18–21 (in Chinese).
3. Wu, Y. (2011). Thinking of development and utilization of coal resources in Xinjiang. *Xinjiang Social Science BBS, 02*, 38–43 (in Chinese).
4. Huang, W. H., et al. (2010). Distribution features of coal for coking resource in china and deep part potential analysis. *Coal Geology of China, 22*(5), 4–6. (in Chinese).
5. Zhang, N. (2013). Xinjiang coking coal supply and demand outlook. *Scientific and Technological Information, 20*, 469 (in Chinese).
6. Hu, D. S. (2012). Solving coal resources problem in Xinjiang Bayi Iron & Steel Co. Ltd. China Metallurgical News 2012-8-16 ($C0_2$) (in Chinese).
7. Patrick, J. W., Reynolds, M. J., & Shaw, F. H. (1973). Development of optical anisotropy in vitrains during carbonization. *Fuel, 52*(3), 198–204.
8. Brooks, J. D., Taylor, G. H., et al. (1968). Chemistry and physics of carbon; (04), p. 243.
9. Barranco, Richelieu, Patrick, John, Snape, Colin, et al. (2007). Impact of low-cost filler material on coke quality. *Fuel, 86*(03), 2179–2185.
10. Qian, Z., Marsh, H., & Clarke, D. E. (1985). Bonding of solid additives in cokes from coals: A microscopy study. *Fuel, 64*(1), 125–128.
11. Valentina, Z. V. (2006). Influence of polyethylene terephthalate on the carbonisation of bituminous coals and on the modification of their electric and dielectric properties. *Fuel, 85*(12), 1652–1665.
12. Choudhury, S. B., Brahmachari, B. B., Dwivedi, S. R., et al. (1996). Solvent-refined coal from high-ash non-coking coals and washery middlings for use in metallurgical coke making Part 1. Production, testing and characterization. *Fuel Processing Technology,* 47, 203–213.
13. Peng, C. H. (2005). Cokemaking test for coal blending of anthracite. *Coal Chemical Industry, 6*, 47–53.

14. GB/T 212-2008. (2008). *Proximate analysis of coal* (pp. 1–14). Beijing: Standards Press of China (in Chinese).
15. Li, M., Zeng, F., Chang, H., et al. (2013). Aggregate structure evolution of low rank coals during pyrolysis by in-stu X-ray diffraction. *International Journal of Coal Geology, 116–117*, 262–269.
16. Lu, L., et al. (2001). Quantitative X-ray diffraction analysis and its application to various coals. *Carbon, 39*, 1821–1833.
17. Dinnebier, R. E., & Billinge, S. J. L. (2008). Powder diffraction theory and practice. *The Royal Society of Chemistry*.
18. Chen, C., Gao, J. S., & Yan, Y. J. (1998). Observation of the type of hydrogen bonds in coal by FTIR. *Energy & Fuels, 12*, 446–449.
19. Cai, M. F., & Smart, R. B. (1994). Comparison of seven west Virginia coals with their N-methyl-2-pyrrolidinone-soluble extracts and residues.1. Diffuse reflectance infrared Fourier transform spectroscopy. *Energy & Fuels, 8*, 369–374.
20. Lin, X. C., Wang, C. H., Keiko, I., et al. (2014). Insight into the function group transformation of a Chinese brown coal during slow pyrolysis by combining various experiments. *Fuel, 118*, 258–260.
21. Sonibare, O. O., Haeger, T., & Foley, S. F. (2010). Structural characterization of Nigerian coals by X-ray diffraction. *Raman and FTIR Spectroscopy, Energy, 35*, 5347–5353.
22. Xin, X., Zhag, G., Rogers, H., et al. (2014). Effects of annealing on microstructure and microstrength of metallurgical coke. *Metallurgical and Materials Transactions B: Process Metallurgy and Materials Processing Science, 45*, 106–112.
23. Siddiqui, M. N., Ali, M. F., & Shirokoff, J. (2002). Use of X-ray diffraction in assessing the aging pattern of asphalt fractions. *Fuel, 81*, 51–58.
24. Manoj, B., & Kunjomana, A. G. (2012). Study of stacking structure of amorphous carbon by X-ray diffraction technique. *International Journal of Electrochemical Science, 7*, 3127–3134.
25. Nyathi, M. S., Clifford, C. B., & Schobert, H. H. (2013). Characterization of graphitic materials prepared from different rank Pennsylvania anthracites. *Fuel, 114*, 244–250.

Investigating the Dissolution Characteristics of Strontium Sulfide

İbrahim Göksel Hizli, Ayşegül Bilen, Raşit Sezer, Emre Yilmaz, Selim Ertürk and Cüneyt Arslan

Abstract Celestite is the main source of strontium containing chemicals and metallic strontium. Unlike other sulfate containing minerals, celestite is not soluble in water. Further processing of $SrSO_4$ requires conversion to water-soluble strontium sulfide (SrS). In this study, solubility of SrS in distilled water is investigated. SrS used in experimental study was obtained by roasting celestite at 1050 °C in a lab-scale rotary furnace. Experiments were carried out to determine the effects of time (90, 120, and 150 min), solid/liquid (S/L) ratio (1/5, 1/10, and 1/25 w/v) and temperature (25, 55, 75, and 95 °C). Stirring speed (500 rpm) was constant. XRD results of leach cakes showed that increasing temperature increases dissolution rate and efficiency. However, increasing S/L ratio decreases the leaching efficiency.

Keywords Celestite · Strontium sulfide · Leaching

Introduction

Strontium metal is obtained from the naturally found celestite ($SrSO_4$) and strontianite ($SrCO_3$) minerals. The strontium percentage is higher in strontianite, but the high calcium content limits its industrial usage. Hence, despite the lower strontium percentage of celestite, it has higher commercial value [1].

Strontium minerals are the main source of many strontium-based compounds such as strontium carbonate ($SrCO_3$), strontium nitrate ($Sr(NO_3)_2$), strontium chloride ($SrCl_2$), strontium hydroxide ($Sr(OH)_2$), and strontium oxide (SrO). The

İ.G. Hizli · A. Bilen (✉) · R. Sezer · E. Yilmaz · S. Ertürk · C. Arslan
İstanbul Technical University, Istanbul, Turkey
e-mail: aysegul.bilen92@gmail.com

İ.G. Hizli
İstanbul University, Istanbul, Turkey

R. Sezer
Karadeniz Technical University, Trabzon, Turkey

© The Minerals, Metals & Materials Society 2017

M.A. Meyers et al. (eds.), *Proceedings of the 3rd Pan American Materials Congress*,
The Minerals, Metals & Materials Series, DOI 10.1007/978-3-319-52132-9_54

most common usage of strontium carbonate is the production of colored television tubes and magnetic materials. Apart from that, it can be used for zinc refining, strong magnets, and as a lubricator for bearings when converted to strontium ferrite [2].

For the production of metallic strontium and other strontium-based compounds, the most important step is obtaining strontium carbonate from celestite. Production of strontium carbonate is carried out by two common methods. In the black ash method, the celestite mineral is roasted at 1100–1200 °C to obtain strontium sulfide, which is soluble in water. By this process, reduced ore is solubilized in water at 80–95 °C. Then, dissolved strontium is precipitated by carbonating agents such as sodium carbonate, ammonium carbonate, ammonium bicarbonate or carbon dioxide [3]. The other method for obtaining strontium carbonate is direct conversion. The ore should be purified at high rates before this process starts. In this refining process, calcium and iron compounds are removed from the ore by washing with HCl or H_2SO_4. Refined ore then reacts with sodium carbonate or ammonium carbonate to give strontium carbonate [4]. This method is also called the *double decomposition process*.

Several studies sought to diminish the disadvantages of high cost and excess process steps. De Buda and co-workers studied an ore containing less than 80% $SrSO_4$, and used HCl to dissolve. They precipitated $SrSO_3$ with 99% efficiency from the solution by adding ammonium carbonate to a $SrCl_2$ charged solution [5]. Erdemoglu and Canbazoglu [3] roasted strontium sulfide and dissolved it in water. They produced $SrCO_3$ from the strontium solution and also investigated the effect of Na_2CO_3 and CO_2 on speed and efficiency. According to their study, Na_2CO_3 has lower efficiency than the CO_2 as a carbonating agent, but the precipitation was faster. Aydogan et al. [6] investigated the solubility of celestite in $BaCl_2$-containing acidic solution. They discovered that increasing the temperature, which was held between 40 and 80 °C, increased the dissolution efficiency. In addition to this, it was observed that in the presence of NaCl, dissolution of celestite efficiency increased.

Experimental Procedure

Black ash, used for dissolution experiments, was produced by two-hour-long roasting, carried out at 1050 °C with a mixture of celestite ore and coke. XRD pattern of the roasting product is shown in Fig. 1.

Experiments took place in a water-jacketed reactor which provided heating by circulating hot water. The system was mixed by a mixer from above up to 1000 rpm. During the experiment, 10 ml samples were taken after 90 and 120 min, and diluted to 100 ml. After 150 min, the remaining liquid was filtered to a flask by filter paper. Then, 250 ml of solution was diluted to 500 ml and stocked. Strontium

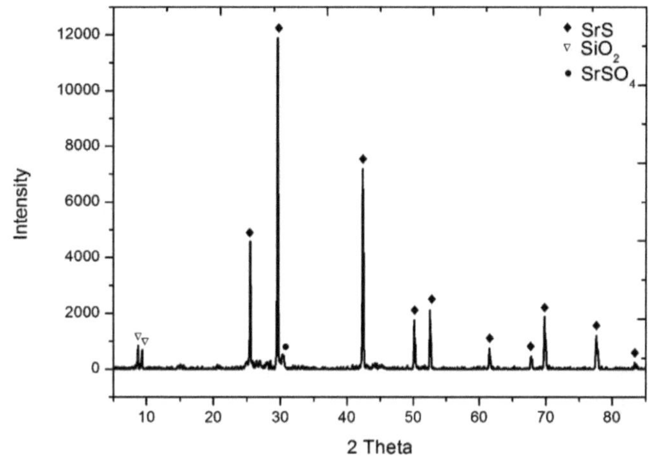

Fig. 1 XRD pattern of the *black* ash

Table 1 Parameters of the experiments

Temperature	25, 55, 75, 95 °C
S/L ratio	1/5, 1/25, 1/50
Time	90, 120, 150 min

concentrations of the samples were analyzed by Atomic Absorption Spectrometer (AAS), and filtered solids were chemically characterized by an X-Ray Diffractometer.

Parameters investigated in this study are listed in Table 1. Volume of the solution and the mixing rate were held at 250 ml and 500 rpm, respectively. Production conditions of the black ash (powders obtained after two hours of roasting at 1050 °C) were the same for all experiments.

Results and Discussion

Weights of the black ash which was added before the experiment and the leach residue obtained after the experiment can be seen in Table 2.

Effect of S/L Ratio

To investigate the effect of S/L ratio on efficiency, three different S/L ratios were used: 1/5, 1/25, and 1/50. Leach residues were analyzed by XRD. Leaching efficiencies were calculated by using the weighting results. The results are shown in Fig. 2.

Table 2 Experimental conditions and dissolution efficiencies of Sr

S/L ratio	Temperature (°C)	Initial weight (g)	Final weight (g)	Sr dissolution efficiency (%)
1/5	25	50	44.48	11.08
	55		37.84	24.35
	75		32.03	35.97
	95		20.04	59.94
1/25	25	10	7.34	26.75
	55		5.80	42.17
	75		5.30	47.11
	95		4.92	50.89
1/50	25	5	3.06	39.04
	55		2.81	44.02
	75		2.75	45.22
	95		2.39	52.39

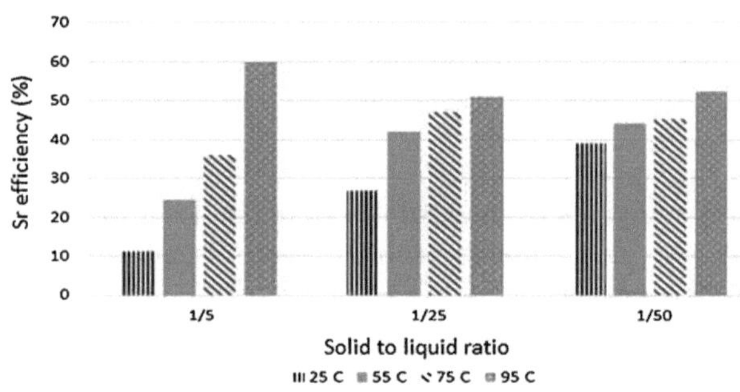

Fig. 2 Sr dissolution efficiency versus S/L ratio

Sr dissolved more efficiently at 75 and 95 °C. The effect of S/L ratio was investigated by the XRD peaks of leach residues of those experiments. The patterns can be seen in Figs. 3 and 4. According to the results, as the solid amount increases, undissolved SrS in the leach residue also increases. This can be seen from the XRD peaks. Occurrence of $Sr(OH)_2$ is also observed by leaching at high temperatures (95 °C), which precipitated from the solution during the cooling period before filtration. Thus, $Sr(OH)_2$ was found in the chemical analysis of the leach residue.

Effect of Temperature

To investigate the effect of temperature, experiments were conducted at 25, 55, 75, and 95 °C. The difference between the weight of black ash before the experiment

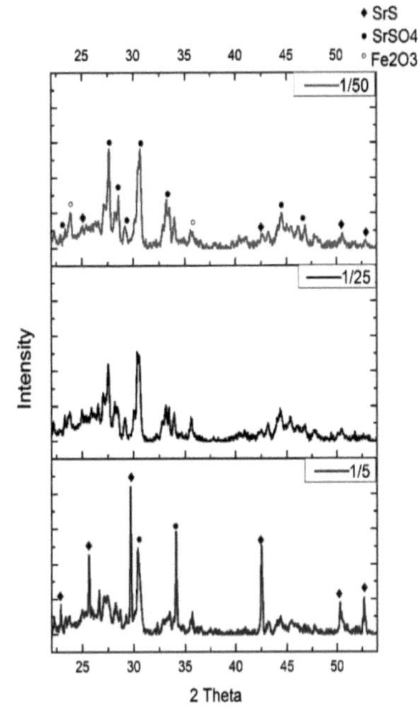

Fig. 3 XRD patterns for different S/L ratios (at 75 °C)

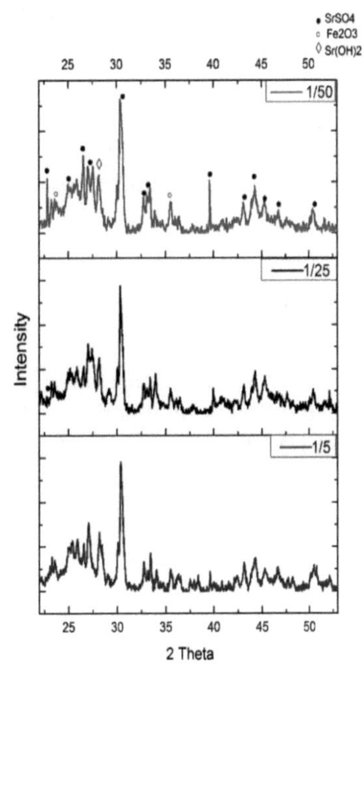

Fig. 4 XRD patterns for different S/L ratios (at 95 °C)

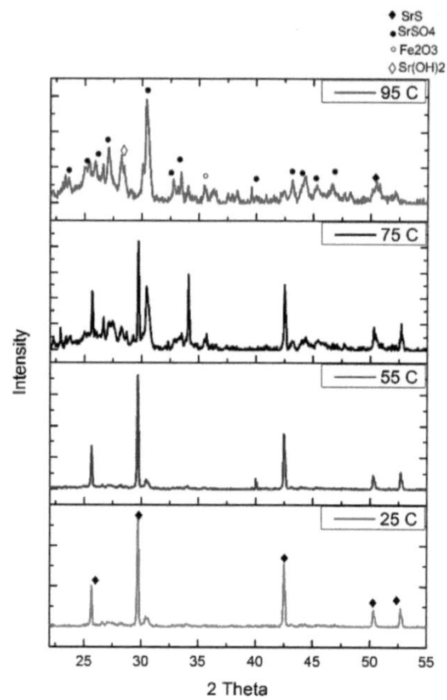

Fig. 5 XRD patterns of the leach residues (1/5 S/L ratio)

and that of the leach residue gives the efficiency of dissolution. Figure 2 also shows this relationship with regard to the temperature. As can be seen, as the temperature increases, efficiency increases.

Figure 5 shows the XRD patterns of the experiments carried out at 25 and 55 °C. Leach residues contained SrS. Leach residues of the experiments done at 75 and 95 °C has unreacted celestite from the black ash.

Effect of Time

In order to investigate the effect of time, leaching experiments were conducted at three different time periods; 90, 120, and 150 min. Leaching temperature of 95 °C and 1/5 S/L ratio were kept constant. Dissolution efficiency was calculated by using AAS results. The change in Sr dissolution efficiency versus time is shown in Fig. 6. The best dissolution efficiency is obtained by leaching for 150 min.

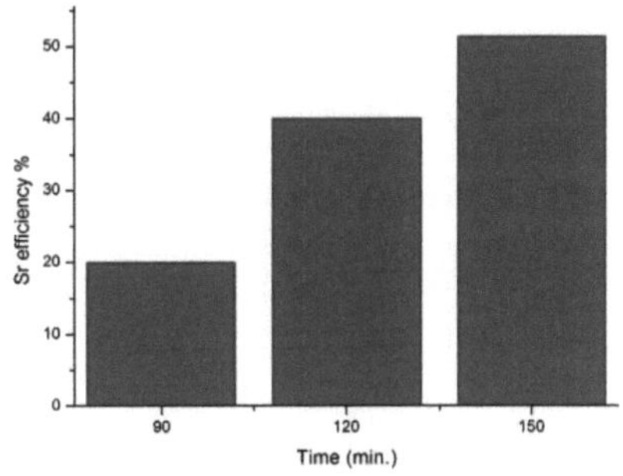

Fig. 6 Sr dissolution efficiency versus time

Conclusion

In this work, dissolution of black ash was studied. Black ash was obtained by roasting celestite and coke. Thus, the effects of S/L ratio, time and temperature on the dissolution efficiency of strontium were investigated. As the S/L ratio increased, dissolution efficiency of strontium decreased. This result was supported by the XRD analysis of leach residues. Increasing temperature noticeably affected the dissolution of strontium. However, strontium dissolved as $Sr(OH)_2$ in leaching experiments conducted at 95 °C and precipitated upon cooling the leachate and ended up in the leach cake; this decreased the dissolution efficiency of strontium. Extending the leaching periods increased the effectiveness. Finally, the optimum conditions were found to be 95 °C, 1/50 S/L ratio and 150 min leaching period.

Acknowledgements The authors greatly acknowledge the financial support provided by Scientific and Research Council of Turkey (TUBITAK) under the Grant No. 115M631 and would like to thank to Barit Maden Türk Ltd. for technical and materials provision.

References

1. Habashi, F. (1997). *Handbook of extractive metallurgy* (Vol. 4, pp. 2329–2336). New York: Wiley-VCH. ISBN 3-527-28792-3.
2. Carrillo, P. F. R., Uribe, S. A., & Castillejos, E. A. H. (1995). A laboratory study of the leaching of celestite in a Pachuca tank. *Minerals Engineering, 8*(4/5), 495–509.
3. Erdemoglu, M., & Canbazoglu, M. (1998). The leaching of SrS with water and the precipitation of $SrCO_3$ from leach solution by different carbonating agents. *Hydrometallurgy, 49*, 135–150.
4. Erkalfa, H., Kalafatoglu, E., Tozun, R., & Ozkan, O. (1981). Ferrit Malzemelerde Kullanılan $BaCO_3$, $SrCO_3$ ve $MnCO_3$'ın Yerli Hammaddelerden Üretim Yöntemleri (pp. 27–39). Marmara Scientific and Industrial Research Institute.
5. De Buda, F. (1987). *Method for recovery and conversion of strontium sulfate to strontium carbonate from low to medium grade celestite ores*. U.S.A. Patent, 4,666,688.
6. Aydogan, S., Erdemoglu, M., Arar, A., Ozkan, G. A., & Ucar, A. (2006). Dissolution kinetics of celestite ($SrSO_4$) HCl solution with $BaCl_2$. *Hydrometallurgy, 84,* 239–246.

Kinetic Study on the Leaching of Vanadium-Bearing Converter Slag with Dilute Sulfuric Acid

Junyi Xiang, Qingyun Huang, Xuewei Lv and Chenguang Bai

Abstract Calcification roasting-sulfuric acid leaching process was used to extract vanadium from converter slag. The kinetics of the extraction process based on both unmilled and ball milled slags in dilute sulfuric acid solution (pH = 2.5 ± 0.2) were investigated. The results showed that the extraction of vanadium was a rapid process with particularly a considerable proportion of vanadium rapidly dissolved in the first 10 min. The kinetics analysis showed that the diffusion through product layer was rate-determining step in the course of the dissolution of vanadium. The linear relationship between the rate constant and the inverse square of the particle diameter also illustrated this phenomenon. Mechanical activation can significantly accelerate the vanadium leaching efficiency, and decrease the corresponding apparent activation energy from 9.94 to 7.63 kJ/mol.

Keywords Vanadium · Converter slag · Kinetic · Leaching · Sulfuric acid

Introduction

Recycling of metallurgical wastes has become very important during the past decade due to the reinforcement of environmental regulations. Vanadium-bearing converter slag was one of the most important metallurgical wastes, as well as the most important raw material for the production of vanadium oxide [1]. Although sodium-roasting-water leaching process is the most widely used technique to extract vanadium from converter slag, the emission of corrosive gases and the dumping sodium containing wastewater are restricted more and more seriously [2–4].

J. Xiang · X. Lv (✉) · C. Bai
School of Materials Science and Engineering,
Chongqing University, Chongqing 400044, China
e-mail: lvxuewei@163.com

Q. Huang
School of Materials and Metallurgical Engineering,
Chongqing University of Science and Technology, Chongqing 401331, China

© The Minerals, Metals & Materials Society 2017
M.A. Meyers et al. (eds.), *Proceedings of the 3rd Pan American Materials Congress*,
The Minerals, Metals & Materials Series, DOI 10.1007/978-3-319-52132-9_55

Calcification roasting-acid leaching process appears to be a practical clean production method for extraction of vanadium from converter slag, with less gas emission, less harmful wastewater and little pollution [5].

Despite the industrial importance of the calcification roasting-acid leaching process for converter slag dissolution for producing V_2O_5 there has been surprisingly little published on the kinetics and mechanisms of the dissolution of calcification roasted converter slag in sulfuric acid. The dissolution of converter slag by calcification roasting-acid leaching process can be summarized by reactions as follow [6]:

Calcification roasting:

$$FeV_2O_4 + 5/4O_2 + 2CaCO_3 = Ca_2V_2O_7 + 1/2Fe_2O_3 + 2CO_2 \tag{1}$$

Leaching (pH 2.5)

$$Ca_2V_2O_7 + 2H_2SO_4 = H_4V_2O_7 + 2CaSO_4 \tag{2}$$

Leaching (pH < 0.5)

$$Ca_2V_2O_7 + 3H_2SO_4 = (VO_2)_2SO_4 + 2CaSO_4 + 3H_2O \tag{3}$$

When leaching at pH < 0.5, CaV_2O_7 dissolved in the sulfuric acid solution as VO_2^{2+}. The leaching behavior and kinetics of the calcification roasted converter slag in sulfuric acid (1–27%) have been investigated [7]. However, CaV_2O_7 dissolved in the sulfuric acid solution as $V_2O_7^{4-}$ when leaching at pH 2.5, The leaching of calcification roasted converter slag in sulfuric acid solution at pH 2.5 has been studied by lots of workers [5, 8–10]. They are mainly focus on the effect of roasting and leaching parameters. However, the kinetics of the leaching is less studied. The present study has been undertaken in order to clarify the details of the mechanism of the leaching of calcification roasted converter slag with dilute sulfuric acid at pH 2.5. The mechanical activation on the leaching of converter slag also has been investigated. The process conditions studied include mechanically activated sample or not, leaching temperature, particle size of the roasted sample and leaching time.

Experimental

Vanadium-bearing converter slag used in this study was kindly provided by Panzhihua Iron and Steel Group Corp. (China). The slag was ground, sieved and magnetic separated. The non-magnetic part was used in this study. 100 g of non-magnetic part was mixed with 12.3 g of calcium carbonate and divided into two parts. One part was introduced into a planetary ball mill (Retsch PM 100,

Table 1 Chemical analysis of the roasted sample, wt%

Compound	V_2O_5	TFe	TiO_2	SiO_2	MnO	CaO	MgO
wt%	13.20	27.35	12.73	12.84	6.39	8.56	2.55

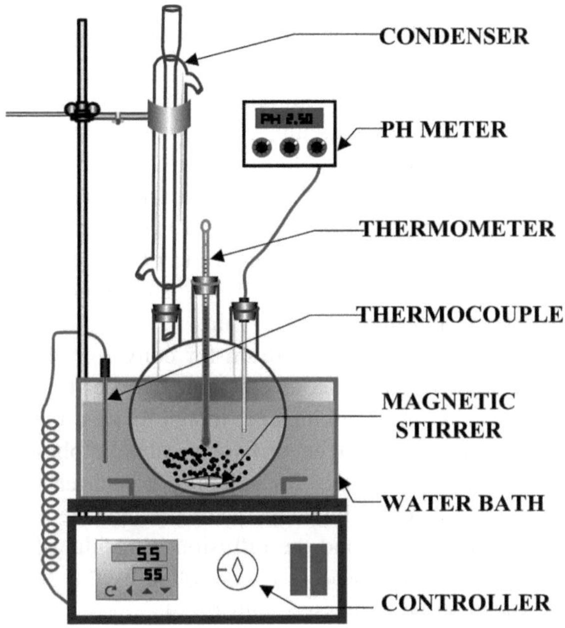

Fig. 1 Schematic diagram of the leaching device

Germany) for mechanical activation and the other part did not. The mechanical activation experiments were conducted at a ball to ore weight ratio of 5:1, rotation rate of 400 rpm, and lasting 80 min. Then both parts were separately oxidation roasted in a muffle furnace at 800 °C for 2 h. After roasting, the samples were cooled down to room temperature in the air, then milled and screened by sieves in different meshes. The chemical analysis of the roasted sample was shown in Table 1.

The roasted samples were leached in dilute sulfuric acid solution in the temperature range of 30–55 °C for periods of up to 60 min. As shown in Fig. 1, a three-necked round-bottomed flask equipped with a water condenser, a thermometer and a pH meter was used for leaching. The flask was heated by a thermostatically water bath. During leaching, 15 g roasted sample was mixed with 300 ml of solution. In the entire leaching process, sulfuric acid concentration of 15% was used to maintain the acidity of the leaching solution at 2.5 ± 0.2. In all cases, 150 rpm rotating speed was needed to keep a good mixing effect. Solution samples of 2 ml were withdraw at a regular time during a run and analyzed the vanadium content by ferrous ammonium sulfate titration method.

Results and Discussion

The effect of temperature on the vanadium extraction from converter slag was investigated in the temperature range 30–55 °C. The samples with the particle size finer than 75 μm were used. Figures 2 and 3 show the relation between vanadium extraction and time at various reaction temperatures for unmilled and milled slag, respectively.

It was shown that both temperature and time have a significant influence on the leaching ratio of vanadium. The leaching rate of vanadium is very fast at the initial stages then decreases with time at all the temperatures investigated. The higher the temperature, the faster the rate of leaching at the initial stages.

In addition, the leaching ratio of vanadium reached to about 75–80% at the first 10 min. However, the leaching ratio of vanadium only increased to about 85–90% with the extent of time to 60 min. This indicates that the extraction of vanadium mainly occurs in the first 10 min.

The extraction of vanadium for converter slag in a dilute H_2SO_4 solution is a solid-liquid heterogeneous reaction. In order to identify the reaction mechanism of this process, the shrinking core model for reaction control under the assumption that the sample is a homogeneous spherical solid phase. Such model is quite extensively used in kinetic studies for leaching converter slag with sulfuric acid [3, 7, 11]. Three rate-determining steps were studied: the diffusion in liquid film, the surface chemical reaction, and the diffusion in residual layer.

If the leaching process is controlled by the diffusion in liquid film, the leaching kinetics can be expressed with Eq. 4:

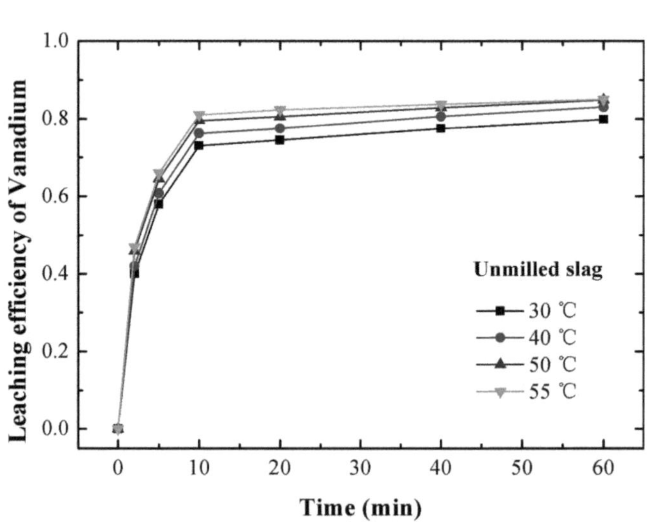

Fig. 2 Relation between vanadium extraction and time at various reaction temperatures (unmilled slag)

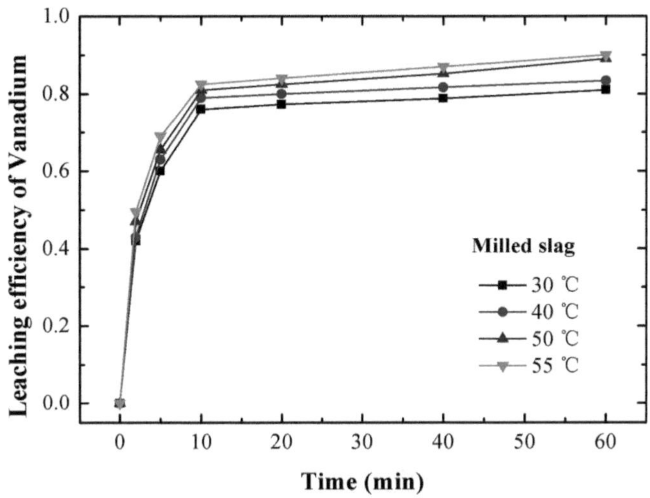

Fig. 3 Relation between vanadium extraction and time at various reaction temperatures (milled slag)

$$1 - (1 - a)^{\frac{2}{3}} = \frac{Mk_cC_At}{dr} = k_1t \tag{4}$$

If the leaching process is controlled by the surface chemical reactions, the leaching kinetics can be expressed with Eq. 5:

$$1 - (1 - a)^{\frac{1}{3}} = \frac{Mk_cC_At}{dr} = k_1t \tag{5}$$

If the leaching process is controlled by the diffusion in residual layer, the leaching kinetics can be expressed with Eq. 6:

$$1 + 2(1 - a) - 3(1 - a)^{\frac{2}{3}} = \frac{6uMDC_At}{dr^2} = k_3t \tag{6}$$

In Eqs. 4–6, a is fraction of vanadium dissolved at time t; M is the molecular weight of the solid reactant; k_c is the chemical rate constant (cm/min); C_A is the concentration of sulfuric acid (mol/m^3); r is the initial radius of the solid reactant; k_1 is the rate constant of the external diffusion; k_2 is the rate constant of the surface chemical reactions; k_3 is the rate constant of the internal diffusion.

To confirm the rate-controlling step and kinetics of the extraction of vanadium from the unmilled and milled slags in dilute sulfuric acid solution, Eqs. 4–6 were applied to the experimental data obtained at the first 10 min. The slops of fitted straight lines passing through the origin equalled to the rate constants. Table 2 shows the obtained rate constants and correlation coefficients of regression. It was showed that only Eq. 6 gives perfectly straight lines but Eqs. 4 and 5 do not, meaning that the diffusion in residual layer was the rate-determining step for the

Table 2 The summary of model fitting results

Parameter	$1-(1-x)^{2/3}$		$1-(1-x)^{1/3}$		$1-3(1-x)^{2/3}+2(1-x)$	
	K	R^2	K	R^2	K	R^2
Unmilled sample						
30 °C	0.0533	0.802	0.0330	0.864	0.0285	0.995
40 °C	0.0565	0.803	0.0355	0.870	0.0322	0.996
50 °C	0.0593	0.779	0.0381	0.857	0.0363	0.992
55 °C	0.0609	0.779	0.0395	0.860	0.0385	0.992
Milled sample						
30 °C	0.0562	0.807	0.03525	0.874	0.0320	0.997
40 °C	0.0595	0.814	0.03805	0.885	0.0360	0.997
50 °C	0.0609	0.783	0.03946	0.863	0.0384	0.994
55 °C	0.0623	0.752	0.04084	0.839	0.0407	0.982
Particle size						
100–150 μm	0.02886	0.747	0.01595	0.777	0.008	0.977
75–100 μm	0.03865	0.754	0.02228	0.799	0.0147	0.984
50–75 μm	0.05273	0.779	0.03257	0.844	0.0282	0.993
<50 μm	0.06333	0.762	0.04184	0.851	0.0422	0.988

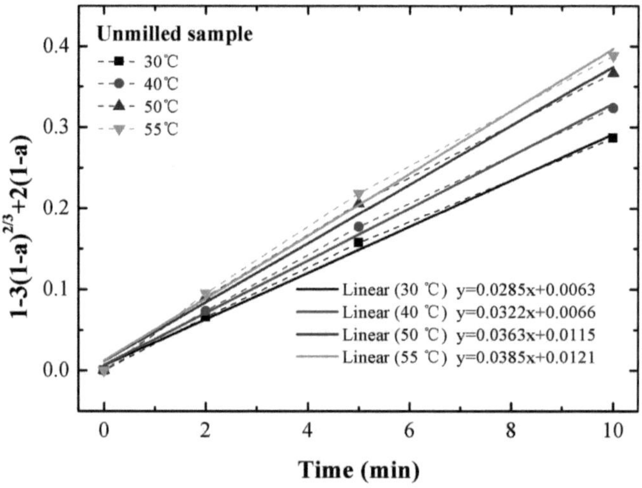

Fig. 4 Plots of $1 + 2(1-a) - 3(1-a)^{2/3}$ versus time at various reaction temperature (unmilled sample)

dissolution of vanadium in both unmilled and milled slags. The plots of the data from Figs. 2 and 3 in first 10 min for temperature range 30–55 °C drawn according Eq. 3 are depicted in Figs. 4 and 5, respectively.

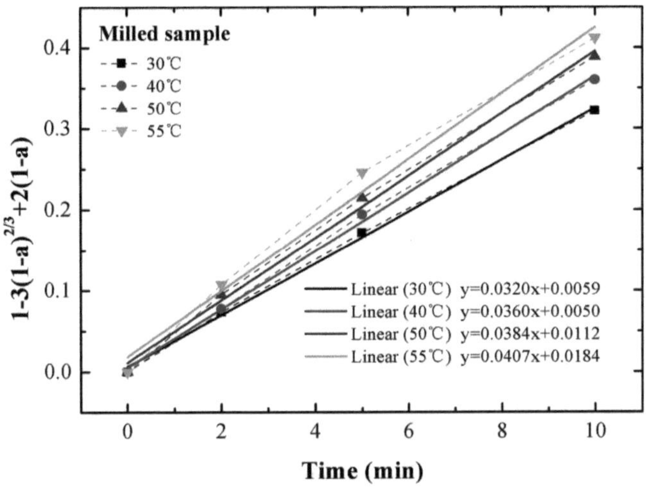

Fig. 5 Plots of $1 + 2(1 - a) - 3(1 - a)^{2/3}$ versus time at various reaction temperature (milled sample)

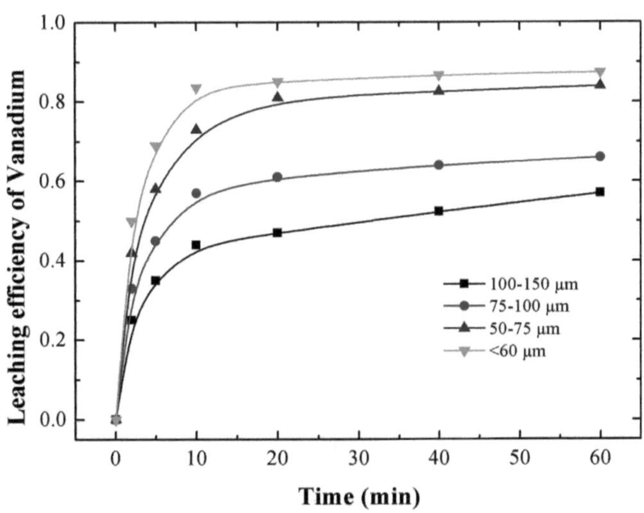

Fig. 6 Relation between vanadium extraction and time at various particle size (unmilled slag)

In order to identify the rate-determining step of the extraction of vanadium, the effect of particle size on the leaching efficiency of vanadium was carried out with four particle sizes, −60, 50–75, 75–100, 100–150 μm. As shown in Fig. 6, the leaching efficiency of vanadium directly increased with the decrease of particle size. Based on the experimental date in the first 10 min, a plot of $1 + 2(1 - a) - 3$

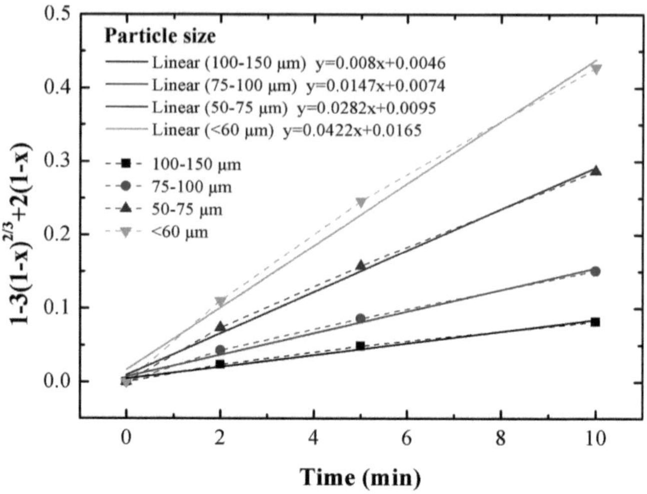

Fig. 7 Plots of $1 + 2(1 - a) - 3(1 - a)^{2/3}$ versus time at various particle size (unmilled sample)

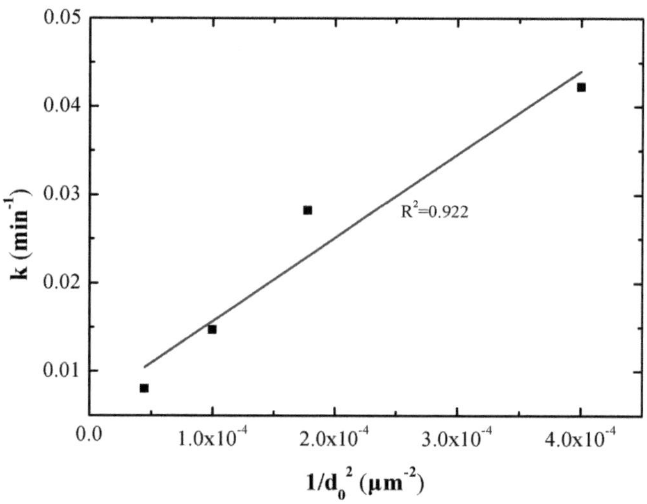

Fig. 8 A plot of the rate constant versus $1/d_0^2$ (d_0 the initial particle diameter of samples)

$(1 - a)^{2/3}$ versus time at various particle size are depicted in Fig. 7. The calculated apparent rate constants are plotted versus the inverse square of the initial particle diameter d_0, as shown in Fig. 8. The linear relationship between the rate constant and $1/d_0^2$ also indicates that the production layer diffusion is the rate-determining step.

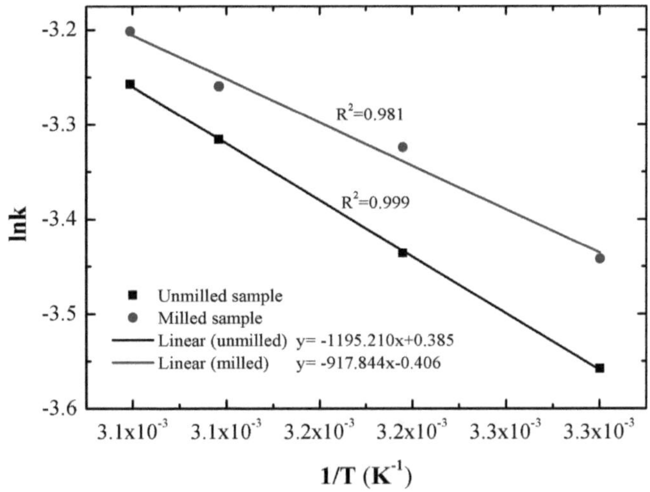

Fig. 9 Arrhenius plot for leaching of unmilled and milled slags

In order to obtain the activation energy of the reaction, Arrhenius type equation was used. The Arrhenius type equation can be represented simply as Eq. 7:

$$\ln k = \ln k_0 - \frac{E}{RT} \tag{7}$$

The reaction rate constants (k) at various temperature for unmilled and milled samples were obtained from the linearized plots (Figs. 4 and 5), summarized in Table 2. Then plotted against temperature according to the Eq. 7, as shown in Fig. 9. The slop of the line is equal to −E/R, from which we can calculate the apparent activation energy. The apparent activation energy for unmilled sample and milled sample were calculated as 9.94 and 7.63 kJ/mol, respectively.

Conclusions

The following conclusions can be drawn from this study.

(1) The extraction of vanadium was a rapid process with particularly a considerable proportion of vanadium rapidly dissolved in the first 10 min.
(2) The results indicated that, under the experimental condition, the shrinking core model for spherical particles is applicable to dilute sulfuric acid leaching of converter slag.
(3) The mechanism of leaching indicated that, the diffusion through the product layer was the rate-determining step for the dissolution of vanadium in both unmilled and milled slags. The apparent activation energy for leaching of

unmilled and milled slags were calculated as 9.94 and 7.63 kJ/mol, respectively.

Acknowledgements This work was supported by the Natural Science Foundation of China [grant numbers 5140, 4047] and the Basic and Frontier Research Program of Chongqing [grant numbers cstc2014jcyjA50011].

References

1. Moskalyk, R., & Alfantazi, A. (2003). Processing of vanadium: A review. *Minerals Engineering, 16,* 793–805.
2. Li, H.-Y., et al. (2016). Selective leaching of vanadium in calcification-roasted vanadium slag by ammonium carbonate. *Hydrometallurgy, 160,* 18–25.
3. Aarabi-Karasgani, M., Rashchi, F., Mostoufi, N., & Vahidi, E. (2010). Leaching of vanadium from LD converter slag using sulfuric acid. *Hydrometallurgy, 102,* 14–21.
4. Voglauer, B., Grausam, A., & Jörgl, H. P. (2004). Reaction-kinetics of the vanadium roast process using steel slag as a secondary raw material. *Minerals Engineering, 17,* 317–321.
5. Zhang, J., et al. (2015). Mechanism of vanadium slag roasting with calcium oxide. *International Journal of Mineral Processing, 138,* 20–29.
6. Kozlov, V. A., & Demidov, A. E. (2000). Chemical principles of a technology for making pure vanadium pentoxide. *Metallurgist, 44,* 428–433.
7. Yang, Z., et al. (2014). Leaching kinetics of calcification roasted vanadium slag with high CaO content by sulfuric acid. *International Journal of Mineral Processing, 133,* 105–111.
8. Fu, Z. (2014). Experimental research on vanadium extraction by calcified roasting and acid leaching. *Iron Steel Vanadium Titanium,* 1–6.
9. Li, X., Zhang, R., & Xie, B. (2012). Effects of vanadium-containing slag calcium roasting conditions on leaching rate of vanadium. *The Chinese Journal of Process Engineering,* 54–58.
10. Yin, D., et al. (2012). Influencing factors of calcified roasting and thermal analysis to the process of vanadium slag produced from Pangang. *Metal Mine,* 91–94.
11. Zhang, G., et al. (2015). Effects of microwave roasting on the kinetics of extracting vanadium from vanadium slag. *JOM Journal of the Minerals Metals and Materials Society, 68,* 577–584.

Leaching of Spent Ni–Mo Hydrodesulphurization (HDS) Catalyst in Oxalic Acid Solutions

Sedat Ilhan

Abstract In this work, the effect of crushing, temperature, oxalic acid concentration and stirring speed on the leaching of spent Ni–Mo HDS catalyst was investigated. Spent Ni–Mo HDS catalyst was roasted at 500 °C under dynamic air atmosphere for 4 h prior to leaching experiments. It was found that roasted spent Ni–Mo HDS catalyst included mainly Al_2O_3, MoO_3, $NiMoO_4$ and $AlPO_4$. Leaching experiments were performed using 5 g of uncrushed roasted Ni–Mo HDS catalyst, 1 L of 0.25, 0.50, 0.75 and 1 M $H_2C_2O_4$ solutions, 25, 40, 55 and 70 °C temperatures and 200, 250 and 300 rpm stirring speeds. Leaching behavior of Mo, Al, P and Ni was determined. It is seen that $H_2C_2O_4$ can be used successfully for the leaching of roasted HDS catalyst. More than 90% of the molybdenum and 80% of Nickel is extracted from the experiments carried out using 300 rpm stirring speed, 1 M $H_2C_2O_4$ and 40 °C.

Keywords Leaching · Hydrodesulphurization catalyst · Mo recovery · Ni recovery

Introduction

It is important to recycle valuable metals from secondary sources to meet growing demand on valuable metals since production of metals from primary sources are more expensive, time consuming and hazardous to environment. Spent hydrodesulphurization (HDS) catalysts include mainly Mo, Al, Ni, Co, S and C compounds with different composition and are very important secondary sources for Mo, Ni and Co. In petroleum refining industry hydrodesulphurization (HDS) catalysts are extensively used for the removal of sulfur. Main elements used in HDS catalysts are Mo, Ni or Co and Al. MoS_2 active phase with Ni or Co

S. Ilhan (✉)
Department of Metallurgical and Materials Engineering, Faculty of Engineering, Istanbul University, Avcilar, 34320 Istanbul, Turkey
e-mail: ilhans@istanbul.edu.tr

© The Minerals, Metals & Materials Society 2017
M.A. Meyers et al. (eds.), *Proceedings of the 3rd Pan American Materials Congress*, The Minerals, Metals & Materials Series, DOI 10.1007/978-3-319-52132-9_56

promoters supported on γ-Al_2O_3. During the hydrodesulphurization process C, S and elements as V, Fe, As, Zn accumulated on the catalyst surface and the hence catalysts are deactivated gradually and discharged as hazardous waste [1, 2].

Spent catalysts are very important secondary sources because of not only their large amounts and enormous economic values, but also because of the environmental concerns if disposed off. Thus the most interesting solution to utilize spent HDS waste is recovery of metals from economic and environmental aspects. Various methods including pyro metallurgical (direct smelting [3], soda ash roasting [4]), hydrometallurgical [5–8] or pyro-hydrometallurgical [9–11] have been investigated for the recovery of valuable metals from spent HDS catalyst in literature. High energy consumption, difficulties in the control of reactions, side reactions and wasting of reagents are main drawbacks of direct processing of spent HDS catalysts. Use of a combination of pyro metallurgical and hydrometallurgical methods is one of the most effective methods for the recovery of valuable metals from spent HDS catalysts. In the first step spent catalysts are roasted at about 500 °C for the elimination of S, C and oil. In the second step roasted HDS catalysts are leached in acidic or basic medium [12].

Although different leaching reactants have been used for the recovery of valuable metals from roasted HDS catalysts in literature, there is no study carried out using $H_2C_2O_4$. In this study $H_2C_2O_4$ is used as a leaching agent due to its good chelating and oxidizing properties. The effect of temperature, oxalic acid concentration, stirring speed and crushing on the leaching of spent Ni–Mo HDS catalyst was investigated.

Materials and Methods

$H_2C_2O_4$ solutions were prepared using anhydrous $H_2C_2O_4$ (Fluka). Spent Ni–Mo HDS catalyst was obtained from Turkish Petroleum Refineries Co. (TUPRAS). Spent Ni–Mo HDS catalyst was roasted at 500 °C for 4 h in air atmosphere in a tube furnace (LENTON LTF 75) for the elimination of S, C and oil. CHNS elemental analysis instrument (Thermo Finnigan Flash EA 1112), X–ray Fluorescence Spectroscopy (XRF) (PANalytical Axios Minerals) and X-ray Powder Diffractometer (XRD) (Rigaku D/Max-2200, Cu–K_α monochromatic X-ray) instruments were used for the analysis of composition and phase.

Leaching experiments were performed for 3 h in water heated, jacketed borosilicate glass reactor system (HWS DN 100) using 5 g of roasted Ni–Mo HDS catalyst, 1 L of $H_2C_2O_4$ solutions with 0.25, 0.5, 0.75 and 1 M concentrations, 25, 40, 55 and 70 °C temperatures and 200, 250 and 300 rpm stirring speeds. Details of the experimental set-up mentioned elsewhere [13]. 1 L of $H_2C_2O_4$ solution was filled to the reactor and heated to experimental temperature. When isothermal conditions were obtained, roasted Ni–Mo HDS catalyst was added to the reactor and stirred at constant stirring speed. Solutions were taken from the reactor at certain time intervals and dissolved elements were analyzed quantitatively by

Inductively Couple Plasma-Optic Emission Spectroscope (ICP-OES) (Spectro Ciros Vision). When experiment completed, solid leach residue was filtered through a Gooch crucible, washed with distilled water, dried in oven (BINDER ED 115) at 105 °C for 24 h.

Results and Discussion

Characterization

CHNS analysis of the spent Ni–Mo HDS catalyst is shown in Table 1. It is seen from Table 1 that total amount of volatile compounds is 25.71 wt% of the total mass.

Result of XRF analysis carried out for the determination of roasted Ni–Mo HDS catalyst shows that roasted HDS catalyst mainly includes Al, Mo, Ni and P compounds (Table 2).

XRD analysis of roasted Ni–Mo HDS catalyst indicates that Al_2O_3 (ICDD Card No: 050-0741), $NiMoO_4$ (ICDD Card No: 012-0348), MoO_3 (ICDD Card No: 005-0508) and $AlPO_4$ (ICDD Card No: 003-025-0581) phases form after roasting process (Fig. 1).

Table 1 CSHN analysis of spent Ni–Mo HDS catalyst

Element	S	C	H	N
(wt%)	14.13	10.43	0.8	0.35

Table 2 Elemental analysis of roasted Ni–Mo HDS catalyst

Element	Al	Mo	Ni	P
(wt%)	27.692	20.238	5.267	4.331

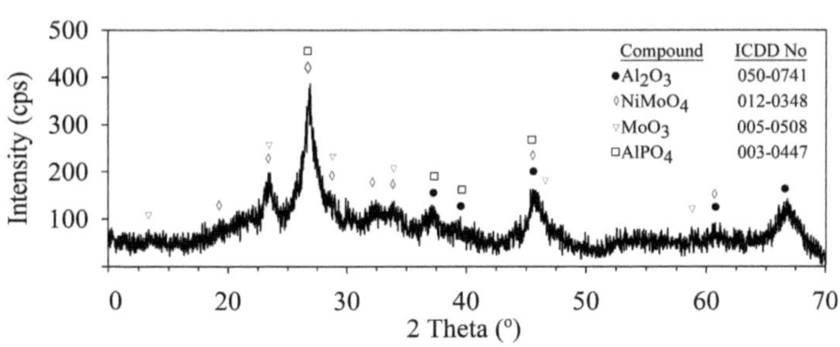

Fig. 1 XRD diagram of roasted Ni–Mo HDS catalyst

Leaching Experiments

Experiments carried out for the determination of the effect of crushing on the leaching rate of roasted Ni–Mo HDS catalyst. 5 g of uncrushed and crushed roasted Ni–Mo HDS catalyst, the highest oxalic acid concentration (1 M), temperature (70 °C) and stirring speed (300 rpm) were used in the experiment. Because there is no significant difference in fractional conversion of Mo (X_{Mo}) values obtained from the experiments carried with crushed and uncrushed roasted Ni–Mo HDS catalyst (Fig. 2), leaching experiments were carried out by using uncrushed roasted Ni–Mo HDS catalyst.

Leaching experiments were carried out using 5 g of uncrushed roasted Ni–Mo HDS catalyst, 1 L of 1 M $H_2C_2O_4$ solution, 70 °C and 200, 250 and 300 rpm stirring speeds to determine the effect of stirring speed on the leaching of spent Ni–Mo HDS catalyst (Fig. 3). It is seen from Fig. 3 that 300 rpm stirring speed is enough to eliminate the effect of liquid film layer resistance. Thus, leaching experiments were performed using 300 rpm stirring speed.

Leaching experiments were carried out using 5 g of uncrushed roasted Ni–Mo HDS catalyst, 1 L of 0.25, 0.5, 0.75 and 1 M $H_2C_2O_4$ solutions, 25, 40, 55 and 70 °C and 300 rpm stirring speed to determine the effect of temperature and $H_2C_2O_4$ concentration on leaching of roasted Ni–Mo HDS catalyst. It is seen from Fig. 4a–d that leaching rate of Mo increases by increasing temperature. In addition total conversion is obtained from the experiments carried out at 55 and 70 °C. On the other hand increase in $H_2C_2O_4$ concentration increases the X_{Mo} slightly. Thus it is not possible to mention a significant effect of concentration on the leaching of Mo.

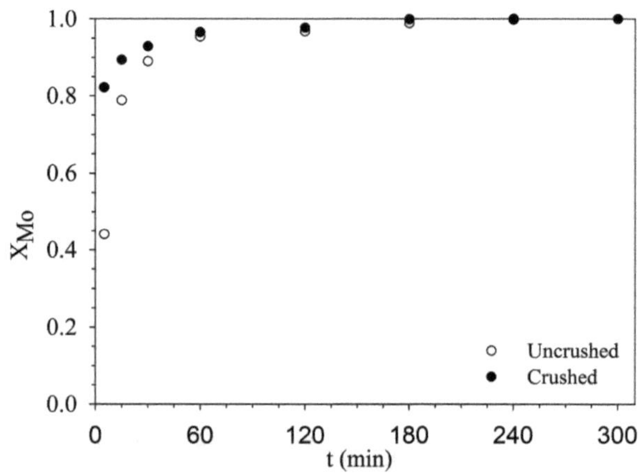

Fig. 2 Effect of crushing on the leaching of Mo in roasted Ni–Mo HDS catalyst

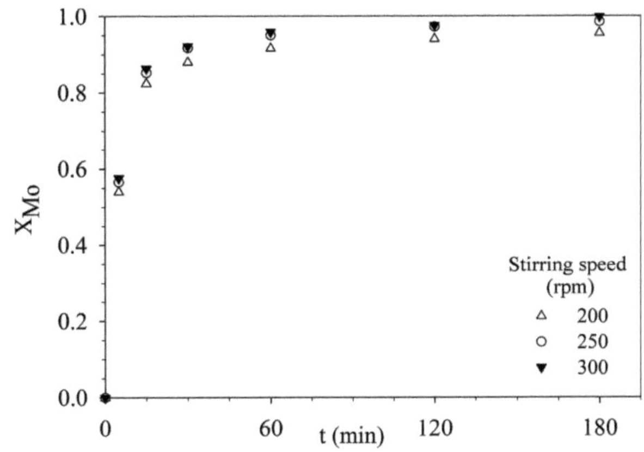

Fig. 3 Effect of stirring speed on the leaching of Mo in roasted Ni–Mo HDS catalyst

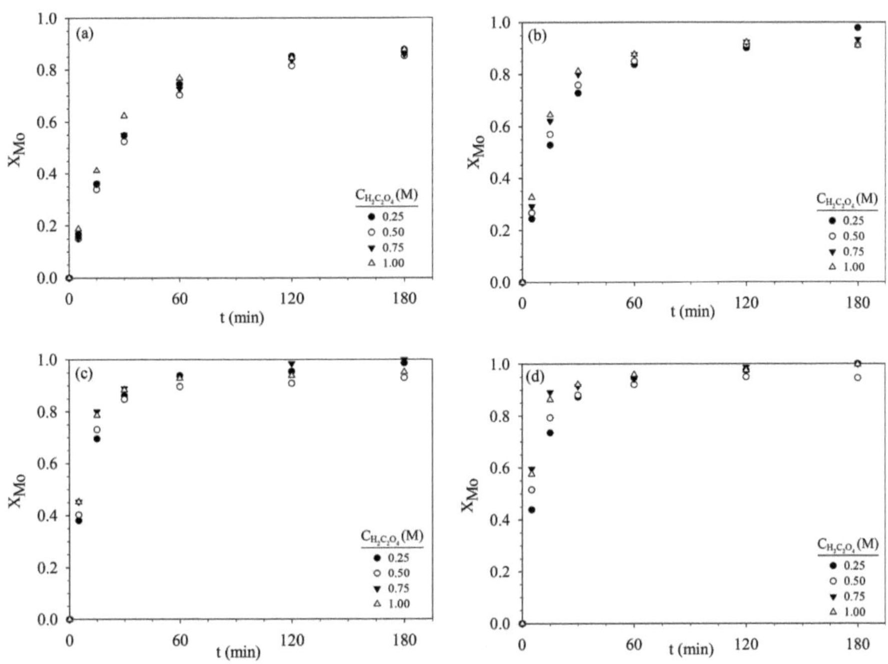

Fig. 4 X_{Mo} − t diagrams obtained at different temperatures (25 (**a**), 40 (**b**), 55 (**c**) and 70 °C (**d**))

Fractional conversion of Ni (X_{Ni}) increases rapidly at the beginning of leaching experiments. While X_{Ni} decreases after a maximum conversion in the experiments carried out at high temperatures (55 and 70 °C), increase in X_{Ni} continues slightly in the experiments carried out at low temperatures (25 and 40 °C) (Fig. 5a–d). X_{Ni} decreases by increasing $H_2C_2O_4$ concentration.

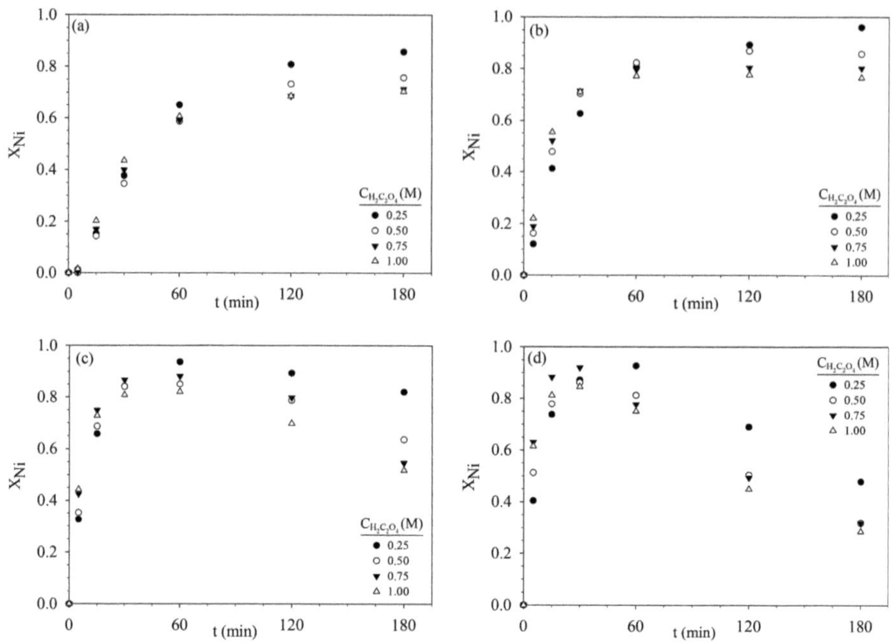

Fig. 5 $X_{Ni} - t$ diagrams obtained at different temperatures (25 (**a**), 40 (**b**), 55 (**c**) and 70 °C (**d**))

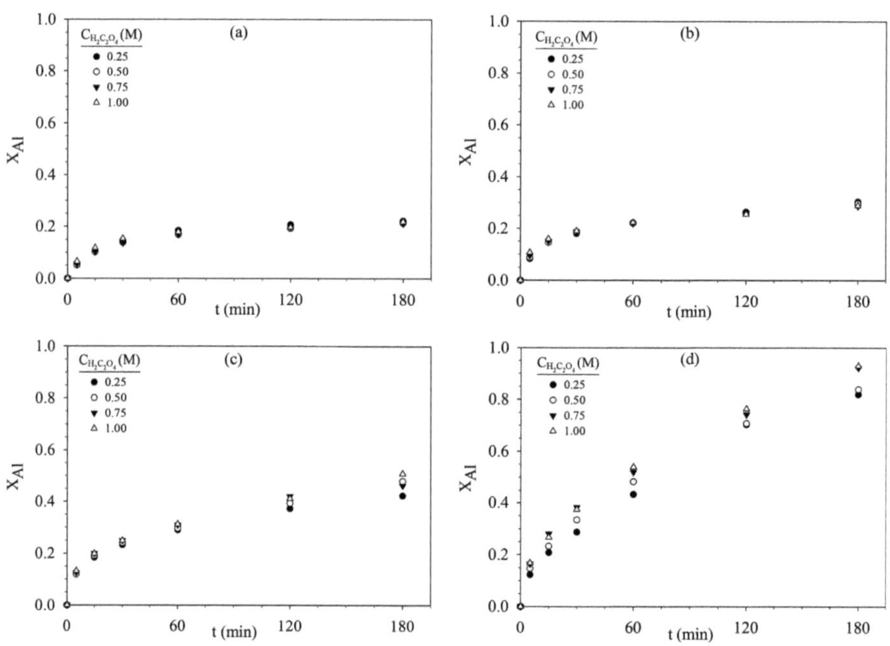

Fig. 6 $X_{Al} - t$ diagrams obtained at different temperatures (25 (**a**), 40 (**b**), 55 (**c**) and 70 °C (**d**))

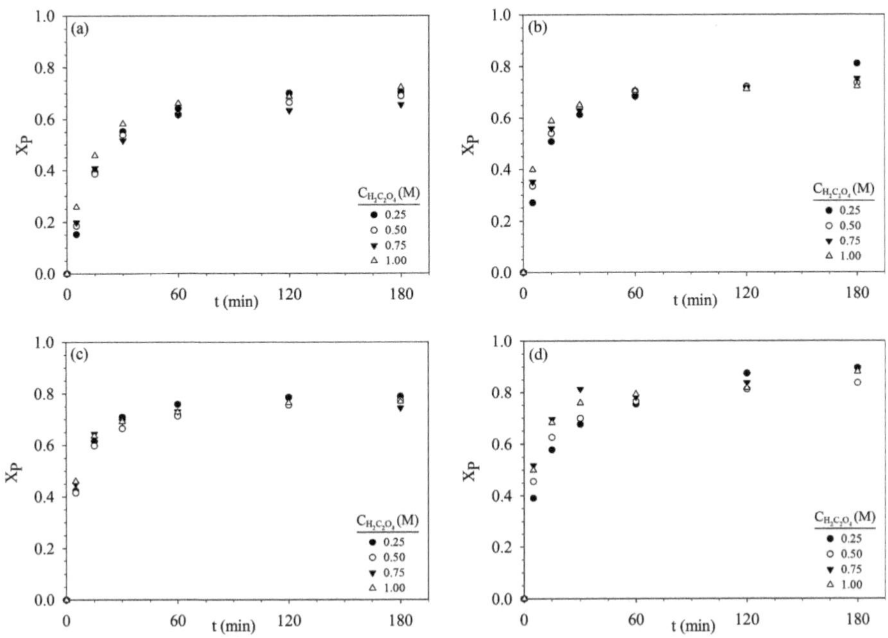

Fig. 7 $X_P - t$ diagrams obtained at different temperatures (25 (**a**), 40 (**b**), 55 (**c**) and 70 °C (**d**))

Fractional conversion of Al (X_{Al}) strongly dependent on the temperature (Fig. 6a–d). X_{Al} increases rapidly by increasing experimental temperatures from 25 to 70 °C. On the other hand increase in $H_2C_2O_4$ concentration has no considerable effect on the leaching rate of Al.

While fractional conversion of P (X_P) is slightly dependent on temperature (Fig. 7a–d), $H_2C_2O_4$ concentration has no significant effect on the leaching rate of P.

Conclusions

- C, S and residual oil are eliminated successfully at 500 °C by roasting under dynamic air atmosphere for 4 h.
- It is found that crushing has no significant effect on the leaching of roasted Ni–Mo HDS catalyst.
- It is determined that leaching behavior of Ni is different than other elements.
- Optimum leach conditions are determined as 40 °C, 1 M $H_2C_2O_4$ concentration and 300 rpm stirring speed. 92% of Mo, 86% of Ni, 30% of Al and 73% of P are leached in these experimental conditions.
- Change in temperature results in a rapid change in the leaching behavior of Ni and Al. High experimental temperatures (over 40 °C) should not be chosen in case to avoid obtaining higher Al and P extraction rates.

References

1. Dufresne, P. (2007). Hydroprocessing catalysts regeneration and recycling. *Applied Catalysis, A: General, 322,* 67–75.
2. EPA, U. (2001). Hazardous waste management system; identification and listing of hazardous waste: Spent catalysts from dual-purpose petroleum hydroprocessing reactors. In Protection E, Agency (Eds.), *Federal register 2001* (pp. 35379–35384).
3. Howard, R. A., & Barnes, W. R. (1991). *Process for recovering valuable metals from spent catalysts*. Google Patents.
4. Kar, B. B., Datta, P., & Misra, V. N. (2004). Spent catalyst: Secondary source for molybdenum recovery. *Hydrometallurgy, 72,* 87–92.
5. Siemens, R. E., Jong, B. W., & Russell, J. H. (1986). Potential of spent catalysts as a source of critical metals. *Conservation and Recycling, 9,* 189–196.
6. Hyatt, D. E. (1987). *Value recovery from spent alumina-base catalyst*. Google Patents.
7. Lai, Y. C., Lee, W. J., Huang, K. L., & Wu, C. M. (2008). Metal recovery from spent hydrodesulfurization catalysts using a combined acid-leaching and electrolysis process. *Journal of Hazardous Materials, 154,* 588–594.
8. Rabah, M. A., Hewaidy, I. F., & Farghaly, F. E. (1997). Recovery of molybdenum and cobalt powders from spent hydrogenation catalyst. *Powder Metallurgy, 40,* 283–288.
9. Park, K. H., Mohapatra, D., & Nam, C. W. (2007). Two stage leaching of activated spent HDS catalyst and solvent extraction of aluminium using organo-phosphinic extractant, Cyanex 272. *Journal of Hazardous Materials, 148,* 287–295.
10. Toyabe, K., Kirishima, K., Shibayama, H., & Hanawa, H. (1995). *Process for recovering valuable metal from waste catalyst*. Google Patents.
11. Valverde, I. M., Jr., Paulino, J. F., & Afonso, J. C. (2008). Hydrometallurgical route to recover molybdenum, nickel, cobalt and aluminum from spent hydro treating catalysts in sulphuric acid medium. *Journal of Hazardous Materials, 160,* 310–317.
12. Barik, S. P., Park, K. H., Parhi, P. K., Park, J. T., & Nam, C. W. (2012). Extraction of metal values from waste spent petroleum catalyst using acidic solutions. *Separation and Purification Technology, 101,* 85–90.
13. Ilhan, S., Kalpakli, A. O., Kahruman, C., & Yusufoglu, I. (2013). The use of oxalic acid as a chelating agent in the dissolution reaction of calcium molybdate. *Metallurgical and Materials Transactions B, 44,* 495–505.

Novel Adsorbent from Iron Ore Concentration Tailings for Toxic Cationic Dye Removal from Water

Yongmei Wang, Alejandro López-Valdivieso, Teng Zhang, Teza Mwamulima and Changsheng Peng

Abstract A novel adsorbent (FB-mZVI) was prepared, using tailings from iron ore concentration plants, fly ash and bentonite, to remove crystal violet (CV) and methylene blue (MB) from polluted water. The adsorbent was prepared as cylindrical-shape pellets to easily separate it from the water after the adsorption process. The statistical orthogonal method was used to evaluate the factors determining the synthesis of FB-mZVI. Batch experiments revealed that the optimum mass ratio of fly ash, bentonite, iron ore tailings and palm kernel shell was 2:2:1:1. And in the synthesis of FB-mZVI adsorbent, the optimum reaction temperature was 800 °C, the reaction time was 10 min and the heating rate was 10 °C/min. SEM Characterization revealed that the adsorbent was highly porous and constituted by Fe^0 particles finely dispersed on an aluminosilicate matrix. The adsorbent showed to have a superior adsorption capacity for CV and MB when compared to other conventional adsorbents. The adsorption kinetics and isotherms of CV and MB on FB-mZVI adsorbents could be expressed by the pseudo-second-order model and Langmuir isotherm.

Keywords FB-mZVI · Crystal violet (CV) · Methylene blue (MB) · Removal · Water

---

Y. Wang · T. Zhang · T. Mwamulima · C. Peng (✉)
College of Environmental Science and Engineering,
Ocean University of China, Qingdao 266100, People's Republic of China
e-mail: cspeng@ouc.edu.cn

Y. Wang · A. López-Valdivieso
Laboratorio de Quimica de Superficie, Instituto de Metalurgia,
Universidad Autónoma de San Luis Potosí, Av. Sierra Leona 550,
San Luis Potosí C.P. 78210, Mexico

C. Peng
Key Laboratory of Marine Environmental Science and Ecology,
Ministry of Education, Qingdao 266100, People's Republic of China

© The Minerals, Metals & Materials Society 2017
M.A. Meyers et al. (eds.), *Proceedings of the 3rd Pan American Materials Congress*,
The Minerals, Metals & Materials Series, DOI 10.1007/978-3-319-52132-9_57

Introduction

Water pollution by toxic cationic dyes has become a serious issue all over the world due to increase in population and the fast industrialization [1, 2]. Crystal violet (CV) and Methylene blue (MB) is widely known as the highly toxic cationic dyes existing in effluents drained from industries such as fabrics, leather, paper, cosmetics, polymers [3]. These toxic cationic dyes discharged from different industries are of specific concern, since the dyes are generally non-biodegradable, toxic, mutagenic, and carcinogenic [4, 5], thereby posing a great threat to human health and ecological environment. Therefore, toxic cationic dyes must be removed from wastewater during the water treatment to safeguard the aquatic life [6, 7]. During the different technologies to purify the polluted water, adsorption was one of the superior approach, because of its simplicity, flexibility, low-cost and economical feasibility of design as well as high efficacy [4, 6, 8].

Iron ore tailings, a kind of mineral waste, which were generated from the iron ore mining industry [9]. Owing to the fast development of steel production, huge amount of iron ore tailing were generated and disposed as waste in landfills, quarries, rivers, oceans, etc. [10]. China produces millions of iron ore tailings every year, however, only a few of them were reused to improve agricultural soils and produce cement [11]. It is of vital importance to find new ways to utilize these IOT effectively to realize the resource conservation and environmental protection.

Recently, iron-based porous materials were widely used as good adsorbents to remove toxic dyes from wastewater [12, 13]. Nanoscale zero valent iron (nZVI) with unique redox potential, large reaction sites and high surface area was widely used to remove toxic dyes from aqueous solution [14]. In previous studies, different matrices was used to overcome the restrictions such as easy being agglomerate and oxidized. Montmorillonite modified by iron could adsorb and degradate crystal viole effectively [15]. Three-dimensional graphene decorated by iron nanoparticles could degrade azo dye rapidly and efficiently [16].

In this study, fly ash and bentonite-supported microscale zero valent iron (FB-mZVI) was synthesized using iron ore tailings, palm kernel shell, fly ash and bentonite. Palm kernel shell was used as the reductant to reduce the iron ore tailings to get the microscale zero valent iron (mZVI). And fly ash and bentonite were used as matrices to support mZVI to increase the removal efficiency for CV and MB from aqueous solution. The statistical orthogonal method was used to evaluate the synthesis factors determining the adsorbent preparation. Likewise, the adsorption kinetics and isotherms of the adsorbent were also studied.

Materials and Methods

Materials and Chemicals

After dried in an oven for 24 h at 105 °C, Iron ore tailings (Wu Han Province, China), palm kernel shell, fly ash and bentonite were sieved through a 500 mesh screen prior to use in experiments. Analytical grade crystal violet (CV) ($C_{25}H_{30}ClN_3$, Mw: 407.98 g/mol, λmax = 582 nm) and methylene blue (MB) ($C_{16}H_{18}ClN_3S$, Mw: 319.86 g/mol, λmax = 664 nm) were provided by Beijing Chemical Reagents Company, Beijing, China. The chemical reagents in this study were all analytical grade and distilled water was used in all experiments.

Synthesis of MZVI and FB-MZVI Adsorbents

For the direct iron oxide reduction, iron ore tailings were reduced by palm kernel shell to get mZVI. Batch experiments were carried out at different temperatures, amount of reductant, reaction times, and heating rate to assess the effect of the reduction yield. The reduction yield (ω) was calculated by Eq (1).

$$\eta = \frac{\mathrm{m}_1 M_1}{2m_2 M_2 \omega} \times 100 \tag{1}$$

where m_1 (mg) is the mass of mZVI; m_2 (mg) is the mass of iron ore tailings; M_1 is molecular weight of Fe_2O_3; M_2 is molecular weight of Fe; ω is the percentage of Fe_2O_3 in iron ore tailings.

Based on the direct iron reduction data, the orthogonal experiment of three factors three levels was design to optimize the conditions to prepare the FB-mZVI adsorbent. Using Orthogonal Design Assistant, the experiment design was developed with the ratio of iron ore tailings, palm kernel shell, fly ash and bentonite; reaction temperature; reaction time and heating rate as variables.

After iron ore tailings, palm kernel shell, fly ash and bentonite were mixed, they were compressed into a model of elongated cylinder-type pellets with some water. First dried in air, then the materials were dried in a electro-thermostatic blast oven at 60 °C for 24 h, after which these granules were calcined at a certain temperature for some minutes and a given heating rate in an atmosphere controlled sintering furnace under anaerobic conditions to prepare the FB-mZVI adsorbent.

Characterization of FB-MZVI Adsorbents

The FB-mZVI adsorbents texture were characterized using a Joel JSM-6610 LV scanning electron microscope. The chemical components of this adsorbents before and after adsorption were determined by XRD analysis (08 Advance Davinci, Bruker, λ Cu = 1.5406 Å). FT-IR spectrophotometer (Tensor 27, Bruker) was used to analyze the structure changes of CV and MB.

Adsorption Experiments

0.5 mg of adsorbent was mixed with 100 mL of CV and MB aqueous solution (800 mg L^{-1}) in 250 mL glass conical flasks. The flasks were shaken in a vibrator with a constant speed of 120 rpm at 30 °C. 0.2 mL aqueous solution was withdraw from the flasks at 1, 3, 5, 7, 9, 24, 29, 36, 56 and 72 h to determine the concentration of CV and MB. This was done using a UV-vis spectrophotometer at a wavelength of 582 nm for CV and 664 nm for MB. The adsorption capacity (Qt) and removal efficiency (R) of CV and MB were obtained by Eqs (2) and (3).

$$Q_t(\mathrm{mg/g}) = \frac{(C_0 - C_t)V}{m} \quad (2)$$

$$\mathrm{R}(\%) = \frac{C_0 - C_t}{C_0} \times 100 \quad (3)$$

where C_0 (mg L^{-1}) is the initial concentration, and C_t (mg L^{-1}) is the concentration at reaction time t, V (L) is the volume of aqueous solution, m (g) is the adsorbent mass.

Results and Discussion

Adsorbent Preparation

Direct Iron Oxide Reduction

Using the technology of direct reduction of iron oxide to metallic iron, palm kernel shell as a reductant to reduce the iron ore tailings to get zero valent iron. Figure 1 showed the reduction yield under different reaction temperature. As noted, the reduction yield slowly increased with temperature up to 700 °C, then, it sharply increased until a plateau was reached above 800 °C. At 800 °C, the reduction yield achieved at 42%. In the iron ore tailing samples, iron is mainly as hematite (Fe_2O_3) and goethite (FeOOH) together with quartz (SiO_2). When the reaction temperature

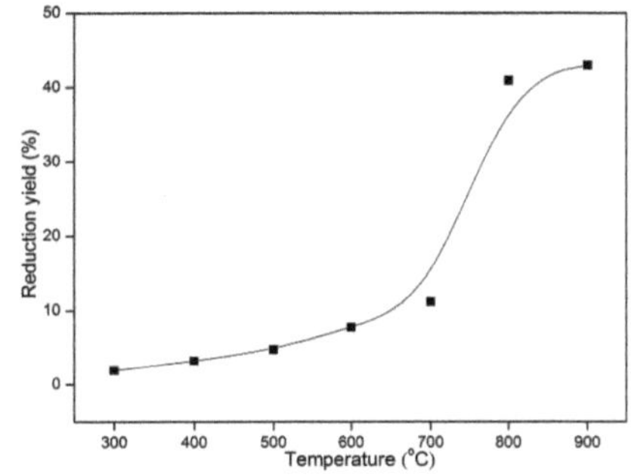

Fig. 1 Effect of temperature on the reduction yield

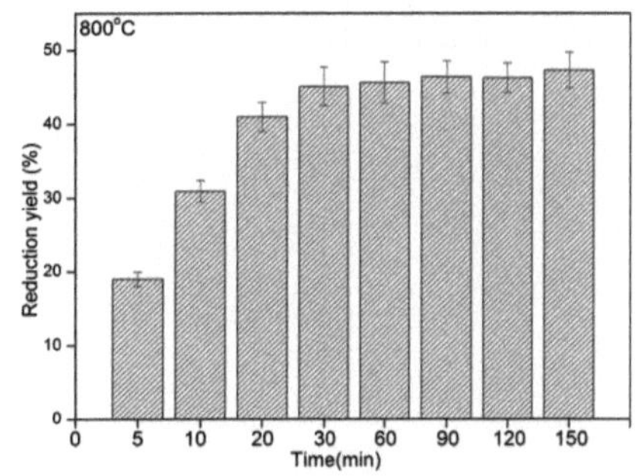

Fig. 2 Effect of time on the reduction yield

was below 700 °C, the reduction yield was below 10%, which can be accounted for by the dehydration of goethite, which turns to hematite [17]. At 800 °C hematite reduced to zero valence iron [18]. Accordingly, 800 °C could be chosen as the optimum temperature.

The effect of time on the reduction yield at 800 °C was shown in Fig. 2. The reduction yield increased with time until a plateau is reached after 30 min, which indicated that the reduction reaction was finished. Therefore, 30 min could be selected as the optimum reaction time high-efficiency and low-cost.

Figure 3 showed the effect of dosage of reducing agent to the direct reduction iron process. The highest reduction yield was 43%, when the ratio of palm kernel shell and iron ore tailing was 1:1. Below a ratio of 1:1, the reduction yield increased with the ratio increasing. However, when the ratio is higher than 1:1, the increase of palm kernel shell dosage decreased the reduction yield, which might be deduced that ferrite changed to cementite [19]. Accordingly, the optimum ratio of palm kernel shell and iron ore tailing was chosen as 1:1.

Figure 4 showed the effect of heating rate on the reduction yield. Increasing the heating rate, the reduction yield increased. It might be deduced that carbonyl in palm kernel shell cleaved to reducing gas (H_2, CO) quickly, which could promote the reduction reaction. The reduction yield reached 47% when the heating rate was 10 °C/min. Therefore, 10 °C/min was chosen as the optimum condition.

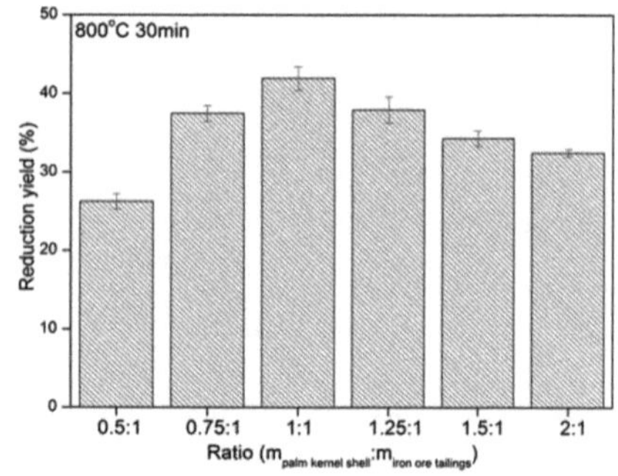

Fig. 3 Effect of amount of reductant on the reduction yield

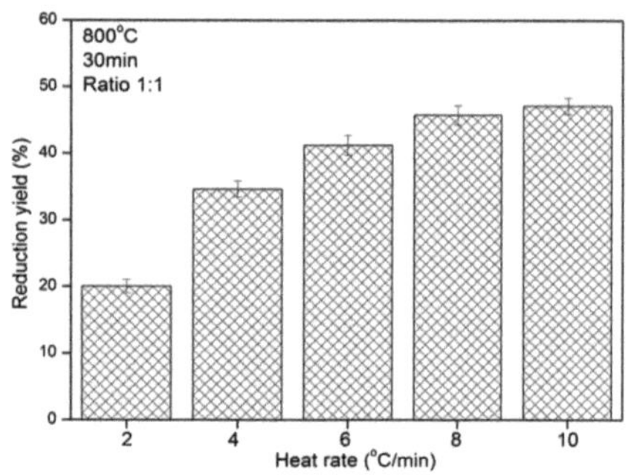

Fig. 4 Effect of heating rate on the reduction yield

Orthogonal Analyses

A three-factor, three-level orthogonal experiment was designed to optimize the conditions of preparation of FB-mZVI adsorbents [20]. The influence of the different reaction factors on the removal efficiencies of CV by FB-mZVI adsorbents is depicted in Table 1.

Orthogonal Design Assistant software was used to establish the condition of preparation of FB-mZVI adsorbents [21], as shown in Table 2. The higher the range and the F-value, the greater the influence of the factor on FB-mZVI adsorbents preparation. Compared with the ranges and F-values in Table 2, factors influencing FB-mZVI adsorbent preparation were: reaction time > temperature > ratio.

The higher the K value, the higher the removal efficiency, and the better the level of the factor. The K values from K_1 to K_3, which was from the lowest to the highest. The highest K was K_1 for the factor of temperature, which represented the level 800 °C was the optimum temperature. The change in K values for the factor of temperature and reaction time, indicated that CV removal decreased with temperature and reaction time increasing. Therefore, the optimal conditions for preparation of FB-mZVI adsorbents were 800 °C, 10 min, and a ratio of fly ash, palm kernel shell, iron ore tailing and bentonite of 2:2:1:1.

Table 1 Orthogonal experimental design and the results obtained from the full 3^3 factorial experiment matrix

	Factors			
	Temperature (°C)	Ratio	Reaction time(min)	Removal efficiency (%)
1	800	1:1.5:1:1	10	91.45
2	800	2:2:1:1	30	74.8
3	800	3:2.5:1:1	60	68.75
4	850	1:1.5:1:1	30	61
5	850	2:2:1:1	60	55.65
6	850	3:2.5:1:1	10	95.85
7	900	1:1.5:1:1	60	41.8
8	900	2:2:1:1	10	93.4
9	900	3:2.5:1:1	30	52.45
K_1	235	194.25	280.7	T = 635.15 T^2 = 403,415.52 P = 44,823.95
K_2	212.5	223.85	188.25	
K_3	187.65	217.05	166.2	
K_1^2	55,225	37,733.06	78,792.49	
K_2^2	45,156.25	50,108.82	35,438.06	
K_3^2	35,212.52	47,110.70	27,622.44	
Q	45,197.92	44,984.19	47,284.33	
S	373.98	160.25	2460.38	

Table 2 Range analysis and variance analysis of the orthogonal test

Factors	Range analysis				Variance analysis			
	K_1	K_2	K_3	Ranges	S_A	DOF	F-value	F critical values
Temperature	235	212.5	187.65	47.35	373.98	2	4.02	$F_{0.05}$ (2,2) = 19
Ratio	194.25	223.85	217.05	29.6	160.25	2	1.72	$F_{0.10}$ (2,2) = 9
Reaction time	280.7	188.25	166.2	114.5	2460.38	2	26.43	

Adsorbent Characterization

As shown in Fig. 5a, the FB-mZVI adsorbents were elongated cylinder-type pellets with high compressive strength. The texture of the adsorbent is presented in the SEM photomicrograph shown in Fig. 5b, c. Volatile CO and hydrogen generated by the organic matter (cellulose, hemicellulose and lignin) in the palm kernel shell through a biomass pyrolysis chemical reaction formed highly porous on the surface of FB-mZVI adsorbents [22], shown in Fig. 5b. In Fig. 5c, the bright spots were

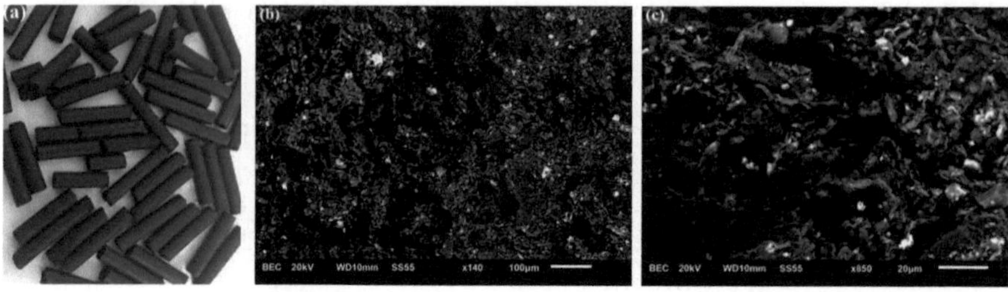

Fig. 5 Images of FB-mZVI adsorbent. **a** Photo; **b** SEM (×140); **c** SEM (×850)

mZVI particles. It could be found that the matrix, fly ash and bentonite, was used to support the mZVI particles, which dispersed throughout the matrix and attached to the matrix.

Figure 6 showed the X-ray diffractogram of FB-mZVI adsorbents before and after adsorbing CV and MB. SiO_2, Fe_3O_4 and Fe^0 to be the major components in the FB-mZVI adsorbents, shown in Fig. 6a. As shown in Fig. 6b, c, the spectrum peak of Fe^0 was $2\theta = 44.90$, which decreased after CV and MB adsorption compared to the FB-mZVI adsorbents. The XRD patterns of FB-mZVI adsorbents after adsorbing CV and MB revealed the presence of Fe_2O_3 ($2\theta = 35.68$), which leads to suggest that Fe^0 underwent oxidation with the adsorption process as has been proposed by Ponder et al. [23].

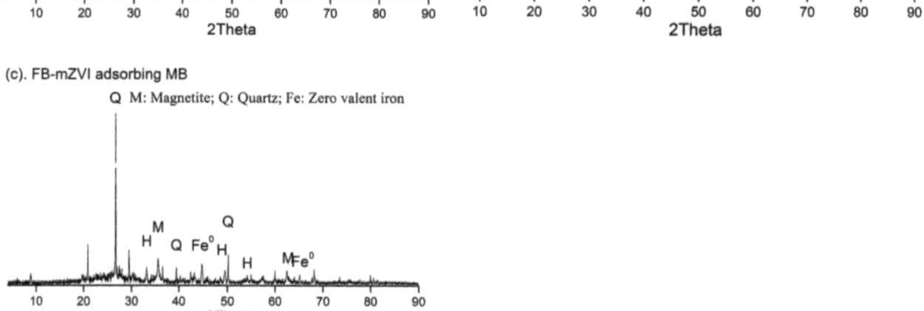

Fig. 6 X-ray diffractogram of FB-mZVI adsorbents before and after adsorbing CV and MB, **a** FB-mZVI adsorbent, **b** FB-mZVI adsorbent after adsorbing CV, **c** FB-mZVI adsorbent after adsorbing MB

Removal of CV and MB

Adsorption Kinetics

The FB-mZVI adsorbent was used to remove CV and MB in aqueous solution. As shown in Fig. 7, the adsorption of CV and MB onto FB-mZVI adsorbent was initially fast and then gradually decreased at 30 °C. When reaction time increased to 36 h, the MB adsorption capacity achieved 24 mg/g. The FB-mZVI adsorbent efficiently removed CV and MB, being the removal of CV much higher than that of MB. When reaction time was 36 h, the CV adsorption capacity reached 52 mg/g, significantly higher than MB adsorption capacity.

The pseudo-first-order kinetic model represented by Eq. (3) and the pseudo-second-order kinetics equation represented by Eq. (4) were examined to correlate the time and adsorbent adsorption density for CV and MB.

$$\ln(Q_e - Q_t) = \ln Q_e - k_1 t \tag{4}$$

$$\frac{\mathrm{t}}{Q_t} = \frac{1}{Q_e} t + \frac{1}{K_2 Q_e^2} \tag{5}$$

The best-fit model was selected based on both the correlation coefficient (r^2) and the calculated Qe value. The values of the parameters (Qe, K_1, K_2) and the respective correlation coefficients (r^2 values) for CV and MB adsorption were calculated from Origin software data fitting for the pseudo-first-order model and plot of t/Qt versus t for the pseudo-second-order model as shown in Fig. 8a. Table 3 shows that the pseudo-second-order kinetic model provided nearly a perfect match between the theoretical and experimental data for CV and MB

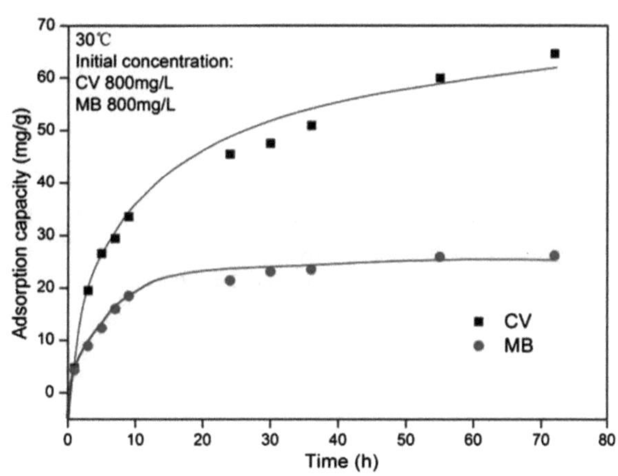

Fig. 7 Adsorption kinetics of CV and MB on FB-mZVI, respectively

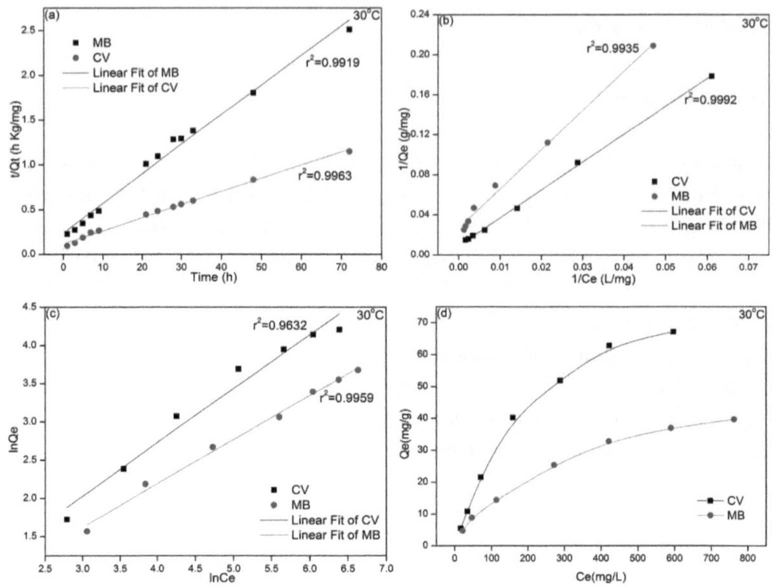

Fig. 8 Linear fit and adsorption isotherms of CV and MB adsorption by FB-mZVI adsorbents, **a** the pseudo-second-order kinetics equation for CV and MB, **b** Langmuir isotherm for CV and MB, **c** Freundlich isotherm for CV and MB, **d** adsorption isotherms at 30 °C, respectively

Table 3 Parameters obtained from pseudo-first-order kinetics model, pseudo-second-order kinetics model for FB-mZVI adsorbents

	CV			MB		
	Q_e (mg/g)	K_1 (min^{-1})/K_2 (mg min)	r^2	Q_e (mg/g)	K_1 (min^{-1})/ K_2(mg min)	r^2
Pseudo-first-order kinetics model	57.63	0.104	0.9446	24.99	0.151	0.8973
Pseudo-second-order kinetics equation	68.74	0.0018	0.9963	30.65	0.0043	0.9919

adsorption kinetics (r^2 all greater than 0.99 for CV and MB). Accordingly, it can be concluded that the adsorption mechanisms of CV and MB contained both physical and chemical contributions [20, 24, 25].

Adsorption Isotherms

Langmuir isotherm represented by Eq. (6) and Freundlich isotherm given by Eq. (7) were used to correlate the equilibrium concentration and adsorbent adsorption density for CV and MB. The values of the parameters (Qe, K_1, K_2) and the respective correlation coefficients (r^2 values) for CV and MB adsorption were calculated from plot of $1/Q_e$ versus $1/C_e$ for the Langmuir isotherm model shown in Fig. 8b and plot of $\ln Q_e$ versus $\ln C_e$ for Freundlich isotherm model as shown in Fig. 8c.

$$\frac{1}{Q_e} = \frac{1}{bQ_m} \cdot \frac{1}{Ce} + \frac{1}{Q_m} \tag{6}$$

$$\ln Q_e = \frac{1}{n} \ln C_e + \ln K_f \tag{7}$$

The Langmuir adsorption isotherm represented a better fit of the experimental data, due to the relatively better r^2 values (all greater than 0.99 for CV and MB), compared with that of the Freundlich isotherm, as shown in Fig. 8b, c. This might be due to homogenous distribution of active sites on the microspheres adsorbent surface, since the Langmuir equation assumes that the surface is homogeneous and adsorption proceeds up to a monolayer. Then it could be said that adsorption of MB and CV on the adsorbent took place within a monolayer region.

UV-Vis Analysis

UV-vis spectra of the CV and MB aqueous solutions before and after adsorption are shown in Fig. 9a, b, respectively. As noted in Fig. 9a, a new band at 582 nm appears in the spectrum after adsorption, indicating that a new compound is formed upon adsorption of CV. This is an evidence that CV is chemisorbing on the FB-mZVI explaining its higher adsorption in comparison to MB.

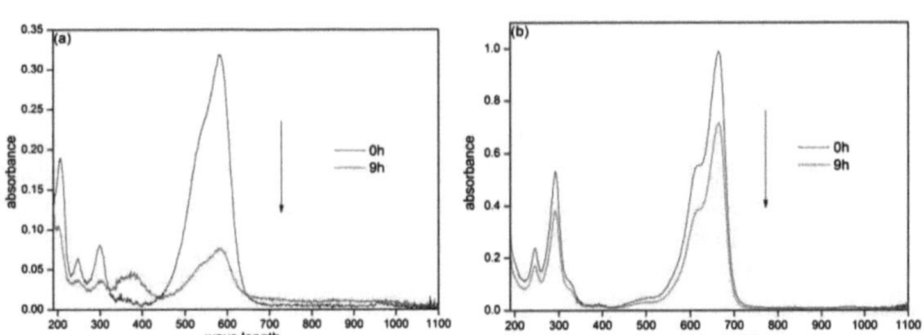

Fig. 9 UV-vis spectra of dye solutions, **a** CV and FB-mZVI adsorbent, **b** MB and FB-mZVI adsorbent, before and after adsorption for 9 h

Conclusions

FB-mZVI adsorbent was successfully prepared from iron ore tailings, palm kernel shell, fly ash and bentonite and applied for the removal of CV and MB from water. The new adsorbent was successfully synthesized at 800 °C, treatment time of 10 min, heating rate was 10 °C/min and an optimum ratio of fly ash, bentonite, iron ore tailings and palm kernel shell of 2:2:1:1. The FB-mZVI adsorbent efficiently removed CV and MB, being the removal of CV much higher than that of MB. The CV adsorption capacity was 52 mg/g, while that for MB it was 24 mg/g at 30 °C. The adsorption kinetics and isotherms of CV and MB on FB-mZVI adsorbents could be expressed by the pseudo-second-order model and Langmuir isotherm.

Acknowledgements This work was supported by the State Key Laboratory of Environmental Criteria and Risk Assessment (No. SKLECRA2013FP12) and the Public science and technology research funds projects of ocean, China (201105020). The authors would like to thank the anonymous reviewers for their recommendations and comments. Also, Y. Wang would like to thank the State Scholarship Fund organized by the China Scholarship Council for offering her the scholarship during her joint Ph.D. studying in Mexico.

References

1. Fan, Y., Liu, H. J., Zhang, Y., & Chen, Y. (2015). Adsorption of anionic MO or cationic MB from MO/MB mixture using polyacrylonitrile fiber hydrothermally treated with hyper-branched polyethylenimine. *Journal of Hazardous Materials, 283,* 321–328.
2. Phetphaisit, C. W., Yuanyang, S., & Chaiyasith, W. C. (2016). Polyacrylamido-2-methyl-1-propane sulfonic acid-grafted-natural rubber as bio-adsorbent for heavy metal removal from aqueous standard solution and industrial wastewater. *Journal of Hazardous Materials, 301,* 163–171.
3. Cottet, L., Almeida, C. A. P., Naidek, N., Viante, M. F., Lopes, M. C., & Debacher, N. A. (2014). Adsorption characteristics of montmorillonite clay modified with iron oxide with respect to methylene blue in aqueous media. *Applied Clay Science, 95,* 25–31.
4. Rasalingam, S., Peng, R., & Koodali, R. T. (2015). An insight into the adsorption and photocatalytic degradation of rhodamine B in periodic mesoporous materials. *Applied Catalysis, B: Environmental, 174,* 49–59.
5. Gao, Y. W., Guo, Y. Z., & Zhang, H. (2016). Iron modified bentonite: Enhanced adsorption performance for organic pollutant and its regeneration by heterogeneous visible light photo-fenton process at circumneutral pH. *Journal of Hazardous Materials, 302,* 105–113.
6. Soon, A. N., & Hameed, B. H. (2013). Degradation of acid blue 29 in visible light radiation using iron modified mesoporous silica as heterogeneous photo-fenton catalyst. *Applied Catalysis, A: General, 450,* 96–105.
7. Patra, A. S., Ghorai, S., Ghosh, S., Mandal, B., & Pal, S. (2016). Selective removal of toxic anionic dyes using a novel nanocomposite derived from cationically modified guar gum and silica nanoparticles. *Journal of Hazardous Materials, 301,* 127–136.
8. Cho, D. W., Chon, C. M., Kim, Y., Jeon, B. H., Schwartz, F. W., Lee, E. S., et al. (2011). Adsorption of nitrate and Cr(VI) by cationic polymer-modified granular activated carbon. *Chemical Engineering Journal, 175,* 298–305.

9. Osinubi, K. J., Yohanna, P., & Eberemu, A. O. (2015). Cement modification of tropical black clay using iron ore tailings as admixture. *Transportation Geotechnics, 5,* 35–49.
10. Shettima, A. U., Hussin, M. W., Ahmad, Y., & Mirza, J. (2016). Evaluation of iron ore tailings as replacement for fine aggregate in concrete. *Construction and Building Materials, 120,* 72–79.
11. Zheng, J., Gao, Z., He, H., Yang, S., & Sun, C. (2016). Efficient degradation of acid orange 7 in aqueous solution by iron ore tailing fenton-like process. *Chemosphere, 150,* 40–48.
12. Wang, J. Q., Liu, Y. H., Chen, M. W., Xie, G. Q., Louzguine-Luzgin, D. V., Inoue, A., et al. (2012). Rapid degradation of azo dye by Fe-based metallic glass powder. *Advanced Functional Materials*. doi:10.1002/adfm.201103015
13. Zhang, C., Zhang, H., Lv, M., & Hu, Z. (2010). Decolorization of azo dye solution by Fe-Mo-Si-B amorphous alloy. *Journal of Non-Crystalline Solids, 356,* 1703–1706.
14. Kerkez, D. V., Tomašević, D. D., Kozma, G., Bečelić-Tomin, M. R., Prica, M. D., Rončević, S. D., et al. (2014). Three different clay-supported nanoscale zero-valent iron materials for industrial azo dye degradation: A comparative study. *Journal of the Taiwan Institute of Chemical Engineers*. http://dx.doi.org/10.1016/j.jtice.2014.04.019
15. Guz, L., Curutchet, G., Torres, R. M., Sánchez, & Candal, R. (2014). Adsorption of crystal violet on montmorillonite (or iron modified montmorillonite) followed by degradation through fenton or photo-fenton type reactions. *Journal of Environmental Chemical Engineering, 2,* 2344–2351.
16. Wang, W., Cheng, Y., Kong, T., & Cheng, G. (2015). Iron nanoparticles decoration onto three-dimensional graphene for rapid and efficient degradation of azo dye. *Journal of Hazardous Materials, 299,* 50–58.
17. Rashid, R. Z. A., Salleh, H. M., Ani, M. H., Yunus, N. A., Akiyama, T., & Purwanto, H. (2014). Reduction of low grade iron ore pellet using palm kernel shell. *Renewable Energy, 63,* 617–623.
18. El-Hussiny, N. A., & Shalabi, M. E. H. (2011). A self-reduced intermediate product from iron and steel plants waste materials using a briquetting process. *Powder Technology, 205,* 217–223.
19. Mohsenzadeh, M. S., & Mazinani, M. (2016). On the yield point phenomenon in low-carbon steels with ferrite-cementite microstructure. *Materials Science and Engineering A, 673,* 193–203.
20. Shi, L. N., Zhang, X., & Chen, Z. L. (2011). Removal of chromium (VI) from wastewater using bentonite-supported nanoscale zero-valent iron. *Water Research, 45,* 886–892.
21. El Hajjouji, H., Ait Baddi, G., Yaacoubi, A., Hamdi, H., Winterton, P., Revel, J. C., et al. (2008). Optimisation of biodegradation conditions for the treatment of olive mill wastewater. *Bioresource technology, 99,* 5505–5510.
22. Guo, D., Zhu, L., Guo, S., Cui, B., Luo, S., Laghari, M., et al. (2016). Direct reduction of oxidized iron ore pellets using biomass syngas as the reducer. *Fuel Processing Technology, 148,* 276–281.
23. Ponder, S. M., Darab, J. G., & Mallouk, T. E. (2000). Remediation of Cr(VI) and Pb (II) aqueous solutions using supported, nanoscale zero-valent iron. *Environmental Science and Technology, 4,* 2564–2569.
24. Chen, Z., Wang, T., Jin, X., Chen, Z., Megharaj, M., & Naidu, R. (2013). Multifunctional kaolinite-supported nanoscale zero-valent iron used for the adsorption and degradation of crystal violet in aqueous solution. *Journal of Colloid and Interface Science, 398,* 59–66.
25. Frost, R. L., Xi, Y., & He, H. (2010). Synthesis, characterization of palygorskite supported zero-valent iron and its application for methylene blue adsorption. *Journal of Colloid and Interface Science, 341,* 153–161.

Preliminary Analysis of the Application of Sensor Based Sorting on a Limestone Mine in the Region Caçapava do Sul, Brazil

Evandro Gomes dos Santos, Régis Sebben Paranhos, Carlos Otavio Petter, Aaron Young and Moacir Medeiros Veras

Abstract This work seeks to evaluate the effect of the sorting of calcium/magnesium carbonates (limestone ores) through the use of X-ray and CCD camera sensors. Collection of the 17 samples used for this study was performed after primary crushing at a size range between 76 and 152 mm (3–6″). This work seeks to evaluate the effect of the sorting of calcium/magnesium carbonates (limestone ores) through the use of X-ray and CCD camera sensors. Collection of the 17 samples used for this study was performed after primary crushing at a size range between 76 and 152 mm (3–6″). Sampling was intended to represent all of the possible variations of the ore and waste currently being extracted. For initial characterization, samples were washed and subjected to analysis by a handheld X-ray fluorescence spectrometer. Sensor based sorting tests were performed at the Federal University of Rio Grande do Sul (UFRGS) using a COMEX lab sorter. The lab sorter is comprised of a conveyor belt, mounted sensors, a data processing system, and a pneumatic flap system for the physical separation of the samples. Results of the X-ray separation showed that of the 12.2 kg of samples, 10.5 kg were sorted into product and only 1.7 kg were separated as waste. However, within the waste fraction, only 54.5% was considered waste, and likewise 7% of the product fraction was considered waste. When using the CCD camera, 100% of the ore fraction and 100% of the waste fractions were considered to be sorted correctly. While it is important to note that this is only a preliminary study analyzing the viability of implementation of the automatic sorting system, these preliminary results obtained with the use of both sensors were considered satisfactory.

E.G. dos Santos (✉) · R.S. Paranhos
Federal University of Pampa, Caçapava do Sul, Brazil
e-mail: eg.evandrogomes@gmail.com

R.S. Paranhos
e-mail: regis.paranhos@unipampa.edu.br

C.O. Petter · A. Young · M.M. Veras
Mineral Processing Laboratory,
Federal University of Rio Grande do Sul, Porto Alegre, Brazil

M.M. Veras
Federal Institute of Amapá, Macapá, Brazil

© The Minerals, Metals & Materials Society 2017

M.A. Meyers et al. (eds.), *Proceedings of the 3rd Pan American Materials Congress*,
The Minerals, Metals & Materials Series, DOI 10.1007/978-3-319-52132-9_58

Keywords Limestone · Sensor-Based seperation · Sorting

Introduction

Limestone mining in the municipality of Caçapava do Sul (RS, Brazil) began in the mid-twentieth century and today is one of the most important industrial activities in the southern region of Brazil. As mentioned by Sampaio and Almeida [1], the extraction and marketing of limestone is present all over the world, with this mineral being responsible for 1–3% of the earth's crust. These rocks may have, according to their composition, application in a wide range of products, ranging from construction and agriculture to industries including plastics, ornaments, paint, glass etc. [2].

The use of the ore mined in Caçapava do Sul, Brazil, was initially limited to the manufacturing of lime for civil construction, and involved a completely manual mining process. Beginning in the 1950s, the advent of ground limestone for use as a corrective of soil acidity and increased demand for raw materials for the realization of large civil works caused regional limestone ore extraction and processing systems to modernize, with operations becoming almost fully mechanized before the 1970s. Although the studied industry accounts for 70% of the corrective soil acidity limestone production and 25% of the state of the lime market, manual sorting remains included in the beneficiation processes [2]. This manual sorting stage, due to the geological characteristics of the deposit, plays an important role in the quality of the final products, because although there is a careful selection process by mining equipment, the ROM material that reaches the crusher contains approximately 5% waste.

Automated sorting technology has its first research dated from the 1950s [3]. However, only since the 1980s, with the development of larger amounts of computer processing power has it been made possible in practice [4]. Today, measurement methods using sensors are used to classify different types of materials, using measurement systems that recognize contrasts of color, shape, density etc. In cases where these features do not provide sufficient contrast in order to obtain classification accurately, measurement methods that assess the materials based on their molecular or atomic structure can be used [5]. According to Salter [6], the use of Sensor-Based Sorting (SBS) can result in many benefits, such as reduced operating and capital costs of transport and increased processing recovery. In this sense, the present study aims to evaluate the effect of the selection of limestone ore through the X-ray sensors and CCD camera.

Materials and Methods

Samples

The limestone samples from the mine were collected from the ROM ore immediately out of the primary crusher. The particle size of the material ranges between 76 and 152 mm (3–6″). Features of formation of the limestone occurrences in the Caçapava do Sul region of Brazil are very specific due to the wide range of geological processes that the limestone formations were submitted. This factor makes sampling a complex task, considering the number of lithologies found.

Although the term limestone is more common, and will be used throughout this work, the bentonite deposits of carbonate of the region are actually dolomitic marbles. These marbles (approximate age 750 Ma) are part of the Vacacaí Metamorphic Complex, surrounding the Caçapava do Sul Granitic Suite (approximate age 580 Ma) [7]. The main types of waste encountered within the formation are syngenetic intercalated pelitic mafic intrusions and granitic apophyses. The ore color is also variable, covering shades of white, gray and pink.

In total, 17 samples were collected, seeking to cover in the most coherent way possible, all existing ore and waste variants. Samples were washed and subjected to analysis on a portable X-ray fluorescence spectrometer (Fig. 1), from which was generated a table containing various percentages of chemical elements. The equipment used is manufactured by Bruker Corporation, model S1 TURBOSD.

For the calculation of Neutralizing Power values (NP), which indicates the potential ability of the limestone to neutralize the soil acidity, the following equation was used [8]: $ECaCO_3\% = (CaO\% \times 1.79) + (MgO\% \times 2.48)$, where $ECaCO_3\%$ is the NP and 1.79 and 2.48 are the neutralization capacity indices relative to $CaCO_3$ for these oxides.

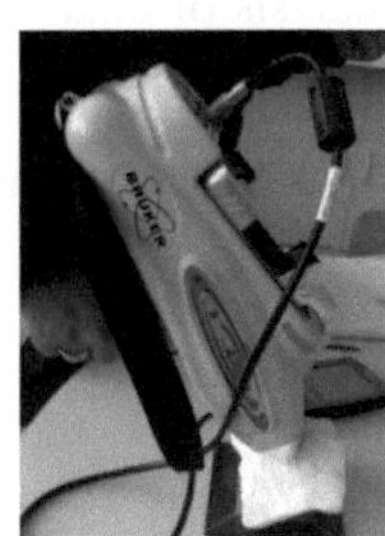

Fig. 1 Limestone sample being analyzed with Bruker S1 TURBOSD portable X-ray fluorescence spectrometer

Equipment of Classification (Sorter)

The sorter used was developed by the Comex company and consists of a conveyor belt, mounted sensors, a data processing system and a pneumatic system responsible for the separation (Fig. 2). The unit has three types of sensors: laser, CCD camera (charge-coupled device) and X-rays. Whereas the objective of this study was to perform a separation based on chemical characteristics and not on particle shape, only the optical and X-ray sensors of the equipment were employed. In using any of these sensors, the samples are placed on the conveyor belt which feeds the equipment and led to the detection zone. Reaching this point, through dedicated software, the data obtained in the sample readings with the chosen sensor are processed and calibration of the equipment may be performed. During calibration the parameters that characterize the material as product or waste are set, ie, the fraction of material which have free passage to reach the pneumatic valves and to be separated. Typically, the ratio of the difference in size between the highest and lowest particle should be at most 3:1 [9].

Identification of samples through X-ray is based on the recognition of the atomic density found within the particles. It is based on the principle that different atomic densities represent different classes of materials to be separated [10]. The software implemented in the equipment produces a range of colors based on the interaction between X-rays and the material undergoing identification, from which images are generated and used in the calibration.

The optical sensor operates within the visible light range of the electromagnetic spectrum (wavelengths between 390 and 780 nm). This sensor converts the light intensities captured in measurable electronic signals [11]. From these images, using RGB color scale parameters, it is possible to detect the visible contrast between different materials and calibrate the equipment accordingly.

After each test performed, the samples were removed from the equipment, already separated into fractions, which could be used to generate tables with the mass balance obtained. MS Excel spreadsheets were used for processing and analysis of data.

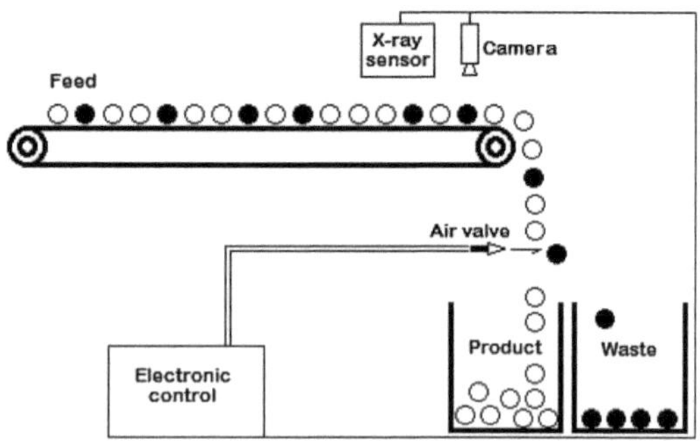

Fig. 2 Functional diagram of the sorter

Results and Analysis

The first characterization results obtained using the portable X-ray fluorescence spectrometer are shown in Table 1. Of interest were only the values that change in everyday mining operations (amounts in yellow). Of these, the most commonly used parameter is the NP obtained, which governs the allocation to be given to each material. Material with NP greater than 50% is considered ore (values highlighted in green), and material with NP less than 50% is considered waste (values highlighted in red) and sent to waste piles.

Table 1 Results of sample analysis of limestone and waste obtained from portable X-ray fluorescence spectrometer (Color table online)

ID	Class	Description	Kg	SiO_2	MgO	CaO	PN
1	Product	Pink Limestone	0,617	7,54	20,40	40,20	122,55
2	Product	Mixed Limestone	0,616	9,67	22,45	36,90	121,73
3	Waste	Metabasite	0,478	84,50	1,79	2,92	9,67
4	Product	White Limestone	0,699	0,83	21,90	40,10	126,09
5	Product	Pink Limestone	0,684	0,00	12,00	44,60	109,59
6	Product	Mixed Limestone	0,801	0,00	24,40	45,40	141,78
7	Product	White Limestone	0,624	2,52	25,70	43,50	141,60
8	Product	Pink Limestone	0,669	0,00	12,50	21,60	69,66
9	Product	Gray Limestone	0,884	0,00	7,01	52,70	111,72
10	Waste	Granite	0,610	65,50	4,78	2,80	16,87
11	Product	White Limestone	0,485	3,32	28,80	43,60	149,47
12	Product	Mixed Limestone	0,587	28,00	18,40	33,90	106,31
13	Product	Mixed Limestone	0,682	13,70	25,00	35,70	125,90
14	Product	Mixed Limestone	1,416	1,90	14,90	38,80	106,40
15	Product	Pink Limestone	0,625	5,99	22,50	68,00	177,52
16	Waste	Diabase	0,661	42,10	19,20	0,83	49,10
17	Product	Gray Limestone	1,012	3,59	24,20	58,20	164,19

Calibration of the automatic sorting equipment, in concept, sought to separate the samples numbered 3, 10 and 16 from all of the other samples (Fig. 3). These samples, besides presenting NP lower than the cut-off values for ore, also have high values of SiO_2, which may cause a decrease in the levels of the end products and premature wear to comminuting equipment.

As one can see, the sorter was fed with a total weight of samples equal to 12.2 kg. After screening of the material, 10.5 kg were separated as product and 1.7 kg as the waste, however, 54.5% of the waste fraction is composed of product and over 7% of the product is waste. This result shows a fairly effective separation in the product portion, but an inefficient sorting of waste, which could lead to the huge amount of waste, considering real values of production in industry (Fig. 4).

As mentioned previously, test of the automatic sensor based sorting process employed two types of sensors: Optical and X-rays. Table 2 shows the results obtained with the X-ray sensor.

Fig. 3 Samples positioned within the sorter for calibration

Fig. 4 Samples taken after the sorting process using X-rays. Note the presence of white limestone with the waste (*top tray*) and diabase *black color* next to the product (*lower tray*)

Table 2 Sorting results from the use of the X-ray sensor

	Limestone		Waste		Total	
	Kg	%	Kg	%	Kg	%
Feed	10.4	85.6	1.8	14.4	12.2	100.0
Product	9.8	80.7	0.7	5.3	10.5	86.0
Waste	0.6	4.9	1.1	9.1	1.7	14.0

Fig. 5 Samples taken from the sorter after the sorting process through optical sensor (CCD camera). Note the presence of only waste material in the *top tray* and only product on the *bottom tray* (Color figure online)

Table 3 Sorting results from the use of the optical sensor (CCD camera)

	Limestone		Waste		Total	
	Kg	%	Kg	%	Kg	%
Feed	10.4	85.6	1.8	14.4	12.2	100.0
Product	10.4	85.6	0.0	0.0	10.4	85.6
Waste	0.0	0.0	1.8	14.4	1.8	14.4

One possibility for the poor performance obtained initially, is that the samples have a heterogeneous format (although within the quoted range limit of 3:1). This shape creates, at certain points, a false high density print due to the thickness of the sample thus it may happen that thick low density samples become equal to thin high density samples as they pass the X-ray sensor. This hypothesis can be seen in Fig. 5, where there is a difference in the upper tray among the size of the ore sample (white) with respect to waste samples (pink and black).

Noteworthy, analyzing the data, is the high efficiency achieved using the CCD camera to give 100% recovery of the ore fraction and 100% disposal of the waste fraction (Fig. 5). In the same image, the existing similarity between the sample of pink granite (top tray) and pink limestone samples is observable (lower tray). This result, in a way, brings surprise, since the existence of colors so close was thought to restrict conditions for sorting through optics.

Turning to the second test, the same group of samples was subjected to analysis by optical sensor (CCD camera). The results are presented in Table 3.

Conclusion

Taking into account that this is a preliminary study assessing the feasibility of implementation of the automatic sorting system, the initial results obtained using both methods were satisfactory, and noteworthy was the optical method, which had 100% efficiency.

The question of the use of the X-ray sensor, as already discussed, deserves attention in the sequence of studies, because it may be possible that the particle size and shape may influence the sorting results significantly. At the same time, without the efficiency values of the current manual sorting used at the mine, it is hard to qualify the effectiveness of the sensor.

To continue this work means to increase the reliability of the same, new samples must be analyzed, seeking to ensure that the largest possible parts of the lithological combinations are studied. Also tests should be performed with material of a smaller particle size, in attempt to resolve the problem raised in relation to counterfeit density found by the X-ray, and thus increase the efficiency of the method.

Acknowledgements The authors would like to thank The Federal University of Pampa (UNIPAMPA) and the Federal University of Rio Grande do Sul (UFRGS), which, with the Re-Maud project (Capes/Cofecub, Brazil), made possible the realization of this study. Also we offer a thanks to company, for the transfer of its industrial area to carry out this study and providing all of the necessary subsidies.

References

1. Sampaio, J. A., & Almeida, S. L. M. (2005). *Rochas & Minerais Industriais: Usos e Especificações*. Rio de Janeiro: CETEM. 350 pp.
2. Ministry of Mines and energy—MME. (2009). *Perfil do Limestone*. Brasilia: MME. 56 p.
3. Gülcan, E., Gulsoy, Ö. Y., & Ergün, L. S. (2014). Comparison of color and NIR camera in sorting of lignites. In: *Sensor-Based Sorting, International Conference and Exhibition 2014*. Alemanha, RWTH Aachen University. CD Sensor-Based Sorting.
4. Pretz, T. (2012). Preface. In: *Sensor-Based Sorting, International Conference and Exhibition, 2012*. Alemanha, RWTH Aachen University. CD Sensor-Based Sorting.
5. Feierabend, A., Bohling, C., & Cordts, L. (2014). Using LIBS spectroscopy for analysis of raw materials and production waste. In: *Sensor-Based Sorting, International Conference and Exhibition, 2014*. Alemanha, RWTH Aachen University. CD Sensor-Based Sorting.
6. Salter, J. (1991, novembro de). Sorting in the minerals industry: Past, present and future. *Minerals Engineering, 4*, 779–796.
7. CPRM—SERVIÇO GEOLÓGICO DO BRASIL. (2000). Programa Levantamentos Geológicos Básicos do Brasil. Cachoeira do Sul, Folha SH.22-Y-A. Estado do Rio de Grande do Sul. Escala 1:250.000. Organizado por Carlos Alfredo Porcher e Ricardo da Cunha Lopes. Brasília.
8. Primavesi, A. C., & Primavesi, O. (2004). Características de corretivos agrícolas. *Documentos, 37*, 28 p (São Carlos: Embrapa Pecuária Sudeste).
9. Ergün, L. Ş., Gülsoy, Ö. Y., & Gülcan, E. (2014). In: *Sensor-Based Sorting, International Conference and Exhibition, 2014*. Alemanha, RWTH Aachen University. CD Sensor-Based Sorting.
10. Jong, T., & Dalmijn, W. (2003). Dual energy X-ray transmission imaging: applications in metal processing. In: M. E. Schlesinger (Ed.), *EPD Congress*.
11. Holst, G. C., & Lomheim, T. S. (2011). *CMOS/CCD Sensors and Camera Systems* (2nd ed.). Winter Park: JCD Publishing.
12. Gülcan, E., Gulsoy, Ö. Y., & Ergün, L. S. (2014). Comparison of color and NIR camera in sorting of lignites. In: *Sensor-Based Sorting, International Conference and Exhibition, 2014*. Alemanha, RWTH Aachen University. CD Sensor-Based Sorting.

Preparation of High Grade Industrial Copper Compound from a Nigerian Malachite Mineral by Hydrometallurgical Process

Alafara A. Baba, Ruth O. Sanni, Abdulrahman Abubakar, Rafiu B. Bale, Folahan A. Adekola and Abdul G.F. Alabi

Abstract Increasing demands for copper and copper sulphate with diverse industrial applications has prompted the development of a low-cost and eco-friendly technique as a substitute for conventional ore treatments by reduction-roasting route, requiring high energy consumption. In this study, hydrometallurgical treatment of a Nigerian malachite through acid leaching and solvent extraction was investigated. The effects of leachant concentration, reaction temperature and particle size were examined. The rate of ore dissolution increased with increasing H_2SO_4 concentration, temperature and decreasing particle size. At optimal leaching conditions, 96.2% of the ore reacted with 2.0 mol/L H_2SO_4 solution at 75 °C within 120 min. The calculated activation energy was 25.41 kJ/mol, supporting the proposed diffusion controlled mechanism for the dissolution process. The leach liquor at optimal conditions was further beneficiated to obtain high grade industrial copper sulphate using a combinations of solvent extraction and precipitation methods, leading to 98% process efficiency.

A.A. Baba (✉) · R.O. Sanni · F.A. Adekola
Department of Industrial Chemistry, University of Ilorin, P.M.B 1515, Ilorin 240003, Nigeria
e-mail: baalafara@yahoo.com

A. Abubakar
Department of Chemistry, University of Ilorin, P.M.B 1515, Ilorin 240003, Nigeria

R.B. Bale
Department of Geology and Mineral Sciences, University of Ilorin, P.M.B 1515, Ilorin 240003, Nigeria

A.G.F. Alabi
Department of Materials and Metallurgical Engineering, University of Ilorin, P.M.B 1515, Ilorin 240003, Nigeria

A.G.F. Alabi
Department of Material Science and Engineering, Kwara State University, P.M.B 1530, Malete, Nigeria

© The Minerals, Metals & Materials Society 2017
M.A. Meyers et al. (eds.), *Proceedings of the 3rd Pan American Materials Congress*, The Minerals, Metals & Materials Series, DOI 10.1007/978-3-319-52132-9_59

Keywords Malachite mineral · Nigeria · Leaching · Solvent extraction · Copper · Copper sulphate

Introduction

Copper is a metal with high thermal and electrical conductivity with wide array of uses as a conductor of heat and electricity, building material and as a constituent of various metal alloys [1]. Also, copper metal and its compounds, such as copper sulphate ($CuSO_4$) are used in anti-biofouling, antibacterial applications and correction of copper deficiency in the ecosystem [2, 3]. Consequently, the increased demand for copper metal supplied from industry leads to the fast use of the best and easiest accessible resources, which in turn inspires the innovations for new eco-friendly solutions to enable processing of poor deposits or recovery of metal from industrial waste. This calls for high need of rich copper ore like malachite mineral and its design for economical method of the ore processing.

Malachite ($CuCO_3{\cdot}Cu(OH)_2$) is a copper carbonate mineral more common than azurite. It is typically associated with copper deposits in limestone, and is an important source of industrial copper production, apart from its use as ornament and mineral pigment due to its ease of impurity removal [4]. In the past years, sulphide ores have widely been used in copper production for pyrometallurgical operations because they are easily separated from gangues and concentrated by conventional flotation techniques. For example, in differential flotation, different concentrates obtained are of poor quality with low metal recovery; this makes the continuous use of conventional method of these ores very difficult, expensive and makes them difficult to commercialize [5]. Conversely, the depletion of high-grade copper sulphide ores has led to focusing on the extraction of copper from oxidized copper ores in various carbonate including malachite mineral [6].

Due to the aforementioned draw-backs, the metal value is preferably extracted directly from the low grade ores even waste dumps through hydrometallurgical process [7–9]. Hydrometallurgical operations for metal extraction and purification include leaching, solvent extraction and electrowinning [6]. Apart from ease of impurities removal, solubility and good metal complex formation, the hydrometallurgical approach by Cyanex 272 is found to be economical, eco-friendly and simplicity in design operation and was adopted in this study [7, 10]. Therefore, the purpose of this study was to investigate leaching-cum-solvent extraction of copper from Malachite mineral of Nigeria origin for possible production of an industrial copper compound in Sulphuric acid media with Cyanex® 272 through hydrometallurgical method. Vast deposits of malachite ore are abundant with less exploration in Nigeria and this study would no doubt contribute to the solid mineral development of the country.

Materials and Method

The copper carbonate hydroxide mineral malachite ($Cu_2CO_3(OH)_2$) used for this study was obtained from the Department of Geology and Mineral Sciences, University of Ilorin, Ilorin—Nigeria. The mineral was sourced from Sabon-Gari in Zamfara State, Nigeria. The fractions of the sample used was generated by comminution with acetone-rinsed mortar and pestle and sieved into three different particle sizes (−63 + 45, −90 + 63, and −112 + 90 μm) with the aid of ASTM standard sieve. All experiments were performed only with the −63 + 45 μm fraction, unless otherwise stated [11]. Characterization of the raw ore and selected products at optimal conditions were carried out by X-ray fluorescence, X-ray diffraction, Scanning electron microscopic, Atomic absorption spectroscopy and Fourier Transform-Infra-red spectroscopic techniques.

Leaching Tests

Leaching experiments were carried out in a 600 mL glass reactor equipped with a mechanical stirrer. 100 mL of H_2SO_4 (0.1–2.5 mol/L) solutions were introduced into the reactor and brought to heating to the desired temperature with agitation [12]. The acid concentration with highest dissolution was used for further optimization studies such as reaction temperature and particle size variations. In all experiments, the fraction of the ore dissolved was evaluated from the initial difference in weight of the amount dissolved or undissolved at various time intervals up to 120 min, after oven-drying at about 60 °C. The post leaching residual product at 75 °C in 2 mol/L H_2SO_4 solution was then analyzed by X-ray diffraction and Scanning Electron Microscope. The appropriate Shrinking Core Model (SCM) was used to obtain the dissolution kinetic parameters for better understanding of the copper extraction conditions. The characterizations of the leach residues before and after leaching at optimal conditions were accordingly carried out.

Solvent Extraction/Beneficiation Tests

Equal volumes (25 mL) of the aqueous (leach liquor) and organic phases were taken in a stoppered bottle (125 mL) and shaken for 30 min. After equilibration and phase separation, the concentration of Cu^{2+} ion in the organic phase was calculated from the difference between its concentration in the aqueous phase before and after extraction. The percent of total copper extracted was quantitatively evaluated. Prior to actual total copper extraction by CYANEX 272 extractant, the iron impurity from the pregnant solution was removed by precipitation using H_2O_2 and NaOH at pH 3.00, at 27 ± 2 °C. Pure copper solution obtained after these operations was beneficiated as copper sulphate pentahydrate [13, 14].

Results and Discussion

Characterization Studies

The malachite mineral, as analysed by X-ray fluoresence (XRF), is primarily composed of 61.21 wt% Cu, 14.26 wt% Fe and 4.46 wt% Al. Also, the mineralogical examination by X-ray diffraction (XRD) showed the principal occurrence of malachite ($CuCO_3{\cdot}Cu(OH)_2$: 046-0858) and quartz (α-SiO_2: 005-0490).

Leaching Results

The investigations of leaching studies showed that the malachite mineral dissolution in H_2SO_4 solution increases with increasing acid concentration, temperature and with decrease in particle diameter at a moderate stirring. The leaching was carried out under the following conditions:

(i) **H_2SO_4 concentration variation**: (0.1–2.5 mol/L), 55 °C, 5–20 min, −63 + 45 μm: Increasing the concentration apparently increases the ore dissolution with diffusional characteristics of the solution to the inner core as the acid concentration increases up till 2.0 mol/L H_2SO_4 solution. However, drastic decrease in dissolution was observed as the concentration increased to 2.5 mol/L solution, where 91.2% dissolution was attained. Possible reason for the decreasing ore dissolution might be due to the precipitation phenomenon at higher acid concentration [11]. Hence, at 2.0 mol/L H_2SO_4 solution, the amount of the ore dissolution reached 96.2% during 120 min leaching.

(ii) **Reaction temperature**: 27–75 °C, 2.0 mol/L H_2SO_4, 5–120 min, −63 + 45 μm: Increasing the reaction temperature from 27 to 75 °C appreciably increases the ore dissolution from 34.1 to 96.2% within 120 min.

(iii) **Particle size variation**: The dissolution results for particle size variations −63 + 45, −90 + 63, and −112 + 90 μm gave 96.2, 73.6 and 62.5%, respectively in 2.0 mol/L H_2SO_4 solutions within 120 min and at 75 °C temperature. The unleached residue constituting about ≈4% of the input was analysed by XRD and found to contain silica (α-SiO_2: 14-1085). The kinetics data evaluated by the shrinking core model was found to be the diffusion controlled mechanism. The reaction order and activation energy, E_a, for the dissolution process determined from the appropriate kinetic and Arrhenius plots were 0.480 ≈ 1/2 (half order) and 25.41 kJ/mol, respectively, supporting the proposed dissolution mechanism.

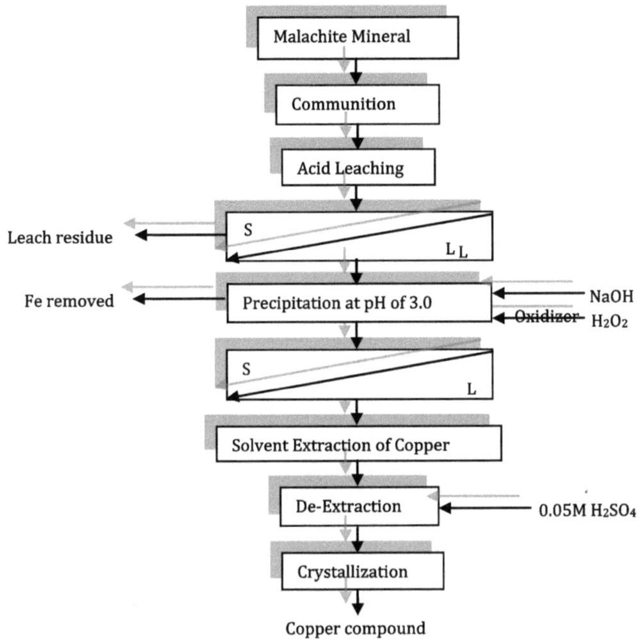

Fig. 1 A hydrometallurgical processing flow chart for a Nigerian malachite mineral to obtain high grade Industrial copper compound [15]

Solvent Extraction/Beneficiation Studies

Solvent extraction of malachite leach liquor containing predominantly 3900.0 mg/L Cu and 580 mg/L Fe was carried out by 0.15 mol/L CYANEX 272 in kerosene at 27 ± 2 °C for 30 min contact. The combination of precipitation and goethite methods were adopted in the purification and removal of iron to achieve extraction and beneficiation efficiencies of about 98% pure copper. Further purification through crystallization to obtain high grade industrial copper compound was done. The low cost and eco-friendly operational flow-chart summarizing the analytical procedures for the copper extraction and beneficiation routes for malachite mineral processing of Nigeria origin is summarized in Fig. 1.

Conclusion

In this study, leaching, solvent extraction and precipitation have been utilized to purify the resulting solution from the sulphuric acid-leaching of Sabon-Gari, Zamfara (Nigeria) malachite mineral by CYANEX® 272 extractant. The efficiency of the various techniques used in this study was >98%. This study affirmed the possibility of extracting pure copper for the production of high grade copper sulphate by hydrometallurgical processing. Hence, the use of Nigerian sourced malachite minerals for our industries would enable the country to generate more revenue apart from that earned from crude oil exploration. Therefore, this study enhances Nigeria's development and her economic sustainability.

References

1. McHenry, C. (1992). *The New Encyclopedia Britannica 3* (15th ed., p. 612). Chicago: Encyclopedia Britannica, Inc. (ISBN 085-229553-7).
2. Courtade, M., Vesinet, L., & Ramel, G. (1999). Bordeaux mixture process for its manufactureand cupric fungicidal compositions containing it. *USP, 5*(958), 438.
3. Andras, S., Tamas, D., Tamasne, M., Laszlone, K., Istvan, R., & Zoltan, S. (2009). Bordeaux mixture suspension and process for the preparation thereof. *EP, 2*(063), 713.
4. Gettens, R. J., & Fitzhugh, E. W. (1993). In Roy, A. (ed.), *Malachite and Green Verditer, in Artists' Pigments. A Handbook of Their History and Characteristics* (Vol. 2, pp. 183–202). Oxford: Oxford University Press.
5. Hyvarinen, O., Hamalainen, M., & Leimala, R. (2002). *Outokumpu Hyro-copper Process: A Novel Concept in Copper Production* (pp. 19–23). Montreal, Canada: Chloride Metallurgy.
6. Liu, Z., Yin, Z., Hu, H., & Chen, Q. (2012). Dissolution kinetics of malachite in ammonia/ammonium sulphate solution. *Journal of Central South University of Technology, 19*, 903–910.
7. Baba, A. A., Ayinla, I. K., Adekola, F. A., Bale, R. B., Ghosh, M. K., Rout, P.C., et al. (2015). Extraction and purification of copper from a Nigerian chalcopyrite ore leach liquor by dithizone in kerosene. *Solvent Extraction Research and Development*, Japan, 22(2), 135–146.
8. Hiroyoshi, N., Miki, H., & Hirajima, T. (2001). Enhancement of chalcopyrite leaching by ferrous ions in acidic ferric sulfate solution. *Hydrometallurgy, 60,* 185–197.
9. Edward, E. Marouf. (1985). *New Developments in Hydrometallurgy* (pp. 22–25). San Francisco, CA: American Mining Congress Mining Convention.
10. Baba, A. A., Adekola, F. A., & Bale, R. B. (2009). Development of a pyro- and hydro-metallurgical route to treat spent zinc-carbon batteries. *Journal of Hazardous Materials, 171*(1–3), 838–844.
11. Baba, A. A., Ayinla, I. K., Adekola, F. A., Bale, R. B., Ghosh, M. K., Alabi, A. G. F., et al. (2013). Hydrometallurgical application for treating a Nigerian chalcopyrite ore in chloride media: Part I. Dissolution kinetic assessment. *Journal of Minerals, Metallurgy, and Materials, 20,* 1021–1028.
12. Bingol, D., & Canbazoglu, M. (2004). Dissolution kinetics of malachite in sulphuric acid. *Hydrometallurgy, 72,* 159–165.
13. Kokes, H., Morcali, M. H., & Acma, E. (2014). Dissolution of Copper and iron from Malachite ore and Precipitation of Copper Sulfate Pentahydrate by Chemical Process. *Engineering Science and Technology, An International Journal, 17,* 39–44.
14. Sole, K. C., & Brent Hiskey, J. (1995). Solvent extraction of copper by Cyanex 272, Cyanex 302 and Cyanex 301. *Hydrometallurgy, 37*, 129–147.
15. Sanni, O. Ruth (2016) *Preparation of high grade industrial copper compound from a nigerian malachite mineral by hydrometallurgical process* (M.Sc. Thesis) Department of Industrial Chemistry, University of Ilorin, Ilorin-240003, Nigeria (pp. 1–138).

Process of Improving the Flotation Using Ultrasonic Bombardment

Erivelto L. Souza, Orimar B. Reis, Denise F. Pereira, Luiz C. Borges and Jeisa F.P. Rodrigues

Abstract The quality of the iron ore, over the decades of exploitation lowered, due to the reduction of the rich ores found at the beginning of the exploration. It resulted in ores with lower grades and more mixed with the gangue. Iron ore is processed after comminution and grinding, to have appropriate particle size and release the stages of flotation, which is divided generally in Rougher steps, Cleaner, Scavenger and Re-Cleaner. Flotation is a process that uses an ore slurry with additions of flocculants, surfactants, and other products. These substances are fundamental to the operation. However, it compromises of the reuse of the water. The proposal presented in this paper introduces a pre-treatment process of the pre-flotation material, still in Rougher step, which will increase its efficiency in such way that only one Cleaner step would be sufficient to generate a concentrate ready to be considered flotation of the final product.

Keywords Hematite flotation · Ore recuperation · Ultrasonic treatment

E.L. Souza (✉)
DTECH/CAP/UFSJ, Ouro Branco, MG, Brazil
e-mail: souza.erivelto@ufsj.edu.br

O.B. Reis · L.C. Borges
IFMG—OP, Ouro Preto, MG, Brazil
e-mail: orimar.reis@ifmg.edu.br

L.C. Borges
e-mail: jeisapr@yahoo.com.br

D.F. Pereira
QTEC, Ouro Branco, MG, Brazil
e-mail: pereira.denisede@gmail.com

J.F.P. Rodrigues
PPGEM—UFOP, Ouro Branco, MG, Brazil
e-mail: jeisapr@yahoo.com.br

© The Minerals, Metals & Materials Society 2017

M.A. Meyers et al. (eds.), *Proceedings of the 3rd Pan American Materials Congress*, The Minerals, Metals & Materials Series, DOI 10.1007/978-3-319-52132-9_60

Introduction

Mining in the world, on increasing expansion, it has been characterized by the need to improve the productivity to attend a growing demand worldwide. In contrast natural resources, non-renewable, like the deposits of considerable levels of the elements of interest are being depleted. This fact encouraged an exploration of reserves where the levels are lower and have more complex minerals such as the exploitation of marginal levels of minerals (with Fe content lower and higher concentrations of silica and contaminants). Thus, the search for a more efficient technology and ways to continue to meet the specifications of an increasingly demanding market with greater competitiveness, which requires a quality increasing the product obtained from the poor ore, encourages the search for gravity concentration processes, magnetic and/or more efficient flotation. The purpose of this work is the application of ultrasound techniques in the feeds of the steps of a flotation. This efficiency was measured by conducting tests in the laboratory with the feed and reject pulps of a flotation iron ore, consisting of columns rougher, cleaner and scavenger, using an ultrasonic bombardment.

Objectives

General Objective

Evaluate the use of ultrasonic bombardment in iron ore pulps subjected to the steps rougher, cleaner and scavenger of a flotation, to improve the efficiency of these steps and reduce the generation of impacting residual wastes in environment, optimizing this flotation.

Specific Objectives

- To evaluate the characteristics of the ore in order to obtain the information needed to process efficiency evaluation, which will be observed in the tailings flotation of rougher and cleaner columns;
- Conducting tests with pulps of flotation steps before and after the bombardment of ultrasound with the aim of increasing the useful material recovery efficiency (Fe) in each stage;
- Demonstrate through benchmarking the performance of ultrassom in the pulps that feed each stage of flotation;
- Judging the results obtained with tests carried out in steps flotation using ultrasonic bombardment.

Justification

Even with the improvement of the steps of a flotation columns, the factor that determines the efficiency of steps rougher cleaner and guarantees the quality of the final product, is the degree of liberation that the particles are before each step. Even improving the stages, while the metallic and non-metallic particles are aggregated, it generates waste with still considerable iron content, and the potential for recovery.

Thus, this project is linked to seeking the best way to treat the pulps prior to each stage to promote better individualization of particles of those pulps, so that stages of flotation have better performance. Thus it promotes greater reductions in disposal of materials generation thus demonstrating the best use of the process, reducing the ore reuse cost and less waste which will result in less wear of the environment.

Theoretical Framework

Influence of Mud Layers in the Flotation Process

The particles between 100 and 10 μm are considered thin, between 10 and 1 μm ultrathin, below 1 μm colloids. Sludge is the ultrafine particles and colloids and comprise the part that serve as ligands of metal and non-metal particles [1–3].

The major problem nowadays the flotation process is the presence of sludge in the recovery of minerals, ores, which brings a detrimental effect on the process efficiency. Step responsible for the elimination of the natural ultrafine particles or generated in the comminution process due to classification and further desliming performed by classifying cyclones [3].

The surface properties of superfine influence the creation of sludge coatings on other particles (slimes coating), alter the consistency of the foam blocking the surface of minerals, interfering with the bubble-mineral contact and make ineffective the performance of reagents, and contaminate the concentrate [1, 4].

To obtain a good efficiency of release of particles, the ore is deslimed with addition of dispersants, which generates costs, and compromises the used water when discarded into the environment, Fig. 1 [5, 6].

Mass Balance in the Process of Flotation

The principle Lavoisier determines that a concentration process should have a mass that enters identical to the mass exiting However, due to losses sprinkling and natural evaporation occurring small losses, however, they can be neglected for its small influence on the process and a whole. As showed in Fig. 2 [7, 8].

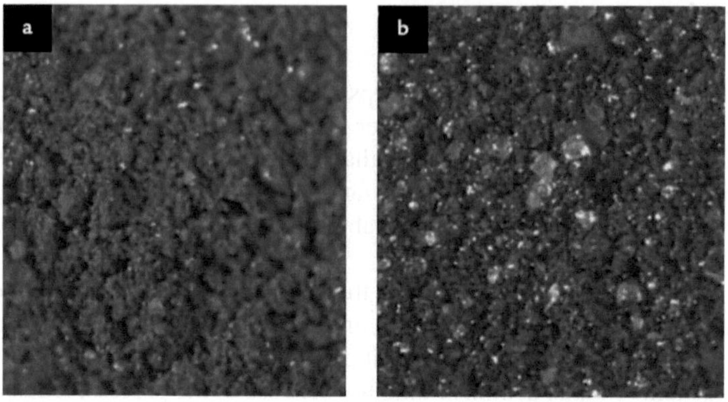

Fig. 1 Material before and after desliming **a** shows deslimed without addition of dispersing reagent; **b** sample deslimed with addition of dispersing reagent [4]

Fig. 2 Mass relationship balance in flotation cell [10]

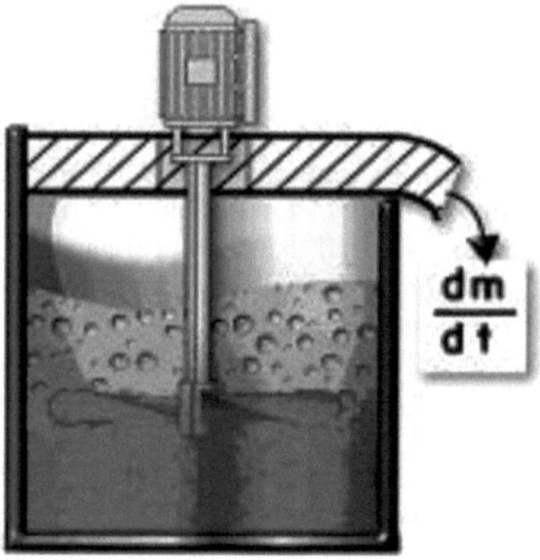

However, cumulatively these losses represent a factor of consumption of water. With an increased process efficiency, a smaller number of steps is required for achieving the expected results, which is explained when considering the volume of material and number of steps that an ore concentration process contains [4, 9].

The degree of approach to equilibrium define the metallurgical process results, the circuit type, the number of stages, the circulating or recycling cleaning fillers, etc. The process will be more or less efficient, both in metallurgical results as in production costs, as its degree of approach to equilibrium. In two-phase systems (such as pulp and foam) when these are not in balance, a spontaneous change is produced by macromolecular migration causing the system to the equilibrium state, where changes cease [10].

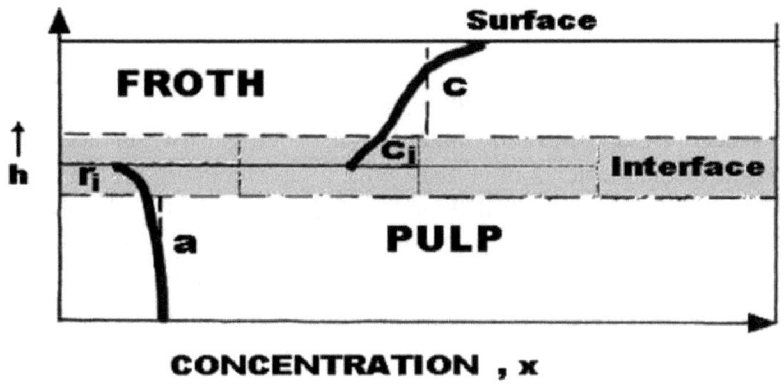

Fig. 3 Interface concentration profile [10]

The mass flow of the substance of our interest $\left(\dot{M} = \text{mass/time}\right)$ can be defined by the following expression (Fig. 3 shows the Interface concentration profile) [10].

$$\dot{M} = K_T.A_c.\nabla c$$

K_T Transfer coefficient, kg/m^2 s;
A_C Contact area, m^2;
∇c Composition gradient (a − r_i), %;
a Average content of metal component in the mass fed into the pulp, %;
r_i Average content of the metal component in the interface between the solute (pulp) and the solvent (air), %.

Operating Flotation Properties

Defining Flotation as a separation phenomenon with mass macromolecular transfer, we analyze the following aspects mentioned in this table, such as: mass transfer contact between the phases, transfer mechanisms, gradient and driving forces in the phases involved and operational aspects, the conventional form of engineering processes, such as chemical engineering operations are studied [5, 10].

Flotation is a separation operation of heterogeneous mixtures of components from particles suspended in the aqueous phase. This allows a separate component other is different level of hydrophobicity on the surface of particles of different substances. At this point the greater individualization of the particles that make up the pulp is critical to the efficiency of the process [8, 10].

Gradient and Driving Forces

The particles collide with air bubbles when activated by hydrophobic energy naturally or induced by appropriate reagents, adsorbed on the surface, decrease the surface tension of the particles in contact with a gas, focusing on the liquid surface. The average particle velocity will depend on its degree of hydrophobicity if the bubble adhering faster and stronger and more stable manner is rapidly conducted to the surface. The particles are more hydrophobic when they are released form impurities (particles greater individualization content), and are better transported when they have a certain weight and size compatible with the size of bubbles, which is the transport vehicle. In phase pulp, by mechanical or pneumatic agitation, is rapidly produced the bubble/particle contact and, to maintain the uniformity, is required for this phase (density, mineralogical distribution, grain size, etc.). The macromolecular transfer speed is very fast, mainly in mechanical cells [4, 10].

Table 1 shows the steps compound the characteristics of flotation, and Fig. 4 shows the phenomena it the flotation [10].

Table 1 Mass transfer occurring in flotation [10]

Item	Characteristics
Phenomenon	Separation and particle concentration
Transport	Particles transfer
Mechanism	Forced migration
Transport gradient	Hydrophobicity of the recovered particles
Operation	Natural gradient
Main phenomenon	Natural + operating
Phases	Heterogeneous (pulp, foam)

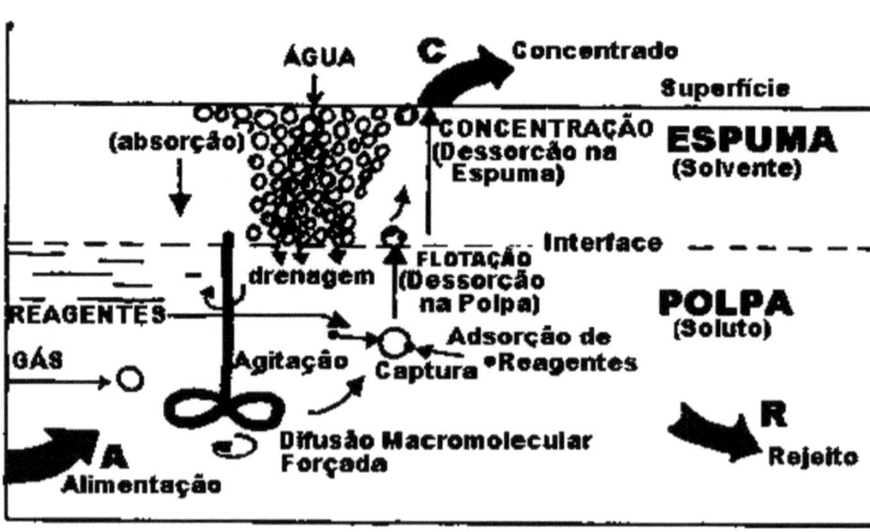

Fig. 4 Transportation mechanisms in a flotation cell [10]

Methodology

Materials and Equipment

The following equipment was used in the laboratorial experiment (as showed in Figs. 5, 6 and 7).

Fig. 5 Beakers with various scales

Fig. 6 Ultrasonic bath vessel

Fig. 7 100× electronic magnifier

- Beakers;
- Ultrasonic vessel;
- Electronic magnifier 100×;
- Poor content milled iron ore;
- Rougher step feed iron ore.

Methods and Procedures

The feeding of the flotation was blended to homogenize. After this step was separated into two storage containers. The first container was subjected to an ultrasonic bombardment, the second was directed to a flotation cell counter (Figs. 8, 9, 10 and 11).

Fig. 8 Poor content milled iron ore

Fig. 9 Iron ore pulp (**a**), after 6 h of decantation (**b**)

Fig. 10 Using the ultrasound vessel to accelerate decantation

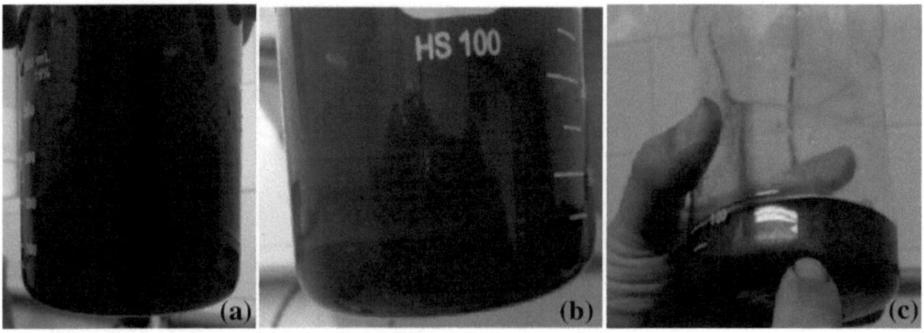

Fig. 11 Iron ore pulp after 180 s of ultrasound decantation: **a** agitated pulp; **b** decanted after 180″ ultrasound; **c** dewatered

Results

After Ultrasound Processing

The two pulps after simple washing step were subjected to rougher flotation step in the process. The pulp subjected to ultrasound showed a significant improvement in metal content analysis. The pulp scavenger stage reject the rougher stage of the ultrasound untreated pulp was also subjected to a ultrasound process. It was possible to notice the change in the aspect of it, how the particles showed release after the ultrasound (Figs. 12 and 13).

Fig. 12 Iron ore flotation feed without ultrasound process. 20× amplifier. Can be perceived as a homogeneous mass

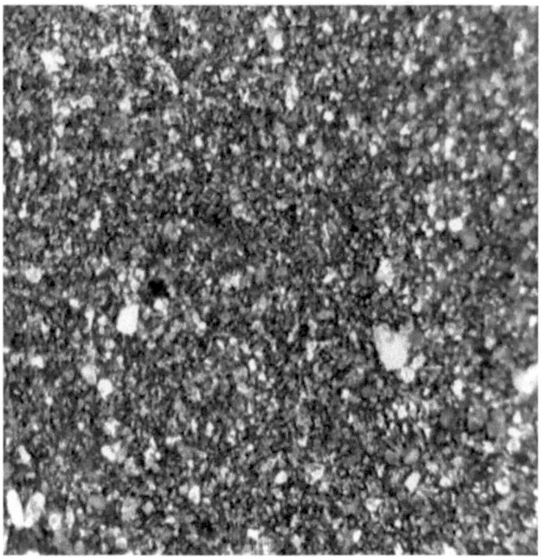

Fig. 13 Iron ore flotation feed after ultrasound process. 20× amplifier. Can be perceived the identity of particles

We can perceive that in Fig. 15 the particles are significantly cleaner, in comparison of the particles of Fig. 14.

After Chemical Analysis

The two pulps was submitted to chemical analysis to found the effect of the ultrasound process on iron ore pulp. Table 2 shows the chemical analysis of feed rougher stage, without the application of ultrasonic bombardment. Table 3 shows the same feed rougher stage after it is exposed to ultrasound. The two phases were

Fig. 14 Scavenger feed without ultrasound process. 100× amplifier

Fig. 15 Scavenger feed after ultrasound process. 100× amplifier

Table 2 Food rougher step flotation before ultrasound bombardment and dewatered

Compound	Fe	SiO_2	Al_2O_3	P	Mn	CaO	MgO	TiO_2	Total
%	49.37	26.64	1.00	0.034	0.119	0.027	0.037	0.144	98.722

Table 3 Food rougher step flotation after ultrasound bombardment and dewatered

Compound	Fe	SiO_2	Al_2O_3	P	Mn	CaO	MgO	TiO_2	Total
%	61.42	10.2	0.69	0.039	0.061	0.019	0.023	0.254	99.202

Table 4 Residue of rougher step without ultrasound bombardment

Compound	Fe	SiO_2	Al_2O_3	P	Mn	CaO	MgO	TiO_2	Total
%	48.68	28.29	0.35	0.024	0.51	0.026	0.023	0.013	99.056

Table 5 Residue of rougher after ultrasound bombardment

Compound	Fe	SiO_2	Al_2O_3	P	Mn	CaO	MgO	TiO_2	Total
%	49.37	26.64	1.00	0.034	0.119	0.027	0.037	0.144	98.722

dewatered for analysis. Table 4 shows the residue of the rougher stage, which was processed without the use of ultrasound, analyzed after dewatering. Table 5 shows the same residue of the rougher stage that had their food treated with ultrasound, after being dewatered.

Conclusão

Although ultrasound had not intended to concentrate but to release the particles, the release process after dewatering, allowed the concentration of the metal content for allowing the fine that comprise the colloids be removed Dewatering. After the ultrasonic treatment, a concentration effect due to the dissociation of fine particles are composed of predominantly nonmetallic levels, enabled a consequent concentration.

In the waste could be observed the effect of ultrasonic treatment. Once the particles were almost entirely released with its promoted individualization, the result of the waste of rougher stage stands out by having lower levels of metal content, as well as higher levels of non-metallic.

Acknowledgements My sincere thanks to FAPEMIG and CNPq, without which I would not have completed this work, DTECH/CAP/UFSJ, Gorceix Foundation, IFMG-OP, and to my colleagues, partners in this work.

References

1. Chaves, A. P., & Leal, L. S. (2006). *Tratamento de minérios* (3rd ed., pp. 411–455). Rio de Janeiro: Luz, A.B., CETEM/ MCT, cap 10.
2. Souza, E. L. (2006). *Estudo da Recuperação do Conteúdo Metálico Presente em Resíduos Mínero-Metalúrgicos Através do Uso do Ultra-som*. Ph.D. Thesis, REDEMAT/UFOP.
3. Guimarães, G. C., Lima, R. M. F., & de Oliveira, M. L. M. (2014). *Flotação de diferentes tipologias de minério fosfático de Tapira/MG, usando o coletor óleo de soja hidrogenado* (Vol. 57, no. 3). MINERAÇÃO: Revista Escola de Minas; REM: Ouro Preto. ISSN 0370-4467. July/Sept. 2004. Disponível em: http://www.scielo.br/scielo.php?script=sci_arttext&pid=S0370-44672004000300011. Acessado em 03 de mai. de 2014.

4. Yovanovic, A. P. (1973). Engenharia da Concentração de Massa por Flotação. MODELO OPERACIONAL - Novo Fundamento Teórico dos Processos Minerais. U. del Norte, Chile.
5. Totou, A. R. (2011). *Efeito da Dispersão em Polpas de Minérios Itabiríticos*. REM: R. Esc. Minas, Ouro Preto, *64*(2), 227–232, abr. jun./ 2011.
6. Santos, E. P. (2010). *Alternativas para o Tratamento de Ultrafinos de Minério de Ferro da Mina do Pico/MG por Flotação em Coluna* (Master Dissertation, UFRGS – Escola de Engenharia. Postgraduate Program of Mining, Metallurgical and Materials Engineering – PPGE-3M).
7. Souza, E. L., & Pedrosa, C. C. (2012). *Recuperação do Rejeito da Flotação Reversa de Minério de Ferro com a Utilização de Bombardeamento de Ultrassom* (Work of Course Conclusion, FATEC/UNIPAC – Lafaiete).
8. Magriotis, R. S. (1995). *Efeito do Tipo de Amina na Flotação Reversa de um Minério Itabirítico* (Master Dissertation in Metallurgical and Mine Engineering, Escola de Engenharia, UFMG).
9. Nascimento, D. R. (2010). *Flotação Aniônica de Minério de Ferro* (Master Dissertation, PPGEM. UFOP).
10. Gontijo, C. D. F. (2010). *Aumento da seletividade da flotação reversa de minério de ferro com a utilização de polieletrólitos*. (11° Simpósio Brasileiro de Minério de Ferro – ABM, Belo Horizonte – MG).

Production of Strontianite from Celestite Ore in Carbonate Media

İbrahim Göksel Hizli, Ayşegül Bilen, Raşit Sezer, Selim Ertürk and Cüneyt Arslan

Abstract Celestite ore ($SrSO_4$) is used as raw material for the production of strontium compounds. The aim of this study is to develop a new and efficient strontianite ($SrCO_3$) production method. Strontianite was obtained from domestic celestite concentrate with direct conversion method in which celestite reacts with CO_3^{2-} and NH_4^+ containing solution. Thus, the effect of particle size, time, ammonia concentration on the conversion was investigated. $(NH_4)_2CO_3$ solution was used as carbonate agent. Because the solution is decomposed at 58 °C, experiments were performed low temperatures. Celestite ore was converted to strontianite between 3 and 6 h depending on ammonia concentration and celestite mass. Characterization of celestite ore, products and reaction residues were carried out by DSC, TG, XRD, and AAS analytical techniques.

Keywords Direct conversion · Celestite · Strontianite

Introduction

Celestite is the most common mineral of strontium and its compounds. Metallic strontium and strontium containing compounds used frequently in industry; specialty glasses, pyrotechnics, modifiers in aluminum industry, etc. The most popular strontium compound is strontianite. It can be obtained by two methods, Black Ash and Direct Conversion. In the Black Ash method; celestite ore is reduced with coal

İ.G. Hizli (✉)
İstanbul University, Istanbul, Turkey
e-mail: gokselhizli@gmail.com

İ.G. Hizli · A. Bilen · R. Sezer · S. Ertürk · C. Arslan
İstanbul Technical University, Maslak, Turkey

R. Sezer
Karadeniz Technical University, Trabzon, Turkey

© The Minerals, Metals & Materials Society 2017

M.A. Meyers et al. (eds.), *Proceedings of the 3rd Pan American Materials Congress*, The Minerals, Metals & Materials Series, DOI 10.1007/978-3-319-52132-9_61

at high temperatures (1273–1473 K). The reduction product is strontium sulfide, which is soluble in water. This solution reacts with carbonating agents to produce strontium carbonate, *strontianite*. Na_2CO_3, $(NH_4)_2CO_3$, and CO_2 are suitable agents to precipitate strontianite [1–4]. In Direct Conversion method; celestite ore reacts with carbonate ion-containing solutions [1, 3, 5–10]. On the other hand, the Black Ash method needs high heating energy, and undesirable pollutants are released. In the Direct Conversion method, high energy is not needed, as temperatures of just 353 K are sufficient. Moreover, the Direct Conversion method requires shorter time periods (3–6 h), and no undesirable pollutants are released.

Iwai and Toguri [6] studied strontianite production with direct conversion from celestite in Na_2CO_3 solution. They found that the conversion was dominated by the surface reaction. It explains the diffusion of carbonate ions through the pores of strontianite layer. Castillejos and Uribe [7] also studied celestite conversion in Na_2CO_3 solution. They found out the conversion rate increased with increasing temperature, Na_2CO_3 concentration, and decreasing particle size. Kobe and Deiglmeier [1] studied the conversion of celestite to strontianite with 98% efficiency in carbonate ion-containing solution at 368 K for 2 h. De Buda [8] obtained strontianite without impurities, by using a two-step purification process with an HCl solution to extract impurities. The Direct Conversion method has been studied extensively. The most effective results were obtained when ammonia was utilized to assure a full conversion [11].

Previous studies showed that strontianite can be obtained from celestite very efficiently, only when the right reactants are used.

Ammonium carbonate ($(NH_4)_2CO_3$) is commercially available as a mixture of NH_4HCO_3 and NH_4COONH_2. Dissolution of NH_4HCO_3 and hydrolysis of NH_4COONH_2 take place as shown in reactions 1 and 2, respectively. Hydrogen ions, produced by the dissociation of ammonium ions (reaction 3) are used up by the carbonate and bicarbonate ions (reactions 4 and 5).

$$NH_4HCO_3 \rightarrow NH_4{}^+ + HCO^{3-} \tag{1}$$

$$NH_4COONH_2 + H_2O \rightarrow NH_4{}^+ + CO_3{}^{2-} \tag{2}$$

$$NH_4{}^+ \rightarrow NH_3 + H^+ \tag{3}$$

$$HCO_3{}^- + H^+ \rightarrow H_2CO_3 \tag{4}$$

$$CO_3{}^{2-} + H^+ \rightarrow HCO_3{}^- \tag{5}$$

In this study, the effects of celestite mass, and volume of ammonia on conversion efficiency were investigated. First of all, prepared a solution which is containing $NH_4{}^+$, $HCO_3{}^-$, dissolved NH_3 and $CO_3{}^{2-}$ ions.

Previous studies mainly focused on lab-scale investigations of converting celestite to strontium carbonate, in small amounts. This study, however, aims at determining the possibility of transforming celestite to strontium carbonate in large scale. The results of this study, therefore, should bring some light on forming an alternative method to Black Ash method, which requires high temperatures, longer time periods, and multiple process steps.

Materials and Method

Celestite concentrate utilized in this study was kindly provided by Barit Maden Türk Ltd. Quantitative chemical analysis of this concentrate is given in Table 1. Celestite concentrate was crushed, ground and sieved before being used.

Conversion experiments were performed in a water-jacketed borosilicate glass reactor. The reactor's four necks were used for inserting propeller, cooler, thermometer and sampler (Fig. 1).

Table 1 Chemical composition of celestite concentrate fractions (wt%)

Compound	Content (wt%)
$SrSO_4$	95.5
$CaSO_4$	3
$BaSO_4$	0.5
SiO_2	0.5
Fe_2O_3	0.5

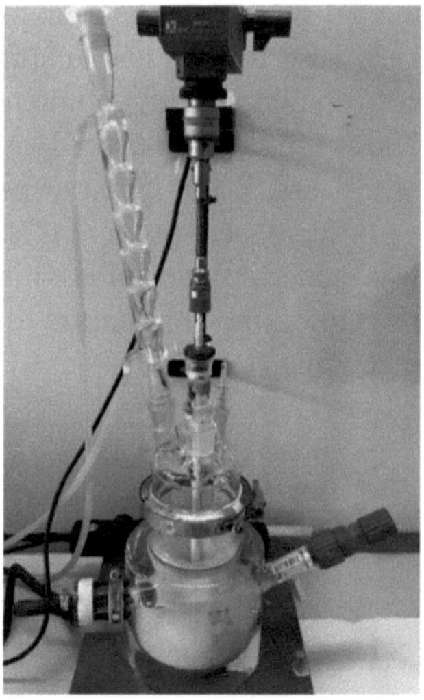

Fig. 1 Experimental setup

Previous studies showed that the conversion ratio increased with decreasing particle size. Thus, −54 μm particles in 50, 100, and 200 g batches were used in experiments. Furthermore; the conversion ratio increased with increasing temperature. Since ammonium carbonate starts decaying at 331 K, experiments were performed at 323 K to stay at a safe temperature range.

In addition, stoichiometric amounts of ammonium carbonate were used, as indicated by the reaction of celestite with ammonium carbonate solution;

$$SrSO_4 + CO_3^{2-} \rightarrow SrCO_3 + SO_4^{2-} \quad (6)$$

Ammonium carbonate dissolved in distilled water and different volumes of ammonia (2, 5, and 10%) were added. Samples were taken from the mixture after 3, 4, 5, and 6 h, without disturbing the reaction. Specimens were analyzed by XRD.

Results and Discussion

Three to six-hour long reactions of celestite concentrate (all −54 μm in size and with weights of 50, 100, and 200 g) were performed with the addition of 2, 5, and 10 wt% ammonia. Stoichiometry and reaction time are important for full conversion of celestite to strontianite [11].

XRD patterns of the products are shown in Figs. 2 and 3. XRD peaks must be investigated to understand the exact conditions at which the reactions take place, showing the conversions with regard to the amount of ammonia and time. It can be concluded from the results that for a complete conversion, the amount of ammonia and time must be prolonged as the amount of celestite concentrate increases.

While 5% ammonia and a three-hour period is adequate for full conversion when the 50 g batch was used, they are not sufficient for 100 g sample. Six hours is necessary for full conversion when the amount of ammonia is kept the same. The conversion was not complete since the amount of ammonia was not enough when a 200 g sample was used. However, utilizing 10% ammonia solved this problem and at the end of the six-hour period, conversion was complete.

It has been proven that conversion was not quite possible if no ammonia was added. When ammonia was added however, according to the Lé Chatelier principle, the carbonating effect of ammonium carbonate continues throughout the reaction.

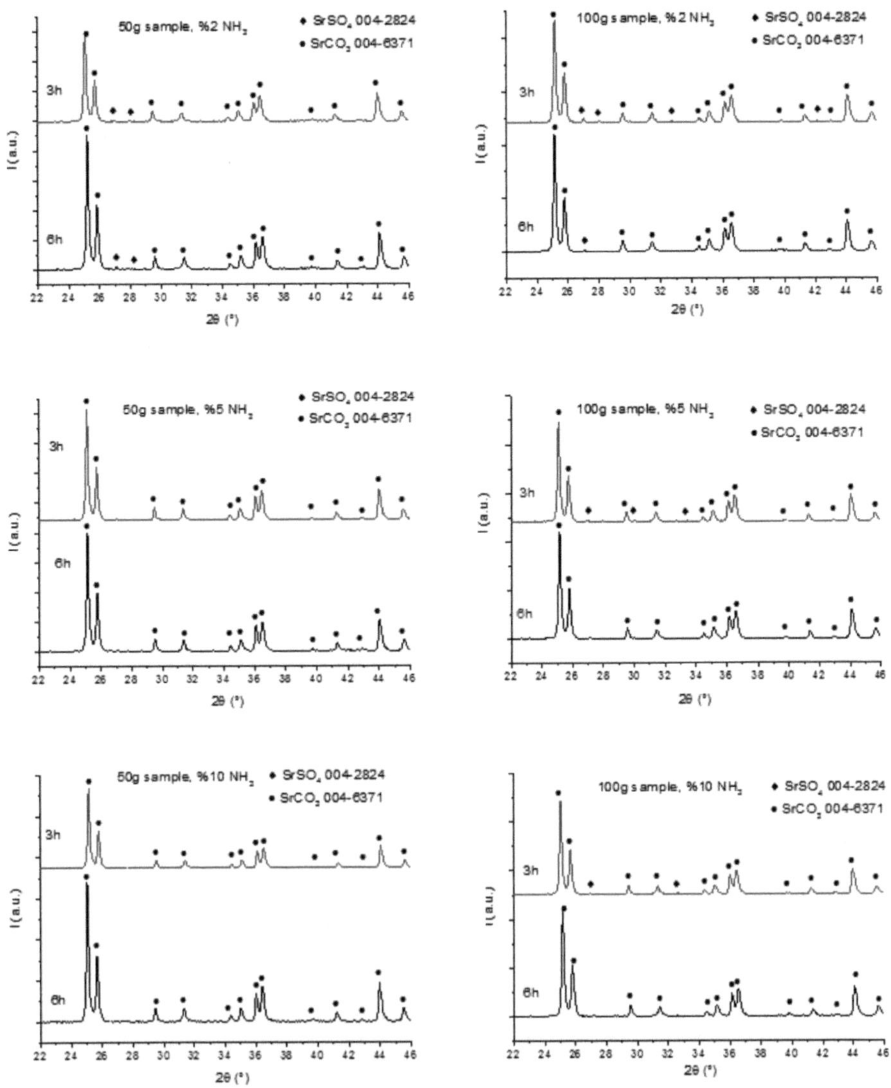

Fig. 2 XRD patterns of three- and six-hour reaction products of 50 and 100 g celestite samples with 2, 5, and 10% NH_3 additions

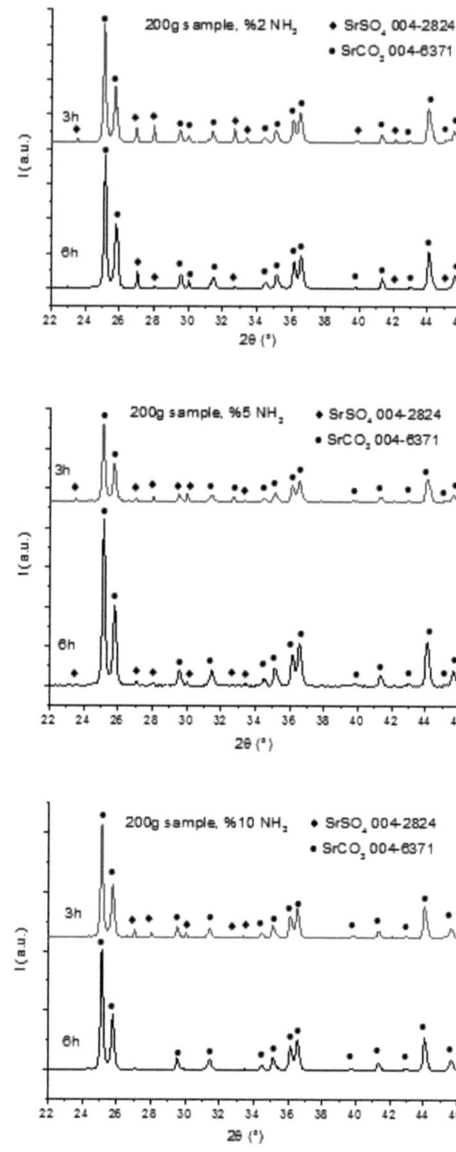

Fig. 3 XRD patterns of three- and six-hour reaction products of 200 g celestite samples with 2, 5, and 10% NH_3 additions

Conclusion

Effects of the amount of initial material, time, and the amount of ammonia were investigated on the precipitation of celestite as strontium carbonate from carbonating agent containing solutions.

It was demonstrated that ammonium carbonate alone was not effective in the production of strontium carbonate from celestite ore, and thus ammonia in stoichiometric amounts was needed.

It can be assumed that time is as effective as the ammonia amount in the conversion process, which suggests that the reaction is diffusion controlled.

It is now safe to accept a highly efficient method exists to produce strontium carbonate on an industrial scale.

Acknowledgements The authors greatly acknowledge the financial support provided by the Scientific and Research Council of Turkey (TUBITAK) under Contract 115M631 and would like to thank Barit Maden Türk Ltd. for technical and materials support.

References

1. Kobe, A. K., & Deiglmeier, N. J. (1943). Conversion from strontium sulfate by metathesis with alkali carbonate solution. *Industrial and Engineering Chemistry, 35*, 323–325.
2. Erdemoğlu, M., & Canbazoğlu, M. (1998). The leaching of SrS with water and the precipitation of $SrCO_3$ from leach solution by different carbonating agents. *Hydrometallurgy, 49*(1), 135–150.
3. Owusu, G., & Litz, J. E. (2000). Water leaching of SrS and precipitation of $SrCO_3$ using carbon dioxide as the precipitating agent. *Hydrometallurgy, 57*(1), 23–29.
4. Sonawane, R., et al. (2000). Effect of a catalyst on the kinetics of reduction of celestite ($SrSO_4$) by active charcoal. *Metallurgical and Materials Transactions B, 31*(1), 35–41.
5. Chiang, J. S., & Goldstein, D. (1983). *Preparation of strontium carbonate*. Google Patents.
6. Iwai, M., & Toguri, J. (1989). The leaching of celestite in sodium carbonate solution. *Hydrometallurgy, 22*(1–2), 87–100.
7. Castillejos, A., & Uribe, A. (1996). The direct conversion of celestite to strontium carbonate in sodium carbonate aqueous media. *Hydrometallurgy, 40*(1), 207–222.
8. De Buda, F. (1987). *Method for recovery and conversion of strontium sulfate to strontium carbonate from low and medium grade celestite ores*. Google Patents.
9. Bingol, D., Aydogan, S., & Gultekin, S. S. (2010). Neural model for the leaching of celestite in sodium carbonate solution. *Chemical Engineering Journal, 165*(2), 617–624.
10. Zoraga, M., & Kahruman, C. (2014). Kinetics of conversion of celestite to strontium carbonate in solutions containing carbonate, bicarbonate and ammonium ions, and dissolved ammonia. *Journal of Serbian Chemical Society, 79*, 345–359.
11. Zoraga, M., Kahruman, C., & Yusufoglu, I. (2016). Conversion kinetics of $SrSO_4$ to $SrCO_3$ in solutions obtained by dissolving/hydrolyzing of equimolar amounts of NH_4HCO_3 and NH_4COONH_2. *Hydrometallurgy, 163*, 120–129.

Reduction Kinetics and Characterization Study of Synthetic Magnetite Micro Fines

Saikat Kumar Kuila, Ritayan Chatterjee and Dinabandhu Ghosh

Abstract The present work deals with the characterization of pure magnetite microfines (<5 μm) and its hydrogen reduction. The structural and morphological properties of magnetite powder were analyzed by X-ray diffraction (XRD), X-ray photoelectron spectroscopy (XPS), Fourier transform infrared spectroscopy (FTIR), nitrogen adsorption analysis by five points isotherm BET method, and scanning electron microscopy along with energy dispersive X-ray spectroscopy (SEM-EDS). The hydrogen reduction of the magnetite powder was carried out in a thermogravimetric analyzer (TGA) under a steady flow of hydrogen or hydrogen-argon mixture (to produce different partial pressures of hydrogen). The variables studied were reduction temperature (973–1273 K), hydrogen partial pressure (0.25–1 atm) and sample bed height (0.184–0.68 cm). The apparent activation energy was obtained as 22 kJ mol^{-1}. The rate equations developed for the reaction system under study were applied to determine the rate controlling step. The reduction was found to be rate controlled by diffusion through the stagnant gas film enclosed above the sample inside the crucible. The true activation energy was calculated to be 9 kJ mol^{-1}.

Keywords Magnetite fines · Characterization · Hydrogen reduction · Activation energy

---

S.K. Kuila · R. Chatterjee (✉) · D. Ghosh
Department of Metallurgical and Materials Engineering,
Jadavpur University, Kolkata 700032, India
e-mail: ritayanchatterjee@gmail.com; ritayanc@research.jdvu.ac.in;
ritayan@hithaldia.in

S.K. Kuila
Department of Metallurgical and Materials Engineering,
Indian Institute of Technology, Kharagpur 721302, West Bengal, India

R. Chatterjee
Department of Applied Sciences,
Haldia Institute of Technology, Haldia 721657, West Bengal, India

© The Minerals, Metals & Materials Society 2017
M.A. Meyers et al. (eds.), *Proceedings of the 3rd Pan American Materials Congress*,
The Minerals, Metals & Materials Series, DOI 10.1007/978-3-319-52132-9_62

Introduction

The blast furnace has dominated the iron extraction industry over the years even though it, in recent times, confronts a challenge on the issue of environmental pollution caused by its exit gas (BF gas) enriched in carbon dioxide. While a greater CO_2/CO ratio in the exit gas is coke saving, it is conducive to greenhouse effect. In this context, finding alternative ways to treat iron ores is a major aim of research in process metallurgy. The direct reduction of iron oxides has drawn much interest as an alternative iron-making route. Iron oxides have been reduced by pure hydrogen, or by gas mixtures like hydrogen–nitrogen, hydrogen–carbon monoxide in search of a less polluting extraction process [1–8]. Jeong et al. [1] investigated the reductions of goethite, hematite and magnetite in H_2–H_2O gas mixtures. In the case of magnetite (Fe_3O_4) reduction, they reported an activation energy of 11 kJ mol^{-1}. The small activation energy, large reduction rate and the maintenance of particle size even under high temperature were observed. Ramadan et al. [2] studied isothermal and non-isothermal reduction of nano-sized (25–30 nm) synthetic magnetite in H_2–N_2 atmosphere. They reported a two-step reduction of Fe_3O_4, $Fe_3O_4 \rightarrow FeO \rightarrow Fe$, above 723 K, and a single-step reduction of FeO(OH) ($FeO(OH) \rightarrow Fe_3O_4$) below 723 K. They explained the two-step reduction on the basis of two consecutive pathways in which the composition of the intermediate product was found to be controlled by the composition of reducing gas atmosphere and the characteristics of the starting material. Kuila et al. [3], who reduced magnetite concentrate by hydrogen above 973 K, also reported a two-step reduction process. They reported activation energies of 42 and 55 kJ mol^{-1} for the two successive steps. Pineau et al. [4] investigated hydrogen reduction of Fe_3O_4 in a wide temperature range of 483–1223 K. They found three different (apparent) activation energies at three temperature segments: 200 kJ mol^{-1} below 523 K, 71 kJ mol^{-1} between 523 and 663 K, and 44 kJ mol^{-1} above 663 K. The change in activation energy at 523 K was explained by the removal of hydroxyl group and point defects present in the magnetite. A mathematical modeling of the experimental data suggested a diffusion-controlling kinetics for the reduction. Liu et al. [6] studied the solid-state reduction kinetics of pre-oxidized vanadium-titanium magnetite concentrate using coal as the reductant at 1223–1373 K. They calculated the apparent activation energy as 67 kJ mol^{-1}, and claimed that the reduction, which went up to 94.08%, followed the shrinking core model. Wang and Sohn [9] studied the high temperature reduction kinetics of magnetite concentrates in the temperature range 1423–1673 K. The nucleation and growth kinetics was found to be appropriate to describe the reduction rate of fine concentrate particles.

The main objective of the present work was to study the reduction kinetics of synthetic magnetite by hydrogen. The variables studied were reduction temperature (973–1273 K), partial pressure of hydrogen (0.25–1 atm) and sample bed height (0.184–0.68 cm). More specifically, it was intended to find the rate-controlling mechanism consistent with the reaction geometry used in the thermogravimetric set-up (TGA) in the study. In this arrangement, the sample is taken in a cylindrical crucible and the reducing gas flows over it at a sufficiently high rate, maintaining the inlet gas composition at the top of the crucible and leaving an entrapped gas film over the sample bed inside the crucible. Another purpose of the present work was to characterize the starting oxide because the properties of the specific reactant solid is likely to have some impact on the reduction mechanism [2]. The structural and morphological properties of magnetite powder were analyzed by X-ray diffraction (XRD), X-ray photoelectron spectroscopy (XPS), Fourier transform infrared spectroscopy (FTIR), nitrogen adsorption analysis by five points isotherm BET method, and scanning electron microscopy along with energy dispersive X-ray spectroscopy (SEM-EDS).

Experimental

Magnetite (Fe_3O_4) powder of particle size less than 5 μm and with 95% purity, supplied by Sigma-Aldrich (China), was used as the starting material for the reduction study. A brief characterization of the starting material was carried out as follows. An X-ray diffraction (XRD) study was done by a Rigaku Ultima X-ray diffractometer working in Bragg-Brentano geometry using Cu-Kα radiation. A continuous scan mode with 1° per minute scan rate between 10° and 80° was applied with the sample width of 0.02 mm. A Fourier-transform infrared spectroscopic (FTIR) study in the range of 1400–400 cm^{-1} was obtained adopting the KBr pellet technique, using an IR Prestige 21, 200V CE, Shimadzu FTIR spectroscope. Approximately 2 mg of the powder was thoroughly mixed with 200 mg of spectroscopic-grade KBr and pressed into pellets to be subjected for recording the spectrum. The chemical composition of the starting material was verified by X-ray photo electron spectroscopic (XPS) study using a PHI 5000 Versa Probe II, ULVAC–PHI, INC spectroscope, with an incident Al-Kα radiation of energy 1486.6 eV. The morphology was studied by scanning electron microscopy (SEM) using a Zeiss EVO 60 microscope and the elemental analysis of the scanned area was obtained by energy dispersive X-ray spectroscopy (EDS). The surface area was made conducting by a Polaron SC7620 sputter coater with the gold-palladium target. The specific surface area of the sample was measured with the five point BET nitrogen isotherms obtained by varying relative pressure using a Nova 1000E surface area analyzer. The relative pressure (p/p_0) was maintained between 0.053 and 0.3. The density value was measured by a Quanta Chrome Ultrapye 1200 E pycnometer.

The magnetite powder was reduced by 99.9% pure hydrogen gas (supplied by Indian Refrigeration Stores, Kolkata). Argon gas with the same purity and from the same supplier was used for purging purpose, before and after the reduction. In addition, argon was mixed with hydrogen for studying the effect of partial pressure. The reduction was carried out in a thermogravimetric analyzer (TGA) (model: Okay, make: Bysakh, Kolkata). The analyzer was made up by assembling a furnace with a Pt–Rh heating element, an impervious reaction tube (internal diameter,

2.5 cm) made of alumina, a balance with an accuracy of 0.1 mg and a cylindrical alumina crucible, with 0.88 cm height and 0.72 cm internal diameter, for holding the sample. The temperature was controlled within ±2 K by a Pt–13%Rh/Pt (R type) thermocouple. About 180 mg of the magnetite powder was taken inside the alumina crucible, which formed a shallow bed inside the crucible. The crucible was placed in the TGA at room temperature and its initial mass was recorded by the in-built balance. Next, the temperature of the furnace was raised at a rate of 10 K per minute to the selected reduction temperature in an argon atmosphere. When the temperature was attained, argon was stopped and hydrogen gas was started passing at the selected flow rate, recorded by a flowmeter (rotameter). Gases were made free from oxygen, carbon dioxide, and moisture by passing through an alkaline pyrogallol solution and through the columns of silica gel and molecular sieve, respectively. The weight of the sample was recorded as a function of time at regular intervals (typically, 1 min) by an in-built software attached to the apparatus. The fractional removal of oxygen (*X*) at each instant *t* was calculated from the sample weight at the time, using the relation

$$X = \frac{W_0 - W(t)}{W_0 \times 0.2764} \tag{1}$$

where W_0 is the initial sample weight and $W(t)$ is the sample weight at time *t*. The maximum possible weight loss, which corresponds to the fractional removal of 1, is obtained when all oxygen (O) in Fe_3O_4, which is 0.2764 g per 1 g oxide, is removed.

Results and Discussion

A. *Characterization of Magnetic Powder*

The X-ray diffraction of the starting material (magnetite powder) is presented in Fig. 1. The most intense peak was observed at $2\theta = 35.38°$. The other peak positions were obtained at 18.29°, 30.24°, 37.04°, 43.02°, 53.57°, 57.04°, 62.59°, 71.21°, 74.05° and 79.06°. These values were compared with the JCPDS standard values (JCPDS 89-0691) which confirmed that the cubic Fe_3O_4 is the only crystalline phase present in the starting material [10]. The corresponding planes (*hkl*) are shown on the peaks in the figure. The Fourier-transform infrared spectrum in percentage transmission for magnetite is shown in Fig. 2. An absorption due to O–H (νOH) is observed at the wave number 3430 cm^{-1} [11]. Two absorptions at 2920 and 2850 cm^{-1} unveiled the presence of hydrocarbon species (νCH) [12]. A close-up of the Fe–O vibration region with wave numbers ranging from 400 to 650 cm^{-1} is shown in the inset. Absorptions at 583 and 410 cm^{-1} are the signature peaks of Fe–O bond vibration in Fe_3O_4 lattice [11]. Peaks obtained at 628 and 446 cm^{-1} indicated the presence of oxyhydroxy species [11, 13].

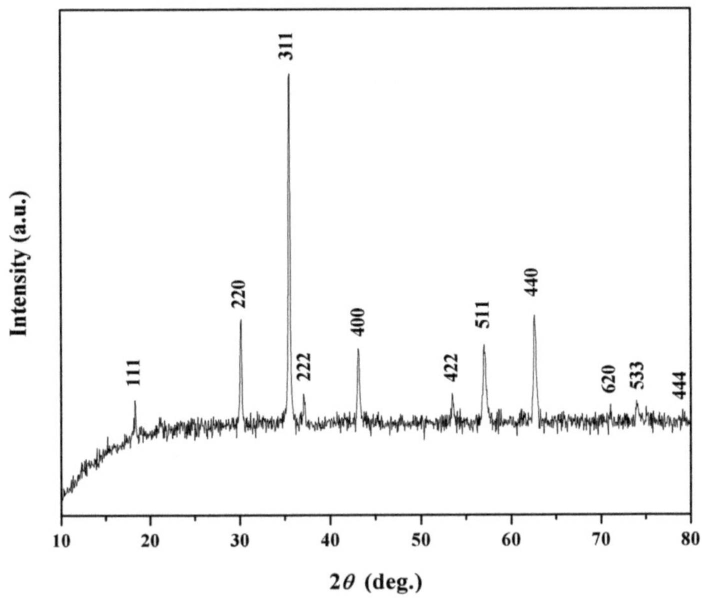

Fig. 1 XRD pattern of the starting material powder (synthetic magnetite)

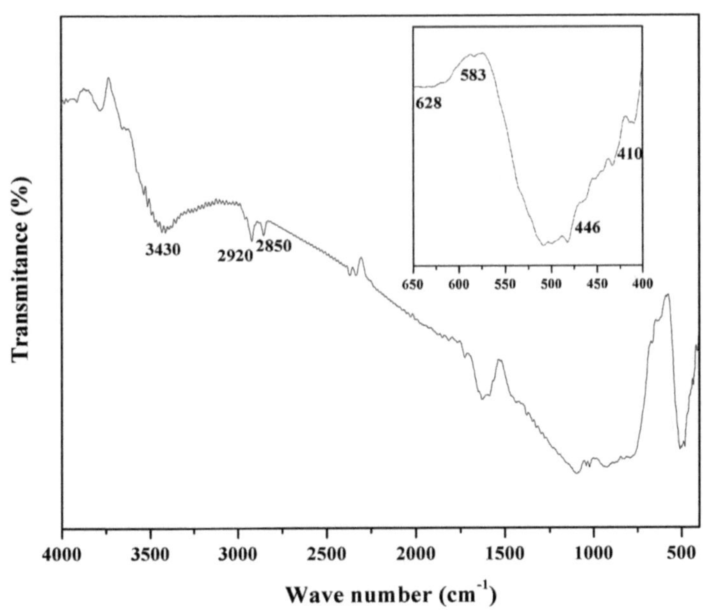

Fig. 2 FTIR spectra of synthetic magnetite; the spectra in the range 400–650 cm^{-1} is shown enlarged in the *inset*

The chemical compositions and bonding of Fe_3O_4 was studied by X-ray photo electron spectroscopy. The spectra are shown in Fig. 3. It shows two peaks at about 710 and 723.8 eV, which occur due to Fe 2p3/2 and Fe 2p1/2 respectively. There is

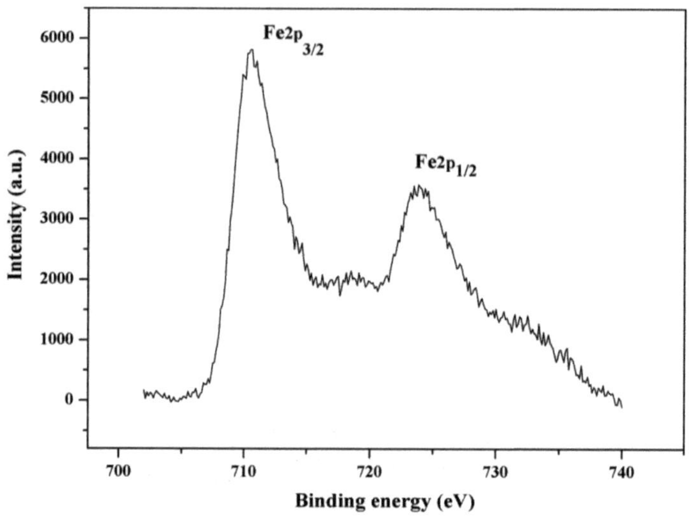

(a)

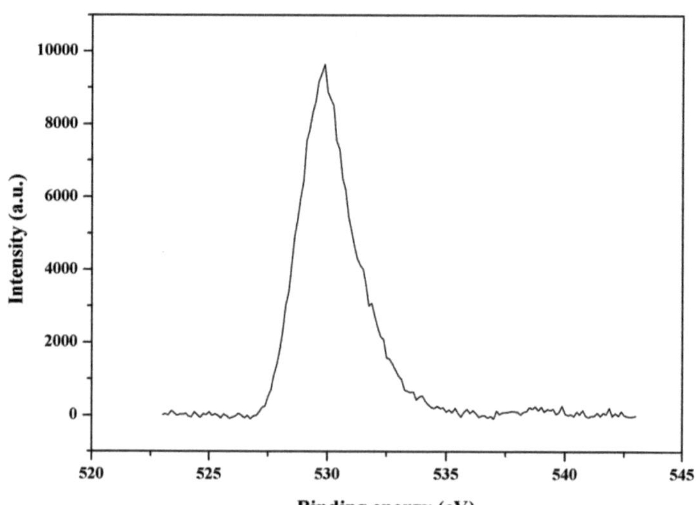

(b)

Fig. 3 High resolution XPS spectra of synthetic magnetite showing the binding energy of **a** Fe 2p3/2 and Fe 2p1/2, and **b** O 1s no other oxidation state except Fe^{3+}, confirming that Fe_3O_4 is the only iron oxide present in the starting material. Single peak at 529.8 eV is due to O 1s.

The density of magnetite powder was in the range of 4.8–5.1 g cm^{-3} as provided by the supplier's specifications. For the purpose of cross-checking, the density was measured by pycnometry and the average of five measurements yielded 4.786 g cm^{-1}, with ±2% error. The pore size distribution (*P*) and the surface area (*S*) were measured by the nitrogen adsorption technique using the Langmuir adsorption model and considering homogeneous surface and monolayer coverage. The following two relations were used:

$$\frac{P}{X_A(p_0 - p_A)} = \frac{1}{X_M C}\left[1 + \frac{p_A}{p_0}(C - 1)\right] \quad (2)$$

$$S = \frac{X_M N_A A_0}{W_P M_A} \quad (3)$$

where, in Eq. (2), p_A is the partial pressure of adsorbate (N_2); p_0, the saturation pressure of adsorbate; X_A, the amount of gas adsorbate under pressure; X_M, the monolayer capacity of the powder; and C, a constant related to adsorption enthalpy. In Eq. (3), N_A is Avogadro's number; M_A, the molar weight of adsorbate; A_0, the adsorption cross section of the adsorbing species; and W_P, the weight of the powder sample taken. The porosity and the (specific) surface area thus calculated were 6.1% and 10.453 m^2 g^{-1}, respectively. The high surface area value indicates a good surface reactivity. The SEM image of the magnetite powder and the EDS spectra are shown in Fig. 4a, b respectively. The elemental compositions are given in Table 1. The SEM image shows that the particle size is nearly uniform and the elemental weight percentages are in rough agreement with the weight percentages calculated from stoichiometry.

B. *Hydrogen Reduction of Magnetite*

The reduction of Fe_3O_4 with H_2 takes place in two successive stages, the standard free energy changes [14] of which are as follows:

$$Fe_3O_4\,(s) + H_2\,(g) = 3FeO\,(s) + H_2O\,(g);\ \Delta G^\circ = 70160 - 92.25T\ \text{J mol}^{-1} \quad (4)$$

$$FeO\,(s) + H_2\,(g) = Fe\,(s) + H_2O\,(g);\ \Delta G^\circ = 16240 - 8.45T\ \text{J mol}^{-1} \quad (5)$$

The overall reaction from the above two stages is

$$Fe_3O_4\,(s) + 4H_2\,(g) = 3Fe\,(s) + 4H_2O\,(g) \quad (6)$$

The reaction assembly used in this reduction study, as described in the experimental section, was comprised of a cylindrical alumina crucible containing the powder magnetite sample placed in the vertical furnace of the TGA set-up. The reactant gas (H_2 or H_2–Ar mixture) entered the furnace through the bottom of

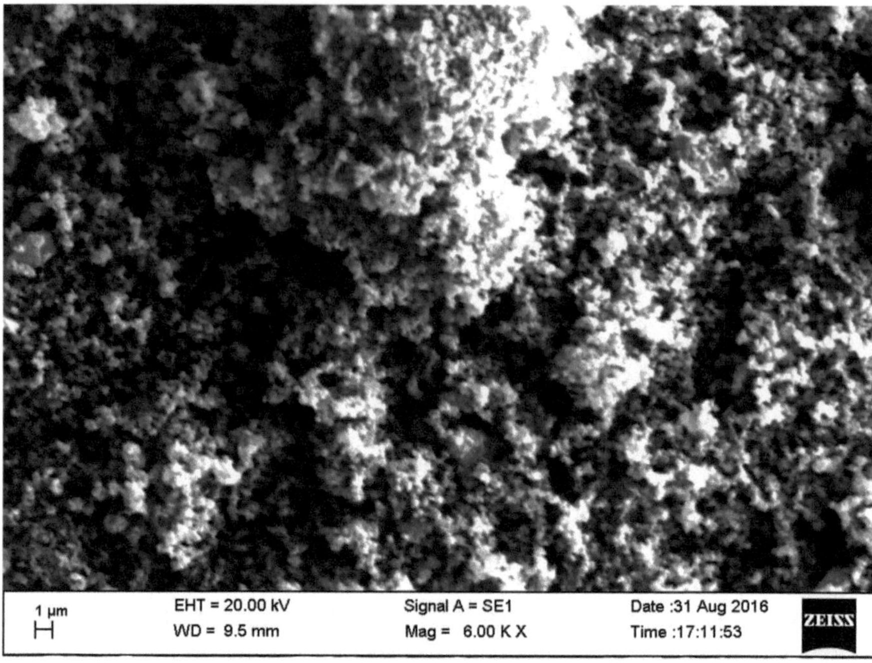

(a)

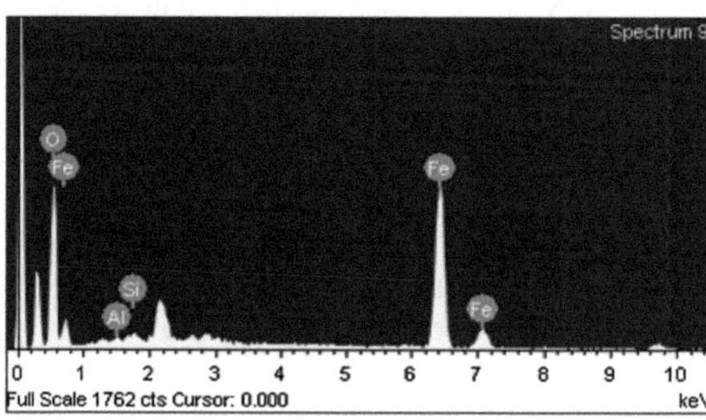

(b)

Fig. 4 **a** SEM micrographs (magnification 6000) and **b** corresponding EDS spectra, of synthetic magnetite

Table 1 Elemental analysis of the synthetic magnetite by EDS

Weight percentage			
Fe	O	Al	Si
66.56	32.22	0.98	0.25

the furnace tube and flew past the top of the crucible leaving a stagnant gas film above the solid within the crucible. A schematic diagram of the assembly can be found elsewhere [15]. Three kinetic steps are typically involved in this type of gas-solid reduction. Due to the chemical reaction, there will be a concentration gradient developed across the gas film and the counter diffusion of H_2 and H_2O through the gas film will constitute the first kinetic step, namely, the gas film diffusion (mass transfer) step. The rate equation for the gas film diffusion controlled kinetics is given by

$$X = \frac{C.D_{H_2-H_2O}(x_{H_{2in}} - x_{H_{2eq}})}{(z_2 - z_1^0)z_1^0.\rho_m}.t = \frac{D_{H_2-H_2O}(C_{H_{2in}} - C_{H_{2eq}})}{(z_2 - z_1^0)z_1^0.\rho_m}.t \tag{7}$$

where X is the fractional removal of oxygen from Fe_3O_4 at time t; $D_{H_2-H_2O}$ is the inter-diffusion coefficient of H_2 and H_2O; C is the total concentration;$C_{i_{in}}$ and $C_{i_{eq}}$ are the concentrations of species i at inlet and at equilibrium, respectively; $x_{i_{in}}$ and $x_{i_{eq}}$ are the mole fractions of species i at inlet and at equilibrium, respectively; z_2 is the crucible height; z_1^0 is the initial bed height; and ρ_m is the molar density of Fe_3O_4. At any given time, there will be a formation of reduced iron layer of certain thickness covering the unreacted oxide. Diffusion through this porous product (or the pores initially present in the sample, or both) will constitute the second kinetic step, namely, the pore diffusion step. The rate equation for the pore diffusion controlled kinetics is given by

$$X^2 = \frac{2C.D_{pore}(x_{H_{2in}} - x_{H_{2eq}})}{z_1^{02}\rho_m}.t = \frac{2D_{pore}(C_{H_{2in}} - C_{H_{2eq}})}{z_1^{02}\rho_m}.t \tag{8}$$

where D_{pore} is the pore diffusivity. The third kinetic step is the chemical reaction, typically given by Eq. (6), taking place at the interface of produced iron and unreacted iron oxide, namely, interfacial reaction step. The rate equation for the interfacial reaction controlled kinetics is given by

$$X = \frac{C.k_f'\left[x_{H_{2in}} - \frac{x_{H_2O_{in}}}{K}\right]t}{\rho_m.z_1^0} = \frac{k_f'\left[C_{H_{2in}} - \frac{C_{H_2O_{in}}}{K}\right]t}{\rho_m.z_1^0} \tag{9}$$

where k_f' is the 1st-order (forward) rate constant; K, equilibrium constant for the reaction given in Eq. (5). The detailed derivation of Eqs. (7)–(9) can be found elsewhere [15].

The flow rate of inlet gas above 0.4 L min^{-1} was found to have no effect on the reduction of the magnetite powder. Accordingly, the gas flow rate of 0.4 L min^{-1} was used in studying the effects of other variables on the reduction. However, it is to be noted that the use of the flow rate of 0.4 L min^{-1} does not eliminate the resistance due to H_2 and H_2O counter-diffusion through the gas film enclosed between the top of the sample and the top of the crucible. The flow rate of 0.4 L min^{-1} merely maintains the inlet hydrogen concentration ($C_{H_{2_{in}}}$) at the top of crucible.

1. *Effect of temperature*

Fractional removal of oxygen from magnetite are plotted, in Fig. 5, against reduction time for four different temperatures (973, 1073, 1173 and 1273 K) with the hydrogen flow rate of 0.4 L min^{-1}. The fractional conversion (X) went up to 0.9. In Fig. 5, the conversion is clearly linear with respect to t at each temperature. The linearity, according to Eqs. (7) and (9), is to be observed at a given temperature in either of the gas film diffusion (mass transfer) and the interfacial reaction controlled kinetics, in view of the terms accompanying t being constants at a given temperature and pressure. However, these two rate controlling mechanisms can be distinguished by considering the (i) effect of the partial pressure of hydrogen, (ii) effect of the initial bed height (z_1^0), and (iii) magnitude of the activation energy. The good surface reactivity obtained by five point BET method implies a fast chemical reaction at the gas-solid interface and the chemical reaction is therefore unlikely to be rate controlling. The apparent rate of the reduction (k_a) at each temperature was calculated from the slope of the conversion plot considering two points from each straight line at $t = 0$ and at $t = 12$ min (Fig. 5), and is given in the second column of Table 2. An Arrhenius plot was drawn with the apparent rates and is shown in Fig. 6. The apparent activation energy (E_a) obtained from the plot is 22 kJ mol^{-1}.

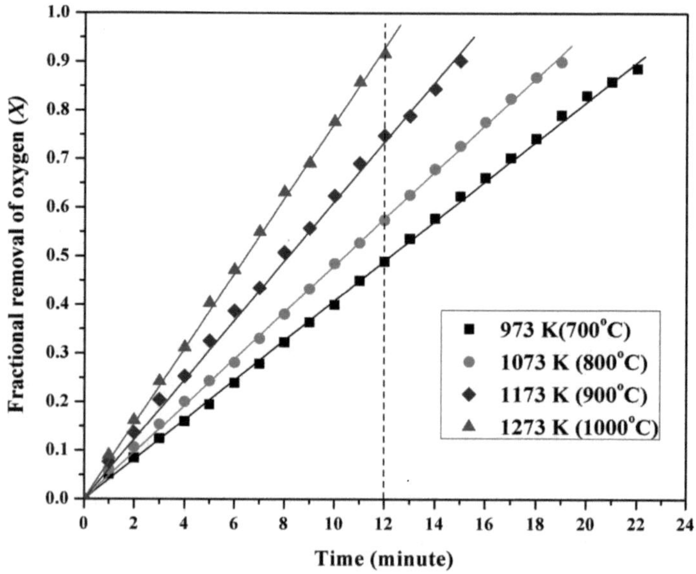

Fig. 5 The plot of fractional removal of oxygen versus time for the reduction of synthetic magnetite at different temperatures

Table 2 Apparent reaction rate and calculated binary gas (H_2–H_2O) diffusivity at different temperatures

Temperature (K)	k_a (s^{-1})	$D_{H_2-H_2O}$ ($m^2\ s^{-1}$)
973	4.083×10^{-2}	5.914×10^{-5}
1073	4.808×10^{-2}	6.747×10^{-5}
1173	6.126×10^{-2}	8.439×10^{-5}
1273	7.733×10^{-2}	1.062×10^{-4}

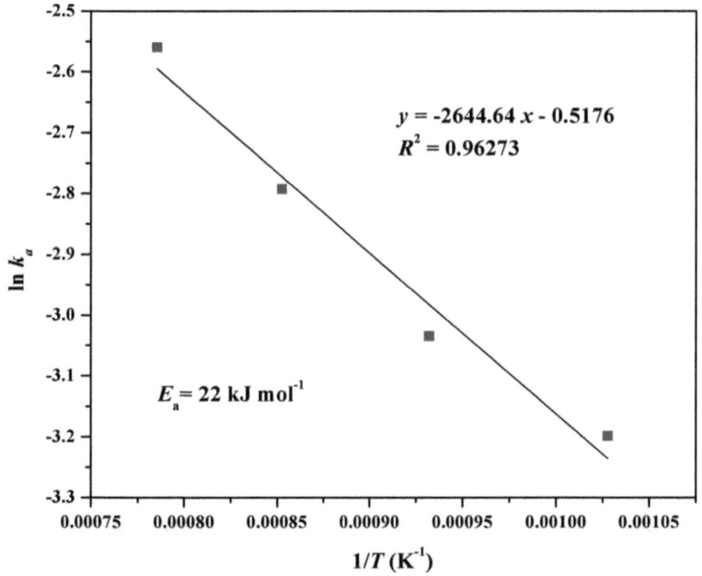

Fig. 6 Arrhenius plot based on the apparent reaction rates at different temperatures, yielding the apparent activation energy

The small value of the apparent activation energy suggests that the mass transfer step is rate controlling. Typically, the gas film diffusion is associated with a low activation energy between 1.65RT and 2RT depending on the temperature [16], producing at 1173 K, for example, a value of about 20 kJ mol^{-1}. R is the universal gas constant. To determine the true activation energy on the basis of the binary gas diffusivity ($D_{H_2-H_2O}$), the diffusivity value for each temperature is calculated by using the following relation, obtained by rearranging of Eq. (7). The values are listed in the third column of Table 2.

$$D_{H_2-H_2O} = [X.(z_2 - z_1^0)z_1^0.\rho_m/(C_{H_{2in}} - C_{H_{2eq}}).t]\ m^2\ s^{-1}$$

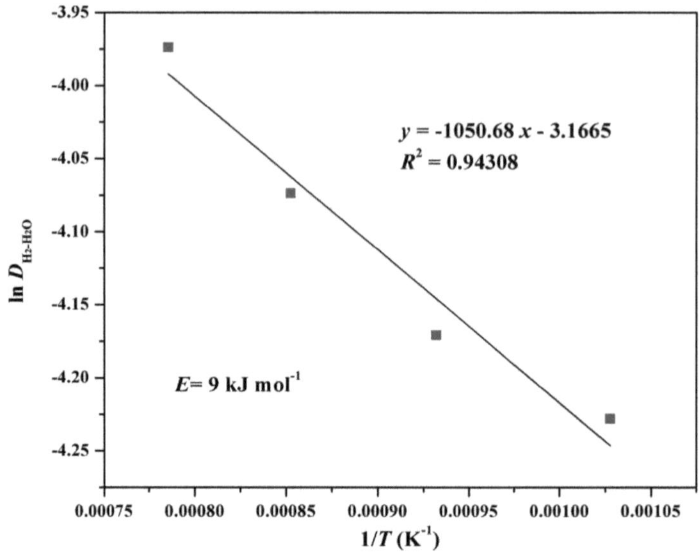

Fig. 7 Arrhenius plot based on the binary gas diffusivities at different temperatures, yielding the true activation energy

The following values for the different terms were used for the calculation: t = 12 min; ρ_m = 16412.37 mol m^{-3}; $z_2 = 8.8 \times 10^{-3}$ m; $z_1^0 = 3.7 \times 10^{-3}$ m; $C_{H_{2in}} = p_{H_{2in}}/R.T = 1/(0.082 \times 10^{-3}).T$ mol m^{-3}. The values of $C_{H_{2eq}}$ calculated using the $\Delta G°$ given in Eq. (5) and the values were 9.2588, 7.9274, 6.8687 and 6.0163 mol m^{-3} at 973, 1073, 1173, and 1273 K, respectively. The plot of ln $D_{H_2-H_2O}$ versus 1/T is shown in Fig. 7 and the true activation energy calculated from the plot is 9 kJ mol^{-1}. It may be noted that this value is in close agreement with the finding of Jeong et al. [1].

2. *Effect of hydrogen partial pressure*

Fractional removal of oxygen versus time is plotted in Fig. 8 for the runs carried out under four partial pressures of hydrogen, namely 1, 0.75, 0.5, and 0.25 atm (where 1 atm = 101.3 kPa) at a fixed temperature of 1173 K. These partial pressures were made by mixing H_2 and Ar in proper ratios keeping the total flow rate fixed at 0.4 L min^{-1}. For example, the hydrogen partial pressure of 0.5 atm was generated by mixing hydrogen flowing at a rate of 0.2 L min^{-1} with argon flowing at a rate of 0.2 L min^{-1}. It may be noted that the maximum fractional removal decreases from 0.9 to 0.35 as p_{H_2} decreases from 1 to 0.25 atm. Now, X is proportional to $\left(x_{H_{2in}} - x_{H_{2eq}}\right)$ for a given t, according to Eq. (7). This means that the X versus t plot for a smaller $x_{H_{2in}} - x_{H_{2eq}}$, and hence approximately for a smaller $x_{H_{2in}}$, should lie below that for a greater $x_{H_{2in}}$, which is observed in Fig. 8. For a more quantitative agreement, X is plotted against $\left(x_{H_{2in}} - x_{H_{2eq}}\right)$ for a given time

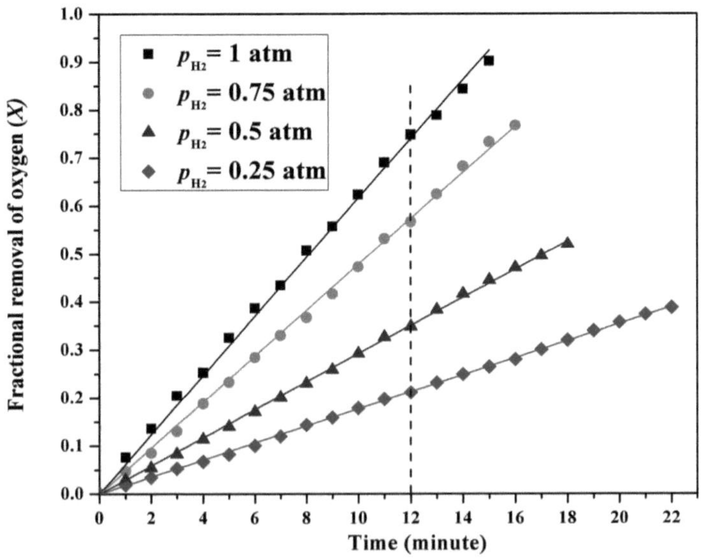

Fig. 8 The plot of fractional removal of oxygen versus time for the reduction of synthetic magnetite using different partial pressures of hydrogen

(t = 12 min) and checked if their relation is linear. This result is shown in Fig. 9 and the linearity of the plot is evident. It should be noted that the plot is required to satisfy the requirement of passing through the point $X = 0$ corresponding to $x_{H_{2in}} - x_{H_{2eq}} = 0$. This lends further support to the previous deduction that gas film diffusion is the rate controlling step. Here, the values of $x_{H_{2in}}$ are known directly from the experimental conditions. The values of $x_{H_{2eq}}$ are found by thermodynamic calculation. To illustrate, in the case of $x_{H_{2in}} = 0.25$ (corresponding to $p_{H_{2in}} = 0.25$ atm), the following three relations are set up: $p_{H_2O}/p_{H_2} = 0.51$ (from the equilibrium constant of Reaction [5]), $p_{H_2S} + p_{H_2} + p_{Ar} = 1$(since the total pressure is 1 atm); and $0.75/0.25 = (n_{Ar}/n_{H_2})_{in} = (n_{Ar}/n_{H_2})_{eq} = [n_{Ar}/(n_{H_2} + n_{H_2O})]_{eq} = [p_{Ar}/(p_{H_2} + p_{H_2O})]_{eq}$ (from the mass balance of H_2 and Ar). From these three relations, one gets, $p_{H_{2eq}} = 0.165$ atm and $x_{H_{2in}} - x_{H_{2eq}} = 0.085$.

3. *Effect of sample bed height*

Fractional removal of oxygen versus time is plotted in Fig. 10, for the runs carried out with four different initial bed heights (0.184, 0.37, 0.43, and 0.68 cm) of the synthetic magnetite powder, keeping the hydrogen flow rate fixed at 0.4 L min^{-1} and the temperature fixed at 1173 K. The bed height was varied by varying the initial mass of the sample from 0.09 to about 0.27 g. It may be noted that in the experiments in which the bed height was not a variable, the initial mass was taken as 0.18 g which corresponds to 0.37 cm bed height. The extent of reduction remarkably decreases with the decreasing bed height from 0.68 to 0.43 cm which is

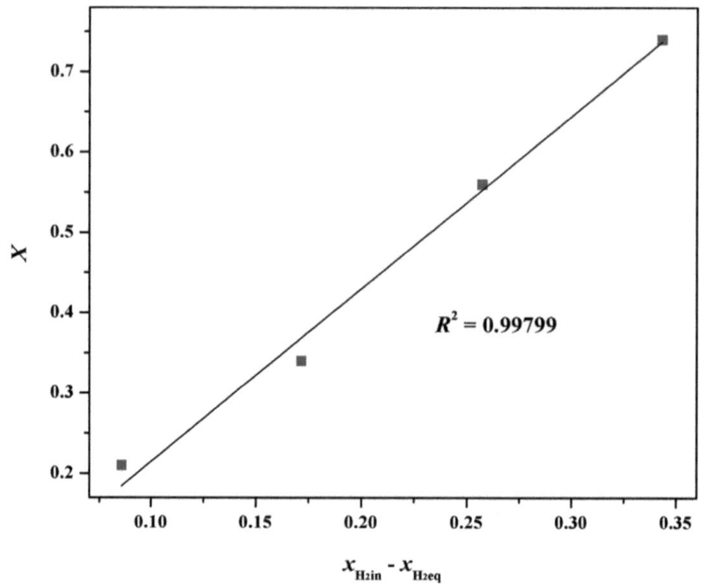

Fig. 9 Linearity check on the plot of X versus $x_{H_{2_{in}}} - x_{H_{2_{eq}}}$

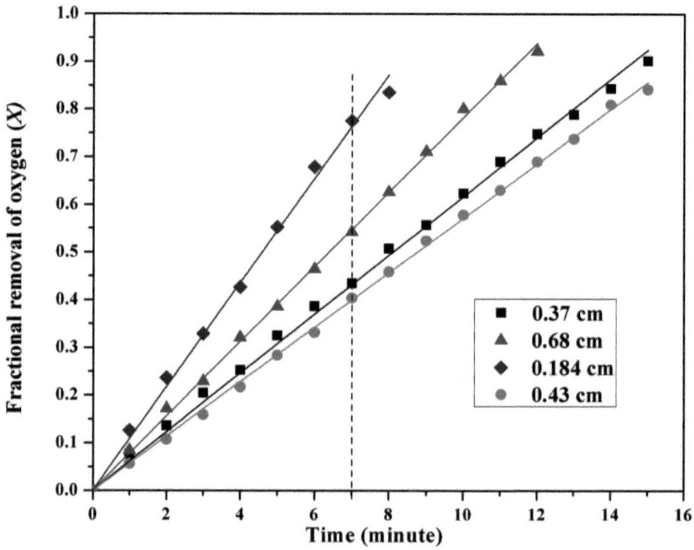

Fig. 10 The plot of fractional removal of oxygen versus time for the reduction of synthetic magnetite with different sample bed heights

nearly half of the crucible height, and then it starts to increase with decreasing bed height towards 0.184 cm. The minimum rate of reduction at a bed height of 0.43 cm, which is half the crucible height (0.88 cm), is very much characteristic of the gas film diffusion control for the current reaction geometry as can be mathematically established from Eq. (7). To explain, for a given time, the first derivative dX/dz_1^0 becomes zero at $z_1^0 = 1/2z_2$ and the second derivative d^2X/dz_1^{02} is positive at $z_1^0 = 1/2z_2$. Consequently, the plot of fractional removal of oxygen versus time (X vs. t) for $z_1^0 = 1/2z_2$ should be the lowest-lying one among the X versus t plots for different bed heights (z_1^0), which is seen in Fig. 10. This result, once again, confirms that the rate of the reduction was controlled by the gas film counter diffusion (mass transfer) of H_2 and H_2O. And the possibility of rate control either by the interfacial reaction or by pore diffusion is ruled out. For a more quantitative agreement, X is plotted against $1/(z_2 - z_1^0)z_1^0$ for a given time (t = 7 min) to check for the linearity. The result is shown in Fig. 11 and the increment of X with increasing $1/(z_2 - z_1^0)z_1^0$ is clear. It should be noted that the plot is drawn satisfying the requirement of passing through the point $X = 0$ corresponding to $1/(z_2 - z_1^0)z_1^0 = 0$, although this point is not shown in the figure. With the correlation coefficient (R^2) of 0.97973, the plot can be considered approximately linear.

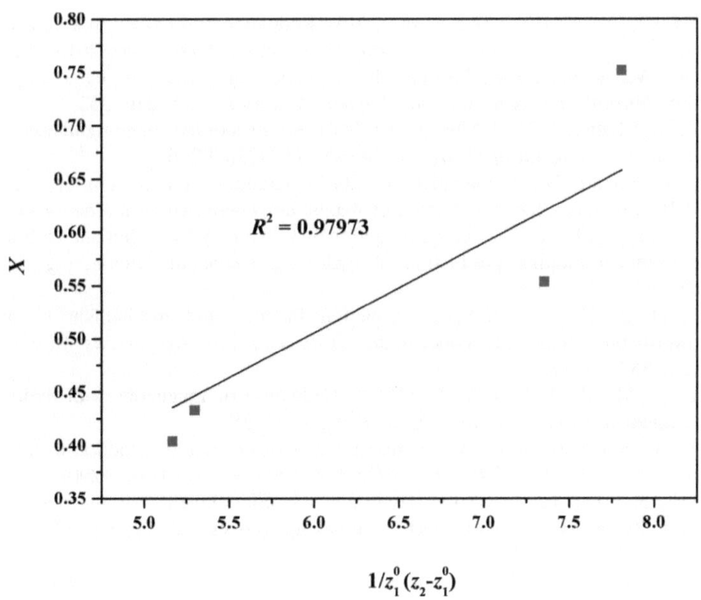

Fig. 11 Linearity check on the plot of X versus $1/(z_2 - z_1^0)z_1^0$

Conclusions

In the present work, hydrogen reduction of pure magnetite powder (<5 μm) was carried out in a thermogravimetric analyzer (TGA) under a steady flow of hydrogen or hydrogen–argon mixture (to produce different partial pressures of hydrogen). The variables studied were reduction temperature (973–1273 K), hydrogen partial pressure (0.25–1 atm) and sample bed height (0.184–0.68 cm). The rate equations developed for the reaction system under study were applied to determine the rate controlling step. The reduction was found to be rate controlled by counter diffusion of H_2–H_2O through the stagnant gas film enclosed above the sample inside the crucible. The rate controlling mechanism was established by the calculation of the activation energy (9 kJ mol^{-1}) and by studying the effects of hydrogen partial pressure and sample bed height.

The structural and morphological properties of magnetite powder were also analyzed by X-ray diffraction (XRD), X-ray photoelectron spectroscopy (XPS), Fourier transform infrared spectroscopy (FTIR), nitrogen adsorption analysis by five points isotherm BET method, and scanning electron microscopy along with energy dispersive X-ray spectroscopy (SEM-EDS).

References

1. Jeong, M. H., Lee, D. H., & Bae, J. W. (2015). Reduction and oxidation kinetics of different phases of iron oxides. *International Journal of Hydrogen Energy, 40*, 2613–2620.
2. Ramadan, W., et al. (2014). Particle characteristics and reduction behavior of synthetic magnetite. *Journal of Magnetism and Magnetic Materials, 355*, 246–253.
3. Kuila, S. K., Chatterjee, R., & Ghosh, D. (2016). Kinetics of hydrogen reduction of magnetite ore. *International Journal of Hydrogen Energy, 41*, 9256–9266.
4. Pineau, A., Kanari, N., & Gaballah, I. (2007). Kinetics of reduction of iron oxides by H_2. Part II. Low temperature reduction of magnetite. *Thermochimica Acta, 456*, 75–88.
5. Liu, W., et al. (2014). Kinetics of the reduction of Wüstite by hydrogen and carbon monoxide for the chemical looping production of hydrogen. *Chemical Engineering Science, 120*, 149–166.
6. Liu, S., et al. (2014). Solid-state reduction kinetics and mechanism of pre-oxidized vanadium-titanium magnetite concentrate. *Transactions of Nonferrous Metals Society of China, 24*, 3372–3377.
7. Al-Kahtany, M. M., & Rao, Y. K. (1980). Reduction of magnetite with hydrogen. Pt. 1: Intrinsic kinetics. *Ironmaking and Steelmaking, 7*, 49–58.
8. Kim, W. H., et al. (2013). The retardation kinetics of magnetite reduction using H_2 and H_2-H_2O mixtures. *International Journal of Hydrogen Energy, 38*, 4194–4200.
9. Wang, H., & Sohn, H. Y. (2013). Hydrogen reduction kinetics of magnetite concentrate particles relevant to a Novel Flash ironmaking process. *Metallurgical and Materials Transactions B, 44*, 133–145.
10. International Center for Diffraction Data, 12 Campus Boulevard, Newton Square, PA, 19073–3273.
11. Gadsden, J. A. (1975). *Infrared spectra of minerals and related inorganic compounds*. London: Butterworths.

12. Degen, I. A. (1997). *Tables for characteristic group frequencies for the interpretation of infrared and Raman spectra*. Harrow: Acolyte Publications.
13. Šaric, A., et al. (1998). Microstructural properties of Fe–oxide powders obtained by precipitation from $FeCl_3$ solutions. *Materials Science and Engineering B, 56,* 43–52.
14. Kubaschewski, O., Alcock, C. B., & Spencer, P. J. (1993). *Materials thermochemistry*. New York: Pergamon Press.
15. Chatterjee, R., & Ghosh, D. (2015). Kinetics of hydrogen reduction of chalcopyrite concentrate. *Metallurgical and Materials Transactions B, 46,* 2692–2705.
16. Bird, R. B., Stewart, W. E., & Lightfoot, E. N. (1960). *Transport Phenomena* (p. 527). New York, NY: Wiley.

Study on Leaching Valuable Elements from Bayan Obo Tailings

Bo Zhang, Xiangxin Xue, Xiaowei Huang, He Yang and Jianxin Han

Abstract After extracting rare earths minerals, iron minerals and fluorite from Bayan Obo ore, niobium and scandium which are riched in the mineral processing tailings coexist with the remaining rare earth. In order to recovery these valuable elements, the tailings was disposed with the method of activating roasting-acid leaching. The tailings was roasted with NaCl–$Ca(OH)_2$ under a temperature of 900 °C for 1.5 h. Weight percentages of NaCl-to-tailings and $Ca(OH)_2$-to-tailings are 10% and 20% respectively. Then the roasted ore experienced two leaching stages from "hydrochloric acid pre-leaching at 90 °C for 1.5 h" to "intensified sulfuric acid leaching at 300 °C for 1 h". The results show that, the leaching rates of niobium, scandium and rare earth in roasted ore could reach 86.80, 97.42 and 97.94% respectively under the conditions above. Moreover, the radioactivity per unit mass of leaching residue was reduced to 745 Bq/kg and environmentally friendly.

Keywords Mineral processing tailings · Niobium · Scandium · NaCl–$Ca(OH)_2$ · Leaching rates

Introduction

Bayan Obo ore is a deposit which mainly contains iron, rare earths, niobium and other valuable elements [1–4]. A series of problems has been faced and solved by technical personnel in Baotou Steel. At present, mineral separation technologies about iron minerals and rare earths mineral are very mature [5–9]. However, niobium concentration with niobium grade 4.20% and recovery 28.25% can be obtained by conventional flotation process [10, 11]. Due to that unsatisfactory result, much work of niobium flotation was made continuously. Li et al. recycled rare earths and niobium from mineral processing tailings by the method of flotation-magnetic

B. Zhang (✉) · X. Xue · X. Huang · H. Yang · J. Han
Shenyang, Liaoning, China
e-mail: neuyanghe@163.com

© The Minerals, Metals & Materials Society 2017
M.A. Meyers et al. (eds.), *Proceedings of the 3rd Pan American Materials Congress*,
The Minerals, Metals & Materials Series, DOI 10.1007/978-3-319-52132-9_63

separation-gravity separation. They obtained niobium concentrate with a grade of 1.66% and a recovery of 35.58% [12]. An ore dressing experiment on extracting niobium from eschynite was made by Beijing General Research Institute of Mining and Metallurgy. Grade and recovery of niobium concentrate in this process are 15.12 and 57.46% respectively [13]. In view of low grades, various types and small sizes, it is very difficult to extract niobium from Chinese niobium minerals with flotation. Niobium minerals are soluble in hydrofluoric acid and heating sulfuric acid under suitable conditions. As a result hydrometallurgical is used for extracting niobium by more and more people. They also may be decomposed by sodium hydroxide and chlorine decomposition. Wu et al. leached niobium from tantalum-niobium bearing minerals with sulfuric acid and hydrogen peroxide. 81% of the niobium leaching rate is achieved [14]. The scandium has no independent minerals in Bayan Obo ore and exists as isomorphism in other minerals. Method of beneficiation is not suitable for the effective extraction of scandium. Li et al. confirmed that rare earth elements including scandium can be transformed into solutions by roasting the target minerals in concentrated sulfuric acid at 250–300 °C and leaching with water [15]. In addition, low-content niobium and scandium existing in Bayan Obo tailings are hardly to be separated out by flotation. So hydrometallurgy is a good selection for extracting the valuable elements from Bayan Obo tailings. In this paper, the process of activating roasting- acid leaching will be studied to extract niobium, scandium and rare earth from tailings.

Experimental

Materials

Niobium and scandium are concentrated in tailings after iron minerals, rare earth minerals and fluorite are separated out from Bayan Obo ore. The tailings above are known simply as "Bayan Obo tailings" and selected as materials in this study. Chemical composition of tailings is listed in Table 1. Table 2 is the specific mineral composition of tailings.

Comparing with Bayan Obo ore, contents of niobium and scandium have tripled in Bayan Obo tailings. Percentage of niobium and scandium in tailings reach 0.36% and 0.03%. Furthermore, rare earth which is not separated off takes a percentage of 2.14% in tailings. It is worthy extracting these valuable elements from the tailings. Tables 3 and 4 are distributions of Nb_2O_5 and Sc_2O_3 in tailings respectively.

Table 1 The chemical composition of Bayan Obo tailings (%)

Na_2O	K_2O	MgO	CaO	BaO	MnO_2	SiO_2	TiO_2	ThO_2	Al_2O_3	FeO
2.25	0.93	5.14	11.68	3.15	1.50	33.26	1.08	0.021	2.24	2.38
Sc_2O_3	REO	P_2O_5	Nb_2O_5	F	S	TFe	Magnetic iron		Burning loss	
0.030	2.14	0.87	0.36	3.14	2.15	15.74	0.50		7.81	

Table 2 The mineral composition of Bayan Obo tailings (%)

Magnetite	Hematite	Siderite	Ilmenite	Pyrite	Nigrine	Rutile	Ilmenorutile
0.61	13.56	0.01	0.26	3.13	0.01	0.16	0.42
Bastnasite	Monazite	Allantite	Aeschynite	Pyrochlore	Baotite	Niobite	Fergusonite
0.97	0.49	0.16	0.24	0.11	0.02	0.10	0.02
Pyrolusite	Rhodochrosite	Fluorite	Apatite	Barite	Wollastonite	Feldspar	Amphibole
0.14	0.02	1.93	1.26	2.71	0.31	5.77	15.04
Pyroxene	Mica	Serpentine	Chlorite	Quartz	Dolomite	Calcite	Others
16.38	7.62	0.32	0.40	9.37	13.16	3.87	1.43

Table 3 Distribution of Nb_2O_5 in tailings

Minerals	Minerals contents in tailings (%)	Nb_2O_5 content in each mineral (%)	Percentages of Nb_2O_5 in tailings (%)	Nb_2O_5 occupancy of each mineral (%)
Niobite	0.10	73.67	0.076	21.11
Ilmenorutile	0.42	10.34	0.043	11.94
Eschynite	0.24	30.58	0.073	20.28
Pyrochlore	0.11	67.94	0.075	20.83
Others	99.13	0.094	0.093	25.84
Totals	100.00		0.36	100.00

Table 4 Distribution of Sc_2O_3 in tailings

Minerals	Minerals contents (%)	Sc_2O_3 content in each mineral (ppm)	Percentages of Sc_2O_3 in tailings (ppm)	Sc_2O_3 occupancy of each mineral (%)
Hematite	13.56	102	13.83	4.61
Bastnasite	0.97	442	4.29	1.43
Monazite	0.49	240	1.18	0.39
Fluorite	1.93	50	0.97	0.32
Apatite	1.26	300	3.78	1.26
Pyroxene, Amphibole	31.42	528	165.90	55.30
Mica	7.62	70	5.33	1.78
Dolomite, Calcite	17.03	68	11.58	3.86
Quartz, Feldspar	15.14	20	3.03	1.01
Barite	2.71	104	2.82	0.94
Niobite	0.10	400	0.40	0.13
Ilmenorutile	0.42	1400	5.88	1.96
Eschynite	0.24	230	0.55	0.18
Pyrochlore	0.11	202	0.22	0.07
Others	7.00	1146	80.25	26.76
Totals	100.00		300.00	100.00

Distributions of niobium and scandium in tailings show that, more than 70% of the niobium exists in the form of niobite, ilmenorutile, eschynite and pyrochlore. Scandium which distributes in pyroxene and amphibole accounts for 55.3% of the total content. According to characteristic of Bayan Obo ore, existing forms of rare earth are mainly bastnasite and monazite. So the minerals mentioned above are determined to be target minerals for extracting niobium, scandium and rare earth.

Experiment Theory

The method of activating roasting-acid leaching is applied in this study. Calcium hydroxide is chosen as decomposer and activator in roasting process. Figure 1 is XRD diagram of the roasted products of niobium concentrate and $Ca(OH)_2$. The calcium hydroxide could transform niobium minerals into acid-soluble minerals. Calcium hydroxide is conductive to the decomposition of silicate minerals such as pyroxene and amphibole at high temperature. Bastnasite and monazite could be decomposed into rare earth oxide effectively under the action of $Ca(OH)_2$ [16]. On the face of it, activating roasting is extremely advantageous for leaching niobium and scandium further. Furthermore, sodium chloride could improve chemical reaction rate through accelerating mass transfer depending on its low-melting point. So the added sodium chloride will reduce required temperature of decomposition reaction. That practice not only saves energy consumption, but also prevents rare earth from losing activity under a high roasting temperature. Then roasted ore experiences hydrochloric acid leaching and sulfuric acid leaching successively. Hydrochloric acid is used for leaching rare earth and scandium to prevent rare earth forming deposits with Na^+ and SO_4^{2-}. In order to guarantee leaching rate of niobium, hydrochloric acid leaching residue is leached in sulfuric acid for a intensified leaching process.

The chemical reactions mentioned in Eqs. 1–4 describe chemical reactions between rare earth minerals and added $Ca(OH)_2$. Figure 1 is XRD diagram of roasted products of niobium concentrate with $Ca(OH)_2$. It is clearly that the roasted products are mainly many Ca-containing compounds, such as $Ca((Ti_{0.8}Fe_{0.1}Nb_{0.1})O_3)$ and $Ca(Ti_{0.4}Fe_{0.3}Nb_{0.3})O_3$.

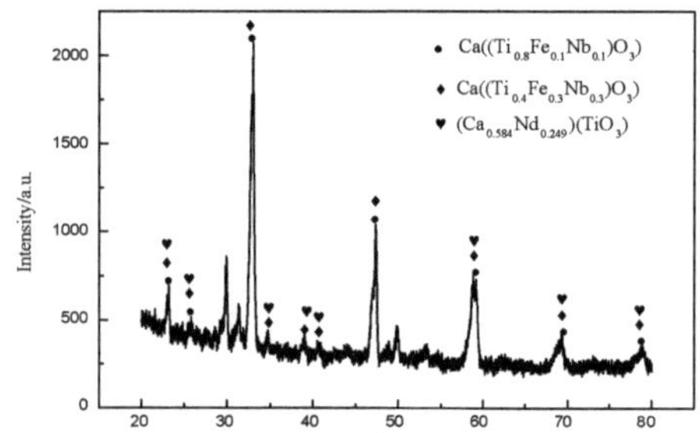

Fig. 1 XRD diagram of roasted products of niobium concentrate and $Ca(OH)_2$

$$CeCO_3F(s) = CeOF(s) + CO_2(g) \quad (1)$$

$$2CeOF(s) + Ca(OH)_2(s) = CaF_2(s) + Ce_2O_3(s) + H_2O(g) \quad (2)$$

$$2CePO_4(s) + 3Ca(OH)_2(s) = Ca_3(PO_4)_2 + Ce_2O_3(s) + 3H_2O(g) \quad (3)$$

$$3Ca_3(PO_4)_2(s) + CaF_2(s) = 2Ca_5(PO_4)_3F(s) \quad (4)$$

Experiment Method

The tailings are mixed with NaCl–$Ca(OH)_2$ according to a certain percentage. Then the mixture is disposed with the method of activating roasting in furnace. The roasted ore is grinded and leached by 6 mol/L hydrochloric acid at 90 °C for 1.5 h. The leaching residue and liquid are separated through filtering. After this step, the leaching residue is leached by 18 mol/L sulfuric at 300 °C for 1 h. A new leaching residue and liquid are obtained. At last, contents of valuable elements in leaching residue and leaching liquid are determined with ICP.

Effect of Adding Amount of NaCl on Leaching Rates of Valuable Elements

As a low-melting compound, sodium chloride is added into reactants to intensity mass transfer at low temperature. It is showed in Fig. 2 that, leaching rates of rare earth and scandium rise as adding amount of sodium chloride increases. The more amount of sodium chloride is added in tailings, the stronger effect of mass transfer is. Then the decomposition effect of $Ca(OH)_2$ on Sc-containing silicate (pyroxene, amphibole) intensifies accordingly. Decomposition reactions between rare earth minerals and $Ca(OH)_2$ become severe, too. However, the rule is different for the leaching process of niobium minerals. Adding excessive sodium chloride at the roasting process is detrimental to leaching of niobium. When the percentage of NaCl-to-tailings is 10%, the sodium chloride is beneficial for the leaching of all valuable elements. In that case, leaching rates of niobium, scandium and rare earth are 83.7, 89.94 and 96.50% respectively.

Effect of Adding Amount of $Ca(OH)_2$ on Leaching Rates of Valuable Elements

This experiment explored variation trends of objective elements leaching rates as the adding amount of $Ca(OH)_2$ increases. It can be known from Fig. 3 that a small

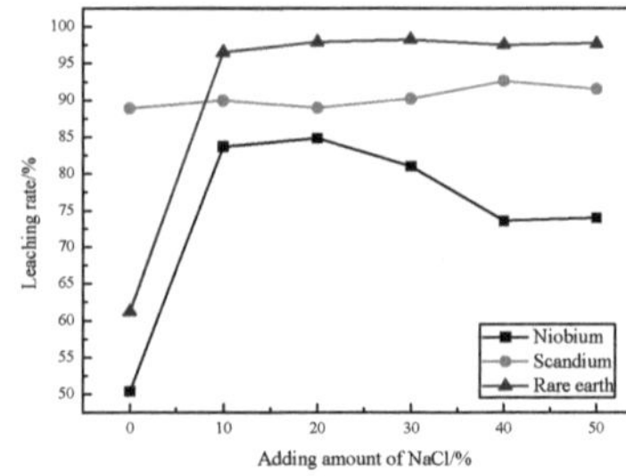

Fig. 2 The relationship between adding amount of NaCl and leaching rate

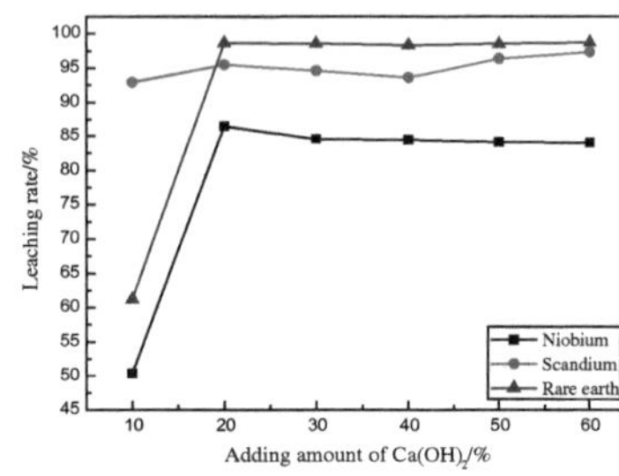

Fig. 3 The relationship between adding amount of $Ca(OH)_2$ and leaching rate

amount of $Ca(OH)_2$ could play roles of decomposer, activator. When the weight percentage of $Ca(OH)_2$-to-tailings is 20%, leaching rates of niobium, scandium and rare earth are 86.53, 95.47 and 98.61% respectively. Increasing additive amount of $Ca(OH)_2$ continuously cannot improve leaching rates of valuable elements obviously. Overmuch $Ca(OH)_2$ not only increases cost, but also absorbs much heat with the decomposition of $Ca(OH)_2$. The best additive amount of $Ca(OH)_2$ is 20%.

Effect of Calcination Time on Leaching Rates of Valuable Elements

Sufficient calcination time is a requisite for high conversion rate of reactant. The mixed ore was roasted in muffle furnace at 800 °C for 0.5, 1, 1.5, 2 and 2.5 h. Weight percentages of $Ca(OH)_2$-to-tailings and NaCl-to-tailings are still 20% and 10% respectively. Leaching rates of each element under different roasting time are showed in Fig. 4. When the roasting time is less than 1.5 h, leaching rates of scandium stay in low values which are lower than 85%. If roasting time is extended to 1.5 h or more, the index increased by ten percent and reached above 97.42%. Roasting time is equally important to leaching rates of niobium, too. Niobium minerals cannot be transformed into acid soluble compounds completely in a short calcination time. According to the change trend of niobium minerals leaching rate in Fig. 4, the shortest transformation time is determined as 1.5 h. However, calcination

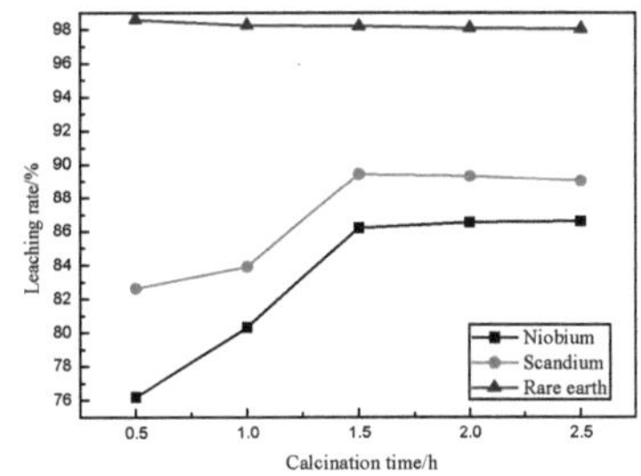

Fig. 4 The relationship between calcination time and leaching rate

time has little effect on the leaching rates of rare earth. Its leaching rate has reached 98.14% when roasting time is 0.5 h. In consideration of the leaching rates of niobium, scandium and rare earth, 1.5 h is thought to be the best calcination time.

Effects of Calcination Temperature on Leaching Rates of Valuable Elements

Certain amount of tailings is mixed with NaCl–$Ca(OH)_2$. Weight percentages of NaCl-to-tailing and $Ca(OH)_2$-to-tailngs are 10% and 20% respectively. The mixture was calcined in muffle furnace at different temperatures for 1.5 h. The leaching rates of niobium, scandium and rare earth are showed in Fig. 5. It is obvious that the leaching rate of niobium is less than 80% when calcination temperature is lower than 800 °C. In contrast, leaching rates of rare earth is higher than 90%, which is relatively high. When calcination temperature is higher than 800 °C, leaching rate of niobium is close to 90%. This phenomenon is related to the transmission rate of the reactant. The transmission rate of the substance in solid is slower than that in liquid. The added sodium chloride plays a role of liquid mass transfer when the calcination temperature approaches melting point of NaCl (801 °C). The same change rule emerges at the leaching process of scandium. The leaching rates of scandium increased over 90% only when the calcination temperature exceeded 800 °C. In the experimental temperature range, leaching rates of the rare earth is almost constant. A better calcination temperature is determined to be 900 °C.

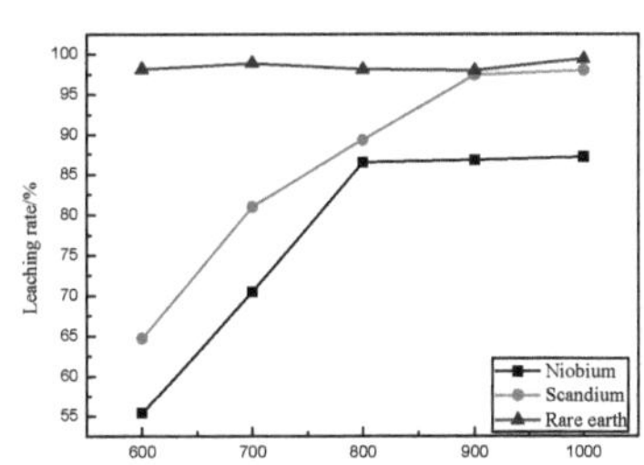

Fig. 5 The relationship between calcination temperature and leaching rate

Radio activity per unit mass of acid-leached residue was tested because of the presence of thorium in Bayan Obo tailings. The test results showed that per unit mass of acid-leached residue had been reduced to 745 Bq/kg.

Conclusions

(1) Elements analysis of Bayan Obo tailings shows that niobium and scandium contents all triple comparing that in Bayan Obo ore. In the tailings, niobium mainly exists in niobite, ilmenorutile, eschynite and pyrochlore. More than half of the scandium which has no independent minerals exists in pyroxene and amphibole.
(2) Under the action of $Ca(OH)_2$, niobium minerals are transformed into Ca-containing compounds. The niobium in new generated compounds could be leached out by sulfuric acid effectively.
(3) As a decomposer, activator and fluxing agent, NaCl–$Ca(OH)_2$ was mixed with Bayan Obo tailings. Percentages of $Ca(OH)_2$-to-tailings and NaCl-to-tailings were 20% and 10%. The mixture was roasted in muffle furnace at 900 °C for 1.5 h. Then the roasted ore was leached by hydrochloric acid and sulfuric acid for 1.5 and 1 h step by step. Leaching temperatures of hydrochloric acid process and sulfuric acid process are 90 and 300 °C. At this condition, leaching rates of niobium, scandium and rare earth reach 86.80, 97.42 and 97.94%, respectively.
(4) Because there is 0.021% thorium existing in the Bayan Obo tailings, it is necessary to evaluate radioactivity of the leaching residue. The test result showed that per unit mass of acid-leached residue had been reduced to 745 Bq/kg. The acid-leached residue is environmentally friendly.

References

1. Yu, L., Liu, J., & Wang, Z. C. (2007). New progress in comprehensive utilization of tailings in Bao Steel's concentrator. *Multipurpose Utilization of Mineral Resources, 3,* 32–34.
2. Cai, Z. L., Cao, M. L., Che, L. P., Yu, Y. F., & Hong, H. Y. (2009). Study on the beneficiation process for recovering rare-earth from the LIMS tailings of HLMS rougher concentrate after magnetizing roasting in Baogang concentrator. *Metal Mine, 7,* 155–157.
3. Wu, B., Shang, H., & Wen, J. K. (2013). Leaching of niobium from low-grade refractory tantalum-niobium bearing minerals. *Chinese Journal of Rare Metals, 37*(5), 791–797.
4. EI-Hussaini, O. M., & Mahdy, M. A. (2002). Sulfuric acid leaching of Kab Amiri niobium-tantalum bearing minerals, Central Eastern Desertm. Egypt. *Hydrometallurgy, 64,* 219–229.
5. Wang, W. W., Pranlol, Y., & Cheng, C. Y. (2011). Metallurgical processes for scandium recovery from various resources: A review. *Hydrometallurgy, 108,* 100–108.

6. Makanyire, T., Jha, A., & Sutcliffe, S. (2016). Kinetics of hydrochloric acid leaching of niobium from TiO_2 residues. *International Journal of Mineral Processing, 64,* 219–229.
7. Rodriguez, M. H., Rosales, G. D., Pinna, E. G., & Suarez, D. S. (2015). Extraction of niobium and tantalum from ferrocolumbite by hydrofluoric acid pressure leaching. *Hydrometallurgy, 156,* 17–20.
8. Liao, C. S., Jia, J. T., Zhang, Y., & Xu, G. (2011). Extraction of scandium from ion-adsorptive rare earth deposit by naphthenic acid. *Journal of Alloys and Compound, 323–324,* 833–837.
9. Zhang, B., Liu, C. J., Li, C. L., & Jiang, M. F. (2014). A novel approach for recovery of rare earth and niobium from Bayan Obo tailings. *Minerals Engineering, 65,* 17–23.
10. Sun, X. Q., Ji, Y., Guo, L., & Chen, J. (2011). A novel ammonium ionic liquid based extraction strategy for separating scandium from yttrium and lanthanides. *Separation and Purification Technology, 81,* 25–30.
11. Xu, Y. H., Liu, H. J., Meng, Z. J., Cui, J. G., Zhao, W. Y., & Li, L. C. (2012). Decomposition of bastnasite and monazite mixed rare earth minerals calcined by alkali liquid. *Rare Earths, 30* (2), 155–158.
12. Gibson, C. E., Kelebek, S., & Ghamirian, M. A. (2015). Niobium oxide mineral flotation: A review of relevant literature and the current state of industrial operations. *International Journal of Mineral Processing, 137,* 82–97.
13. Xu, Y. H., Liu, H. J., Cui, J. G., Meng, Z. J., Zhao, W. Y., & Li, L. C. (2012). Techniques for clean smelting and resource comprehensive recycle of Baotou rare earth concentrates. *Journal of Chinese Social Rare Earths, 30*(5), 632–635.
14. Kuzmin, V. I., Pashkov, G. L., Lomaev, V. G., Voskresenskaya, E. N., Kuzmina, V. N. (2012). Combined approaches for comprehensive processing of rare earth metal ores. *Hydrometallurgy, 129–130,* 1–6.
15. Li, L. C., et al. (2007). *Extraction and separation of rare earth* (p. 1). Chi Feng: Inner Mongolia Science and Technology Press.
16. Wu, W. Y., Bian, X., Wu, Z. Y., Sun, S. C., & Tu, G. F. (2007). Reaction process of monazite and bastnaesite mixed rare earth minerals calcined by $CaO\text{-}NaCl\text{-}CaCl_2$. *Transactions of Nonferrous Metals Society of China, 17,* 864–868.

Study on Thermal Decomposition and Oxidation Characteristics of Iron Ores

Qingfeng Kang, Jianliang Zhang, Donghui Liu, Zhengjian Liu and Jie Yan

Abstract Thermal decomposition and oxidation characteristics of three typical iron ores were studied with differential thermal analysis and XRD techniques. The results displayed that DSC curve of limonite showed a significant endothermic arc in the range of 605–1150 °C. It was thought that the decomposition of crystal water in limonite increased the porosity and surface area of limonite, which improved the adsorption capacity of limonite. DSC curve of hematite showed a weak endothermic arc in the range of 312–652 °C. DSC curve of magnetite showed a significant exotherm arc in the range of 162–915 °C. It was thought that lots of heat released when Fe_3O_4 in magnetite was oxidized.

Keywords Iron ore · Thermal decomposition oxidation characteristics · Optimization of ore blending · Sintering

Introduction

The formula is $nFe_2O_3{\cdot}mH_2O$ (n = 1–3, m = 1–4); this is a cost-effective iron ore but not conducive to reducing coke rates during sintering. Hematite (Fe_2O_3) is conducive to the formation of calcium ferrite for the sintering process. Magnetite (Fe_3O_4) can be oxidized to generate heat, which is conducive to reducing coke rates during sintering.

Q. Kang · J. Zhang (✉) · D. Liu · Z. Liu
School of Metallurgical and Ecological Engineering,
University of Science and Technology Beijing, Beijing 100083, China
e-mail: jl.zhang@ustb.edu.cn

Q. Kang
e-mail: 18810517311@163.com

J. Yan
Hebei Iron & Steel Group; Chengde Iron and Steel Co.,
Chengde 067002, Hebei, China

© The Minerals, Metals & Materials Society 2017
M.A. Meyers et al. (eds.), *Proceedings of the 3rd Pan American Materials Congress*,
The Minerals, Metals & Materials Series, DOI 10.1007/978-3-319-52132-9_64

Extensive research on limonite and magnetite iron ore fines has been done. Sato et al. reported that the decomposition of crystal water in limonite increased the surface area of limonite. Wang et al. [1, 2] reported that pre-granulation technology can help to reduce coke rates and increase productivity. Takayuki et al. [3] reported that the surface characteristics of limonite have a great influence on the compressive strength of green pellets. Pan et al. [4] reported that the decomposition of crystal water in limonite is completed in the range of 200–400 °C. Li et al. [5] reported that decomposition of crystal water in limonite is a slow-fast-slow process. Wu et al. [7] reported that the assimilation of Brazilian and Australian ores is lower and higher than the optical zone respectively, but their characteristics are complementary.

However, the study of thermal decomposition oxidation characteristics of different iron ores at higher heating rates and wider temperature ranges is less common. Therefore, to efficiently utilize different iron ore powder and further reduce fuel consumption in the sintering process, thermal decomposition oxidation characteristics of three typical iron ores at higher heating rates and wider temperature ranges are studied with differential thermal analysis and XRD techniques, for optimization of ore blending during sintering.

Materials and Experiment

Materials

Iron ore fines used in this experiment come from a steel plan. The drying process of iron ore fines is conducted out in a DHG-9140A electric heated blast dry box. The drying temperature is 105 °C, for two hours. Iron ore fines after drying were ground to 0.074 μm or less, and then subjected to chemical analysis. Iron ore fines A1, B1 and C1, are limonite, hematite and magnetite; their chemical composition is shown in Table 1.

Experimental

Iron ore fines after drying are ground to 0.074 μm or less, and then subjected to differential thermal analysis. Sample masses of 10 mg were loaded into the corundum crucible. The differential thermal analysis experiments were conducted in

Table 1 The chemical composition of the three iron ore fines (mass fraction)

Materials	TFe	CaO	SiO_2	Al_2O_3	FeO	MgO	LOI
A1	57.30	0.02	6.02	1.48	0.22	0.06	10.37
B1	59.45	0.20	11.22	1.14	0.28	0.27	2.01
C1	64.97	0.19	8.54	0.20	25.44	0.48	–

static air conditions. The temperature was increased from ambient to 1200 °C at 4 °C/min, and then held for four minutes. TG, DTG, and DSC curves were measured in the differential thermal analysis experiments.

Iron ore fines after drying were ground to −0.074 μm, and subjected to XRD experiments.

Results and Discussion

Analysis of Thermal Decomposition Characteristics of Limonite(A1)

Type of reaction can be judged according to the DSC-TG curve, as shown in Table 2.

Thermal analysis and XRD curves of A1 iron ore fines as shown in Figs. 1, 2, 3, and 4. As shown in Fig. 1, the TG curve of A1 iron ore fines in the vicinity of 287 °C illustrates significant weight loss. As shown in Fig. 2, the DSC curve of A1 iron ore fines in the vicinity of 287 °C signifies an obvious endothermic reaction. Combined with Table 2, it can initially determine that the reaction types of taking place are parse, sublimation, decomposition or evaporation. To further determine the type of reaction in the vicinity of 287 °C, XRD of A1 iron ore fines at room temperature was performed. As shown in Fig. 3, FeO(OH) is the main mineral component of A1 iron ore fines and a small amount of Fe_2O_3 and $V_{0.875}Fe_{0.125}O(OH)$ are included. Since $V_{0.875}Fe_{0.125}O$ (OH) content is small, its thermal effect can be ignored. So it can judge that the decomposition reaction of FeO(OH) mainly occurred in the vicinity of 287 °C.

As shown in Fig. 2, the DSC curve of A1 iron ore fines appears a significant exotherm convex arc between 605–1150 °C and that indicates that an exothermic reaction occurred at the appropriate temperature. Combined with Table 2, it can initially determine that the reaction types of taking place are oxidation, absorbing, solid phase transition or crystal transition. The XRD detection and analysis results

Table 2 Type of reaction can be judged according to the DSC-TG curve

DSC		TG		Reaction process
Endothermic	Exothermic	Weightlessness	Weight gain	
√		√		Desorption or parse
√		√		Sublimation
√		√		Evaporation
√		√		Decomposition
	√		√	Oxidation
	√		√	Absorbing
√	√			Solid phase transition
√	√			Crystal transition

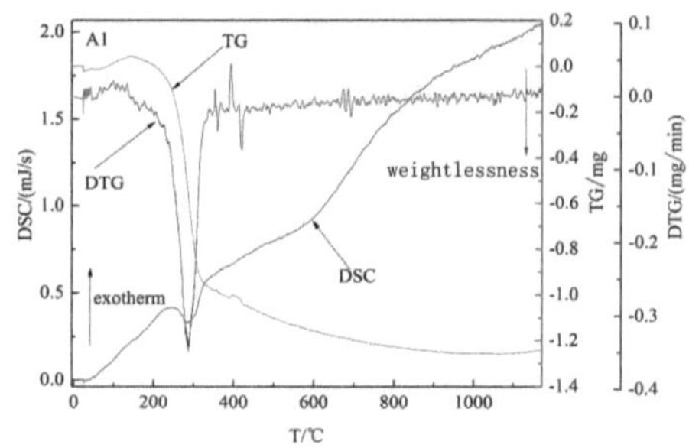

Fig. 1 DSC-TG-DTG curves of A1 iron ore fines

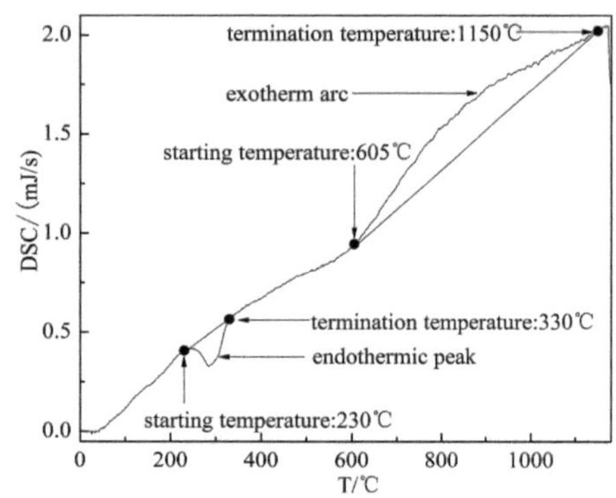

Fig. 2 DSC curve of A1 iron ore fines

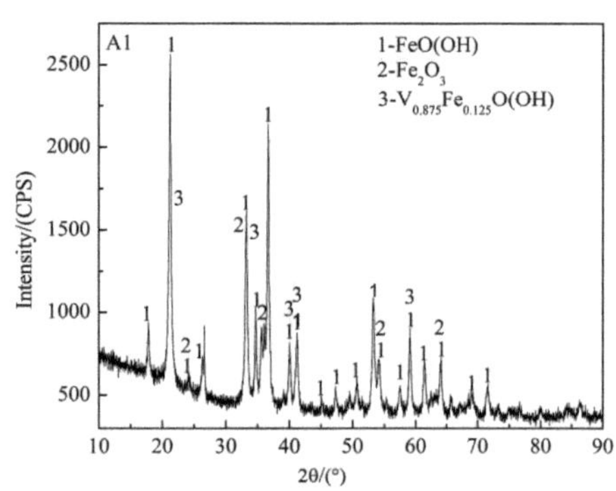

Fig. 3 XRD of A1 iron ore fines

indicate that adsorption reaction mainly occurred between 605–1150 °C. As shown in Fig. 4, weight loss rate and weight loss conversion rate of A1 iron ore fines at 600 °C are 12.003% and 0.922, which indicates that most of the crystal water in the A1 iron ore fines in the vicinity of 600 °C has been removed. When the crystal water of A1 iron ore fines is removed, the porosity and specific surface area increase and then the adsorption capacity increases. In addition, the influence of crystal water decomposition endothermic gradually reduces. Therefore, the DSC curve of A1 iron ore fines has a significant exotherm convex arc between 605–1150 °C.

Fig. 4 The weight loss of A1 iron ore fines

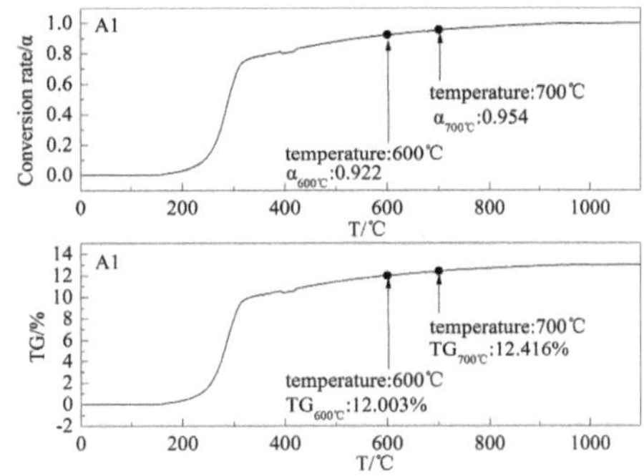

When the ratio of the limonite is too high, over-melting of the sintering bed will occur, resulting in sinter fuel consumption increase and lower sinter strength [4]. As mentioned above, the increase of the porosity and specific surface area of A1 fines is the main reason of the over-melting of the sintering bed. As shown in Fig. 4, the weight loss rate and weight loss conversion rate of A1 iron ore fines at 700 °C are 12.416% and 0.954, which indicates that most of the crystal water in the A1 iron ore fines in the vicinity of 700 °C were decomposed. However, a small part of the crystal water still has not been removed in the lower temperature zone and decomposed in the high-temperature combustion zone, which also increased the fuel consumption. Therefore, the crystal water of limonite should be decomposed as far as possible in the lower temperature zone. It is seen by the Gibbs energy function:

$$Fe_2O_3 \cdot H_2O \rightarrow Fe_2O_3 + H_2O_{(g)}$$
$$R\ln P_{H_2O} = -\frac{a}{T} - b$$

As mentioned above, FeO(OH) is the main mineral component of A1 iron ore fines and a small amount of Fe_2O_3 and $V_{0.875}Fe_{0.125}O(OH)$ are included. The decomposition reaction of FeO(OH) mainly occurred in the vicinity of 287 °C and the DSC curve of A1 iron ore fines in the vicinity of 287 °C shows an endothermic peak. When the crystal water of A1 iron ore fines is removed, the porosity and specific surface area increase and then the adsorption capacity increases. In addition, the influence of crystal water decomposition endothermic gradually reduces. Therefore, the DSC curve of A1 iron ore fines has a significant exotherm convex arc between 605–1150 °C. In addition, the increase of the porosity and specific surface area of A1 iron ore fines is the main reason of the over-melting of the sintering bed.

Analysis of Thermal Decomposition Characteristics of Hematite (B1)

Thermal analysis and XRD curves of B1 iron ore fines as shown in Figs. 5, 6, 7, and 8. As shown in Fig. 6, the weight loss rate of B1 iron ore fines in the vicinity of 287 °C is not obvious and the DSC curve of B1 iron ore fines in the vicinity of 287 °C has no obvious peak. However, the DSC curve of B1 iron ore fines appeares a slight exotherm convex arc between 312–652 °C. Combined with Table 2, it can initially be determined that the reaction types taking place are parse, sublimation, decomposition or evaporation. To further determine the type of reaction in the vicinity of 287 °C, XRD of B1 iron ore fines at room temperature was performed. As shown in Fig. 7, Fe_2O_3 and $Fe_{7.332}O_{11}$ H_2O are the main mineral component of B1 iron ore fines. Similarly with A1 iron ore fines, it can determine that the adsorption process of B1 iron ore fines occurred in the vicinity of 287 °C.

As shown in Fig. 8, the weight loss rate and weight loss conversion rate of B1 iron ore fines at 312 °C are 1.1025% and 0.442. Moreover, the crystal water in the B1 iron ore fines is much less than the crystal water in the A1 iron ore fines and the endothermic effect of crystal water decomposition can be ignored. However, when the crystal water of B1 iron ore fines is removed, the porosity and specific surface area increase and then the adsorption capacity increase. Therefore, the DSC curve of B1 iron ore fines appeares a slight exotherm convex arc between 312–652 °C.

Fig. 5 DSC-TG-DTG curves of B1 iron ore fines

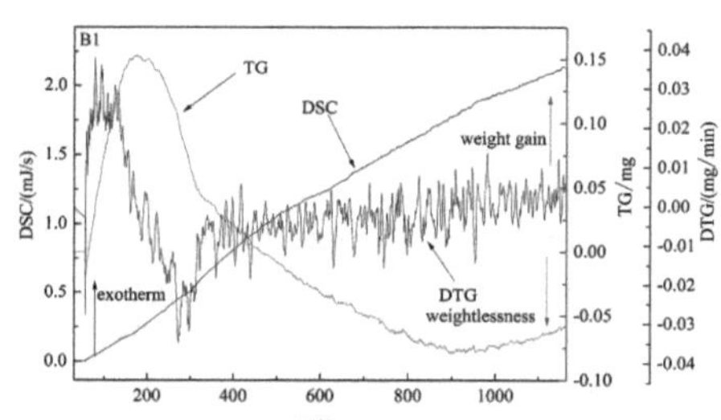

Fig. 6 DSC curve of B1 iron ore fines

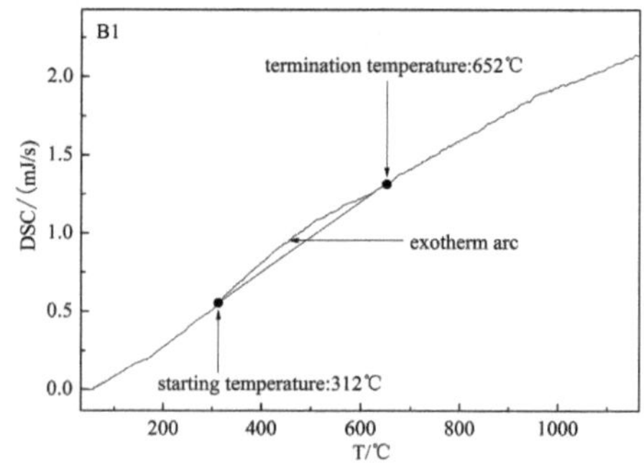

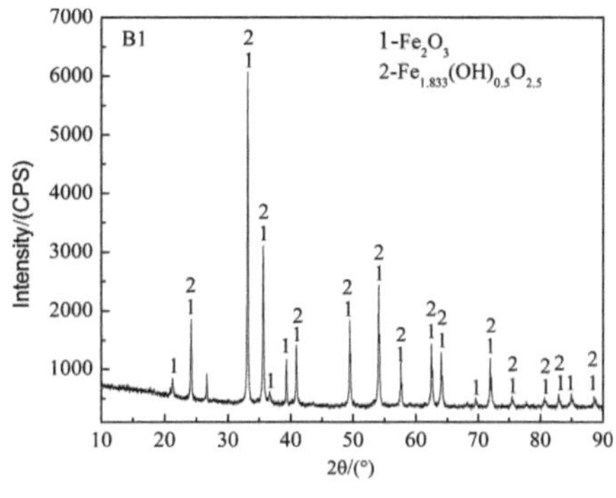

Fig. 7 XRD of B1 iron ore fines

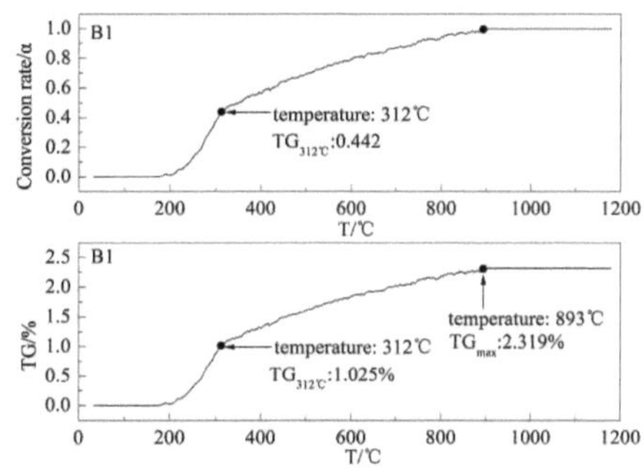

Fig. 8 The weight loss of B1 iron ore fines

As mentioned above, Fe_2O_3 and $Fe_{7.332}O_{11}$ H_2O are the main mineral components of B1 iron ore fines. The crystal water in the B1 iron ore fines is much less than the crystal water in the A1 iron ore fines and the endothermic effect of crystal water decomposition can be ignored. In addition, the porosity, specific surface area and adsorption capacity increase. Therefore, the DSC curve of B1 iron ore fines has a slight exotherm convex arc between 312–652 °C.

Analysis of Thermal Oxidation Characteristics of Magnetite (C1)

Thermal analysis and XRD curves of C1 iron ore fines as shown in Figs. 9, 10, 11, and 12. As shown in Fig. 9, the TG curve of C1 iron ore fines between 162–915 °C illustrates a significant weight gain phenomenon. As shown in Fig. 10, the DSC curve of C1 iron ore fines has a significant exotherm convex arc between 162–915 °C, which indicates that an exothermic reaction occurred at the appropriate temperature. Combined with Table 2, it can initially determine that the reaction types of taking place are oxidation, absorbing, solid phase transition or crystal transition. To further determine the type of reaction between 162–915 °C,

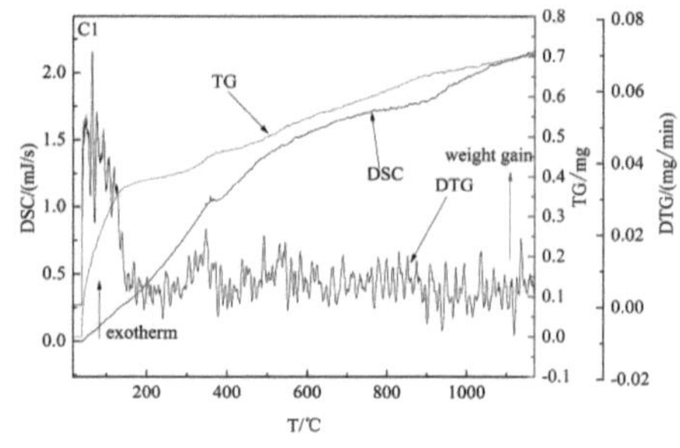

Fig. 9 DSC-TG-DTG curves of C1 iron ore fines

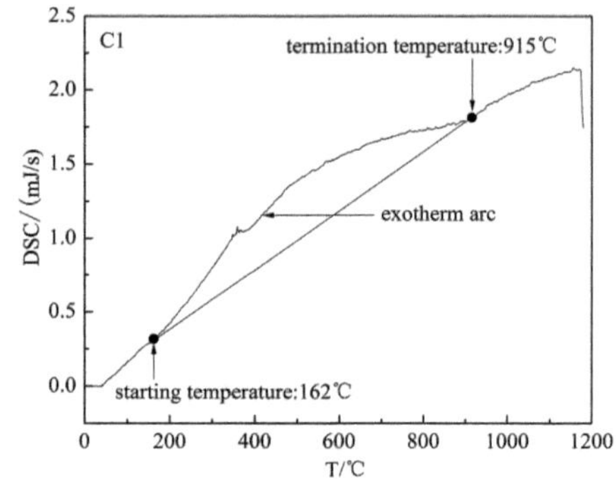

Fig. 10 DSC curve of C1 iron ore fines

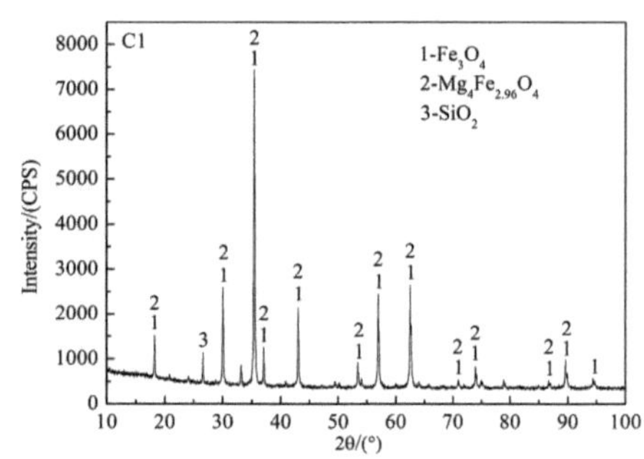

Fig. 11 XRD of C1 iron ore fines

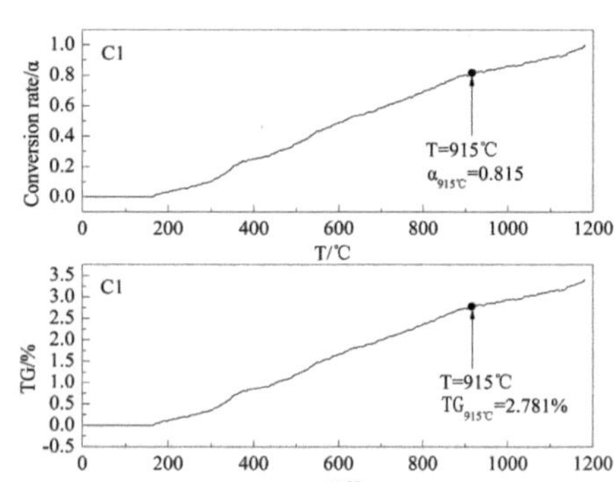

Fig. 12 The weight loss of C1 iron ore fines

XRD of C1 iron ore fines at room temperature was performed. As shown in Fig. 11, Fe_3O_4 is the main mineral component of C1 iron ore fines and a small amount of $Mg_4Fe_{2.96}O_4$ and SiO_2 are included. So it can judge that the oxidation reaction of Fe_3O_4 mainly occurred between 162–915 °C.

A large amount of heat is released when C1 iron ore fines are oxidized. Therefore, the DSC curve of C1 iron ore fines has a significant exotherm convex arc between 162–915 °C. As shown in Fig. 12, weight gain rate and weight gain conversion rate of C1 iron ore fines at 915 °C are 2.781% and 0.815, which indicates that most of the C1 iron ore fines are oxidized at 915 °C. In addition, a large amount of heat is released when C1 iron ore fines are oxidized, which is conducive to reducing sintering fuel consumption.

Conclusion

(1) The decomposition reaction of FeO(OH) mainly occurred in the vicinity of 287 °C and the DSC curve of A1 iron ore fines in the vicinity of 287 °C illustrates a endothermic peak. When the crystal water of A1 iron ore fines is removed, the porosity and specific surface area increase and then the adsorption capacity increases. In addition, the influence of endothermic crystal water decomposition gradually reduces. Therefore, the DSC curve of A1 iron ore fines has a significant exotherm convex arc between 605–1150 °C. In addition, the increase of the porosity and specific surface area of A1 iron ore fines is the main reason of the over-melting of the sintering bed.

(2) The crystal water in the B1 iron ore fines is much less than the crystal water in the A1 iron ore fines, and the endothermic effect of crystal water decomposition can be ignored. In addition, the porosity, specific surface area and adsorption capacity increase. Therefore, the DSC curve of B1 iron ore fines has a slight exotherm convex arc between 312–652 °C.

(3) A large amount of heat is released when C1 iron ore fines are oxidized. Therefore, the DSC curve of C1 iron ore fines has significant exotherm convex arc between 162–915 °C.

References

1. Yang, L. X., & Witchard, D. (1998). Sintering of blends containing magnetite concentrate and hematite or/and goethite ores. *ISIJ International, 38*(10), 1069–1076.
2. Yang, L. X., & Davis, L. (1999). Assimilation and mineral formation during sintering for blends containing magnetite concentrate and hematite/pisolite sintering fines. *ISIJ International, 39*(3), 239–245.
3. Maeda, T., Kikuchi, R., Ohno, K., et al. (2013). Effect of particle size of iron ore and coke on granulation property of quasi-particle. *ISIJ International, 53*(9), 1503–1509.

4. Pan, W., Wu, K., Wang, W., et al. (2013). Thermal decomposition characteristics of limonite and their influences on the sintering process. *Journal of Northeastern University, 34*(9), 1277–1281.
5. Li, J., Liu, W., Li, Y., et al. (2011). Applied research on thermal decomposition of limonite with crystal water by thermal analysis technology. *Iron and Steel, 46*(5).
6. Zhang, J., Shu, B., Che, X., et al. (2011). Experimental study of assimilation of several domestic and foreign iron ore fines. *Chinese Journal of Process Engineering, 11*(1), 97–102.
7. Wu, S., Dai, Y., Fei, Y., et al. (2010). Optimization of ore blending during sintering based on complementation of high temperature properties. *Journal of University of Science and Technology Beijing, 6,* 719–724.

The Direct Leaching of Micro-disseminated Gold Concentrate by Bromide Process and the Characterization of Leaching Products

Chao Li, Hongxu Li and Qiankun Jing

Abstract Gold is an important industry and strategic reserves of metal, but cyanide leaching process have adverse to the environment. In this paper, an absolutely non-cyanide bromide gold-leaching system was puts forward to leaching a kind of micro-disseminated gold concentrate from China. The influence of different reagent concentration and temperature on the bromide leaching process were tested in this research, and various testing methods as X-ray diffraction (XRD), scanning electron microscopy, infrared spectroscopy, and laser particle size analysis were carried to characterize the leaching residue. Results show that the leaching rate of Au is increasing as the concentration of leaching agent and oxidant rising; temperature rise will accelerate the leaching reaction rate, gold-leaching efficiency after 8 h can reach 97.93%; XRD and infrared spectroscopy testing indicate that the mineral structure is destroyed completely, the desulfurization rate of S element can reach 96%.

Keywords Environment-friendly · Gold concentrate leaching · Brominating process

Introduction

As the classic method of hydrometallurgy, cyanide process has been occupying the leading position in the gold industry for more than one hundred years [1]. At present, 75% of the global sodium cyanide is used on gold industry, among which

C. Li · H. Li (✉) · Q. Jing
Beijing Key Lab of Green Recycling and Extraction of Metals,
University of Science and Technology, Beijing,
30# Xueyuan Road, Beijing 100083, China
e-mail: lihongxu@ustb.edu.cn

H. Li · Q. Jing
School of Metallurgical and Ecological Engineering, University of Science and Technology Beijing, 30# Xueyuan Road, Beijing 100083, China

© The Minerals, Metals & Materials Society 2017
M.A. Meyers et al. (eds.), *Proceedings of the 3rd Pan American Materials Congress*,
The Minerals, Metals & Materials Series, DOI 10.1007/978-3-319-52132-9_65

the most is consumed by the gold mining industries in North America, South America, Russia, Middle Asia, Austria, South Africa and China. But cyanide has acute toxicity that can pollute the environment, thus it is imperative to develop the clean and highly-effective gold-leaching technology [2].

Numerous researches about new leaching reagents include the halide, thiourea, thiosulfate and thiocyanate have been conducted to search for a non-cyanide hydrometallurgical systems for gold extraction [3–9]. Among all the methods, bromination can be less interfered by impurity metal ions, has a better performance in choosing gold, more stable coordination compounds and less pollution to the environment. Bromination characterized by high leaching speed, non-toxic, strong adaptability to the changes of PH, low cost of environmental protection facilities, high recovery rate of gold and almost equal amount of medicament cost with that of cyaniding. Besides, leaching reagents can be reused, which is corresponding with the advocated "green metallurgy". As to extract gold from the refractory gold ore, bromination also better than cyaniding method, with a higher leaching efficiency.

Materials and Experiment

Properties of Materials

The raw minerals is flotation gold concentrate from Hebei province in China, which have a gold content of 141.5 g/t. It belongs to micro-disseminated type sulfur refractory gold ore, the results of XRD and petrographic analysis of the concentrate shows the major gold minerals are native gold, petzite and kustelite, mostly are wrapped in pyrite or as growth in mineral fissures in the shape of granular and long strips. Part of coarse native gold is included in albite and potassium feldspar. The major metallic minerals are hematite, pyrite, limonite, chalcopyrite, galena and trace of tellurium. The primary element content of gold concentrate as shown in Table 1.

Leaching Experiments

Before the experiments, gold concentrate was ground to more than 95% of −0.074 mm by ball grinder and flour mill. The flow diagram of experiments as shown in Fig. 1.

Table 1 Primary element content of gold concentrate

Element	Au (g/t)	Ag (g/t)	Si	Fe	S	Al	K	Na
Content (%)	141.5	43.9	46.47	17.35	5.25	12.83	7.65	3.51
Element	Mg	Ba	Sr	Ti	Pb	Mn	Cu	Zn
Content (%)	0.72	0.46	0.37	0.25	0.20	0.19	0.18	0.13

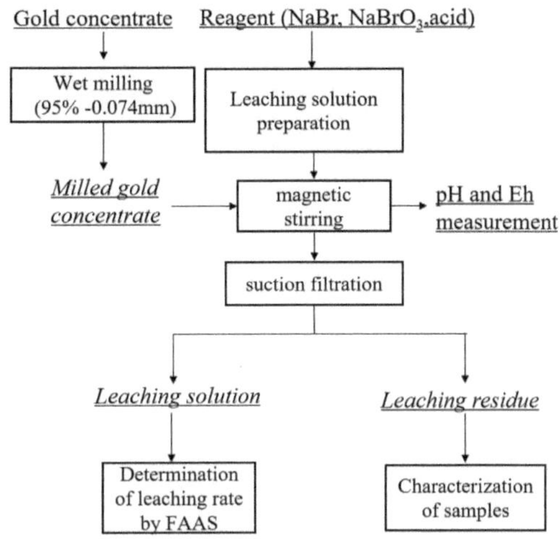

Fig. 1 The flow diagram of bromide leaching experiments

Characterization Method

Fire assaying method was adopted in this experiment to analyze the content of gold and silver in mineral, and the flame atomic absorption spectroscopy method and the ICP-AES method were used to trace and monitor the changes of the leaching rates of minerals. With the representations, before and after the hydrometallurgy, of mineral compositions of minerals and mineral waste residue, the gold occurrence states, the surface appearances, the grain compositions and the mental valence states, the structure changes of the minerals after leaching and the leaching performance can be analyzed.

Results and Discussion

Experimental Results

As can be seen from Fig. 2, leaching rate increased with reagent concentration. The increase of the concentration of NaBr contributes to the leaching reaction of gold. When the concentration of NaBr is 0.1942 mol/L, the leaching rate can reach 64% after 4-h stirring, while the leaching rate was lower with insufficient leaching agent after the equal time span of stirring, and the gold concentration was just 0.25 μg/mL in blank test with the leaching rate of 5%, which shows that the concentration of the leaching agent generates a large influence to the leaching result.

When the NaBr is sufficient, higher concentration of the oxidizing agent brings stronger oxidability of the solution and contributes to higher the oxidation potential, which helps the leaching in theory. As shown in the Fig. 3, under the high

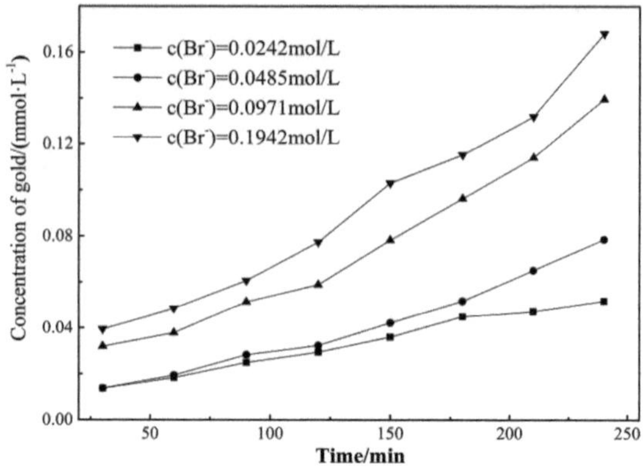

Fig. 2 Variation of gold leaching quantity over time in different NaBr concentration (L-S ratio 10:1, oxidant concentration is 15 g/L, initial pH = 3, room temperature)

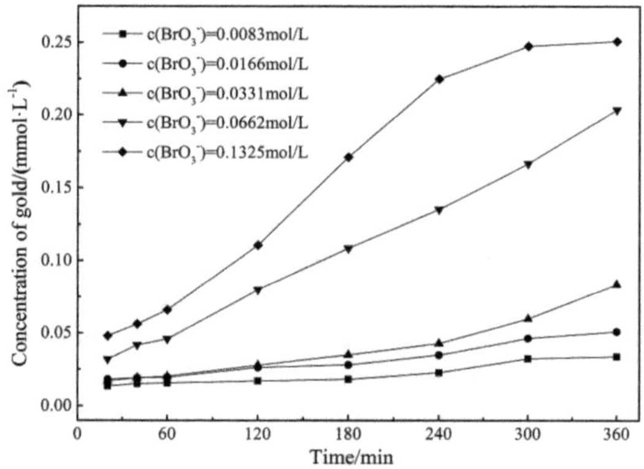

Fig. 3 Gold leaching quantity changes over time in different $NaBrO_3$ concentration

concentration of oxidizing agent, higher gold leaching rates can be achieved, which shows that the increase of the concentration of oxidizing agents is in favor of gold leaching and can shorten the time of leaching process.

As shown in the Fig. 4, temperature has a big influence to leaching efficiency, that is, more leaching amount of gold with higher temperature. In normal temperature, after 8 hours' leaching, the leaching rate can reach as much as 97%. Here, the environment temperature mainly influence the dynamic condition rather than the thermodynamic condition of the leaching process, higher temperature contributing to leaching chemical reactions and diffusion processes thus shortening the time before the leaching course gets into equilibrium state.

XRD Analysis

Compare the X-ray diffraction patterns (Fig. 5) and chemical composition (Table 2) of gold concentrate before and after leaching, it can be seen obviously that the

Fig. 4 Leaching quantity changes over time at different temperature

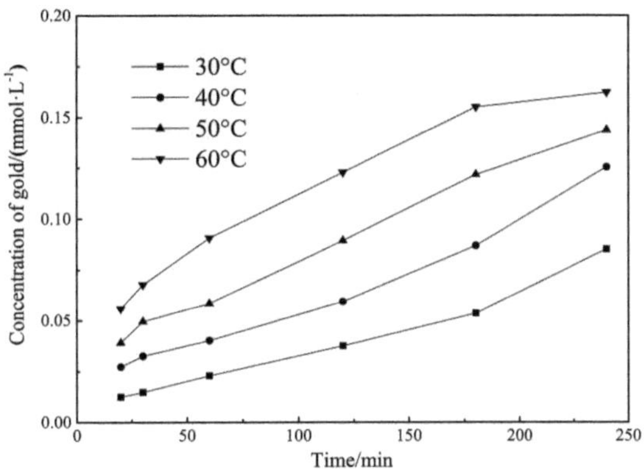

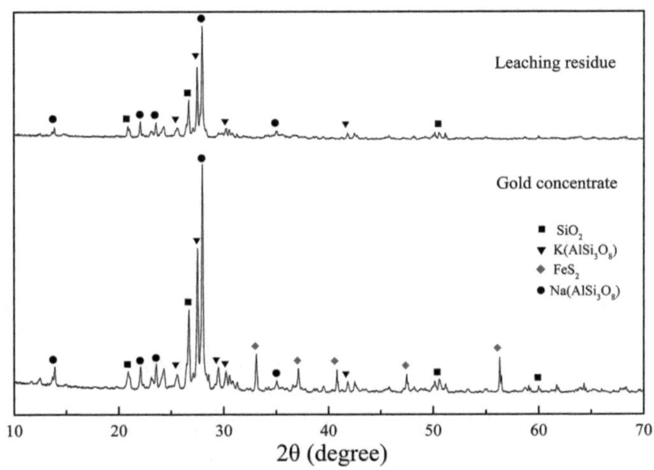

Fig. 5 XRD diffraction result of oxidation leaching residue in gold concentrate and leaching residue by bromine leaching method

Table 2 Chemical composition of leach production

Element	Au (g/t)	Ag (g/t)	Si	Fe	S	Al	K	Na
Content (%)	5.30	2.80	52.16	18.79	0.23	13.19	8.95	3.58
Element	Mg	Ba	Sr	Ti	Pb	Br	Cu	Zn
Content (%)	0.53	0.55	0.31	0.34	0.15	0.24	0.03	0.03

diffraction peaks of pyrite disappear completely, and the peaks of gangue components have no significant change. It means that pyrite is decomposed completely, the micro-fine gold wrapped in sulfide minerals can be exposed and react with the leaching agent, sequentially the stable complex ion was formed in aqueous solution. The main oxidation and leaching reactions that may occurred shows as follows:

$$2FeS_2 + 15HBrO + 7H_2O \rightarrow 2Fe(OH)_3 + 4H_2SO_4 + 15HBr \quad (1)$$

$$2Au + 3HBrO + 3H^+ + 5Br^- \rightarrow 2[AuBr_4]^- + 3H_2O \quad (2)$$

In the leaching process, Fe being oxidized to Fe^{3+} is good for the gold leaching process. We can see from the comparison with the Fig. 1 that the desulfurization degree can reach as high as 96%. And the content of the Pb and Zn, which existing in the form of sulphide ore, was reduced. There may exist little Br, which might be caused by bromide attaching at the surface of the leaching residues.

SEM Characterization

The samples were characterized by scanning electron microscopy for observing the morphology changes of minerals before and after gold leaching. The photograph of gold-bearing minerals and leaching residue by scanning electron microscope are showed in Figs. 6 and 7.

According to the SEM images, the surface appearance of gold concentrate changed remarkably after leaching in bromine system. Before leaching, the surface of the concentrate is smooth and in regular shapes. While after leaching, the grain sizes of the leaching residues decreased obviously, with rough surface and traces of erosion, it illustrate the mineral structure was destroyed completely. It also can be seen that the particle surface is porous, it is helpful for the diffuse process.

The porous structure makes the minerals touching the gold easy, and it is enable $[AuBr_4]^-$ to diffuse into the solution, which thus can improve the leaching efficiency, the best gold leaching efficiency in this experiment can be up to 97.93%.

Fig. 6 Surface morphology of gold concentrate before leaching

Fig. 7 SEM photograph of leaching residue after leaching

Table 3 Variation of particle size distribution and specific surface area before and after leaching

Sample	X10 (μm)	X50 (μm)	X90 (μm)	BET (m^2/cm^3)
Gold concentrate	1.490	2.914	6.089	2.765
Leaching residue	1.532	2.700	5.117	2.860

Laser Particle Size Analysis

Laser particle analyzer was adopted to study the distribution intervals of the particle sizes of the materials and the leaching residues before and after the leaching, the results as following in Table 3. We can see from the table that the data of the leaching residues of size X50 and X90 are smaller than that of the leaching materials, and in the process the grain sizes reduced after bromination leaching, specific surface area increased and the distribution of the leaching residues more centralized, which indicates that the higher monomer dissociation degree of the minerals after leaching can increase the action area between the leaching solution and the minerals, thus helps to the leaching.

Conclusion

Bromide leaching system is a stable gold leaching system with advantages of low pollution, no harmful gas, fast reaction speed, short process and the reagents is also easy to get, has a great potential in industrial application.

The influence of different reagent concentration and temperature on the bromide leaching process were tested in this research, a variety of modern analysis methods as X-ray diffraction (XRD), scanning electron microscopy, and laser particle size analysis to characterize the leaching residue.

Research results show that the leaching dynamics condition within the concentration range of this experimental study is good, the leaching rate of Au is increasing as the concentration of leaching agent and oxidant rising; temperature rise will accelerate the leaching reaction rate, gold-leaching rate measured after 8 h can reach 97.93%; the mineral structure is destroyed completely, the prolapse rate of S element more than 96%, which implements the one-step direct leaching of gold concentrate.

Acknowledgements The authors gratefully acknowledge the financial support of the National Science Foundation of PRC for the Research Project (2012-51234008).

References

1. Manivannan, S., & Ramaraj, R. (2012). Synthesis of cyclodextrin-silicate sol–gel composite embedded gold nanoparticles and its electrocatalytic application. *Chemical Engineering Journal, 210,* 195–202.
2. Wai, W., Eugene, L., Mujumdar, A. S. (2009). Gold extraction and recovery processes. *Minerals, Metals and Materials Technology Centre (M3TC).*
3. Senanayake, G. (2004). Gold leaching in non-cyanide lixiviant systems: Critical issues on fundamentals and applications. *Minerals Engineering*, *17*(6), 785–801.7.
4. Baghalha, M. (2012). The leaching kinetics of an oxide gold ore with iodide/iodine solutions. *Hydrometallurgy, 113–114,* 42–50.
5. Mensah-Biney, R. K., Reid, K. J., & Hepworth, M. T. (1994). Kinetics of gold loading from bromide solution onto an anion exchange resin. *Minerals Engineering, 7*(7), 865–887.
6. Gönen, N., Körpe, E., Yıldırım, M. E., et al. (2007). Leaching and CIL processes in gold recovery from refractory ore with thiourea solutions. *Minerals Engineering, 20*(6), 559–565.
7. Feng, D., & van Deventer, J. S. J. (2010). The effect of iron contaminants on thiosulphate leaching of gold. *Minerals Engineering, 23*(5), 399–406.
8. Syed, S. (2006). A green technology for recovery of gold from non-metallic secondary sources. *Hydrometallurgy, 82*(1–2), 48–53.
9. Vukcevic, S. (1996). A comparison of alkali and acid methods for the extraction of gold from low grade ores. *Minerals Engineering, 9*(10), 1033–1047.

Working Experience on the New WOX Washing and Leaching Plant at ZGH Boleslaw S.A., Poland

Angel Selke, Leszek Stencel, Mirosław Fatyga, Bogdan Pieczonka and Łukasz Zięba

Abstract The new facility for washing and leaching of 70,000 tpy of WOX (zinc waelzoxide) with a high content of chlorine and fluorine, was put into service at the end of March 2013. The process technology of removing chlorine and fluorine is made of two stages of continuous washing. The first step—is a counter-current washing of WOX at atmospheric pressure and the second stage is the high pressure washing. Second stage washing is carried out at the pressure of about 2 bar and temperature 120 °C, in an autoclave. The next step of the process is leaching of washed WOX with spent electrolyte coming as a feedback from existing Cell House. After separation and filtration the zinc sulphate solution is directed to the existing neutral leaching section. The final level of chlorine and fluorine are Cl < 0.005%, and F < 0.01%.

Keywords Hydrometallurgy · Zinc · Waelzoxide · Cl-and F-Removal

Project Background

ZGH "Bolesław" S.A. is a producer of high-quality SHG zinc with an output of more than 80,000 t/year zinc in 2015, producing mainly from its own sulphide concentrates. Since 2009, ZGH "Bolesław" S.A. has been processing zinc oxides, which initially constituted 3.5% share in the zinc production (in the first year) and currently reach up to 30%. The zinc oxides used for the production come from a sister company, Bolesław Recycling, where are produced in two batch streams:

A. Selke (✉)
ingenium GmbH, Schoene Aussicht 8c/2, 61348 Bad Homburg, Germany
e-mail: selke@ingenium-online.com

L. Stencel · M. Fatyga · B. Pieczonka · Ł. Zięba
ZGH Bolesław S.A., ul. Kolejowa, 3732-332 Bukowno, Poland

© The Minerals, Metals & Materials Society 2017
M.A. Meyers et al. (eds.), *Proceedings of the 3rd Pan American Materials Congress*, The Minerals, Metals & Materials Series, DOI 10.1007/978-3-319-52132-9_66

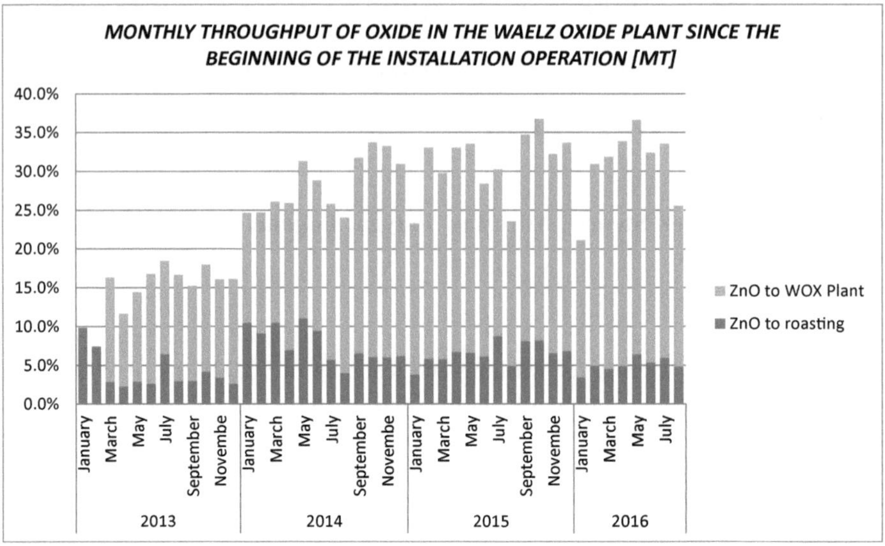

Fig. 1 Actual waelzoxide distribution of zinc oxide production since start up of the new WOX-plant at ZGH Bolesław S.A

- The stream of waste slurry (including the slurry from the neutral leaching): the oxides from this stream are directed to the zinc production at HC "Miasteczko Śląskie" for its processing into the IS-Furnace.
- The stream of steel dust (EAFD): prior to 2013 these oxides were directed to de-chlorination, and then to roasting in the fluidized-bed furnace at the existing roaster in ZGH Bolesław.

After the installation of the new Waelzoxide De-Chlorination and De-Fluorination Facility, and its start-up in March 2013, the oxides recovered from steel dust (EAFD) are now mainly processed at the Waelzoxide Plant through a modern washing and leaching technology and then routed directly to the existing neutral leaching section, where the stream is directly mixed with zinc sulphate solution obtained during the calcine leaching. A portion of the waelzoxide is still continuously directed to roasting furnace.

The actual waelzoxide distribution from the two technological process directions of the zinc production is shown in Fig. 1.

Waelzoxide Plant Technology

In 2011, the Board of Directors of ZGH "Bolesław" S.A. decided to build a new Waelzoxide Treatment Plant based on the long-term proven technology used by the Electrolytic Zinc Plant at Datteln, Germany. The contractor for the process design was ingenium GmbH, Bad Homburg, Germany (Fig. 2).

Fig. 2 Waelzoxide treatment plant at ZGH Bolesław S.A.—pressure washing

ingenium GmbH was also responsible for the commissioning and start-up of the Waelzoxide Treatment Plant during in the period of January 2013 to March 2013.

The commissioning and start-up of the Waelzoxide Treatment Plant was successfully completed and afterwards the performance tests proved the design data of the WOX-Plant. To date, the zinc oxides at ZGH "Bolesław" S.A. are processed in two technological process routes directions. The first process route direction is the newly built Waelzoxide Treatment Plant, and the second direction is fluidized-bed roasting together with sulphide concentrate.

The main process route is processing the waelzoxides into in the new Waelzoxide Treatment Plant, which share in the production of electrolytic zinc increases gradually.

The nominal throughput capacity of the Waelzoxide Plant is 70,000 t/year, and the planned target share of zinc oxide in the production of electrolytic zinc from this process route is estimated to achieve at least 40%.

The new Waelzoxide Plant at ZGH "Bolesław" S.A. was built on the basis of the technology and equipment purchased from Datteln (approx. 30% of the equipment), including two autoclaves with 150 m^3 capacity each. These autoclaves are the heart of the whole oxides washing technology. At a pressure of 1.5–2.0 bar and a temperature of 120–130 °C, the washing process of fluorine takes place more quickly and efficiently. ingenium GmbH, drawing up the process design, relied on the experience of the Datteln Plant, where the process was examined and improved for several years.

The original basic process design of the pressure washing was updated by using ingenium's and ZGH "Bolesław" S.A. engineers' experience. Thus, the process design has been improved and is different from the original basic one used by the Electrolytic Zinc Plant Ruhr-Zink GmbH Datteln, Germany.

The main differences in relation to the original process design are:

- Counter-current washing: in the first stage of the oxide washing
- Filtration on a vacuum filter belt: in a second stage of the oxide washing
- Thickening on the lamella filter: in the third stage of oxide leaching.

These and other modifications are the base for reliable operation of the whole system, which now is more stable, and achieves on continuous basis better results regarding chlorine and fluorine removal.

Brief Process Description

The whole process works continuously and consists of the following main steps:

Preparation of Oxide Suspension

The so-called zinc oxide pre-washing is continuously carried out in a reactor, into which process water, waelzoxide and Na_2CO_3 solution are mixed. The process is conducted at atmospheric conditions with the aim to remove mainly chlorine from the waelzoxide.

Counter-Current Washing

The suspension thickening of de-chlorinated waelzoxide (after pre-washing) with the aim of phase separation, and so-called counter-current washing of zinc oxide. The underflow from the thickener is directed to the filtration facility, while the clear overflow containing the removed chlorides and fluorides are sent to the effluent treatment plant.

Filtration

The filtration of the de-chlorinated zinc oxide suspension takes place onto a vacuum drum filter equipped with cake washing facilities. Fresh process water is used in order to improve the pre-washing efficiency.

Pressure Washing

The zinc oxide pressure washing process carried out in a four-chamber horizontal autoclave. The aim of the process is to remove mainly fluorine and the remaining amount of chlorine from the zinc oxide.

Filtration

The filtration of the de-fluorinated zinc oxide suspension after the pressure washing step takes place onto a horizontal vacuum belt filter, where the filter cake is efficiently counter-current washed and dried in order to obtain low fluorine contents into the filter cake remaining humidity.

Leaching the Purified Zinc Oxide

The leaching of the purified zinc oxide after removal of chlorine and fluorine is carried out with spent electrolyte from the existing cellhouse. The leaching process of the zinc oxides achieves an efficiency up to 99.5%.

Thickening and Filtration

The process of thickening the acidic suspension obtained from the purified zinc oxide leaching, is carried out in an inclined plate thickener (Lamella thickener).

The thickener overflow containing the zinc sulphate solution is directed to the existing calcine leaching section while the underflow is filtered on a chamber-membrane filter press. The resulting filter cake with a low zinc content and a high lead content is directed to a lead recovery plant.

Achievements (Results)

From the start-up of the Waelzoxide Plant up to date approx. 80,000 t of the oxide from EAFD have been processed, in accordance with the attached chart (Fig. 3).

The oxide throughput to the WOX-Plant mainly depends on the EAF dust availability and production of waelzoxide at Bolesław Recycling Plant. As the chart shows, an increase in the oxides throughput is gradual, and the share in the zinc oxide production from this process route grows according to Fig. 1.

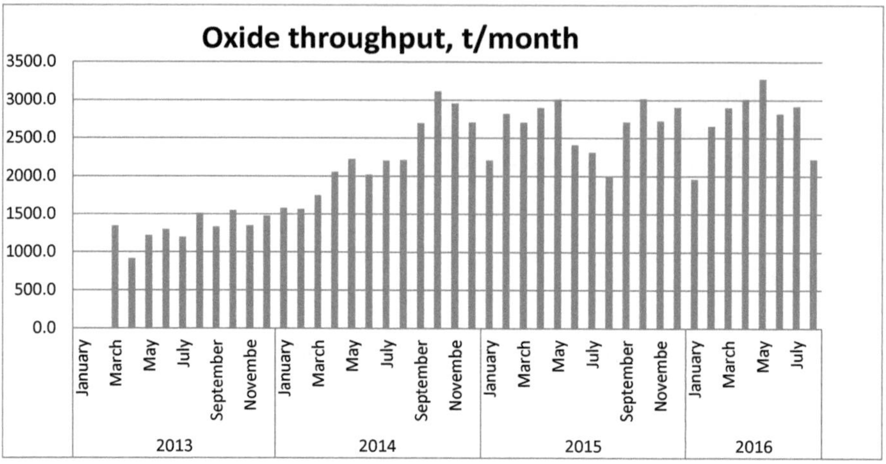

Fig. 3 Monthly throughput of oxide in the Waelzoxide plant since the beginning of the installation operation

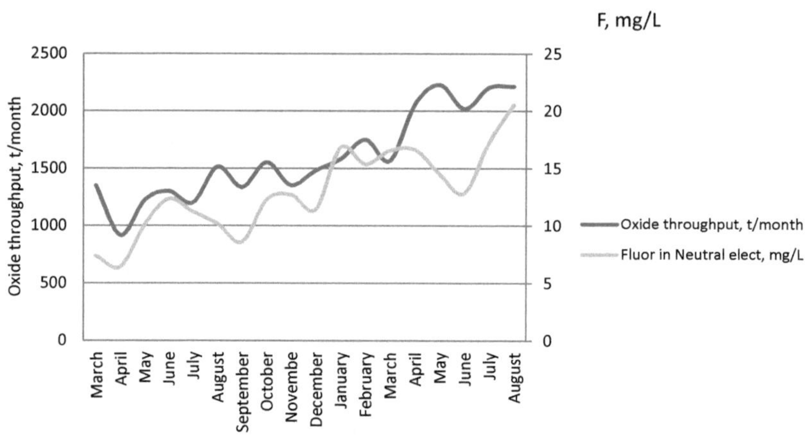

Fig. 4 Relation of fluorine concentration in the electrolyte versus the amount of processed oxide in the Waelzoxide plant, since the beginning of the production

It should be noted that since the moment of the first introduction of washed zinc oxide into existing zinc sulphate solution circuit, the fluorine content in the circulating electrolyte is continuously monitored.

As expected, (a simulation of the fluorine content increase depending on the amount and the content of fluorine in the filter cake was previously conducted), the increase in the fluorine content in the circulating electrolyte was observed and followed the amount of fluorine introduced with the oxide. In the chart (Fig. 4), the relation of the fluorine content in circulating electrolyte against the amount of the oxide processed by the Waelzoxide Plant is shown.

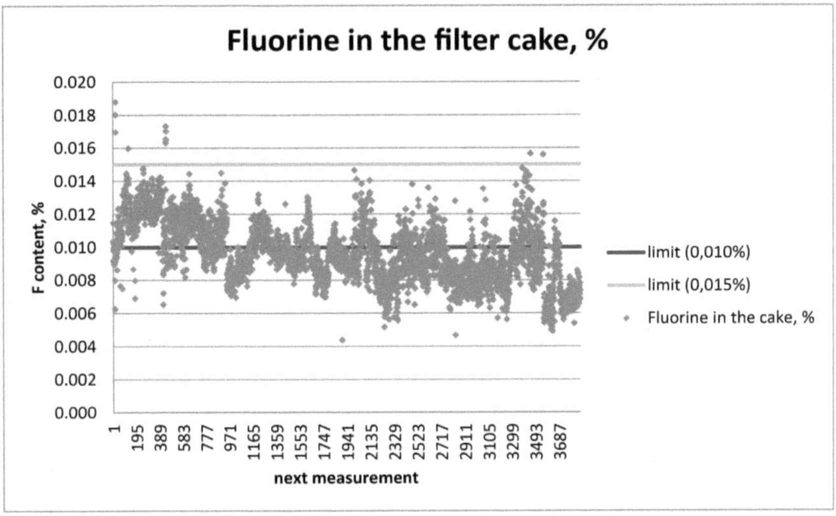

Fig. 5 Fluorine content in the filter cake from the belt filter since the start of production of the Waelzoxide plant

Zinc hydro-metallurgists know very well how important the concentration monitoring of fluorine in the circulating electrolyte is. As a reminder—at the high fluorine concentration in the electrolyte, adhesion of zinc to an aluminum cathode occurs and the adhesive strength depends on the fluorine concentration, and in the industrial practice it should not exceed 30 mg/l (15 mg/l at AZSA, Spain). However, it has to be mentioned, that in the Electrolytic Zinc Plant at Datteln, Germany where the cathode stripping was mechanically carried out of the 3.5 m^2. Cathodes, the electrolysis process was carried out at maximum levels of 60 mg/l fluorine.

Currently, the level of fluorine in the circulating electrolyte at ZGH "Bolesław" S.A. is 20–25 mg/l, and shows a stable trend.

An extremely important thing is that the fluorine content in the washed zinc oxide should be kept at the lowest. After implementing the new technology of oxide washing, fluorine and chlorine levels of **<0.01** and **<0.005%** respectively in the filter cake produced in the new Waelzoxide Plant are achieved.

The chart under Fig. 5 represents the fluorine content in the filter cake being obtained since the beginning of the production at the Waelzoxide Plant.

Conclusions

- The Waelzoxide Plant at ZGH "Bolesław" S.A. has been working smoothly and continuously since March 2013; the waelzoxide throughput has been continuously increased and depends basically on the untreated raw material (waelzoxide) availability, which is determined by the potential raw material suppliers.

- The Waelzoxide Plant nominal capacity is 70,000 t/year, what in a consequence allows to increase the share of zinc oxide in the production up to 50% instead of the planned 40% (the Datteln case).
- The purity of the washed zinc oxide is at the level of **0.01% F** and below, what allows to increase the share of the oxide in the electrolytic zinc production.
- ZGH "Bolesław" S.A. intends to continue to maintain the two technological process routes (zinc calcine and zinc waelzoxide), what can result in an increase of the waelzoxide share in the production up to 50%.

Part VIII
Nanocrystalline and Ultra-fine Grain Materials and Bulk Metallic Glasses

Continuous Dynamic Recovery in Pure Aluminium Deformed to High Strain by Accumulative Press Bonding

Sajjad Amirkhanlou, Mostafa Ketabchi, Nader Parvin and Fernando Carreño

Abstract Microstructural evolutions of AA1050 aluminium alloy, prepared by accumulative press bonding, have been investigated by means of sophisticated analytical tools, including STEM and EBSD. The results revealed that continuous dynamic recovery was dominant mechanism in grain refinement, and resulted in formation of nano/ultrafine grains with average diameter of 450 nm in pressing direction and 320 nm in transverse direction. By increasing strain during APB process, the mean misorientation angle and the fraction of high angle boundaries increased and reached a saturation value of $\sim 35°$ and $\sim 78\%$, respectively.

Keywords Nano/ultrafine structured materials · Severe plastic deformation (SPD) · Accumulative press bonding (APB) · Scanning transmission electron microscopy (STEM) · Electron backscattered diffraction (EBSD)

S. Amirkhanlou (✉) · M. Ketabchi · N. Parvin
Department of Mining and Metallurgical Engineering,
Amirkabir University of Technology, Tehran, Iran
e-mail: sajjad.amirkhanlou@brunel.ac.uk

M. Ketabchi
e-mail: ketabchi@aut.ac.ir

N. Parvin
e-mail: nparvin@aut.ac.ir

S. Amirkhanlou
BCAST, Brunel University London, London UB8 3PH, UK

F. Carreño
Department of Physical Metallurgy, CENIM-CSIC,
Av. Gregorio del Amo 8, 28040 Madrid, Spain
e-mail: carreno@cenim.csic.es

© The Minerals, Metals & Materials Society 2017
M.A. Meyers et al. (eds.), *Proceedings of the 3rd Pan American Materials Congress*,
The Minerals, Metals & Materials Series, DOI 10.1007/978-3-319-52132-9_67

Introduction

Over recent years, efforts have been made by scientists around the world for manufacturing bulk materials with submicron or even nanosized grains, since the grain refinement plays a significant role on improvement of the mechanical properties [1, 2]. There are several techniques for processing metals with ultrafine grained (UFG, 100–1000 nm), and nanocrystalline (NC, less than 100 nm) structures [3, 4]. For producing UFG and NC materials two different methods are being employed and these are recognized as the bottom-up and the top-down approaches [5, 6]. The top-down approach includes severe plastic deformation (SPD) processing techniques [7, 8]. SPD can be explained as deformation to large strains usually below recrystallization temperature without intermediate thermal treatments that can result in UFG/NC structures [9]. Up to now, a number of SPD techniques such as equal channel angular pressing (ECAP) [10], high pressure torsion (HPT) [11] and accumulative roll bonding (ARB) [12] have been developed. The accumulative press bonding (APB) is a novel severe plastic deformation method proposed by the present authors [13–15]. The basic goal of APB is to impose an extremely high plastic strain on the material, which results in structural refinement and improvement of mechanical properties. The objective of the present study is to assess the microstructural changes of AA1050 pure aluminium during the APB process.

Experimental

As-received commercial AA1050 aluminium strips with the dimensions of 100 mm × 50 mm × 1.5 mm were annealed at 623 K for 1 h and used as raw materials. In the APB process, most attention should be paid on surface preparation in order to obtain a sufficient metallurgical bond. It is necessary to remove any contamination from the surfaces of the two strips [16–18]. The preparation processes for specimens included degreasing in an acetone bath and, subsequently, scratch brushing with a steel crimped wire wheel brush attachment (0.4 mm wire diameter) mounted in electric drill (AEG SBE 600R) with peripheral speed of 2800 rpm. In order to prevent contamination and thick alumina layer formation, the press bonding process carried out immediately after surface preparation. Figure 1 illustrates the schematic of APB process.

The press bonding was performed by using laboratory hydraulic press machine (Toni Technik Baustoffprüf systeme GmbH). Two sheets were stacked and bound tightly. Thickness reduction per cycle was 50%, which is applied in a single pass and is equivalent to a von Mises strain, $\varepsilon_m = 0.8$. Then, the press bonded sheet was cut in half and stacked. The stacked sheets were pressed again with the same reduction ratio and the same procedure was repeated up to 14 cycles at ambient temperature. The microstructure evaluations of the aluminum sheets by various

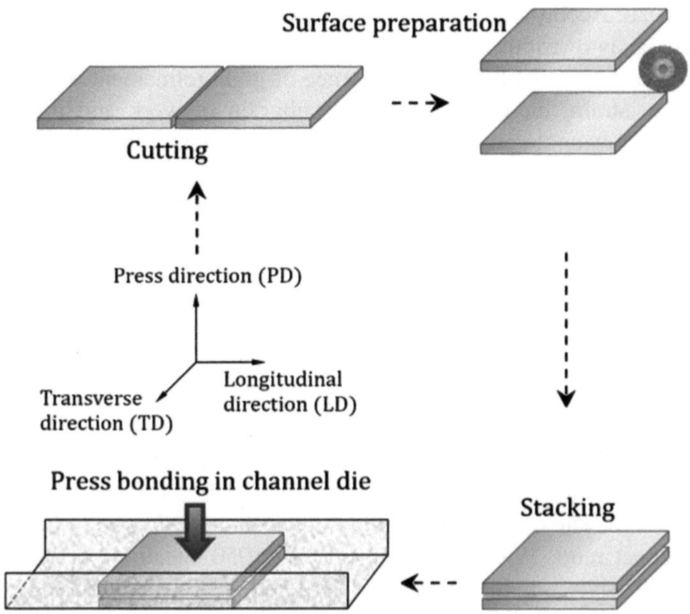

Fig. 1 Schematic illustration of accumulative press bonding (APB) process

APB cycles were carried out by field-emission scanning transmission electron microscopy (FE-STEM, Hitachi S-4800) operating at 30 kV. STEM characterization was performed on the longitudinal direction-transverse direction (LD-TD) and longitudinal direction-press direction (LD-PD) of the sheets. The microstructure of the some specimens was also characterized by transmission electron microscopy (TEM, JEOL JEM 2000) operating at 200 kV. Due to further investigation, electron backscatter diffraction (EBSD) analysis was also carried out by JEOL JSM 6500 F field emission gun scanning electron microscope (FE-SEM) operating at 20 kV with a working distance of 15 mm and tilt angle of 70°. For TEM, EBSD and STEM investigations, 3 mm diameter disks were prepared from the samples and were thinned using a twin-jet electro-polishing tool and a 30% nitric acid and 70% methanol solution at 11 V and −28 °C.

Results and Discussion

In examining the effects of APB process on the microstructural evolution of pure aluminium, STEM observations were carried out in both LD-TD and LD-PD planes. Figure 2 illustrates the microstructures of the specimens in the press direction, i.e. LD-TD planes. Figure 2a shows that the average grain size after first cycle is 1.5 μm while after three cycles the grain size becomes finer and reaches to

1.2 μm, Fig. 2b. According to Fig. 2f, the specimen after 14 cycles, was filled with the homogenously distributed ultrafine grains with average grain size of ~450 nm. Overall, Fig. 2 shows that there is a sequence of grain refinement when the number of cycles, i.e. strain, increases. First, the subgrain network and dislocation cell structures with low angle grain boundaries formed. Then, with the progress of APB due to multi directional slip [19, 20], the aforementioned structures became finer and the misorientation at dislocation cell structures and low angle boundaries of subgrains increased gradually. Ultimately, an ultrafine grained structure with high angle grain boundaries was fabricated. For further analyses, STEM observations were focused on the specimens processed by 2, 5, 10 and 14 cycles in the LD-PD planes. Unlike the LD-TD planes, the entire structure was determined by a lamellar structure, which was subdivided by two types of boundaries; lamellar boundaries almost parallel to the longitudinal direction and short transverse boundaries interconnecting the lamellar boundaries [21]. As shown in Fig. 3a, structure consisting of primarily lamellar grains elongated in the longitudinal direction. Grains with an aspect ratio of less than 1.5, grain aspect ratio (GAR < 1.5), were approximately 9% after 2 passes. The overall structure has an average grain size of 720 nm, with a 3.4 mean aspect ratio. When analyzing the AA1050 specimen processed by five cycles, Fig. 3b shows a further refined structure that is built predominantly of lamellar grains. However, the mean aspect ratio of the observed structure was specified to be 2.5, thereby demonstrating that grain fragmentation of the elongated grains happened during the additional three cycles.

It was also observed that the percentage of grains with GAR < 1.5 increased to 19% within the structure. The average grain size determined for the whole structure was 580 nm in this case. As shown in Fig. 3c, after 10 cycles the microstructure becomes finer and the average grain size reaches 510 nm and the fraction of grains with GAR < 1.5 increased to 48%.

Fig. 2 STEM micrographs of AA1050 pure aluminium processed by APB after different cycles, taken from the LD-TD planes; **a** 1, **b** 3, **c** 5, **d** 7, **e** 10 and **f** 14 cycles

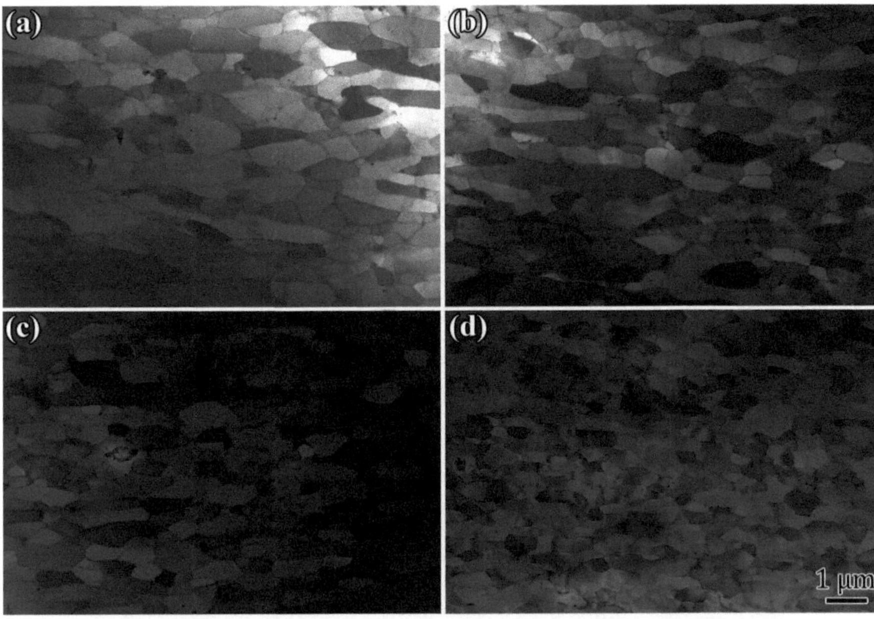

Fig. 3 STEM micrographs of AA1050 pure aluminum processed by APB after different cycles, taken from the LD-PD planes; **a** 2, **b** 5, **c** 10 and **d** 14 cycles

The final specimen analyzed by STEM, shown in Fig. 3d, was processed by 14th APB cycle. The microstructure was mostly filled with equiaxed grains. The average grain size attained was ~320 nm, 67% were determined to show GAR < 1.5. It should be noted that when the number of APB cycle increased, the grains became more equiaxed. As seen in Fig. 3, the longitudinal section microstructures represented elongated grains after the first few cycles, but for further numbers of cycles, these elongated grains are converted into finer equiaxed grains. The main reasons for these changes can be attributed to grain subdivision mechanisms and the generation and arrangement of dislocations during severe plastic deformation. Figure 4 displays STEM micrographs of AA1050 processed by APB after 14th cycle in high magnification. It is obvious that after 14th cycle, the microstructure considerably becomes finer and a certain amount of nanograins, i.e. grains less than 100 nm, are also obtained. It should be noted that after 14 cycles, almost 10% of all grains were less than 100 nm. These grains can significantly improve the mechanical properties of the materials [22, 23].

Figure 5 shows the OIM/EBSD maps obtained by EBSD measurement of the specimens at different APB cycles, respectively. In Fig. 5, high-angle grain boundaries (HAGBs) of which misorientation angles are higher than 15°, are depicted by black lines, while low-angle grain boundaries (LAGBs) of which misorientation angles are 2°–15°, are depicted by red lines. All these images were recorded on LD-TD planes. For each specimen the fraction of high angle

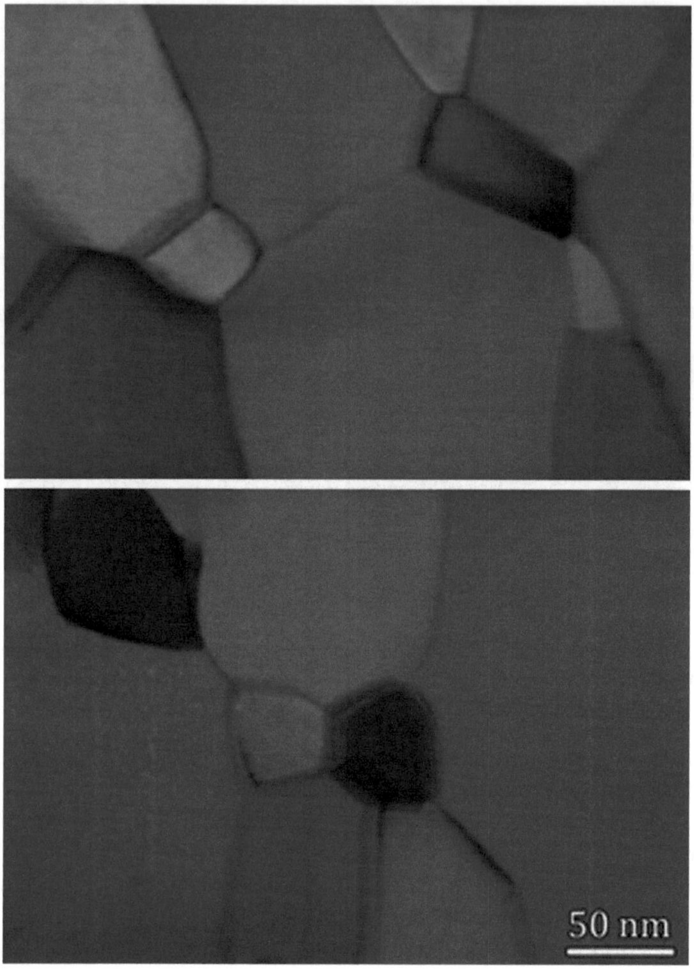

Fig. 4 STEM micrographs of AA1050 pure aluminium processed by APB after 14 cycles, formation of nanograins

boundaries (f_{HAB}), and mean misorientation angle ($\bar{\theta}$) are represented in the figure. The fraction of high angle grain boundaries and the mean misorientation angle of boundaries for the annealed specimen are 90% and 41°, respectively. As shown in Fig. 5b, after the first APB cycle, f_{HAB} and $\bar{\theta}$ are very low, 15% and 4°, respectively. But these values increase to 63% and 27° after 5 cycles. This is a clear indication that the misorientation between the subgrains increases and a balanced distribution of low-angle and high-angle boundaries is obtained. The mean misorientation angle increases by increasing strain during APB process and reaches a saturation value of about 34° after 10 cycles and 35° after 14 cycles. On the other hand, f_{HAB} is 73 and 78% after 10 and 14 cycles, respectively. The EBSD analysis

Fig. 5 Grain boundary maps of AA1050 pure aluminium processed by APB after different cycles, taken from the LD-TD planes; **a** initial aluminium, **b** 1, **c** 3, **d** 5, **e** 10 and **f** 14 cycles

shows that by continuing APB process low angle grain boundaries turn into high angle boundaries and the fraction of high angle grain boundaries increases. Also, the mean misorientation angle of the boundaries increases and the grain size decreases. Conventional dislocation theory [12, 24–27] explains the main mechanism that can contribute to this evolution. The conventional dislocation theory explains that at the primary stage of deformation, a very high dislocation density is created, which leads to the fabrication of an intragranular structure including cells with thick cell walls and low angles of misorientation. The thickness of the cell walls is decreased by increasing the strain. These walls convert to grain boundaries, and finally arrays of ultrafine grains with high angle nonequilibrium grain boundaries are formed.

The grain refinement mechanisms during APB process can be listed as follows: (I) severe shear deformation: This mechanism influences initially, and specially, the microstructure just below the surface due to the friction between the strips and the press anvil. This shear deformation considerably increases the equivalent strain and causes grain refinement. [28–30], (II) continuous dynamic recovery: this mechanism is dominant phenomenon in FCC materials with high stacking fault energy, especially aluminium. It has been suggested [12, 31] that the continuous dynamic recovery takes place in the UFGs and the processes of ultra-fine grain subdivision, where recovery to form clear sub-boundaries and short-range grain boundary migration can be observed [32–35].

Conclusions

The APB process, as new severe plastic deformation method, has been carried out successfully up to 14 cycles at ambient temperature on commercial AA1050 pure aluminium. The main results are summarized as follows: The APB process led to decreasing the grain size of AA1050 strip from 40 μm to the nano/ultrafine grain structure with average grain size of 450 nm in LD-TD planes and 320 nm in LD-PD planes as shown in STEM micrographs. EBSD analysis shows that the fraction of high angle boundaries (f_{HAB}) and mean misorientation angle ($\bar{\theta}$) increased by increasing strain during APB process and reached the saturation values of 78% and 35°, respectively. The APB refining mechanism was continuous dynamic recovery characterized by subdivision to ultrafine grains by severe deformation and progressive recovery to form clear and highly misoriented ultrafine grains.

Acknowledgements The authors acknowledge Alfonso García Delgado and Antonio Tomás López for their assistance with the SEM, STEM, TEM and EBSD at CENIM-CSIC, and financial support from CICYT (Spain) under program MAT2012-38962-C03-01, and the Ministry of Science, Research and Technology of Iran.

References

1. Pande, C., & Cooper, K. (2009). Nanomechanics of Hall-Petch relationship in nanocrystalline materials. *Progress in Materials Science, 54*, 689–706.
2. Estrin, Y., & Vinogradov, A. (2013). Extreme grain refinement by severe plastic deformation: A wealth of challenging science. *Acta Materialia, 61*, 782–817.
3. Furukawa, M., et al. (1996). Microstructural characteristics of an ultrafine grain metal processed with equal-channel angular pressing. *Materials Characterization, 37*, 277–283.

4. Valiev, R., & Langdon, T. (2006). Developments in the use of ECAP processing for grain refinement. *Reviews on Advanced Materials Science, 13*, 15–26.
5. Meyers, M. A., Mishra, A., & Benson, D. J. (2006). Mechanical properties of nanocrystalline materials. *Progress in Materials Science, 51*, 427–556.
6. Saxl, I., et al. (2009). Grain and subgrain boundaries in ultrafine-grained materials. *Materials Characterization, 60*, 1163–1167.
7. Pippan, R., et al. (2010). Saturation of fragmentation during severe plastic deformation. *Annual Review of Materials Research, 40*, 319–343.
8. Cepeda-Jiménez, C., et al. (2011). Influence of the supersaturated silicon solid solution concentration on the effectiveness of severe plastic deformation processing in Al–7 wt% Si casting alloy. *Materials Science and Engineering A, 528*, 7938–7947.
9. Cherukuri, B., Nedkova, T. S., & Srinivasan, R. (2005). A comparison of the properties of SPD-processed AA-6061 by equal-channel angular pressing, multi-axial compressions/forgings and accumulative roll bonding. *Materials Science and Engineering A, 410*, 394–397.
10. Iwahashi, Y., et al. (1998). The process of grain refinement in equal-channel angular pressing. *Acta Materialia, 46*, 3317–3331.
11. Zhilyaev, A. P., & Langdon, T. G. (2008). Using high-pressure torsion for metal processing: Fundamentals and applications. *Progress in Materials Science, 53*, 893–979.
12. Saito, Y., et al. (1999). Novel ultra-high straining process for bulk materials—Development of the accumulative roll-bonding (ARB) process. *Acta Materialia, 47*, 579–583.
13. Amirkhanlou, S., et al. (2013). Accumulative press bonding; A novel manufacturing process of nanostructured metal matrix composites. *Materials and Design, 51*, 367–374.
14. Amirkhanlou, S., et al. (2014). Structural evaluation and mechanical properties of aluminum/tungsten carbide composites fabricated by continual annealing and press bonding (CAPB) process. *Metallurgical and Materials Transactions B, 45*, 1992–1999.
15. Amirkhanlou, S., et al. (2015). Homogeneous and ultrafine-grained metal matrix nanocomposite achieved by accumulative press bonding as a novel severe plastic deformation process. *Scripta Materialia.*
16. Amirkhanlou, S., et al. (2011). Manufacturing of high-performance A1356/SiCp composite by CAR process. *Materials and Manufacturing Processes, 26*, 902–907.
17. Amirkhanlou, S., et al. (2011). Fabrication and characterization of Al/SiCp composites by CAR process. *Materials Science and Engineering A, 528*, 4462–4467.
18. Jamaati, R., et al. (2011). CAR process: A technique for significant enhancement of as-cast MMC properties. *Materials Characterization, 62*, 1228–1234.
19. Park, K.-T., et al. (2001). Microstructural characteristics and thermal stability of ultrafine grained 6061 Al alloy fabricated by accumulative roll bonding process. *Materials Science and Engineering A, 316*, 145–152.
20. Jamaati, R., et al. (2012). Investigation of nanostructured Al/Al_2O_3 composite produced by accumulative roll bonding process. *Materials and Design, 35*, 37–42.
21. Jamaati, R., et al. (2014). Fabrication of nano/ultra-fine grained IF steel via SPD processes: A review. *Transactions of the Indian Institute of Metals, 67*, 787–802.
22. Valiev, R. Z. (1997). Structure and mechanical properties of ultrafine-grained metals. *Materials Science and Engineering A, 234*, 59–66.
23. Zhang, Y., Tao, N., & Lu, K. (2008). Mechanical properties and rolling behaviors of nano-grained copper with embedded nano-twin bundles. *Acta Materialia, 56*, 2429–2440.
24. Valiev, R. Z., Korznikov, A. V., & Mulyukov, R. R. (1993). Structure and properties of ultrafine-grained materials produced by severe plastic deformation. *Materials Science and Engineering A, 168*, 141–148.
25. Valiev, R. Z., Islamgaliev, R. K., & Alexandrov, I. V. (2000). Bulk nanostructured materials from severe plastic deformation. *Progress in Materials Science, 45*, 103–189.
26. Valiev, R. (2004). Nanostructuring of metals by severe plastic deformation for advanced properties. *Nature Materials, 3*, 511–516.
27. Zehetbauer, M. J., & Zhu, Y. T. (2009). *Bulk nanostructured materials.* Wiley.

28. Shaarbaf, M., & Toroghinejad, M. R. (2008). Nano-grained copper strip produced by accumulative roll bonding process. *Materials Science and Engineering A, 473*, 28–33.
29. Eizadjou, M., Manesh, H. D., & Janghorban, K. (2009). Microstructure and mechanical properties of ultra-fine grains (UFGs) aluminum strips produced by ARB process. *Journal of Alloys and Compounds, 474*, 406–415.
30. Pasebani, S., & Toroghinejad, M. R. (2010). Nano-grained 70/30 brass strip produced by accumulative roll-bonding (ARB) process. *Materials Science and Engineering A, 527*, 491–497.
31. Lee, S. H., et al. (2002). Microstructures and mechanical properties of 6061 aluminum alloy processed by accumulative roll-bonding. *Materials Science and Engineering A, 325*, 228–235.
32. Amirkhanlou, S., et al. (2011). High-strength and highly-uniform composites produced by compocasting and cold rolling processes. *Materials and Design, 32*, 2085–2090.
33. Amirkhanlou, S., et al. (2011). Refinement of microstructure and improvement of mechanical properties of Al/Al_2O_3 cast composite by accumulative roll bonding process. *Materials Science and Engineering A, 528*, 2548–2553.
34. Ardakani, M. R. K., Amirkhanlou, S., & Khorsand, S. (2014). Cross accumulative roll bonding-A novel mechanical technique for significant improvement of stir-cast Al/Al_2O_3 nanocomposite properties. *Materials Science and Engineering A, 591*, 144–149.
35. Kamali Ardakani, M. R., et al. (2014). Application of compocasting and cross accumulative roll bonding processes for manufacturing high-strength, highly uniform and ultra-fine structured Al/SiCp nanocomposite. *Materials Science and Engineering A, 592*, 121–127.

Effects of Natural Aging and Post-processed Heat Treatment on the Microstructure and Mechanical Properties of Friction Stir Processed Al-7B04

Y. Chen, H. Ding and J.Z. Li

Abstract 2 mm thick Al-7B04 alloy sheets under T4 and O tempers were subjected to friction stir processing (FSP). The microstructure, Vickers hardness and tensile properties of the stir zone were characterized. The results show that FSP led to the formation of full recrystallized microstructure, the average grain size in the stir zone (SZ) was about 2 μm. Besides, the stir zone produced using the base metal under T4 temper was softened, while a significant strengthening was observed when the base metal temper was O. Furthermore, both the natural aging (NA) and post-processed heat treatment (PPHT) were also conducted on FSP samples. And it indicates that NA was beneficial to strengthening the SZ regardless of the initial base metal temper. On the contrary, PPHT only contributed to the strengthening of SZ using the base metal under T4-temper.

Keywords Friction stir · Aluminum alloy · Natural aging · Post-processed heat treatment · Mechanical properties

Introduction

Friction stir processing (FSP) is an adaptation of friction stir welding (FSW), a process invented at TWI, UK in 1991 [1, 2]. The concept of FSP is remarkably simple: a rotating tool is plunged into a fixed workpiece and traverses along the line

Y. Chen · H. Ding (✉)
School of Materials Science and Engineering, Northeastern University,
Shenyang 110819, People's Republic of China
e-mail: hding2013@163.com

Y. Chen
e-mail: crainy11@126.com

J.Z. Li
AVIC Beijing Aeronautical Manufacturing Technology Research Institute,
Beijing 100024, People's Republic of China

© The Minerals, Metals & Materials Society 2017
M.A. Meyers et al. (eds.), *Proceedings of the 3rd Pan American Materials Congress*,
The Minerals, Metals & Materials Series, DOI 10.1007/978-3-319-52132-9_68

of interest [3]. In FSP, the workpiece experiences intense plastic deformation at elevated temperatures, which will lead to a stir zone (SZ) with recrystallized and fine-grained microstructure. At present, the effects of processing parameters on the microstructure and mechanical properties in the SZ have been extensively investigated. And it was reported that the evolution of the SZ during FSP was affected by not only the critical process parameters (including the tool geometry, rotation speed, traverse speed and external cooling conditions) but also the initial base material temper [4–6].

Al-7B04 is a precipitation-hardened Al–Zn–Mg–Cu aluminum alloy. It has been widely applied to produce aircraft parts due to its light weight and high strength. Over the years, different aging tempers, such as natural aging-T4 and annealing-O, have been developed for this kind of alloys. Based on our previous study [7], the initial base metal temper has a great influence on the microstructure and mechanical properties of friction stir processed Al-7B04. And after FSP, natural aging (NA) and post-processed heat treatment (PPHT) are always applied to enhance the mechanical properties of SZ further. However, effects of NA and PPHT on the microstructure and mechanical properties of friction stir processed Al-7B04 with different initial base metal tempers have not been well reported so far. Therefore it is necessary to do further research on this topic. In the present work, effects of NA and PPHT on the microstructure and mechanical properties of friction stir processed Al-7B04 under different initial base metal temper were systematically investigated.

Experimental

Rolled 7B04 aluminum alloy sheets of 2 mm were applied in this study, and the nominal chemical compositions of this alloy are listed in Table 1. Two heat treatments (T4 and O) were adopted for Al-7B04 plates before FSP. Natural aged T4-temper was conducted by solution treatment at 480 °C, soaking for 80 min followed by water quenching and aging at ambient temperature for 60 days. Annealed O-temper comprised of initial heating to 400 °C, soaking for 1 h, followed by furnace cooling to 150 °C and air cooling down to ambient temperature. The plates were friction stir processed parallel to the rolling direction. A H13 steel tool with a concave shoulder of 10 mm in diameter and a threaded and tapered pin (diameters of the root and head are 4 and 2.7 mm, respectively) was used. The FSP was performed with a tool rotation speed of 800 rpm and a tool traverse speed of 200 mm/min. The FSP samples using base metal under T4 and O tempers were designated in brief form as T4-sample and O-sample. Both the natural aging (NA)

Table 1 Nominal chemical compositions (wt%) of the studied Al-7B04 aluminum alloy

Zn	Mg	Cu	Mn	Fe	Si	Cr	Al
6.0	2.2	1.6	0.3	0.1	<0.1	0.1–0.25	Bal

and post-processed heat treatment (PPHT) were also conducted on FSP samples. The FSP samples were kept under room temperature for one year, and PPHT was carried out at 165 °C for a soaking period of 6 h.

Microstructure of the cross-section of FSP sample was characterized using an Olympus DSX-500 optical microscopy (OM) and transmission electron microscopy (TEM). The samples for OM were ground and polished and then etched using Keller's reagent (190 ml water, 2 ml hydrofluoric acid, 3 ml hydrochloric acid and 5 ml nitric acid). Thin foils for TEM observation, cut from the base metal and the center of SZ using an electrical-discharge machine, were prepared by jet electro-polishing with a solution of 70% methanol and 30% nitric acid at −30 °C and 19 V. The TEM study was performed on a Tecnai $G^2$20 TEM.

The mechanical properties of the FSP samples were evaluated by Vickers hardness and tensile tests. The Vickers hardness measurement was carried out along the centerline of the cross-section of FSP sample with a distance between neighboring measured points of 0.5 mm under a load of 50 g for 10 s. Room-temperature tensile tests were conducted on a universal testing machine with strain rate of 10^{-3} s^{-1}. The tensile specimens were cut along the process direction with a gauge length of 20 mm and width of 4 mm which covered only SZ.

Results

Microstructure Under PPHT State

Figure 1 shows the low-magnification OM images of the cross-section of the FSP samples using different initial base metal tempers under PPHT state. A distinct SZ is evident in each FSP sample and no defects such as cavity, tunnel and "zigzag line" [8] are found. Based on our previous study [7], the shapes of the SZ are

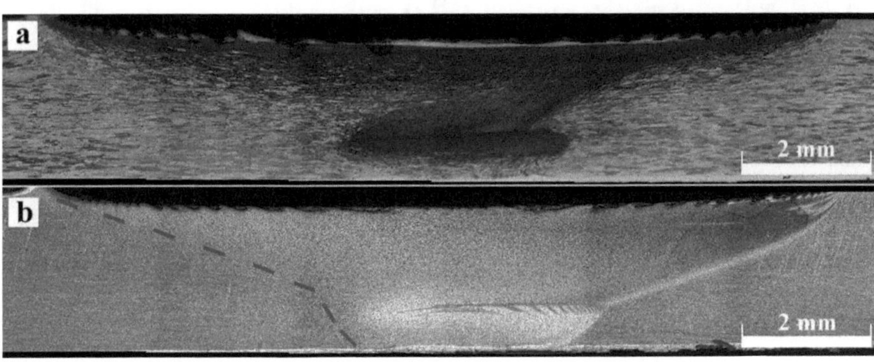

Fig. 1 Low-magnification OM of transversal cross-section of FSP samples produced by base metal under **a** T4 and **b** O tempers

apparently dependent on the initial BM temper. Under PPHT state, the shapes of the SZ keep similar to those under as-processed state. And it can be seen that both the SZs of T4- (Fig. 2a) and O-samples (Fig. 2b) are unsymmetrical, besides, the SZ of O-sample is larger than that of T4-sample. The microstructure of the BM under T4 and O tempers are shown in Fig. 2. The difference in the morphology of grains is nonsignificant between T4 and O tempers, where all the microstructure primarily consists of elongated grains due to the inherent from rolling (Fig. 2a, b). With a T4 heat treatment, the strengthening precipitates are mainly Guinier-Preston (GP) zones [9, 10], however, GP zones are hard to discern here due to their fine features and coherent relationship to the matrix (Fig. 2c). And because of the low cooling speed of annealing, the strengthening precipitates η (η′) under O temper grow and become much coarser than those under T4 tempers (Fig. 2d). The microstructure developed in the center of SZ using different initial base metal tempers under PPHT state are shown in Fig. 3. Comparing with the original banded structure, the evolved microstructure is dominated by completely recrystallized fine

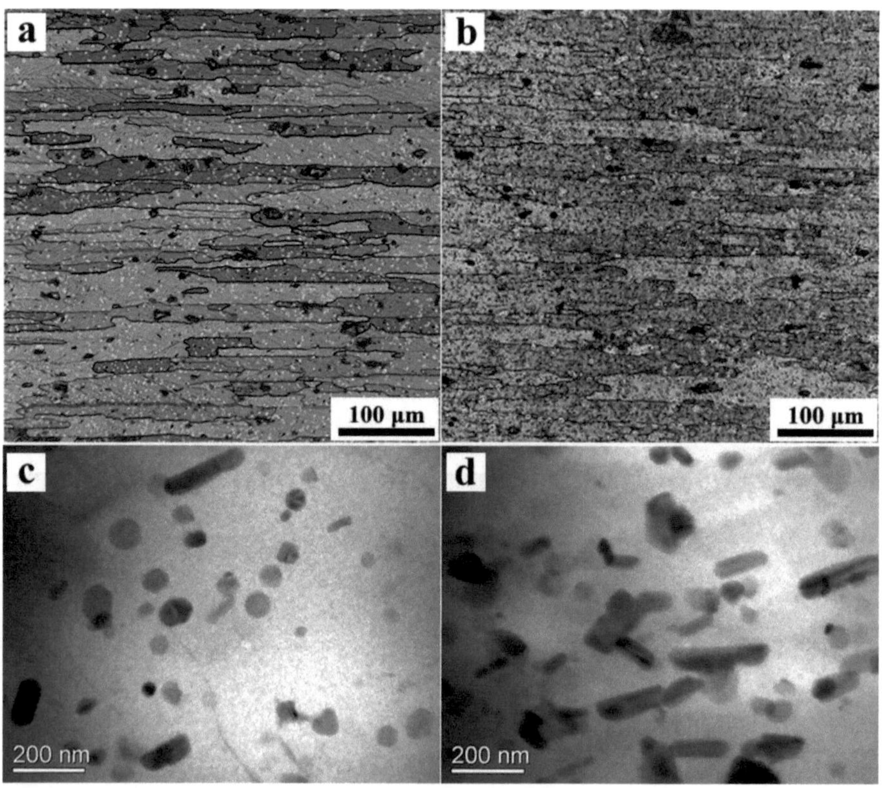

Fig. 2 Microstructure of base metal under different tempers OM micrographs of **a** T4 and **b** O tempers, TEM images of **c** T4 and **d** O tempers

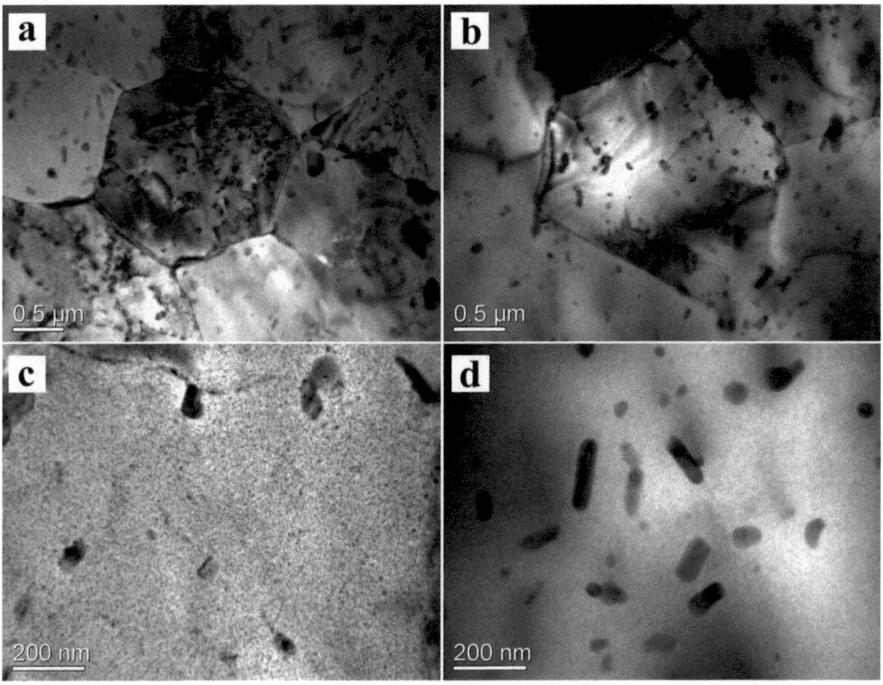

Fig. 3 Grain morphology of **a** T4- and **b** O-samples, precipitation distribution of **c** T4- and **d** O-samples

and equiaxed grains which are resulted from the severe plastic deformation and thermal exposure during FSP, and the average grain size of T4- and O-sample is similar which is around 2 μm (Fig. 3a, b). The evolution of the precipitates during FSP for T4- and O-sample is different. For T4-samples, a large amount of fine η (η′) re-precipitated during PPHT due to its relatively high solution effect (Fig. 3c) [6], however, no fine η (η′) was observe and only coarse precipitates remained in the O-sample (Fig. 3d).

Mechanical Properties Under NA and PPHT States

The hardness profiles measured across the cross-section of plates using different initial base metal tempers under as-FSP state are presented in Fig. 4. After FSP, the hardness distribution varies dependent on the initial base metal temper (Fig. 4a). The low hardness zone of T4-sample is SZ, and the softening of hardness in this region is not much. Differently, the hardness profiles of O-sample exhibits a reverse "U" shape where the SZ is significantly strengthened, and the hardness of SZ is

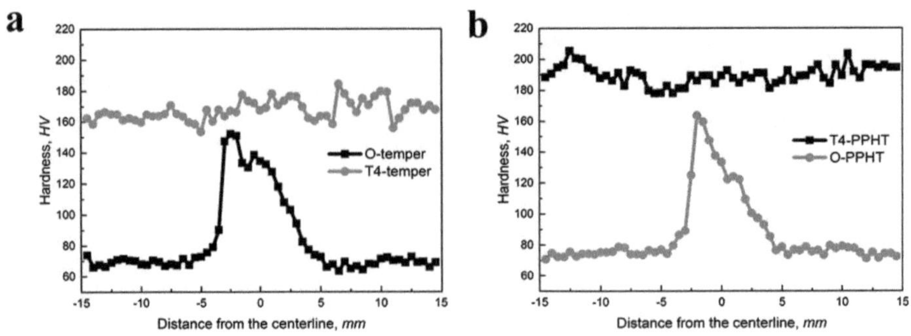

Fig. 4 Hardness distribution across the cross-section of plates under **a** As-FSP and **b** NA states

Table 2 Tensile properties of BM and FSP samples under As-FSP and NA states

Samples	YS (MPa)	UTS (MPa)	Elongation (%)
T4-BM	325	520	23.3
O-BM	110	220	20.0
T4-sample (As-FSP)	320	495	12.0
O-sample (As-FSP)	145	390	10.7
T4-sample (NA)	415	538	10.6
O-sample (NA)	310	425	11.1

twice than that of BM under O temper. A 165 °C for 6 h PPHT was conducted for all FSP samples, and the hardness under PPHT state was shown in Fig. 4b, the hardness in the SZ of T4-samples increases after PPHT due to the re-precipitation of fine η (η′), by contrary, and the increment in the hardness after PPHT of O-sample is negative.

The results of the tensile tests are listed in Table 2. Under as-FSP state, compared with the tensile strength of respective BM, the SZ of T4-samples are softened while an enhancement is observed in O-sample. It can be indicated that the softening of T4-sample was attributed to the dissolution of strengthening precipitates, where the increase in the strength of the grain refinement could not offset the decrease of the precipitation strengthening [11]. Differently, for O-sample, the mechanical properties of the SZ were enhanced by grain refinement. Based on the result of hardness, the PPHT is not suitable for strengthening the SZ of O-tempers. Therefore, the NA is selected to enhance the tensile properties of FSP samples. After one year NA, the GP zones entirely re-precipitate, and the tensile strength of both T4- and O-samples are increased.

Conclusions

In summary, FSP was an effective method to produce fine-grained Al-7B04, and the microstructure and mechanical properties were affected by the initial base metal temper. Due to the difference in solution effect, the post-processed heat treatment strengthened the stir zone of T4-sample while softened that of O-sample. Differently, the stir zones of both T4- and O-samples could be strengthened by natural aging.

References

1. Mishra, R. S., & Ma, Z. Y. (2005). Friction stir welding and processing. *Materials Science and Engineering R, 50*, 1–78.
2. Cui, G. R., Ma, Z. Y., & Li, S. X. (2009). The origin of non-uniform microstructure and its effects on the mechanical properties of a friction stir processed Al-Mg alloy. *Acta Materilia, 19*, 5718–5729.
3. Liu, H. J., & Feng, X. L. (2013). Effect of post-processing heat treatment on microstructure and microhardness of water-submerged friction stir processed 2219-T6 aluminum alloy. *Materials and Design, 47*, 101–105.
4. Chen, Y., Ding, H., Li, J. Z., et al. (2016). Influence of multi-pass friction stir processing on the microstructure and mechanical properties of Al-5083 alloy. *Materials Science and Engineering A, 650*, 281–289.
5. Sharma, C., Dwivedi, D. K., & Kumar, P. (2015). Influence of pre-weld temper conditions of base metal on microstructure and mechanical properties of friction stir weld joints of Al-Zn-Mg alloy AA7039. *Materials Science and Engineering A, 620*, 107–119.
6. Malopheyev, S., Kulitskiy, V., Mironov, S., et al. (2014). Friction-stir welding of an Al-Mg-Sc-Zr alloy in as-fabricated and work-hardened conditions. *Materials Science and Engineering A, 600*, 159–170.
7. Chen, Y., Ding, H., Cai, Z. H., et al. (2016). Effect of initial base metal temper on microstructure and mechanical properties of friction stir processed Al-7B04 alloy. *Materials Science and Engineering A, 650*, 396–403.
8. Chen, Y., Ding, H., Li, J. Z., et al. (2015). Effect of welding heat input and post-welded heat treatment on hardness of stir zone for friction stir-welded 2024-T3 aluminum alloy. *Transactions of Nonferrous Metals Society of China, 25*, 2524–2532.
9. Stiller, K., Warren, P. J., Hansen, V., et al. (1999). Investigation of precipitation in an Al-Zn-Mg alloy after two-step ageing treatment at 100 and 150 °C. *Materials Science and Engineering A, 1*, 55–63.
10. Fuller, C. B., Mahoney, M. W., Calabrese, M., et al. (2010). Evolution of microstructure and mechanical properties in naturally aged 7050 and 7075 Al friction stir welds. *Materials Science and Engineering A, 527*, 2233–2240.
11. Feng, X. L., Liu, H. J., & Babu, S. S. (2011). Effect of grain size refinement and precipitation reactions on strengthening in friction stir processed Al-Cu alloys. *Scripta Materialia, 12*, 1057–1060.

Evaluation of the Hardening and Softening Effects in Zn–21Al–2Cu with as Cast and Homogenized Microstructure Processed by Equal Channel Angular Pressing

J.L. Hernández-Rivera, E.E. Martínez-Flores, E. Ramírez Contreras, J. García Rocha, J.J. Cruz Rivera and G. Torres-Villaseñor

Abstract In this work it is reported the evolution of the microstructure of the as-cast and homogenized Zn–21Al–2Cu samples, after 2 and 6 passes in an equal channel angular pressing (ECAP) at room temperature. A homogenization treatment for 24 h at 350 °C was applied in the as-cast samples and then they were deformed. An annealing heat treatment was made in all samples after ECAP process. One of the main results is that the homogenized and deformed samples showed a uniform fine-grained microstructure after annealing, while as-cast samples without annealing presented only some regions with fine-grained microstructure. The micro segregation level was higher in the as-cast samples in contrast to the homogenized ones even after annealing. Vickers microhardness measurement on

J.L. Hernández-Rivera (✉)
CONACYT-Instituto de Metalurgia, Universidad Autónoma de San Luis Potosí, Sierra Leona No. 550, Z.C., 78210 San Luis Potosí, S.L.P., Mexico
e-mail: jlhernandezri@conacyt.mx

E.E. Martínez-Flores · E. Ramírez Contreras · J. García Rocha · J.J. Cruz Rivera
Instituto de Metalurgia-Facultad de Ingeniería, Universidad Autónoma de San Luis Potosí, Sierra Leona No. 550, Z.C., 78210 San Luis Potosí, S.L.P., Mexico
e-mail: emartine@uaslp.mx

E. Ramírez Contreras
e-mail: emmanuel.ramirez@alumnos.uaslp.edu.mx

J. García Rocha
e-mail: garcrojo@uaslp.mx

J.J. Cruz Rivera
e-mail: jdjcr35@uaslp.mx

G. Torres-Villaseñor
Instituto de Investigaciones en Materiales, Universidad Nacional Autónoma de México, Z.C., 04519 México, D.F., Mexico
e-mail: gtorres@unam.mx

© The Minerals, Metals & Materials Society 2017
M.A. Meyers et al. (eds.), *Proceedings of the 3rd Pan American Materials Congress*,
The Minerals, Metals & Materials Series, DOI 10.1007/978-3-319-52132-9_69

samples after deformation is smaller than the original material indicating a softening, however after the annealing treatment the microhardness increased, indicating that there was a slight hardening of the material.

Keywords Severe plastic deformation · Work softening · Annealing hardening

Introduction

The Zn–Al eutectoid alloy has been extensively studied, because it has a low melting point, good castability, wear resistance, low forming temperature, high strain rate sensitivity and high plastic deformability because presents the superplasticity phenomenon [1, 2]. This alloy forms a lamellar microstructure during solidification, so that to form a fine equiaxed-grained microstructure both, stages of plastic deformation and annealing treatments are required [2, 3]. Numerous thermo-mechanical treatments have been studied for eutectoid alloy, these have enabled to obtain fine-grained microstructure which is required for this alloy can be deformed superplastically, most of these treatments first considered homogenizing the as cast alloy, followed of multistage of rolling [3] or extruding [2] and then different annealing treatments. An atypical phenomenon has been observed in Zn–Al eutectoid alloy as well as alloys modified with 0.3 and 2% by weight copper, this phenomenon consists in a work-softening and anneal-hardening of these materials [2–4]. For potential applications, it is of interest find grain refinement processes that allow to obtain a fine grain microstructure from as cast alloy, with fewer processing steps. The aim of this work is to study the effect of extrusion in constant angular channel (ECAP) on the microstructure of an Zn–Al eutectoid alloy as well as Zn–22%Al–2%Cu, comparing the effect in both as-cast and homogenized microstructure. We assessed whether these alloys show the work softening and annealing hardening effect, observed for thermomechanical processes involving rolling and extrusion. The results are compared with those obtained by Yang et al. [3] for the Zn–Al eutectoid alloy and the eutectoid modified with 0.3% by weight copper. The effect of copper content on atypical behavior of these alloys and the advantages of processing by ECAP for obtaining a fine grained microstructure were studied.

Experimental Procedure

A frequency induction furnace was used to produce the Zn–22Al–2.5Cu alloy by melting the corresponding quantities of Zn (99.99%), Al (99.99%) and Cu (99.96%). The melt was poured into a graphite crucible in air and casting it into cylindrical bars of 19 mm in diameter and 35 mm in length. After that, some bars were subjected to a homogenization treatment at 350 °C for 24 h in air atmosphere. Cast and homogenized samples were subjected to an equal channel angular

Fig. 1 Tools used during the ECAP process

extrusion (ECAP) in a die with two cylindrical channels of an equal diameter (15.8 mm). The die consisted of two channels intersecting at an angle (φ) of 90° and an outer angle (ψ) of 36° (Fig. 1). A FEM simulation was carried out using the DEFORM 3D software. A similar geometry of the die and sample were employed. An experimental flow curve obtained from compression test of the alloy was loaded to the software. All samples were extruded by 2 and 6 passes with a ram velocity of 5 mm/min and by using B_c route.

The lubricant used was MoS_2 and it was applied in both channels for each pass. The process was carried out in a universal testing machine Shimadzu AG-I 600 kN at room temperature. Load-displacement curves were registered during the ECAP process. Then, deformed samples were annealed at 270 °C for 30 min. Longitudinal sections of annealed and deformed samples were ground and polished in order to characterize their microstructure in a JEOL 6610 LV scanning electron microscope. Vickers hardness was evaluated in a Shimadzu HMV-G21DT micro-hardness testing machine using 1.96 N for 15 s. At least 8 values were taken for each specimen to obtain an average.

Results and Discussion

Effect of SPD on Microstructure and Microhardness

Figure 2 exhibits the cast microstructure at low and high magnification. For example in Fig. 2a it can be seen the dendritic pattern composed by the zinc rich phase (η) and aluminum rich phase (α). The former can be seen in a bright contrast and the latter in a grey contrast. In the incise b of the same Fig. 2, it can be distinguished two zones (with laminar and granular morphology) product of the eutectoid reaction surrounded by η phase. Figure 2c, d shows the same dendritic pattern with some preferred orientation visible from left to right, which can be

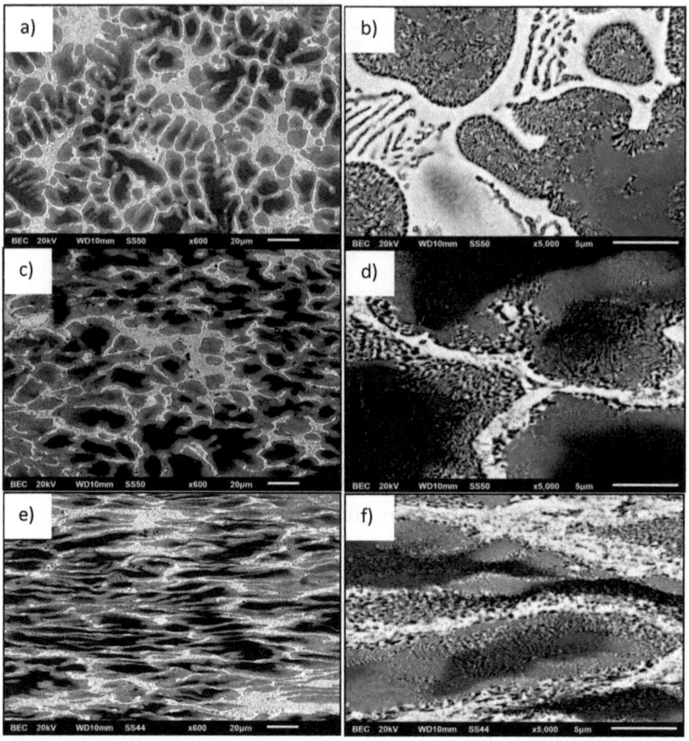

Fig. 2 SEM micrographs at low and high magnification of the samples with initial cast morphology, **a**, **b** without deformation, **c**, **d** with 2 passes, **e**, **f** with 6 passes

attributed to the 2 passes of ECAP that were applied to this sample. In Fig. 2e, f it can be noted that the deformation pattern is more accentuated in the microstructure, however the microsegregation level is still very significant. It is also evident that bright and grey areas exhibit a more fine grained morphology inside. This change can be associated to a combination of both mechanisms: mechanical twist effect of the α phase and dynamic recrystallization of η phase [3].

Figure 3 shows the homogenized microstructure of the samples. It can be noted (incises a and b) that a laminar morphology results but there are also some areas of η phase that remain with the original dendritic morphology. In the incises b and c, the microstructure corresponding to 2 passes of ECAP is shown. The micrograph of the incise d exhibits how the lamellar microstructure is now distorted as a result of the application of the severe deformation. In some areas, the initial morphology is very clear, however in other regions, the distance between the layers of both phases are no more distinguishable, a fact that is proposed as an evidence of the mechanical twist effect caused by ECAP. Micrographs corresponding to 6 passes of ECAP are presented on Fig. 3e, f. As it can be seen, there are still some areas of η phase which is in the initial dendritic morphology, while the majority of the

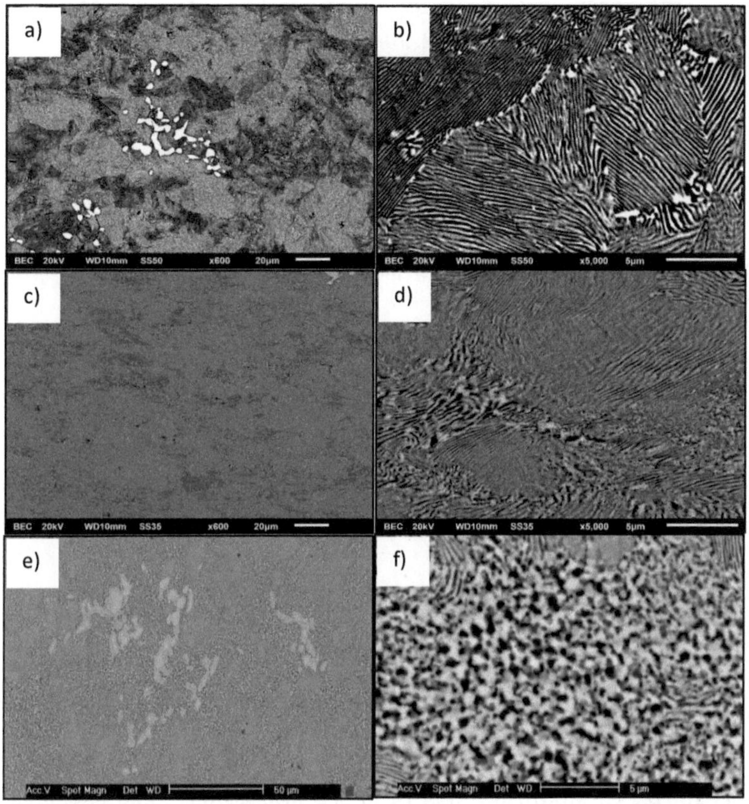

Fig. 3 SEM micrographs at low and high magnification of the samples with homogenized morphology, **a**, **b** without deformation, **c**, **d** with 2 passes, **e**, **f** with 6 passes

microstructure shows a fine granular morphology of α and η. As it has been already mentioned before, it is believed that this microstructure is the result of the mechanical twist effect of the aluminum rich phase and the dynamic recrystallization of zinc rich phase. Yang et al. [3] have demonstrated by means of DSC runs that the recrystallization temperature of the α phase was 337 °C and the corresponding to η phase was −12 °C. Consequently, it is proposed that the former phase is fractured and mixed along the microstructure while the latter experiments dynamic recrystallization during severe plastic deformation at room temperature. This is the explanation about the fine microstructure shown in Fig. 3f. It is important to note that even for 6 passes, there are still some areas that remain with a lamellar morphology due to the deformation heterogeneity as it is shown in Fig. 4. In this case, it may be necessary to subject the sample to longer homogenization treatments and/or more passes of ECAP in order to see if these areas disappeared with higher strain.

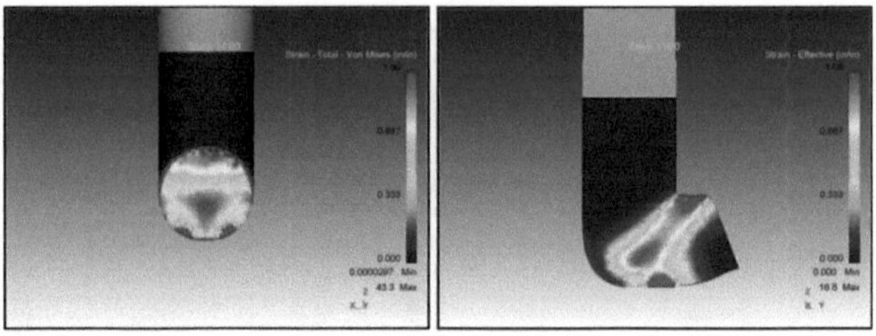

Fig. 4 Strain effective contour maps for 1 pass of ECAP obtained by finite element method simulation

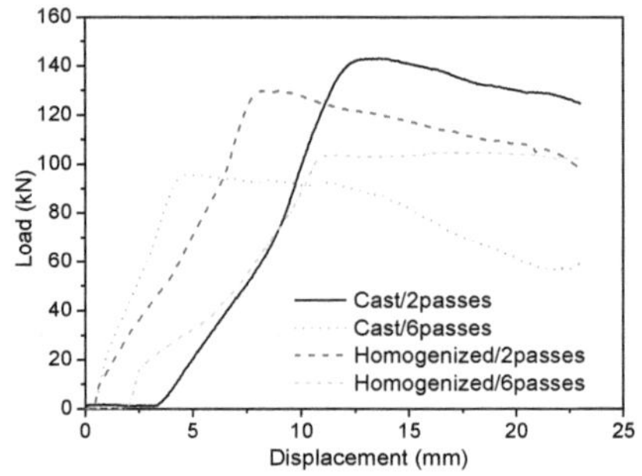

Fig. 5 Load-displacement curves for 2 and 6 passes of ECAP in the cast and homogenized state

El-Danaf et al. [5] have recently reported a similar work softening phenomenon in an eutectic Pb-Sn alloy subjected to ECAP processing. They concluded that a combination or fracture of eutectic crystals and recrystallization of α and β phases lead to a change in morphology from lamellar to fine granular after 4 passes of ECAP. On Fig. 5 it is presented the behavior of load during the ECAP process for 2 and 6 passes for the cast and homogenized samples. In both cases it can be noted that there is a significant reduction in the required load for the process. For example, in the case of the cast samples there is a reduction from 140 to 95 kN while for homogenized samples the reduction is from 132 to 100 kN. The reductions observed can be associated to the recrystallization of η phase as has been already mentioned in the precedent section. It is proposed that there is a major drop in load in the cast samples in comparison to the homogenized ones ought to the higher microsegregation in the former.

Figure 6 exhibits the behavior of the microhardness of the cast and homogenized samples as they are subjected to several passes of ECAP. These are compared with results from the literature [3] on which Zn–22Al and Zn–22Al–0.3Cu alloys microhardness values are reported for lower strain values. As can be seen the general tendency is a decrement of the microhardness as the strain increases.

It is important to point out that in this work it has been confirmed that the softening experimented during deformation is still valid for higher values of deformation in comparison to the values reported in the literature. It is also

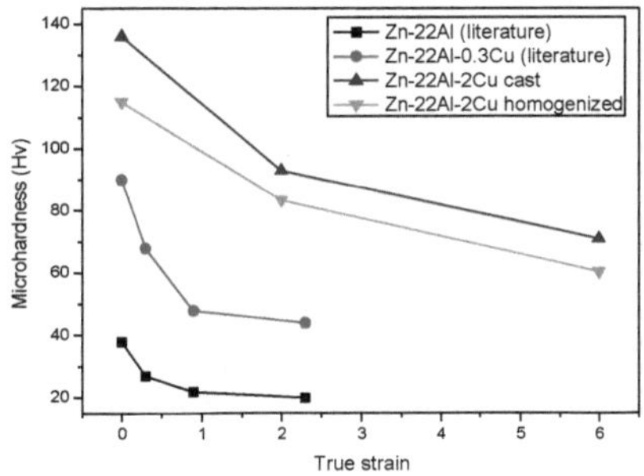

Fig. 6 Microhardness graphs for the alloy studied in this work compared with values from the literature

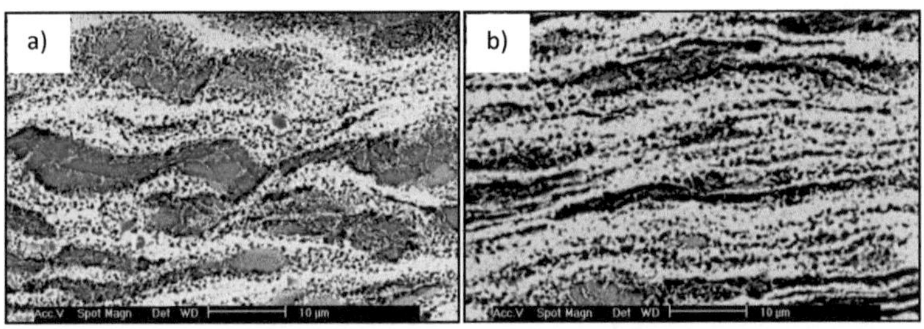

Fig. 7 SEM micrographs of the cast samples subjected to 6 passes of ECAP for **a** 30 min and **b** 120 min of annealing at 275 °C

important to mention that the Cu content has a significant effect in the hardness of the alloy since our values are higher in both cases: cast and homogenized compared to the ones reported in [3]. The general phenomenon observed in the microhardness of the deformed samples supports the hypothesis about the recrystallization of the Zn rich phase during the ECAP process. It is proposed that the higher hardness values exhibited for the cast specimens are due to the presence of a higher microsegregation in comparison with the homogenized samples.

Evolution of Microstructure and Microhardness During Annealing

Figure 7 shows the microstructure of samples with 6 passes of ECAP when they are subjected to 30 and 20 min at 275 °C. As it is observed, the microsegregation level is still present in the entire sample (Fig. 7a, b). Also it is evident that there are some areas on which there exists a microstructure with fine grains. On the other hand when the homogenized sample with 6 passes of ECAP is annealed for the same time at 275 °C the resultant microstructures are shown in Fig. 8. It is evident that for 30 min the microstructure has fine grains with a size close to 1 μm (Fig. 8a).

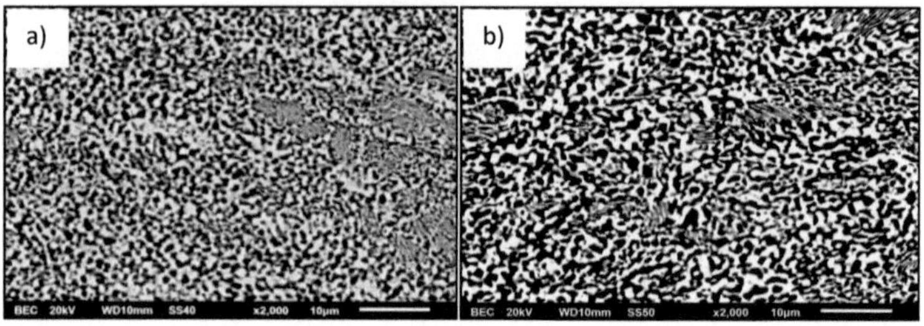

Fig. 8 SEM micrographs of the homogenized samples subjected to 6 passes of ECAP for **a** 30 min and **b** 120 min of annealing at 275 °C

Fig. 9 Comparison of the hardness behavior of the experimental samples and literature values during the annealing time

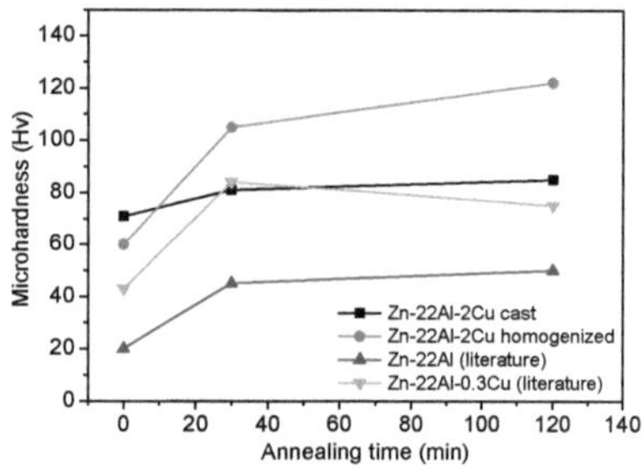

Also there are still some areas with lamellar morphology that did not transform to fine grains. This result is associated to the heterogeneity in deformation present in the sample even for 6 passes. Figure 8b shows the corresponding microstructure for 120 min of annealing. It is clear that the fine grains have coarsened and also zones with laminar morphology remained in the microstructure.

Figure 9 shows the microhardness behavior of the cast and homogenized samples (with 6 passes of ECAP) annealed for 30 and 120 min at 275 °C. These results are compared to the ones presented in Ref. [3]. It can be noted that the hardness increases in both samples as the time is higher. However the increment is significantly higher in the case of the homogenized sample in comparison to the cast one. This result can be attributed to the initial morphology in the samples. We have already demonstrated that even for 6 passes of ECAP the cast sample possessed a mixed morphology of dendritic and granular arrays and also the microsegregation level is evident while the homogenized sample showed fine grain morphology with some isolated islands of lamellar microstructure (Fig. 3f). According to [3] as annealing time is higher there is an increment in the high angle boundaries quantity in the microstructure which in turn lead to the reduction of the heterogeneous nucleation sites of dynamic recovery which causes the hardening observed.

In our case, we proposed that the increment in the relative amount of high angle boundaries is higher in the homogenized sample in comparison to the cast sample, however a deeper study about the quantification of this parameter is carried out at present to prove this assumption. Zhang et al. [2] have associated this anneal hardening phenomenon to a phase transformation of Zinc-rich phase (α_2) to equilibrium phases Al-rich α_1 and the Zinc-rich η. However it is stated that this mechanism is not operative in the system studied because the phase α_2 is improbable to form during the slow cooling conditions that prevail during cast and homogenization processes.

Conclusions

It was demonstrated that work softening in an Zn–22Al–2Cu is valid even for true strains close to 6 in both samples: cast and homogenized.

It was found that homogenized samples showed a fine grain microstructure (less than 1 μm), while in the case of cast sample the microsegregation pattern and zones with initial dendritic microstructure remained after 6 passes of ECAP.

It was found that during the annealing of cast and homogenized samples with 6 passes of ECAP, there was an annealing hardening phenomenon which was attributed to the increment in the relative amount of high angle boundaries in both samples; however it was observed that this increment was more significant in the case of the homogenized samples.

References

1. Zhu, Y. H., & Man, H. C. (1997). Influence of extrusion temperature on structure of eutectoid Zn-Al alloy. *Materials and Manufacturing Processes, 12,* 1149–1162.
2. Zhang, Y., Yang, L., Zeng, X., Zheng, B., & Song, Z. (2013). The mechanism of anneal-hardening phenomenon in extruded Zn-Al alloys. *Materials and Design, 50,* 223–229.
3. Yang, C. F., Pan, J. H., & Lee, T. H. (2009). Work-softening and anneal-hardening behaviors in fine-grained Zn-Al alloys. *Journal of Alloys and Compounds, 468,* 230–236.
4. Ramos Azpeitia, M., Martinez Flores, E. E., & Torres, V. G. (2012). Superplastic behavior of Zn-Al eutectoid alloy with 2% Cu. *Journal of Materials Science, 47,* 6206–6212.
5. El-Danaf, E. A., Khalil, K. A., & Soliman, M. S. (2012). Effect of equal-channel angular pressing on superplastic behavior of eutectic Pb-Sn alloy. *Materials and Design, 34,* 235–241.

Part IX
Steels

Effect of Titanium Sulfide Particles on Grain Size in Low Carbon Steel

Yuan Wu, Bowen Peng, Fangjie Li, Shaobo Zheng and Huigai Li

Abstract The effect of titanium sulfide particles on the grain size characteristic of low carbon steel was analyzed. Optical microscope (OM), scanning electron microscope (SEM) were used to characterize the grain size and particles. The result showed that grain size increased from 19.95 to 60.56 μm after heat treatment. The particles were mainly titanium sulfide in the size range of 0.2–0.8 μm and the volume fraction decreased significantly from 0.0084 to 0.0023%. The thermodynamic calculation resulted that these particles were dissolved during heat treatment. The pinning force of grain boundary and the driving force of grain growth were calculated. Based on experimental results and theoretical calculations, titanium sulfide particles with diameter from 0.2 to 0.8 μm and volume fraction of 0.0084% would be sufficient to inhibit the ferrite grain growth.

Keywords Low carbon steel · Grain size · Particles · Pin effect · Heat treatment

Introduction

It is well known that controlling grain size has been recognized as a viable approach to obtain desired properties [1]. Moreover, inclusions are known to play an important role in grain size because they can pin dislocations and grain boundaries [2, 3]. Additionally, the type and the size of the precipitates are related to the pinning force, and the grain growth is determined by the balance of the driving force and the pinning force. When the driving force exceeds the pinning force, grain growth can take place [4].

Y. Wu · B. Peng · F. Li · S. Zheng · H. Li (✉)
State Key Laboratory of Advanced Special Steel & Shanghai Key Laboratory of Advanced Ferrometallurgy & School of Materials Science and Engineering, Shanghai University, Shanghai 200072, China
e-mail: lihuigai@shu.edu.cn

© The Minerals, Metals & Materials Society 2017

M.A. Meyers et al. (eds.), *Proceedings of the 3rd Pan American Materials Congress*, The Minerals, Metals & Materials Series, DOI 10.1007/978-3-319-52132-9_70

The element titanium acts with sulfur in steel and forms various kinds of precipitates such as $Ti_4C_2S_2$, TiS, TiS_2, Ti_2S, etc. Many studies have been carried out on Ti-added ultra-low carbon steels and found that small amount of alloying elements Ti can retarding the grain growth through the precipitation of second phases [5]. Many researchers have also studied the effect of hundreds of nanometers particles on the grain size during heat treatment. Zhou et al. [6] established that the particles in the size range of 0.3–1.2 μm could effectively pinned the grain boundary and strongly hindered the grain growth. Nakayama et al. [7] found that AlN in the size of less than 0.5 μm in semi-processed non-oriented electrical steel inhibited the grain growth at the final annealing. Ti is commonly used as alloying element to improve the properties and it has strong chemical affinity for sulfide in steels [8]. However, the pinning effect of titanium sulfides in the size range of 0.1–1 μm on grain size in low carbon steel is rarely reported.

The aim of the present study, therefore, was to investigate the effect of titanium sulfides in the size range of 0.1–1 μm on grain size in low carbon steel during heat treatment. The results indicate that the dissolution of titanium sulfides particles in the size range of 0.2–0.8 μm during heat treatment will be effective in facilitating grain growth.

Experiment Procedure

The sample S1 was obtained after stress relief annealing of low carbon steel sheet. The stress relief annealed steel sheet was then annealed in 1023 K for about several hours, cooled in the furnace to room temperature. The sample taken from this procedure was named as sample S2. The carbon content of steel is ultra-low, it also contains Si, S, P, N, Al, Ti and other elements.

For grain analysis and particle characterization, samples were cut from original steel. Then the samples (10 mm × 10 mm × 0.35 mm) were prepared by grinding and polishing. The sample was etched by 3% Nital for microstructure observation by using LEICA DM6000 M light microscope. For volume fraction calculation of particles in the size of 0.1–1 μm, 75 micrographs were continuously taken at 3000× magnification using Hitachi SU1510 scanning electron microscope (SEM). For the purpose of determining the types of the particles in the size of 0.1–1 μm, the particles were chemical extracted with nonaqueous solvent and then collected on a filter with pore size of 3, 1 and 0.1 μm successively. The 0.1 μm film were examined with JSM-6700F cold field emission scanning electron microscope (FE-SEM) equipped with energy dispersive spectrometry (EDS) to analyze the chemistry composition of the particles.

Results

Grain Growth Behavior

Both steel samples, S1 and S2, exhibite a ferrite microstructure as shown in Fig. 1. Sample S1 has a finer grain than sample S2, and sample S2 has a more uniform distribution than sample S1. The mean grain sizes of samples S1 and S2 measured by linear intercept method are 19.95 and 60.56 μm, respectively.

Particle Characterization

Figure 2 represents the number of particles per unit area (N_A) in detected size range of 0.1–1 μm in samples S1 and S2, the results were grouped with interval of 0.1 μm. The particles detected are nearly spherical, therefore the size represents the

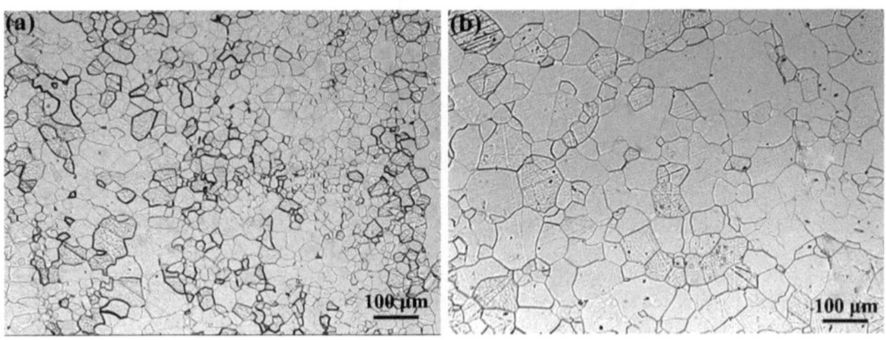

Fig. 1 Optical microstructures of samples: **a** sample S1 and **b** sample S2

Fig. 2 The number of particles per unit area in detected size range of 0.1–1 μm in samples S1 and S2

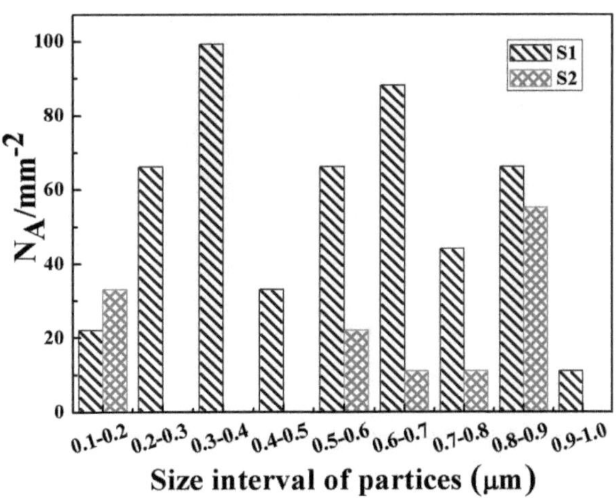

diameter of the particles. It was shown that fewer particles in this size range were detected in sample S2. As for particles in size range of 0.2–0.8 μm, the value of N_A decrease significantly from 397 to 55. The volume fraction (f_v) of the particles can also be calculated by using Dehoff equations [9] based on the date obtained from 75 SEM micrographs captured continuously at a magnification of 3000×. The f_v was calculated to be 0.0084 and 0.0023% for sample S1 and S2, respectively.

Figures 3 and 4 represents the typical particles extracted by electrolysis from samples. It can be found that the particles in sample S1 were mainly titanium sulfide with a nearly spherical shape as shown in Fig. 3a. Figure 3b shows a typical EDS

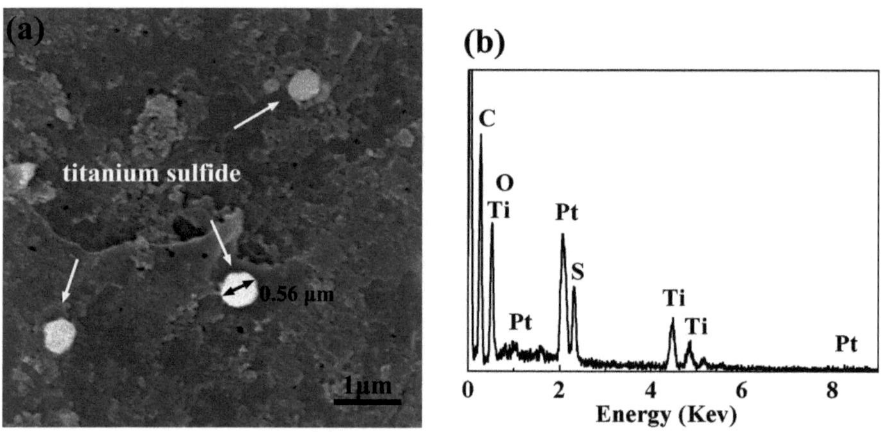

Fig. 3 Typical particles observed by FE-SEM in sample S1: **a** nearly spherical titanium sulfide in size of around 0.5 μm, **b** the EDS spectrum of titanium sulfide particle

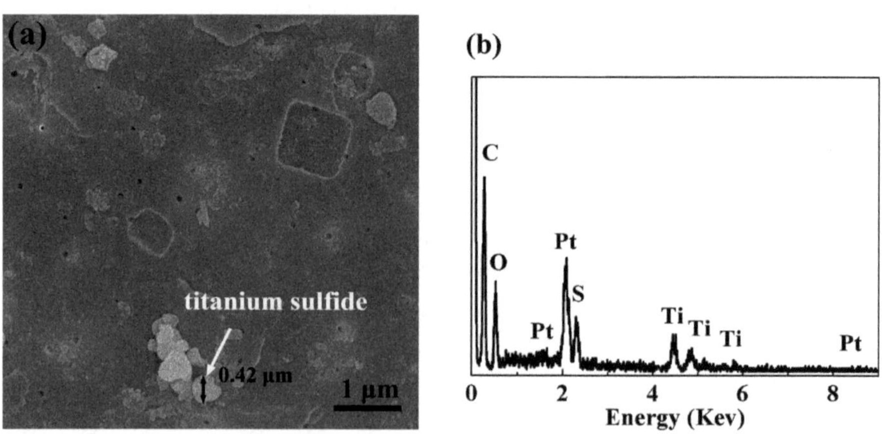

Fig. 4 Typical precipitates observed by FE-SEM in sample S2: **a** some nearly spherical titanium sulfide particles in size of around 0.5 μm. **b** the EDS spectrum of titanium sulfide particle

Fig. 5 The atomic ratios of Ti and S of titanium sulfides

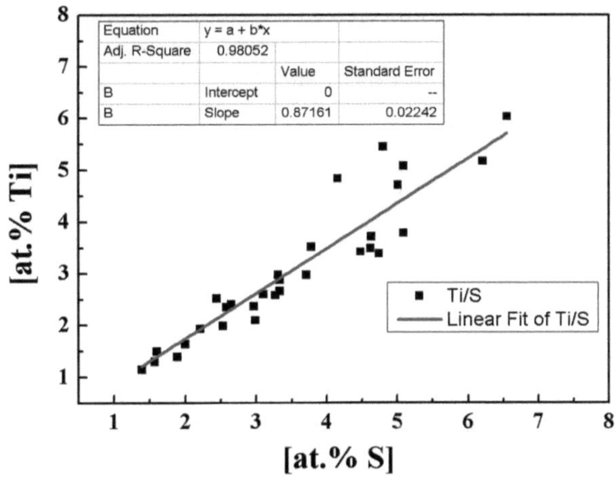

spectrum from such particles, the Ti/S atomic ratio is close to one. In sample S2, FE-SEM analysis shows that few titanium sulfide particles were found as shown in Fig. 4a. Figures 3b and 4b show the typical EDS spectrum of titanium sulfide particles from Figs. 3a and 4a, respectively. And the Ti/S atomic ratio is close to one. The C, O and Pt peaks result from the filter membrane.

The atomic ratios of Ti and S were investigated in order to deduce the type of titanium sulfide particles in samples. The result were represented in Fig. 5. It shows obviously that the atomic ratios are between 0.7 and 1.1, and most of them are close to 1.0.

Discussions

Thermodynamics Calculation

Based on the analysis of the atomic ratios of Ti and S, it can be deduced that titanium sulfide particles observed in sample S1 are TiS particles. The assumption for the thermodynamics calculation is that titanium sulfide particles in this study are TiS particles. And then the effect of heat treatment temperature on TiS particles was analyzed. The corresponding equilibrium solubility product formula of TiS is [10]:

$$\log([\%\mathrm{Ti}][\%\mathrm{S}]) = \frac{-13975}{\mathrm{T}} + 5.43 \quad (1)$$

where T is the temperature (K), the mass percent of Ti and S in the steel are 0.00108 and 0.0008% respectively.

The relationship between the amount of precipitation of sulfur and the titanium is expressed in Eq. 2. The concentrations of titanium and sulfur at lower temperature were expressed in Eqs. 3 and 4.

$$\Delta[\%S] = \frac{32}{48}\Delta[\%Ti] \quad (2)$$

$$Ti_2 = Ti_1 - \Delta Ti \quad (3)$$

$$S_2 = S_1 - \Delta S \quad (4)$$

where $\Delta[\%S]$ and $\Delta[\%Ti]$ are the amount of sulfur and the titanium respectively, S_1 and Ti_1 are the concentration of sulfur and the titanium when the temperature is T_1, and S_2, Ti_2 are the concentration of sulfur and the titanium when the temperature is dropped to T_2.

$$\lg\left[\frac{32}{48}Ti_2^2 + \left(S_1 - \frac{32}{48}Ti_1\right)Ti_2\right] = \frac{-13975}{T} + 5.43 \quad (5)$$

Equation 5 is derived from the Eqs. 2–5. According to Eq. 5, the amount of TiS particle and the value of [%Ti], [%S] can be calculated at different temperatures, as shown in Fig. 6. On the one hand, based on the chemical composition of present experimental steel, the TiS particles are precipitated in the temperature ranged from 964 to 1216 K. On the other hand, soaking time is long enough for the dissolution of TiS particles. It can be deduced that the TiS particles may dissolve during the heat treatment in the condition of 1023 K in this study. This also could explain the fact that there were few titanium sulfide particles detected in sample S2. Therefore, heat treatment can cause dissolution of titanium sulfide particles, which could be responsible for the grain growth.

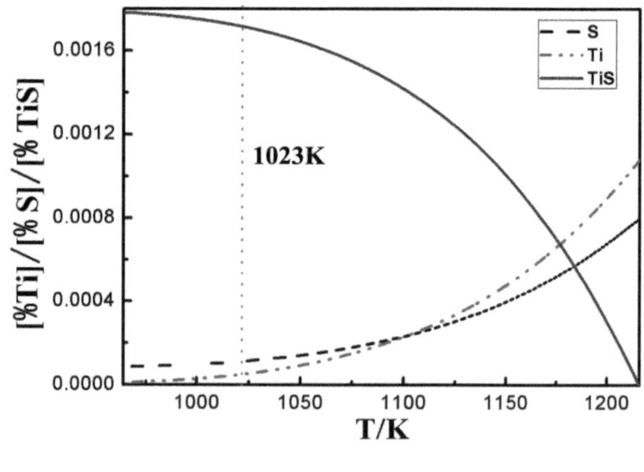

Fig. 6 Variation of [%Ti] and [%S] and amount of TiS precipitation at different temperatures

Pinning Force and Driving Force

In order to further analyze the influence of titanium sulfide particles on the grain growth during heat treatment, the pinning force of particles and the driving force of grain growth were discussed. The assumption for the theoretical calculation is that all particles in size range from 0.2 to 0.8 μm are titanium sulfide particles. For calculation of the pinning force of particles, the following type of equation [11] has often been used:

$$F_p = \frac{3\gamma f^{\frac{2}{3}}}{\pi r} \tag{6}$$

where F_p is the pinning force, γ is grain boundary energy, taking 0.8 J/m^2 [12, 14], r is mean radius of particles, f is the volume fraction of particles in the size of 0.2–0.8 μm. According to Eq. 6, the larger the f, the greater the pinning force. F_p was calculated to be 6.3×10^4 Pa in this study, where r is 0.50 μm and f is 0.0084%. The driving force of grain growth is expressed as the following type of equation [13]:

$$F_d = \left(\frac{3}{2} - \frac{2}{Z}\right)\left(\frac{\gamma}{D}\right) \tag{7}$$

where F_d is driving force, D is the average radius of the grains and Z is the size advantage (the ratio of the maximum grain size to the average grain size), taken as 2 [14]. The F_d was calculated to be 4×10^4 Pa, where D is 9.98 μm of sample S1, the value of the γ is the same as the above. It should be noted that the pinning force caused by titanium sulfide particles was larger than the driving force of grain growth. That is to say, the titanium sulfide particles in the size of 0.1–1 μm played a role in hindering the grain growth.

Conclusions

(1) The particles in the size range of 0.2–0.8 μm are mainly spherical titanium sulfide. Thermal calculation indicated that titanium sulfide could dissolve during the heat treatment.
(2) The driving force for grain growth is 4×10^4 Pa. It is less than the pinning force, which was calculated to be 6.3×10^4 Pa. Titanium sulfide particles with diameter from 0.2 to 0.8 μm and volume fraction of 0.0084% would be sufficient to inhibit the ferrite grain growth.
(3) The grain size is significantly increased from 19.95 to 60.56 μm after the heat treatment, and this phenomenon is related to the dissolution of titanium sulfide particles during heat treatment.

Acknowledgements This work was supported by 973 Project (No.2011CB012902) and National Natural Science Foundation of China (No. U1460103). The authors also wish to express their grateful thanks to the Instrumental Analysis & Research Center of Shanghai University for the instrument supports.

References

1. Estrin, Y., & Vinogradov, A. (2013). Extreme grain refinement by severe plastic deformation: a wealth of challenging science. *Acta Materialia, 61*(3), 782–817.
2. Langdon, T. G. (2013). Twenty-five years of ultrafine-grained materials: Achieving exceptional properties through grain refinement. *Acta Materialia, 61*(19), 7035–7059.
3. Apps, P. J., Bowen, J. R., & Prangnell, P. B. (2003). The effect of coarse second-phase particles on the rate of grain refinement during severe deformation processing. *Acta Materialia, 51*(10), 2811–2822.
4. Gladman, T. (2004). *Grain size control* (p. 183). London: Maney.
5. Tao, X., Gu, J., & Han, L. (2014). Carbonitride dissolution and austenite grain growth in a high Cr ferritic heat-resistant steel. *ISIJ International, 54*(7), 1705–1714.
6. Zhou, B., et al. (2016). In-situ observation of grain refinement in the simulated heat-affected zone of high-strength low-alloy steel by Zr-Ti combined deoxidation. *Metal and Material International, 22,* 267–275.
7. Nakayama, T., & Honjou, N. (2000). Effect of aluminum and nitrogen on the magnetic properties of non-oriented semi-processed electrical steel sheet. *Journal of Magnetism and Magnetic Materials, 213*(1), 87–94.
8. Mitsui, H., et al. (2009). Phase equilibria in FeS-XS and MnS-XS (X = Ti, Nb and V) systems. *ISIJ International, 49*(7), 936–941.
9. Dehoff, R. T., & Rhines, F. N. (1968). *Quantitative microscopy* (p. 128). New York: McGraw-Hill Book Company.
10. Yang, X., et al. (1996). Solubility products of titanium sulphide and carbosulphide in ultra-low carbon steels. *ISIJ International, 36*(10), 1286–1294.
11. DeArdo, A. J., Ratz, G. A., & Wray, P. J. (1982). Thermomechanical processing of microalloyed austenite: Proceedings of the International Conference on the Thermomechanical Processing of Microalloyed Austenite. *Metallurgical Society of AIME.*
12. Chapa, M., et al. (2002). Influence of Al and Nb on optimum Ti/N ratio in controlling austenite grain growth at reheating temperatures. *ISIJ International, 42*(11), 1288–1296.
13. Gladman, T. (1966). *Proceedings of the Royal Society of London, A294,* 298.
14. Anna, K., et al. (2015). Effect of Nb microalloying on reversion and grain growth in a high-Mn 204Cu austenitic stainless steel. *ISIJ International, 55*(10), 2217–2224.

Evolution of Austenite Dislocation Density During Hot Deformation Using a Physical Dynamic Recrystallization Model

Peng Zhou and Qingxian Ma

Abstract A new model to predict the dislocation density evolution of 30Cr2Ni4MoV steel during hot deformation was proposed in this study. Hot compression of 30Cr2Ni4MoV steel was carried out on Gleeble 1500 at different temperatures and strain rates. A series of flow curves was obtained and the experimental dislocation density evolution was derived from the experimental flow curves. Based on the obtained flow curves, the dependences of yield stress, critical stress and strain of dynamic recrystallization and the saturation stress on temperature and strain rate were determined. Two sets of dislocation density equation were derived from the experimental flow curves: (I) a dislocation density relation describing the grains in which dynamic recovery took place only; and (II) an average dislocation density expression pertaining to the recrystallized grains. All the parameters needed for the determination of the dislocation equations were calculated and expressed as a function of strain, temperature and strain rate. A physically realistic and practical kinetics model of dynamic recrystallization was determined with the aid of the above relations. Finally, the dependence of the dislocation density on strain, deformation temperature and strain rate was determined and the predicted results agreed well with the experimental results.

Keywords Austenite · Dislocation density · Dynamic recrystallization · Kinetics model

Introduction

Dynamic recrystallization (DRX), a method for the structure control and mechanical properties improvement during hot forging, is extensively studied in the past years. It is classified into continuous dynamic recrystallization (cDRX) [1, 2] and

P. Zhou · Q. Ma (✉)
Department of Mechanical Engineering,
Tsinghua University, Beijing 100084, China
e-mail: maqxdme@mail.tsinghua.edu.cn

© The Minerals, Metals & Materials Society 2017
M.A. Meyers et al. (eds.), *Proceedings of the 3rd Pan American Materials Congress*,
The Minerals, Metals & Materials Series, DOI 10.1007/978-3-319-52132-9_71

discontinuous dynamic recrystallization (dDRX) [3, 4] based on the different nucleation mechanisms [5, 6]. For the metals with low to medium stacking fault energy, the dDRX is the main mechanism when they are subjected to plastic deformation at high temperature. The plastic deformation of metal at high temperature is achieved by dislocation moving. Both the stress and the microstructure evolution are affected by the dislocation density evolution. Kocks and Mecking proposed the K-M model [7] that is capable of describing the work hardening behavior of the metal. The dislocation density can be expressed as a function of strain, deformation temperature and strain rate. Based on the K-M model, Estrin further presented a unified phenomenological description of work hardening dislocation density [8]. With the aid of the flow curves obtained from the experiment, an improved method to model the dislocation density evolution pertaining to the grains in which the dynamic recovery takes place alone was put forward by Jonas et al. [9]. The equation describing the mean dislocation density of the grains in which dDRX has taken place is established by Quelennec et al. [10]. Considering the dDRX nucleates at the critical strain, the average dislocation density pertaining the recrystallized grains begins at the critical strain of dDRX. With the aid of the modeling dislocation density and the experimental dislocation density derived from the experimental flow curves, the recrystallization fraction caused by recrystallization can be determined and expressed in the form of Avrami equation.

For the alloys that will undergo phase transition from austenite to martensite when it is cooled to room temperature, it is difficult to detect the dislocation density of the austenite at high temperature with transmission electron microscope. Considering the relationship between dislocation density and stress, it is possible to establish the dislocation density model based on the obtained stress-strain data from experiments.

In this study, a new approach is proposed from the view point of dislocation density for the determination of recrystallization fraction. Dislocation density model for austenite deformed at high temperature is established as a function of strain, deformation temperature and strain rate.

Modeling Method

General Equation Describing the Dislocation Density

When the metals are compressed at high temperature, once the DRX occurs, the microstructure is composed of work hardening grains and recrystallized grains. So the dislocation density, ρ, can be expressed using the simple rule of mixtures as follows:

$$\rho = (1 - X)\rho_{wh} + X\overline{\rho}_{rex}, \tag{1}$$

Here, ρ is the dislocation density of the deformed alloy at a certain strain, X the volume fraction of DRX, ρ_{wh} the dislocation density pertaining to the work hardening grains and $\overline{\rho}_{rex}$ the average dislocation density of the grains in which DRX has taken place.

Calculation of the Volume Fraction of DRX

In order to determine the dislocation density model as expressed in Eq. 1, the dynamic recrystallization volume fraction X is necessary. In this study, the Avrami equation is employed to describe the volume fraction of DRX as follows:

$$X = 1 - \exp(-kt^n), \tag{2}$$

$$\lg\ln[1/(1 - X)] = \lg k + n\lg t. \tag{3}$$

Here, k is the Avrami constant and t the time exponent which can be expressed as:

$$t = (\varepsilon - \varepsilon_c)/\dot{\varepsilon}, \tag{4}$$

where ε is the true strain, ε_c the critical strain of DRX and $\dot{\varepsilon}$ is the strain rate.

From Eq. 1, it can be readily derived that the volume fraction of DRX can be expressed as:

$$X = \frac{\rho_{wh} - \rho}{\rho_{wh} - \overline{\rho}_{rex}}. \tag{5}$$

As long as the ρ, ρ_{wh} and $\overline{\rho}_{rex}$ at different deformation conditions can be obtained, the Xs can be calculated by Eq. 5. By regression analysis, the values of k and n can be obtained from the plots of $\lg\ln[1/(1 - X)]$ with respect to $\lg t$ (Eq. 3) at different deformation conditions. Then the dependences of k and n on the temperature and strain rate can be determined.

Modeling of Work Hardening Dislocation Density (ρ_{wh} and $\overline{\rho}_{rex}$)

Accurate description of ρ_{wh} and $\overline{\rho}_{rex}$ are also necessary for modeling the dislocation density as expressed in Eq. 1. The dependence of dislocation density on strain can be expressed as [9]:

$$d\rho/d\varepsilon = k_1 - k_2\rho, \tag{6}$$

Here, k_1 is the athermal work-hardening rate and independent on the strain, k_2 the recovery rate activated thermally, which can be expressed as a function of temperature and strain rate.

As the details of integration described in equations A1–A6 in Ref [9], the ρ_{wh} during the course of work hardening, only DRV occurrence, can be expressed as follows:

$$\rho_{wh} = \rho_{sat} - (\rho_{sat} - \rho_0)\exp[-k_2(\varepsilon - \varepsilon_0)], \tag{7}$$

Here, ρ_{sat} is the dynamic recovery saturation dislocation density corresponding to the saturation stress resulted from dynamic recovery (DRX) alone, ρ_0 the yield dislocation density at which point the stress corresponds to the yield stress (σ_0) and the strain equals to the yield strain (ε_0).

The relationship between ρ and σ is according to [7]:

$$\rho = \sigma^2/(M\alpha\mu b)^2, \tag{8}$$

Here, M is the Taylor factor, α a material constant, μ the shear modulus and b the Burgers vector. In this study, M and α are constant and set equal to 3 and 0.5, respectively. b [11] and μ [12] are temperature dependence parameters. So the dependence of ρ_0 on σ_0 and ρ_{sat} on σ_{sat} can be described as:

$$\rho_0 = \sigma_0^2/(M\alpha\mu b)^2, \tag{9}$$

$$\rho_{sat} = \sigma_{sat}^2/(M\alpha\mu b)^2, \tag{10}$$

where σ_{sat} is the dynamic recovery saturation stress.

In order to describe the evolution of the average dislocation density of the grains in which DRX has taken place, the modified formalism of Eq. 7 is adopted as expressed in Eq. 11 [10]. ρ_c is the critical dislocation density for the initiation of DRX, which corresponds to the critical stress for the initiation of DRX (σ_c) in the flow curve, and ε_c is the critical strain of DRX. k_2' is the average recovery rate of the recrystallized grains and the relationship between k_2' and k_2 is described in Eq. 12.

$$\overline{\rho}_{rex} = \rho_c - (\rho_c - \rho_0)\exp\left[-k_2'(\varepsilon - \varepsilon_c)\right], \tag{11}$$

$$k_2' = k_2\frac{\rho_{sat}}{\rho_c} = k_2\frac{\sigma_{sat}^2}{\sigma_c^2}. \tag{12}$$

With the aid of Eq. 8, the relationship between ρ_c and σ_c can be described as:

$$\rho_c = \sigma_c^2/(M\alpha\mu b)^2. \tag{13}$$

Determination of Characteristic Stresses and Strains of DRX ($\sigma_0, \sigma_c, \sigma_{sat}, \varepsilon_0$ and ε_c)

From the Eqs. 7 and 11, it can be seen that the $\sigma_0, \sigma_c, \sigma_{sat}, \varepsilon_0$ and ε_c are necessary for the final description of ρ_{wh} and $\overline{\rho}_{rex}$. All the flow curves obtained from experiments are firstly fitted with a seventh order polynomial and smoothed from the yield strain (ε_0) [9]. The plot of work hardening rate ($\theta = d\sigma/d\varepsilon$) with respect to stress, derived from the flow curve, is employed to determine the values of ε_c, σ_c and σ_{sat} [9]. In order to obtain the accurate value of the critical stress for DRX (σ_c), the very accurate method proposed by Poliak and Jonas in 1996 [13] is used on account of it's widely application without the limitation of strain rate [14] and testing mode, i.e. tension, compression and torsion [15].

Recovery Rate Parameters Calculation (k_2 and k_2')

The k_2 can be obtained by replotting the experiment stress-strain data in the form of $\theta\sigma$ versus σ^2 as described in details in Ref. [16] as follows:

$$\theta\sigma = 0.5k_2\sigma_{sat}^2 - 0.5k_2\sigma^2. \tag{14}$$

As long as the plot is linear, k_2 can be directly obtained from the slope, 0.5 k_2. After the determination of σ_{sat}, σ_c and k_2, the k_2' can be obtained with the aid of the relationship between k_2' and k_2 as described in Eq. 12.

Modeling Procedure

The aim of the dislocation density model derived from the experimental stress-strain data is to predict the dislocation density evolution at any strain rates and temperatures. So the dislocation density model should be expressed as a function of strain rate, temperature and strain, $\rho = f(\dot{\varepsilon}, T, \varepsilon)$. In other words, the parameters mentioned above, i.e. σ_0, σ_{sat}, σ_c, ε_0, ε_c, k_2, k_2', n and k, should be expressed functions of temperature and strain rate.

In the following section, the dependences on temperature and strain rate of the above parameters will be determined.

Table 1 Design of experiment: deformation temperature and strain rate

Temp. (°C)	Strain rate (s^{-1})
1200, 1150	0.01, 0.1, 0.25, 0.5, 1
1100	0.01, 0.1, 0.25, 0.5
1050	0.01, 0.1, 0.25
1000	0.01, 0.1
950	0.01

Material and Experiment

The single pass compression deformation was carried out on a Gleeble-1500 thermal mechanical simulation tester and the material used in this study was turbine rotor steel 30Cr2Ni4MoV cut from 600 t ingot with a composition of 0.28C–0.02Mn–0.01Si–0.003P–0.003S–1.72Cr–0.41Mo–3.63Ni–0.11 V–(bal.)Fe, all values given in wt%.

Cylinder specimens for hot compression were machined with size of Φ 8 × 12 mm. In order to minimize the friction between the punch head and the specimens head, Ta pieces were used in the course of compression deformation. The samples were firstly heated to 1473 K with a rate of 10 K/s and held for 5 min for austenization and then the temperature was decreased to the deformation value followed by a 1 min preservation to eliminate the temperature grade at different positions of the cylinder sample. The details of the design of experiment is shown in Table 1.

Parameters Calculation

Experimental Dislocation Density

A selection of the dislocation density curves, derived from flow curves with the aid of Eq. 8, are shown in Fig. 1. It is obviously that the behavior of the dislocation density is affected by the deformation temperature and strain rate, i.e. the Zener-Hollomon parameter expressed as follows:

$$Z = \dot{\varepsilon} \exp(Q/RT), \tag{15}$$

where T is the absolute temperature, R the gas constant that equals to 8.314 J mol^{-1} K^{-1} in this study and Q is the activation energy of the deformation. The deformation activation energy Q can be calculated to be equals to 375 kJ mol^{-1} using the method described by Ref. [17].

The Calculation Results of the Characteristic Stresses and Strains

Figures 2 and 3 show the dependences of σ_0, σ_c, σ_{sat} and ε_c on Z, respectively, and these relationships are described by Eqs. 16–19.

$$\sigma_0 = 2.053 * Z^{0.34}, \tag{16}$$

$$\sigma_c = 0.725 * Z^{0.148}, \tag{17}$$

$$\sigma_{sat} = 1.689 * Z^{0.127}, \tag{18}$$

$$\varepsilon_c = 5.26 \times 10^{-4} * Z^{0.18}. \tag{19}$$

Work Hardening Parameters

In the case of the k_2 calculation, a series of $\theta\sigma - \sigma^2$ plots converted from flow curves are shown in Fig. 4. The slope, 0.5 k_2, can be obtained by linear fitting of the data between yield strain and critical strain. The values of k_2' can be calculated

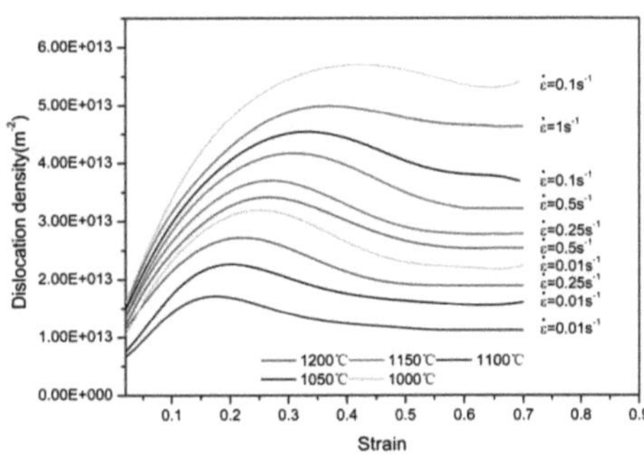

Fig. 1 Experimental dislocation density derived from flow curves

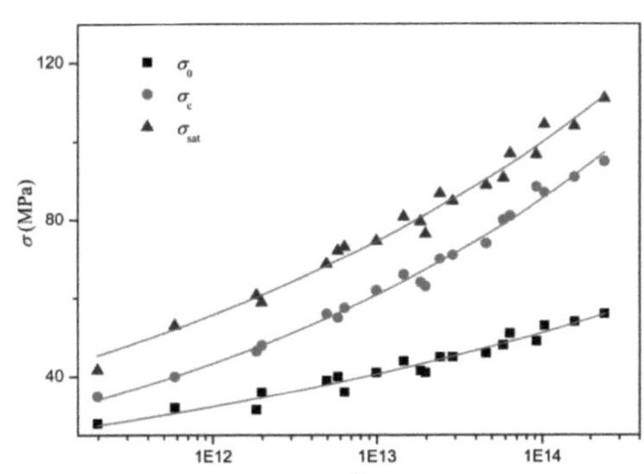

Fig. 2 Dependence of σ_0, σ_c and σ_{sat} on Z

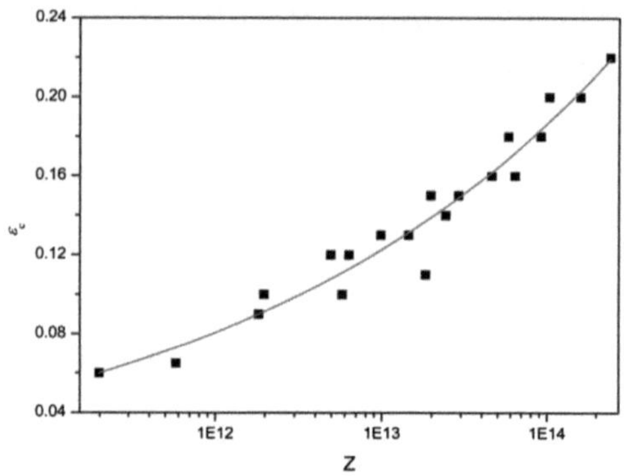

Fig. 3 Dependence of the critical strain of DRX, ε_c, on Z

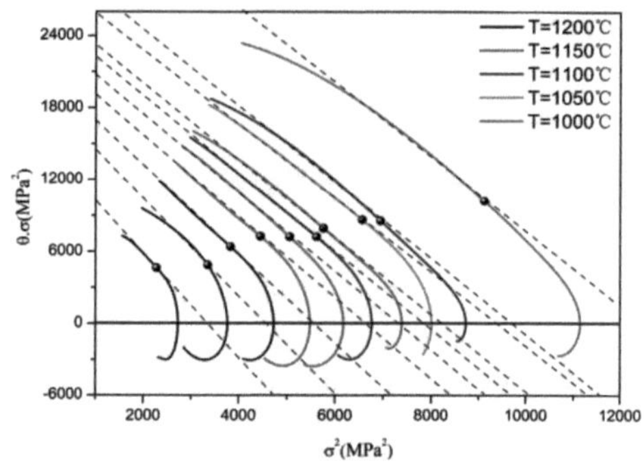

Fig. 4 A selection of $\theta\sigma - \sigma^2$ curves derived from the flow curves

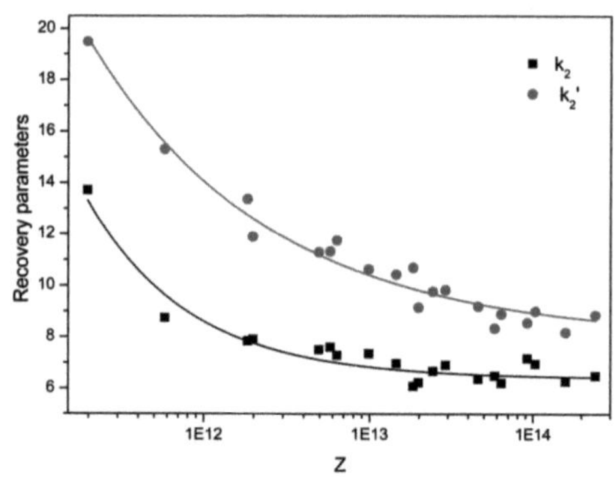

Fig. 5 Dependences of work hardeing paremeters, k_2 and k_2', on Z

by Eq. 12. The dependences of k_2 and k_2' on Z are plotted in Fig. 5 and they can be expressed in Eqs. 20 and 21 as follows:

$$k_2 = 6.37 + 5.6 \times 10^8\, Z^{-0.7}, \tag{20}$$

$$k_2' = 7.96 + 3.85 \times 10^5\, Z^{-0.4}. \tag{21}$$

Avrami Kinetics Parameters Determination

The last two parameters for the determination of dislocation density model are n and k in Avrami equation (Eq. 2). With the aid of the ρ_{wh} and $\overline{\rho}_{rex}$, the Xs can be obtained at different deformation conditions under consideration in this study. From the plots of lgln[1/(1 − X)] with respect to lg$t$ (Fig. 6), we can calculate the values of n and k.

Figure 7 illustrates the dependence of Avarmi exponent, n, on Z and the relationship is expressed in Eq. 22.

$$n = 10.26 * Z^{-0.04}. \tag{22}$$

In the case of the dependence of k on deformation temperature and strain rate, it is much more complex in comparison with n. In this study, the t_{50}, half time of complete dynamic recrystallization, is employed which can be expressed as follows [9].

$$t_{50} = Ad_0^{-v}\exp(Q_{drx}/RT) * \dot{\varepsilon}^{-q}, \tag{23}$$

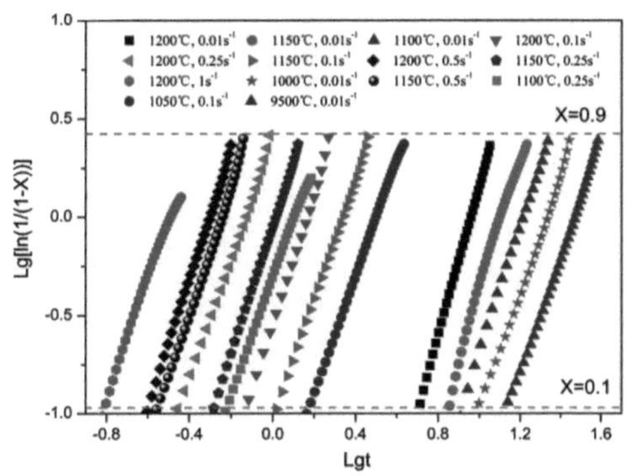

Fig. 6 A selection of lgln[1/(1 − X)] − lg$t$ plots at different deformation conditions

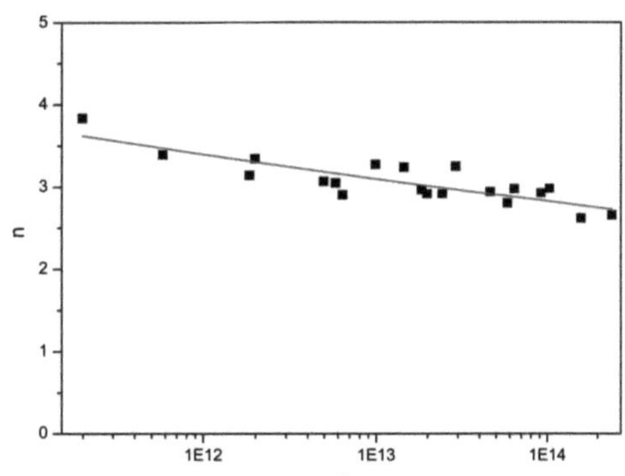

Fig. 7 The dependence of Avrami exponent, n, on Z

here, A is the material constant, d_0 the initial grain size, v the grain size exponent, q the strain rate exponent and Q is the activation energy associated with DRX mentioned above.

Considering the $Ad_0^{-v}\exp(Q_{drx}/RT)$ in Eq. 23 in this research is constant at a certain temperature, t_{50} can be considered as a power function of strain rate, $\dot{\varepsilon}$, at different deformation temperatures and the pre-exponential factor of the function is $Ad_0^{-v}\exp(Q_{drx}/RT)$. Figure 8a shows the t_{50} dependence on strain rate at different deformation temperatures. And then the dependence of $Ad_0^{-v}\exp(Q_{drx}/RT)$ on temperature can be obtained as illustrated in Fig. 8b. Figure 8c shows the

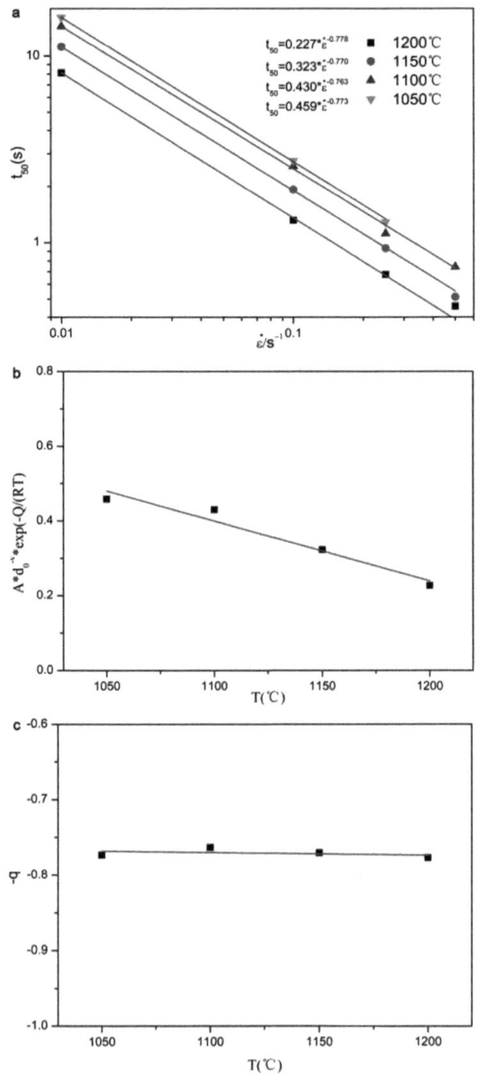

Fig. 8 **a** The dependence of t_{50} on strain rate at different temperatures, **b** the dependence on T of $Ad_0^{-v}\exp(Q_{drx}/RT)$, **c** the dependenc of $-q$ on T

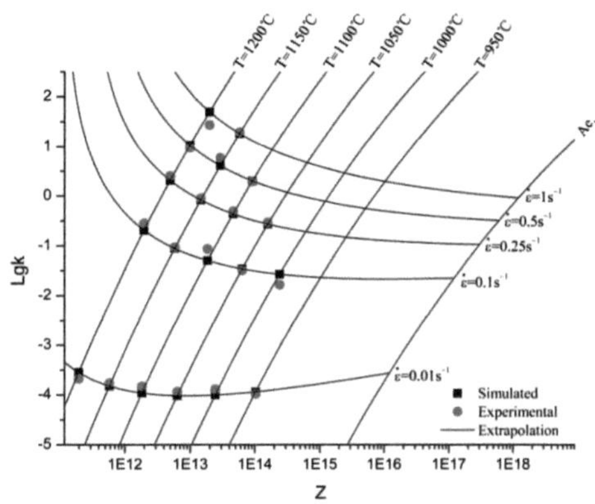

Fig. 9 Comparison between the values of simulated lg$k$ and experimental lg$k$ and the line in the diagram represent the extrapolation of the values of simulated lg$k$

dependence of $-q$, the exponent of t_{50} as a function of strain rate, on temperature. By linear regression, the dependence of $d_0^{-v}\exp(Q_{drx}/RT)$ and $-q$ on temperature can be expressed as follows:

$$Ad_0^{-v}\exp(Q_{drx}/RT) = -0.0016 * T + 2.1629, \tag{24}$$

$$-q = -3.71 \times 10^{-5} * T - 0.73. \tag{25}$$

Combination of Eqs. 24 and 25, the k can be expressed as Eq. 26:

$$k = \frac{-\ln 0.5}{t_{50}^{n}}. \tag{26}$$

In order to verify the accuracy of the simulated k, the comparison between the values of the experimental lg$k$ and predicted lg$k$ is illustrated in Fig. 9. The predicted values of lg$k$ agree well with the experimental values of lg$k$.

Till now, all the parameter needed for the modeling of the dislocation density evolution are determined.

Dislocation Density Modeling Validation

With the aid of the parameters obtained in Sect. "Parameters Calculation", the model of the austenite dislocation density can be expressed as a function of strain, temperature and strain rate. So it can predict the dislocation density at any given deformation conditions. It should be pointed out that the dislocation density is equal to ρ_{wh} when the strain is less than the critical strain of DRX. Figure 10 shows some examples of the comparison between the predicted and experimental dislocation density curves. It is obviously that these two results consist with each other well which means that the determined model can predict the dislocation density at any given deformation conditions.

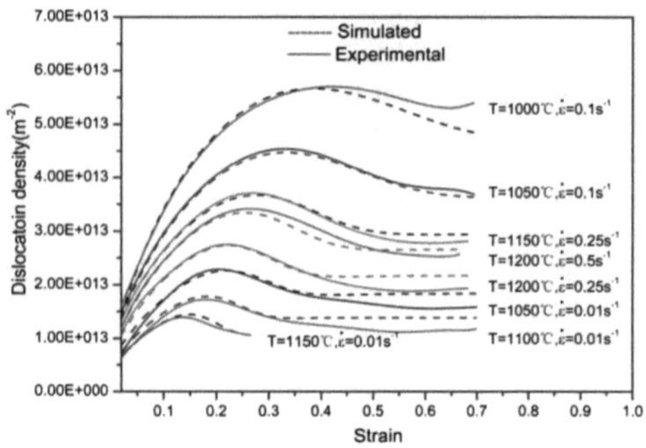

Fig. 10 Comparison between simulated and experimental dislocation density at different deformation conditions

Conclusion

Hot compression deformation of 30Cr2Ni4MoV steel is carried out on Gleeble 1500 and the flow curves under different deformation condition are obtained. Based on the stress data, a new method of volume fraction calculation of DRX is proposed from dislocation density point view in this study. Combination of the DRX volume fraction with the Avrami DRX kinetics plot, it leads to the dislocation density model successful establishment. Comparison between the simulated and experimental dislocation density is implemented and these two results consist with each other well.

Acknowledgements The authors gratefully acknowledge financial support from National Basic Research Program of China (2011CB012903).

References

1. Tsuzaki, K., Huang, X. X., & Maki, T. (1996). Mechanism of dynamic continuous recrystallization during superplastic deformation in a microduplex stainless steel. *Acta Mateialia, 44,* 4491–4499.
2. Gourdet, S., & Montheillet, F. (2000). An experimental study of the recrystallization mechanism during hot deformation of aluminium. *Materials Science and Engineering A, 283,* 274–288.
3. Sakai, T. (1995). Dynamic recrystallization microstructures under hot working conditions. *Journal of Materials Processing Technology, 53,* 349–361.
4. Belyakov, A., Miura, H., & Sakai, T. (1998). Dynamic recrystallization under warm deformation of a 304 type austenitic stainless steel. *Materials Science and Engineering A, 255,* 139–147.
5. Sakai, T., Belyakov, A., Kaibyshev, R., Miura, H., & Jonas, J. J. (2014). Dynamic and post-dynamic recrystallization under hot, cold and severe plastic deformation conditions. *Progress in Materials Science, 60,* 130–207.
6. Doherty, R. D., Hughes, D. A., Humphreys, F. J., Jonas, J. J., Juul Jensen, D., Kassner, M. E., et al. (1997). Current issues in recrystallization: A review. *Materials Science and Engineering A, 238,* 219–274.

7. Mecking, H., & Kocks, U. F. (1981). Kinetics of flow and strain-hardening. *Acta Metallurgica, 29,* 1865–1875.
8. Estrin, Y., & Necking, H. (1984). A unified phenomenological description of work hardening and creep based on one-parameter models. *Acta Metallurgica, 32,* 57–70.
9. Jonas, J. J., Quelennec, X., Jiang, L., & Martin, E. (2009). The Avrami kinetics of dynamic recrystallization. *Acta Materialia, 57,* 2748–2756.
10. Quelennec, X., Bozzolo, N., Jonas, J. J., & Logé, R. (2011). A new approach to modeling the flow curve of hot deformed austenite. *ISIJ International, 51,* 945–950.
11. Seki, I., & Nagata, K. (2005). Supersaturation of carbon in austenite during carburization by CO gas. *ISIJ International, 45,* 1536–1542.
12. Ghosh, G., & Olson, G. B. (2002). The isotropic shear modulus of multicomponent Fe-base solid solutions. *Acta Materialia, 50,* 2655–2675.
13. Poliak, E. I., & Jonas, J. J. (1996). A one-parameter approach to determining the critical conditions for the initiation of dynamic recrystallization. *Acta Materialia, 44,* 127–136.
14. Poliak, E. I., & Jonas, J. J. (2003). Critical strain for dynamic recrystallization in variable strain rate hot deformation. *ISIJ International, 43,* 692–700.
15. Poliak, E. I., & Jonas, J. J. (2003). The critical strain for dynamic recrystallization in rolling mills. *Materials Science Forum, 426,* 57–66.
16. Quelennec, X., & Joseph, J. J. (2012). Simulation of austenite flow curves under industrial rolling conditions using a physical dynamic recrystallization model. *ISIJ International, 52,* 1145–1152.
17. Chen, F., Cui, Z. S., & Chen, S. J. (2011). Recrystallization of 30Cr2Ni4MoV ultra-super-critical rotor steel during hot deformation. Part I: Dynamic recrystallization. *Materials Science and Engineering A, 528,* 5073–5080.

A Rapid Heating Method for Press Hardening Processing

Anatolii Andreiev, Olexandr Grydin and Mirko Schaper

Abstract The work presents results of investigation on a new heating method for production of high-strength car body elements. It's proposed to substitute the conventional heating of blanks in gas or electric furnaces through the rapid contact heating. The blank at this process is pressed between two heated plates during few seconds and subsequently quenched in water-cooled dies to obtain high-strength properties due to the martensitic transformation. The influence of heating temperature in the range between 800 and 1000 °C and dwell time from 4 to 16 s on the microstructure and mechanical properties of 1 mm thick sheet of a low alloyed manganese-boron steel was studied. Furthermore, press hardening including common heating in electric furnace at 950 °C during 360 s and quenching in water-cooled dies of the same sheet was performed to compare the resulted microstructure and mechanical properties with the rapid heated and press hardened material.

Keywords Press hardening · Rapid heating · Short austenitization · 22MnB5

Introduction

The contribution of the transportation sector to greenhouse gas emissions equals approximately 14% [1] and, since increased greenhouse gas emissions are linked to increased climate change, efforts to reduce greenhouse gas emissions in an effort to

A. Andreiev (✉) · O. Grydin · M. Schaper
Chair of Materials Science, Paderborn University,
Warburger Strasse 100, 33098 Paderborn, Germany
e-mail: andreiev@lwk.upb.de

O. Grydin
e-mail: olexandr.grydin@uni-paderborn.de

M. Schaper
e-mail: @lwk.uni-paderborn.de

© The Minerals, Metals & Materials Society 2017
M.A. Meyers et al. (eds.), *Proceedings of the 3rd Pan American Materials Congress*,
The Minerals, Metals & Materials Series, DOI 10.1007/978-3-319-52132-9_72

reduce climate change strongly affect the automobile industry. The bulk of CO_2-emissions produced during a car's lifecycle, including manufacturing and recycling/disposal, occurs during the utilization phase (about 85%), of which 36% of these emissions are linked to the weight of the car [2]. The car body and chassis together represent about 60% of the complete weight, and, therefore, they exhibit the highest potential for light-weighting [3]. One common way to manufacture a lightweight car body is the application of materials possessing significantly lower density as compared to steel. Examples include light metals (aluminium or magnesium), fibre reinforced composites (FRC) or metal foams. However, application of these materials can strongly raise the final price of a car, often making it an inappropriate option for low or middle class cars. Development of advanced high strength steels (AHSS), which exhibit higher strength properties compared to common steels and are more cost-effective than the aforementioned light materials, can solve the problem of light-weighting of cars by reducing the wall thickness of car body elements, i.e. crash-relevant elements, while simultaneously raising the safety level for occupants during a crash.

The most typical steel grade of AHSS for manufacturing of crash-relevant elements is low alloyed steel 22MnB5 (1.5528), which exhibits excellent strength properties after quenching. Due to the addition of boron, its critical quenching rate for obtaining a completely martensitic microstructure equals 30 K/s. This allows to produce finished parts in one simultaneous step of forming and cooling by means of press hardening [4]. The press hardening process is a conventional process for the manufacturing of high strength crash-relevant car body elements and consists of the following steps: cutting off blanks from a previous cold or hot rolled sheet, austenitization and subsequent simultaneous forming and quenching in press between water-cooled dies [5].

Commonly, the heating is performed in 40 m long electric or gas through-type furnaces with a heating rate of 12 K/s, whereas higher heating rates are critical for Al–Si coated blanks due to the threat of melting the coating [6]. High investment costs for the aforementioned furnace (which amount to 44% of the total cost of the premises), the large footprint needed to accommodate the furnace, long heating durations (which can cause significant oxidation of uncoated blanks and consume large amounts of energy) and low energy efficiency (energy efficiency in only 55% with an optimum load of the furnace), are the main disadvantages of this heating method [7–9].

Therefore, numerous studies of press hardening during the last decade have been directed to the development of new, rapid heating methods, which could eliminate these disadvantages. All these methods can be divided into three groups: resistance [10], induction [11] and contact heating [9, 12, 13]. The main features, as well as advantages and disadvantages, of these approaches were briefly discussed in [14]. Contact heating seems to be the most optimal according to the results of previous investigations and could be relatively easily implemented in industrial conditions. Furthermore, this method allows car body components to be manufactured with tailored properties due to the simplicity of realizing a temperature gradient along the preheated plates. Preheated plates with the temperature gradient then heat the blank thus producing partially austenized areas during heating, which cannot completely undergo a martensitic transformation during the cooling stage.

It is worth mentioning that rapid heating with short austenitization durations, or so-called "short austenitization", strongly influences the entire austenitization process and consequently the martensitic transformation at cooling, and, thus, the mechanical properties after quenching/press hardening. The investigations of this influence were previously performed by various researchers and were also briefly discussed in [14]. Thus, the most relevant results for the current study are mentioned here.

The effect of heating rates at continuous heating, heating temperatures and soaking durations at isothermal heating on A_{c1} and A_{c3} temperatures, austenite grain size, and hardness values were investigated for low, middle and high carbon steels, as well as alloyed steels, by Orlich et al. [15]. The authors summarized obtained data in continuous and isothermal heating transformation diagrams. In this study, an increase of the heating rate from 0.05 to 300 K/s for steel 20MnCr5 (1.7147), which, compared to 22MnB5 (1.5528), has up to 1 wt% higher chromium content, lead to a rise of both A_{c1} (from 735 to 810 °C) and A_{c3} (from 840 to 910 °C) temperatures. At the same time, at heating rates of 1 and 100 K/s at a constant austenitization temperature of 950 °C, the austenite grain size, according to ASTM classification, was 5 and 9 respectively, whereas hardness was approximately 460 HV1 in all cases.

In another study, Lolla et al. [16] revealed a simultaneous positive effect of rapid heating with a heating rate of approximately 410 K/s and short soaking times of 2 s at 1100 °C with subsequent quenching in water on strength, ductility and toughness of low carbon steel 20NiCrMo2-2 (1.6523).

An investigation performed by Loebbe et al. showed an improvement of ductility and slight decrease of strength properties of steel 22MnB5 (1.5528) after induction rapid heating with heating rate of 100 K/s up to 950 °C and subsequent press hardening compared to conventional heat treated material. They found that an increase of austenitization temperature from 950 to 1100 °C led to a slight decrease of both strength and ductility, whereas the difference in soaking times of 3 and 10 s at austenitization temperature had an insignificant influence on mechanical properties [17].

With the help of resistance heating and subsequent quenching in water-cooled dies, Andreiev et al. studied the effect of austenitization temperature in the range from 830 to 900 °C at short austenitization with a constant heating rate of 160 K/s and a soaking time of 2 s on the strength properties of steel 22MnB5 (1.5528) with different initial microstructures. It was shown that all rapid heat treated samples exhibited higher hardness and tensile strengths than conventional heat treated material, whereby the highest increase of hardness (up to 9%) and of tensile strength (up to 13%) was shown for the lowest investigated austenitization temperature of 830 °C. Furthermore, the steel with an initial microstructure of ferrite, bainite and pearlite showed slightly higher strength properties than that with an initial ferrite-pearlite microstructure [14].

Holzweissig et al. investigated the influence of different heating parameters at contact heating, such as temperature of preheated plates (1000 and 1100 °C) and dwell time of blanks between these plates (from 3 to 15 s), with a heating rate of approximately 220 K/s and subsequent quenching in water-cooled dies on final properties of a quenched blank of steel 22MnB5 (1.5528). The most optimal relation of strength and ductility was obtained at the lowest investigated temperature of 1000 °C and dwell time of 6 s, whereas the lowest dwell time of 3 s was determined as insufficient for the complete austenitization of initial microstructure [12].

Thus, mentioned investigations of short austenitization of steel 22MnB5 (1.5528) or other low carbon steels with similar alloy content showed that, after such treatment, the material exhibits similar, or in some cases better, resultant mechanical properties than the conventional heat treated. However, in previous studies, an improvement of mechanical properties after short austenitization was presented for miniature specimens, and the scatter of obtained results was relatively high. The purpose of this study is to characterize the effect of contact heating on mechanical properties of steel 22MnB5 using standard samples for tensile and bending tests, which are common for industry conditions. In addition, the described positive effect of low austenitization temperatures and dwell times on mechanical properties will also be proven.

Experimental Procedure, Equipment and Materials

Material

In this study, an uncoated sheets of steel 22MnB5 with thickness of 1.0 mm were used. The chemical composition of this material is shown in Table 1.

Furthermore, in Fig. 1, a schematic illustration of the investigated blanks is displayed including the—samples' extraction position. The samples were employed to conduct hardness measurements, tensile tests and bending tests.

Table 1 Chemical composition of the investigated steel 22MnB5

Element (%)	C	Si	Mn	Al	Cr	B	Ti	Nb	V	Fe
22MnB5	0.267	0.28	1.267	0.036	0.139	0.0028	0.023	0.012	<0.001	Balance

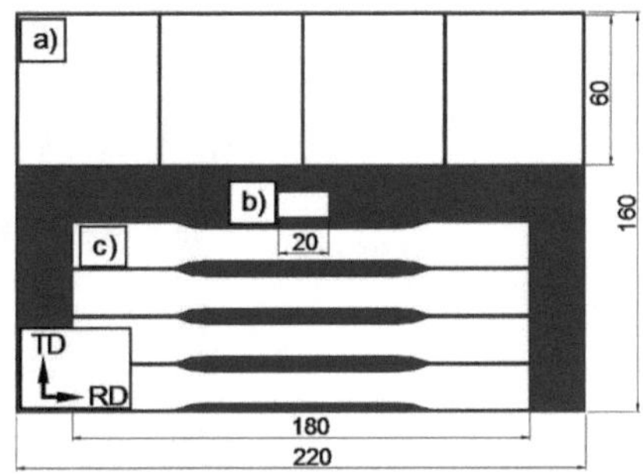

Fig. 1 Extracting location of the samples in transversal direction (TD) and rolling direction (RD): for: *a* bending tests; *b* hardness measurements microstructure characterization; *c* tensile tests

Conventional and Short Austenitization Treatment with Subsequent Quenching

The reference heat treatment with conventional austenitization, which is similar to the common press hardening process, was performed under following conditions: a blank was heated in an electric furnace at a temperature of 950 °C for 360 s. Then, the blank was transported within 4 s to a quenching device with water-cooled dies. Subsequently, it was quenched by closing of dies with a pressure of approximately 0.3 MPa. The cooling speed was obtained by means of temperature measurements with the help of thermocouple type K. In the range from 800 to 400 °C, the cooling speed was determined with almost 35 K/s.

The investigations on the short austenitization process were performed by using a contact heating device. Therefore, the two process parameters, i.e. austenitization temperature and dwell time were varied. The temperatures of the preheated plates were ranging between 800 and 1000 °C with 20 °C steps of, between which blank of 22MnB5 was heated. Four dwell times at 4, 8, 12 or 16 s were chosen. Thereafter, the heated blank was transported within 4 s to the quenching device and cooled as described in the reference heat treatment.

The schematic representation of both performed heat treatment experiments is presented in Fig. 2.

Characterization of Mechanical Properties and Analysis of Material Microstructure

The characterization of the mechanical properties and microstructure observations were carried out for the initial, the conventional austenized, and the short austenized material.

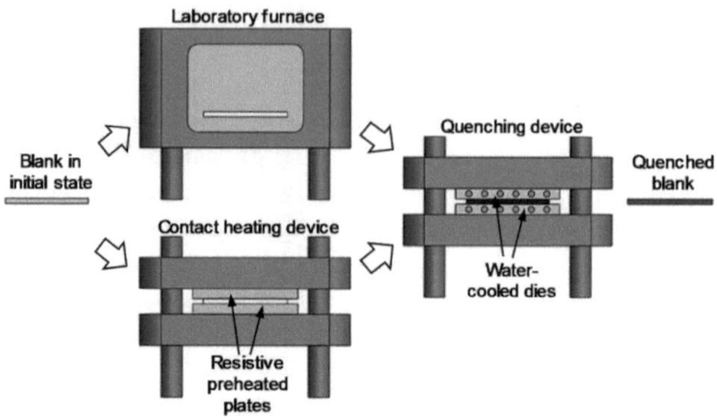

Fig. 2 Schematic representation of the performed heat treatments: conventional (heating in laboratory furnace) and short austenitization (heating in contact heating device) with subsequent quenching in water-cooled dies

Hardness values (HV5) were measured in the cross section of the cold mounted samples along the rolling direction (see Fig. 1b). These samples were further used for the microstructural analysis. The results of the hardness measurements for each material condition are presented as an arithmetical mean consisting of five values.

In addition, tensile tests according to DIN EN ISO 6892 with standard geometry of sample A_{50} were performed to obtain the yield strength ($R_{p0.2}$), the tensile strength (R_m), and the total elongation (e_t) of each material conditions in initial state as well as after conventional and short austenitization with subsequent quenching. As it can be seen from Fig. 1c, four samples for one heat treated condition were tested and thus, the showed below results represent the mean of values for these samples.

Bending angle (α) and maximal force (F_{max}) for the bending tests were conducted to gain knowledge on the materials' crash-behaviour characteristics and were according to the test specification VDA 238-100 after both conventional austenitization and short austenitization with quenching. The presented results of bending tests were also calculated as a mean of values for four tested samples (see Fig. 1a).

For the microstructural analysis, the cross section of the samples in initial, conventionally heat treated and rapidly heat treated conditions were cold mounted, mechanically grinded and polished as well as etched in alcoholic solution with 2% nitric acid. Subsequently, the microstructure of the samples was analysed by employing light microscopy. Furthermore, the microstructure was analysed by using scanning electron microscope. Additional detailed microstructure analysis on the microstructure components in material after short austenitization was performed using transmission electron microscopy.

Results and Discussion

Mechanical Properties

The mechanical properties of the samples in the initial condition are as follows: hardness—155 HV5, yield strength ($R_{p0.2}$)—336 MPa, tensile strength (R_m)—486 MPa and total elongation (e_t)—21%. Thus, the material in advance to the press hardening procedures were comparably soft.

With the help of hardness measurements, it has been revealed, that short austenitization temperatures below 860 °C are not sufficient for a complete austenitization of blank at all investigated dwell times, since the hardness for these cases was lower than 400 HV5. Furthermore, a dwell time of 4 s was solely sufficient for a complete transformation of the initial austenitic microstructure at plate temperatures of 1000 °C.

The generated hardness, tensile, and bending test results of the short austenized samples, excluding the above mentioned parameters, which lead to an incomplete $\alpha \rightarrow \gamma$ transformation, are presented in Fig. 3. The value of each bar (z-axis) corresponds to certain parameters that were investigated at short austenitization treatment: temperature (axis x) and dwell time (axis y). The dark blue colour columns in the illustrated diagram represent the conventional austenitization condition with a subsequent quenching.

The sample after short austenitization with a heat treatment at 920 °C and a dwell time of 8 s (see Fig. 3a) exhibits hardness values of approximately 472 HV5, which represents the minimum of all investigated specimen. The latter condition possesses the same hardness as the conventional heat treated material.

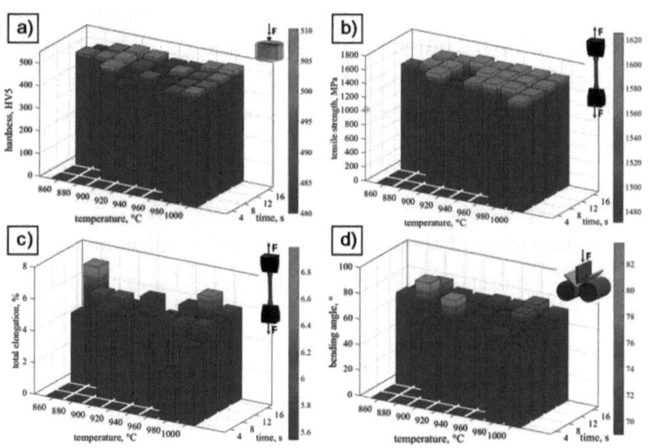

Fig. 3 The mechanical properties of the condition after short austenitization with a subsequent quenching are shown: **a** hardness; **b** tensile strength; **c** total elongation; **d** bending angle

The tensile strength values short austenitization conditions, excluding the temperature of 920 °C with dwell times of 8 and 12 s as well as temperature of 980 °C and dwell time of 4 s, are higher than in conventional austenized and quenched material. The highest hardness and tensile strength were achieved at the two short austenitization temperatures of 900 °C with dwell times of 8 s (517 HV5 and 1624 MPa respectively) and 12 s (512 HV5 and 1612 MPa respectively). Due to the scatter of total elongation values, both short austenitization parameters seem to have a similar effect on the obtained mechanical properties.

The bending angles of the material after the short austenitization are almost the same to those of the conventional austenized and quenched material. Whereas, an increase of dwell time during short austenitization leads to a slight increase of bending angles. The minimal bending angle, i.e. 64°, was obtained at short austenitization condition at a temperature of 900 °C and a dwell time of 8 s. At the same time, after short austenitization at temperature of 880 °C and dwell time of 12 s, the material exhibits the maximal bending angle of 83°. The maximal forces obtained during the bending tests after the short austenitization and quenching are up to 15% higher compared to the maximal forces occurring while testing the conventional heat treated samples (see Table 2).

Mechanical properties of conventional austenized and quenched material (950 °C_360 s) as well as mechanical properties of the favourable parameter sets for the short austenitization are presented in Table 2.

As it can be seen in Table 2, the material after quenching possesses, typical for this steel, high tensile strengths and low uniform elongations. The obtained mechanical properties correspond to the presented in literature values [5]. The strength properties of short austenized material such as hardness, yield and tensile strength at all favorised parameters of short austenitization procedure (Table 2) exceed the conventional austenitization strength properties. In general, the total elongation values of the material after short austenitization and quenching are equally to the conventional heat treated states. Such a difference of strength and ductility values between short austenized and conventional austenized and quenched material are in good alignment to the results of previous author's works [12, 14].

Table 2 Favourable parameters of short austenitization and resultant mechanical properties

Conditions, temperature in °C/dwell time (s)	Hardness, HV5	$R_{p0.2}$ (MPa)	R_m (MPa)	e_t (%)	α (°)	F_{max} (kN)
950 °C/360 s	480 (±6)	1036 (±30)	1471 (±33)	5.58 (±1)	69 (±7)	5.6 (±0.2)
880 °C/12 s	498 (±10)	1117 (±51)	1574 (±19)	5.87 (±1.3)	83 (±6)	6 (± 02)
880 °C/16 s	497 (±13)	1120 (±58)	1580 (±35)	5.09 (±0.7)	76 (±3)	5.7 (±0.2)
900 °C/8 s	517 (±15)	1197 (±29)	1624 (±12)	5.75 (±1.1)	64 (±1)	5.6 (±0.1)
900 °C/12 s	512 (±11)	1147 (±32)	1612 (±11)	5.93 (±0.6)	66 (±3)	6.3 (±0.1)

Concerning the bending angles and the maximal forces during bending tests, further analyses have to be done to profoundly elucidate the influence of the short austenitization parameter sets on the obtained results.

Microstructure

The microstructures of the investigated material after the conventional and the short austenitization with subsequent quenching between water cooled dies are presented in Fig. 4.

The micrograph (Fig. 4b) of the material after the short austenitization at a temperature of 900 °C and dwell time of 4 s shows, that the dwell time was insufficient at this temperature. In this case, only a small part of the initial ferrite-pearlite microstructure transformed in austenite during short austenitization and subsequently in bainite/martensite during quenching. Thus, a high quantity of the untransformed ferrite grains resulted in low hardness values (202 ± 13HV5). In contrast, the 8 s dwell time at the same temperature resulted in a complete transformation of the initial microstructure into austenite. After following quenching, the material exhibits martensite and/or bainite laths, which is significantly finer than martensite laths in material after conventional austenitization and quenching.

A further increase of the temperature and the dwell times lead gradually to a coarsing of the resulting microstructure (see Fig. 4d). It is well-known, that the

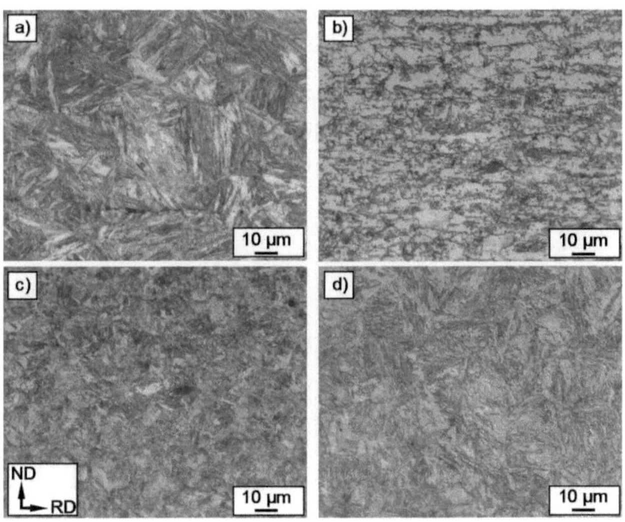

Fig. 4 Optical micrographs: conventional austenitization (**a**) and short austenitization at a temperature of 900 °C and dwell times of 4 s (**b**) and 8 s (**c**) as well as at a temperature of 1000 °C and a dwell time of 16 s (**d**). ND and RD: are accordingly to the normal and rolling direction at previous rolling

austenitization process and the evolving austenite grain sizes are directly influenced by the temperature and the dwell time, whereas higher temperatures and longer dwell times result in a coarsing of the austenite grains. During the following quenching, these grains then transform in martensite/bainite with higher lath, block, and packet sizes.

Detailed investigations on the microstructure morphology of the blanks after conventional and short austenitization treatments were performed using scanning electron microscopy (Fig. 5).

After both heat treatments, the material exhibits a lath-like microstructure. In addition, very fine precipitations were observed in some laths (see Fig. 5a, b). Furthermore, relative big particles can be seen in sample after the short austenitization (see Fig. 5b).

Quantitative analysis of chemical composition of the big particles was carried out with the help of energy-dispersive X-ray spectroscopy (EDS) and showed, that they have higher carbon, chromium and manganese content than parent lath microstructure. Thus, these particles were classified as carbides. Absence of any kind of tempering treatment in the current work excludes the possibility of a carbide formation from the parent martensite lath microstructure. Probably, short dwell times, i.e. few seconds, suppress a diffusion dissolving of the carbides and thus, some of them are presented in steel after short austenitization and quenching [12, 16].

The observed fine precipitations in the samples after short austenitization at the temperature of 900 °C and the dwell time of 8 s were also studied using transmission electron microscope (TEM). Obtained bright and dark field images are presented in Fig. 6.

Figure 6a presents detailed view of precipitations within one lath, which should be fine carbides. These carbides are aligned along more than one habit plane variants and are similar to carbides, which originate at first stage of tempering [18]. However, in the current study, a tempering treatment has not been performed and thus, these carbides point out a presence of so-called "auto-tempered" martensite [19].

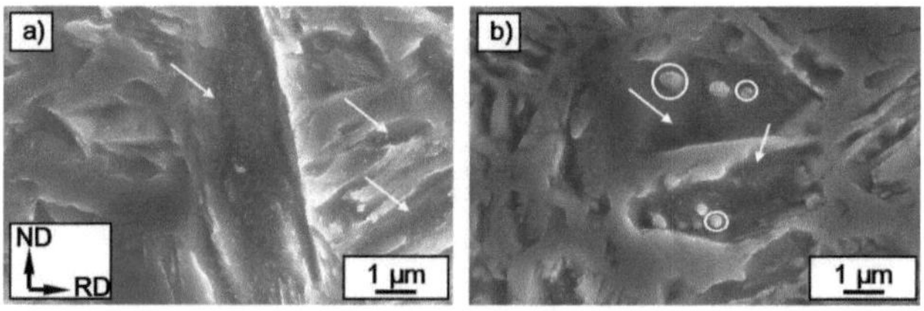

Fig. 5 High-magnification micrographs of the blanks after **a** conventional and **b** short austenitization treatment at a temperature of 900 °C and a dwell time of 8 s. *Arrows* indicate a presence of laths with very fine precipitations and *circles* point out a presence of relative big particles

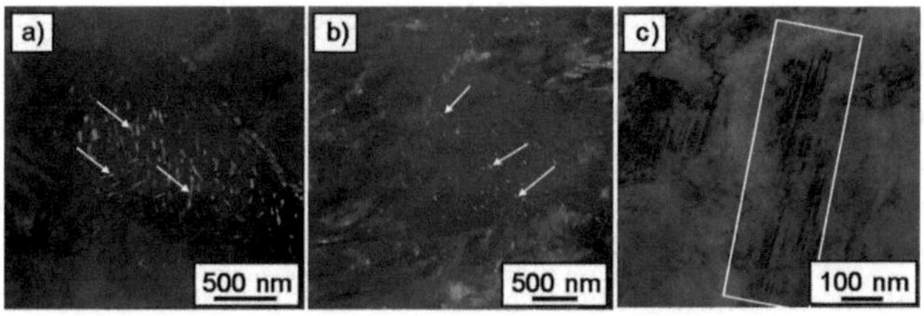

Fig. 6 TEM-images of microstructure after short austenitization at the temperature of 900 °C and the dwell time of 8 s: **a** dark field image: *arrows* indicate differently oriented fine precipitations within one lath, which were previously observed on the SEM-micrographs (see Fig. 5b); **b** dark field image: *arrows* indicate fine precipitations, which are oriented in the same direction; **c** bright field image: *rectangle* indicates observed fine twins

According to previous studies of the microstructure features and hardness of steel 22MnB5 at quenching and/or press-hardening with different cooling rates, the cooling rate of 30 K/s or higher is sufficient to obtain complete martensitic microstructure [20]. Such cooling rate is high enough to suppress any transformation of austenite before reaching the martensite start temperature. However, if the following cooling from the martensite start temperature to ambient temperature is increased up to 300 K/s, it leads to an increase of hardness in comparison with cooling with a rate of 30 K/s in the same temperature range. In this case, the increase of hardness can achieve 50 HV1. Described effect was explained by pronounced auto-tempering of martensite at cooling with lower cooling rates, which lead to a softening of microstructure [20].

In the present study, conventional and short austenitization treatments were performed with the same quenching conditions and therefore, auto-tempered martensite could be observed in both cases.

In Fig. 6b, fine carbides are aligned along one habit plane variant and the angle between them and lath's axis amounts approximately 55–60°. Such positioning of carbides is characteristically for lower bainite.

Observed fine twins in microstructure point out a presence of high-carbon plate martensite (see Fig. 6c).

The formation of presented mixed microstructure consisting of bainite, lath and plate martensite and undissolved carbides cannot be predicted using conventional CCT diagram. Starting microstructure for determining conventional CCT or TTT diagrams is homogeneous austenite, which has uniform distribution of alloying elements. Here, the formation of homogeneous austenite is time and temperature defined process. In case of insufficient austenitization temperatures or times so-called "inhomogeneous" austenite with or without undissolved primary carbides forms in steel [15]. Such inhomogeneous austenite has a nonuniform distribution of alloying elements, first of all carbon. Austenite grains, which originate near the carbide particles of initial ferrite-pearlite microstructure, have higher carbon content than austenite grains or regions, which are remote from carbide particles [12, 16, 21].

In present study, the formation of inhomogeneous austenite at short austenitization is caused by insufficient austenitization times. At following quenching, "carbon-enriched" austenite transforms into low-carbon lath or even high-carbon plate martensite, whereby "carbon-poor" austenite transforms into bainite [16]. Determination of CCT diagram for steel 22MnB5 with starting microstructure of inhomogeneous austenite instead of homogeneous austenite and thus, prediction of its transformation at cooling stage could be the possible direction for further investigation.

Conclusions

The current work introduces a novel process to produce crash relevant car body components. The process includes short austenitization treatment by means of contact heating instead of conventional furnace heating and subsequent press hardening of heated blanks. This promises reduction of investments costs, necessary production spaces, energy consumption, heating duration from 300 to 4–16 s and simultaneous improvement of strength properties.

The effect of different short austenitization parameters, such as plates temperature in the range from 800 to 1000 °C with a step of 20 °C and dwell time between preheated plates of 4, 8, 12 and 16 s with subsequent quenching between two water-cooled dies on hardness, tensile strength, total elongation as well as and bending angle has been investigated. It has been revealed that hardness values of short austenized material at all investigated dwell times below the plates temperature of 860 °C were lower than 400 HV5. Due to this fact, these short austenitization parameters were insufficient for complete austenitization of microstructure and thus, the following martensitic transformation was incomplete. Furthermore, dwell time of 4 s was sufficient for complete transformation of initial microstructure in austenite only at plates temperature of 1000 °C.

The highest hardness and tensile strength were achieved at short austenitization temperature of 900 °C and dwell times of 8 s (517 HV5 and 1624 MPa respectively) and 12 s (512 HV5 and 1612 MPa respectively). Due to scatter of total elongation values, it is difficult to recognise any positive or negative influence of short austenitization parameters on this characteristic.

An increase of dwell time leads to slightly increase of bending angles. The minimal bending angle was obtained at short austenitization temperature of 900 °C and dwell time of 8 s (64°) and the maximum (83°)—at short austenitization temperature of 880 °C and dwell time of 12 s.

Generally, short austenized blanks exhibits up to 8% higher hardness values, up to 15% higher yield strength and up to 11% higher tensile strength than conventional austenized and quenched material. Total elongation and bending angles are almost the same after conventional and short austenitization.

Increase of plates temperature/dwell time at short austenitization lead to coarsing of microstructure. However, even after short austenitization with highest plates temperature of 1000 °C and longest dwell time of 16 s the martensite/bainite laths remain finer than in case of conventional austenitization. Thus, the positive effect of short austenitization on mechanical properties on the one hand is connected with refinement of resulting microstructure. On the other hand, in comparison with conventional heat treatment the microstructure after short austenitization contains not only as-quenched and auto-tempered martensite laths, but also undissolved carbides of initial ferrite-pearlite microstructure, bainite laths and high-carbon plate martensite. This mixed microstructure also influences mechanical properties of quenched component. Therefore, the purpose of future work is to estimate the influence of different components of this mixed microstructure on resulting mechanical properties.

Acknowledgements The authors would like to thank the Ministry of Innovation, Science and Research of the State of North Rhine-Westphalia for the financial support of the scientific work within the scope of the project "Light—Efficient—Mobile".

References

1. IPCC. (2014). Summary for policymakers. In O. Edenhofer, R. Pichs-Madruga, Y. Sokona, E. Farahani, S. Kadner, K. Seyboth, et al. (Eds.), *Climate change 2014: Mitigation of climate change. Contribution of working group III to the fifth assessment report of the intergovernmental panel on climate change*. Cambridge, United Kingdom and New York: Cambridge University Press.
2. VDI. (2014). *VDI-GME-Studie: Werkstoffinnovationen fuer nachhaltige Mobilitaet und Energieversorgung*. Duesseldorf: Verein Deutscher Ingenieure e.V.
3. Kelkar, A., Roth, R., & Clark, J. (2001). Automobile bodies: Can aluminum be an economical alternative to steel? *Journal of the Minerals Metals and Materials Society, 53,* 28–32.
4. Geiger, M., Merklein, M., & Hoff, C. (2005). Basic investigations on the hot stamping steel 22MnB5. *AMR, 6–8,* 795–804.
5. Karbasian, H., & Tekkaya, A. E. (2010). A review on hot stamping. *Journal of Materials Processing Technology, 210,* 2103–2118.
6. Naganathan, A., & Penter, L. (2012). Hot stamping. In T. Altan & A. E. Tekkaya (Eds.), *Sheet metal forming—processes and applications* (pp. 133–156). ASM International.
7. Goetze, U., Zoennchen, S., & Schönherr, J. (2013). Wirtschaftliche Bewertung von Prozesskettenvarianten am Beispiel von Strukturbauteilen. In R. Neugebauer, U. Goetze, & W.-G. Drossel (Eds.), *Energetisch-wirtschaftliche Bilanzierung und Bewertung technischer Systeme – Erkenntnisse aus dem* Spitzentechnologiecluster eniPROD, Tagungsband zum 1. und 2. Methodenworkshop der Querschnittsarbeitsgruppe 1 "Energetisch-wirtschaftliche Bilanzierung" des Spitzentechnologieclusters eniPROD (pp. 375–394). Auerbach: Wissenschaftliche Scripten.
8. Rasera, J. N., Daun, K. J., & D'Souza, M. (2014) *Direct contact heating for hot forming die quenching*. Paper presented at ASME 2014 international mechanical engineering congress and exposition, Montreal, Quebec, Canada, November 14, 2014, p. 8.
9. Kolleck, R., & Veit, R. (2011). *Current and future trends in the field of hot stamping of car body parts*. Paper presented at 3rd International Conference on Steels in Cars and Trucks, Salzburg, Austria, June 5–9, 2011, p. 8.

10. Mori, K., Maki, S., & Tanaka, Y. (2005). Warm and hot stamping of ultra high tensile strength steel sheets using resistance heating. *CIRP Annals—Manufacturing Technology, 54,* 209–212.
11. Kolleck, R., Veit, R., Merklein, M., Lechler, J., & Geiger, M. (2009). Investigation on induction heating for hot stamping of boron alloyed steels. *CIRP Annals—Manufacturing Technology, 58,* 275–278.
12. Holzweissig, M. J., Lackmann, J., Konrad, S., Schaper, M., & Niendorf, T. (2015). Influence of short austenitization treatments on the mechanical properties of low-alloy steels for hot forming applications. *Metallurgical and Materials Transactions A, 46,* 3199–3207.
13. Ploshikhin, V., Prihodovsky, A., Kaiser, J., Bisping, R., Lindner, H., Lengsdorf, C., et al. (2011). New heating technology for the furnace-free press hardening process. Paper presented at Conference "Tools and Technologies for Processing Ultra High Strength Materials", Graz, Austria, September 19, 2011, p. 8.
14. Andreiev, A., Grydin, O., & Schaper, M. (2016). Evolution of microstructure and properties of steel 22MnB5 due to short austenitization with subsequent quenching. *Steel Research International, 87,* 9.
15. Orlich, J., Rose, A., & Wiest, P. (1973). *Atlas for heat treatment of steels* (p. 264). Duesseldorf: Verlag Stahleisen M.B.H.
16. Lolla, T., Cola, G., Narayanan, B., Alexandrov, B., & Babu, S. S. (2011). Development of rapid heating and cooling (flash processing) process to produce advanced high strength steel microstructures. *Materials Science and Technology, 27,* 863–875.
17. Loebbe, C., Hering, O., Hiegemann, L., & Tekkaya, A. E. (2016). Setting mechanical properties of high strength steels for rapid hot forming processes. *Materials, 219,* 19.
18. Krauss, G. (2001). Deformation and fracture in martensitic carbon steels tempered at low temperatures. *Metallurgical and Materials Transactions B, 32B,* 205–221.
19. Speich, G. R., & Leslie, W. C. (1972). Tempering of steel. *Metallurgical Transactions, 3,* 1043–1054.
20. Nishibata, T., & Kojima, N. (2013). Effect of quenching rate on hardness and microstructure of hot-stamped steel. *Journal of Alloys and Compounds, 577S,* 549–554.
21. Grydin, O., Nuernberger, F., Zou, Y., Schaper, M., & Brosius, A. (2014). Formation and properties of mixed ferritic-martensitic microstructures in the air-hardening steel LH800. *Steel Research, 85,* 1340–1347.

New Generation Niobium Bearing Structural Steels for Future Infrastructure Demands

Steven G. Jansto

Abstract The new generation of value-added low carbon niobium (Nb) microalloyed beam, plate and rebar construction steels for both low and high yield strength and energy absorption applications are shifting designers to these new high performance lower cost materials. The civil engineering and end user community demand structural reinforcing bars, shapes, beams and plates with improved energy absorption and fatigue properties. The future market demands better fire and seismic resistance, yield-to-tensile ratio consistency, improved bendability and weldability. These attributes are difficult to obtain from steel producers today with their current higher carbon microalloyed steel approach. However, there is a global shift in motion to low C–Nb bearing construction steels displacing traditional materials. For example, in the construction beam sector and rebar sector improved properties result for 0.02–0.04%Nb in low carbon steel for S355 and S420 beams and for S500 and S600 low carbon reinforcing bars.

Keywords Fatigue · Niobium · Rebar · Seismic · Structural steels

Introduction

Beam, reinforcing bar, plate and other infrastructure structural applications account for over 500 million tonnes of global steel production. With environmental climate change, more severe weather conditions around the globe and increased seismic activity, construction materials experience increased load, stress, strain and fatigue levels. These climatic changes often drive the customer to request improved low temperature toughness, better fracture toughness, improved cyclic fatigue performance, more homogeneous grain size and better resistance to earthquake, typhoon

S.G. Jansto (✉)
CBMM North America, Inc., 1000 Old Pond Road,
Bridgeville, PA 15017, USA
e-mail: jansto@cbmmna.com

© The Minerals, Metals & Materials Society 2017
M.A. Meyers et al. (eds.), *Proceedings of the 3rd Pan American Materials Congress*,
The Minerals, Metals & Materials Series, DOI 10.1007/978-3-319-52132-9_73

and hurricane environments. These mechanical property attributes are primary drivers for the increased use of Nb-containing structural steels the long product global segment.

Chemical and mechanical property requirements are quite diverse to accommodate the end user infrastructure requirements. Nb-bearing steels are being methodically introduced into beams, plates, rail, rebar, shapes and pre-stressed concrete wire rod. These products can now meet the ever increasing end user infrastructure demands for improved mechanical properties and tighter specifications compared to the traditional construction steels produced today. Traditional construction steels have been applied for decades with relatively insignificant mechanical performance enhancements for the end user. Recently, the incorporation of more Nb into numerous steel grades is displacing traditional higher carbon-higher manganese V-bearing structural steels. Within the building, bridge and rebar sector, there is a shift to lower carbon–Nb containing steels which significantly improves toughness, formability, consistency, energy absorption and weldability at a more economical cost of construction than traditional structural steels. Niobium may be a key element in eutectoid (0.80%C) compositions for rail and pre-stressed concrete wire rod for improved fracture toughness in both heavy haul and light rail. In the end, the application of Nb-technology in long products depends upon the design criterion of the end user. On the metallurgical processing side, with additions of Nb for structural steel plate and long product applications, normalizing and heat treatment cycles can be shortened or entirely eliminated for bridge and other infrastructure applications, thereby increasing throughput, productivity, capacity and operational cost reductions between 5 and 10%.

Other steel sectors, such as energy pipelines, have adopted the lower carbon approach years ago. End user performance and cost considerations drove the development of HSLA steels to allow higher operating pressures and gas transmission production rates through the development of lighter wall, higher strength low carbon steel pipelines. Although the introduction of HSLA into the structural sector is being done, the rate of product development and implementation is significantly slower. The technology is already proven; it is the implementation that is necessary. Right now, there is a compelling need to accelerate this adaptation of the Nb-HSLA infrastructure technology within the structural market with the same intensity that has been experienced within the global automotive and energy pipeline sectors.

Niobium Infrastructure Structural Steel Metallurgical Approach

The unique metallurgical attributes that niobium provides in structural steels create the opportunity to successfully meet stringent mechanical, corrosion and elevated temperature demands. Nb-based structural steels were in limited production during the 1980s. Different from the automotive or pipeline segment where carbon levels are typically less than 0.10%, many of the plate structural products still exceed 0.10%C approaching allowable specification maximum carbon levels of 0.22%. Also, a large volume of structural steels are produced within the higher carbon peritectic regions (i.e. 0.11–0.16%C). Some mills choose this higher carbon level approach to achieve strength, but sacrifice toughness, weldability and end user product performance at an increased cost. Mills which have not yet adopted the low carbon–Nb approach are missing out on the inherent advantages Nb provides in yield strength, ductility, toughness and weldability at a lower overall operational cost [1]. Several case applications are presented which exhibit the quality and production advantages of producing Nb-LCLA structural steels at less than 0.10%C. Specific case application examples involve medium and heavy section structural beams for high rise structures such as the Freedom Tower in New York City, wind tower structural supports with improved fatigue and fracture toughness and high carbon eutectoid pre-stressed concrete wire rod. The seismic-resistant rebar sector can also benefit through the adoption of low carbon Nb-containing rebar. This sub-segment of the structural market has the potential to significantly further reduce the carbon footprint and simultaneously improve fatigue and seismic performance.

All of these described applications will translate into less raw materials consumption, lower emissions and less overall energy consumption. With the global infrastructure construction segment, the opportunity exists to further reduce the carbon footprint by an additional 5–10% HSLA usage in the S355 and S420 strength grades. For example, through the application of Nb-microalloyed structural steels, the opportunity exists today to reduce the total weight of a given structure, such as a bridge, compared to a non-microalloyed steel construction design. The cost savings associated with less beam, plate or rebar material and lower construction cost translates into lower overall project cost. The reduction of the carbon footprint through the application of Nb-bearing steels can contribute to the beneficial reduction in CO_2 emissions and energy consumption in that less steel is required to construct the structure.

Background

Over 50% of the structural plate and beam sections currently produced today are intermediate carbon levels from 0.11 to 0.22%. There is a gradual shift at some mills seeking participation in the value-added structural plate and beam segment to produce Nb-bearing structural grades at less than 0.10%C to make lower carbon base alloys for both plate and some long product applications. The benefits are not only improved mechanical properties and functional performance, but also the opportunity to reduce overall steelmaking cost per tonne through improved productivity, reduced diverts and improved product quality [1]. With increasing raw material and energy costs, the focus on processing parameters, such as reheating temperature and cooling rate after hot rolling to achieve improved mechanical properties can result in significant savings. A lower total cost of production may be achieved through low carbon-low alloy (LCLA) chemistries with selective accelerated cooling and better control of reheat furnace temperatures.

Structural Plates and Beams

Successful commercialization has resulted in the production of Nb-bearing beams for the Freedom Tower in New York City replacing V-bearing beams. Significant toughness improvements are realized in beams and plates produced to such specifications as ASTM992, ASTM572, ASTM588, ASTM710, Q345e, and S355 to name a few. The incorporation of Nb technology has significantly improved toughness properties through grain refinement and strategic cooling practices during rolling. The LCLA chemistry is comprised of less than 0.10%C, 0.025–0.035%Nb, less than 0.010%S, less than 0.015%P, less than 1.40%Mn and residuals less than 0.70% (i.e. Cr + Ni + Cu + Mo). The Nb addition refines the grain by 2–3 ASTM sizes, lowers the carbon equivalent by 0.07% and significantly improves the beam toughness compared to V-bearing low C steels as shown in Fig. 1a, b.

Results from various mills in different geographic regions have been congruent. Near net shape cast structural beams containing only a single Nb microalloy exhibit double to triple the Charpy impact strength energy at room temperature compared to a V-only microalloy system at similar sulfur, phosphorous and nitrogen levels and cooling rates as illustrated in Fig. 1. At mill #2 in China, an even higher cooling rate was employed and Charpy values increased more.

A second part of the study investigated a comparison of different cooling rates. Micrographic analysis revealed that the primary microstructural constituents at a

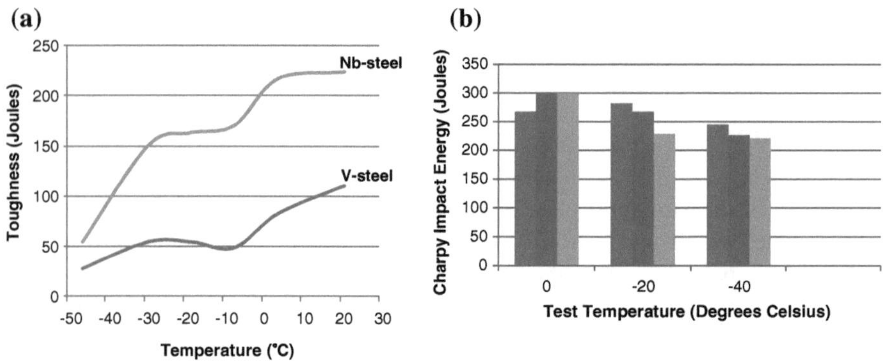

Fig. 1 **a** Charpy V-notch impact strength comparison—Nb versus V at Mill #1 [2]. **b** Charpy V-notch impact strength toughness Nb steel at Mill #2

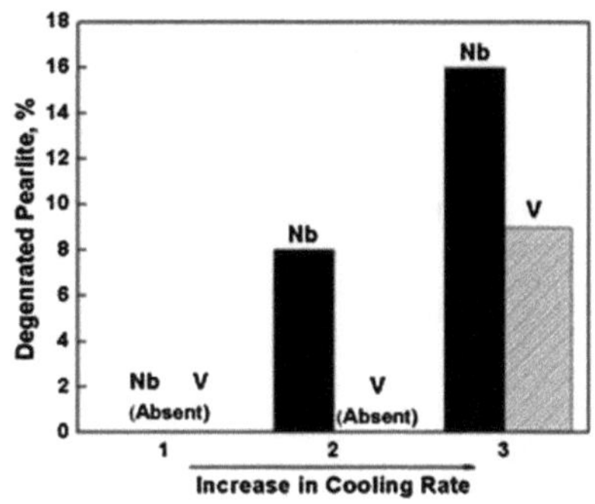

Fig. 2 Percent degenerated pearlite versus cooling rate—Nb and V comparison [3]

low cooling rate were polygonal ferrite and pearlite. At intermediate and high cooling rates, the microstructure consisted of lath-type/bainitic ferrite and degenerated pearlite together with conventional ferrite-pearlite. With an increase in cooling rate, there was an increased tendency towards formation of lath ferrite/bainitic ferrite with a consequential decrease in the conventional ferrite-pearlite microstructure. Figure 2 illustrates the influence of Nb on the transformation to the formation of degenerated pearlite (approaching a bainitic-type microstructure) which contributes to the improved toughness. No degenerated pearlite was observed in the V-bearing steel grade.

Construction Plates for Wind Tower Infrastructure

The wind tower end users require demand greater power generation efficiency which requires construction of towers to higher elevations. Fatigue and fracture toughness limitations of traditional steel higher carbon structural supports moved designers to consider carbon fiber composites. As a result of this threat of carbon fiber composite substitution for the HSLA S355 structural steel supports, a new steel material design was required to halt the threat. With the proven success of the beam applications, the Nb-Low Carbon Low Alloy (Nb-LCLA) hot rolled plate product provided a viable cost effective solution. Table 1 compares the mechanical properties of low C (0.06%C)–Nb (0.03%Nb) versus medium C (0.15%C)–Nb (0.02%Nb) containing wind tower plate for 20 mm plate thickness.

Note the isotropic CVN toughness at 15.5 °C for the low C–Nb compared to the anisotropic toughness behavior of the medium C-V in the transverse direction. A closer analysis of the upper shelf energy difference between the Nb and V is quite remarkable. A significant difference is exhibited in upper shelf CVN energy performance for the Nb LCLA compared to the low carbon V wind tower constructional plate in both directions. The comparison of low carbon–Nb and low carbon-V wind tower construction plates is illustrated in Fig. 3a, b [4, 5].

Table 1 Mechanical property comparison [4]

Steel	Orientation	Yield strength (MPa)	Tensile strength (MPa)	Elongation in 200 mm (%)	CVN @ −15.5 °C (J)
Low C–Nb	L	436	514	29.7	384
	T	450	521	28.1	371
Med C–Nb	L	439	561	21.9	103
	T	442	569	23.3	42
Med C–Nb norm	L	384	528	28.3	243
	T	391	530	27.6	132
ASTM A572-50 ASTM A709-50		345 min	448 min	18 min	34 min LCVN @ −12.2 °C
EN10025-2 S355K2		345 min	469-627	20 min	41 min LCVN @ −20 °C

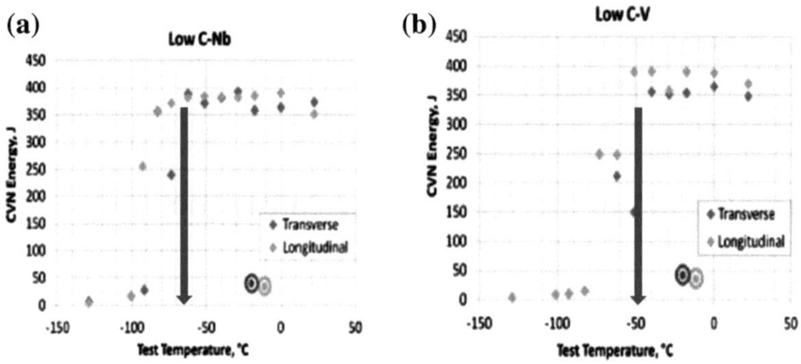

Fig. 3 **a** Low C-Nb transverse and longitudinal impact toughness. **b** Low C-V transverse and longitudinal impact toughness

At −65 °C test temperature, the CVN energy of the Nb wind tower supports is 400 J in both directions compared to V wind tower plate which is only 250 J in the longitudinal direction and 200 J in the transverse direction. With the Nb-containing microstructure, the isotropic properties are excellent with 400 J in both the longitudinal and transverse directions. Based upon the superior isotropic Nb-LCLA, the fatigue and fracture toughness was measured as shown in Table 2.

The weldability of these low C–Nb plates is significantly improved as well with the move from Zone II for the medium carbon which requires preheating into Zone I which does not require any preheating as shown in Fig. 4.

Table 2 Fatigue and fracture toughness comparison [4, 5]

	Endurance limit (MPa)	Fracture toughness (MPa $m^{1/2}$)
Low C–Nb	303	412
Med C–Nb	269	258
Med C-Normalized	245	275
Low C-V	245	Invalid test J integral[a]

[a]Due to anisotropy and microstructure inhomogeneity

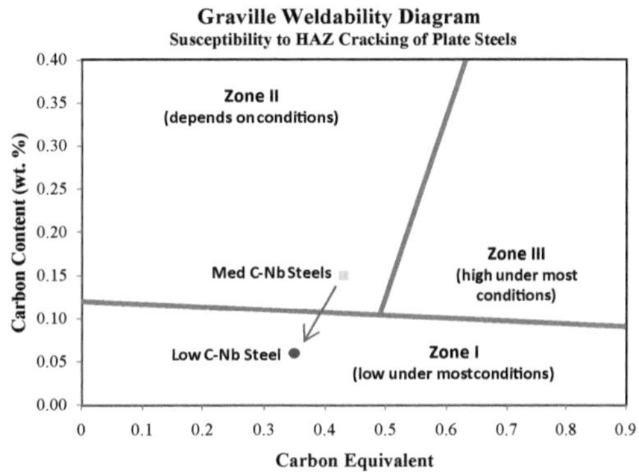

Fig. 4 Graville diagram [4]

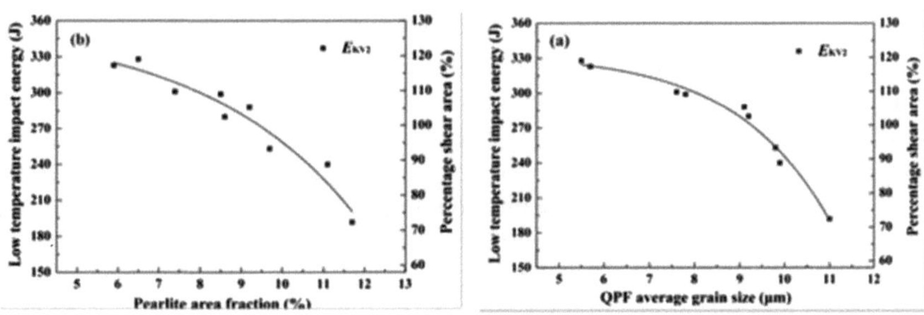

Fig. 5 Effects of **a** pearlite area fraction. **b** QPF grain size on *E*KV2 (−40 °C) [6]

Nb-Low Carbon Weathering Bridge Steel

A high-performance Q370qE-HPS bridge steel in China with low carbon content ($\leq$0.10wt%), Nb microalloying (0.025–0.050wt%) and low carbon equivalent (CEV) ($\leq$0.38%) has been produced using TMCP procedure. The results show that the microstructure consists of fine-grained quasi-polygonal ferrite (QPF), less pearlite and a large number of finely dispersed 5–10 nm Nb-rich precipitates. As shown in Fig. 5a, b, the *E*KV2 (−40 °C) of Q370qE-HPS steel plate increases significantly with the decreasing $d\alpha$ (interlammelar spacing) or P% (pearlite area fraction). These effects can be described quantitatively by the Boltzmann model.

Low Carbon Nb Containing Seismic-Resistant Rebar

Weldable and non-weldable reinforcing steel bars are one of the most important steel products widely applied in civil construction approaching over 240 million tons of usage in 2016. The compelling need for the development of even higher quality rebar for seismic applications is driven by recent catastrophic earthquakes in Haiti, Japan, Peru and China. Therefore, research projects are in progress around the world with a focus on the development of a family of low C–Nb-containing S500 and S600 grades with superior toughness, excellent low temperature energy absorption, fatigue resistance and less yield to tensile ratio variation. The available strength levels of Nb-bearing rebar has increased progressively from S345, S390, S500 and S600 grades. Traditionally, higher strength grades were produced with vanadium. However, recent niobium bearing rebar developments combine clean steelmaking practices of lower carbon steels at the melt shop with selective accelerated and controlled cooling practices at the rolling mill to produce low carbon equivalent high strength and earthquake resistant reinforcing bars. Typical chemistries are shown in Table 3. The goal is to eventually progress to a "New Generation" ultra-low carbon–Nb–B rebar for the most demanding seismic conditions after the successful evaluation of the less than 0.10%C rebar.

The production practices from the melting stage through the crack-free continuous casting of the billets through the hot rolling and accelerated cooling are keys to maximize the niobium effectiveness when producing these high quality, high strength reinforcing bar grades. One major obstacle to improved fatigue and fracture toughness performance involves the traditional Tempcore® produced reinforcing bar which results in a tempered martensite shell and a pearlitic or bainitic core mixed microstructure. This segregated microstructure is old technology and is incongruent with the fundamental goal of a uniform fine grain microstructure to accommodate high fatigue and fracture toughness environments.

There has been limited reliable published research on the impact and toughness properties of rebar. Some fundamental process metallurgy considerations should be incorporated into the production scheme to effectively manufacture homogeneous fine grain size across the entire cross section of S355, S420, S500 and S600 seismic rebars. There is a need to reduce carbon levels of rebar to improve fatigue and fracture behavior is seismic regions. It is initially a two-step process, first moving to 0.20%C and then to 0.10%C. The reduction of sulfur and phosphorous is imperative. The homogeneous fine-grain refinement that results from the Nb addition is critical. Figure 6 schematically captures these three critical success factors in designing and producing consistent high quality rebar with exceptional properties over the currently produced earthquake resistant rebars.

The goal is consistent production of lighter-weight reinforcing bar products at a lower carbon equivalent with improved weldability, higher elongation and better toughness at a lower cost. Low carbon grades with microalloys of Nb, Ti and Mo and the judicious application of accelerated/controlled cooling with or without the Tempcore® Process was compared to study the energy absorptive properties via the

Table 3 Nb-rebar technological development progression to "New Generation" Nb-rebar

	C	Mn	Si	V	Nb	S	P	Cr	Ni	Cu	B	N (ppm)
W/Nb	0.18–0.21	1.10–1.30	0.25 max	–	0.025–0.050	0.015 max	0.020 max	0.20–0.30	0.20 max	0.30 max	–	110 max
LowC + Nb	0.08–0.10	1.10–1.30	0.25 max	–	0.040–0.050	0.007 max	0.020 max	0.20–0.30	0.20 max	0.30 max	–	110 max
New generation[a]	0.03–0.05	0.90–1.20	0.25 max	–	0.10 max	0.007 max	0.015 max	0.20–0.30	0.20 max	0.30 max	0.002	90 max

[a]Not yet industrially produced

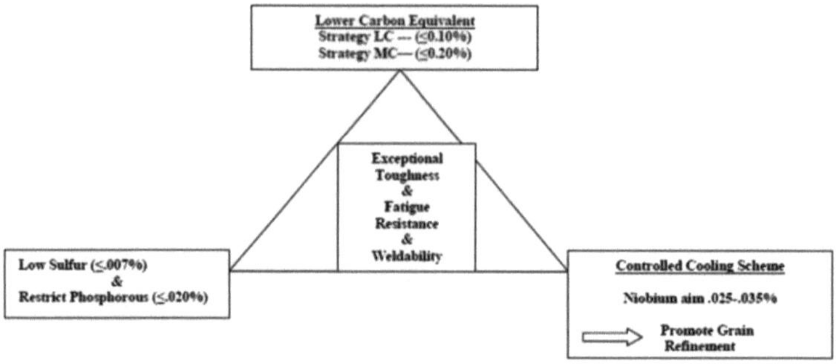

Fig. 6 Ultra-tough seismic-resistant rebar approach

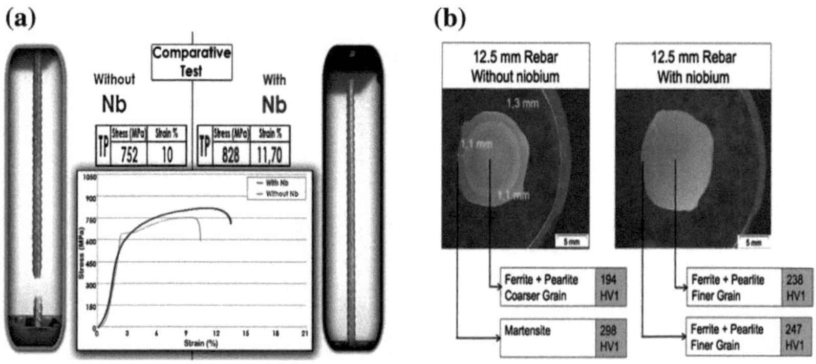

Fig. 7 **a** Air cooled Nb-rebar versus conventional Tempcore® rebar. **b** Tempcore® versus Nb-air cooled rebar microstructure

Table 4 Non-Nb Tempcore® versus air cool Nb rebar

	C	Si	Mn	P	S	Nb	YS (MPa)	TS (MPa)	Elong (%)	TS/YS
Tempcore	0.25	0.15	0.65	0.041	0.034	0.00	620	752	10	1.21
Air-Nb	0.28	0.26	1.00	0.030	0.042	0.02	600	828	13	1.38

stress-strain curve behavior. Figure 7a illustrates the stress-strain curve comparison of a high strength mixed microstructure Tempcore® rebar (0.26%C) compared to an air cooled homogeneous fine grain Nb-containing rebar at similar carbon content. Figure 7b illustrates the homogeneous fine grain microstructure compared to the discontinuous inhomogeneous Tempcore®-produced microstructure [7].

The chemistry and mechanical properties are shown in Table 4.

The homogeneous microstructure of the Nb-containing rebar is fundamentally favorable for fatigue and fracture toughness for earthquake resistant value added rebar.

Conclusions

Several case applications including high rise buildings, wind towers and bridges are presented which exhibit the quality and production advantages of producing Nb-LCLA structural steels at less than 0.10%C compared to the traditional 0.11–0.16%C steels. As a result of documented environmental climate change, more severe weather conditions around the globe and increased seismic activity, upgraded construction materials are required in order to sustain increased load, stress, strain and fatigue levels. Niobium-containing low carbon structural steels can meet this need. Nb-steels promote homogenous fine grain microstructures which improve performance under more severe climatic conditions. The 240 million tonne global rebar market is a huge opportunity as well to apply lower carbon fine grained Nb-containing steels with improved fatigue and fracture toughness performance.

References

1. Jansto, S. (2015, October 4–8). Metallurgical implications during the production of peritectic structural flat and long products. In *Materials Science & Technology Conference*, Columbus, Ohio.
2. Jansto, S. (2010). Niobium-bearing structural steels for the 21st century. *TMS, Niobium Bearing Structural Steels*, 1–26.
3. Shanmugan, S., et al. (2007). Effect of cooling rate on microstructure and mechanical properties of Nb-microalloyed steels. *Materials Science and Engineering A, 460–61,* 335–345.
4. Bodnar, R. E., et. al. (2011). Evaluation of low- and medium-carbon microalloyed plate steels for wind tower applications. In *International Symposium and Recent Developments in Plate Steels* (pp. 139–150), AIST.
5. Garsina, V. (2015, May). *Fracture toughness evaluation of five microalloyed plate steels for wind tower applications (Graduate Master's Thesis).* Illinois Institute of Technology.
6. Wang, H., & Jansto, S. (2016, August 1–5). High performance steels used for railway bridge structures in China. In *9th Pacific Rim International Conference on Advanced Materials and Processing (PRICM9),* Kyoto, Japan.
7. Jansto, S. (2015, July 12–15). Niobium-bearing steel technological developments for long and forged products. In *International Conference on Advances in Metallurgy of Long and Forged Products*, AIST, Vail, Colorado.

Kinetic Study of the Austenite Decomposition During Continuous Cooling in a Welding Steel

Octavio Vázquez-Gómez, Edgar López-Martínez, Alexis Iván Gallegos-Pérez, Heber Santoyo-Avilés, Héctor Javier Vergara-Hernández and Bernardo Campillo

Abstract The kinetics of austenite decomposition during the continuous cooling of a low-carbon welding steel was determined by dilatometric analysis. Based on the measurements, the critical transformation temperatures of austenite decomposition were determined for both ferrite and pearlite. The cooling conditions were established from the Stelmor® controlled cooling process for the manufacturing of wire rod. It was observed that the critical temperatures decrease when the cooling time was reduced as a result of an increased cooling rate in the range of 600–900 °C. Using the decomposition temperatures, a continuous cooling transformation CCT diagram was created and it may be observed that austenite decomposition occurs in two steps. The kinetic parameters for each step were determined and compared with

O. Vázquez-Gómez · A.I. Gallegos-Pérez · H.J. Vergara-Hernández
División de Estudios de Posgrado e Investigación,
Instituto Tecnológico de Morelia, Av. Tecnológico 1500,
58820 Morelia, Michoacán, Mexico

O. Vázquez-Gómez
Consejo Nacional de Ciencia y Tecnología, Av. Insurgentes 1582,
03940 Ciudad de México, Mexico

E. López-Martínez
Universidad del Istmo, Campus Tehuantepec Ciudad Universitaria s/n,
Barrio Santa Cruz 4a Secc., 70760 Sto. Domingo Tehuantepec, Oaxaca, Mexico

H. Santoyo-Avilés
Ternium México S. A. de C. V., Av. Universidad 992,
Cuauhtémoc, 66450 San Nicolás de los Garza, Nuevo León, Mexico

B. Campillo
Facultad de Química, Universidad Nacional Autónoma de México,
Ciudad Universitaria, 04510 Ciudad de México, Mexico

B. Campillo
Instituto de Ciencias Físicas, Universidad Nacional Autónoma de México,
Av. Universidad 1001, Chimalpa, 62209 Cuernavaca, Morelos, Mexico

O. Vázquez-Gómez (✉)
CONACyT-ITM, Av. Tecnológico 1500, Lomas de Santiaguito,
58120 Morelia, Michoacán, Mexico
e-mail: ovazquezgo@conacyt.mx

© The Minerals, Metals & Materials Society 2017
M.A. Meyers et al. (eds.), *Proceedings of the 3rd Pan American Materials Congress*,
The Minerals, Metals & Materials Series, DOI 10.1007/978-3-319-52132-9_74

the Johnson-Mehl-Avrami-Kolmogorov diffusive model. The final microstructure was analyzed using an optical microscope and evaluated by nanoindentation to determine the effect of the cooling rate on the nanohardness of each phase.

Keywords Nanoindentation · Low-carbon steel · Continuous cooling transformation · Dilatometric analysis

Introduction

Low-carbon steel is used to produce electrodes and wire rod. The American Welding Society (AWS) has classified this type of alloy for its application in the electric arc welding process through the A5.18 standard, in which required mechanical properties and chemical compositions are specified.

During the manufacturing process of low-carbon steel wire rod, the wire rod obtains its final mechanical properties via certain processes, such as the Morgan Stelmor® controlled cooling process. The cooling conditions in this process are critical to obtaining the desired final microstructure and mechanical properties. Different studies have demonstrated the transformation kinetics of various types of steel under separate continuous or isothermal cooling conditions [1–7]. These studies have focused on demonstrated the transformation mechanisms that occur in steels and understanding their microstructural behavior when exposed to heat or thermomechanical treatments. However, since an infinite combination of alloys exists, it is impossible to understand the transformation mechanisms for all of these alloys. For this reason, the transformation kinetics must be determined. There are thermal analysis techniques that can be used to study solid-solid phase transformations; for example, dilatometric analysis is a technique that continuously measures, in real time, the dimensional changes of a specimen undergoing a controlled heating and cooling process. Dilatometric analysis compares the internal changes with the phase transformation [7–11]. The purpose of this study is determine the non-isothermal kinetics of austenite decomposition in a welding steel, emulating the cooling cycle that occurs in the Morgan Stelmor® controlled cooling process. An additional goal of this investigation is to determining the effect of cooling rate on the nanohardness of the phases formed during the continuous cooling.

Experimental Procedure

Materials

Cylindrical samples of low-carbon welding steel (AWS ER70S-6) 5 mm in diameter and 15 mm in length were tested. The initial microstructure was composed

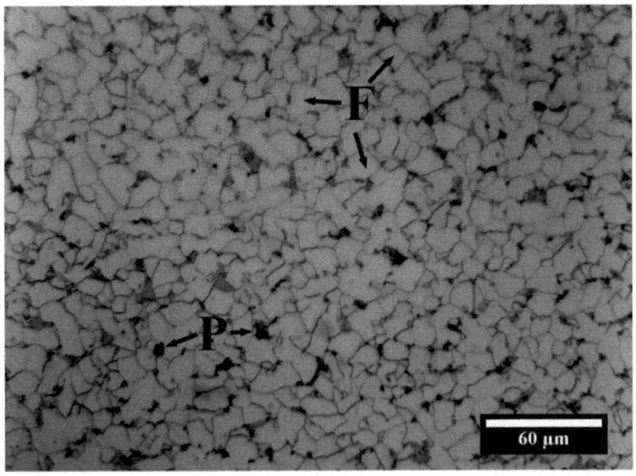

Fig. 1 Initial microstructure of low-carbon steel, where "F" refers to proeutectoid ferrite and "P" refers to pearlite

Table 1 Chemical composition of the low-carbon steel

%C	%Mn	%Si	%P	%S	%Ni	%Cr
0.083	1.455	0.831	0.006	0.004	0.005	0.015

of ferrite and pearlite (volume fraction of ferrite: 0.97 ± 0.009) with a ASTM No. 10 grain size in accordance with the ASTM E112 standard. Figure 1 shows the initial microstructure of the steel, which was revealed using a 3% nitric acid solution in alcohol (Nital 3). The chemical composition of the steel is listed in Table 1.

The lower and upper surfaces of each sample were prepared by grinding them with 600-grit SiC paper (15 μm) and polishing them with 0.5 μm of alumina for subsequent analysis in a vertical model L75-V dilatometer (Linseis, Germany).

Dilatometry: Heating Cycle

The samples were heated to 1150 °C at a constant heating rate of 19 °C min^{-1} in an inert argon atmosphere at 12 psi. The heating conditions were established by the billet reheating process, considering the time it takes a billet to be heated from room temperature to 1150 °C.

Dilatometry: Cooling Cycles

After the heating cycle, the samples were cooled in three stages. In the first stage, the samples were cooled at a fixed cooling rate of 50 °C min^{-1} until they attained a temperature of 1025 °C. The second stage consisted of cooling at 25 °C min^{-1} until the samples reached a temperature of 900 °C, which emulated the cooling conditions during the rolling process prior to the Morgan Stelmor® controlled cooling process. Once a temperature of 900 °C was attained, the samples were cooled at different rates: 1, 5, 15 and 25 °C min^{-1} until they reached a temperature of 50 °C. The argon pressure in the interior of the dilatometer was maintained at 12 psi.

Results and Discussion

Austenite Formation

Figure 2 shows the dilatation strain $(\Delta L/L_o - T)$ at a cooling rate of 19 °C min^{-1}, and this strain is compared with its first derivative with respect to temperature $(d(\Delta L/L_o)/dT - T)$. This figure also shows that the austenite formation occurs in two steps. In the first step, austenite is formed from pearlite decomposition, and in the second step austenite is formed from proeutectoid ferrite [12].

Austenite Decomposition

Figure 3 shows the dilatometric cooling curve at a rate of 1 °C min^{-1}. As in the previous case for the austenite formation, the decomposition of austenite occurs in two steps delimited for the critical temperatures A_{r3}, A_{r1s} and A_{r1f}.

During the continuous cooling, the austenite decomposes in two steps: (1) the transformation of austenite into proeutectoid ferrite at temperatures ranging from A_{r3} to A_{r1s}; and (2) the transformation of austenite into pearlite at temperatures ranging from A_{r1s} to A_{r1f}. The second step maintains the volume fraction of proeutectoid ferrite formed in the first step.

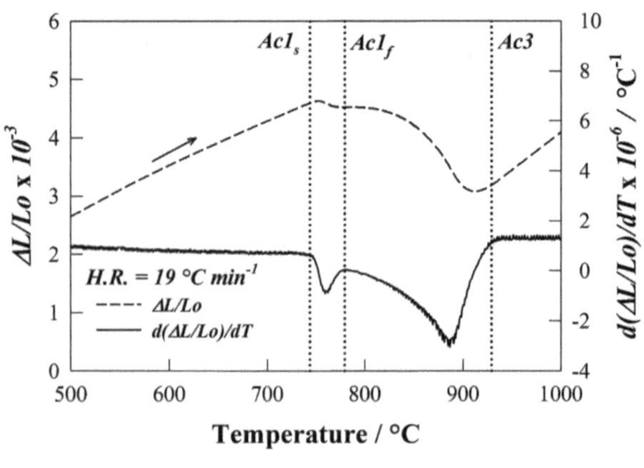

Fig. 2 Dilatometric heating curve of low-carbon steel at a cooling rate of 19 °C min^{-1}. The *discontinuous line* indicates the dilatation strain, and the *continuous line* indicates the change of the dilatation stain with respect to the temperature

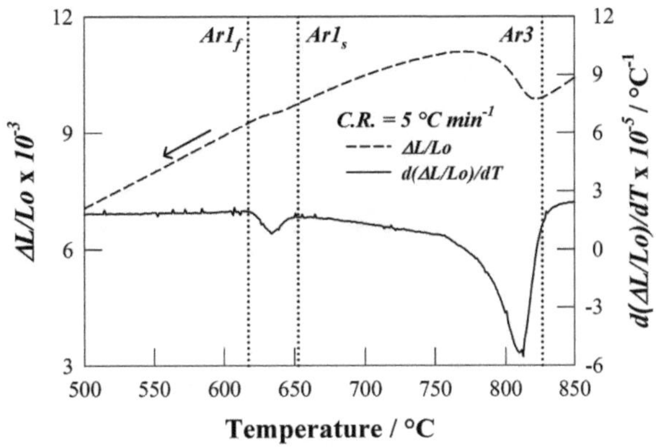

Fig. 3 Dilatometric cooling curve of low-carbon steel at a cooling rate of 5 °C min^{-1}. The *discontinuous line* indicates the dilatation strain, and the *continuous line* indicates the change of the distortion with respect to the temperature

Continuous Cooling Transformation Diagram

Using the decomposition temperatures of the austenite, a continuous cooling transformation (CCT) diagram was created, as shown in Fig. 4. This figure shows the cooling stages prior to the controlled cooling process between 900 and 1150 °C. Once a temperature of 900 °C was attained, the controlled cooling process was initiated. The CCT diagram is composed of two transformation zones, as indicated by the dilatometric curves: (1) the transformation zone of austenite to ferrite (intercritical zone) denoted by F + A; and (2) the transformation zone of austenite to pearlite plus the proeutectoid ferrite transformation zone denoted by F + A + P.

Both zones grow relative to the cooling rate. However, in the pearlitic transformation zone the change is more evident because the zone is larger when the cooling rate is increased above 15 °C min^{-1}. The transformation zones are consistent with the micrographs shown in Fig. 5a, b, which show the final microstructure after the cooling cycle at rates of 1 and 25 °C min^{-1}.

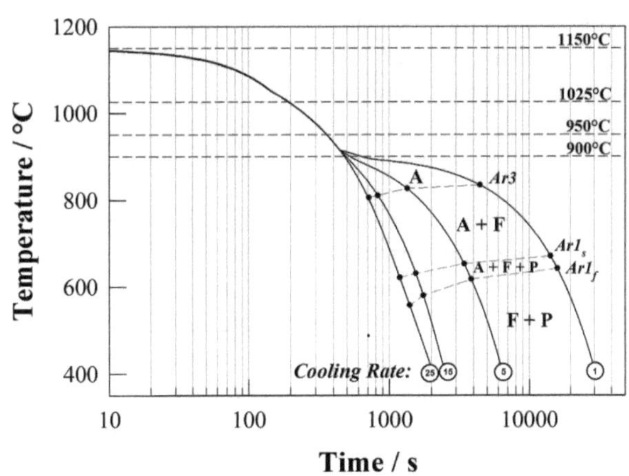

Fig. 4 Continuous cooling transformation diagram of low-carbon steel for a range of cooling rates: 1–25 °C min^{-1}

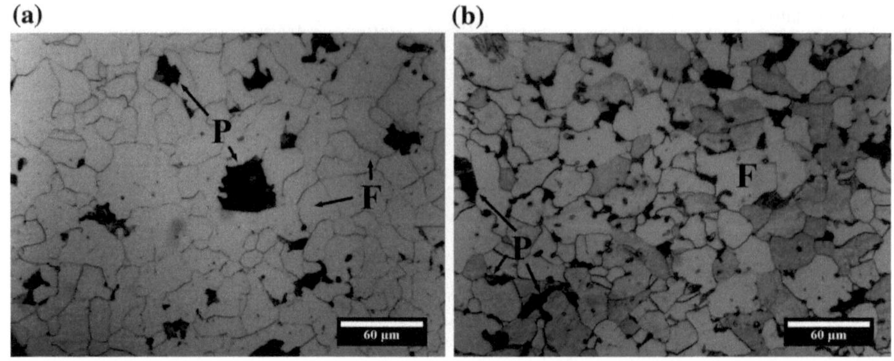

Fig. 5 Final microstructures of the samples tested with different cooling rates: **a** 1 and **b** 25 °C min^{-1}. "F" refers to proeutectoid ferrite and "P" refers to pearlite

Figure 5a corresponds to the microstructure of the sample cooled at a rate of 1 °C min^{-1}, where the volume fraction of proeutectoid ferrite measured was 0.92 ± 0.04 with a ASTM No. 7 grain size. On the other hand, Fig. 5b shows the microstructure of the sample cooled at a rate of 25 °C min^{-1}. In this case, the volume fraction of proeutectoid ferrite slightly increased to 0.94 ± 0.02, and the grain size decreased to ASTM No. 8. In the micrographs shown, the pearlite fraction decreased with the cooling rate. However, the change in the proportion of the phases was not significant, but ASTM grain size increased from No. 7 to 8, which was higher than the initial grain size.

Austenite Decomposition Non-isothermal Kinetics

The kinetics of austenite decomposition was analyzed, and the volume fraction was calculated for each step of the transformation using extrapolation and the intersection lines method and the lever rule. The data obtained were compared with the Johnson-Mehl-Avrami-Kolmogorov diffusive model, which has been used to estimate the kinetics parameters k and n under non-isothermal conditions [1–5, 7, 9]:

$$f_i = 1 - \exp(-k \cdot t^n) \quad (1)$$

where f_i is the volume fraction of each phase, i, k and n are the kinetics parameters and t is the transformation time.

For each decomposition step, the volume fraction of transformed austenite is shown in Fig. 6a, b for cooling rates of 1 and 25 °C min^{-1}. As indicated above, the first decomposition step was associated with the transformation of austenite into proeutectoid ferrite, and the second step was related to the formation of pearlite.

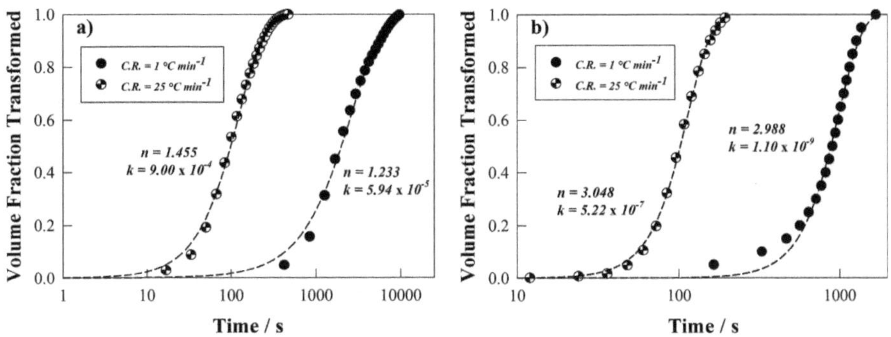

Fig. 6 Volume fraction of transformed austenite for each step of the decomposition: (**a**) Step I, formation of proeutectoid ferrite; (**b**) Step II, formation of pearlite

Figure 6a shows the volume fraction of transformed austenite for the first step of the decomposition as a function of the transformation time and cooling rate. Both curves exhibit similar behavior with respect to the volume fraction of transformed austenite, showing only the effect of the cooling rate on the transformation time, which decreases at higher cooling rates. The discontinuous lines show the adjustment of the transformation curve with the JMAK model, where the kinetic parameters k and n can be observed. In this figure, the parameter k increases by one order of magnitude from 10^{-4} to 10^{-5}, which indicates that the transformation rate increases directly proportional to the cooling rate and the transformation interval occurs at lower temperatures, as the CCT diagram shows (Fig. 4). The parameter n increased slightly from 1.23 to 1.45. These values are consistent with a preferential nucleation in grain boundaries (i.e., the nucleation mode remains virtually unchanged and only the nucleation rate is affected), which increases when the cooling rate is increased due to the driven force transformation defined as a difference in temperature per unit time.

Figure 6b shows the volume fraction of transformed austenite for the second decomposition step as a function of the transformation time and cooling rate. As in the first step, both curves exhibit similar behavior in terms of the difference in transformation time, which decreases as the cooling rate increases. The observations are similar to those shown in the first austenite decomposition step. However, the change in the parameter k is more pronounced since it changes its value by two orders of magnitude from 10^{-9} to 10^{-7}, which indicates that the transformation rate increases directly proportional to the cooling rate. On the other hand, the parameter n remains virtually unchanged from 2.98 to 3.05 when the cooling rate is changed from 1 to 25 °C min^{-1}; however, the value is considerably higher, which indicates that the nucleation mode is similar in both cases and independent of the cooling rate. Based on the value of n, we can estimate that pearlite nucleates predominantly in the corners of the proeutectoid ferrite grains.

Nanoindentation Analysis

Figure 7 shows the force-displacement curve of the nanoindentation tests for the samples that were cooled at 1 and 25 °C min^{-1}. Two features can be observed in this figure: (1) the behavior of the force-displacement curve and (2) the maximum penetration. For both of the cooling rates, proeutectoid ferrite not exhibit significant changes in these features. However, differences can be noted in the behavior and maximum penetration of pearlite.

Figures 8 and 9 show the results of the nanohardness measurements of the samples tested with a cooling rate of 1 and 25 °C min^{-1}, respectively. These figures show that the nanohardness of proeutectoid ferrite (pearlite) is approximately 3 (5) GPa. It can also be noted that the pearlite measurement data have higher levels of dispersion and that the nanohardness measurements of the microconstituents are affected by the grain boundaries. Figure 8 shows that the nanohardness of pearlite is approximately 4 GPa for measurements taken near the grain boundary of the proeutectoid ferrite and approximately 5 GPa for measurements taken further from the grain boundary. Figure 9 shows that the nanohardness of pearlite is approximately 5 GPa even when the measurement is taken near the grain boundary of the proeutectoid ferrite. These results indicate that the hardening mechanism via phase transformation has a more pronounced effect on nanohardness than the hardening mechanism via the grain boundary for the pearlite cooled at a cooling rate of 25 °C min^{-1}. Furthermore, the nanohardness determined near the interception of three proeutectoid ferrite grains is higher than the nanohardness measured further from the grain boundary (Fig. 9). López-Martínez et al. [13] noted that the contribution

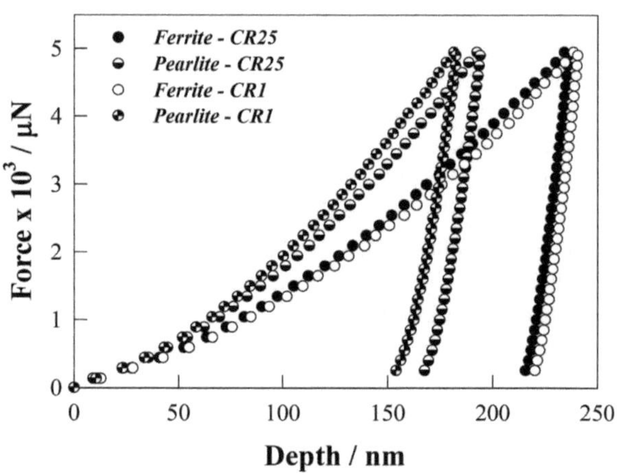

Fig. 7 Force-displacement curves of proeutectoid ferrite and pearlite at different cooling rates: 1 and 25 °C min^{-1}

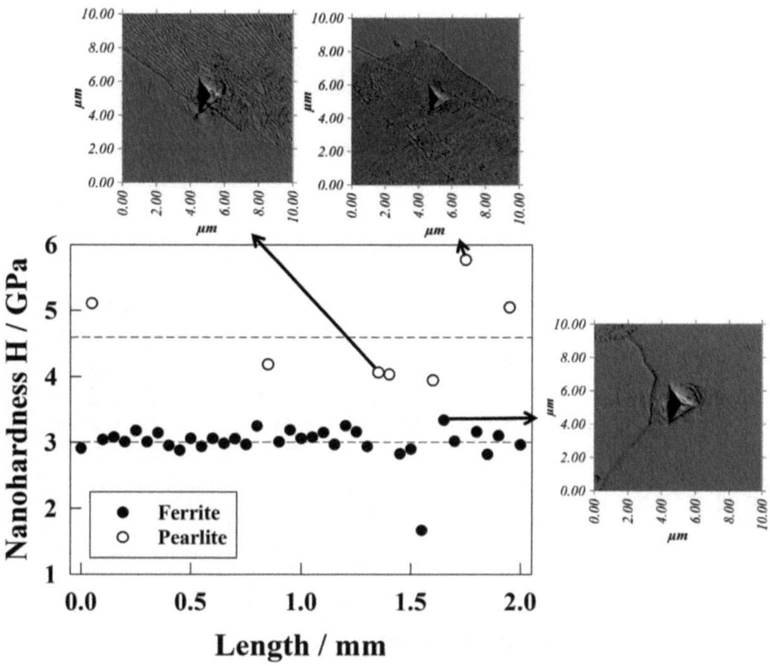

Fig. 8 Nanohardness of the microconstituents as a function of position from the center of the sample. Cooling rate: 1 °C min^{-1}

of the grain boundary to the nanohardness can be upwards of 24% of phase nanohardness.

Figure 10 shows the nanohardness of the microconstituents as a function of the cooling rate. In the case of proeutectoid ferrite, it can see that the nanohardness is not a function of the cooling rate since there is no significant difference in phase nanohardness. The variances shown in the error bars in this phase are primarily due to the effect at the grain boundary; this statement can be validated using the hardness measurements. For cooling rates of 1 and 25 °C min^{-1}, the hardness measurements are 42 and 47 HRA, respectively. Furthermore, it can also be seen in Fig. 10 that the nanohardness of pearlite has a greater effect on the cooling rate than proeutectoid ferrite. As in the case of proeutectoid ferrite, the error bars of the pearlite nanohardness correspond to the effect of the grain boundary where the effect is greater in pearlite formed at a cooling rate of 1 °C min^{-1}.

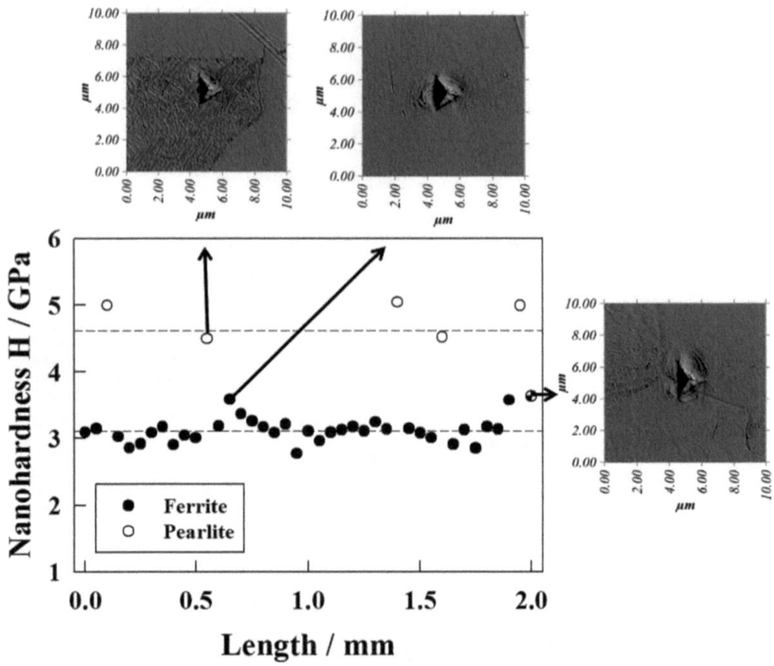

Fig. 9 Nanohardness of the microconstituents as a function of position from the center of the sample. Cooling rate: 25 °C min^{-1}

Fig. 10 Nanohardness of the microconstituents as a function of cooling rate

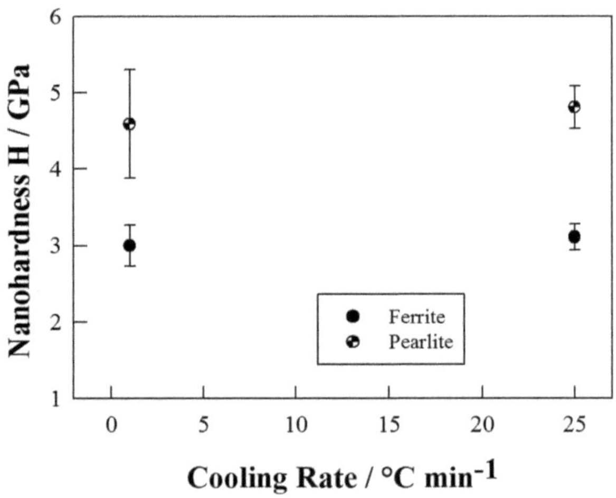

Conclusion

Austenite decomposition in a low-carbon welding steel (AWS ER70S-6) occurs in two steps. During cooling, austenite decomposes into proeutectoid ferrite and pearlite. The final microstructure remains virtually independent of the cooling rate decreased the grain size; an increase of the cooling rate results in a smaller grain size. On the other hand, kinetic non-isothermal analysis was conducted and revealed that during the continuous cooling in the first step the inter-critical zone decreased directly proportional to the cooling rate, thereby increasing the value of k while the value of n remained constant. This finding suggests there were no significant changes in the nucleation mode. Something similar occurs in the formation of pearlite, in which the parameter k considerably increases compared with the first step and does so in accordance with the cooling rate. Furthermore, the parameter n exhibited values close to 3, which indicates that the nucleation of pearlite occurs at the grain corners. However, the parameter n remained unaffected by the cooling rate. Finally, based on nanoindentation tests one can see that proeutectoid ferrite does not contribute to an increase in hardness given an increase in the cooling rate.

Acknowledgements The authors would like to thank the National Council of Science and Technology of Mexico (CONACyT) for the support received through grant No. 256843 and also to PAPIIT-UNAM (grant IN115616), PRODEP (grant UNISTMO-PTC-108) for their financial support.

References

1. Pham, T. T., Hawbolt, E. B., & Brimacombe, J. K. (1995). Predicting the onset of transfomation noncontinuous cooling conditions: Part I. *Metallurgical and Materials Transactions A, 26A,* 1987–1992.
2. Pham, T. T., Hawbolt, E. B., & Brimacombe, J. K. (1995). Predicting the onset of transfomation noncontinuous cooling conditions: Part II. *Metallurgical and Materials Transactions A, 26A,* 1993–2000.
3. Hawbolt, E. B., Chau, B., & Brimacombe, J. K. (1983). Kinetics of austenite-pearlite transformation in eutectoid carbon steel. *Metallurgical Transactions A, 14A,* 1803–1815.
4. Umemoto, M., Horiuchi, K., & Tamura, I. (1983). Pearlite transformation during continuous cooling and its relation to isothermal transformation. *Transactions of the Iron and Steel, Institute of Japan, 23,* 690–695.
5. Militzer, M., Pandi, R., & Hawbolt, E. B. (1996). Ferrite nucleation and growth during continuous cooling. *Metallurgical and Materials Transactions A, 27A,* 1547–1556.
6. Somani, M. C., Porter, D. A., & Karjalainen, L. P. (2014). On various aspects of decomposition of austenite in a high-silicon steel during quenching and partitioning. *Metallurgical and Materials Transactions A, 45,* 1247–1257.
7. López-Martínez, E., et al. (2015). Effect of initial microstructure on austenite formation kinetics in high-strength experimental microalloyed steels. *International Journal of Minerals, Metallurgy and Materials, 22*(12), 1304–1312.
8. Farrar, R. A., & Zhang, Z. (1993). Experimental verification of the continuous-cooling transformation diagram produced by the dilatometry-metallography method. *Journal of Materials Science Letters, 12,* 1606–1611.

9. García de Andrés, C., et al. (1998). Modelling of kinetics and dilatometric behavior of non-isothermal pearlite-to-austenite transformation in an eutectoid steel. *Scripta Materialia, 39*(6), 791–796.
10. García de Andrés, C., et al. (2002). Application of dilatometric analysis to the study of solid-solid phase transformations in steels. *Materials Characterization, 48*(1), 101–111.
11. San Martín, D., Rivera-Díaz-del-Castillo, P. E. J., & García-de-Andrés, C. (2008). In situ study of austenite formation by dilatometry in a low carbon microalloyed steel. *Scripta Materialia, 58*(10), 926–929.
12. Batista-Lopes, M. M., & Barros-Cota, A. (2014). A study of isochronal austenitization kinetics in a low carbon steel. *REM. Revista Escola de Minas, 67*(1), 61.
13. López-Martínez, E., et al. (2016). Mechanical characterization of the welding of two experimental HSLA steels by microhardness and nanoindentation tests. *Metals and Materials International, 22*(6), 987–994.

Microstructural Evolution in Microalloyed Steels During Thermomechanical Rod Rolling

Lijia Zhao, Robert L. Cryderman and John G. Speer

Abstract Steel rods are hot-rolled at high strains and strain rates with a subsequent controlled cooling process to influence the microstructure. The microstructure and mechanical properties of the hot-rolled rods are controlled to produce high-strength fasteners in the cold heading process without subsequent heat treatment. In the present study, simulations of rod rolling by torsional deformation and controlled time-temperature schedules were conducted to examine the effects of thermomechanical processing parameters and microalloying additions on the microstructure evolution and mechanical properties of low-carbon steel rods. Transformation and precipitation behaviors during the thermomechanical process were investigated and related to the increased strength in the steel rods.

Keywords Rod rolling · Dynamic transformation · Grain refinement · Microalloying

Introduction

Current high strength steel fasteners of grade 8.8 (i.e., 800 MPa tensile strength with minimum 0.8 yield strength/tensile strength ratio) and above are produced from a variety of steels in the form of rod and wire, including plain carbon, medium carbon, low alloy steels, etc. As illustrated in Fig. 1, the traditional method for manufacturing high strength fasteners involves two thermal treatments: (i) softening annealing prior to cold deformation (drawing, extrusion and heading) and (ii) heat treatment (quench and temper) after cold deformation to obtain the required strength levels. Cost reductions in terms of alloy content, rod/wire processing and component heat treatment would be achieved if high strength fasteners could be

L. Zhao (✉) · R.L. Cryderman · J.G. Speer
Department of Metallurgical and Materials Engineering,
Colorado School of Mines, Advanced Steel Processing
and Products Research Center, Golden, CO 80401, USA
e-mail: zhao@mines.edu

© The Minerals, Metals & Materials Society 2017
M.A. Meyers et al. (eds.), *Proceedings of the 3rd Pan American Materials Congress*,
The Minerals, Metals & Materials Series, DOI 10.1007/978-3-319-52132-9_75

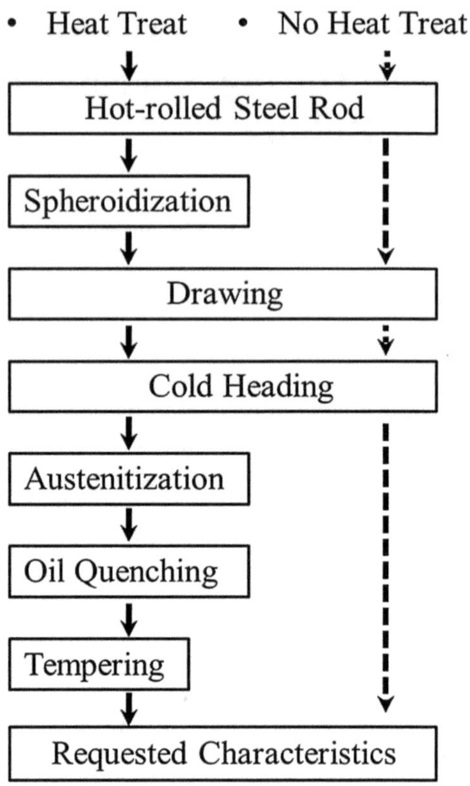

Fig. 1 Schematic manufacturing process evolution of cold deformed fasteners

directly produced from as-rolled rods, without the need for a subsequent hardening heat treatment. When softening (e.g. spheroidization) and hardening (e.g. quench and temper) heat treatments of the rods are eliminated, it becomes very important to tailor the strength properties of the hot-rolled rods to meet the final strength requirement after cold drawing and cold heading. Based on the current industrial infrastructures for rod rolling, the production of high strength fasteners without any heat treatment could be possible by the development of microalloyed steels [1–6] through transformation, precipitation strengthening and work hardening (by cold deformation) [7–10].

Rod rolling is a high-strain and high-strain rate process. The accumulated true strain applied on the rods could be up to 10 with a strain rate as high as 100 s^{-1}. The deformation of austenite at such a high strain and strain rate can generate a large amount of heat which is hard to dissipate at such a short deformation time, resulting in the occurrence of adiabatic heating which could accelerate the coarsening of austenite grains and subsequently transformed phases. The present study discusses the adiabatic heating phenomenon during rod rolling and its effect on the final microstructure and mechanical property of the steel rods. A new strategy for strengthening hot-rolled steel rods is then proposed to offset the negative effect of adiabatic heating, aiming at fabrication of ultrafine grained steel rods to achieve high strength and high toughness (ductility) simultaneously. Finally, the effects of microalloying additions and thermomechanical processing parameters are examined.

Experimental Procedure

A microalloyed steel (0.049C-1.37Mn-0.013Ti-0.07Nb) and a reference steel (0.052C-1.15Mn) were utilized in the present study. It has been well recognized that the addition of Titanium (Ti) and Niobium (Nb) can not only refine austenite grains prior to hot-rolling, but also help suppress recrystallization of austenite during hot-rolling. Therefore, microstructural differences between the two kinds of steels could be expected. Torsion samples with a gauge length of 14.4 mm and a diameter of 7.2 mm were machined from industrially melted and rolled steel plates for simulating the industrial rod rolling on the Gleeble® 3500. The torsion tests are described as follows: the torsion samples were soaked at 1100 °C for 20 min, and naturally cooled to 1050 °C to simulate roughing to finishing deformation passes, and then cooled by nitrogen to 920, 830 or 730 °C to simulate final V-block pass. After final V-block deformation, the samples were cooled at a rate of 50 °C/s to 800 °C (in the case of 920 or 830 °C). Stelmor® cooling simulation was conducted from 800 to 500 °C at different cooling rates ranging from 5 to 20 °C/s, followed by natural cooling to room temperature. In the case of V-block temperature of 730 °C, Stelmor® cooling simulation was conducted immediately after the deformation.

After testing, each torsion sample was cut through the center of the gauge section and metallographically prepared by grinding and polishing with diamond suspension down to 1 μm. Microstructure characterization and Vickers microhardness testing were conducted according to ASTM E-92 at 72.4 pct of the sample radius, representing the effective radius upon which strain calculations were based. Twenty measurements were taken per sample for the hardness testing. The mechanically polished samples were etched using 3 pct Nital. Scanning electron microscopy (SEM) observation was conducted with a JEOL 7000 field-emission SEM (FESEM).

True stress-true strain curves for the deformation passes were calculated according to the recorded torsion and torque data during the Gleeble® simulation. Shear strain was calculated from θ, the rotation angle of the sample in rad, using

$$\gamma = \frac{r\theta}{L} \tag{1}$$

where γ is shear strain, r is the effective radius of the torsion sample (2.61 mm), and L is the gauge length (14.4 mm). Shear stress, τ, in MPa was calculated using

$$\tau = \frac{32Tr}{\pi D^4} \quad (2)$$

where T represents measured torque in Nm and D is the effective diameter of the torsion sample (5.22 mm). The Von Mises yielding criterion was used to relate shear to normal stress and strain, assuming a true torsion stress state with τ_{xy} as the only nonzero stress component [11]. The simplified conversions are provided for true strain, according to

$$\varepsilon = \frac{\gamma}{\sqrt{3}} \quad (3)$$

And for true stress, σ, according to

$$\sigma = \sqrt{3}\tau \quad (4)$$

Results and Discussion

Figure 2a shows the temperature variation with time during the thermomechanical simulation on the Gleeble® 3500. Adiabatic heating phenomenon was observed during most of the deformation steps, as marked by solid blue circles. Another evidence of the adiabatic heating can be revealed in Fig. 2b, which shows true stress-true strain curves of the torsion specimens during the thermomechanical process. The softening of true stress-true strain curves (especially the sharp decrease of flow stress) is believed to be due to dynamic recrystallization of austenite and adiabatic heating during the high-strain deformation at high strain rates.

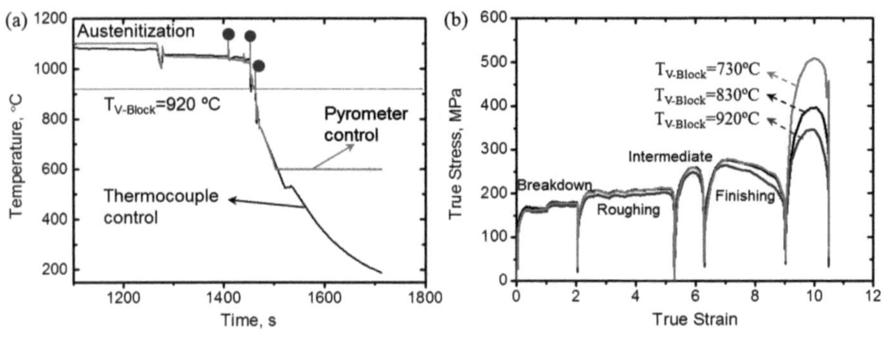

Fig. 2 Temperature profile (**a**) and experimental true stress-true strain curves (**b**) for the hot torsion specimens deformed at different temperatures during the thermomechanical simulation (Color figure online)

The adiabatic heating can accelerate the coarsening of recrystallized austenite, and consequently, coarsening the transformed phases. Figure 3 shows SEM images of the reference steel deformed at different temperatures from 920 to 730 °C at the V-block pass with subsequent cooling at a rate of 20 °C/s to 500 °C followed by natural cooling to room temperature. The microstructures consist of mainly ferrite phase with a small fraction of martensite and granular bainite. Industrially, the deformation temperatures at the V-block pass are usually above 800 °C. The microstructures obtained at temperatures of 830 °C (Fig. 3a) and 920 °C (Fig. 3b) are mainly composed of coarse transformed ferrite grains due to adiabatic heating. The ferrite grains were greatly refined by decreasing the V-block temperature to 730 °C, where dynamic transformation of austenite to ferrite may occur during the deformation of austenite. As shown in Fig. 3c, the hot-rolled microstructure consists of a small amount of ultrafine grained ferrite (grain size of 1–2 μm) and ultrafine martensite particles (particle size of 1–2 μm). It should be noted that the ultrafine ferrite grains at 1–2 μm level were observed in spite of the occurrence of adiabatic heating.

Figure 4 shows SEM images of the Ti/Nb-alloyed steel deformed at 920 °C (a), 830 °C (b) and 730 °C (c) at the V-block pass with subsequent cooling at a rate of 20 °C/s from 800 °C (a, b) or 730 °C (c) to 500 °C followed by natural cooling to

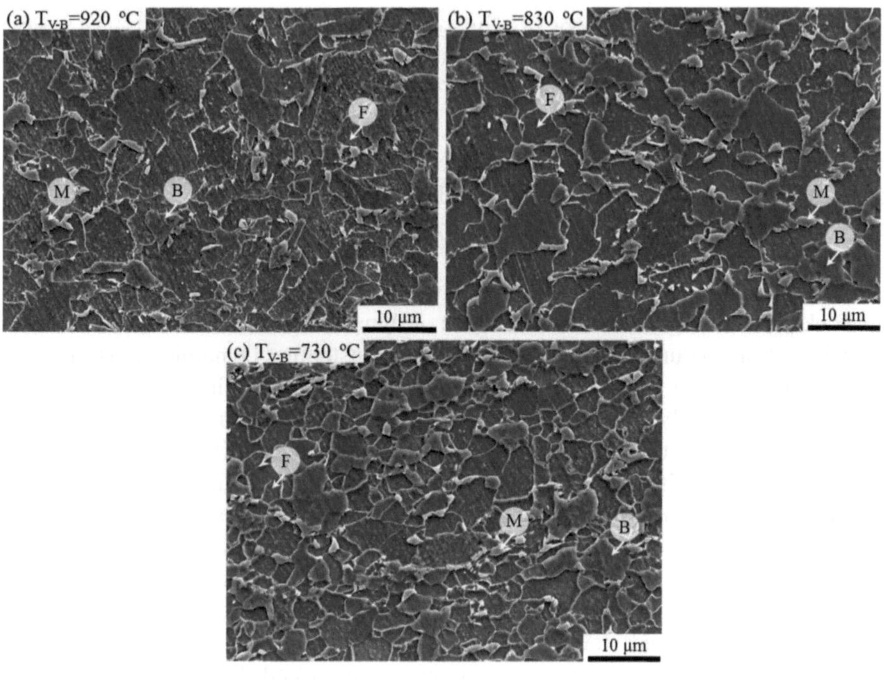

Fig. 3 FESEM images of reference steel specimens deformed to simulate V-block strain at **a** 920 °C, **b** 830 °C and **c** 730 °C with subsequent cooling at a rate of 20 °C/s from **a**, **b** 800 °C or **c** 730–500 °C followed by natural cooling to room temperature

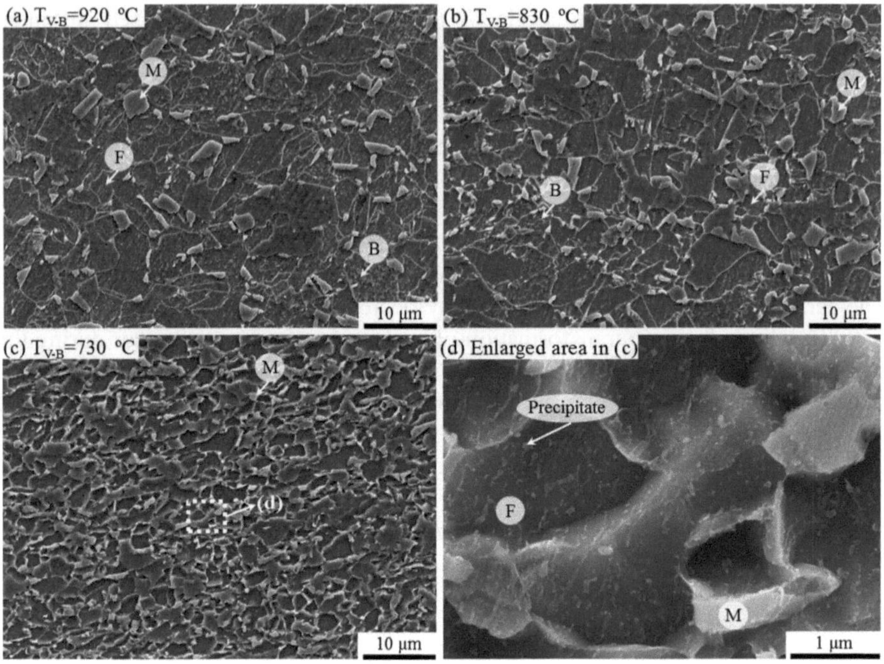

Fig. 4 FESEM images of the Ti/Nb microalloyed steel specimens deformed to simulate V-block strain at **a** 920 °C, **b** 830 °C and **c** 730 °C with subsequent cooling at a rate of 20 °C/s from **a**, **b** 800 °C or **c** 730–500 °C followed by natural cooling to room temperature

room temperature. Similar to the reference steel shown in Fig. 3, grain refinement of ferrite is achieved by decreasing the deformation temperature. Nano-sized precipitates were formed in the ultrafine ferrite grains (Fig. 4d), which could increase the hardness of ferrite.

The present study confirms that lowering the deformation temperature at V-block pass could lead to the occurrence of dynamic transformation, that is, ferrite transformation occurs during deformation of austenite. Dynamic transformation was first reported in the pioneering work by Yada et al. [12] in the mid-1980s and has been extensively investigated especially in plate and sheet steels ever since [13, 14]. Dynamic transformation is considered to be effective for ferrite grain refinement. The ferrite grains could be refined to around 1–4 μm after rapid cooling (usually by water quenching) in the thermomechanical process. The present study shows a possibility that grain refinement by dynamic transformation can be achieved during rod rolling even at a cooling rate of 20 °C/s or lower.

Figure 5 summarizes the variation of ferrite grain size and Vickers hardness of the Ti/Nb microalloyed steel and reference steel processed at different simulated V-Block deformation temperatures. It is again confirmed that grain refinement of ferrite can be significantly enhanced by decreasing the deformation temperature and increasing the cooling rate for both steels (Fig. 5a). It is suggested that many lattice

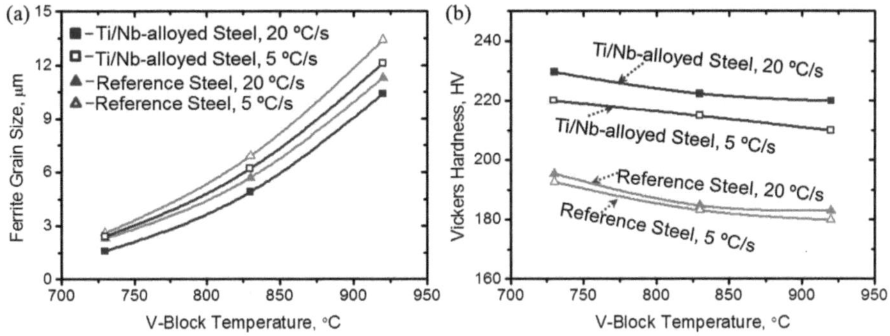

Fig. 5 Variation of ferrite grain size (**a**) and Vickers hardness (**b**) of the Ti/Nb microalloyed steel and reference steel deformed at various temperatures at the V-block pass with subsequent cooling at a rate of 20 °C/s or 5 °C/s followed by natural cooling to room temperature

defects and deformation microstructures introduced into austenite by deformation at lower temperatures could greatly increase the nucleation density for ferrite and lead to an early impingement between ferrite grains nucleated and grown, which is the main cause for the formation of ultrafine ferrite grains. Grain refinement of ferrite by decreasing the deformation temperature at V-block pass results in an increase of Vickers hardness for both steels (Fig. 5b). Fast cooling rate is in favor of improving the hardness of the steels, especially for the microalloyed steel. The difference in hardness between the two kinds of steels indicates the precipitation strengthening effect on the microallyed steel. The improvement of hardness in Fig. 5 suggests a promising way to strengthen the hot-rolled steels, i.e., combination of grain refinement strengthening (by dynamic transformation) and precipitation strengthening (by Ti/Nb-containing precipitates).

The Vickers hardness of Ti/Nb microalloyed steel processed at various thermomechanical conditions (Fig. 5) falls into a range of 210–230 HV, which corresponds to a tensile strength range of 675–740 MPa. It should be noted that these tensile strengths are achieved in the hot-rolled steel rods, that is, prior to the cold drawing and heading. Basically cold drawing and heading increases dislocation density in the rods, which can help improve the strength by 300 MPa or higher [15]. Therefore, it is expected that the hot-rolled steels with hardness of 210–230 HV could be suitable for the production of fasteners with strength properties meeting grade 10.9 requirements.

Conclusion

Simulations of rod rolling by torsional deformation and controlled time-temperature schedules were conducted using the Gleeble® 3500 to examine the effects of thermomechanical processing parameters and microalloying additions on the microstructure evolution and mechanical properties of low-carbon steel rods. The main conclusion is drawn as follows:

(1) High-strain deformation at high strain rates during rod rolling led to the occurrence of adiabatic heating, which greatly accelerated the coarsening of recrystallized austenite grains, resulting in coarse transformed microstructures with low strength.
(2) A new strategy for strengthening hot-rolled steel rods was proposed based on dynamic transformation mechanism, by which the grain refinement of ferrite was significantly enhanced, resulting in a microstructure consisting of ultrafine ferrite with nano-sized precipitates and ultrafine martensite particles.
(3) Hardness of the hot-rolled steel rods was improved by grain refinement and fast cooling. Microalloying addition of Ti and Nb was confirmed to improve the hardness of simulated rolled rods. Vickers hardness of the thermomechanically processed (Ti + Nb)-microalloyed steel were in the range of 210–230 HV, corresponding to a tensile strength range of 675–740 MPa (prior to the cold drawing and heading), which could be suitable for the production of fasteners with strength properties meeting grade 10.9 requirements.

Acknowledgements The authors gratefully acknowledge the continued support of the sponsors of the Advanced Steel Processing and Products Research Center, an industry/university cooperative research center at Colorado School of Mines.

References

1. Hua, Z. H. (2014). Preparation process of bainite structure steel for high-strength fastener. *China Patent, CN103789621 A*.
2. Cao, J., et al. (2010). Non-hardened bainite cold heading steel for fastener and manufacturing method thereof. *China Patent, CN101880826 A*.
3. Pritchard, S. B., & Trowsdale, A. J. (2002). Dual phase steel-high strength fasteners without heat treatment. *Corus Construction and Industrial, U.K.*, 1–10.
4. Pierson, G. (1992). Dual phase steels for non-heat treated high tensile bolts. *Wire Journal International, 6*, 53–57.
5. Trowsdale, A. J., & Pritchard, S. B. (2002). *Dual phase steel—High strength fasteners without thermal treatment* (Report from Corus Construction and Industrial, U.K., 1–11).
6. Enginner, S., & Huchtemann, B. (1996). Review and development of microalloyed steels for forgings, bars and wires. In C. J. Van Tyne, G. Krauss, & D. K. Matlock (Eds.), *Proceedings of Fundamentals and Applications of Microalloying Forging Steels Symposium* (pp. 61–78), TMS.
7. Enginner, S., Lukas, S., & Wittek, R. (1990). Precipitation hardening ferritic-pealitic steel for cold-deformation. *Thyssen Edelstahl Technische Berichte, 16*, 20–25.
8. Krauss, G., & Banerji, S. K. (1987). *Fundamentals of microalloying forging steels*. Warrendale, PA: TMS-AIME.
9. Jeong, D. H., & Lee, H. C. (1992). *Microstructure and mechanical properties of Mn-Mo, B bainitic steel*. M.A Thesis, Seoul National University, Seoul.
10. Bhadeshia, H. K. D. H. (1998). Alternatives to the ferrite-pearlite microstructures. *Materials Science Forum, 284–286*, 39–50.

11. Richardson, G. J., Hawkins, D. N., & Sellars, C. M. (1985). *Worked examples in metalworking* (pp. 19–23). London, England: The Institute of Metals.
12. Yada, H., Matsumura, Y., & Nakajima, K. (1984). *U.S. Patent, 4,466,842.*
13. Hurley, P. J., Hodgson, P. D., & Muddle, B. C. (1999). Analysis and characterisation of ultra-fine ferrite produced during a new steel strip rolling process. *Scripta Materialia, 40,* 433–438.
14. Adachi, Y., Xu, P. G., & Tomota, Y. (2008). Crystallography and kinetics of dynamic transformation in steels. *ISIJ International, 48,* 1056–1062.
15. ArcelorMittal. (2010). Low-carbon precipitation-strengthened steel for cold heading applications. *Europe Patent, EP2199422 A1.*

Modeling of Metal-Slag Mass and Momentum Exchanges in Gas-Stirred Ladles

Marco Ramírez-Argáez and Carlos González-Rivera

Abstract Ladle refining plays a key role in achieving the quality of the steel. Specifically the metal-slag mass exchange is studied through a scaled water physical model in which thymol, a solute, is added to the water (steel) and silicon oil (slag) picks up the thymol, while the ladle is agitated with the central gas injection and samples of water were taken to track the thymol concentration with time with a UV-visible spectrophotometer. Also, a mathematical model was developed and solved with the CFD code Fluent Ansys to represent the fluid flow and the mass transfer phenomena through the solution of the continuity, the turbulent momentum conservation and species mass conservation equations. A good agreement between the measured and the computed results regarding the thymol concentration evolution in water was found so the model was validated and it may be used to study metal-slag exchanges in the steel ladle.

Keywords Gas/stirred ladle · Mathematical model · Physical model · Slag-metal exchange

Introduction

Ladle steelmaking plays a key role to produce high quality steel grades. Independently on the steelmaking route, the most important step, which defines the quality of the steel, is the secondary refining in the Ladle Furnace (LF), where important refining operations such as deoxidation, desulphurization, chemical composition adjustment, uniformity of composition and temperature and elimination or modification of the non-metallic inclusions take place. However, opacity of

M. Ramírez-Argáez (✉) · C. González-Rivera
Facultad de Química, UNAM, Avenida Universidad 3000, Ciudad Universitaria, C.P. 04510 Coyoacán, Cd. México, Mexico
e-mail: marco.ramirez@unam.mx

© The Minerals, Metals & Materials Society 2017

M.A. Meyers et al. (eds.), *Proceedings of the 3rd Pan American Materials Congress*, The Minerals, Metals & Materials Series, DOI 10.1007/978-3-319-52132-9_76

liquid steel during operation makes plant trials difficult to be performed. Therefore, process engineering through physical and mathematical models are excellent tools to understand, to control, and to optimize the process.

In the LF not only the liquid steel is present, but also a basic slag layer that protects the steel from the environment avoiding re-oxidation of the steel and heat losses through radiation but also the slag absorbs the oxides and sulfides coming from the deoxidation and desulphurization of the steel among other metal-slag exchanges. In order to accelerate the refining processes, the steel bath is stirred by injecting argon gas through a porous plug located at the bottom of the ladle. Then, the LF is a very complex three-phase fluid flow system with momentum, heat and mass transport phenomena.

Hundreds of works have been published on the LF as Mazumdar and Guthrie [1], Mazumdar and Evans [2] and Irons et al. [3] have pointed out in their reviews on LF. All of the works reported so far are in one of the following categories: (a) Mathematical modeling, (b) Physical modeling, and (c) a combination between physical and mathematical studies.

Regarding physical modeling, the majority of models use water in the role of steel and air in the role of argon by satisfying similarity criteria such as the dynamic similarity through the modified Froude number and the kinematic similarity is assumed to be met since the kinematic viscosities of water at 20 °C and of liquid steel at 1600 °C are the same. With the physical models, the two-phase fluid flow structure has been characterized in the plume region, fluid flow patterns of the liquid, gas holdup, mixing correlations, spout characteristics at the free surface, mass transfer and particle dynamics. All these studies from the last century to the beginning of the current century only considered the presence of two phases. Recently, it has been recognized that the slag phase needs to be considered for a realistic representation of the fluid flow system since this phase affects the fluid flow phenomena by dissipating around 10–20% of the stirring energy in the LF [4] and also since without the presence of this phase the metal-slag exchanges can not be modeled. However, only a few works on physical modeling have involved a slag layer, such as the work by Amaro et al. [5] who reported a mixing correlation under the present of a slag layer of different layer size and viscosity or the work by Peranandhanthan and Mazumdar [6] who reported the size of the slag opening at the free surface (called slag "eye"). Even fewer works on mathematical modeling for three-phase systems are found in literature and most of them have not enough quality to be considered as real contributions.

Several mathematical models have been used to investigate the flow and mixing phenomena in gas-stirred ladle systems. These models have been classified into three types: Quasi-single or pseudo-single phase models, Lagrangian-Eulerian two-phase models (Lagrangian) and Eulerian-Eulerian two-phase models (Eulerian). Detailed reviews of related simulations and model studies are found in literature [1–3].

The presence of the top slag layer involves a three-phase system (liquid steel/liquid slag/gas). Mendez et al. [7] analyzed in detail the effects of drag and non-drag forces as well as the critical role of the top slag layer on fluid flow. The relative importance of each force depends on the size of the bubbles and the degree of turbulence. Drag forces are originated by pressure, friction forces act in a direction opposite to bubble motion, virtual mass forces due to acceleration of the liquid displaced by the bubbles, lift forces due to the unsymmetrical pressure distribution, and turbulent dispersion forces. Their simulations proved the importance to include drag and non-drag forces to properly represent fluid flow. In addition to this, they also indicate that is mandatory to include the top slag layer in order to obtain a more realistic representation of the industrial process. Other three-phase models can be found elsewhere [8–11].

In the current work a three-phase mathematical model was developed to describe the fluid flow in a cylindrical gas-stirred ladle in steelmaking with metal slag exchange. The model was validated in a 1/17th water-oil physical model of a 250 tons ladle from an industrial plant in Mexico. In the physical model thymol (2-isopropyl-5-methylphenol, playing the role of a solute such as sulfur) is added to the water (playing the role of steel) and silicon oil (playing the role of slag) picks up the thymol, while the ladle is agitated with a central injection of gas. Also, fluid dynamics validation of the CFD model was achieved by comparing the numerical results against PIV measurements of liquid flow patterns and turbulence. The final goal is to set a robust three-phase mathematical model to study, mixing and metal-slag exchanges in a realistic manner.

Theory and Experiment

Mathematical Model

The model was developed and cast into the commercial CFD code Fluent Ansys to represent the fluid flow phenomena and the slag metal/mass transfer through the solution of the continuity equation, the turbulent momentum conservation equations and the species mass conservation equation.

The three-phase fluid flow model representing a typical gas-stirred steel ladle is based on the following assumptions: (a) all phases are incompressible and Newtonian fluids; (b) turbulence in the flow may be considered to be isotropic; (c) the flow field is considered to be isothermal, i.e. there are no thermal gradients; (d) the gas stirred ladle shows azimuthal symmetry; (e) turbulence is considered to be present only in the liquid phases (fluid flow inside the gas bubbles is assumed to be laminar but turbulence affects the motion of the gas phase so turbulence is indirectly present in the average phase conservation equations through the turbulent coefficients of gas dispersion); (f) bubbles are considered to be at constant size, Dp, for a given gas flow rate according to the correlation used in [12], i.e. no coalescence nor disintegration of bubbles are considered and the effect of pressure on the bubble size is also neglected.

The mathematical model is based on the Eulerian frame of reference including a multiphase approach where pressure and turbulence fields are shared among phases whereas phasic equations are solved for continuity and momentum equations.

The continuity equation for phase q can be written as:

$$\frac{\delta}{\delta t}(\alpha_q \rho_q) + \nabla \cdot (\alpha_q \rho_q \overrightarrow{u_q}) = 0 \tag{1}$$

where α_q, ρ_q and u_q are the volume fraction, density and velocity of the qth phase respectively. The momentum balance for the qth phase is:

$$\frac{\delta}{\delta t}(\alpha_q \rho_q \overrightarrow{u_q}) + \nabla \cdot (\alpha_q \rho_q \overrightarrow{u_q}\overrightarrow{u_q}) = -\alpha_q \nabla P + \nabla \overline{\tau_q} + \alpha_q \rho_q \vec{g} + \overrightarrow{F}_{drag,q} + \overrightarrow{F}_{lift,q} + \overrightarrow{F}_{VM,q} \tag{2}$$

where τ_q, P and g are the qth phase stress-strain tensor, pressure and gravity acceleration respectively. $F_{drag,q}$ is the drag force, $F_{lift,q}$ is the lift force and $F_{VM,q}$ is the virtual mass force acting on the qth phase. The drag force between phases can be described as:

$$\overrightarrow{F}_{drag,q} = \frac{3f\mu_q}{L} A_i (\overrightarrow{u_p} - \overrightarrow{u_q}) \tag{3}$$

where A_i is the interfacial area, L and μ_q are the diameter and viscosity of qth phase respectively, f is the drag coefficient. In this work, the symmetric drag model was used.

Taking into account that the gas stirred ladle under study shows axial symmetry. the governing equations can be written and solved in two dimensions: radial and axial.

In gas-stirred ladle, the species in liquid steel, such as aluminum, silicon, manganese, etc., can be transferred simultaneously by molecular diffusion, convection, and turbulent velocity fluctuations, and they can be removed from or produced into liquid steel due to chemical reactions at the slag–metal interface. In the current model, these mass exchanges can be represented in the following form, which represents the species mass conservation equation:

$$\frac{\delta}{\delta t}\left(\alpha_l \rho \frac{[\text{pct } Y_i]}{100}\right) + \nabla \cdot \left(\alpha_q \rho \overrightarrow{u_l} \frac{[\text{pct } Y_i]}{100}\right) = \nabla \cdot \left(\alpha_l \frac{\mu_t}{Sc_t}\left(\nabla \frac{[\text{pct } Y_i]}{100}\right)\right) + S_i \tag{4}$$

where [*pctYi*] is the local mass fraction of species i in liquid steel, Sct is the turbulent Schmidt number (default value is 0.7). S_i is the source term of mass transfer equation of species i due to chemical reactions at the slag–metal interface and in slag eyes.

At the slag–metal interface, the overall slag–metal reaction rate, S_i, is controlled by mass transport processes in slag and metal phases and can be written as

$$S_i = \alpha_l \rho_l k_{eff} 100 \frac{A_{cell}}{V_{cell}} \left\{ [\text{pct } Y_i] - \frac{(\text{pct Yi})}{L_i} \right\} \tag{5}$$

where the (*pctYi*) indicates the mass fraction of species *i* in slag phase, and L_i is the equilibrium constant or the partition coefficient that states the ratio of thymol dissolved in the slag over the thymol dissolved in water at the thermodynamic equilibrium. A_{cell} and V_{cell} are the bubble-liquid interface and volume for each cell of the domain. Finally, k_{eff}, is the mass transfer coefficient that depends on the stirring conditions prevailing in the ladle and specifically close to the water-oil interface.

To account for the turbulence, the standard k-epsilon model [13] was employed, which solves two additional conservation equations: one for turbulent kinetic energy, k, and other for its dissipation rate, epsilon. Boundary conditions sets zero velocities at the bottom and lateral walls while standard wall functions are used to interpolate the turbulent core to the laminar region near the walls. The symmetry axis is a zero flux boundary for all the transported quantities. At the nozzle only gas is injected (gas volume fraction equal to one) at a constant velocity given by the gas flow rate and the nozzle diameter. Finally, the top boundary is opened to the atmosphere and gas is allowed to scape at this boundary. Since the problem was solved in transient form, initial conditions must be set, which involved a static system with a layered structure where water is present at the bottom followed by a layer of small thickness of oil at the top of the water and finally at the top of the oil a layer of air is present.

In the present work, the CFD model of the system shown schematically in Fig. 1a and in its discretized 2D representation in Fig. 1b, was solved using the commercial computational fluidynamics software Fluent ANSYS 14.5.

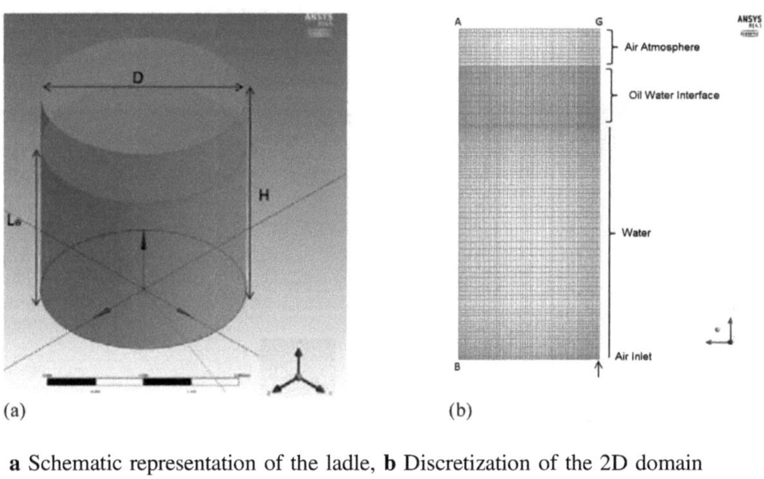

(a) (b)

Fig. 1 **a** Schematic representation of the ladle, **b** Discretization of the 2D domain

Solution was done numerically in transient form using a delta time of 1×10^{-4} s and the 2D mesh shown in Fig. 1a with 17,100 nodes was employed. The mesh shows fine nodes in the liquid-liquid and liquid-gas interfaces to resolve accurately the surface deformations and to avoid numerical diffusion and the final mesh was a result of a long mesh sensitivity study resulting in this optimized mesh. Every simulation considered took 15 s of real time to reach a steady state conditions in the flow patterns and in the distributions of phases and every calculation took around 4 days of computer time to converge the final solution in a personal computer with 8 GB in RAM and a clock speed of 3.4 GHz.

Physical Model

Physical Model Procedure for Hydrodynamics

Figure 2 shows a photograph of the measurements (Fig. 2a) and a scheme of the experimental setup (Fig. 2b). The arrangement involves a 1/17th scaled acrylic physical model of a 140 ton LF containing water and silicon oil and a PIV equipment Dantec Dynamics® operated in the single frame with a time interval between pulses of 100 μs and using 30% of the total power of the laser of 10 mJ. Images were captured with a High Speed camera Speedsense M320 with a resolution of 1920 pixels × 1200 pixels. A total of 500 pair of photos were captured and statistically processed with the software DynamicStudio version 4.0 to get the final results on flow patterns and turbulent intensity. The model used a gas flow rate of 2.85 l/min and a layer of slag of 6.6 mm and the central longitudinal plane (containing the nozzle) was used to make the measurements. Physical properties of all the three phases simulated can be consulted in Table 1.

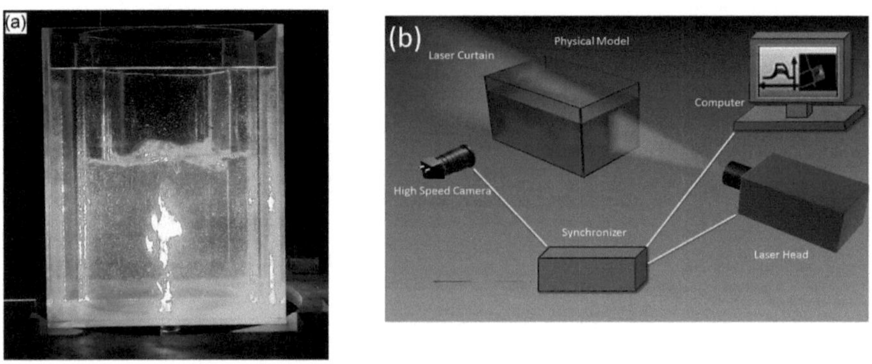

Fig. 2 **a** Photograph of a measurement with PIV; **b** Scheme of the experimental setup

Table 1 Physical properties of the phases present in the system

Property/substance	Water	Silicon oil	Air
Density [kg/m^3]	1000.0	800.0	1.23
Kinematic viscosity [m^2/s]	1×10^{-6}	1×10^{-4}	1.45×10^{-5}

Table 2 Physical dimensions and operating parameters in the ladle and model

	Diameter [m]	Height [m]	Liquid column [m]	Slag layer [m]	Q [m^3/min]
Industrial ladle	3.226	3.736	2.802	0.112	0.2
Physical model	0.189	0.219	0.164	0.0065	0.00285

Physical Model for Mass Transfer Experiment of Slag-Metal Exchange

A 1/17th scale water model of a 250 ton steel ladle was employed to investigate slag/metal exchange (see Fig. 2a). In this, physical dimensions and operating parameters in the model ladle were deduced from those of the full-scale system on the basis of geometric, dynamic and kinematic similarity. For ready reference these are summarized in Table 2.

In the scaled water physical model, thymol (2-isopropyl-5-methylpheno) plays the role of a solute such as sulfur and is added to the water (which plays the role of steel) and silicon oil (playing the role of slag) picks up the thymol, while the ladle is agitated with the central injection of gas.

At the beginning of an experiment, the required amounts of thymol-water solution and silicon oil, according with the geometrical quantities shown in Table 2, were sequentially and quietly poured into the ladle, then the gas injection was started. Thymol-water solution samples were taken by introducing a syringe into a hole located at the bottom of the ladle, at a radial distance of 1/3 of the inner ladle radius from the symmetry axis. Samples of 1.5 ml were taken at 1, 3, 5, 15, 30, 60, 90, 120 and 150 min. The solute content in water as a function of time was determined with the aid of a Spectrophotometer Shimadzu UV-2600, previously calibrated to measure thymol content in water in the range of interest. The associated calibration curve is shown in Fig. 3.

Results and Discussion

Figure 4 shows a comparison of the liquid velocity field measured (left) and computed (right). It is notorious the excellent agreement between predicted and measured velocity fields which shows only half the measured plane for symmetry reasons. The velocity filed shows the very well known structure where at the center the liquid is directed upwards by the drag that the bubbles exert on the liquid. When the liquid reaches the free surface there is a deformation called "spout" which is measured and predicted accurately and the flow is directed radially from the center

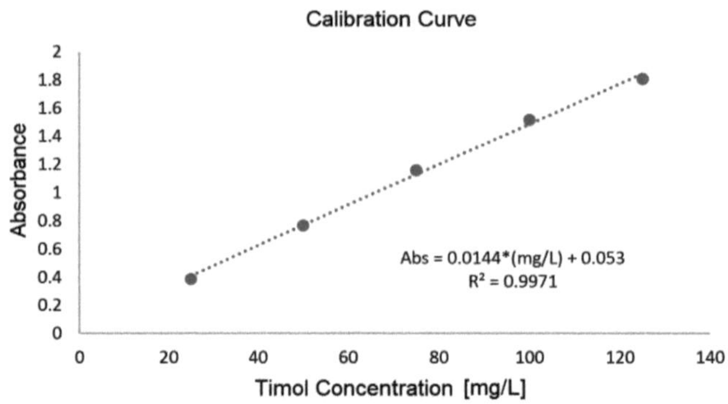

Fig. 3 Calibration curve for Thymol content determination

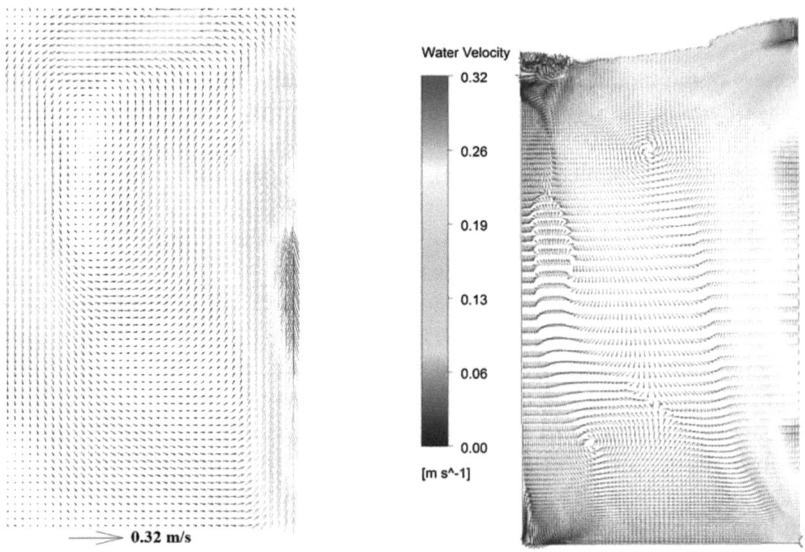

Fig. 4 Measured (*left*) and predicted (*right*) velocity fields

to the lateral walls and when the fluid reaches the lateral wall turns down creating a circulating pattern with the eyes of the circulation located at the top of the LF. The lower region of the LF has low motion. Maximum liquid velocity of 0.32 m/s is found in both experiments and simulations.

Figure 5 shows the predicted (right) and measured (left) turbulent kinetic energy fields. Again, the agreement in the k fields is qualitatively excellent but quantitatively just reasonable. It is known that the k-epsilon turbulence model is probably not the best choice for a three-phase problem and especially for bubbly flows. However, the turbulence model included a term to account for the extra turbulence promoted by the bubbles. This result suggests future work to test several turbulence

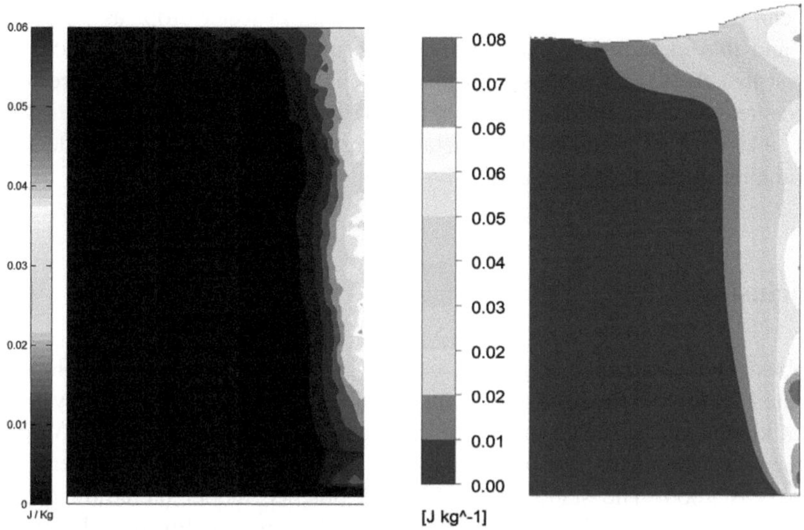

Fig. 5 Measured (*left*) and predicted (*right*) turbulent kinetic energy fields

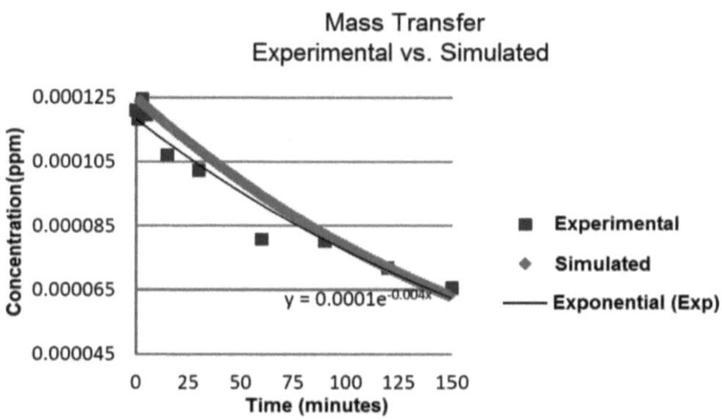

Fig. 6 Evolution of thymol concentration measured and computed

models to determine which one of them gives the best description of the measured (real) intensities. The plume is the region where turbulence is concentrated and especially close to the "spout" region in the LF according to both measurements and simulations. In general, the model overestimates turbulence in comparison with the measured values of k.

Typical results of thymol content evolution in water as a function of time are shown in Fig. 6. In this figure the plot of thymol content in water against time is shown, including the experimental determinations, shown as discrete squares and the exponential function resulting from the numerical fitting of these points, as well as the thymol content evolution predicted by the model. The good fitting of the experimental points to an exponential function indicates that thymol elimination follows a kinetics behavior controlled by mass transfer.

It can be seen, as expected, that thymol content in water drops as a function of time and that experimental and simulated results are very similar which in turn suggest that the model proposed in this work is able to capture the more relevant features related to the fluidynamic behavior of the three phase system studied in this work, and may be used to study metal-slag exchanges in the steel ladle such as deoxidation or desulphurization.

Conclusions

A robust mathematical model of a centric gas-stirred ladle was developed and validated in this work against PIV measurements of the liquid velocity and turbulent kinetic energy fields including three-phases (water-oil and gas). Agreement between measurements and predictions in both velocities and turbulent kinetic energies is good. The standard k-epsilon turbulence model including a term to account for the extra turbulence is able to reproduce reasonably well the measured turbulent structure of the system but slightly overestimate it, and therefore, it is necessary to study which of the turbulent models available agrees better to the experimental measurements.

A physical water model where thymol is a solute, water represents steel and silicon oil represents slag, was employed to investigate slag/metal exchanges by tracking the thymol concentration with time in order to compare experimental results with predictions of the model at a given gas flow rate.

There is a good agreement between the measured and the computed results, regarding the thymol concentration evolution in water and consequently the mathematical model was validated regarding the mass species metal-slag exchanges, confirming that mass transfer at the water-oil interface controls the kinetics of removal of thymol from water, just as it happens in the slag-metal reactions in a industrial ladle furnace.

Finally, our model may be used to study in a realistic way other phenomena of interest in the ladle such as mixing, heat transfer, inclusion dynamics, etc.

Acknowledgements Authors would like to thank Project PAPIIT IN114815 from DGAPA-UNAM for financial support. Technical assistance provided by A. Ruiz, A. Amaro and C. Atlatenco is also acknowledged.

References

1. Mazumdar, D., & Guthrie, R. I. L. (1995). The physical and mathematical modelling of gas stirred ladle systems. *ISIJ International, 35*(1), 1–20.
2. Mazumdar, D., & Evans, J. W. (2004). Macroscopic models for gas stirred ladles. *ISIJ International, 44*(3), 447–461.

3. Irons, G., Senguttuvan, A., & Krishnapisharody, K. (2015). Recent advances in the fluid dynamics of ladle metallurgy. *ISIJ International, 55*(1), 1–6.
4. Mazumdar, D., & Guthrie, R. I. L. (2010). Modeling energy dissipation in slag-covered steel baths in steelmaking ladles. *Metallurgical and Materials Transactions B, 41B*(5), 976–989.
5. Amaro-Villeda, A. M., Ramírez-Argáez, M. A., & Conejo, A. N. (2014). Effect of slag properties on mixing phenomena in gas-stirred ladles by physical modeling. *ISIJ International, 54*(1), 1–8.
6. Peranandhanthan, M., & Mazumdar, D. (2010). Modeling of slag eye area in argon stirred ladles. *ISIJ International, 50*(11), 1622–1631.
7. Méndez, C. G., Nigro, N., Cardona, A., Begnis, S. S., & Chiapparoli, W. P. (2002). Physical and numerical modelling of a gas stirred ladle. *Mecanica Computacional, 21,* 605–619.
8. Liu, H., Qi, Z., & Xu, M. (2011). Numerical simulation of fluid flow and interfacial behavior in three-phase argon-stirred ladles with one plug and dual plugs. *Steel Research International, 82,* 440–458.
9. Sulasalmi, P., Visuri, V. V., Kärnä, A., & Fabritius, T. (2015). Simulation of the effect of steel flow velocity on slag droplet distribution and interfacial area between steel and slag. *Steel Research International, 86,* 212–222.
10. Lou, W., & Zhu, M. (2015). Numerical simulation of slag-metal reactions and desulfurization efficiency in gas-stirred ladles with different thermodynamics and kinetics. *ISIJ International, 55*(5), 961–969.
11. Lou, W., & Zhu, M. (2014). Numerical simulation of desulfurization behavior in gas-stirred systems based on computation fluid dynamics-simultaneous reaction model (CFD–SRM) coupled model. *Metallurgical and Materials Transactions B, 45B,* 1706–1722.
12. López-Gutiérrez, A. (2015). MSc Thesis, Universidad Nacional Autónoma de México.
13. Launder, B. E., & Spalding, D. B. (1974). The numerical computation of turbulent flows. *Computer Methods in Applied Mechanics and Engineering, 3*(2), 269–289.

Study on Adjustment and Optimization of LF Refining Slag of Spring Steel 55SiCrA

Chao Gu, Yan-ping Bao, Lu Lin, Min Wang, Li-hua Zhao and Zi-xuan Wu

Abstract The spring steel 55SiCrA produced in a domestic steel factory is deoxidized by Si with "100 t EAF-LF-VD-CC". The fluoric content (CaF_2 = 17.41–22.15%) and the binary basicity (R = 1.9–3.0) of applied LF refining slag are high. To avoid environmental pollution caused by fluorine, meanwhile guarantee the speed of melting slag and smelting effect on molten steel, the structure of refining slag was adjusted. A new fluorine-free slag system was obtained by changing slagging process, adjusting adding amounts of slag-making materials, such as lime, calcium carbide, silicon carbide and so on. This new fluorine-free slag was applied in industrial tests. The results showed that this new fluorine-free slag could meet smelting demands well. The average mass fraction of oxygen content of spring steel 55SiCrA billets could decreased from 11×10^{-6} to 9×10^{-6} and the plasticity of inclusions also changed better. All compositions of observed inclusion in billets were within the 1400 °C liquidus in Al_2O_3-SiO_2-CaO-MgO-MnO phase diagram.

Keywords Spring steel 55SiCrA · LF refining slag · Fluorine-free

Introduction

With rapid development of national economy, the quality requirements for spring steel gradually improved. Refining slag plays an important role in the smelt of spring steel. Slags with different compositions are applied in different deoxidation methods [1–5]. Correlational researches [6–9] showed that appropriate refining slag could help desulfuration and deoxidation, protect refractory lining of furnace and absorb inclusions in steel. Besides, different refining slags also have significant

C. Gu · Y. Bao (✉) · L. Lin · M. Wang · L. Zhao · Z. Wu
State Key Laboratory of Advanced Metallurgy,
University of Science and Technology Beijing, Beijing 100083, China
e-mail: gu_chao2011@sina.com

© The Minerals, Metals & Materials Society 2017

M.A. Meyers et al. (eds.), *Proceedings of the 3rd Pan American Materials Congress*,
The Minerals, Metals & Materials Series, DOI 10.1007/978-3-319-52132-9_77

effects on the category, property, and size of inclusions [10, 11]. Slags with low basicity are normally used for steel deoxidized by silicon to decrease the spherical inclusions in steel, such as calcium-aluminates [12, 13].

Industrial tests were conducted in one factory in China to confirm a kind of LF refining slag fit for spring steel 55SiCrA deoxidized by silicon, which has reasonable slugging speed and refining results and could decrease environmental pollution caused by CaF_2 addition in slag. Though adjusting slagging process and compositions of refining slag, a new fluoride-free refining slag system was obtained. The total oxygen content, inclusions, and so on of spring steel 55SiCrA smelted by the new slag were analyzed and compared with steel smelted by the previous slag. Theoretical and practical guidance on LF refining slag for spring steel 55SiCrA deoxidized by Si was provided.

Industrial Tests

Optimization of Slagging Process

The industrial trails of spring steel 55SiCrA were carried out in a steel plant in China, and the production route is 100 t electric arc furnace (100 t EAF) → Ladle Furnace (LF) → VD → Continuous Casting (CC). The composition requirement of spring steel 55SiCrA is shown in Table 1.

According to thermochemical calculation of equilibrium contents of O, Si, and Al in slag-steel system with FactSage 6.4 and correlational researches, proper control range of refining slag for spring steel 55SiCrA is SiO_2: 32–42%, CaO: 41–51%, Al_2O_3: 2–4%, MgO: 9–11%, basicity R is around 0.9–1.5. There is no pretreatment of hot metal for 55SiCrA in this plant, thus the desulfurization capacity of refining slag should be desirable. Considering too low basicity will affect the desulfurization capacity of refining slag, the basicity of refining slag in the industrial tests was chosen to be the upper limit of optimization range, which was around 1.4–16. The objective compositions of slag of the industrial tests are shown in Table 2.

Table 1 Composition requirement of spring steel 55SiCrA/mass%

C	Si	Mn	P	S	Ni	Cr
0.51–0.59	1.20–1.60	0.50–0.80	$\leq$0.025	$\leq$0.025	$\leq$0.35	0.50–0.80

Table 2 Objective compositions of slag of the industrial tests

CaO/%	SiO_2/%	MgO/%	Al_2O_3/%	CaF_2/%	R
46–51	32–36	9–11	2–4	0	1.4–1.6

The deoxidation and slagging process was adjusted as endpoint carbon content of EAF was controlled less than 0.15%; over-oxidation decreased; quantity of slag was controlled; 3 kg/t refining slag and 2 kg/t lime were added for slagging when tapping; 16 kg/t low aluminum ferrosilicon and 5 kg/t mid-carbon ferromanganese were applied for deoxidation and alloying. Ferrosilicon was added for deoxidation in earlier stage of LF refining according to the deoxidation condition. Lime, refining slag, and other slag charge were added to make white slag with reducibility (Specific addition quantities were according to calculated objective composition). Meanwhile slag thickness was guaranteed and 0.4 kg/t carborundum and a small amount of calcium carbide were added in batches for diffusible deoxidation. After deoxidation and slagging, high carbon ferro-chrome and carbon powder were added for content adjustment of C and Cr in molten steel. Electrifying time was guaranteed more than 20 min and soft blow time was guaranteed more than 10 min.

Research Methods

Two heats industrial tests were conducted after adjustment of refining slag of spring steel 55SiCrA. Steel samples were taken with a pail-sampler (diameter 60 mm, height 100 mm) in 6 positions including after EAF, middle of LF, after LF, after breaking vacuum of VD, after VD, and tundish. Steel samples of billet were also taken. Cylinder samples ($\phi 5 \times 50$ mm) and metallographic specimens ($10 \times 10 \times 20$ mm) were machined respectively from the pail samplers as well as billets. Cylinder samples were used to analyze total oxygen content with infrared-absorption method. Metallographic specimens were polished and used to analyze inclusions in steel with scanning electron microscope (SEM) and the compositions of inclusions were analyzed by energy disperse spectroscopy (EDS). Besides, the slag samples were grinded to pass 200-mesh sieve and the compositions were analyzed by X-ray fluorescence (XRF).

Results and Discussion

Compositions of LF Refining Slag

Compositions of LF refining slags of 3 heats before adjustment and 2 heats after adjustment were compared (Table 3). As shown in Table 3, the basicity of refining slag decreased from 1.9–3.0 to 1.5–1.6. The content of CaF_2 in refining slag also decreased sharply.

According to objective compositions of refining slag, the slagging process was eventually adjusted as below: addition amount of lime decreased about 2 kg/t, addition amount of ferrosilicon decrease about 0.5 kg/t; addition amount of fluorite

Table 3 Composition comparison of refining slag before and after adjustment

	Al_2O_3/%	CaO/%	MgO/%	MnO/%	SiO_2/%	CaF_2/%	R
Before optimization	4.47	45.79	6.92	0.11	24.04	17.41	1.9
	2.33	51.64	5.85	0.11	15.97	22.15	3.2
	2.49	51.97	6.39	0.15	17.33	19.54	3.0
After optimization	3.91	51.87	8.62	0.20	32.67	<0.1	1.6
	2.76	44.40	6.47	0.17	29.98	13.95	1.5

decreased from 0.5–1.5 kg/t to 0–0.3 kg/t. With this new slag, smelting performance of slag was improved, furnace burden consumption decreased, and cost of production also decreased.

Change of Total Oxygen Content in Whole Production Route

Average total oxygen content in whole production route of 2 heats industrial tests were shown in Fig. 1. Total oxygen content in molten steel after EAF was high (55×10^{-6}). The effect of LF refining was obvious. Total oxygen content after LF

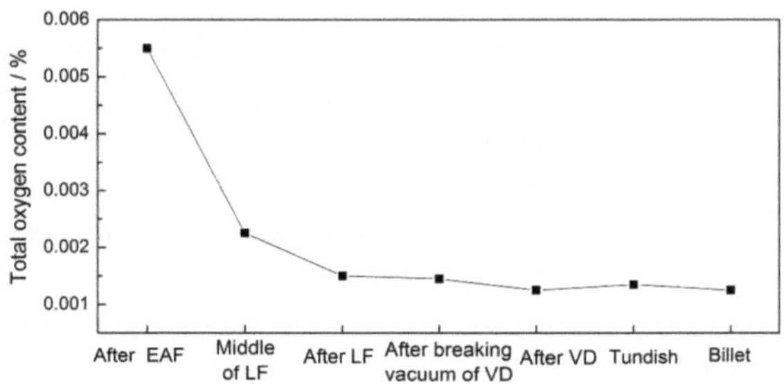

Fig. 1 Change of total oxygen content in whole production route

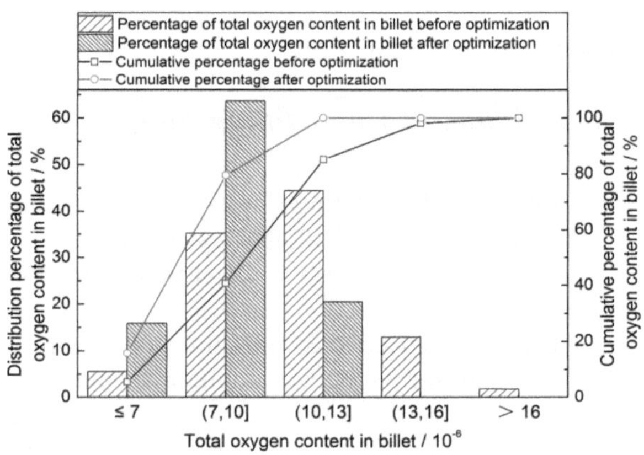

Fig. 2 Distribution change of mass fraction of total oxygen content in billets before and after optimization

decreased to 15×10^{-6}. From VD refining to billet, the total oxygen content barely changed. In the billet, the average total oxygen content was 12.5×10^{-6}.

Total oxygen contents in billets of 44 heats of spring steel 55SiCrA after optimization and 54 heats before optimization were analyzed. The distribution was shown in Fig. 2. Cumulative percentages of total oxygen content in billets after optimization were all larger than those before optimization. The percentage of total

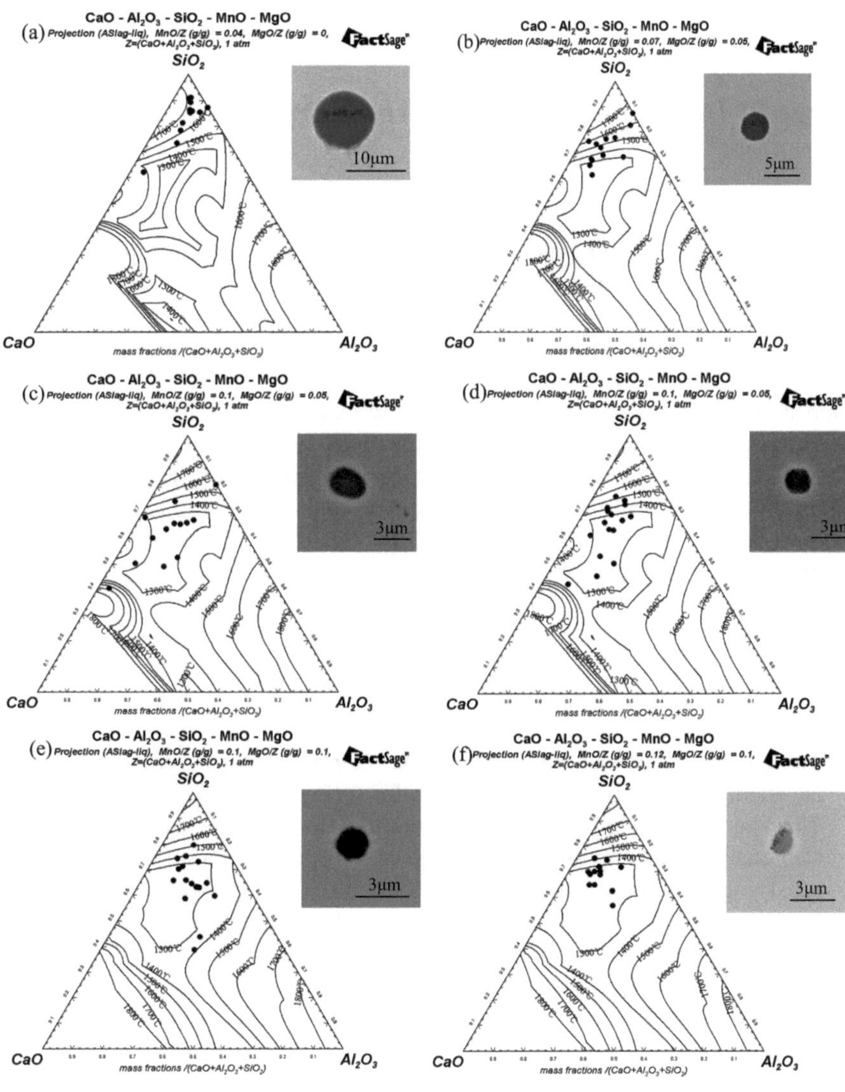

Fig. 3 Melting-point changes of inclusions during melting in industrial tests in Al_2O_3-SiO_2-CaO-MgO-MnO phase diagram (**a** After EAF, **b** After LF, **c** After VD vacuum breaking, **d** After VD, **e** Tundish, **f** Billets)

oxygen content in billets in the range of 10×10^{-6} to 13×10^{-6} were the largest before optimization. After optimization, the percentages of total oxygen content in billets in the ranges of $\leq 7 \times 10^{-6}$ and 7×10^{-6} to 10×10^{-6} increased sharply. The largest percentage of total oxygen content in billets changed to the range of 7×10^{-6} to 10×10^{-6}. The percentage of total oxygen content in billets less than 10×10^{-6} increased from 40.7 to 77.3%. No total oxygen contents were larger than 13×10^{-6}. Besides, the average total oxygen content after optimization decreased from 11×10^{-6} to 9×10^{-6}.

Change of Inclusion Plasticity in Whole Production Route

The observed inclusions in metallographic specimens taken in after EAF, middle of LF, after LF, after breaking vacuum of VD, after VD, tundish, and billet were shown in Al_2O_3-SiO_2-CaO-MgO-MnO pentabasic phase diagram (Fig. 3). And the typical morphologies of inclusions were also shown in Fig. 3.

As shown in Al_2O_3-SiO_2-CaO-MgO-MnO phase diagram, the melting point of inclusions gradually decreased with smelting process. After EAF, most of inclusions were in high-melting-point area (over 1600 °C). Through LF refining, the compositions of inclusions moved towards low-melting-point area gradually. Most inclusions were within 1500 °C liquidus after breaking vacuum of VD. From VD refining to tundish and billet, inclusions fluctuated around 1500 °C liquidus. In VD refining process, several inclusions were still in high-melting-point area; in billet, all observed inclusions were within 1400 °C liquidus. The effect of optimized refining slag on plasticity control of inclusions in spring steel 55SiCrA was ideal.

Conclusions

(1) Though changing slagging process, the compositions of refining slag were adjusted. Fluoride-free refining slag fit for spring steel 55SiCrA was obtained. This slag could help protect the environment from fluorine as well as decrease slag charge consumption and save the cost of production on the condition of satisfying smelting results.
(2) Deoxidation effect of refining was obvious after optimization. The percentage of total oxygen content in billets less than 10×10^{-6} increased from 40.7 to 77.3%. The average total oxygen content after optimization decreased from 11×10^{-6} to 9×10^{-6}. Total oxygen contents decreased sharply, which reached optimization result.
(3) Effect of the optimized refining slag on plasticity of inclusions was obvious. In billet, all observed inclusions were within 1400 °C liquidus. The plasticity of inclusions improved substantially.

Acknowledgements This research was supported financially by the National Natural Science Foundation of China (No. 51404022) and State Key Laboratory of Advanced Metallurgy Foundation (No. 41614014).

References

1. Zhao, D. W., et al. (2013). Composition optimization of LF refining slag in high cleanliness Al-killed steel. *Steelmaking, 29*(2), 9–13.
2. Ma, W. J., et al. (2014). Influence of slag composition on bearing steel cleanness. *Ironmaking and Steelmaking, 41*(1), 26–30.
3. Oshiro, T., et al. (1989). Verbesserung der dauerhalt-barkeit von ventilfederdraht. *Stahl und Eisen, 21*(10), 1011–1015.
4. Suito, H., & Inoue, R. (1996). Thermodynamics on control of inclusions composition in ultraclean steels. *ISIJ International, 36*(5), 528–536.
5. Onoe, T., et al. (1987). Shape control of inclusions for steel tire cord (development in ladle arc refining). *Transactions of ISIJ, 27,* B249.
6. Ge, Y. Z., et al. (2013). Best composition of CaO-Al_2O_3 slag system for refining of 20CrMnTiH steel. *Iron and Steel, 48*(10), 23–27.
7. Wu, C., et al. (2013). Effect of refining slag with different basicities on inclusions in spring steel. *Journal of Wuhan University of Science and Technology, 36*(4), 254–257.
8. Jiang, M., et al. (2010). Laboratory study on evolution mechanisms of non-metallic inclusions in high strength alloyed steel refined by high basicity slag. *ISIJ International, 50*(1), 95–104.
9. Reis, B. H., Bielefeldt, W. V., & Vilela, A. C. F. (2014). Efficiency of inclusion absorption by slags during secondary refining of steel. *ISIJ International, 54*(7), 1584–1591.
10. Liu, F. Y., Xu, Z. G., & Shang, M. (2013). A study on 80 t converter-ladle furnace refining slag for high quality GCr15 bearing steel steelmaking. *Special Steel, 34*(1), 25–27.
11. Yan, P. C., et al. (2013). Effect of the CaO-Al_2O_3-Based top slag on the cleanliness of stainless steel during secondary metallurgy. *Metallurgical and Materials transactions B, 44B* (10), 1105–1119.
12. Deng, Z. Y., et al. (2013). Effect of basicity on deoxidation capability of refining slag. *Journal of Iron and Steel Research International, 20*(2), 21–26.
13. Chen, S. H., et al. (2013). Industrial application of desulfurization using low basicity refining slag in tire cord steel. *Journal of Iron and Steel Research International, 20*(1), 26–33.

Tempering Response of Bainitic and Martensitic Microstructures

Igor Vieira and Emmanuel De Moor

Abstract The tempering response of fully martensitic microstructures has been well characterized. However, bainitic microstructures may also be found in quenched industrial materials and the present study investigates the tempering response of fully martensitic and fully bainitic microstructures. Specific thermal cycles were developed to generate both microstructures in a boron added 0.17 wt pct carbon steel. The tempering response was assessed through dilatometry and microstructural characterization was conducted using scanning electron microscopy and Mössbauer spectroscopy. The dilatometric analysis of the tempering response of the martensitic microstructures provided information about retained austenite decomposition and cementite precipitation whereas bainitic microstructures showed a less sensitive dilatometric response during tempering likely due to the low amount of carbon in solution and absence of retained austenite as measured by Mössbauer spectroscopy.

Keywords Tempering · Dilatometry · Mössbauer spectroscopy

Introduction

High strength heavy gauge plate steels develop different cooling rates throughout the thickness of the material during quenching. Several grades do not fully harden to martensite resulting in mixed microstructures of martensite and bainite, which are subsequently tempered. Most published relationships for the tempering response of high strength steels apply to fully martensitic microstructures [1–3]. Therefore, it is desirable to better describe the tempering of bainitic microstructures in order to allow tempering optimization for materials with mixed microstructures.

I. Vieira (✉) · E. De Moor
Advanced Steel Processing and Products Research Center,
Colorado School of Mines, 1500 Illinois St., Golden, CO 80401, USA
e-mail: ivieira@mines.edu

© The Minerals, Metals & Materials Society 2017
M.A. Meyers et al. (eds.), *Proceedings of the 3rd Pan American Materials Congress*,
The Minerals, Metals & Materials Series, DOI 10.1007/978-3-319-52132-9_78

Non-isothermal analysis has been used extensively to describe solid state transformation kinetics including tempering [4–7]. Different properties such as hardness, electrical resistivity and magnetization are susceptible to changes during tempering, which allows analysis of transformation kinetics. Specific length also varies during tempering and can be measured using dilatometry which is the technique utilized in this work.

Martensite tempering proceeds in various stages and consists of the precipitation of transition carbides; decomposition of retained austenite into ferrite and cementite; precipitation of cementite, and finally the formation of alloy carbides, also known as secondary hardening [8]. Mössbauer spectroscopy is suitable to assess and quantify the formation of transition carbides and cementite as well as to quantify retained austenite. Most of the studies involving this technique are focused on quench and tempered steels [9, 10]. It should be noted that Mössbauer spectroscopy is able to identify iron carbides, but not alloy carbides.

The present study was developed to contrast the tempering response of fully martensitic and fully bainitic steels through non-isothermal tempering in a dilatometer and microstructural characterization in order to develop an understanding of tempering of mixed bainitic/martensitic microstructures.

Experimental Procedure

A 0.17C-1.23Mn-0.29Si wt pct (0.76C-1.24Mn-0.57Si atomic pct) steel was received as laboratory produced hot rolled plate. Non-isothermal analysis was performed using a Thermal Analysis (TA) Quenching 805L Dilatometer. Cylindrical samples measuring 4 mm in diameter and 10 mm in length were machined at a distance of 0.5 cm from the plate surface for dilatometry. Coupons approximately 2 × 2 × 0.5 cm^3 were also sectioned from the plate to be used for Mössbauer spectroscopy and scanning electron microscopy (SEM).

The bainitic samples were isothermally transformed and the holding temperature was determined based on the bainite start (B_s) and martensite start (M_s) temperatures calculated using the following equations [11, 12]:

$$\begin{aligned} B_S[^\circ C] = 830 &- 270[\text{wt pct C}] - 90[\text{wt pct Mn}] - 37[\text{wt pct Ni}] \\ &- 70[\text{wt pct Cr}] - 83[\text{wt pct Mo}] \end{aligned} \tag{1}$$

$$\begin{aligned} M_S[^\circ C] = 539 &- 423[\text{wt pct C}] - 30.4[\text{wt pct Mn}] - 12.1[\text{wt pct Cr}] \\ &- 17.7[\text{wt pct Ni}] - 7.5[\text{wt pct Mo}] - 7.5[\text{wt pct Si}] \end{aligned} \tag{2}$$

amounting to 430 and 674 °C, respectively. Thus, the temperature of 475 °C was chosen as the isothermal bainitic holding temperature.

Martensitic samples were austenitized at 997 °C for 10 min and quenched. Dilatometry specimens were helium quenched while coupons were austenitized in a box furnace and water quenched. The bainitic samples were also austenitized at 997 °C for 10 min, quenched to 475 °C, isothermally held at that temperature for 10 min, and finally quenched to room temperature. The quenching steps were performed using helium gas quenching in the dilatometer. The coupons were austenitized in a box furnace, transferred to molten salts at adequate temperature, held at temperature, and finally water quenched. In addition, the martensitic samples were tempered at 500 °C for one hour.

For the non-isothermal tempering, martensitic and bainitic samples were heated from 100 to 700 °C with heating rates of 10, 20, 50, and 100 °C/min. Plots of the derivative of change in length (L) with respect to time (t) versus temperature (T) were used to identify phase transformations occurring during heating.

Microstructural characterization was performed using SEM on samples etched with 2 pct Nital. The martensitic, bainitic and tempered martensitic coupons were evaluated through Mössbauer spectroscopy. The arrangement for the Mössbauer spectrometer consists of a Co57 γ-ray source that is oscillated by a drive with velocity varying from −6.5 to 6.5 mm/s, a collimator used to filter the γ-rays, a sample and a multichannel analyzer (MCA). The Mössbauer spectrum is obtained by passing a beam of γ-rays emitted from the radioactive source through a thin absorber which is the sample being analyzed. A large surface area and proper thickness of the sample facilitated data fitting, providing a more accurate evaluation. A 4 mm (5/32 in.) collimator was used. The γ-rays passing through the sample are detected, amplified, and stored in the MCA. Each channel stores the total number of γ-rays transmitted by the absorber and, once the MCA is connected to the drive, total numbers are related to a particular source velocity [13]. Sample preparation required mechanical grinding to a thickness between 100 and 150 μm followed by final thinning to a thickness between 20 and 40 μm using a solution of hydrofluoric acid.

Results and Discussion

Scanning Electron Microscopy

Figure 1 shows the microstructures of the as-quenched martensite, tempered martensite, and bainite. In Fig. 1a, the as-quenched microstructure shows the packets of martensite laths and a considerable amount of transition carbides formed during the quenching process due to auto-tempering. In the tempered condition, Fig. 1b, the presence of a larger amount of carbides is observed. At high tempering temperatures, such as 500 °C, cementite is the stable iron carbide [2]. Figure 2 presents the bainitic microstructure with two different morphologies of bainite, lower (indicated by the white arrow) and upper (indicated by the black arrow). Significant carbide precipitation is observed.

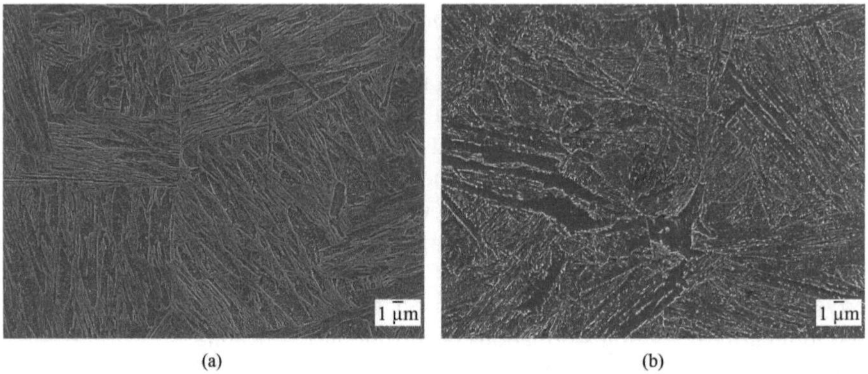

Fig. 1 SEM micrographs of the **a** as-quenched martensite and **b** martensite tempered at 500 °C

Fig. 2 SEM micrograph of the bainitic microstructure. *White arrow* indicates lower bainite and *black arrow* indicates upper bainite

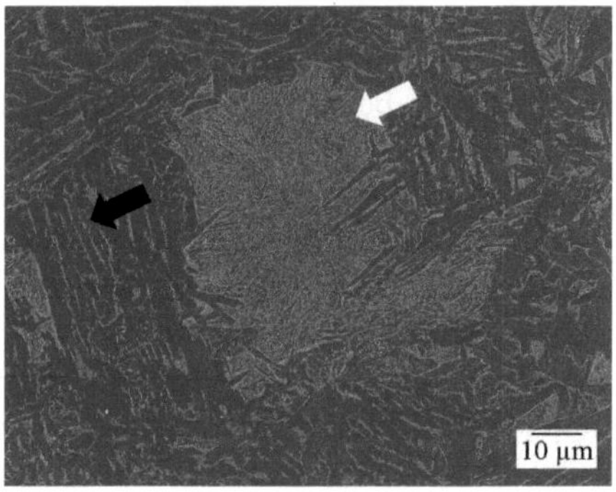

Dilatometry

The change in specific length induced in the fully martensitic sample by non-isothermal tempering at a heating rate of 100 °C/min and the derivative of change in length with respect to time (dL/dt) are shown in Fig. 3a. It is observed that the change in length does not vary in a perfect linear manner with respect to temperature. The slight variation of slope from 100 to approximately 300 °C as well as the variation above 400 °C is likely associated with the change in expansion coefficient with temperature. Between 300 and 400 °C, a significant drop occurs in the dL/dt curve indicating the precipitation of cementite [14, 15]. The carbon diffusion out of martensite produces an overall reduction in the specific length, which is detected in the non-isothermal approach as a reduction in the rate of change in length with temperature. Figure 3b shows the dL/dt curves versus temperature for different heating rates. As the heating rate decreases, the drop associated with the precipitation of cementite becomes less sharp. The rate of the decrease in dL/dt is reduced between 340 and 400 °C for the heating rate of 10 °C/min. This

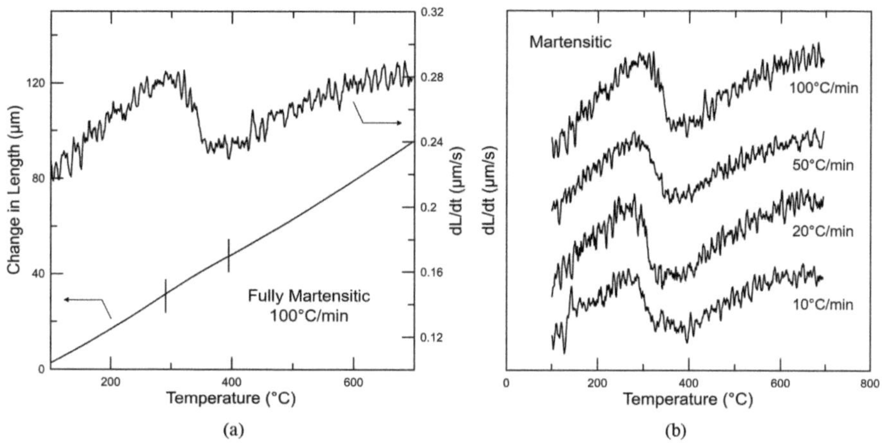

Fig. 3 **a** Dilation curve for a fully martensitic specimen heated at a rate of 100 °C/min and corresponding derivative of change in length with respect to time. **b** Derivative of change in length with respect to time versus temperature for martensitic samples and heating rates of 100, 50, 20, and 10 °C/min

change can be attributed to the decomposition of retained austenite [15]. The transformation of retained austenite into ferrite and cementite promotes an increase in change in length as observed. This transformation has been reported to overlap with the formation of cementite, agreeing with the observed results in this study [14].

The bainitic specimens were also submitted to non-isothermal analysis. Figure 4a shows dL/dt versus temperature for the different heating rates used. No significant variations are demonstrated with respect to temperature and among the various heating rates suggesting that no major transformation takes place during tempering of a fully bainitic steel. Figure 4b compares non-isothermal results for bainite and martensite submitted to the same heating rate. Smoothing of the curves was performed in order to facilitate comparison between the responses of bainite and martensite to tempering. The curve associated with bainite exhibits an increase in dL/dt in the temperature range proposed whereas the martensitic sample shows a clear drop associated with cementite formation.

Mössbauer Spectroscopy

Mössbauer spectroscopy was conducted for the martensitic and bainitic samples as well as for the martensitic sample tempered at 500 °C. Figure 5 shows the obtained spectra for the three different conditions. Each spectrum was fitted with multiple subspectra to account for martensite/ferrite, η-carbide (stoichiometric = η_S and

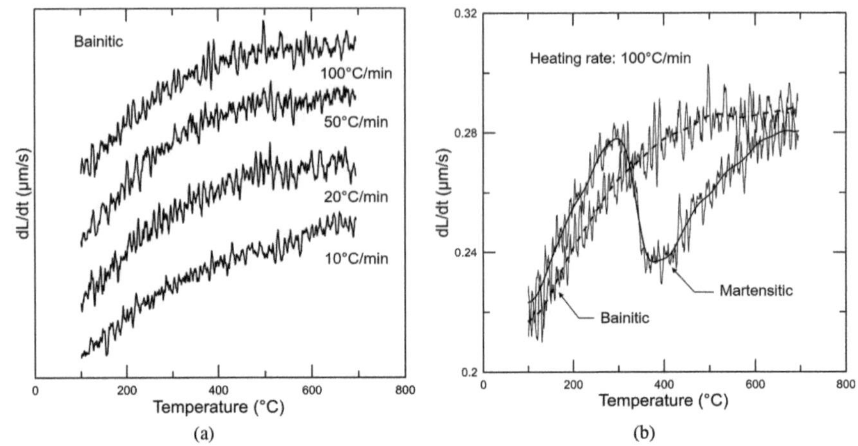

Fig. 4 **a** Derivative of change in length with respect to time versus temperature for bainitic samples for heating rates of 100, 50, 20, and 10 °C/min. **b** Comparison between dL/dt obtained for martensitic and bainitic samples for a heating rate of 100 °C/min

non-stoichiometric = η_{NS}) [16], cementite (θ), retained austenite (A(0) = Fe atoms in austenite without carbon as nearest neighbor and A(C) = Fe atoms in austenite with carbon as nearest neighbor), and a non-magnetic iron phase (P). The stick diagrams displayed in Fig. 5 indicate the resonance associated with each phase. The components attributed to martensite/ferrite were removed for clarity.

Table 1 shows the Mössbauer spectral parameters for η and θ carbides. The values found for η_S are consistent with results presented in the work of Pierce et al. [16] for a steel with 0.38 wt pct C, with an isomer shift (IS) of 0.16 ± 0.05 mm/s, quadrupole shift (QS) of −0.09 ± 0.1 mm/s, and a magnetic hyperfine field (B_{hf}) of 17.4 ± 0.6 T. The signal for the η_{NS} was weak and therefore the parameters were fixed during the fitting based on the results from Pierce et al. [16] for a steel with 0.38 wt pct C. The parameters found for θ carbide were also in agreement with results found in literature [17], with an IS of 0.18 mm/s, a QS of 0.02 mm/s, and a B_{hf} of 20.7 T. The magnetic hyperfine field was the only parameter that presented a significant difference. Schaaf et al. [18] demonstrated that the presence of Mn in the lattice can be responsible for a reduction in the magnetic field found for cementite.

Table 2 presents the atomic percentage of iron in the various phases found in each condition. The martensitic sample showed a total of approximately 0.73 atomic pct Fe as η carbide. Assuming no partitioning of substitutional elements takes place and that η carbide corresponds to Fe_2C, approximately 0.36 atomic pct carbon is found as η carbide, which confirms the auto-tempering of the material during cooling, as expected for a steel with a M_S temperature of 430 °C [19]. The specimen still shows 2.18 atomic pct Fe in retained austenite. When tempered at 500 °C, the only carbide found was θ, which consumed the majority of the carbon available, around 0.61 atomic pct C. In this condition, a non-magnetic phase was

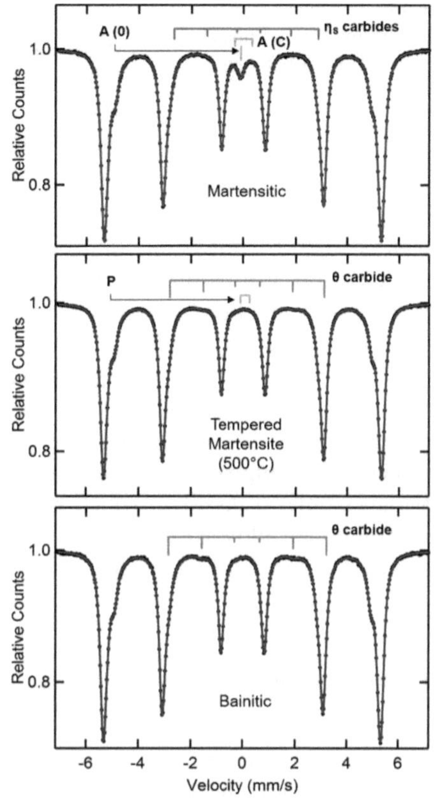

Fig. 5 Mössbauer spectra for the as-quenched martensite, tempered martensite, and bainite. The stick diagrams represent the resonance attributed to retained austenite, η carbide, θ carbides and a paramagnetic phase. The stick diagrams are not to scale and components associated with martensite and ferrite have been removed for better visualization

Table 1 Mössbauer spectral parameters determined for η and θ carbides

Carbide	Parameter	Values
η_S	IS (mm/s)	0.15 ± 0.05
	QS (mm/s)	−0.10 ± 0.05
	B_{hf} (T)	17.6 ± 0.4
η_{NS}	IS (mm/s)	0.13
	QS (mm/s)	−0.36
	B_{hf} (T)	25.2
θ	IS (mm/s)	0.17 ± 0.02
	QS (mm/s)	0.02 ± 0.03
	B_{hf} (T)	18.5 ± 0.2

IS isomer shift, *QS* quadrupole splitting, and B_{hf} magnetic hyperfine field

also found. The phase is believed to be $(Fe,Mn)_3C$. Schaaf et al. [18] demonstrated that in the presence of more than 12 atomic pct Mn, $(Fe,Mn)_3C$ can be produced, which is a paramagnetic phase. The parameters found for the phase in Schaaf's work are similar to those observed for the P phase. At 500 °C, the temperature is

Table 2 Atomic percentages of iron for various phases detected by Mössbauer spectroscopy

Condition	η_S	η_{NS}	θ	γ_{ret}	P
Martensitic	0.60	0.13	–	2.18	–
Tempered martensite (500 °C)	–	–	1.84	–	0.15
Bainitic	–	–	2.26	<0.1	–

high enough to allow some diffusion of Mn and promote Mn enrichment in small particles of θ. With a substantial amount of the carbon found as cementite and some in the non-magnetic phase, only around 0.1 atomic pct C is available for formation of alloy carbides or in solution. Finally, the bainitic material presented 2.26 atomic pct Fe in cementite, consuming approximately 0.75 atomic pct C and leaving only 0.01 atomic pct C in solution. In addition to the low amount of carbon in solid solution, cementite observed in bainitic microstructures is usually coarser than cementite in tempered martensitic microstructures [20]. The coarser cementite takes a longer time to dissolve, which reduces the rate of replacement of cementite by substitutional alloy carbides [20]. Thus, secondary hardening is expected to be slower for steels with a bainitic microstructure.

Conclusions

Non-isothermal tempering was conducted for martensitic and bainitic microstructures at different heating rates using dilatometry. Martensite demonstrated substantial sensitivity to the technique allowing the identification of retained austenite decomposition and an estimation of the range of temperature for cementite precipitation. The bainitic microstructure demonstrated less sensitivity during tempering, which is likely attributed to the low amount of carbon in solution and absence of retained austenite prior to tempering based on Mössbauer spectroscopy analysis.

Acknowledgements The support of the sponsors of the Advanced Steel Processing and Products Research Center, an industry-university cooperative research center at the Colorado School of Mines, and the Roberto Rocca Education Program is gratefully acknowledged. Tata Steel is also gratefully acknowledged for providing the material.

References

1. Irvine, K., & Pickering, F. (1960). The tempering characteristics of low-carbon low-alloy steels. *Journal of the Iron and Steel Institute, 194,* 137–153.
2. Speich, G. R., & Leslie, W. C. (1972). Tempering of steel. *Metallurgical Transactions, 3,* 1043–1054.

3. Crafts, W., & Lamont, J. L. (1948). *Secondary hardening of tempered martensitic alloy steels* (pp. 471–512).
4. Mittemeher, E. J., Cheng, L., Schaaf, P. J., Brakman, C. M., & Korevaar, B. M. (1988). Analysis of nonisothermal transformation kinetics; tempering of iron-carbon and iron-nitrogen martensites. *Metallurgical Transactions A, 19,* 925–932.
5. Mittemeijer, E. J. (1992). Review—Analysis of the kinetics of phase transformations. *Journal of Materials Science, 27,* 3977–3987.
6. Viswanathan, U. K., Kutty, T. R. G., & Ganguly, C. (1993). Dilatometric technique for evaluation of the kinetics of solid-state transformation of maraging steel. *Metallurgical Transactions A, 24,* 2653–2656.
7. Leiva, J. A. V., Morales, E. V., Villar-Cociña, E., Donis, C. A., & Bott, I. S. (2010). Kinetic parameters during the tempering of low-alloy steel through the non-isothermal dilatometry. *Journal of Materials Science, 45,* 418–428.
8. Krauss, G. (2015). *Steels—Processing, Structure, and Performance* (pp. 373–403). Materials Park, OH: ASM International.
9. Williamson, D. L., Nakazawa, K., & Krauss, G. (1979). A study of the early stages of tempering in an Fe-1.2 Pct alloy. *Metallurgical Transactions A, 3,* 1351–1363.
10. Williamson, D. L., Schupmann, R. G., Materkowski, J. P., & Krauss, G. (1979). Determination of small amounts of austenite and carbide in hardened medium carbon steels by Mössbauer spectroscopy. *Metallurgical Transactions A, 10,* 379–382.
11. Steven, W., & Haynes, A. G. (1956). The temperature of formation of martensite and bainite in low-alloy steels. *Journal of the Iron and Steel Institute, 183,* 349–359.
12. Andrews, K. W. (1965). Empirical formulae for the calculation of some transformation temperatures. *Journal of the Iron and Steel Institute, 203,* 721–727.
13. Huffman, G. P., & Huggins, F. E. (1992). The use of Mössbauer spectroscopy in the analysis of the raw materials and products of the steel industry. In *AIP Conference Proceedings* (Vol. 84, pp. 150–156).
14. Morra, P. V., Böttger, A. J., & Mittemeijer, E. J. (2001). Decomposition of iron-based martensite: A kinetic analysis by means of differential scanning calorimetry and dilatometry. *Journal of Thermal Analysis and Calorimetry, 64,* 905–914.
15. De Moor, E., Lacroix, S., Samek, L., Penning, J., & Speer, J. G. (2006). Dilatometric study of the quench and partitioning process. In *3rd International Conference on Advanced Structural Steels* (pp. 1–6).
16. Pierce, D. T., Coughlin, D. R., Williamson, D. L., Clarke, K. D., Clarke, A. J., Speer, J. G., et al. (2015). Characterization of transition carbides in quench and partitioned steel microstructures by Mössbauer spectroscopy and complementary techniques. *Acta Materialia, 90,* 417–430.
17. Le Caer, G., Dubois, J. M., & Senateur, J. P. (1976). Etude par Spectrométrie Mössbauer des Carbures de Fer Fe_3C et Fe_5C_2. *Journal of Solid State Chemistry, 19,* 19–28.
18. Schaaf, P., Wiesen, S., & Gonser, U. (1992). Mössbauer study of iron carbides: Cementite $(Fe,M)_3C$ (M = Cr, Mn) with various manganese and chromium contents. *Acta Metallurgica et Materialia, 40,* 373–379.
19. Jung, M., Lee, S. J., & Lee, Y. K. (2009). Microstructural and dilatational changes during tempering and tempering kinetics in martensitic medium-carbon steel. *Metallurgical and Materials Transactions A: Physical Metallurgy and Materials Science, 40,* 551–559.
20. Bhadeshia, H. (2001). *Bainite in steels transformations, microstructure and properties* (pp. 91–92). London: IOM Communications Ltd.

The Research on the Relationship Between Gas Movement Behaviors and Circulation Flow of the Molten Steel in RH

Jia-liang Xu, Yan-ping Bao, Li-hua Zhao, Min Wang, Lu Lin, Ya-di Li and Xing-le Fan

Abstract Circulating kinetic energy has important influences on process of rapid decarburization rate in RH. A physical model of 300 t RH in 1:6 ratio is established to study the relationship between gas behaviors and circulating kinetic energy in this paper. The studies showed that the circulating kinetic energy of liquid steel has the exponential relationship with the bubble velocity and gas volume fraction (17–26%). However, the proportion that the circulating kinetic energy accounted for the total energy is less than 1%, when the void fraction in the up leg is lower than 22% or the bubble velocity less than 0.5 m/s. Therefore, in order to avoid appearing lowest energy value and take advantage of gas-driven energy, must satisfy threes below requests. Ensuring the void fraction is greater than 22%. The vacuum chamber bubble residence time is around 0.15 s. And the bubble rising velocity is not less than 0.5 m/s. Satisfying the above conditions can optimize flow effect of molten steel and improve the decarbonization rate.

Keywords Gas behaviors · Circulation flow · Circulating kinetic energy · Void fraction

Introduction

As the demand for steel products with superior physical and chemical properties, the production of interstitial free (IF) steel with ultra-low carbon content has been increased.

J. Xu · Y. Bao (✉) · L. Zhao · M. Wang · L. Lin · Y. Li · X. Fan
State Key Laboratory of Advanced Metallurgy,
University of Science and Technology Beijing, Beijing 100083, China
e-mail: xujialiangsmile@gmail.com

© The Minerals, Metals & Materials Society 2017
M.A. Meyers et al. (eds.), *Proceedings of the 3rd Pan American Materials Congress*,
The Minerals, Metals & Materials Series, DOI 10.1007/978-3-319-52132-9_79

RH degasser is the key instrument in the process of the refining technology for ultra-low carbon steel, especially in how to get a faster decarburization rate. During the process of RH refining, the molten steel has the Circular flow between the vacuum chamber and the ladle, so it is important to understand the effect of fluid flow on the ladle [1–6].

As the RH vacuum degassing circulation process is conducted at high temperature, it is difficult to observe and measure the circulation flow process. With the rapid development of physical simulation, the complex flow can be solved using physical model [7–11]. Many researchers had gotten many results about the steel flow in the ladle, such as the reaction rate with plug flow is faster than that with perfectly mixed flow and that when a dead zone exists the reaction rate is slower than that with perfectly mixed flow. No dead zone exists that can disturb the decarburization rate in the process, and the shape of the ladle has little effect on the decarburization rate [12, 13].

However, these studies concerning the effect of fluid mixing and the mixing phenomena occurring in the RH ladle didn't be precisely investigate the energy that is carried by the molten steel that has the lower carbon content coming from the vacuum chamber. In addition, the circulating energy of flow steel in ladle is less studied and got little attention.

In this paper, the main purpose is to present a directly relationship between the circulating energy and the bubble behaviors in the RH degasser. The hydrodynamic calculations and the physical model will be bonded to get the bubble velocity, void fraction in the up leg, and the proportion of circulating energy.

Theoretical Analysis

The up leg of RH circulation reactor is a regular circular tube and the gas incidence to the circular tube is well-regulated. The status of gas-liquid two phase flow inside will be influence by the gas fraction. Meanwhile it will also affect the state of air in the vacuum.

The relationship between the void fraction and the liquid velocity can be derived through the empirical formula 2, and the relationship in up and down leg flow velocity of liquid is shown in the formula 1 [14, 15].

The void fraction is solved by using mathematics software and the analysis of the relationship between circulating kinetic energy and the void fraction in up leg is depend on it.

$$v_l' = \frac{v_l}{1-\beta} \tag{1}$$

$$\beta = 0.647\left(\frac{v_l'}{v_l' + v_{\mathrm{a}}}\right)^{0.68} v_{\mathrm{a}}^{0.34} \tag{2}$$

$$\beta = 0.647\left(\frac{v_l}{v_l + \frac{4Q_g}{\pi d_0^2} \times (1-\beta)}\right)^{0.68}\left(\frac{4Q_g}{\pi d_0^2}\right)^{0.34} \quad (3)$$

v_l the liquid velocity in down leg, m s^{-1};
v_l' the liquid velocity in up leg, m s^{-1};
Q_g the amount of injecting gas, m^3 s^{-1};
d_0 inner diameter of snorkel, m;
v_a superficial gas velocity, m s^{-1};
β the void fraction in up leg, %.

Experiments

A 1/6 scale model of an RH degasser made of plastic was used for water experiments. Figure 1 shows a schematic representation of the RH degasser and the controlling system used for physical modeling and experimentation. The TDS-100H Handheld flowmeter was used to measure the water velocity in the down leg.

This model is introduced in detail in Table 1. Geometric similarity, kinematic similarity, vacuum similarity are satisfied between physical model and the prototype (Table 2).

The physical simulation must not only keep Geometric parameters of model are similar to the prototype, but also keep model has the same similar criteria with the proto. When Re > 5000, the flow phenomena is same in metallurgical.

In this paper, the Re of model and prototype are both above 5000 and the Fr is same. So Similar motion and dynamic similarity of this paper can be guaranteed.

In order to research all the experimental circulating kinetic energy, samples are simultaneously taken in different refining parameters. The water level in the model chamber in the vacuum chamber is ranged from 3 to 11 mm and the blowing flow is changed from 1500 to 4500 L/min.

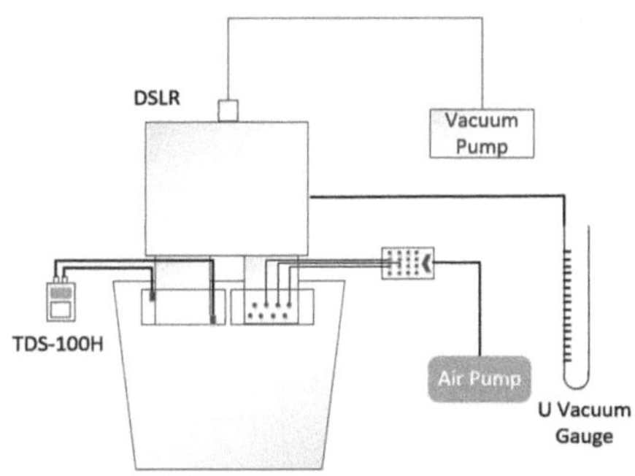

Fig. 1 A schematic representation of the RH degasser and the controlling system

Table 1 Geometric parameters of model and prototype

Item (m)	Prototype	Model
Height of ladle	4.20	0.70
The up diameter of ladle	3.92	0.65
The down diameter of ladle	3.64	0.61
the depth of the liquid steel in ladle	3.92	0.65
Inner diameter of vacuum chamber	2.52	0.42
Inner diameter of snorkel	0.75	0.13
Outer diameter of snorkel	1.50	0.25
Length of snorkel	1.65	0.28

Table 2 Physical parameters of model and prototype

Item	Prototype	Model
Density of air (kg m^{-3})	1.784	1.29
Density of liquid (kg m^{-3})	7000	1000
Liquid viscosity (Pa s)	0.0064	0.0009
Surface tension (N m^{-1})	1.5	0.06

Results and Discussion

The Relationship About the Circulating Kinetic Energy of Liquid Steel, Bubble Velocity and the Void Fraction in up Leg

The Effect of the Bubble Rising Velocity

The circulating kinetic energy for 300 t RH is shown in Fig. 2 changing with the bubble velocity in the vacuum room rising.

As shown in Fig. 2, the circulating kinetic energy of liquid steel is increased exponentially as the increasing of bubble rising velocity (within the range of 0.36–0.60 m/s), and the formula is $y = 0.08343 - 0.1552x + 0.6982x^2$ that the square of correlation coefficient R^2 can get above 0.96. Because the bubble is the

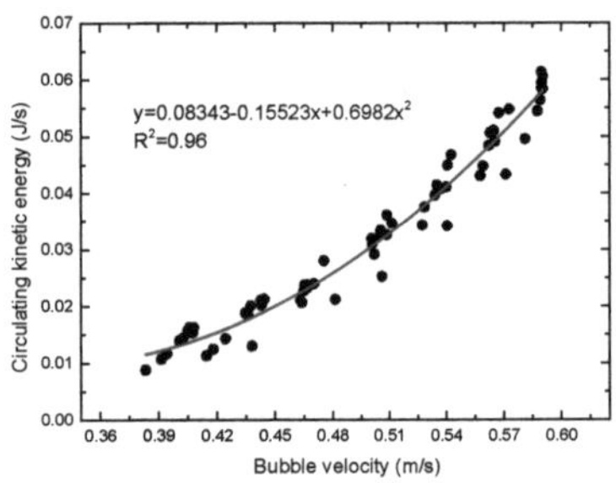

Fig. 2 The relationship between the circulating kinetic energy and the void fraction in up leg

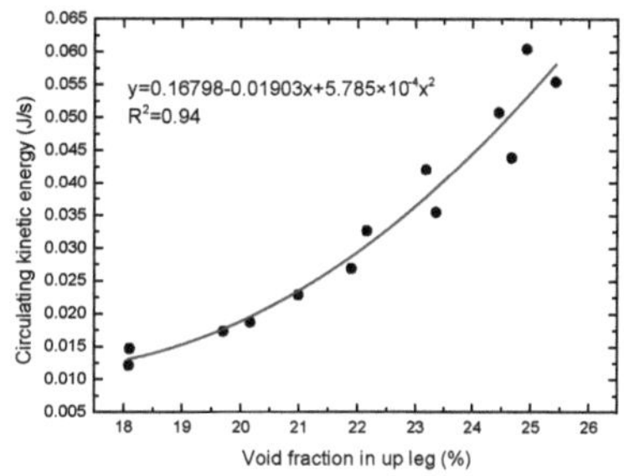

Fig. 3 The relationship between the circulating kinetic energy and the bubble rising velocity

power source in RH reaction system, when the increasing of the bubble velocity, more energy will be brought into steel, then the circulating kinetic energy will be larger than before.

The Effect of Void Fraction in up Leg on the Circulating Kinetic Energy

The void faction in up leg has a similar rule with the bubble rising velocity, and the formula is $y = 0.16798 - 0.01903x + 5.785 \times 10^{-4}x^2$, the square of correlation coefficient R^2 can get above 0.94. More bubbles can bring more energy (Fig. 3).

This phenomenon is up to the mustard that the bubble afford the power of the molten steel flow in RH. W is mainly influence by the blowing value can prove the conclusion in the below equation that obtained by Tie [8]. The greater the blowing flow, the greater the void faction, so the void fraction in up leg has a favourable effect on the circulating kinetic energy.

$$W = 2\rho g H Q_g \tag{4}$$

W the total energy from gas to the liquid, $J\ s^{-1}$;
ρ density of the liquid, $kg\ m^{-3}$;
H the depth of air injection, m.

The Proportion of Circulating Kinetic Energy Linked to Behavior of Bubbles

Bubble Residence Time in the Vacuum Chamber

Figure 4 shows the movements of bubbles from the entrance to the liquid surface in the vacuum chamber. The residence time can be calculated depend on it.

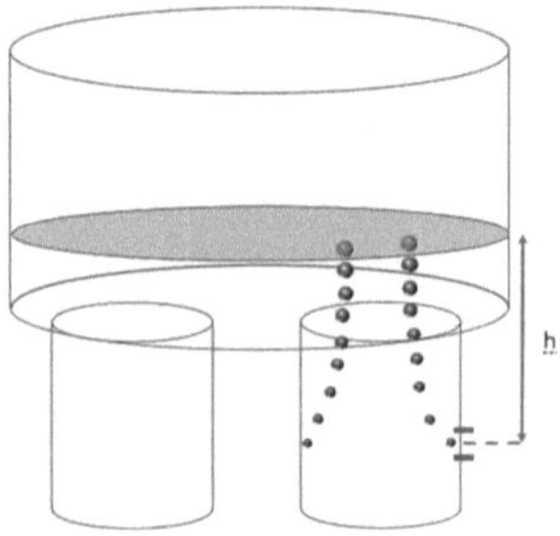

Fig. 4 The sketch map of bubble movements

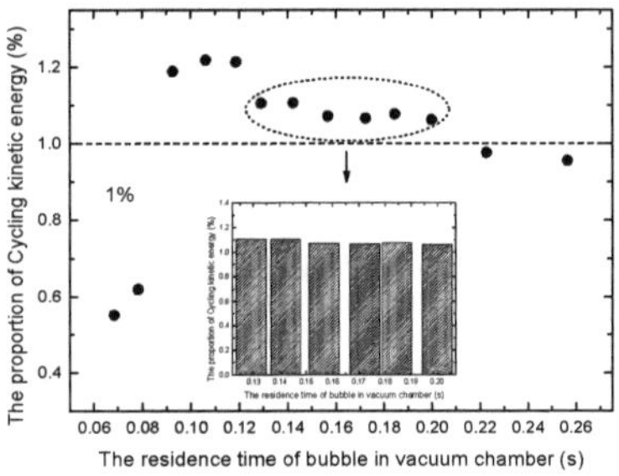

Fig. 5 Experimental data and fitted relational curves between proportion of circulating kinetic energy and bubble residence time

As shown in Fig. 5. As bubble residence time in the vacuum chamber is changed from 0.07 to 0.26 s, the proportion of cycling kinetic energy has the lowest value and the value of remaining the same is existed. When bubble residence time is very small under 0.09 s, the bubbles do not have enough time in exchange energy with the steel, so the energy can't transfer between the bubble and steel efficiently.

The proportion of cycling kinetic energy can reach the high level about 1.2%, but the situation doesn't insist in a big range of bubble residence time. It's just about 0.3 s. However the proportion of cycling kinetic energy still above 1% when bubble residence time in 0.13–0.20 s, and it is in a stabilization. More energy is used for the steel circulation.

The controlling of residence time is difficult and unstable, so making it keep in 0.1–0.2 s and choosing the 0.15 s as the mark point is helpful to get the best result.

The residence time is too long to afford more efficient energy to transfer because of the low bubble velocity.

The Influence of Void Fraction in up Leg to the Proportion of Circulating Kinetic Energy

The proportion of circulating kinetic energy is above 1%, when the void fraction is bigger than 22% and sill increase as it lifts in Fig. 6a.

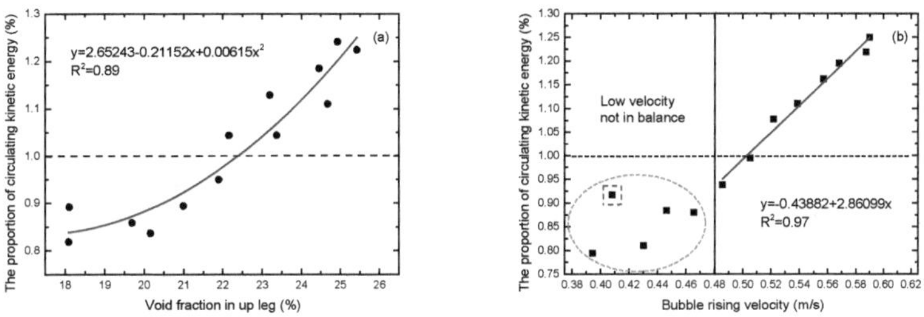

Fig. 6 Important parameters for the proportion of circulating kinetic energy

The Relationship Between the Bubble Velocity and the Proportion of Circulating Kinetic Energy

When bubble rising velocity is less than 0.50 m/s, the proportion of circulating kinetic energy is not very stable such as the point in Fig. 6b because of the flow pattern in the vacuum room. There is a liner relation between bubble rising velocity and the proportion of circulating kinetic energy, as the bubble rising velocity is bigger than 0.48 m/s until the 0.60 m/s. The relative coefficient is more than 0.99.

Conclusions

(1) The circulating kinetic energy of liquid steel increase exponentially as the bubble velocity and the void fraction in up leg rising and the formula is $y = 0.16798 - 0.01903x + 5.785 \times 10^{-4}x^2$ (within the range of 18–26%). Its $R^2 > 0.94$ (correlation coefficient).

(2) When choosing the 0.15 s as the mark residence time and controlling it keep in 0.1–0.2 s, a good refining results can be obtained.

(3) To realize the proportion of circulating kinetic energy is above 1%, void fraction is greater than 22%, the vacuum chamber bubble residence time is around 0.15 s and the bubble rising velocity of not less than 0.5 m/s.

Acknowledgements This work was supported by Nature Science Foundation of China (No. 51404022).

References

1. Wang, X., & Wang, W. (2009). Numerical simulation of circulating flow field structure in RH-refining process. *Journal of Northeastern University (Natural Science), 30*(10), 1481–1484.
2. Ai, X. G., et al. (2009). A study on water model for circulation flow rate optimization in RH degassing refining process. *Special Steel, 30*(3), 1–3.

3. Geng, D. Q., Lei, H., & He, J. C. (2012). Simulation on flow field and mixing phenomenon in RH degasser with ladle bottom blowing. *Ironmaking and Steelmaking, 39*(6), 431–438.
4. Li, C. W., et al. (2011). Mechanism of surface and splash droplet decarburization of RH treatment for ultra-low carbon steel. *Steelmaking, 27*(5), 35–42.
5. Zhang, L., et al. (2009). Numerical simulation of liquid steel flowing in RH refining process and its application. *Journal of University of Science and Technology Beijing, 31*(7), 821–825.
6. Hu, H. T., & Wei, J. H. (2005). Mathematical modeling of molten steel flow during RH refining process: Mathematical model of the flow. *Journal of Anhui University of Technology, 22*(4), 630–634.
7. Jing, Y. G., Xu, H. H., & Zhu, M. Y. (2000). Physical modeling of vacuum degassing process in RH reactor. *Steelmaking, 16*(5), 39–42.
8. Zhu, W. J., et al. (2014). Study on circulation characteristics in 150 t RH refining unit by water modeling. *Steelmaking, 30*(5), 38–42.
9. Ono, K., Yanagida, M., Katoh, T., Okamoto, T. (1981). The circulation rate of RH degassing process by water model experiment. *Electric Furnace Steel, 52*(3), 149–157.
10. Kuwabara, T., Umezawa, K., Mori, K., & Watanabe, H. (1988). Investigation of decarburization behavior in RH-reactor. *Transactions ISIJ, 28,* 305–314.
11. Seshadri, V., Souza Costa, S. L. (1986). Cold model studies of RH degassing process. *Transactions ISIJ, 26,* 133–138.
12. Kleimt, B., Kohle, S., Ponten, H. J., Matissik, W., & Schewe, D. (1993). Dynamic modelling and control of vacuum circulation process. *Ironmaking and Steelmaking, 20*(5), 390–395.
13. Kato, Y., et al. (1993). Fluid flow in ladle and its effect on decarburization rate in RH degasser. *ISIJ International, 33*(10), 1088–1094.
14. Kuwabara, Kamata, C., & Ito, K. (1995). Cold model experiments on the application of gas lift pump to the transportation of molten metal. *ISIJ International, 35*(7), 859–865.
15. Olivieri, G., et al. (2007). Local and global hydrodynamics in a two-phase internal loop airlift. *Chemical Engineering Science, 62*(24), 7068–7077.

Author Index

© The Minerals, Metals & Materials Society 2017

M.A. Meyers et al. (eds.), *Proceedings of the 3rd Pan American Materials Congress*,
The Minerals, Metals & Materials Series, DOI 10.1007/978-3-319-52132-9

Subject Index

© The Minerals, Metals & Materials Society 2017

M.A. Meyers et al. (eds.), *Proceedings of the 3rd Pan American Materials Congress*,
The Minerals, Metals & Materials Series, DOI 10.1007/978-3-319-52132-9

MIX
Papier aus verantwortungsvollen Quellen
Paper from responsible sources
FSC® C105338
FSC
www.fsc.org

If you have any concerns about our products,
you can contact us on
ProductSafety@springernature.com

In case Publisher is established outside the EU,
the EU authorized representative is:
Springer Nature Customer Service Center GmbH
Europaplatz 3, 69115 Heidelberg, Germany

Printed by Libri Plureos GmbH
in Hamburg, Germany